Common Problems in Low- and Medium-Energy Nuclear Physics

NATO ADVANCED STUDY INSTITUTES SERIES

A series of edited volumes comprising multifaceted studies of contemporary scientific issues by some of the best scientific minds in the world, assembled in cooperation with NATO Scientific Affairs Division.

Series B: Physics

RECENT VOLUMES IN THIS SERIES

Volume 39 – Hadron Structure and Lepton–Hadron Interactions – *Cargèse* 1977
edited by Maurice Lévy, Jean-Louis Basdevant, David Speiser, Jacques Weyers, Raymond Gastmans, and Jean Zinn-Justin

Volume 40 – Kinetics of Ion–Molecule Reactions
edited by Pierre Ausloos

Volume 41 – Fiber and Integrated Optics
edited by D. B. Ostrowsky

Volume 42 – Electrons in Disordered Metals and at Metallic Surfaces
edited by P. Phariseau, B. L. Györffy, and L. Scheire

Volume 43 – Recent Advances in Group Theory and Their Application to Spectroscopy
edited by John C. Donini

Volume 44 – Recent Developments in Gravitation – *Cargèse* 1978
edited by Maurice Lévy and S. Deser

Volume 45 – Common Problems in Low- and Medium-Energy Nuclear Physics
edited by B. Castel, B. Goulard, and F. C. Khanna

Volume 46 – Nondestructive Evaluation of Semiconductor Materials and Devices
edited by Jay N. Zemel

Volume 47 – Site Characterization and Aggregation of Implanted Atoms in Materials
edited by A. Perez and R. Coussement

Volume 48 – Electron and Magnetization Densities in Molecules and Crystals
edited by P. Becker

Volume 49 – New Phenomena in Lepton-Hadron Physics
edited by Dietrich E. C. Fries and Julius Wess

This series is published by an international board of publishers in conjunction with NATO Scientific Affairs Division

A Life Sciences B Physics	Plenum Publishing Corporation London and New York
C Mathematical and Physical Sciences	D. Reidel Publishing Company Dordrecht and Boston
D Behavioral and Social Sciences	Sijthoff International Publishing Company Leiden
E Applied Sciences	Noordhoff International Publishing Leiden

Common Problems in Low- and Medium-Energy Nuclear Physics

Edited by

B. Castel
Queen's University
Kingston, Ontario, Canada

B. Goulard
University of Montreal
Montreal, Quebec, Canada

and

F. C. Khanna
Chalk River Nuclear Laboratories
Chalk River, Ontario, Canada

PLENUM PRESS • NEW YORK AND LONDON
Published in cooperation with NATO Scientific Affairs Division

Library of Congress Cataloging in Publication Data

Nato Advanced Study Institute, Banff, Alta., 1978.
Common problems in low and medium energy nuclear physics.

(NATO advanced study institutes series: Series B, Physics: v. 45)
"Lectures presented at the Nato Advanced Study Institute/1978 Banff Summer Institute on Nuclear Theory, held in Banff, Canada, August 21–September 1, 1978."
Includes index.
1. Particles (Nuclear physics)–Congresses. 2. Nuclear physics–Congresses. I. Castel, Boris, II. Goulard, B. III. Khanna, F. C. V. Title: Low and medium energy nuclear physics. VI. Series.
QC793.N37 1978 539.7 79-17039
ISBN 978-1-4684-8952-1 ISBN 978-1-4684-8950-7 (eBook)
DOI 10.1007/978-1-4684-8950-7

Lectures presented at the NATO Advanced Study Institute/1978 Banff Summer Institute on Nuclear Theory, held in Banff, Canada, August 21–September 1, 1978.

Softcover reprint of the hardcover 1st edition 1979
A Division of Plenum Publishing Corporation
227 West 17th Street, New York, N.Y. 10011

Preface

The 1978 Advanced Study Institute in Nuclear Theory devoted to common problems in Low and Intermediate Energy Nuclear Physics was held at the Banff Centre in Alberta, Canada from August 21 through September 1, 1978.

The present volume contains the text of 25 lectures and seminars given at the Institute and illustrates the directions that nuclear physicists are taking in the evolution toward a unified picture of low, medium and high energy phenomena.

Recent attempts at unifying the weak and electromagnetic interaction in particle physics have led naturally to question their role in nuclei. The success of the quark model at interpreting the new resonances in high energy physics makes it imperative to consider their role in dealing with nuclear physics problems at the microscopic level. Is our present knowledge of the nuclear potential consistent with recent experimental evidence at low and medium energy and can it correlate meaningfully nuclear and pion physics phenomena? These are some of the fundamental questions debated in this book attempting to offer a consistent picture of the nuclear system as it emerges using the electromagnetic, weak and strong interaction probe.

The lectures and seminars forming the present volume have been divided into four sections dealing with a) the weak interaction, b) quarks and nuclear structure, c) physics of electrons, protons and kaons, and finally d) pion physics.

The Institute's Organizing Committee gratefully acknowledges the financial support of the NATO Scientific Affairs Division and the co-sponsors, Atomic Energy of Canada Ltd. and the National Research Council. For the preparation of these Proceedings the Committee is particularly indebted to Mrs. Irene High.

B. Castel
B. Goulard
F.C. Khanna

Contents

PART III: PHYSICS OF ELECTRONS, PROTONS AND KAONS

PART IV: PION PHYSICS

Part I

The Weak Interation

WEAK NEUTRAL CURRENTS IN NUCLEAR PROCESSES

H. Primakoff

Department of Physics, University of Pennsylvania

Philadelphia, Pennsylvania, 19104, U.S.A.

I. INTRODUCTION

Some five years have now passed since the discovery at CERN and at FNAL of muonless neutrino-induced events, events whose interpretation strongly suggested the existence of a {weak neutral current (WNC)} x {weak neutral current (WNC)} interaction. At present, such a {WNC} x {WNC} is not only firmly established with its structure essentially determined, but its relationship to the {weak charged current (WCC)} x {weak charged current (WCC)} interaction and to the electromagnetic {EM} interaction is also rather well understood.

As regards examples of processes which could be caused by the {WNC} x {WNC} interaction we first mention the decays

$$H_a \rightarrow H_b + \nu_e + \bar{\nu}_e, \; H_b + \nu_\mu + \bar{\nu}_\mu, \ldots . \qquad (1)$$

where ν_e (ν_μ) and $\bar{\nu}_e$ ($\bar{\nu}_\mu$) are electron (muon)-type neutrinos and antineutrinos and H_a and H_b are hadrons, e.g., H_a is a nucleus in an excited state ($[A,Z]^*$) and H_b is the same nucleus in the ground state ($[A,Z]$) or H_a is a positive kayon (K^+) and H_b is a positive pion (π^+). Actually, such processes have never been observed - in the nuclear case because the {EM} - caused decay $H_a \rightarrow H_b + \gamma$ (or $H_a \rightarrow H_b + \gamma + \gamma$, $H_b + e^- + e^+, \ldots$) proceeds much faster than any {WNC} x {WNC} - caused decay and in the $K^+ \rightarrow \pi^+$ case, where the decays $K^+ \rightarrow \pi^+ + \gamma + \gamma$, $\pi^+ + e^- + e^+, \ldots$ are quite slow (since $K^+ \rightarrow \pi^+ + \pi^0 \rightarrow \pi^+ + \gamma + \gamma$, $K^+ \rightarrow \pi^+ + \rho^0 \rightarrow \pi^+ + e^+ + e^-, \ldots$ involves one weak and two electromagnetic vertices), because the {WNC} x {WNC} interaction appears incapable of effecting transitions between hadronic states of different strangeness ($S(K^+) = 1$, $S(\pi^+) = 0$). On the other hand, ν-induced elastic and inelastic scattering off nucleons and elastic scattering off electrons (muonless events)

$$
\begin{aligned}
&\nu_\mu + p(n) \to \nu_\mu + p(n) \\
&\nu_\mu + p(n) \to \nu_\mu + \Delta^+(\Delta^0);\ \Delta^+(\Delta^0) \to p(n) + \pi^0,\ n(p) + \pi^+(\pi^-) \\
&\nu_\mu + p(n) \to \nu_\mu + p(n) + \pi^- + \pi^+,\ \nu_\mu + p(n) + \pi^0 + \pi^0 \\
&\qquad\qquad \vdots \\
&\nu_\mu + e^- \to \nu_\mu + e^- \qquad (2)
\end{aligned}
$$

has now been extensively studied by experiment and successfully interpreted as a manifestation of the {WNC} x {WNC} interaction. As is seen from Eqs.(1) and (2), the {WNC} x {WNC} lepton-hadron interaction is distinguished from the {WCC} x {WCC} lepton-hadron interaction by charge conservation in the lepton and in the hadron sectors *separately* (compare, e.g., $\nu_\mu + p \to \nu_\mu + \Delta^+$ with $\nu_\mu + p \to \mu^- + \Delta^{++}$ or $^{12}C^* \to {}^{12}C + \nu_e + \bar{\nu}_e$ with $^{12}B \to {}^{12}C + e^- + \bar{\nu}_e$) so that the {WNC} x {WNC} interaction is assumed to be mediated by a *neutral* vector boson ($W^0 \equiv Z$) analogous to the mediation of the {WCC} x {WCC} interaction by a *charged* vector boson (W^+ or W^-) (compare, e.g., $\nu_\mu + p \to \nu_\mu + Z + p \to \nu_\mu + \Delta^+$ with $\nu_\mu + p \to \mu^- + W^+ + p \to \mu^- + \Delta^{++}$ or $^{12}C^* \to {}^{12}C + Z \to {}^{12}C + \nu_e + \bar{\nu}_e$ with $^{12}B \to {}^{12}C + W^- \to {}^{12}C + e^- + \bar{\nu}_e$).

Further manifestations of the {WNC} x {WNC} interaction are a consequence of its *parity-violating character*. Thus the cross sections of processes such as

$$
\begin{aligned}
&e^- + p \to e^- + p,\ e^- + \text{hadrons (nucleon + pions + ...)} \\
&e^- + d \to e^- + d,\ e^- + \text{hadrons (two nucleons + pions....)} \\
&p + p \to p + p,\ \text{hadrons (two nucleons + pions + ...)} \\
&p + d \to p + d,\ \text{hadrons (three nucleons + pions + ...)} \\
&\qquad \vdots \qquad\qquad (3)
\end{aligned}
$$

are expected to depend on the helicity of the incident e^- because of the interference between the parity-violating {WNC} x {WNC} and parity-conserving {EM} scattering amplitudes (e.g., $e^- + p \to e^- + Z + p \to e^- + \Delta^+$ and $e^- + p \to e^- + \gamma + p \to e^- + \Delta^+$), and on the helicity of the incident p because of the interference between the parity-violating [{WNC} x {WNC} + {WCC}x{WCC}] and parity-conserving [{Strong} + {EM}] scattering amplitudes (e.g., $p + p \to p(n) + Z(W^+) + p \to p(n) + \rho^0(\rho^+) + p \to \Delta^+ + \pi^0 + p \to \Delta^+ + \Delta^+$ and $p + p \to \Delta^+ + \pi^0(\gamma) + p \to \Delta^+ + \Delta^+$) - such an incident-particle helicity dependence has now been definitely observed in $e^- + d \to e^- +$ hadrons. Finally, the parity-violating {WNC} x {WNC} interaction produces "wrong-parity admixtures" to bound electron states in atoms and bound nucleon states in nuclei (in the latter case, the parity-violating {WCC} x {WCC} interaction also contributes). Such wrong-parity admixtures reveal their presence, in the atomic cases by the dependence of the atom's index of refraction on photon helicity, and in the nuclear cases by the incomplete forbiddenness of parity-violating α-decays ($[A,Z;\ J^P=0^-,\ 1^+,2^-,\ldots]^* \to [A,Z;\ J^P=0^+] + [^4\mathrm{He};\ J^P=0^+]$) and the nonvanishing photon circular polarization and photon momentum - nuclear spin correlation in γ-decays ($[A,Z]^* \to [A,Z] + \gamma$).

In concluding this introduction we remark that the {WCC} x {WCC} interaction can produce the same effects in second order as the {WNC} x {WNC} interaction in first order, e.g., $\nu_\mu + p \to \mu^- + W^+ + p \to \nu_\mu + W^- + W^+ + p \to \nu_\mu + W^- + \Delta^{++} \to \nu_\mu + p$ in comparison with $\nu_\mu + p \to \nu_\mu + Z + p \to \nu_\mu + p$. However the corresponding second-order {WCC} x {WCC} amplitudes are very much smaller than those required to fit the experimental data so that first-order {WNC} x {WNC} amplitudes are essential for a quantitative description of the various processes in Eqs.(2) and (3).

II. THE "STANDARD" MODEL FOR THE {WNC} x {WNC} AND {WCC} x {WCC} INTERACTIONS

We proceed to describe the "standard" model for the {WCC} x {WCC} and {WCC} x {WCC} interactions (Fermi; Feynman and Gell-Mann; Cabibbo; Weinberg, Salam; Glashow, Iliopoulos, and Maiani). This model is in good agreement with all available experimental information.

The Lagrangian density of the standard model is expressed in terms of spinor fields for the participating fermions(f): leptons(ℓ) and quarks(q) - $\nu_e, \ell; \nu_\mu, \mu; \nu_\tau, \tau; u,d; c,s; t,b$ and vector fields for the participating bosons - W^+, W^-, Z, γ. We have

$$\mathcal{L} = -\frac{1}{4}\left\{(\partial_\rho W_\lambda^{(k)} - \partial_\lambda W_\rho^{(k)})(\partial_\rho W_\lambda^{(h)} - \partial_\lambda W_\rho^{(h)}) + (\partial_\rho X_\lambda - \partial_\lambda X_\rho)(\partial_\rho X_\lambda - \partial_\lambda X_\rho)\right\}$$

$$-(\bar{\nu}_e \gamma_\lambda \partial_\lambda \nu_e) - (\bar{e}\gamma_\lambda \partial_\lambda e) + ig_1 W_\lambda^{(k)} I_\lambda^{(k)}(\nu_e, e) + ig_2 X_\lambda Y_\lambda(\nu_e, e)$$

$$+\begin{pmatrix} \nu_e \to \nu_\mu \\ e \to \mu \end{pmatrix} + \begin{pmatrix} \nu_e \to \nu_\tau \\ e \to \tau \end{pmatrix}$$

$$-(\bar{u}\gamma_\lambda \partial_\lambda u) - (\bar{d}'\gamma_\lambda \partial_\lambda d') + ig_1 W_\lambda^{(k)} I_\lambda^{(k)}(u,d') + ig_2 X_\lambda Y_\lambda(u,d') +$$

$$+\begin{pmatrix} u \to c \\ d' \to s' \end{pmatrix} + \begin{pmatrix} u \to t \\ d' \to b' \end{pmatrix} \qquad (4)$$

where

$$W_\lambda^{(1)} = \frac{1}{\sqrt{2}} (W_\lambda^{(+)} + W_\lambda^{(-)}), \; W^{(2)} = \frac{i}{\sqrt{2}} (W_\lambda^{(+)} - W_\lambda^{(-)}),$$

$$W_\lambda^{(3)} = Z_\lambda \cos\theta_W + A_\lambda \sin\theta_W, \; X_\lambda = -Z_\lambda \sin\theta_W + A_\lambda \cos\theta_W$$

$$g_1 = e/\sin\theta_W = \sqrt{4\pi\alpha}/\sin\theta_W, \; g_2 = e/\cos\theta_W = \sqrt{4\pi\alpha}/\cos\theta_W, \; \alpha = 1/137$$

θ_W is the Weinberg angle ($[\sin^2\theta_W] \cong 0.20 - 0.24$-see below).

γ_λ are Dirac matrices with $\gamma_5 \equiv \gamma_1\gamma_2\gamma_3\gamma_4$

$d' \equiv d\cos\theta + s\sin\theta$, $s' \equiv -d\sin\theta\cos\theta' + s\cos\theta\cos\theta' - b\sin\theta'$,

$b' \equiv -d\sin\theta\sin\theta' + s\cos\theta\sin\theta' + b\cos\theta'$

θ and θ' are quark mixing angles — $\theta \equiv \theta_{\text{Cabibbo}} = \sin^{-1}0.22$,

$10^{-3} < \theta' \lesssim 10^{-1}$

$$I_\lambda^{(k)}(\nu_e, e) = |\bar{\nu}_{e;L}\bar{e}_L| \gamma_\lambda \frac{\tau(k)}{2} \begin{vmatrix} \nu_{e;L} \\ e_L \end{vmatrix}$$

$$I_\lambda^{(k)}(u, d') = |\bar{u}_L \bar{d}'_L| \gamma_\lambda \frac{\tau(k)}{2} \begin{vmatrix} u_L \\ d'_L \end{vmatrix}$$

$$Y_\lambda(\nu_e, e) = |\bar{\nu}_{e;L}\bar{e}_L| \gamma_\lambda \left(\frac{Q(\nu_e) + Q(e)}{2}\right) \begin{vmatrix} \nu_{e;L} \\ e_L \end{vmatrix} + Q(\nu_e)(\bar{\nu}_{e;R}\gamma_\lambda \nu_{e;R})$$

$$+ Q(e)(\bar{e}_R\gamma_\lambda e_R)$$

$$= -I_\lambda^{(3)}(\nu_e, e) + Q(\nu_e)(\bar{\nu}_e\gamma_\lambda\nu_e) + Q(e)(\bar{e}\gamma_\lambda e)$$

$$Y_\lambda(u, d') = |\bar{u}_L \bar{d}'_L| \gamma_\lambda \left(\frac{Q(u) + Q(d')}{2}\right) \begin{vmatrix} u_L \\ d'_L \end{vmatrix} + Q(u)(\bar{u}_R\gamma_\lambda u_R) +$$

$$+ Q(d')(\bar{d}'_R\gamma_\lambda d'_R) = -I^{(3)}_\lambda(u,d') + Q(u)(\bar{u}\gamma_\lambda u) + Q(d')(\bar{d}'\gamma_\lambda d')$$

$$f_{L,R} \equiv \left(\frac{(1\pm\gamma_5)}{2}\right)f, \bar{f}_{L,R} \equiv \left\{\left(\frac{1\pm\gamma_5}{2}\right)f\right\}^\dagger \gamma_4 = \bar{f}\left(\frac{1\pm\gamma_5}{2}\right)$$

with $f = \nu_e, e; \nu_\mu, \mu; \nu_\tau, \tau; u, d'; c, s'; t, b' \equiv f_1, \ldots, f_{12}$

$\tau^{(k)}$ are weak-isospin Pauli matrices: $\tau^{(1)} = \begin{vmatrix} 0 & 1 \\ 1 & 0 \end{vmatrix}, \tau^{(2)} = \begin{vmatrix} 0 & -i \\ i & 0 \end{vmatrix}, \tau^{(3)} = \begin{vmatrix} 1 & 0 \\ 0 & -1 \end{vmatrix}$

$Q(f)$ is charge of f: $Q(\nu_e)=0, Q(e)=-1, Q(u)=\frac{2}{3}, Q(d') = -\frac{1}{3}$ (5)

and where a particularly simple version of quark mixing ($b \not\rightleftarrows u$, $b \not\rightleftarrows c$, $c \not\rightleftarrows t$) has been assumed. We note that the interaction between the fermion currents and the boson fields in $\mathcal{L}$ is invariant under rotation of weak isospin - this follows since $I^{(k)}_\lambda(\nu_e,e)$, $I^{(k)}_\lambda(u,d')$, ..., and $W^{(k)}_\lambda$ are weak isovectors while $Y_\lambda(\nu_e,e)$, $Y_\lambda(u,d')$, ..., and X_λ are weak isoscalars. We further note that $\mathcal{L}$ is invariant under the gauge transformation

$$W^{(k)}_\lambda \to W^{(k)}_\lambda + \partial_\lambda \Omega^{(k)}, \quad X_\lambda \to X_\lambda + \partial_\lambda \Sigma$$

$$\begin{vmatrix} \nu_{e;L} \\ e_L \end{vmatrix} \to \exp\left(ig_1 \frac{\tau^{(k)}}{2}\Omega^{(k)} + i g_2\left(\frac{Q(\nu_e)+Q(e)}{2}\right)\Sigma\right)\begin{vmatrix} \nu_{e;L} \\ e_L \end{vmatrix}$$

$$\begin{vmatrix} u_L \\ d'_L \end{vmatrix} \to \exp\left(ig_1 \frac{\tau^{(k)}}{2}\Omega^{(k)} + i g_2\left(\frac{Q(u)+Q(d')}{2}\right)\Sigma\right)\begin{vmatrix} u_L \\ d'_L \end{vmatrix}$$

$$f_R \to \exp\left(i g_2 \, Q(f)\Sigma\right) f_R \qquad (6)$$

which gauge invariance would be destroyed if boson or fermion mass terms ($m_W^2(W^{(+)}_\lambda W^{(+)}_\lambda + W^{(+)}_\lambda W^{(+)}_\lambda)$, $m_Z^2 Z_\lambda Z_\lambda$, $m_f \bar{f}f$) were simply inserted into $\mathcal{L}$. On the other hand, if the mass terms in question are generated by the Higgs mechanism (see below) the gauge invariance is not affected - this last fact is crucial since the renormalizability of the unified theory of weak and electromagnetic interactions based on $\mathcal{L}$ depends on its gauge invariance. In addition, using

Eq.(5), we get

$$I_\lambda^{(3)}(\ell) = I_\lambda^{(3)}(\nu_e,e)+I_\lambda^{(3)}(\nu_\mu,\mu)+I_\lambda^{(3)}(\nu_\tau,\tau) = \frac{1}{2}\Big\{(\bar{\nu}_{e;L}\gamma_\lambda \nu_{e;L})-(\bar{e}_L\gamma_\lambda e_L)$$
$$+ (\bar{\nu}_{\mu;L}\gamma_\lambda\gamma_{\mu;L})-(\bar{\mu}_L\gamma_\lambda\mu_L)+(\bar{\nu}_{\tau;L}\gamma_\lambda\nu_{\tau;L})-(\bar{\tau}_L\gamma_\lambda\tau_L)\Big\}$$

$$I_\lambda^{(3)}(q) = I_\lambda^{(3)}(u,d')+I_\lambda^{(3)}(c,s')+I_\lambda^{(3)}(t,b') = I_\lambda^{(3)}(u,d)+I_\lambda^{(3)}(c,s)$$
$$+I_\lambda^{(3)}(t,b)$$
$$= \frac{1}{2}\Big\{(\bar{u}_L\gamma_\lambda u_L)-(\bar{d}_L\gamma_\lambda d_L)+(\bar{c}_L\gamma_\lambda c_L)-(\bar{s}_L\gamma_\lambda s_L)+(\bar{t}_L\gamma_\lambda t_L)-(\bar{b}_L\gamma_\lambda b_L)\Big\}$$

$$I_\lambda^{(3)}(f) = I_\lambda^{(3)}(\ell) + I_\lambda^{(3)}(q)$$

$$Y_\lambda(\ell) = Y_\lambda(\nu_e,e)+Y_\lambda(\nu_\mu,\mu)+Y_\lambda(\nu_\tau,\tau) = -I_\lambda^{(3)}(\ell)$$
$$+\Big\{Q(\nu_e)(\bar{\nu}_e\gamma_\lambda\nu_e)+(Q(e)(\bar{e}\gamma_\lambda e)$$
$$+ Q(\nu_\mu)(\bar{\nu}_\mu\gamma_\lambda\nu_\mu)+Q(\mu)(\bar{\mu}\gamma_\lambda\mu)$$
$$+ Q(\nu_\tau)(\bar{\nu}_\tau\gamma_\lambda\nu_\tau)+Q(\tau)(\bar{\tau}\gamma_\lambda\tau)\Big\}$$
$$= -I_\lambda^{(3)}(\ell) + J_\lambda^{em}(\ell)$$

$$Y_\lambda(q) = Y_\lambda(u,d')+Y_\lambda(c,s')+Y_\lambda(t,b') = Y_\lambda(u,d)+Y_\lambda(c,s)+Y_\lambda(t,b)$$
$$= -I_\lambda^{(3)}(q) + \Big\{Q(u)(\bar{u}\gamma_\lambda u)+Q(d)(\bar{d}\gamma_\lambda d)$$
$$+ Q(c)(\bar{c}\gamma_\lambda c)+Q(s)(\bar{s}\gamma_\lambda s)$$
$$+ Q(t)(\bar{t}\gamma_\lambda t)+Q(b)(\bar{b}\gamma_\lambda b)\Big\}$$
$$= -I_\lambda^{(3)}(q) + J_\lambda^{em}(q)$$

$$Y_\lambda(f) = Y_\lambda(\ell) + Y_\lambda(q) = -(I_\lambda^{(3)}(\ell)+I_\lambda^{(3)}(q))+(J_\lambda^{em}(\ell)+J_\lambda^{em}(q))$$

$$= -I_\lambda^{(3)}(f) + J_\lambda^{em}(f)$$

where $J_\lambda^{em}(\ell)$ and $J_\lambda^{em}(q)$ are, respectively, the electromagnetic currents of all the leptons and all the quarks. Eq. (7), together with Eqs.(4) and (5) shows that $\mathcal{L}$ does not possess couplings of the quark currents to the neutral-vector-boson and photon fields that are off-diagonal in quark flavor (i.e. does not possess couplings $\sim (\bar{d}_{L,R}\gamma_\lambda s_{L,R}+\bar{s}_{L,R}\gamma_\lambda d_{L,R})Z_\lambda$ or $(\bar{d}\gamma_\lambda s+\bar{s}\gamma_\lambda d)A_\lambda$, etc.) so that neither the {WNC} x {WNC} nor the {EM} interactions obtained from $\mathcal{L}$ can effect transitions between hadronic states of different strangeness, charm, bottomness, or topness.

Eqs. (4), (5), and (7) permit explicit expression of $\mathcal{L}$ in terms of $W_\lambda^\pm$, Z_λ and A_λ. We have

$$\mathcal{L} = -\frac{1}{4}\left\{(\partial_\rho W_\lambda^{(k)}-\partial_\lambda W_\rho^{(k)})(\partial_\rho W_\lambda^{(k)}-\partial_\lambda W_\rho^{(k)})+(\partial_\rho X_\lambda-\partial_\lambda X_\rho)(\partial_\rho X_\lambda-\partial_\lambda X_\rho)\right\}$$

$$-\sum_{n=1}^{n=12}(\bar{f}_n\gamma_\lambda\partial_\lambda f_n)+ig_1W_\lambda^{(k)}I_\lambda^{(k)}(f)+ig_2X_\lambda(-I_\lambda^{(3)}(f)+J_\lambda^{em}(f))$$

$$= -\frac{1}{4}\left\{(\partial_\rho W_\lambda^{(+)}-\partial_\lambda W_\rho^{(+)})(\partial_\rho W_\lambda^{(-)}-\partial_\lambda W_\rho^{(-)})+(\partial_\rho W_\lambda^{(-)}-\partial_\lambda W_\rho^{(-)})x(\partial_\rho W_\lambda^{(+)}-\partial_\lambda W_\rho^{(+)})\right.$$

$$\left.+(\partial_\rho Z_\lambda-\partial_\lambda Z_\rho)(\partial_\rho Z_\lambda-\mu_\lambda Z_\rho)+(\partial_\rho A_\lambda-\partial_\lambda A_\rho)(\partial_\rho A_\lambda-\partial_\lambda A_\rho)\right\}$$

$$-\sum_{n=1}^{n=12}(\bar{f}_n\gamma_\lambda\partial_\lambda f_n)+i\left(\frac{e}{2\sqrt{2}\sin\theta_W}\right)\left(W_\lambda^{(-)}J_\lambda^{(-)}(f) + W_\lambda^{(+)}J_\lambda^{(+)}(f)\right)$$

$$-i\left(\frac{e}{2\sin\theta_W\cos\theta_W}\right)Z_\lambda J_\lambda^{(0)}(f) + i\,e\,A_\lambda J_\lambda^{em}(f)$$

where

$$J_\lambda^{(-)}(f)=2\left(I_\lambda^{(1)}(f)-iI_\lambda^{(2)}(f)\right)$$

$$=\{(\bar{e}\Gamma_\lambda\nu_e)+(\bar{\mu}\Gamma_\lambda\nu_\mu)+(\bar{\tau}\Gamma_\lambda\nu_\tau)+(\bar{d}'\Gamma_\lambda u)+(\bar{s}'\Gamma_\lambda c)+(\bar{b}'\Gamma_\lambda t)$$

$$J_\lambda^{(+)}(f)=2\left(I_\lambda^{(1)}(f)+iI_\lambda^{(2)}(f)\right)$$

$$=\{(\bar{\nu}_e\Gamma_\lambda e)+(\bar{\nu}_\mu\Gamma_\lambda\mu)+(\bar{\nu}_\tau\Gamma_\lambda\tau)+(\bar{u}\Gamma_\lambda d')+(\bar{c}\Gamma_\lambda s')+(\bar{t}\Gamma_\lambda b')\}$$

$$J_\lambda^{(0)}(f)=2\left(I_\lambda^{(3)}(f)-\sin^2\theta_W J_\lambda^{em}(f)\right)=\frac{1}{2}\Big\{(\bar{\nu}_e\Gamma_\lambda\nu_e)-(\bar{e}\Gamma_\lambda e)+(\bar{\nu}_\mu\Gamma_\lambda\nu_\mu)-(\bar{\mu}\Gamma_\lambda\mu)$$

$$+(\bar{\nu}_\tau\Gamma_\lambda\nu_\tau)-(\bar{\tau}\Gamma_\lambda\tau)+(\bar{u}\Gamma_\lambda u)-(\bar{d}\Gamma_\lambda d)+(\bar{c}\Gamma_\lambda c)-(\bar{s}\Gamma_\lambda s)+(\bar{t}\Gamma_\lambda t)-(\bar{b}\Gamma_\lambda b)\Big\}$$

$$-2\sin^2\theta_W J_\lambda^{em}(f)$$

$$J_\lambda^{em}(f)=\Big\{Q(\nu_e)(\bar{\nu}_e\gamma_\lambda\nu_e)+Q(e)(\bar{e}\gamma_\lambda e)+Q(\nu_\mu)(\bar{\nu}_\mu\gamma_\lambda\nu_\mu)$$

$$+Q(\mu)(\bar{\mu}\gamma_\lambda\mu)+Q(\nu_\tau)(\bar{\nu}_\tau\gamma_\lambda\nu_\tau)+Q(\tau)(\bar{\tau}\gamma_\lambda\tau)$$

$$+Q(u)(\bar{u}\gamma_\lambda u)+Q(d)(\bar{d}\gamma_\lambda d)+Q(c)(\bar{c}\gamma_\lambda c)+Q(s)(\bar{s}\gamma_\lambda s)$$

$$+Q(t)(\bar{t}\gamma_\lambda t)+Q(b)(\bar{b}\gamma_\lambda t)$$

$$\Gamma_\lambda\equiv\gamma_\lambda(1+\gamma_5)$$

$$d'=d\cos\theta+s\sin\theta,\ s'=-d\sin\theta\cos\theta'+s\cos\theta\cos\theta'-b\sin\theta',$$
$$b'=-d\sin\theta\sin\theta'+s\cos\theta\sin\theta'+b\cos\theta'$$

$$Q(\nu_e)=Q(\nu_\mu)=Q(\nu_\tau)=0,\quad Q(e)=Q(\mu)=Q(\tau)=-1;$$
$$Q(u)=Q(c)=Q(t)=2/3,\ Q(d)=Q(s)=Q(b)=-1/3. \qquad (9)$$

We emphasize that the {EM}-interaction term in the $\mathcal{L}$ of eqs.(8) and (9), i.e. $A_\lambda J_\lambda^{em}(f)$, is of the form required by the gauge invariance of QED, i.e. the $\mathcal{L}$ of Eqs.(8) and (9) is invariant under $A_\lambda \to A_\lambda + \partial_\lambda \chi$, $f \to \exp(i\,e\,Q(f)\chi)f$. Clearly, this last gauge invariance is achieved for a given Θ_W, i.e. for a given relation between $W_\lambda^{(3)}$, X_λ and Z_λ, A_λ, by a proper choice in terms of Θ_W of the weak isovector and weak isoscalar coupling constants in Eq.(8) ($g_1 = e/\sin\Theta_W$ and $g_2 = e/\cos\Theta_W$).

We next consider the Higgs mechanism for the generation of boson and fermion masses. Introducing a weak isodoublet of scalar-fields $\begin{vmatrix}\phi^{(0)}\\ \phi^{(-)}\end{vmatrix}$ (the so-called Higgs' fields) we postulate an additional term in $\mathcal{L}$ which is invariant under rotation of weak isospin and under the gauge transformation of Eq.(6) (with $\begin{vmatrix}\phi^{(0)}\\ \phi^{(-)}\end{vmatrix}$ transforming like $\begin{vmatrix}\nu_{e;L}\\ e_L\end{vmatrix}$), viz.

$$\mathcal{L}(H) = -\left\{\left(-\partial_\lambda + i\,g_1 W_\lambda^{(k)}\frac{\tau^{(k)}}{2} + i\,g_2 X_\lambda\left(\frac{Q(H^o)+Q(H^-)}{2}\right)\right)\begin{vmatrix}\phi^{(0)}\\ \phi^{(-)}\end{vmatrix}\right\}^\dagger$$

$$\times\left\{\left(-\partial_\lambda + i g_1 W_\lambda^{(k)}\frac{\tau^{(k)}}{2} + i g_2 X_\lambda\left(\frac{Q(H^o)+Q(H^-)}{2}\right)\right)\begin{vmatrix}\phi^{(0)}\\ \phi^{(-)}\end{vmatrix}\right\}$$

$$-\left\{g_{\nu_e H}\left(|\bar{\nu}_{e;L}\bar{e}_L|\begin{vmatrix}\phi^{(0)}\\ \phi^{(-)}\end{vmatrix}\nu_{e;R}\right) - g_{eH}\left(|\bar{\nu}_{e;L}\bar{e}_L|\begin{vmatrix}\phi^{(+)}\\ \phi^{(0)}\end{vmatrix}e_R\right) + \text{herm.conj.}\right\}$$

$$-\left\{\begin{pmatrix}\nu_e \to \nu_\mu\\ e \to \mu\end{pmatrix}\right\} - \left\{\begin{pmatrix}\nu_e \to \nu_\tau\\ e \to \tau\end{pmatrix}\right\}$$

$$-\left\{g_{uH}\left(|\bar{u}_L\bar{d}_L|\begin{vmatrix}\phi^{(0)}\\ \phi^{(-)}\end{vmatrix}u_R\right) - g_{dH}\left(|\bar{u}_L d_L|\begin{vmatrix}\phi^{(+)}\\ \phi^{(0)}\end{vmatrix}d_R\right) + \text{herm. conj.}\right\} -$$

$$-\left\{\begin{pmatrix} u\to c \\ d\to s \end{pmatrix}\right\} - \left\{\begin{pmatrix} u\to t \\ d\to b \end{pmatrix}\right\} - \frac{\kappa^2}{4}\left(\left|\begin{matrix}\phi^{(0)} \\ \phi^{(-)}\end{matrix}\right|^{\dagger}\left|\begin{matrix}\phi^{(0)} \\ \phi^{(-)}\end{matrix}\right| - \frac{v^2}{2}\right)^2$$

$$\left|\begin{matrix}\phi^{(-)} \\ -\phi^{(0)}\end{matrix}\right| = C \exp\left(\frac{i\pi}{2}\tau^{(2)}\right)\left|\begin{matrix}\phi^{(0)} \\ \phi^{(-)}\end{matrix}\right|$$

$$Q(H^o) = 0, \; Q(H^-) = -1 \tag{10}$$

where κ and v are constants; the significance of v is that the value of $\left|\begin{matrix}\phi^{(0)} \\ \phi^{(-)}\end{matrix}\right|$ considered as a classical field which minimizes. The last term in $\mathcal{L}(H)$ is $\left|\begin{matrix}\phi^{(0)} \\ \phi^{(-)}\end{matrix}\right| = +v/\sqrt{2}$ or $-v/\sqrt{2}$. Thus, the vacuum expectation value of the quantum field $\phi^{(0)}$ is $+v/\sqrt{2}$ or $-v/\sqrt{2}$ and choosing one of these ("spontaneous" symmetry breaking) we can write

$$\phi^{(0)} = \frac{v}{\sqrt{2}} + \Phi^{(0)} \tag{11}$$

where $\Phi^{(0)}$ is a quantum field with vacuum expectation value equal to zero which destroys and creates H^{o} (analogous to the destruction of H^- and creation of H^+ by $\phi^{(-)}$ and the destruction of H^+ and creation of H^- by $\phi^{(+)}$). Substitution of Eq.(11) into Eq.(10) and use of Eq.(5) yields

$$\mathcal{L}(H) = -\frac{1}{2}\left(\frac{v^2}{4} g_1^2\right)\left(W_\lambda^{(1)}W_\lambda^{(1)} + W_\lambda^{(2)}W_\lambda^{(2)} + \left(W_\lambda^{(3)} - \frac{g_2}{g_1} X_\lambda\right)^2\right)$$

$$-\sum_{n=1}^{n=12}\left(\frac{v}{\sqrt{2}} g_{f_n H}\right)\bar{f}_n f_n - \frac{1}{2}(\kappa^2 v^2)\Phi^{(0)}\Phi^{(0)} + \mathcal{L}'(H)$$

$$= -\frac{1}{2}\left(\frac{v^2 e^2}{4\sin^2\theta_W}\right)\left(W_\lambda^{(+)}W_\lambda^{(-)} + W_\lambda^{(-)}W_\lambda^{(+)}\right) - \frac{1}{2}\left(\frac{v^2 e^2}{4\sin^2\theta_W \cos^2\theta_W}\right) Z_\lambda Z_\lambda$$

$$-\sum_{n=1}^{n=12}\left(\frac{v}{\sqrt{2}} g_{f_n H}\right)\bar{f}_n f_n - \frac{1}{2}(\kappa^2 v^2)\Phi^{(0)}\Phi^{(0)} + \mathcal{L}'(H)$$

where $\mathcal{L}'(H)$ consists of terms which couple the scalar-boson fields $\Phi^{(0)}$ and $\phi^{(\pm)}$ to the vector-boson fields $W_\lambda^{(k)}, X_\lambda$ (or $W_\lambda^{(\pm)}$, Z_λ, A_λ)

and of terms cubic and quartic in $\Phi^{(0)}$ and $\phi^{(\pm)}$. Then, since any boson and fermion mass terms in $\mathcal{L}$ have to be of the form

$$-\frac{1}{2}m_W^2\left(W_\lambda^{(+)}W_\lambda^{(-)}+W_\lambda^{(-)}W_\lambda^{(+)}\right)-\frac{1}{2}m_Z^2 Z_\lambda Z_\lambda-\frac{1}{2}m_\gamma^2 A_\lambda A_\lambda$$

$$-\sum_{n=1}^{n=12} m_{f_n}\bar{f}_n f_n-\frac{1}{2}m_{H^o}^2\Phi^{(0)}\Phi^{(0)}$$

we see that the presence of $\mathcal{L}(H)$ in $\mathcal{L}$ introduces mass terms with masses given by

$$m_W=v\left(\frac{e}{2\sin\theta_W}\right),\ m_Z=v\left(\frac{e}{2\sin\theta_W}\right)\left(\frac{1}{\cos\theta_W}\right)=\left(\frac{1}{\cos\theta_W}\right)m_W,\ m_\gamma=0,$$

$$m_f=v\left(\frac{g_{f_nH}}{\sqrt{2}}\right)=\left(\frac{\sqrt{2}\sin\theta_W g_{f_nH}}{e}\right)m_W,\ m_{H^o}=v\kappa=\left(\frac{2\sin\theta_W\kappa}{e}\right)m_W;$$

$$e=\sqrt{4\pi\alpha},\ m_{\nu_e}<10^{-4}m_e,\ m_{\nu_\mu}<10^{-4}m_e,\ m_\mu=207\ m_e,\ \text{etc.} \qquad (14)$$

We emphasize that the fact that the photon remains massless after the spontaneous symmetry breaking of Eq.(11) (which endows $W^\pm$ and Z with mass) ensures that $\mathcal{L}$ as described by Eqs.(8)-(14) respects the gauge invariance of QED. We also note, in view of Eq.(14), that a determination of θ_W and m_W (see below) will permit immediate calculation of m_Z and g_{f_nH}.

Cut will leave m_{H^o} unspecified since nothing can be said at this stage regarding the numerical value of κ/e. However, m_{H^o} is believed to be much smaller than m_W($m_W \cong 80$ GeV - see below) but greater than a few GeV. Whatever its mass, H^o is coupled most strongly to the heaviest leptons and quarks (Eq. (14)) so that if, e.g., $m_T>m_{H^o}$($T=[b\bar{b}]$ with $m_T\cong 10$ GeV) then a decay mode of T is $T=[b\bar{b}]\to b+\bar{b}+\gamma$, $b+\bar{b}+H^o\to\gamma+H^o$; $H^o\to\tau^++\tau^-$, $c+\bar{c}$ with $\Gamma(T\to\gamma+H^{(0)})/\Gamma(T\to\mu^++\mu^-)\approx e^2g^2_{f_bH}/e^4=\left(\frac{m_b}{m_W}\right)^2/2\sin^2\theta_W\approx 10^{-2}$ which is not impossibly small. As a general feature, the $\mathcal{L}$ of Eqs.(8)-(14) is completely determined if α and m_{f_n}, m_W, m_Z, and m_{H^o} are given; this is obvious from an examination of the equations in question and the recollection that the quark mixing angles are presumably specified in terms of the ratios of the various quark masses.

We conclude this account of the standard model for the {WNC} x {WNC} and {WCC} x {WCC} **interactions by recording the** {WNC} x {WNC} and {WCC} x {WCC} interaction Hamiltonian density corresponding to the Lagrangian density of Eqs.(8)-(14), viz.

$$\mathcal{H} = \left(\frac{4\pi\alpha/8}{\sin^2\theta_W m_W^2}\right)\frac{1}{2}\left\{J_\lambda^{(-)}(f)\,\frac{\delta_{\lambda\rho}-\partial_\lambda\partial_\rho/m_W^2}{1-\partial\bullet\partial/m_W^2}\,J_\rho^{(+)}(f)\right.$$

$$\left.+\,J_\lambda^{(+)}(f)\,\frac{\delta_{\lambda\rho}-\partial_\lambda\partial_\rho/m_W^2}{1-\partial\bullet\partial/m_W^2}\,J_\rho^{(-)}(f)\right\}$$

$$+\left(\frac{4\pi\alpha/4}{\sin^2\theta_W\cos^2\theta_W m_Z^2}\right)\frac{1}{2}\left\{J_\lambda^{(0)}(f)\,\frac{\delta_{\lambda\rho}-\partial_\lambda\partial_\rho/m_W^2}{1-\partial\bullet\partial/m_W^2}\,J_\lambda^{(0)}(f)\right\} \tag{15}$$

where the weak charged currents $J_\lambda^{(\pm)}(f)$ and the weak neutral current $J_\lambda^{(0)}(f)$ **are given in** Eq.(9). Introducing the universal weak-interaction coupling constant $G/\sqrt{2}$ (known from the analysis of the observed rates of processes caused by the {WCC} x {WCC} interaction, e.g.,($\mu^+ \to e^+ + \nu_e + \bar{\nu}_\mu$) we have the identification

$$\frac{G}{\sqrt{2}} = \frac{4\pi\alpha/8}{\sin^2\theta_W m_W^2}\,, \quad G = 1.026 \times 10^{-5}/m_{\text{proton}}^2. \tag{16}$$

Thus, remembering also Eq.(14) for the relation between m_W and m_Z, Eq.(15) becomes

$$\mathcal{H} \cong \frac{G}{\sqrt{2}}\left\{\frac{J_\lambda^{(-)}(f)J_\lambda^{(+)}(f)+J_\lambda^{(+)}(f)J_\lambda^{(-)}(f)}{2} + J_\lambda^{(0)}(f)\,J_\lambda^{(0)}(f)\right\} \tag{17}$$

in the approximation of $|q^2| \ll m_W^2$ where q^2 is the squared four-momentum transfer for the process under consideration.

III: Applications of the Standard Model for the {WNC} x {WNC} and {WCC} x {WCC} Interactions to Various Nuclear Processes

A) Scattering of Neutrinos and Antineutrinos off Nuclei

We first treat elastic scattering of neutrinos and antineutrinos off protons

$$\nu_\mu(\bar{\nu}_\mu) + p \rightarrow \nu_\mu(\bar{\nu}_\mu) + p. \tag{18}$$

Here, the relevant part of the $\mathcal{H}$ of Eqs.(17) and (9) is

$$\mathcal{H}_{\nu,q} = \frac{G}{\sqrt{2}}\, 2\left(\frac{1}{2}\, \bar{\nu}_\mu \Gamma_\lambda \nu_\mu\right) J_\lambda^{(0)}(q)$$

$$J_\lambda^{(0)}(q) = 2\left(I_\lambda^{(3)}(q) = \sin^2\theta_W J_\lambda^{em}(q)\right) = \mathcal{J}_\lambda^{(3)} + \mathcal{J}_{5;\lambda}^{(3)} + \mathcal{J}_\lambda^{(0)} + \mathcal{J}_{5;\lambda}^{(0)} - 2\sin^2\theta_W J_\lambda^{em}(q)$$

$$\mathcal{J}_\lambda^{(3)} = \frac{1}{2}\left\{(\bar{u}\gamma_\lambda u) - (\bar{d}\gamma_\lambda d)\right\}$$

$$\mathcal{J}_{5;\lambda}^{(3)} = \frac{1}{2}\left\{(\bar{u}\gamma_\lambda\gamma_5 u) - (\bar{d}\gamma_\lambda\gamma_5 d)\right\}$$

$$\mathcal{J}_\lambda^{(0)} = \frac{1}{2}\left\{(\bar{c}\gamma_\lambda c) - (\bar{s}\gamma_\lambda s) + (\bar{t}\gamma_\lambda t) - (\bar{b}\gamma_\lambda b)\right\}$$

$$\mathcal{J}_{5;\lambda}^{(0)} = \frac{1}{2}\left\{(\bar{c}\gamma_\lambda\gamma_5 c) - (\bar{s}\gamma_\lambda\gamma_5 s) + (\bar{t}\gamma_\lambda\gamma_5 t) - (\bar{b}\gamma_\lambda\gamma_5 b)\right\}$$

$$J_\lambda^{em}(q) = \frac{2}{3}(\bar{u}\gamma_\lambda u) - \frac{1}{3}(\bar{d}\gamma_\lambda d) + \frac{2}{3}(\bar{c}\gamma_\lambda c) - \frac{1}{3}(\bar{s}\gamma_\lambda s) + \frac{2}{3}(\bar{t}\gamma_\lambda t) - \frac{1}{3}(\bar{b}\gamma_\lambda b) \tag{19}$$

where, as indicated by the notation, $\mathcal{J}_\lambda^{(3)}$, $\mathcal{J}_{5,\lambda}^{(3)}$, $\mathcal{J}_\lambda^{(0)}$, and $\mathcal{J}_{5,\lambda}^{(0)}$ are, respectively, a polar strong isovector, an axial strong isovector, a polar strong isoscalar, and an axial strong isoscalar. Matrix elements of $\mathcal{J}_\lambda^{(0)}$ and $\mathcal{J}_{5;\lambda}^{(0)}$ between nucleon (nuclear) states are small because nucleons (nuclei) are very largely composed of u and d quarks. We therefore consistently neglect such matrix elements though it would, of course, be of considerable interest to obtain reliable quantitative information regarding the form factors which characterize, e.g.,

$$<p,\vec{p}_f|\,\mathcal{J}_\lambda^{(0)}\,|p,\vec{p}_i> \qquad \text{and} \qquad <p;\vec{p}_f|\,\mathcal{J}_{5,\lambda}^{(0)}|p,\vec{p}_i>$$

It remains to calculate $<p,\vec{p}_f|\,\mathcal{J}_\lambda^{(3)}\,|p,\vec{p}_i>$, $<p,\vec{p}_f|\,\mathcal{J}_{5;\lambda}^{(3)}|p,\vec{p}_i>$ and $<p,\vec{p}_f|J_\lambda^{em}(q)|p,\vec{p}_i>$. First of all, $<p,\vec{p}_f|J_\lambda^{em}(q)|p,\vec{p}_i>$ is characterized by form factors known from the QED analysis of experimental data on $e^- + p \to e^- + p$. Further, since $J_\lambda^{em}(q) = \mathcal{J}_\lambda^{(3)} + \frac{1}{6}\{(\bar{u}\gamma_\lambda u) + (\bar{d}\gamma_\lambda d)\} + \frac{2}{3}\{(\bar{c}\gamma_\lambda c) + (\bar{t}\gamma_\lambda t)\} - \frac{1}{3}\{(\bar{s}\gamma_\lambda s) + (\bar{b}\gamma_\lambda b)\} = \mathcal{J}_\lambda^{(3)} + [J_\lambda^{em}]_{isosc.}$,

$$<p,\vec{p}_f|\,\mathcal{J}_\lambda^{(3)}\,|p,\vec{p}_i> = \frac{1}{2}\,\{<p,\vec{p}_f|J_\lambda^{em}(q)|p,\vec{p}_i> - <n,\vec{p}_f|J_\lambda^{em}(q)|n,\vec{p}_i>\} \tag{20}$$

with $<n,\vec{p}_f|J_\lambda^{em}(q)|n,\vec{p}_i>$ characterized by form factors known from the QED analysis of experimental data on $e^- + n \to e^- + n$ (i.e., $e^- + d \to e^- + n + p_{spectator}$). Finally

$$<p,\vec{p}_f|\,\mathcal{J}_{5;\lambda}^{(3)}|p,\vec{p}_i> = \frac{1}{2}\,<n,\vec{p}_f|\,\mathcal{J}_{5;\lambda}^{(-)}|p,\vec{p}_i>$$
$$\mathcal{J}_{5;\lambda}^{(-)} = (\bar{d}\gamma_\lambda\gamma_5 u) \tag{21}$$

with the form factors characterizing $<n,\vec{p}_f|\,\mathcal{J}_{5;\lambda}^{(-)}|p,\vec{p}_i>$ known from the analysis of experimental data on $\mu^- + p \to \nu_\mu + n$ and $\nu_\mu + n \to \mu^- + p$ (i.e., $\nu_\mu + d \to \mu^- + p + p_{spectator}$) using the $\mathcal{H}$ of Eqs. (17), (16) and (9). Thus all the form factors characterizing

$<p;\vec{p}_f|J_\lambda^{(0)}(q)|p;\vec{p}_i>$ are known and the differential cross sections $\frac{d\sigma}{dq^2}(\nu_\mu(\bar{\nu}_\mu) + p \to \nu_\mu(\bar{\nu}_\mu) + p)$ can be calculated in terms of the parameter $\sin^2\theta_W$ entering into the $\mathcal{H}_{\nu,q}$ of Eq. (19) - the best fit to the experimental data (obtained at $E_\nu(E_{\bar{\nu}}) \cong 0.5 - 2.0$ GeV, $0.40\ (\text{GeV/c})^2 < q^2 < 0.90\ (\text{GeV/c})^2$, q^2/q_{m_p} = kinetic energy of recoil proton) corresponds to

$$\sin^2\theta_W = 0.22 \pm 0.04 \tag{22}$$

This value is consistent with that obtained from the comparison of experimental data for $\sigma(\nu_\mu(\bar{\nu}_\mu) + p(n) \to \nu_\mu(\bar{\nu}_\mu) + \text{hadrons})$ and $\sigma(\nu_\mu(\bar{\nu}_\mu) + p(n) \to \nu_\mu(\bar{\nu}_\mu) + \pi^\pm + \text{hadrons})$ with theoretical expressions calculated by means of the quark-parton model. Eqs. (22), (16), and (14) predict

$$m_W = \left[\frac{4\pi\alpha/8}{G\sqrt{2}}\right]^{1/2} / \sin\theta_W = 79.5 \text{ GeV} \quad ,$$

$$m_Z = m_W/\cos\theta_W = 90.0 \text{ GeV} \tag{23}$$

It is clear that a complete justification of the $\mathcal{L}$ of Eqs. (8) -(14) awaits the experimental confirmation of these predictions and the observation of $H^{(0)}$.

We next consider the elastic scattering of neutrinos and antineutrinos from nuclei [A,Z] with A > 1; for the sake of simplicity we confine the discussion to spinless nuclei such as ^{12}C or ^{208}Pb. Observation of this elastic scattering will, unfortunately, be extraordinarily difficult since it will be necessary to detect an unexcited recoil nucleus, accompanied by no other particles, and with a kinetic energy = $q^2/2Am_p$.

For the calculation of the differential cross sections $\frac{d\sigma}{dq^2}(\nu_\mu(\bar{\nu}_\mu) + [A,Z] \to \nu_\mu(\bar{\nu}_\mu) + [A,Z])$ we again use the $\mathcal{H}$ of Eq. (19), neglect $<[A,Z;0^+];\vec{p}_f| \mathcal{J}_\lambda^{(0)}|[A,Z;0^+];\vec{p}_i>$, note that $<[A,Z;0^+];\vec{p}_f| \mathcal{J}_{5;\lambda}^{(3)} + \mathcal{J}_{5;\lambda}^{(0)}|[A,Z;0^+];\vec{p}_i>$ vanishes on the basis of the rotational and inversion symmetry of $|[A,Z;0^+];\vec{p}_i(\vec{p}_f)>$, and parametrize $<[A,Z;0^+];\vec{p}_f| \mathcal{J}_\lambda^{(3)} - 2\sin^2\theta_W J_\lambda^e(q)|[A,Z;0^+];\vec{p}_i>$ as

$$<[A,Z;0^+];\vec{p}_f| \mathcal{J}_\lambda^{(3)} - 2\sin^2\theta_W J_\lambda^{em}(q)|[A,Z;0^+];\vec{p}_i>$$

$$= \frac{i^{-1}(p_f+p_i)_\lambda}{2Am_p}[\tfrac{1}{2}(Z-N)F^{isov}(q^2) - 2\sin^2\theta_W ZF^{em}(q^2)] \qquad (24)$$

$$q^2 = (p_i-p_f)^2 = 2Am_p(K.E)_{nucleus} = 2E_\nu^2(1-\cos\theta_{\nu_f,\nu_i}),N=A-Z$$

where the isovector form factor, and the electromagnetic (electric charge) form factor are each normalized to unity at $q^2 = 0$ and where the parametrization is consistent with $\partial_\lambda J_\lambda^{em}(q) = 0$ and $\partial_\lambda \mathcal{J}_\lambda^{(3)} = 0$. Eqs. (24) and (19) yield

$$\frac{d\sigma}{dq^2}(\nu_\mu(\bar{\nu}_\mu)+[A,Z] \to \nu_\mu(\bar{\nu}_\mu)+[A,Z])$$

$$= \frac{G^2}{2\pi}\left(1 - \frac{q^2}{4E_\nu^2}\right)[\tfrac{1}{2}(Z-N)F^{isov}(q^2) - 2\sin^2\theta_W ZF^{em}(q^2)]^2$$

$$\cong \frac{G^2}{2\pi}\left(1 - \frac{q^2}{4E_\nu^2}\right)[\tfrac{1}{2}(Z-N)-2\sin^2\theta_W Z]^2 (F^{em}(q^2))^2 \qquad (25)$$

and so provide a model-independent determination of $\sin^2\theta_W$ since $F^{em}(q^2)$ can be found from a QED analysis of experimental data on $e^- + [A,Z] \to e^- + [A,Z]$. As a numerical example

$$\sigma(\nu_\mu + {}^{12}C \to \nu_\mu + {}^{12}C) \cong \frac{G^2}{\pi} E_\nu^2[12\sin^2\theta_W]^2 = 0.9 \times 10^{-40}\ \mathrm{cm}^2$$

$$\langle(\mathrm{K.E.})_{\mathrm{nucleus}}\rangle = 2E_\nu^2/2.12\, m_p = 75\ \mathrm{keV} \qquad (26)$$

for $E_\nu = 30$ MeV (ν_μ from stopped $\pi^+ \to \mu^+ + \nu_\mu$) and $\sin^2\theta_W = 0.22$. $\sigma(\nu_\mu + {}^{12}C \to \nu_\mu + {}^{12}C)$ initially increases with E_ν as E_ν^2 and eventually becomes constant for $E_\nu \gtrsim 3$ [radius of ${}^{12}C]^{-1} \cong 3(12^{1/3}/m_\pi)^{-1} = 180$ MeV. On the other hand, this increase of $\sigma(\nu_\mu + {}^{12}C \to \nu_\mu + {}^{12}C)$ with E_ν is accompanied by an increase in the cross sections for the various inelastic processes with E_ν (e.g., $\sigma(\nu_\mu + {}^{12}C \to \nu_\mu + {}^{11}C + n)$ and the optimum signal-to-noise ratio may actually be obtained at a fairly small E_ν).

Finally, we consider inelastic scattering of neutrinos and antineutrinos from nuclei and, as a typical example, treat

$$\nu_\mu(\bar{\nu}_\mu) + {}^{12}C(0^+;I=0,E^*=0) \to \nu_\mu(\bar{\nu}_\mu) + {}^{12}C^*(1^+;I=1,E^*=15.1\ \mathrm{MeV})$$

$${}^{12}C^*(1^+;I=1,E^*=15.1\ \mathrm{MeV}) \to {}^{12}C(0^+;I=0,E^*=0) + \gamma \qquad (27)$$

where E* is the nuclear excitation energy and where the 15.1 MeV γ would be detected. In this case, the $\mathcal{H}$ of Eq. (19) requires the calculation of

$$\langle {}^{12}C^*(1^+;I=1);\vec{p}_f|J_\lambda^{(0)}(q)|{}^{12}C(0^+;I=0);\vec{p}_i\rangle = \langle {}^{12}C^*(1^+;I=1);\vec{p}_f|\mathcal{J}_\lambda^{(3)}$$

$$+ \mathcal{J}_{5;\lambda}^{(3)} - 2\sin^2\theta_W J_\lambda^{em}(q)|{}^{12}C(0^+;I=0);\vec{p}_i\rangle = \langle {}^{12}C^*(1^+;I=1);\vec{p}_f|\mathcal{J}_{5;\lambda}^{(3)}$$

$$+ (1-2\sin^2\theta_W)J_\lambda^{em}(q)|{}^{12}C(0^+;I=0);\vec{p}_i\rangle \qquad (28)$$

with the second equality a consequence of the fact that $\langle I=1|J_\lambda^{em}(q) - \mathcal{J}_\lambda^{(3)}|I=0\rangle = \langle I=1|[J_\lambda^{em}(q)]_{\mathrm{isosc}}|I=0\rangle = 0$.

Further, the form factors characterizing $\langle {}^{12}C^*(1^+;I=1);\vec{p}_f|J_\lambda^{em}(q)|{}^{12}C(0^+;I=0);\vec{p}_i\rangle$ are known from the QED analysis of experimental data on $e^- + {}^{12}C(0^+;I=0,E^*=0) \to e^- + {}^{12}C^*(1^+;I=1,E^*=15.1\ \mathrm{MeV})$ and ${}^{12}C^*(1^+;I=1,E^*=15.1\ \mathrm{MeV}) \to {}^{12}C(0^+;I=0,E^*=0) + \gamma$ while

$$<^{12}C^*(1^+;I=1);\vec{p}_f|\mathcal{J}^{(3)}_{5;\lambda}|^{12}C(0^+;I=0);\vec{p}_i>$$

$$=\frac{1}{\sqrt{2}}<^{12}B(1^+;I=1);\vec{p}_f|\mathcal{J}^{(-)}_{5;\lambda}|^{12}C(0^+;I=0);\vec{p}_i>$$

with the form factors characterizing $<^{12}B(1^+;I=1);\vec{p}_f|\mathcal{J}^{(-)}_{5;\lambda}|^{12}C(0^+;I=0);\vec{p}_i>$ known from the analysis of experimental data on $\mu^- + {}^{12}C \rightarrow \nu_\mu + {}^{12}B$ and ${}^{12}B \rightarrow {}^{12}C + e^- + \nu_e$ using the $\mathcal{H}$ of Eqs. (17), (16), and (9). Taking, for the sake of a numerical example, again the case of relatively small E_ν, i.e., $E_\nu << m_p$, we have

$$|<^{12}C^*(1^+;I=1);\vec{p}_f|J^{em}_\lambda(q)|^{12}C(0^+;I=0);\vec{p}_i>|$$

$$\cong\left|\frac{1}{\sqrt{2}}\varepsilon_{\lambda\rho\kappa\sigma}\xi_\rho\frac{q_\kappa}{2mp}\frac{Q_\sigma}{2.12\cdot m_p}F^{(M)}(q^2=0)\right| << 1$$

$$<^{12}C^*(1^+;I=1);\vec{p}_f|\mathcal{J}^{(3)}_{5;\lambda}|^{12}C(0^+;I=0);\vec{p}_i> \simeq \frac{1}{\sqrt{2}}\xi_\lambda F^{(A)}(q^2=0)$$

$$F^{(A)}(q^2=0) = 0.72\ F^{(M)}(q^2=0) = 2.79 \tag{30}$$

where ξ_λ is the polarization four-vector of the spin-1 $^{12}C^*$, $q^2=2E_\nu(E_\nu=15.1\text{ MeV})(1-\cos\theta_{\nu_f,\nu_i})$, and $Q_\sigma=(p_f+p_i)_\sigma$. Combination of Eqs. (28) and (30) yields

$$\sigma(\nu_\mu(\bar{\nu}_\mu) + {}^{12}C(0^+;I=0,E^*=0) \rightarrow \nu_\mu(\bar{\nu}_\mu) + {}^{12}C^*(1^+;I=1,E^*=15.1\text{ MeV}))$$

$$\simeq\frac{G^2}{\pi}3(E_\nu-15.1\text{ MeV})^2\left[\frac{F^{(A)}(q^2=0)}{\sqrt{2}}\right]^2 = 2.5\times10^{-42}\text{ cm}^2 \tag{31}$$

for E_ν=30 MeV and $F^{(A)}(q^2=0) = 0.70$. In a behaviour analogous to that of the cross section in Eq. (26), this cross section initially increases with E_ν as $(E_\nu-15.1\text{ MeV})$ and eventually becomes constant for $E_\nu \gtrsim 3[{}^{12}C\leftrightarrow{}^{12}C^*\text{ transition radius}]^{-1} \simeq 150$ MeV.

As a last example we treat the inelastic scattering process

$$\nu_\mu(\bar{\nu}_\mu) + {}^{12}C(0^+;I=0,E^*=0) \rightarrow \nu_\mu(\bar{\nu}_\mu) + {}^{12}C^*(1^+;I=0,E^*=12.7\text{ MeV})$$

$$^{12}C^*(1^+;I=0,E^*=12.7\text{ MeV}) \rightarrow {}^8Be^* + {}^4He \rightarrow He^4 + He^4 + He^4 \tag{32}$$

Here

$$<^{12}C^*(1^+;I=0);\vec{p}_f|J_\lambda^{(0)}(q)|^{12}C(0^+;I=0;\vec{p}_i> =$$

$$<^{12}C^*(1^+;I=0)\vec{p}_f|\ \mathfrak{J}_\lambda^{(0)} + \mathfrak{J}_{5;\lambda}^{(0)} + [J_\lambda^{em}(q)]_{isosc.}|^{12}C(0^+;I=0)\vec{p}_i>$$

$$\cong <^{12}C^*(1^+;I=0);\vec{p}_f|[J_\lambda^{em}(q)]_{isosc.}|^{12}C(0^+;I=0);\vec{p}_i>$$

$$= \varepsilon_{\lambda\rho\kappa\sigma}\ \xi_\rho\ \frac{q_\kappa}{2m_p}\ \frac{Q_\sigma}{2\cdot 12\cdot m_p}\ G^{(M)}(q^2) \tag{33}$$

with the isoscalar magnetic form factor $G^{(M)}(q^2)$ in this equation estimated to be about 10 times smaller than the isovector magnetic form factor $(F^{(M)}(q^2)/\sqrt{2})$ in Eq. (30). As a result $\sigma(\nu_\mu(\overline{\nu}_\mu) + {}^{12}C(0^+;I=0,E^*=0) \rightarrow \nu_\mu(\overline{\nu}_\mu) + {}^{12}C^*(1^+;I=0,E^* = 12.7\text{ MeV})) \ll \sigma(\nu_\mu(\overline{\nu}_\mu) + {}^{12}C(0^+;I=0,E^*=0) \rightarrow \nu_\mu(\overline{\nu}_\mu) + {}^{12}C^*(1^+;I=1,E^* = 15.1\text{ MeV}))$ not only for E_ν = 30 MeV but for all higher E_ν as well. Thus, if a serious violation of this very strong inequality were found experimentally, it could hardly be explained on the basis of an I = 1 admixture into $|^{12}C^*(1^+;I=0,E^*=12.7\text{ MeV}>$ and would instead imply either a much larger magnitude for $<^{12}C^*(1^+;I=0);\vec{p}_f|\ \mathfrak{J}_{5;\lambda}^{(0)}|\ ^{12}C(0^+;I=0;\vec{p}_i)>$ than normally anticipated or an inadequacy of the expression of the standard model for $J_\lambda^{(0)}(q)$ in the sense of the actual $J_\lambda^{(0)}(q)$ containing a sizeable axial isoscalar $\{u,d\}$-quark current ($\sim$ const $1/2\ \{(\overline{u}\gamma_\lambda\gamma_5 u) + (\overline{d}\gamma_\lambda\gamma_5 d)\}$).

B) Scattering of Electrons off Nuclei-Helicity Dependence of Cross Sections

As already mentioned in the Introduction, the scattering, elastic and inelastic, of electrons from nuclei is expected to depend on the helicity of the incident electrons because of interference between the parity-violating {WNC} x {WNC} and parity-conserving {EM} scattering amplitudes. The scattering process is described by the relevant parts of the {WNC} x {WNC} Hamiltonian density of Eqs. (17) and (9) and the {EM} Hamiltonian density of Eqs. (8) and (9), viz.

$$\mathcal{H}_{e,q} = \frac{G}{\sqrt{2}}\, 2(-\frac{1}{2}\,(\overline{e}\,\Gamma_\lambda e) + 2\sin^2\theta_W(\overline{e}\gamma_\lambda e))\ J_\lambda^{(0)}(q)$$

$$- 4\pi\alpha(\overline{e}\gamma_\lambda e)\ (\frac{1}{-\partial.\partial})\ J_\lambda^{em}(q)$$

$$J_\lambda^{(0)}(q) = \mathfrak{J}_\lambda^{(3)} + \mathfrak{J}_{5;\lambda}^{(3)} + \mathfrak{J}_\lambda^{(0)} + \mathfrak{J}_{5;\lambda}^{(0)} - 2\sin^2\theta_W J_\lambda^{em}(q) \tag{34}$$

with $\mathcal{J}_\lambda^{(3)}$, $\mathcal{J}_{5;\lambda}^{(3)}$, $\mathcal{J}_\lambda^{(0)}$, $\mathcal{J}_{5;\lambda}^{(0)}$, $J_\lambda^{em}(q)$ given in terms of the quark fields u, d, c, s, t, b in Eq. (19). For elastic scattering of electrons off 0^+, I = 0 nuclei ($^4He, ^{12}C, ^{16}O$, ...) the only part of $J_\lambda^{(0)}(q)$ which effectively contributes is $-2\sin^2\theta_W[J_\lambda^{em}(q)]_{isosc.}$ (see the argument leading to Eq. (24)) so that the parity-violating part of $H_{e,q}$ is

$$\mathcal{H}_{e,q}^{P.V.} = \frac{G}{\sqrt{2}}\, 2\, (-\tfrac{1}{2}\,(\bar{e}\gamma_\lambda\gamma_5 e))\, (-\, 2\sin^2\theta_W[J_\lambda^{em}(q)]_{isosc.}) \qquad (35)$$

The corresponding helicity dependence of the cross section, described via the right-handed helicity vs. left-handed helicity asymmetry, $_{R,L}$, is

$$\mathcal{A}_{R,L} \equiv \frac{\frac{d\sigma_R}{dq^2} - \frac{d\sigma_L}{dq^2}}{\frac{d\sigma_R}{dq^2} + \frac{d\sigma_L}{dq^2}} = 2\, \left(\frac{-(\frac{G}{\sqrt{2}})q^2}{4\pi\alpha}\right) 2\sin^2\theta_W = -1.58\text{x}10^{-4}\, \frac{q^2}{m_\mu^2}(2\sin^2\theta_W)$$

$$q^2 = 2E_e^2\, (1 - \cos\theta_{e_f,e_i}) \qquad (36)$$

Thus a measurement of $\mathcal{A}_{R,L}$ would provide a model-independent determination of $\sin^2\theta_W$; the numerical value of $\mathcal{A}_{R,L}$ expected for $\sin^2\theta_W = 0.22$ (see Eq. (22)) and, e.g., $q^2 = 0.2\ m_p^2$ (which is the value of q^2 at the second diffraction maximum in $e^- + {}^{16}O \rightarrow e^- + {}^{16}O$) is $\mathcal{A}_{R,L} = 1.4 \times 10^{-5}$.

Very recently, $\mathcal{A}_{R,L}$ has been determined, in a beautiful experiment of the SLAC-YALE group, for the process $e^- + d \rightarrow e^- +$ hadrons. Here, again retaining only the contributions of the u and d quarks,

$$\mathcal{H}_{e,q}^{P.V.} = \frac{G}{\sqrt{2}}\, 2 \Big[(-\tfrac{1}{2}\,(\bar{e}\gamma_\lambda\gamma_5 e))\, (\mathcal{J}_\lambda^{(3)} - 2\sin^2\theta_W J_\lambda^{em}(q))$$

$$+ (-\tfrac{1}{2}\,(\bar{e}\gamma_\lambda e) + 2\sin^2\theta_W(\bar{e}\gamma_\lambda e))\, (\mathcal{J}_{5;\lambda}^{(3)})\Big]$$

$$\mathcal{J}_\lambda^{(3)} = \tfrac{1}{2}\,\{(\bar{u}\gamma_\lambda u) - (\bar{d}\gamma_\lambda d)\},\quad \mathcal{J}_{5;\lambda}^{(3)} = \tfrac{1}{2}\,\{(\bar{u}\gamma_\lambda\gamma_5 u) - (\bar{d}\gamma_\lambda\gamma_5 d)\},$$

$$J_\lambda^{em}(q) = \tfrac{2}{3}\,(\bar{u}\gamma_\lambda u) - \tfrac{1}{3}\,(\bar{d}\gamma_\lambda d) \qquad (37)$$

whence, using the quark-parton model to calculate the matrix elements of $\mathfrak{J}_{\lambda}^{(3)}$, $\mathfrak{J}_{5;\lambda}^{(3)}$, and $J_{\lambda}^{em}(q)$

$$\mathcal{A}_{R,L} = 2\left(\frac{-\left(\frac{G}{\sqrt{2}}\right)q^2}{4\pi\alpha}\right)\left[\frac{9}{10} - 2\sin^2\theta_W + \frac{9}{5}\left(\frac{1}{2} - 2\sin^2\theta_W\right)\left(\frac{1-(1-y)^2}{1+(1+y)^2}\right)\right]$$

$$= -1.58 \times 10^{-4}\,\frac{q^2}{m_p^2}\left[\frac{9}{10} - 2\sin^2\theta_W) + \frac{9}{5}\left(\frac{1}{2}\ 2\sin^2\theta_W\right)\left(\frac{1-(1-y)^2}{1+(1+y)^2}\right)\right]$$

$$q^2 = 2E_{e;i}\,E_{e;f}\,(1 - \cos\theta_{e_f,e_i})\ ,\quad y \equiv (E_{e;i} - E_{e;f})/E_{e;i} \tag{38}$$

to be compared with the observed value

$$\mathcal{A}_{R,L} = -(1.33 \pm 0.20) \times 10^{-4} \quad \text{for}$$

$$E_{e;i} = 19.4\ \text{GeV},\quad y = 0.21,\quad q^2 = 1.4(\text{GeV})^2 \tag{39}$$

Combination of Eqs. (38) and (39) yields

$$\sin^2\theta_W = 0.20 \pm 0.03 \tag{40}$$

in very good agreement with Eq. (22). Further measurements of $\mathcal{A}_{R,L}$, especially measurements of $\mathcal{A}_{R,L}$ as a function of y, are in progress.

C) "Wrong-Parity Admixtures" to Bound Nucleon States in Nuclei and Bound Electron States in Atoms

We begin the discussion of wrong-parity admixtures by considering very briefly bound electron states in atoms. Here, the relevant part of the Hamiltonian density of Eqs. (17) and (9) which violates parity conservation is given by Eq. (37) and corresponds, in the case of an electron orbiting around a heavy nucleus, to a parity-violating electron-nucleus potential

$$U^{P.V.}_{\text{el-nucleus}} = -\frac{G}{\sqrt{2}}\{\{\gamma_5\}_{\text{elect}}\,[\tfrac{1}{2}\,(Z-N) - 2\sin^2\theta_W Z]\}\,\delta(\vec{r}) \tag{41}$$

where

$$<[A,Z;J^P];\vec{p}_f=0|\,\mathcal{J}_4^{(3)} - 2\sin^2\theta_W J_4^{em}(q)\,|[A,Z;J^P];\vec{p}_i=0>$$

$$= \frac{1}{2}(Z-N) - 2\sin^2\theta_W Z \qquad (42)$$

(compare Eq. (24)) and where

$$|<[A,Z;J^P];\vec{p}_f=0|\,\mathcal{J}_5^{(3)}\,|[A,Z;J^P];\vec{p}_i=0>| \stackrel{\sim}{\sim} J << Z \qquad (43)$$

and can be neglected. The potential $U^{P.V.}_{el-nucleus}(\vec{r})$ admixes wrong-parity components into otherwise pure-parity atomic electron states with the result that electromagnetic transitions which in the pure-parity limit are M1 (E1) acquire a small E1 (M1) component 90° out of phase with M1 (E1) and similarly for higher multipoles. Because of this, the atom in question possesses a (forward) scattering amplitude for photons which depends on the photon helicity so that a collection of such atoms is characterized by different indices of refraction for right and left circularly polarized electromagnetic radiation and consequently rotates the plane of polarization of linearly polarized electromagnetic radiation. The amount of rotation is proportional to $[1/2\ (Z-N) - 2\sin^2\theta_W Z]$ and to the matrix elements of $\{r\}_{elect}\ \delta(\vec{r})$ between the admixed states (in general, both the initial and final states of the electromagnetic transition are admixed)--a reliable and precise calculation of the wavefunctions of the admixed states is quite difficult. In addition, the experiments available on the rotation of the plane of polarization (in bismuth vapour) have given results which are inconsistent among themselves so that it is impossible, at the time of writing, to say whether the implications of the $V^{P.V.}(\vec{r})$ of Eq. (41) are confirmed by observation.

Finally, we discuss wrong-parity admixtures to bound nucleon states in nuclei. In this case the relevant part of the Hamiltonian density of Eqs. (17) and (9) is

$$\mathcal{H}_{q,q} = \frac{G}{\sqrt{2}}\{\frac{J_\lambda^{(-)}(q)J_\lambda^{(+)}(q) + J_\lambda^{(+)}(q)J_\lambda^{(-)}(q)}{2} + J_\lambda^{(0)}(q)J_\lambda^{(0)}(q)$$

$$J_\lambda^{(-)}(q) = \cos\theta(\mathcal{J}_\lambda^{(-)} + \mathcal{J}_{5;\lambda}^{(-)}) + \sin\theta(K_\lambda^{(-)} + K_{5;\lambda}^{(-)})$$

$$J_\lambda^{(+)}(q) = \cos\theta(\mathcal{J}_\lambda^{(+)} + \mathcal{J}_{5;\lambda}^{(+)}) + \sin\theta(K_\lambda^{(+)} + K_{5;\lambda}^{(+)})$$

$$J_\lambda^{(0)}(q) = \mathcal{J}_\lambda^{(3)} + \mathcal{J}_{5;\lambda}^{(3)} + \mathcal{J}_\lambda^{(0)} + \mathcal{J}_{5;\lambda}^{(0)} - 2\sin^2\theta_W J_\lambda^{em}(q)$$

$$\mathcal{J}_\lambda^{(-)} = (\bar{d}\gamma_\lambda u), \quad \mathcal{J}_{5;\lambda}^{(-)} = (\bar{d}\gamma_\lambda\gamma_5 u), \quad K_\lambda^{(-)} = (\bar{s}\gamma_\lambda u),$$

$$K_{5;\lambda}^{(-)} = (\bar{s}\gamma_\lambda\gamma_5 u)$$

$$\mathcal{J}_\lambda^{(+)} = (\bar{u}\gamma_\lambda d), \quad \mathcal{J}_{5;\lambda}^{(+)} = (\bar{u}\gamma_\lambda\gamma_5 d), \quad K_\lambda^{(+)} = (\bar{u}\gamma_\lambda s),$$

$$K_{5;\lambda}^{(+)} = (\bar{u}\gamma_\lambda\gamma_5 s)$$

$$\mathcal{J}_\lambda^{(3)} = \frac{1}{2}\{(\bar{u}\gamma_\lambda u) - (\bar{d}\gamma_\lambda d)\}, \quad \mathcal{J}_{5;\lambda}^{(3)} = \frac{1}{2}\{(\bar{u}\gamma_\lambda\gamma_5 u) - (\bar{d}\gamma_\lambda\gamma_5 d)\},$$

$$\mathcal{J}_\lambda^{(0)} = -\frac{1}{2}(\bar{s}\gamma_\lambda s), \quad \mathcal{J}_{5;\lambda}^{(0)} = -\frac{1}{2}(\bar{s}\gamma_\lambda\gamma_5 s)$$

$$\begin{aligned} J_\lambda^{em}(q) &= \frac{2}{3}(\bar{u}\gamma_\lambda u) - \frac{1}{3}(\bar{d}\gamma_\lambda d) - \frac{1}{3}(\bar{s}\gamma_\lambda s) \\ &= \mathcal{J}_\lambda^{(3)} + \frac{1}{6}\{(\bar{u}\gamma_\lambda u) + (\bar{d}\gamma_\lambda d)\} - \frac{1}{3}\bar{s}\gamma_\lambda s \\ &= \mathcal{J}_\lambda^{(3)} + [J_\lambda^{em}(u,d)]_{isosc.} - \frac{1}{3}\bar{s}\gamma_\lambda s = \\ &= \mathcal{J}_\lambda^{(3)} + [J_\lambda^{em}(q)]_{isosc.} \end{aligned} \tag{44}$$

where we have neglected the contributions of the (heavy) c, b, t quarks and where θ is the Cabibbo angle ($\sin\theta = 0.22$). The parity-violating, strangeness-conserving part of $\mathcal{H}_{q,q}$ is

$$\mathcal{H}^{P.V.;\Delta S=0}_{q,\bar{q}} =$$

$$\frac{G}{\sqrt{2}}\Big\{\frac{1}{2}\cos^2\theta\Big[\mathcal{J}^{(-)}_{\lambda}\mathcal{J}^{(+)}_{5;\lambda}+\mathcal{J}^{(+)}_{5;\lambda}\mathcal{J}^{(-)}_{\lambda}+\mathcal{J}^{(-)}_{5;\lambda}\mathcal{J}^{(+)}_{\lambda}+\mathcal{J}^{(+)}_{\lambda}\mathcal{J}^{(-)}_{5;\lambda}\Big]$$

$$+\frac{1}{2}\sin^2\theta\Big[K^{(-)}_{\lambda}K^{(+)}_{5;\lambda}+K^{(+)}_{5;\lambda}K^{(-)}_{\lambda}+K^{(-)}_{5;\lambda}K^{(+)}_{\lambda}+K^{(+)}_{\lambda}K^{(-)}_{5;\lambda}\Big]$$

$$+\Big[(\mathcal{J}^{(3)}_{\lambda}+\mathcal{J}^{(0)}_{\lambda}-2\sin^2\theta_W(\mathcal{J}^{(3)}_{\lambda}+[J^{em}_{\lambda}(q)]_{isosc.})(\mathcal{J}^{(3)}_{5;\lambda}+\mathcal{J}^{(0)}_{5;\lambda})$$

$$+(\mathcal{J}^{(3)}_{5;\lambda}+\mathcal{J}^{(0)}_{5;\lambda})(\mathcal{J}^{(3)}_{\lambda}+\mathcal{J}^{(0)}_{\lambda}-2\sin^2\theta_W(\mathcal{J}^{(3)}_{\lambda}+[J^{em}_{\lambda}(q)]_{isos.}))\Big]\Big\}$$

(45)

and enables a nucleon to emit a meson ($\pi,\rho,\omega,\ \ldots$) which meson is then absorbed by another nucleon via $\mathcal{H}_{strong}$. This exchange of a meson between the two nucleons gives rise to a parity-violating internucleon potential $U^{P.V.}_{nucl.-nucl.}(\vec{r})$ which admixes wrong-parity components into otherwise pure-parity bound nucleon states in nuclei with the result that, e.g., photons emitted in electromagnetic transitions between the admixed, and so no longer pure-parity, nuclear states have non-vanishing circular polarizations.

We next consider certain important selection rules which govern matrix elements such as $\langle\pi^+n(\pi^-p)|\mathcal{H}^{P.V.;\Delta S=0}_{q,q}|p(n)\rangle$ and so characterize the pion-exchange contribution to $U^{P.V.}_{nucl.-nucl.}(\vec{r})$ - this contribution has the longest range and, unless suppressed by the selection rules, should be dominant. We first write

$$\langle\pi^+n(\pi^-p)|\mathcal{H}^{P.V.;\Delta S=0}_{q,q}|p(n)\rangle =$$

$$\langle\pi^+n(\pi^-p)|g^{(+)}_{\pi np}(\bar{\psi}_p\psi_n\Phi_{\pi^+}+\mathcal{H}\,\Phi_{\pi^-}\bar{\psi}_n\psi_p)$$

$$+g^{(-)}_{\pi np}(\bar{\psi}_p\psi_n\Phi_{\pi^+}-\Phi_{\pi^-}\bar{\psi}_n\psi_p)|p(n)\rangle \qquad (46)$$

where $g^{(+)}_{np}$ and $g^{(-)}_{np}$ are appropriate constants and ψ_p, ψ_n, Φ_{π^+}, Φ_{π^-} are fields that destroy physical protons, physical neutrons, physical positive pions, and physical negative pions ($\Phi_{\pi^+}=(\Phi_{\pi^-})^+$) We then note that g^{+}_{np} must vanish since $(\bar{\psi}_p\psi_n\Phi_{\pi^+}+\Phi_{\pi^-}\bar{\psi}_n\psi_p)$ is odd under CP (odd under P and even under C); on the other hand, $(\bar{\psi}_p\psi_n\Phi_{\pi^+}-\Phi_{\pi^-}\bar{\psi}_n\psi_p)$ is even under CP (odd under P and odd under C) and, moreover, is an isovector, since

$$(\bar{\psi}_p\psi_n\Phi_{\pi^+} - \Phi_{\pi^-}\bar{\psi}_n\psi_p) = \frac{1}{\sqrt{2}}\bar{\psi}_N(\vec{\tau}\times\vec{\Phi}_\pi)^{(3)}\psi_N \tag{47}$$

where

$$\psi_N \equiv \begin{vmatrix}\psi_p\\ \psi_n\end{vmatrix}, \quad \Phi_\pi^{(1)} = \frac{1}{\sqrt{2}}(\Phi_{\pi^+} + \Phi_{\pi^-}), \quad \Phi_\pi^{(2)} = \frac{i}{\sqrt{2}}(\Phi_{\pi^+} - \Phi_{\pi^-}),$$

$$\Phi_\pi^{(3)} = \Phi_{\pi^0} \tag{48}$$

Also

$$<\pi^0 p(n)|\mathcal{H}^{P.V.}_{q,q}|p(n)> = <\pi^0 p(n)|g'^{(+)}_{\pi pp}(\bar{\psi}_p\psi_p\Phi_{\pi^0} + \Phi_{\pi^0}\bar{\psi}_n\psi_n)$$

$$+ g'^{(-)}_{\pi pp}(\bar{\psi}_p\psi_p\Phi_{\pi^0} - \Phi_{\pi^0}\bar{\psi}_n\psi_n)|\pi^0 p(n)> \tag{49}$$

with $g^{(+)}_{\pi pp} = g^{(-)}_{\pi pp} = 0$ since $(\bar{\psi}_p\psi_p\Phi_{\pi^0} \pm \Phi_{\pi^0}\bar{\psi}_n\psi_n)$ is odd under CP (odd under P and even under C). Thus the pion-exchange contribution to $U^{P.V.}_{nucl-nucl}(\vec{r})$, $[U^{P.V.}_{nuc.-nucl}(\vec{r})]_{\pi\text{-exch.}}$, involves π^+ and π^- only and is an isovector arising from the isovector part of $\mathcal{H}^{P.V.;\Delta S=0}_{q,q}$.

We proceed to discuss this isovector part of $\mathcal{H}^{P.V.;\Delta S=0}_{q,q}$ interchanging first, by means of a Fierz transformation, the u and s fields in products of $(K^{(F)}_\lambda + K^{(F)}_{5;\lambda})$ with $(K^{(\pm)}_\lambda + K^{(\pm)}_{5;\lambda})$. This gives

$$(K^{(-)}_\lambda + K^{(-)}_{5;\lambda})\ (K^{(+)}_\lambda + K^{(+)}_{5;\lambda}) = (\bar{s}\gamma_\lambda(1+\gamma_5)u)\ (\bar{u}\gamma_\lambda(1+\gamma_5)s) \rightarrow$$

$$(\bar{s}\gamma_\lambda(1+\gamma_5)s)\ (\bar{u}\gamma_\lambda(1+\gamma_5)u)$$

$$(K^{(+)}_\lambda + K^{(+)}_{5;\lambda})\ (K^{(-)}_\lambda + K^{(-)}_{5;\lambda}) = (\bar{u}\gamma_\lambda(1+\gamma_5)s)\ (\bar{s}\gamma_\lambda(1+\gamma_5)u) \rightarrow$$

$$(\bar{u}\gamma_\lambda(1+\gamma_5)u)\ (\bar{s}\gamma_\lambda(1+\gamma_5)s) \tag{50}$$

so that, using Eq. (45) with $[K^{(-)}_\lambda K^{(+)}_{5;\lambda} + K^{(+)}_{5;\lambda}K^{(-)}_\lambda + K^{(-)}_{5;\lambda}K^{(+)}_\lambda + K^{(+)}_\lambda K^{(-)}_{5;\lambda}]$ transformed according to Eq. (50)

$$[\mathcal{H}_{q,q}^{P.V.;\Delta S=0}]_{isov.} =$$

$$= \frac{G}{\sqrt{2}}\Big\{\frac{1}{2}\sin^2\theta\Big[K_\lambda^{(-)}K_{5;\lambda}^{(+)}+K_{5;\lambda}^{(+)}K_\lambda^{(-)}+K_{5;\lambda}^{(-)}K_\lambda^{(+)}+K_\lambda^{(+)}K_{5;\lambda}^{(-)}\Big]_{transf.}$$

$$+\Big[(1-2\sin^2\theta_W)(\mathcal{J}_\lambda^{(3)}\mathcal{J}_{5;\lambda}^{(0)}+\mathcal{J}_{5;\lambda}^{(0)}\mathcal{J}_\lambda^{(3)})+$$

$$+(\mathcal{J}_\lambda^{(0)}-2\sin^2\theta_W[J^{em}(q)_{isosc.}])\mathcal{J}_{5;\lambda}^{(3)}$$

$$+\mathcal{J}_{5;\lambda}^{(3)}(\mathcal{J}_\lambda^{(0)}-2\sin^2\theta_W[J^{em}(q)]_{isosc.})\Big]\Big\}$$

$$[K_\lambda^{(-)}K_{5;\lambda}^{(+)}+K_{5;\lambda}^{(+)}K_\lambda^{(-)}+K_{5;\lambda}^{(-)}K_\lambda^{(+)}+K_\lambda^{(+)}K_{5;\lambda}^{(-)}]_{transf.} =$$

$$(\bar{s}\gamma s)(\bar{u}\gamma_\lambda\gamma_5u)+(\bar{u}\gamma_\lambda\gamma_5u)(\bar{s}\gamma_\lambda s)+(\bar{s}\gamma_\lambda\gamma_5s)(\bar{u}\gamma_\lambda u)+(\bar{u}\gamma_\eta u)(\bar{s}\gamma_\lambda\gamma_5s) \tag{51}$$

Further, remembering that $|p\rangle \cong |[uud]\rangle$, $|n\rangle \cong |[udd]\rangle$, $|\pi^+\rangle = |[u\bar{d}]\rangle$, $|\pi^-\rangle = |[\bar{u}d]\rangle$, we can, to a reasonable approximation in calculating $\langle\pi^+n(\pi^-p)|[\mathcal{H}_{q,q}^{P.V.;\Delta S=0}]_{isov.}|p(n)\rangle$, drop all terms where $(\bar{s}\gamma_\lambda s)$ or $(\bar{s}\gamma_\lambda\gamma_5 s)$ operate directly on $|p(n)\rangle$ or $|\pi^+n(\pi^-p)\rangle$. The $[\mathcal{H}_{q,q}^{P.V.;\Delta S}]_{isov.}$ of Eq. (51) can then be effectively replaced by

$$[\mathcal{H}_{q,q}^{P.V.;\Delta S=0}]_{isov.} \cong \frac{G}{\sqrt{2}}(-2\sin^2\theta_W)\{[J_\lambda^{em}(u,d)]_{isosc.}\mathcal{J}_{5;\lambda}^{(3)}$$

$$+\mathcal{J}_{5;\lambda}^{(3)}[J_\lambda^{em}(u,d)]_{isosc.}\}$$

$$[J_\lambda^{em}(u,d)]_{isosc.} = \frac{1}{6}\{(\bar{u}\gamma_\lambda u)+(\bar{d}\gamma_\lambda d)\},$$

$$\mathcal{J}_{5;\lambda}^{(3)} = \frac{1}{2}\{(\bar{u}\gamma_\lambda\gamma_5u)-(\bar{d}\gamma_\lambda\gamma_5d)\} \tag{52}$$

It is important to emphasize that in the approximation of Eq. (52) only the {WNC} x {WNC} interaction contributes to $[\ {}_{q,q}^{P.V.;\Delta S=0}]_{isov.}$

It remains to calculate $[U^{P.V.}_{nucl-nucl}(\vec{r})]_{\pi\text{-exch}}$. In view of Eq. (52) and Eqs. (46) - (48) this can be written as

$$U^{P.V.}_{nucl-nucl}(\vec{r})]_{\pi\text{-exch}} = \frac{[\frac{GM_\pi^2}{\sqrt{2}}(-2\sin^2\theta_W)\kappa][\frac{f_{\pi np}}{\sqrt{2}}]}{2\cdot 4\pi} \times$$

$$[(\vec{\sigma}_1+\vec{\sigma}_2)\ (\text{grad}_{\vec{r}_{12}}\ (\frac{e^{-m_\pi r_{12}}}{m_\pi r_{12}}))\ (\vec{\tau}_1 x \vec{\tau}_2)^{(3)}]$$

$$\kappa = m_\pi^{-3} <\pi^+ n(\pi^- p)|\{[J_\lambda^{em}(u,d)]_{isos.}\ \mathcal{J}^{(3)}_{5;\lambda}$$

$$+ \mathcal{J}^{(3)}_{5;\lambda}\ [J_\lambda^{em}(u,d)]_{isoc.}\}|p(n)> \quad (53)$$

where $f_{\pi\pi p}/\sqrt{2} = f_{\pi pp} = \sqrt{4\pi}\ \sqrt{0.08}$ is the strong pion-nucleon-nucleon coupling constant. The matrix element can be (rather crudely) estimated by use of PCAC ($\Phi_{\pi^+} = (a_\pi m_\pi^3)^{-1}\ \partial_\lambda \mathcal{J}^{(-)}_{5;\lambda}$; a_π = pion decay constant = 0.94), current algebra (commutation relations between $\mathcal{J}^{(-)}_{5;\lambda}$ and $\{[J_\lambda^{em}(u,d)]_{isosc.}\ \mathcal{J}^{(3)}_{5;\lambda} + \mathcal{J}^{(3)}_{5;\lambda} + \mathcal{J}^{(3)}_{5;}\ [J_\lambda^{em}(u,d)]_{isosc.}\}$ and suitable comparison, using the quark model, with (experimentally known matrix elements of the type: $m_\pi^{-3} <\pi^- p|\{\mathcal{J}^{(-)}_\lambda K^{(+)}_{5;\lambda} + K^{(+)}_{5;\lambda}\mathcal{J}^{(-)}_\lambda\}|\Lambda>$--this gives $\kappa \approx 2$ which should be of the right order of magnitude.

With $U^{P.V.}_{nucl.-nucl.}(\vec{r})]_{\pi\text{-exch.}}$ determined we can calculate the wrong-parity admixtures to otherwise pure-parity bound nucleon states in nuclei. As a particularly simple example we consider a pair of excited states in ^{18}F: $|a> = |^{18}F(0^-;(I=0) + \delta(I=1), E_a^* = 1.08\text{ MeV})>$ and $|b> = |^{18}F(0^+;I=1,E_b^* = 1.04\text{ MeV})>$ in the pure-parity approximation where δ measures a small I=1 component in $|a>$ induced by the proton-proton Coulomb interaction. The wave functions of these states with inclusion of wrong-parity admixtures arising from $U^{P.V.}_{nucl-nucl.}(\vec{r})]_{\pi\text{-exch.}}$ are

$$|a>_{admix.} = |a> + \frac{|b><b|U|a>}{E_b^*-E_a^*}, \quad |b>_{admix.} = |b> + \frac{|a><a|U|b>}{E_a^*-E_b^*}$$

(54)

the wrong-parity admixtures being especially large here because of the small numerical value of $E_a^* - E_b^*$ (= 0.04 MeV). The 1.08 MeV photon from $|a\rangle_{admix.}$ to the ground state of ^{18}F: $|g\rangle_{admix.} \cong |g\rangle = |^{18}F(1^+;I=0,E_g^* = 0)\rangle$ is then E1 with an M1 admixture and its circular polarization is estimated to be

$$|P_\gamma| = \left|2\left(\frac{\langle b|U|a\rangle}{E_b^*-E_a^*}\right)\frac{||M1[\{0^+;I=1\to 1^+;I=0\}]||}{\delta||E1[\{0^-;I=1\to 1^+;I=0\}]||}\right| \approx 3 \times 10^{-3} \tag{55}$$

to be compared with an experimental upper bound: $P_\gamma = (-0.7 \pm 2.0) \times 10^{-3}$. In view of the difficulty of estimating κ (P_γ is proportional to κ on the basis of Eqs. (53) - (55)) and of evaluating $\langle b|U|a\rangle$, the discrepancy does not seem too serious.

With one exception, all other so far observed wrong-parity-admixture effects in nuclei, e.g., the finite rate for the α-decay: $^{16}O(2^-;I=0) \to {}^{12}C(0^+;I=0) + {}^4He(0^+;I=0)$, are induced by parity-violating internucleon potentials which have a relatively dominant, or at least a relatively important, isoscalar (or isoscalar + isotensor) part; such $\Delta I=0$ (or $\Delta I=0$, $\Delta I=2$) potentials arise from ρ, ω, ... exchange and while of shorter range than the $\Delta I=1$ π-exchange potential (Eq. (53)) are, within that range, larger than the latter. The derivation of these $\Delta I=0$, $\Delta I=2$ ρ, ω, ... exchange potentials from the $\mathcal{H}_{q,q}^{P.V.;\Delta S=0}$ of Eq. (45) is complicated by the fact that the {WCC} x {WCC} and {WNC} x {WNC} interactions make comparable contributions to these potentials; in addition, the calculation of matrix elements of these potentials between the to-be-admixed states of opposite parity is particularly difficult since available nuclear wavefunctions are not well known at internucleon distances = m_ρ^{-1}. In all, definitive general treatments are lacking and much work remains to be done.

BIBLIOGRAPHY

II: The "Standard" Model of the {WNC} x {WNC} and {WCC} x {WCC} Interactions

S. Weinberg, Phys. Rev. Lett. 19 (1967), 1264.

A. Salam, Elementary Particle Physics edit. by N. Svartholm (Stockholm, Sweden 1968), p. 367.

S.L. Glashow, J. Iliopoulos, and L. Maiani, Phys. Rev. D2 (1970), 1285.

S. Weinberg, Phys. Rev. D5 (1972), 1412.

P.W. Higgs, Phys. Rev. Lett. 12 (1964), 132.

E. Abers and B.W. Lee, Phys. Repts. 9 (1973), 1.

M.A.B. Beg and A. Sirlin, Ann. Rev. Nucl. Sci. 24 (1974), 379.

III: Applications of the Standard Model for the {WNC} x {WNC} and {WCC} x {WCC} Interactions

A) Scattering of Neutrinos and Antineutrinos off Nuclei

T.W. Donnelly and R.D. Peccei, Phys. Repts. (to be published).

H.H. Williams et al, Proc. Inter. Confer. on Neutrino Phys., University of Oxford, Oxford, England (July 1978).

T.W. Donnelly and R.D. Peccei, Phys. Lett. 65B (1976), 196.

S.S. Gerštein et al, Sov. J. Nucl. Phys. 22 (1976), 76.

G.J. Gounaris and J.D. Vergados, Phys. Lett. 71B (1977), 35.

B) Scattering of Electrons off Nuclei-Helicity Dependence of Cross Sections

E. Derman, Phys. Rev. D7 (1973), 2755.

G. Feinberg, Phys. Rev. D12 (1975), 3375.

R. Cahn and F. Gilman, Phys. Rev. D17 (1978), 1313.

J.D. Walecka, Nucl. Phys. A285 (1977), 349.

C.A. Prescott et al, Phys. Lett. 77B (1978), 347.

C) "Wrong-Parity Admixtures" to Bound Nucleon States in Nuclei and Bound Electron States in Atoms

L. Wilets, Proc. Inter. Confer. on Neutrino Phys. and Astrophysics, Purdue Univ., Lafayette, Indiana, U.S.A. (May 1978).

E. Fischbach, Proc. Inter. Confer. on Neutrino Phys. and Astrophysics, Purdue Univ., Lafayette, Indiana, U.S.A. (May 1978).

E. Fischbach and D. Tadic, Phys. Repts. 6C (1973), 123.

M. Gari, Phys. Repts. 6C (1973), 317.

M. Gari, J.B. McGrory, and R. Offerman, Phys. Lett. 55B (1975), 277.

B. Desplanques and J. Micheli, Phys. Lett. 68B (1977), 339.

J.F. Donoghue, Phys. Rev. D13 (1976), 2064; D15 (1977), 184.

C.A. Barnes et al, Phys. Rev. Lett. 40 (1978), 840.

NOVEL TOPICS IN π NUCLEAR PHYSICS

T.E.O. Ericson

CERN

Geneva, Switzerland

What is the role of π-nuclear physics? What can we learn from it? Let me first clearly state what it should not be. It is a temptation for some to use physical pions as a spectroscopic tool to investigate nuclear states, which is little more than an extension of Van de Graaff physics to a new particle and to less preciseness. With a few exceptions (mainly photopions and radiative π-capture) this is not a great challenge to our imagination and it is not the true reason for the importance of pions to nuclear physics. The key importance of pions is that it is the most important constituent in the nuclear force; in this way it indirectly influences every single aspect of nuclear physics. Because of this the nature of its interaction becomes of prime importance and so does its connection to nuclear properties. This is in particular so since the pion field may radically change under the right conditions leading for example to a nuclear phase transition. It is therefore these unfamiliar, novel features of the nucleus, its relativistic phenomena, the modification of the nuclear force by the nuclear medium (3 body forces), isobar effects and exchange currents which are the main motivations for π-nuclear physics: the use of physical pions serve to illustrate the effects and elucidate the many possibilities.

In the present talk, I have concentrated on some very recent developments which illustrate this theme: a) the possibility of π-nuclear bound states and size resonances due to a singularity in the optical potential[20]; b) the possibility of nuclear critical opalescence well before the onset of a π-condensate[14]; c) the intimate relation between nuclear axial-currents and the nuclear π-field ("axial locality")[24], and finally d) the possibility of observing the π-field directly using the selective $(\pi^-,2\gamma)$ reaction[34].

I. CAN THERE BE π-NUCLEAR BOUND STATES?[20]

1. The low-energy π-nuclear optical potential, as derived from the πN interaction and used in pionic atoms, results in an approximate wave equation of the Kisslinger type

$$\nabla\cdot(1-\alpha(r))\nabla\Phi(r)+[-Q(r)+k^2]\Phi(r) = 0 \quad . \tag{1}$$

The potential is thus velocity dependent (non-relativistic notation)

$$2\mu V(r) = \nabla\cdot\alpha(r)\nabla+Q(r) \quad . \tag{2}$$

Here $Q(r)$ is a weakly repulsive term, while the velocity-dependent term, $\nabla\cdot\alpha(r)\nabla$, originating from p-wave πN interaction, is attractive. The asymptotic kinetic energy is $E = k^2/2\mu$. Since pions absorb in nuclei, both $\alpha(r)$ and $Q(r)$ are complex with $\mathrm{Im}\, Q(r) < 0$ and $\mathrm{Im}\, \alpha(r) > 0$.

If $\alpha(r) < 1$, the attraction is insufficient to overcome the intrinsic repulsion of the kinetic energy in eq. (1). However, in the region with $\alpha(r) > 1$, the overall "effective" kinetic energy in eq. (1), $\nabla\cdot[1-\alpha(r)]\nabla$ becomes attractive; bound states may then develop even in the presence of the repulsive local potential $Q(r)$. We will here demonstrate that such bound π-nuclear states are very likely for certain nuclei. This prediction is rather insensitive to the exact form of $\alpha(r)$ as long as $\alpha(r) > 1$ in some region of the nucleus.

2. The crucial properties of eq. (1) for the development of bound states appear clearly in the following oversimplified model. Consider the wave function to be confined to a spherical box of radius R so that $\Phi(r) = 0$ for $|r| \geq R$. The potential parameters are constant inside the box: $\alpha(r) = \alpha$ and $Q(r) = Q$. For a bound state of energy $B = 2\mu\kappa^2$ ($k^2 = -\kappa^2$) the wave number K inside the box is given by eq. (1):

$$-(1-\alpha)K^2 - Q - \kappa^2 = 0 \quad . \tag{3}$$

The wave functions are spherical Bessel functions $j_\ell(K_n r)$ of given orbital angular momentum ℓ, for which the eigenvalues are determined by the boundary condition $j_\ell(K_n R) = 0$. Denoting the nth root of $j_\ell(\pi\beta) = 0$ by $\beta_{\ell n}$, the binding energy $B_{n\ell}$ of the nth state of given ℓ obeys therefore the condition

$$B_{n\ell} = -Q/2\mu + (\alpha-1)\pi^2\beta_{\ell n}^2/2\mu R^2 \quad , \tag{4}$$

where $\beta_{0n} = n > 0$, $\beta_{\ell n} \to (n+1/2)$ for a large n and $\ell > 0$. The wave function for a given number of nodes is independent of α and Q: $j_\ell(K_n r) \equiv j_\ell[\pi\beta_{\ell n}(r/R)]$; it is only subject to the condition that this bound state exists, which is the positivity condition for $B_{n\ell}$ in eq. (4).

For a repulsive $Q > 0$ there can be no bound states in eq. (4) for $\alpha < 1$, since $(\alpha-1)$ is negative. The point $\alpha = 1$ is a singular point at which the "effective" kinetic energy term $(\alpha-1)K^2/2\mu$ changes from repulsion to attraction. This provokes a profound change in the system. For $\alpha > 1$ an infinite number of bound states develop. It is energetically favourable to have states with more nodes, so that the level order is the inverse one of that of a normal potential: $B_{1,\ell} < B_{2,\ell} < \ldots$, etc. This is illustrated in fig. 1a. For an attractive $Q < 0$, the situation is slightly different. Since the normal repulsion of the kinetic energy term is weakened by the factor $(1-\alpha)$ as α approaches unity, the attraction in Q can accommodate more and more bound states. In the limit $\alpha = 1$ an infinity of states will bind, all with the same energy $B_{n\ell} = -Q/2\mu$. These states have the usual level order in a potential for $\alpha < 1$, i.e. $B_{1\ell} > B_{2\ell} > B_{3\ell} > \ldots$. At the point $\alpha = 1$, the level order is reversed (see fig. 1b), so that the binding energies have once more the anomalous order $B_{1\ell} < B_{2\ell} < B_{3\ell} < \ldots$. Although $\alpha = 1$ is a singular point, it is remarkable to note that any wave function with a given number of nodes $(n - 1)$ has no singularity at this point.

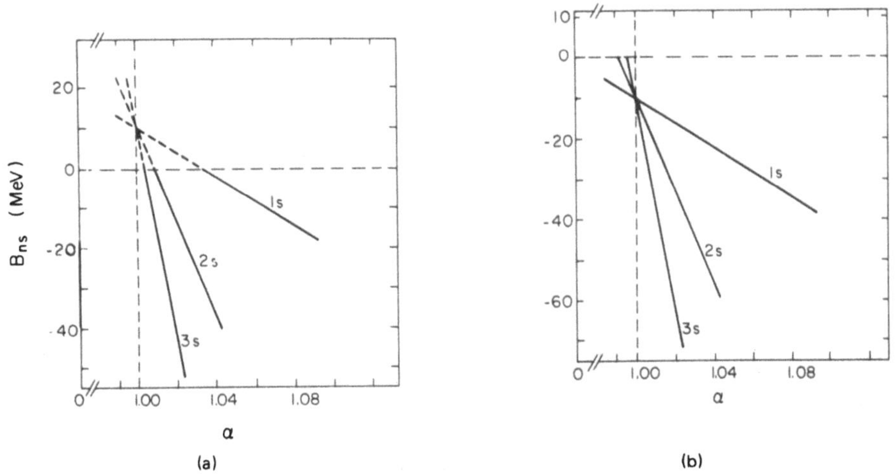

Fig. 1. Binding energies of ns states in the uniform model according to eq. (4). Radius R = 2 fm. (a) Repulsive local potential $Q/2\mu = 10$ MeV; (b) attractive local potential $Q/2\mu = -10$ MeV.

While the uniform model has many features of the general case in which $\alpha(r)$ varies from a positive, finite value near $r = 0$ to zero at $r = \infty$, there are also new features. Multiply first eq. (1) by $\Phi^*(r)$ and make a partial integration. This gives the following relation for the binding energy B:

$$2\mu B \int |\Phi|^2\, dr = \int \{[\alpha(r)-1]|\nabla\Phi|^2 - Q(r)|\Phi|^2\}dr \quad . \qquad (5)$$

It is clear from eq. (5) that in the case of a repulsive $Q(r) > 0$, bound states can only occur if $\alpha(r) > 1$ in some region: only then can the right-hand integral be made positive. Further, we note that for an attractive $Q(r) < 0$, it is only possible to obtain a binding energy of more than $(-Q)_{max}/2\mu$, if $\alpha(r) > 1$ in some region. In both cases this is a generalization of the results of the uniform model. It is physically plausible that whenever $\alpha(r) > 1$ in some region, bound states will develop with the binding energy given by the number of nodes and the size of the region, roughly as stated in condition (4) for the uniform model. The novel feature is, however, the smooth transition from a region of $\alpha(r) > 1$ to one of $\alpha(r) < 1$. This gives rise to a singularity in the wave function as will now be discussed.

4. Bethe[2] pointed out that there should be a pathological behaviour of the wave equation (1) at the point $\alpha(r_0) = 1$, since the wave number becomes infinite there. He did not investigate this point further, although he proposes remedies for infinite nuclear matter[3]. We will give here some important consequences of the singularity. We first transform eq. (1) into a standard wave equation by the substitution[4] $\Phi(\underset{\sim}{r}) = [1-\alpha(\underset{\sim}{r})]^{-1/2}\tilde{\Phi}(\underset{\sim}{r})$:

$$(\nabla^2+k^2)\tilde{\Phi} = \frac{\{[Q(\underset{\sim}{r}) - \alpha(\underset{\sim}{r})k^2] - \frac{1}{2}\nabla^2\alpha(\underset{\sim}{r})\}}{[1-\alpha(\underset{\sim}{r})]}\tilde{\Phi}$$

$$- \frac{1}{4}\frac{[\nabla\alpha(\underset{\sim}{r})]^2}{[1-\alpha(\underset{\sim}{r})]^2}\tilde{\Phi} \quad . \qquad (6)$$

Consider the case of spherical symmetry. Since $1 - \alpha(r) \propto (r - r_0)$ near $\alpha(r) = 1$, one notes immediately that the equation has a potential term singular as $(r - r_0)^{-2}$ and $(r - r_0)^{-1}$. The $(r - r_0)^{-2}$ singularity has the following consequences:

(a) an infinity of bound states for $\alpha > 1$ as demonstrated in sections 2 and 3;

(b) a logarithmic singularity in the wave function at the pole;

(c) an anomalous absorption layer with finite strength at Re $\alpha(r_0) = 1$ even in the limit Im $\alpha \to 0$.

To exhibit the last two features consider the special case of a diffuse plane surface, i.e. $\alpha(\underset{\sim}{r}) = \alpha(x)$ and $Q(\underset{\sim}{r}) = Q(x)$. We take $\alpha(x)$ to vary linearly through the critical region, while all other potential terms are constant there: $\alpha(x) - 1 = \alpha'(x-x_0)$ and $Q(x) = Q$. The most general solution to eq. (1) is then in this case

$$\Phi(x) = f(x) + \ln(x - x_0)g(x)$$

$$\equiv AJ_0\ \{2\ [\frac{k^2-Q}{\alpha'}(x_0 - x)]^{1/2}\}$$

$$+ BN_0\ \{2\ [\frac{k^2-Q}{\alpha'}(x_0 - x)]^{1/2}\} \quad . \qquad (7)$$

Here $f(x)$ and $g(x)$ are regular functions, $J_0(x)$ and $N_0(x)$ are the regular and irregular cylindrical Bessel functions, while A and B are arbitrary constants. The result (7) exhibits point (b) above.

Consider once more the plane surface discussed above with the wave function given by eq. (7) near the singular region. The derivative will then take the form

$$\Phi'(x) \xrightarrow[x\to x_0]{} g(x_0)\ (x-x_0) \quad ,$$

i.e. it becomes very large there.

The absorption generally reflects itself by an imaginary part of the potential. In the limit of a real potential the pole position x_0 moves to the real axis. There will still be absorption in this limit. To see this, consider the imaginary part of eq. (5). The first term in the r.h.s. integrand is then

$$\mathrm{Im}\{[\alpha(x)-1]|\Phi'(x)|^2\} \to \mathrm{Im}\{[\alpha(x)-1]\ \frac{|g(x_0)|^2}{|x-x_0|^2}\ \}$$

$$= \frac{\alpha'\,\mathrm{Im}(x_0)|g(x_0)|^2}{[x-\mathrm{Re}(x_0)]^2 + [\mathrm{Im}(x_0)]^2}$$

$$\xrightarrow[\mathrm{Im}(x_0)\to 0]{} \pi\alpha'|g(x_0)|^2\delta(x-\mathrm{Re}(x_0)) \quad .$$

The consequence is thus that there will be a non-vanishing absorption from this term even in the limit of purely real potential. The origin of the anomaly is an absorptive dipole layer in the region $x = \mathrm{Re}(x_0)$.

5. The singularities of the wave equation (1) reflect two simplifying approximations in the original π-nuclear multiple scattering equations: (i) the use of a vanishing range in the p-wave πN interaction, and (ii) the neglect of the range of the NN pair correlation function. These approximations are not necessary[1]. In the actual description an averaging factor is introduced, so that there is no singularity in the critical region. The mechanism for this is easily seen as follows. Replace $\alpha(\underset{\sim}{x})\underset{\sim}{\nabla}\Phi(\underset{\sim}{x})$ by the averaged hermitian form

$$\alpha(\underset{\sim}{x})\nabla\Phi(\underset{\sim}{x}) \rightarrow \int \alpha(\underset{\sim}{x} + \tfrac{1}{2}\underset{\sim}{\zeta})F(\underset{\sim}{\zeta})\nabla\Phi(\underset{\sim}{x} + \underset{\sim}{\zeta})d\underset{\sim}{\zeta} \quad . \tag{8}$$

Here $F(\zeta)$ is an averaging function which we take spherically symmetrical with a range $\langle\zeta^2\rangle = \int F(\underset{\sim}{\zeta})\zeta^2 d\underset{\sim}{\zeta}$. To leading order in the range parameter $\langle\zeta^2\rangle$, eq. (8) leads to a fourth-order non-singular differential equation.

The nature of this equation is apparent in the previously discussed case of a plane surface with a linear variation $\alpha(x) = \alpha_0 + \alpha_1 x$,

$$\frac{1}{6}\langle\zeta^2\rangle\alpha(x)\,\frac{d^4\Phi(x)}{dx^4} + \frac{1}{3}\alpha_1\langle\zeta^2\rangle\,\frac{d^3\Phi(x)}{dx^3}$$

$$+ \frac{d}{dx}\,[\alpha(x) - 1]\,\frac{d\Phi(x)}{dx} + [k^2 - Q(x)]\Phi(x) = 0 \quad . \tag{9}$$

It is essential to note that the limit $\langle\zeta^2\rangle \rightarrow 0$ is not a smooth one as $\alpha(x) = 1$.

For a uniform medium with $\alpha(x) = \alpha > 1$ and $Q(x) = Q$, a crucial parameter is $\beta = \frac{2}{3}(\kappa^2+Q)\langle\zeta^2\rangle\alpha(\alpha-1)^{-2}$, which determines the magnitude of the fourth-order term as compared to each of the other terms. The eigenfrequencies λ_1 and λ_2 are in this case

$$\lambda_1^2 = \frac{2(\kappa^2+Q)}{(\alpha-1)(1+\sqrt{1-\beta})} \simeq \frac{\kappa^2+Q}{(\alpha-1)} \quad , \qquad \beta < 1 \quad ,$$

$$\lambda_2^2 = \frac{3(\alpha-1)}{\alpha<\zeta^2>}\,[1+\sqrt{1-\beta}] \simeq \frac{6(\alpha-1)}{\alpha<\zeta^2>} \quad , \qquad \beta < 1 \quad , \tag{10}$$

Therefore $\lambda_1^2/\lambda_2^2 \sim \frac{1}{4}\,\beta << 1$.

The normal "long-wavelength" solution λ_1, has a small fourth-order term as required as long as β is small. The high-frequency solution λ_2 has a large such term in this case and must be expected to be highly model dependent. (One notes that for λ_2 it is essential to consider the limits $<\zeta^2> \to 0$ and $(\alpha-1) \to 0$ simultaneously: for $\alpha \neq 1$ the limit is ∞ as $<\zeta^2> \to 0$).

The number of bound states in now finite. For a real uniform potential of radius R this number is approximately obtained from the two conditions: $\kappa^2 > 0$ and $\lambda_1 R = n\pi$ (s-waves only). Neglecting the weak local potential Q, we find for large n:

$$n^2 \lesssim 3(\alpha-1)R^2/(\pi^2<\zeta^2>\alpha) \quad . \tag{11}$$

In addition, the range parameter $<\zeta^2>$ limits the binding energy of the lowest state in eq. (9). Although we have discussed non-relativistic binding, eq. (1) is in fact relativistic with a total π-energy $\omega = \sqrt{k^2+\mu^2}$. The binding energy therefore equals the rest mass for $\omega = 0$ or $\kappa = \mu$. This is a necessary condition for a π condensate[5]. If we neglect the weak local potential Q, we have[†] ±1 for the uniform model (which has no response of the nuclear degrees of freedom) by this condition and by the reality condition $\beta \lesssim 1$ (undamped waves)

$$\alpha > 1 + [\tfrac{2}{3}\,\mu^2<\zeta^2>/\alpha]^{1/2} \tag{12}$$

This demonstrates incidentally that the condition for bound states as discussed here is not the same as the condition for a π-condensate.

6. In actual nuclei bound π-nuclear states, as discussed above, will occur as a general phenomenon provided the decisive condition $Re(\alpha) > 1$ is valid in a sufficiently large region. In its gross features the shape of $\alpha(r)$ is that of the nuclear matter

---

[†]An analogous condition based on a different consideration has been obtained independently by M. Ericson, see section 2.

distribution, although its detailed behaviour is a different one. In fact, microscopic theories for $\alpha(r)$ lead to a non-linear dependence on the nuclear density[1,4,6,7]. For the present considerations these non-linear effects are largely irrelevant, while it is crucial that the effective strength of $\alpha(r)$ of the nucleus is consistent with the large amount of information from pionic atoms. As a consequence we have used $\alpha(r)$ and $Q(r)$ with experimentally determined parameters; these do not vary much from one analysis to another[1,8,9].

The following two conditions favour bound states. Firstly, a high nuclear density near the central region favours bound states of π^-, π^0 and π^+. Secondly, since the p-wave π^-n interaction at low energy is about 10 times the π^-p interaction, it is very favourable for π^--nuclear bound states to have a large neutron excess in this region, while this is unfavourable for π^+ binding, and indifferent for π^0 binding.

For a numerical investigation of favourable cases we have first considered the T = 0 nuclei ^{16}O, ^{28}Si and ^{40}Ca with equal neutron and proton densities; using the experimental charge density for the proton density[10,11] and parametrizing experiments as in Krell and Ericson[1], the maximal value for $\alpha(r)$ of 0.80, 0.88 and 0.83, respectively, is obtained. These nuclei which have higher densities than normal in their central regions, therefore all fail to meet the criterion for bound states; this conclusion a fortiori also extends to the vast majority of other nuclei. We have also considered the doubly magic nucleus ^{208}Pb, for which we optimistically took the neutron and proton distributions to have equal shape[11]. In this case $\alpha(r) \approx 1$ from the centre to nearly 4 fm. Here the possibility of bound states is a more realistic one, since small changes in the parameters or in the detailed shape of $\alpha(r)$ would be sufficient to fulfil the conditions. It is therefore difficult to make a firm prediction, since this is a transition case. As the neutron distribution is more concentrated in the outer region of this nucleus than we have assumed, we feel, however, that chances are that ^{208}Pb just fails to meet the necessary conditions.

These results are a consequence of experimentally constrained values for the "effective" $\alpha(r)$. In terms of a theoretical description based on the πN interactions there are rather important higher-order corrections to $\alpha(r)$ (Lorentz-Lorenz effect[1,6]). If we ignore such effects and simply construct the leading-order π-nuclear optical potential linear in the density, nearly all nuclei would have $\alpha(r) > 1$ in the central region and π-nuclear bound states would be the rule and not the exception; this is not a realistic description, however.

It is at this point there should be a warning, however, to those who investigate the π-nuclear optical model in the continuum: there are a number of examples in which parameters have been used from πN scattering without reducing their effective strength by the L-L effects. Presumably these treatments must face up to singularity as a matter of principle since the strength of the potential is beyond the critical one: results of such studies are at best suspect in the absence of further discussion.

There are nuclei which are expected to surpass the critical density, but unfortunately those I know of are extremely short-lived and unsuitable for targets. An example is ^{32}Na. In this nucleus the 2s neutron shell has just been filled while the 2s proton shell is empty. As a consequence there is a vast surplus of neutrons near the nuclear center with $\rho_n \gtrsim 2\rho_p$; in addition the central mass density is high. In this case $\alpha(0) \gtrsim 1.3$ which is beyond the critical value. In order to see qualitatively what happens in this case we have numerically integrated equation (1) with the potential parameters as above[1]. The spurious singularity at $\alpha(r) = 1$ has been avoided by smoothing the wave-functions in this region. We did this in various ways, and as long as we used wavefunctions with few nodes the results were stable. In this case the 1s state binds barely and has a large width of about 15 MeV. The 2s state binds by about 20 MeV and has 80 MeV width; this state is physically dubious, since it is physically a necessary condition that wave functions are constant over regions of the size of internuclear spacing. As a whole these results confirm qualitatively the results of the uniform model. In particular, the remarkably steep slope of the binding energy with increasing strength of $\alpha(r)$ is confirmed.

The surprisingly large widths of the bound states are noteworthy. The explanation is readily seen in the uniform model with binding energies proportional to $(\alpha-1)$. While $\mathrm{Im}\,\alpha \ll \mathrm{Re}\,\alpha$, $\mathrm{Im}\,\alpha$ is of the same order as $\mathrm{Re}(\alpha-1)$.

The closeness of $\alpha(r)$ to unity in most nuclei has an important practical consequence for the discussion and analysis of experimental data on π-nuclear scattering, pionic atoms, etc. An inadvertent description of such data with $\alpha(r) > 1$ could result simply from a minor variation in parameters. If improperly handled this may lead to unrealistic and unphysical conclusions.

Only a few days ago I received a preprint by Friedman, Gal and Mandelzweig. They observe that a recent analysis of parameters of pionic atoms by Batty et al.[9] give $\alpha > 1$ in a number of nuclei; such as ^{40}Ca, ^{27}Al, etc. In this case even the nuclei which are the source of the parameters must exhibit anomalies, contrary to the case we discussed. This observation is extremely interesting and it shows clearly the urgency of a detailed investigation of the optical model in the region $\alpha \gtrsim 1$.

Friedman, Gal and Mandelzweig attempt an analysis with new parameters deduced, introducing smoothing functions. They then attempt to find bound states. The states they produce in this way have unfortunately very many nodes and are probably not too realistic. As a consequence of this they have also very large widths ($\gtrsim$ 500 MeV) which would make them unobservable. Still the basic observation they make is extremely interesting, since it clearly shows that parameters which could provoke the anomaly are very realistic.

7. Our treatment has implicitly assumed that the nucleus is an inert source of the potential, In fact, the nucleus will respond to the presence of the pion. In particular, the binding energy of the pion is an extremely rapid function of the density and neutron excess in the critical region. As a consequence it is in principle advantageous for the nucleus to adjust to the presence of the pion. This can be done in two ways particularly. The nucleus may be compressed in volume to a higher density ($T = 0$ monopole compression). It may also respond to a π^- by compressing the neutron distribution and dilating the proton distribution ($T = 1$ monopole compression). In the absence of absorption these effects would be important, and bound states would result under more favourable conditions than described here. However, the adjustment time is linked to the monopole frequencies, which are of the same magnitude as the typical absorption width. Consequently, the nuclear response time is at best similar to the absorption time. An important exception to this situation is a π^- in a finite piece of neutron matter, in which case there is no absorption. Even in the absence of an adjustment of the density, the π^- will bind at a density of 0.11 fm^{-3}, i.e. at less than the nuclear density. It is a pity this situation can only be realized in neutron stars.

8. It is clearly quite important both theoretically and experimentally to find out if there are π-nuclear bound states. How can this be done? One thinks immediately about γ-ray transitions, say, from π-atom states to the bound states. Unfortunately the overlap is poor, so the branching ratio is bad and in addition the states are likely to be broad. Similarly one might hope to display them in principle in μ-capture with an important enhancement at high energy or in photoprocesses in nuclei: a bound state would give rise to enhancements below threshold. All of these methods seem difficult to me. However, if there are bound states, there should also be resonances in the low energy π-nuclear scattering, and the resonances could well be there even if the attraction is too weak to bind. The possibility of resonances of this kind was first pointed out in a different context several years ago for

^4He[13]. Experimentally it is therefore quite important to carefully study the energy dependence of π-nuclear scattering over the first 50 MeV in nuclei which are plausible candidates: ^{28}Si and ^{40}Ca are good cases, while heavy elements have the defect that absorption prevents exploration of the central region. The states should be expected to be about 15 MeV wide.

II. NUCLEAR CRITICAL OPALESCENCE

There is at present no evidence that actual nuclei can undergo a phase transition into a π-condensate; in fact the evidence, if any, is negative. At the same time, it has become clear that nuclear density is rather close to the one required for a condensation phenomenon. Indeed, the possibility of π-nuclear bound states as just discussed shows clearly that we are very close indeed to a region in which pions develop quite new phenomena in their behaviour with nuclei. This raises in a very clear way the question whether it is possible to have observable precursor phenomena to a nuclear phase transition. Very recently Delorme and M. Ericson[14] have noticed that the situation in the nucleus is similar to the phase transition in an antiferromagnet: before the onset of the phase transition the $\underline{\sigma}\cdot\underline{\nabla}\Phi$ coupling induces long range correlations between the nuclear spins. These should be physically observable using spin dependent probes. Following their reasoning we will now investigate how this develops physically. As a preparation we will first discuss the nuclear π-field in the static limit qualitatively, and pedagogically so as to display the central physical phenomena and their strong analogy to the theory of e.m. dipoles.

1. Pion Field in the Nucleus in the Static Limit ($\omega = 0$)[15]

In the static limit, the nucleon, which has a permanent axial-dipole moment, radiates a dipole field

$$\phi(\vec{x}) = -\frac{g_r}{8\pi m_N}\,\vec{\sigma}\cdot\vec{\nabla}\,\frac{e^{-m_\pi x}}{x} \tag{13}$$

This form for the field resembles the e.m. dipole field in classical physics: this is the starting point for far-reaching analogies between the two situations which affect very non-trivial phenomena. We will emphasize this analogy strongly in the following. The difference with the electric dipole field $V(x) = -(1/4\pi)\vec{M}\cdot\vec{\nabla}\,1/x$ arises from the finite pion mass, which makes the Yukawa field short-ranged, while the electromagnetic one is long-ranged. This creates quantitative differences between the two cases, but it does not raise any difficulty for the analogies at the conceptual level. The pionic dipole moment corresponding to

$$\vec{M} \quad \text{is} \quad -\frac{g_r}{8\pi m_N}\vec{\sigma} \quad .$$

When several nucleons are assembled to form a nucleus, the simplest assumption takes the nucleonic dipoles to radiate as if they were free, so that the total field

$$\Phi(\vec{x}) = \frac{-g_r}{8\pi m_N}\sum_i \vec{\sigma}_i\cdot\vec{\nabla}\,\frac{e^{-m_\pi|\vec{x}-\vec{x}_i|}}{|\vec{x}-\vec{x}_i|} \tag{14}$$

This is called the impulse approximation. However, it is not correct, since a nucleon submitted to the pionic field of its neighbours responds with the appearance of an induced dipole, which in turn radiates.

We must therefore study the response to an external pion field, the "axial polarizability" of the nuclear system. This concept is in fact quite analogous to the response of a system to an external electric (or magnetic) field, which can be expressed by a structure constant characteristic of the system, the electric polarizability. Indeed the derivation of the π-nuclear optical potential and the Lorentz-Lorenz effect for physical pions is based on this analogy[1,4].

Consider first the response of a (spinless) point nucleon to an incident p-wave pion in the limit $k \to 0$. The incident and scattered pion waves are

$$\phi(r) = \frac{1}{3}\left[(\underset{\sim}{k}\cdot\underset{\sim}{r}) + 3\left(\lim_{k\to 0}(\delta_p/k^3)\right)\frac{\underset{\sim}{k}\cdot\underset{\sim}{r}}{r^3}\right] \tag{15}$$

where the first term is the incident wave and the second is the outgoing scattered wave proportional to the p-wave scattering volume $\lim\,(\delta_p/k^3) = c_0/3$. If we compare this to electric response to a field $\underset{\sim}{E}$ due to the induced dipole $\underset{\sim}{P}$, the electric potential is

$$V(\underset{\sim}{r}) = -\underset{\sim}{E}\cdot\underset{\sim}{r} + \frac{1}{4\pi}\frac{(\underset{\sim}{P}\cdot\underset{\sim}{r})}{r^3} \tag{16}$$

with the induced dipole moment proportional to the applied field

$$\underset{\sim}{P} = a_{el} \cdot \underset{\sim}{E} \quad . \tag{17}$$

From the structure of Eqs. (15) and (16) we see at once that there is a direct correspondence

$$a_{el} \leftrightarrow -4\pi \cdot 3 \cdot (\lim \delta p/k^3) \equiv -4\pi c_o \tag{18}$$

Therefore to the electric-dipole polarizability corresponds the axial-dipole polarizability $-4\pi c_o$.

We can now easily generalize the impulse-approximation for the pion field to include the dipole response (i.e. the pion rescattering) from the individual nucleons. The reasoning follows that for the electric response with scattering due to induced dipoles:

$$\hat{\phi}(\vec{x}) = -\frac{g_r}{8\pi m_N} \Sigma_i \vec{\sigma}_i \cdot \vec{\nabla} \frac{e^{-m_\pi |\vec{x}-\vec{x}_i|}}{|\vec{x}-\vec{x}_i|} + \frac{1}{4\pi} \Sigma_i \vec{p}_i \cdot \vec{\nabla} \frac{e^{-m_\pi |\vec{x}-\vec{x}_i|}}{|\vec{x}-\vec{x}_i|} \tag{19}$$

Here $\vec{p}_i$ denote the induced dipole for the nucleon i. The pion field is an operator in the nucleonic variables (denoted by ^). As compared to the general case there are two simplifications at $\omega = 0$:

a) There is no s-wave response (rescattering) because of soft pion theorems.
b) There is no physical π absorption.

It is convenient to define the spin density $\vec{\sigma}(\vec{x})$ and the polarization density $P(\vec{x})$ between states i and f. These are transition densities as is the pion field itself:

$$\vec{\sigma}(\vec{x}) = \langle f | \Sigma_i \vec{\sigma}_i \; \delta(\vec{x}-\vec{x}_i) | i \rangle$$

$$\vec{\underline{P}}(\vec{x}) = \langle f | \Sigma_i \vec{p}_i \; \delta(\vec{x}-\vec{x}_i) | i \rangle$$

If we now apply the operator $(-\nabla^2 + m_\pi^2)$ to eq. (19) we obtain the Klein-Gordon equation at $\omega = 0$.

$$(-\nabla^2 + m_\pi^2)\, \phi(\vec{x}) = -\frac{g_r}{2m_N} \vec{\nabla} \cdot \vec{\sigma}(\vec{x}) + \vec{\nabla} \cdot \vec{P}(\vec{x}) \tag{20}$$

One recognizes in this equation the axial analogue of the Poisson equation for the electric potential

$$-\nabla^2 V(x) = \rho(x) + \rho_{pol}(x) \tag{21}$$

with $\rho_{pol} = \vec{\nabla}\cdot\vec{P}$. In the axial cases the free charges are given by $-g_r/2m_N\vec{\nabla}\cdot\sigma(x)$ and $\vec{\nabla}\cdot\vec{P}$ are the polarization charges. These last charges are ignored in the impulse approximation.

The field radiated by the induced dipole $\vec{p}_i$ represents also the field scattered by the nucleon i, as is clear from Eq. (15). Therefore Eq. (19) is nothing other than the multiple scattering equation for the virtual pion, which takes into account the rescattering of the virtual pion, a process ignored in the impulse approximation.

We now proceed exactly like in electromagnetism where the induced dipole is

$$\vec{p}_i = a\,\vec{E}_{eff}(x_i) \tag{22}$$

and $\vec{E}_{eff}(x_i)$ is the effective exciting field $\vec{\nabla}\phi(\vec{x},\vec{x}_i)$ at $\vec{x} = \vec{x}_i$. The effective exciting field is the one produced by all the nucleons but the nucleon i itself

$$\hat{\phi}(\vec{x},\vec{x}_i) = -\frac{g_r}{8\pi m_N}\Sigma_{j\neq i}\,\vec{\sigma}_j\cdot\vec{\nabla}\,\frac{e^{-m_\pi|\vec{x}_j-\vec{x}|}}{|\vec{x}_j-\vec{x}|} + \frac{1}{4\pi}\Sigma_{j\neq i}\,\vec{p}_j\cdot\vec{\nabla}\,\frac{e^{-m_\pi|\vec{x}_j-\vec{x}|}}{|\vec{x}_j-\vec{x}|} \tag{23}$$

For this effective field, the summing over $i\neq j$ introduces the effect of pair correlations in a rigorous discussion: when these have short range and are repulsive they are the origin of the Lorentz-Lorenz effect in the usual treatment of the π-optical potential. Here they produce an analogous effect.

If we, therefore, consider a spin in a hole in the medium (simulating short range anti-correlations) we see immediately that the medium is polarized. In terms of $\alpha_o = 4\pi c_o\rho$ the polarization

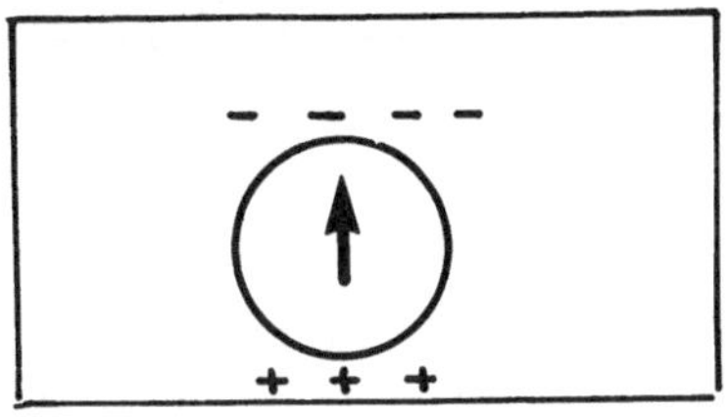

density $\underset{\sim}{P}(\underset{\sim}{x})$ is

$$\underset{\sim}{P} = \alpha_0 \underset{\sim}{\nabla}\phi - \frac{1}{3}\frac{g_r}{2M}\alpha_0 \underset{\sim}{\sigma} - \frac{1}{3}\alpha_0 \underset{\approx}{P} \tag{24}$$

or since $\alpha = \dfrac{\alpha_0}{1+\alpha_0/3}$

$$\underset{\sim}{P} = \alpha\underset{\sim}{\nabla}\phi - \frac{1}{3}\frac{g_r}{2M}\alpha\underset{\sim}{\sigma} \tag{25}$$

If we now introduce this into the Klein-Gordon equation it will read

$$\underset{\sim}{\nabla}(1-\alpha)\underset{\sim}{\nabla}\phi + m_\pi^2\phi = -\frac{g_r}{2M}\underset{\sim}{\nabla}(1 - \frac{1}{3}\alpha)\underset{\sim}{\sigma}(\underset{\sim}{x}) \tag{26}$$

Apart from the usual L-L effect in the propagation we therefore see that it also renormalizes the apparent source strength by $(1 - 1/3\ \alpha)$ [16].

There are various corrections in a more detailed and sophisticated treatment but they do not change the overall picture. However, they change the strength of the L-L effect. This is usually expressed by the parameter ξ which is 1 for an undiluted L-L effect.

The principal corrections are

a) <u>the nucleon size (interaction range)</u>. If the range is larger than the correlation distance the LL effect is wiped out.

b) <u>Pauli correlations</u>[16]. These have longer range than the interaction. They therefore restore part of the L-L effect. In contrast to hard core correlations they are repulsive only in states between equal particles. At nuclear matter density ξ = 0.42 they reduce the correlation effect to about one half.

c) Response to the ρ-field[6]. The axial polarization is more generally

$$\underset{\sim}{P} = \alpha_\pi \underset{\sim}{E}_n + \alpha_\rho \underset{\sim}{E}_\rho \quad .$$

When the joint effect of π and ρ is introduced this brings ξ back to ≈ 1. This result is quite model dependent since the range of ρ is short and its coupling is large and uncertain. Therefore ξ may be as large as 1.5.

2. Nuclear Critical Opalescence[14]

If nuclei had a pion condensate phase these would be characterized by a pion field of non-vanishing expectation value $\langle\phi\rangle \neq 0$. This field would introduce long range order between the nuclear spins by the $\sigma\cdot\nabla\langle\phi\rangle$ coupling. Even before the phase transition the $\underset{\sim}{\sigma}\cdot\underset{\sim}{\nabla}\tilde{\phi}$ coupling will introduce ordering and correlations in the system as fluctuations. Contrary to the correlation phenomena of the condensed phase those due to fluctuations are not direction dependent: slow neutrons show magnetic critical opalescence in scattering from polycrystalline media, reflecting that the short range spin order increases its range to ∞ as the Curie point is approached[17].

The pion condensate introduces spin-correlations very similar to those of antiferromagnets. In this case the peak of the critical scattering with a spin sensitive probe occurs at a finite momentum transfer q_c. As in a magnetic system the local spin order is described by the spin-spin correlation function which is the quantity to study. In the nuclear case we might in addition have the remarkable feature that pion field, itself responsible for the ordering, could be directly accessible to experimentation with luck (for example by π^--nucleus → 2γ + nucleus reaction).

Since the correlations will have range small compared to the radius of a heavy nucleus, we will consider ∞ nuclear matter.

In order to agree with the notations for a magnetic material (susceptibility χ) and so as to display the previously discussed πN range effect explicitly[18] we write the response function α(q) as $-\chi_e(q)\, v^2(q^2)$ with v(q) the πN form-factor. For an assembly of nucleons located at $\underset{\sim}{x}_i$ in the nuclear medium the momentum space pion field is (compare eq. (26)):

$$\phi(\omega=0,\underset{\sim}{q}) = i\,\frac{g_r}{2m_N}\sum_i (1+\chi_e/3)\,\frac{\underset{\sim}{\sigma}_i\cdot\underset{\sim}{q}\; v(q^2)\; e^{i\underset{\sim}{q}\cdot\underset{\sim}{x}_i}}{[m_\pi^2+q^2(1+\chi_e(q)v^2(q^2))]} \qquad (27)$$

While we previously discussed the polarizability χ_e as having its source in the intrinsic excitation of the nucleons, it has an additional important part from low lying nuclear excitations due to spin flip excitations of particle-hole type $[\beta(q^2)]$. In magnetic language we would call the very high excitations of the first type (Δ-mass) "diamagnetism" and the second ones "paramagnetism". The latter are of course quite sensitive to the detailed structure of the medium.

For momentum transfers $q < 2P_{Fermi}$. We have for an ∞ Fermi gas

$$\beta(q^2) = \beta[1 - \frac{1}{12} q^2/P_F^2 + \ldots] \tag{28}$$

with $\beta = \dfrac{-2Mp_F}{\pi^2} \simeq -2.7$ at $\rho = \rho_0$.

The total response is $\chi(q) = -\alpha_0 + \beta(q)$. Since $|\alpha_0| \simeq 0.9$ we have thus $|\beta| > \alpha_0$. The effective response is reduced by the L-L effect as previously discussed[1,6]

$$\chi_e(q) = \frac{\chi(q)}{1-g'\chi(q)} \tag{29}$$

where $g' = 1/3$ to $1/2$[19]. One should note that $\chi(0) \simeq -3.6$ is quite large numerically.

From the expression (27) for the pion field we see that the propagator is renormalized from its free value by $R(q^2)$

$$R = (m_\pi^2 + q^2)/ [m_\pi^2 + q^2(1+\chi_e(q^2)v^2(q^2))] \tag{30}$$

As $\chi_e < 0$ and $v(q^2) \to 0$ for $q^2 \to \infty$, R increases from 1 at $q = 0$, reaching again the value 1 at $q^2 = \infty$. In order to simplify the discussion we use the first order expansion $v^2(q^2) \gtrsim 1 - q^2r^2/3$. This approximation gives the explicit dependence of the critical parameters on the r.m.s. radius r (its accuracy is around 10%). Similarly we expand $\chi_e(q^2)v^2(q^2) \gtrsim \chi_e(1 - q^2s^2/3)$ with $\chi_e \equiv \chi_e(0)$ and $s^2 = r^2 + \beta/[4p_F^2\chi(1 - g'\chi)]$. Above $\rho \gtrsim 0.5\,\rho_0$, s is a slowly varying function of the density which remains close to its limiting value r for infinite density and depends very moderately on g', so that

$$s \simeq r \text{ for } \rho > \rho_0/2 \tag{31}$$

The renormalization factor has therefore schematically the following shape

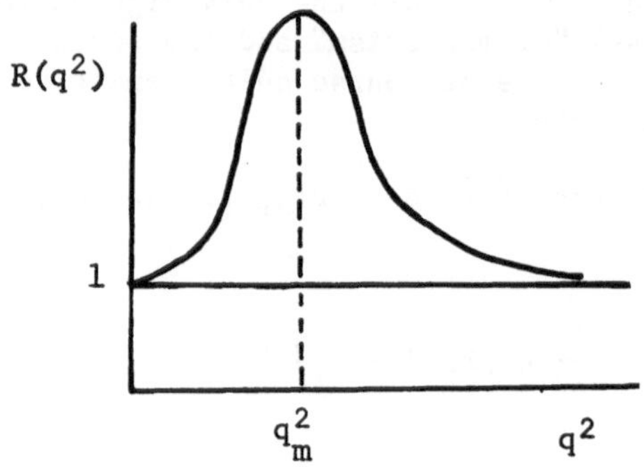

In the quadratic approximation the properties of $R(q^2)$ are easily solved for: The renormalization factor R has then a single maximum at a value q_m of the momentum practically independent of the density: $q_m^2 \approx -m_\pi^2 + \sqrt{3}\, m_\pi/s$. The evolution of R when one varies the density (which fixes the parameters χ_e and s) can be traced from the properties of the poles of the propagator. A double pole is obtained when:

$$(1 + \chi_e)^2 + \frac{4}{3} m_\pi^2 s^2 \chi_e = 0 \tag{32}$$

This equation is equivalent to the two implicit equations in the density (the indices 1 and 2 correspond to the + and - signs):

$$\chi_e^{1,2} = -(1 + \frac{2}{3} m_\pi^2 s^2) \pm \frac{2}{\sqrt{3}} m_\pi s \left(1 + \frac{1}{3} m_\pi^2 s^2\right)^{1/2} \quad . \tag{33}$$

Except in the region of low densities ($\rho/\rho_0 < 0.5$) which are not of interest for our purpose, the variation of the right hand side with the density is very small and may be neglected in practice. As on the other hand the r.h.s. is nearly independent on the Lorentz-Lorenz parameter g', the two polarizabilities χ_e^1 and χ_e^2 are well determined, irrespective of its value. With $c_0 = 0.15\ m_\pi^{-3}$ and $r = 0.48\ m_\pi^{-1}$, we find numerically: $\chi_e^1 \approx -0.57$ and $\chi_e^2 \approx -1.75$.

These two characteristic values define three regions. For $|\chi_e| < |\chi_e^1|$ the two poles are negative and correspond thus to unphysical values of q^2. Between χ_e^1 and χ_e^2, they are complex conjugate. They become real and positive beyond the second characteristic point $\chi_e^2 \equiv \chi_e^c$ which thus corresponds to the condensation threshold†. The double pole q_c^2 is then the critical momentum of the condensation:

$$q_c^2 = -2m_\pi^2/(1 + \chi_e^c) \stackrel{\sim}{\sim} -m_\pi^2 + \sqrt{3}\, m_\pi/s_c \stackrel{\sim}{\sim} q_m^2 \quad . \tag{34}$$

Numerically we get $q_c^2 = 2.6\ m_\pi^2$.

For what concerns the determination of the critical density ρ_c, the exact value of the Lorentz-Lorenz parameter is essential: in order for the polarizability to reach the effective value -1.75, the density has to be $\rho_c/\rho_0 \stackrel{\sim}{\sim} 1.3$ for g' = 1/3 but as large as $\rho_c/\rho_0 \stackrel{\sim}{\sim} 6$ for g' = 0.5. When g' attains values of the order of 0.5, the critical density becomes very sensitive to small changes in the critical polarizability χ_e^c induced by variations of the input data. Beyond the value $g' = |\chi_e^c|^{-1} = 0.57$ the condensation cannot occur at all even for large compressions but this does not prevent the occurrence of critical phenomena. Hence the critical polarizability is well determined while the critical density is very uncertain or may even not exist.

In conclusion the maximum of the renormalization of the pion propagator becomes infinite at the critical density. Below the condensation threshold it can be expressed approximately as:

$$R_m = \chi_e^c/(\chi_e^c - \chi_e) \tag{35}$$

Here χ_e^c is well known whereas χ_e for a given density depends sensitively on g'. For instance, at $\rho = \rho_0$ one finds $R_m = 5$ for g' = 0.5 but a much larger value is attained for g' = 1/3 where χ_e is close to χ_e^c. Away from q_m (for instance $q \lesssim m$), R has a smaller value which is not as sensitive to g'.

This enhancement of the pion field in this momentum range expresses the strong fluctuations of the order parameter. It appears in the subcritical region as a precursor of the Bragg peak corresponding to the condensed field. It is the critical opalescence phenomenon very similar to the one observed for antiferromagnets near the superstructure lines[21]. Our numerical estimate shows that this phenomenon is pronounced at the ordinary density (at least in infinite nuclear matter) even if the critical density is far away or even cannot be reached at all. This may look surprising but our treatment suggests that the expansion parameter should not be $(\rho - \rho_c)$ but instead $(\chi_e - \chi_e^c)$.

---

†A relation similar to eqn. (32) has been derived by Ericson and Myhrer[20] from the condition for a pion-nucleus bound state at $\omega = 0$.

Although experiments always involve a definite momentum (or a definite range of momenta), it is instructive to discuss the behaviour of the field in x-space so as to see how the long range ordering of the spins is established. For simplicity we take one nucleon at $x = 0$ and consider form-factor effects only in the propagator in consistency with the preceding discussion. Anyway our first order expansion of $v(q^2)$ does not allow a proper description of the high momentum components and thus of the close vicinity of the source.

The pion field $\phi(\underset{\sim}{x})$ is expressed in terms of the eigenfrequencies $q_\pm^2$ of the propagator:

$$q_\pm^2 = \{(1 + \chi_e) \pm [(1 + \chi_e)^2 + \frac{4}{3} m_\pi^2 s^2 \chi_e]^{1/2}\}/\frac{2}{3} s^2 \chi_e \quad . \quad (36)$$

One can distinguish two cases according to the value of the density (we exclude the condensation region $\rho > \rho_c$ where our model is not valid).

i) in the low density region $\rho < \rho^1$ (where ρ^1 corresponds to the first characteristic point $\chi_e^1 = -0.57$), the roots are both negative and the field has the form:

$$\phi(\underset{\sim}{x}) \sim (|q_+|^2 - |q_-|^2)^{-1} \underset{\sim}{\sigma}\cdot\underset{\sim}{\nabla}[(e^{-|q_-|x} - e^{-|q_+|x})/x] \quad . \quad (37)$$

At the lowest densities the Yukawa fall-off occurs with the effective mass m'_π in the medium: $|q_-| \simeq m'_\pi = m_\pi/(1+\chi_e)^{1/2}$ and the other Yukawa function has a much shorter range which characterizes the nucleon form factor. When ρ tends to the value ρ_1, both $|q_-|$ and $|q_+|$ converge to $|q_1| = \sqrt{2}\, m_\pi/(1+\chi_e^1)^{1/2}$ and $\phi(\underset{\sim}{x})$ towards $(2|q^1|)^{-1} \underset{\sim}{\sigma}\cdot\underset{\sim}{\nabla}\, e^{-|q^1|x}$.

ii) for an intermediate density $\rho^1 < \rho < \rho_c$, the roots are complex conjugate and the field becomes:

$$\phi(\underset{\sim}{x}) \sim (2\eta\varepsilon)^{-1} \underset{\sim}{\sigma}\cdot\underset{\sim}{\nabla}(\sin \varepsilon x\, e^{-\eta x}/x) \quad (38)$$

where $\varepsilon = |\mathrm{Re}q_\pm|$ and $\eta = |\mathrm{Im}q_\pm|$. Between ρ^1 and ρ_c, ε increases from 0 to q_c and η decreases from $|q^1|$ to 0. The field is characterized by the occurrence of a sinusoidal modulation which announces the ordered phase.

iii) When the critical density is approached, the Yukawa damping disappears and $\phi(\underset{\sim}{x})$ tends to $\phi_c(\underset{\sim}{x}) \sim (2\eta)^{-1} \underset{\sim}{\sigma}\cdot\underset{\sim}{\nabla}(\sin q_c x/q_c x)$. As the field is the agent for spin orientation, the range of the spin ordering becomes infinite with that of the field (Coulomb-like

x^{-1} behaviour) with the modulation sin $q_c x$ characteristic of the condensation. In this theory based on the linear response, the critical exponent for the range of the spin correlation in the disordered phase is 1/2 since η goes to zero as $|\chi_e - \chi_e^c|^{1/2}$.

It is clearly urgent to experimentally probe the spin correlation function of the nucleus and also the π-nuclear field in the neighbourhood of the critical momentum q_m (see ref. 15). It may well be that certain of the principal excited states contributing to the spin-spin correlation function could also show effects from the proximity of a condensate at similar momenta.

Much theoretical effort remains to be done in order to extend the nuclear matter treatment of the field fluctuations to finite systems. Some studies of the condensation problem in finite nuclei have been performed with the conclusion that they do not drastically differ from nuclear matter for what concerns the threshold[22,23]. It is thus natural to hope that the new physical effect that we predict at subcritical densities in infinite systems, namely an enhancement of the pion field near the critical momentum, will survive in actual nuclei.

III. Axial Locality[24,32]

The idea is to provide an explicit construction of the axial current in an extended object (such as a nucleus) and to show that it is naturally linked to the local s-and p-wave interaction of pions in the system. The problem is therefore to establish a possible connection of the axial current with pionic properties in the nuclear medium. It is no surprise for the reader to say that one of the corner-stones in our considerations is the PCAC relation[25].

For the nucleon the axial current is indeed connected to pions in this way for quasi-elastic processes, provided the momentum transfer is so small that multi-meson excitations are not explored. In this limit the "pionic" form factor of the nucleon is nearly a constant. The system can then be regarded as effectively pointlike, provided the long-range pion pole contribution is first separated out (the pseudoscalar form-factor). The axial form-factor follows then from the pion source by the Goldberger-Treiman relation[26].

Extended systems like nuclei differ fundamentally from the nucleon in that the leading excitations are not the multi-meson states. Instead there is characteristically a vast number of states below pion threshold. The nuclear size is the leading scale in this problem; it is not related (at least in any direct manner) to the multi-meson scale. These two scales are distinctly different. If we now choose to consider non-mesic substructures in the nucleus (nucleons, "quasideuterons", etc.) these will characteristically have a mesonic scale rather than a long-range scale like the size of system. This suggests that there is a substantial region of momentum transfers for which the constituents can be considered pointlike in the previous sense. In this region it may therefore be possible to give an explicit program linking the axial current to the pion interaction. Since the local scale parameter must enter into this consideration, we must be prepared to use the local properties of the pion source in the medium in an essential way.

The conventional approach to the study of the axial current in nuclei has implicitly recognized this point. It describes a nucleus as consisting of a set of nucleons in interaction and takes the impulse approximation as a starting point. When the kinematics is close to the quasi-free process on nucleons, this approximation has proved to be quite good (β decay, μ capture). The corrections to this picture involve the interacting meson field and are introduced as "exchange currents"[27]. This amounts to adding a two-nucleon current (predominantly from pion exchange) to the single nucleon current of the impulse approximation. In a recent paper, J. Delorme et al.[16] have shown that, for static nucleons, the axial current can be related to the pionic field in the nucleus in such a way that all modifications to the impulse approximation are identical for the axial current and for the pion source. From this important result it follows that the axial current in this case is a pionic phenomenon. It has in particular the consequence that the renormalization of the axial coupling constant in the nuclear medium is seen as a result of the change in the pion coupling. A close parallel between the behaviour of the electrostatic displacement vector $\vec{D}$ and the axial vector $\vec{A}$ has been established[28], in the sense that both acquire an induced "polarization" coming from the induced "dipoles" in the medium.

While the picture of Delorme et al.[16] can to some extent be generalized also to the nearly static case, it leaves the question of the axial current at large energy transfers open (large on the nuclear scale). In this case one expects the impulse approximation to fail severely as it is known to do in π capture[29]. It is then necessary to take a different viewpoint of the problem. As outlined below, the emphasis will be put on the local distribution of the pionic source in the medium and an explicit reference to how and by what these pionic properties are generated will be avoided.

In the following we will proceed in two steps. We first show that, as a consequence of PCAC, the time component A_0 of the axial current has its matrix elements completely determined by s-wave pion absorption on the system to leading order in the momentum transfer $\vec{q}$. This rigorous result from PCAC determines a piece of the current which contains both pion-pole and pole-free contributions. Its pole-free part would not exist if the current were built from individual nucleon currents in the limit of very massive nucleons. For large values of the energy transfer q_0 (on the nuclear scale), the s-wave pion absorption becomes very important, however, and to speak in terms of the impulse approximation is then meaningless.

When the energy transfer q_0 is large, it is not possible to derive the spatial part of the axial current from PCAC alone, even for a momentum transfer $\vec{q} = 0$. However, since for small systems the pole-free part of this current is closely related to the p-wave pion amplitude, we will conjecture "axial locality". By this we mean that the pole-free axial spatial current $\tilde{\vec{A}}$ is in essence given by the local p-wave interaction in the medium. This local connection is established in terms of the full pionic source at the same point, and it will be discussed in detail below. This conjecture provides a natural basis for a unified discussion of the axial current in terms of local pion properties in the system. As a consequence, it becomes essential to understand the physics of off-shell pions in the nuclear medium in order to quantitatively understand the axial current.

1. Reminder on PCAC

The conservation of electric charge means that the corresponding 4-current is conserved and has the consequence that the electron charge e is a universal unit. Similarly the conservation of the vector current (CVC) has the important consequence that the weak vector charge g_V is a universal quantity unchanged by strong interactions.

Can a similar relation hold for the axial current? The answer is no for two simple reasons. First, $g_A = 1.26$ and not unity for the nucleon, differing from the leptonic value. It misses unity but not by very much. Strict conservation (CAC) is therefore impossible. Even worse, a conserved axial current has the unpleasant feature of making the pion stable. To arrange the pion decay, the assumption is that there is something very special about the pion as compared to other mesons, simply because it is the lightest pseudoscalar meson by quite a margin. In addition the divergence $\partial_\mu A_\mu^{(-)}(x) \neq 0$ is a pseudoscalar of isospin 1 so it has exactly the pion quantum numbers. The PCAC assumption[25] is that the divergence is proportional to the pion field $\phi_\pi(x)$ with a universal constant f_π

$$\text{PCAC:} \quad \partial_\mu A_\mu^{(-)}(x) = f_\pi m_\pi^2 \, \phi_\pi^{(-)}(x) \tag{39}$$

In Eq. (39) we have used a symbolic notation also for the pion field $\phi_\pi(x)$ which represents the transition field $\langle i|\phi_\pi^{(-)}(x)|f\rangle$ between initial and final nuclear states. The constant f_π is given by the pion decay rate.

The PCAC relation is a restriction on the longitudinal component of the axial current. It is therefore immediately clear that PCAC by itself is insufficient in general for a detailed and unique relation between $A_\mu(x)$ and the pion source. As an example: if we go to the limit of $\underset{\sim}{q} = 0$, the PCAC relation uniquely determines $A_0(q_0, \underset{\sim}{q} = 0)$.

The question is then: what can be said about A? The first observation is that PCAC is insufficient to determine $\underset{\sim}{A}$. We have already used PCAC in the determination of A_0; it is not possible to get two unknowns out of one relation. As the saying goes: you can't draw blood from a stone. I will now argue that $\underset{\sim}{A}$ represents the local pion p-wave. To make this seem natural we rewrite the PCAC relation.

2. The Restated PCAC Relation

We can take out the pion pole from the PCAC Eq. (39) by writing $A_\mu = A_\mu^{pole} + \tilde{A}_\mu$, where $\tilde{A}_\mu$ is pole free and A_μ^{pole} = const x $\partial_\mu \phi_\pi(x)$. If we simply recall that the usual Klein-Gordon equation for the pion field with a source is

$$(\partial_\mu^2 + m_\pi^2)\phi_\pi(x) = j_\pi(x) \quad , \tag{40}$$

we obtain immediately the constant by the condition that explicit terms in $\phi_\pi(x)$ should be eliminated.

$$\text{Restated PCAC} \qquad \begin{cases} A_\mu^{pole} = -f_\pi \partial_\mu \phi_\pi(x) \\ \partial_\mu \tilde{A}_\mu = f_\pi j_\pi(x) \quad . \end{cases} \tag{41}$$

In order to see how this new relation can be put to work let us assume we have a "point" nucleus with a source function for s- and p-wave pions

$$j_\pi(\underset{\sim}{x}) = s\partial^{(3)}(\underset{\sim}{x}) - (\underset{\sim}{S}\cdot\underset{\sim}{\nabla})\partial^{(3)}(\underset{\sim}{x}) \quad . \tag{42}$$

Here s and $\underset{\sim}{S}$ are transition matrices with components s_{if} and $\underset{\sim}{S}_{if}$, which carry the nucleus from an initial state i to a final state f on pion absorption.

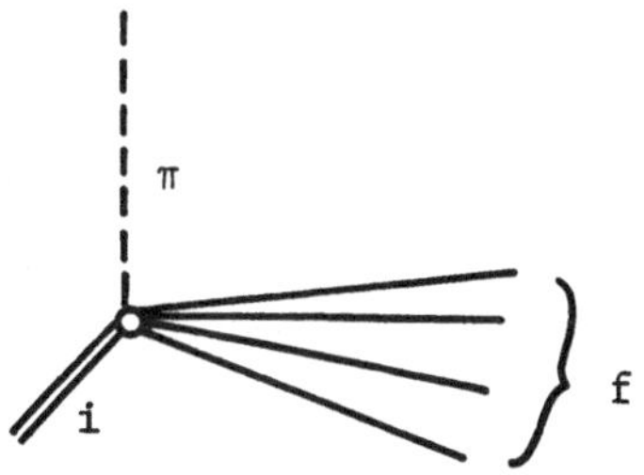

3. Axial Locality Assumption[24]

The question is now how the axial current can be related to the source (42). This is achieved by the following locality hypothesis:

Locality of the axial current (LAC): The pole free part of the axial current $A_\mu(x)$ is determined by the local pion source.

4. Axial Current for Point System

If we apply the LAC hypothesis to point-like s- and p-wave source (32), the result is ($\partial/\partial t = iq_0$):

$$\tilde{A}_0(\underset{\sim}{x}) = -i \frac{f_\pi}{q_0} s\, \delta^{(3)}(\underset{\sim}{x})$$

$$\underset{\sim}{\tilde{A}}(\underset{\sim}{x}) = f_\pi \underset{\sim}{S}\, \delta^{(3)}(\underset{\sim}{x}) \quad . \tag{43}$$

Inside of our hypothesis we note that the time-like part $\tilde{A}_0$ goes with s-wave pions, the space-like part $\underset{\sim}{\tilde{A}}$ goes with p-wave pions. For a pointlike system the reasonableness of the LAC hypothesis is obvious. It would in fact be disturbing to have the axial current associated with points at which there is no matter. On the other hand, even elementary particles are extended objects. A very strict application of the locality principle would then mean that the axial current is associated with s- and p-wave pions in the particle. We do not have to face this question here. As an example, the characteristic distance for μ-capture at large energy transfers is 2 fm and so it is at low energy transfers. Therefore we must ask how point-like the hadronic system is on this scale, thereby avoiding the more difficult question. With this physical interpretation of "locality", we will illustrate our viewpoint with a non-relativistic description of the Goldberger-Treiman relation.

5. The Non-Relativistic Goldberger-Treiman Relation

If we express the non-relativistic axial current at $q_0 = 0$ and choose only the pole free part, the expressions for $\underset{\sim}{A}(x)$ and the source function are

$$\begin{cases} \tilde{\underset{\sim}{A}}(\underset{\sim}{x}) = g_A \underset{\sim}{\sigma}\, \delta^{(3)}(\underset{\sim}{x}) \\ j_\pi(\underset{\sim}{x}) = \frac{ig_r}{2M} (\underset{\sim}{\sigma}\cdot\underset{\sim}{\nabla})\delta^{(3)}(\underset{\sim}{x}) \end{cases} , \tag{44}$$

where g_r is the renormalized πN coupling constant. The PCAC relation with the locality condition (LAC) gives according to Eq. (43)

$$g_A = f_\pi \frac{g_r}{\sqrt{2}M} . \tag{45}$$

This result is the usual Goldberger-Treiman relation[26] which is known to be accurate to 5% without form factor effects[33]. This derivation of the Goldberger-Treiman relation demonstrates that the LAC principle is implicit in the usual applications of PCAC. In the nuclear case the explicit locality statement serves first to separate the contributions to $A_0(x)$ and to $\tilde{\underset{\sim}{A}}(x)$ in a clear way, and secondly (which is its most valuable effect) to determine the axial current inside the extended nucleus.

6. The Extended Nucleus

For an extended system such a simple relation with the global pion amplitudes cannot be expected to be true. In this case an s-wave or p-wave with respect to a local point in the system may even correspond to a very high partial wave with respect to its centre of mass. If one instead takes the attitude that Eqs. (43) express that the (pole free part of the) axial current is given by the local matter density, it becomes reasonable to insist that this should be so also in an extended system. The equivalent relations of Eqs. (43) in this case are particularly easy to visualize, if we think of the extended system as consisting of small, discrete constituents.

The pion source function for local s-and p-wave pions is now described by density functions $s(q_0,x)$ and $\underset{\sim}{S}(q_0,\underset{\sim}{x})$, which contain the <u>full effect</u> of pion rescattering in the medium.

$$j(q_0,\vec{x}) = s(q_0,\vec{x}) - \vec{\nabla}\cdot\vec{S}(q_0,\vec{x}) \tag{46}$$

By the axial-locality hypothesis Eq. (46) gives the explicit form for $A_\mu(q_0,\underset{\sim}{x})$

$$\tilde{A}_0\ (q_0,\vec{x}) = -i\ \frac{f_\pi}{q_0}\ s(q_0,\vec{x}) \tag{47a}$$

$$\tilde{\vec{A}}\ (q_0,\vec{x}) = f_\pi\ \vec{S}(q_0,\vec{x}) \tag{47b}$$

This is the basic result of axial locality. One should note that eqs. (47) are very information-rich. In particular they give back the traditional description of low q_0 process including exchange currents[16].

We emphasize that the conjectures (47) is consistent with PCAC, but it does not follow from it for an extended system. In addition it should be clearly realized that the proposed range of validity of Eqs. (47) is on the nuclear scale of energies and distances: the short range behaviour in the source cannot be explored. The implication of the locality principle is that it provides a well-defined model independent frame for calculation of the nuclear axial current by insisting that it is identical to the local pion current. All that is needed is therefore the correct pion sources $s(\vec{x})$ and $\vec{S}(\vec{x})$ whether these are derived from a particular theoretical model or are extracted from the physical π-nuclear interaction.

7. Analogy With Dielectrics

As an illustration of the content of Eqs. (47) we give the explicit expression for the axial current in a model. Consider a direct absorption (or emission) vertex for a pion in the nuclear medium. We describe this by a local s-and p-wave transition (pseudo) potential

$$j_{direct}(\vec{x}) = \sigma(\vec{x}) - \vec{\nabla}\cdot\vec{\Sigma}(\vec{x}) \tag{48}$$

In addition the pion has a source connected to its rescattering in the nuclear medium. We describe this by a non-local optical potential of the usual type dominated by local s-and p-wave pion scattering

$$2m_\pi V = Q(\vec{x}) + \vec{\nabla}\cdot\alpha(\vec{x})\vec{\nabla} \quad . \tag{49}$$

Although of little importance to our argument, we note that repeated absorption and emission of pions also generates scattering in the medium. Such effects are included into the phenomenological optical potential above to the extent consistent with no double counting[31]. With these interactions the pion field in the medium, associated with total energy q_0, is denoted by $\Phi(q_0,\vec{x})$ and the full pion source functions are given by

$$s(q_0,\vec{x}) = \sigma(\vec{x}) - Q(\vec{x})\,\Phi(q_0,\vec{x})$$

$$\vec{S}(q_0,\vec{x}) = \vec{\Sigma}(\vec{x}) + \alpha(\vec{x})\vec{\nabla}\Phi(q_0,\vec{x}) \quad . \tag{50}$$

From eqs. (41) and (47) the total axial current is constructed as

$$A_0(q_0,\vec{x}) = -i\,\frac{f_\pi}{q_0}\,[q_0^2\Phi(q_0,\vec{x}) + \sigma(\vec{x}) - Q(\vec{x})\Phi(q_0,\vec{x})]$$

$$\vec{A}(q_0,\vec{x}) = f_\pi[-\vec{\nabla}\Phi(q_0,\vec{x}) + \vec{\Sigma}(\vec{x}) + \alpha(\vec{x})\vec{\nabla}\Phi(q_0,\vec{x})] \quad . \tag{51}$$

When written in this form, the axial current shows an interesting analogy with the displacement vector $\vec{D}(\vec{x})$ in an inhomogeneous dielectric. This analogy has been pointed out[16] for the static case $q_0 = 0$. In a medium with a permanent dipole density $\vec{d}(\vec{x})$, the three terms in Eq. (51) would correspond to

$$\vec{D}(\vec{x}) = \vec{E}(\vec{x}) + \vec{d}(\vec{x}) + \vec{P}(\vec{x}) \quad . \tag{52}$$

The contributions originate from the external field $\vec{E} \rightleftarrows -\vec{\nabla}\Phi$, the permanent sources in the medium $\vec{d} \rightleftarrows \vec{\Sigma}$ and the polarization vector $\vec{P} \rightleftarrows +\alpha\,\vec{\nabla}\Phi$ due to the induced dipoles. We see that the same kind of structure remains for the non-static case $q_0 \neq 0$. The time component of the axial current has a completely analogous structure, but it is now associated with s-waves and not p-waves (dipoles). However, this component has no electromagnetic correspondence.

8. Concrete Case Study: μ-capture at Large Energy Transfer ("Strong Interactions of Muons")[30]

Reminder on general properties of μ-capture. Normal μ-capture. The μ-capture in nuclei takes place from the 1s state of the muonic atom. The wave function of the muon is to a good approximation constant over nuclear dimensions, particularly, for light and medium elements. The weak absorption process is the usual V-A interaction for the vector- and axial-currents.

The μ-capture process on the free proton $\mu^- + p \to n + \nu$ has essentially the kinematics $q_0 \simeq 0$; $|\underset{\sim}{q}| \simeq m_\mu = 106$ MeV/c, i.e. $q^2 = -m_\mu^2$. The neutrino carries off nearly all the muon rest mass energy and the neutron is given only a small recoil energy.

Nuclear μ-capture to low excited states ($E^* \ll 106$ MeV) has nearly the same kinematics as for the nucleon with $|\underset{\sim}{q}| \simeq 100$ MeV/c. Since the kinematics corresponds to a quasi-free nucleon process, the conditions are excellent for a description in terms of the impulse approximation (Primakoff theory). For $E^* \lesssim 30$ MeV we can qualitatively understand nuclear μ-capture on this basis sufficiently well to use it for information on nuclear structure.

Breakdown of impulse picture for large E*. For energy transfers larger than about 30 MeV the impulse approximation rapidly breaks down. This region is nearly entirely unexplored both experimentally and theoretically. That is why I have marked this region of "terra incognita" on the figure as it was marked on medieval maps, 'hic sunt dragones' (there are dragons here).

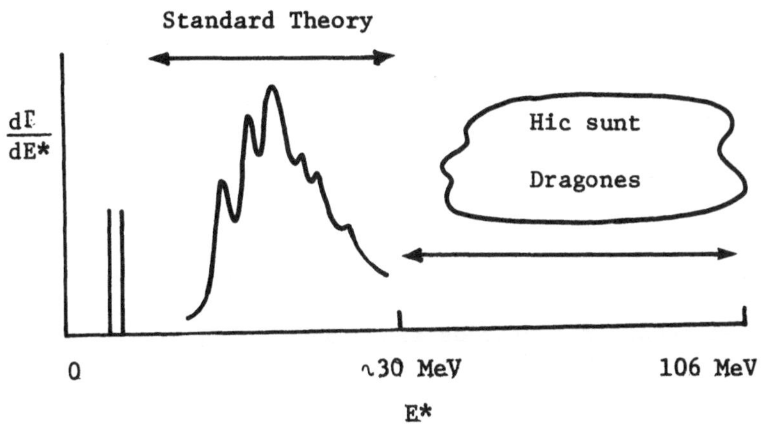

There are a few qualitative experimental indications

a) high-energy neutrons have been obtained with energies up to 60 MeV. They are about 0/00 to % of the total capture rate;

b) high-energy protons have been observed up to 40 MeV and are roughly 10% of the high-energy neutrons;

c) in the limited region of observations both energy spectra are approximately linear on a logarithmic scale.

We can immediately conclude: The high-energy protons cannot result directly from the single nucleon process $\mu^- + p \rightarrow n + \nu$ and are outside the simple impulse approximation.

It is therefore highly likely that the large energy transfer region needs a radical change in theoretical approach.

Kinematics of maximal energy transfer. So as to obtain the purest possible example of large energy transfer, consider the case of μ-capture with no energy going into the neutrino. For $\omega_\nu = 0$ we have

$$q_0 = 106 \text{ MeV } (= m_\mu) \simeq E^*; \; |\underset{\sim}{q}| = 0 \quad ,$$

which is to say that $q^2 = +m_\mu^2$. To an elementary particle physicist the difference $q^2 = \pm m_\mu^2$ may appear rather minor. For nuclei it means a major change of physics. A momentum transfer $|\underset{\sim}{q}| \overset{\sim}{\sim} 100$ MeV/c to a nucleus changes only a form factor, and for light elements it does not even depend very strongly on the nuclear radius, as is well known from electron scattering. An energy transfer of 100 MeV to a nucleus makes great damage and it is quite a traumatic event: it is physically extremely important what amount of energy (q_0) is transferred to the nucleus. This point becomes clearer if we simply compare the transfer of energy-momentum for the absorption of a stopping pion to that of maximal energy transfer in μ-capture

$$q_0^\pi = 139 \text{ MeV} \quad ; \quad |\underset{\sim}{q}| = 0$$

$$q_0^\mu = 106 \text{ MeV} \quad ; \quad |\underset{\sim}{q}| = 0 \quad .$$

The kinematical similarity is striking. We will in the following show that also the physics is closely linked, although non-trivially so. In order to intuitively visualize part of this connection, we choose the pseudoscalar one-pion exchange between a nucleon and the muon (see Fig.).

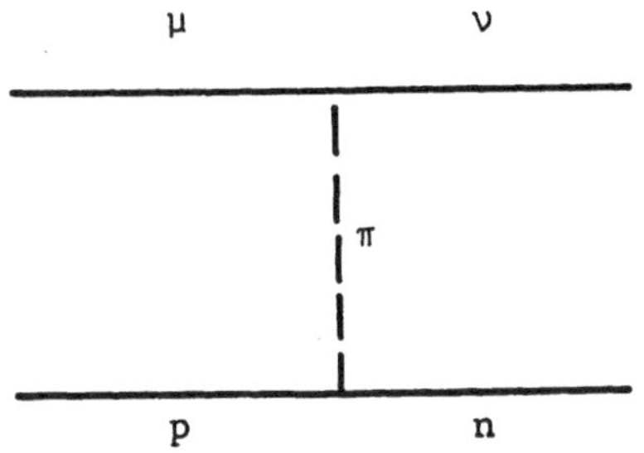

This diagram is usually thought of as the pseudoscalar pion contribution to the nucleon axial current, which then is coupled to the muon. A corresponding pseudoscalar contribution also occurs in the maximal energy transfer μ-capture on a nucleus (see Fig.).

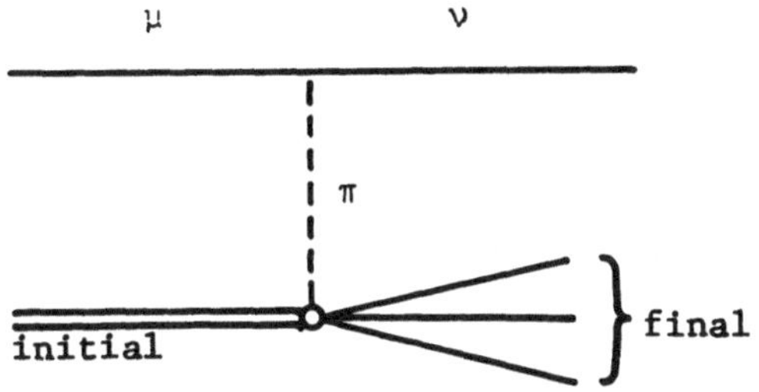

Once more we could try to describe this as a pionic effect in the nuclear axial current. This is, in principle, alright, but it is quite an inefficient way of visualizing the physics. Instead we can look at the upper μνπ vertex as describing the virtual decay of the muon into a pion: in free space the muon surrounds itself by a Yukawa field of pions $e^{-\kappa x}/x$ with $\kappa^2 = m_\pi^2 - m_\mu^2$ or $\kappa^{-1} \simeq 2$ fm. More exactly, the muon will act as a source for the pion field of the nucleus. This will build up a pion wave function which will be absorbed like a near-physical pion by the nucleus. One should at once note that this will not be a simple pole approximation: only if $\kappa R < 1$ will it be possible to describe this process in terms of the external amplitudes in which an incident pion wave falls in on a nucleus. For most nuclei the pion wave is generated well inside nuclear matter by the muon. This changes the physics as we will see shortly.

<u>General expression for μ-capture at $\omega_\nu = 0$.</u> In order to be specific we consider the general expression for μ-capture per unit excitation energy near $\omega_\nu = 0$. This rate $d\Gamma/dE^*$ can symbolically be written

$$\frac{d\Gamma}{dE^*} \propto G^2\psi_\mu(0)^2\omega_\nu^2 \sum_{\text{final states}} \{|\underset{\sim}{V}|^2+|V_0|^2+|\underset{\sim}{A}|^2+|A_0|^2\} \quad . \tag{53}$$

The notation is $G = 10^{-5}/m_p^2$ = weak coupling constant and $\psi_\mu(0)$ is the muon wave at the origin (assumed constant over the nucleus). The central part of Eq. (53) are the symbolic matrix elements V_0, $\underset{\sim}{V}$, ... , from the initial state i to the final state f:

$$V_0 \equiv \langle i|V_0^{(-)}|f\rangle \equiv \int V_0(\underset{\sim}{x})d\underset{\sim}{x} \equiv \int\langle i|V_0(\underset{\sim}{x})|f\rangle d\underset{\sim}{x} \qquad \text{etc.} \tag{54}$$

We will use these notations indiscriminately in the following. In the limit $\omega_\nu = 0$ the transition rate has a particularly simple form when summed over the lepton spin, since there are four incoherent processes: time-like vector V_0 and axial vector A_0 and space-like vector $\underset{\sim}{V}$ and axial vector $\underset{\sim}{A}$. Further the total rate goes to zero as ω_ν^2, so we shall consider the coefficient of this term.

<u>Application of PCAC to μ-capture at $\omega_\nu = 0$.</u> For $\omega_\nu = 0$ we have $\partial/\partial t = iq_0$ and the PCAC relation reads

$$iq_0A_0(\underset{\sim}{x}) - \underset{\sim}{\nabla}\cdot\underset{\sim}{A}(\underset{\sim}{x}) = if_\pi m_\pi^2\phi_\pi(\underset{\sim}{x}) \tag{55}$$

By integration over all space the exact space divergence $\underset{\sim}{\nabla}\cdot\underset{\sim}{A}(\underset{\sim}{x})$ yields zero, so that

$$A_0 = \frac{f_\pi m_\pi^2}{q_0}\,\phi_\pi \tag{56}$$

and

$$\sum_{\text{final states}} |A_0|^2 = \frac{f_\pi^2 m_\pi^4}{q_0^2} \sum_{\text{final states}} |\phi_\pi|^2 \quad . \tag{57}$$

One can therefore directly express the time component of axial μ-capture in terms of pions with $\omega_\pi = m_\mu$. In addition, since $\phi_\pi \equiv \int \phi_\pi(\underset{\sim}{x})d\underset{\sim}{x}$, these pions must be s-wave pions, for all other l-values are suppressed by integration over all space.

There is a simple interpretation of Eq. (57). It corresponds exactly to the muons having produced a source for pions. Since the muon wave function is constant in our approximation, the source is the same everywhere with the constant arranged to unity. The pion produced interacts with the nucleus and gives partial transitions. For a pion with $\omega_\pi \gtrsim m_\pi$ the sum over the partial transitions corresponds to the optical theorem. A very similar relation is valid below threshold but we must first remove the pion pole in Eq. (57):

$$\sum_{\text{final states}} |\phi_\pi|^2 = \sum_f \frac{|\langle i|T_{L=0}(\omega_\pi = m_\mu)|f\rangle|^2}{(m_\pi^2 - m_\mu^2)^2} \equiv \frac{\mathrm{Im}\{i|T_{L=0}(\omega_\pi = m_\mu)|i\}}{(m_\pi^2 - m_\mu^2)^2} \tag{58}$$

Therefore, as long as one is close enough to the physical pions, it is sufficient to know the elastic interaction of the pion with an absorptive interaction. It is not necessary to carry out the complicated sum over the many open final channels.

This is an exact result when $\underset{\sim}{q} = 0$: the summed transitions by the time component of the axial current are described by the exact π-nuclear s-wave "scattering length" extrapolated to $\omega_\pi = m_\mu$.

This remarkably simple relation can be interpreted as due to an external pion source proportional to the muon probability: pions are generated from the muon over the nuclear volume with a strength dictated by PCAC. Since the source has spherical symmetry for $q = 0$, it corresponds to an incident s-wave pion of energy $\omega_\pi = m_\mu$. In complete analogy to the physical pion, the absorption of this pion is given by the imaginary elastic amplitude at $\omega_\pi = m_\mu$.

It would be totally wrong to believe that eq. (58) is the pion pole contribution, with a strength given by the virtual $\mu \to \pi\nu$ amplitude and by the on-shell pion absorption. First, according to PCAC the effective strength of the pion source is m_π^2/m_μ and not m_μ as given by the $\mu \to \pi\nu$ vertex, so that the rate (57) is multiplied by $(m_\pi/m_\mu)^4 \approx 3$. Secondly, the pion amplitude itself is not the physical one and it may differ radically from it. Even if the nucleus is assumed to be pointlike, it will differ, simply because the pion has an imaginary wave number (unrelated to its momentum) $i\kappa$ with $\kappa = (m_\pi^2 - m_\mu^2)^{1/2}$.

Solution for small nucleus $\kappa R < 1$. In the limit of a small nucleus, the π-nucleus s-wave interaction is described by the complex scattering length a_0.

As threshold the pion wave has the form

$$\phi_0(x) = 1 + \frac{a_0}{x} \tag{59}$$

The muon generates an incoming wave λ_μ which is the solution to the inhomogeneous wave equation below threshold

$$[\nabla^2 + (\omega_\pi^2 - m_\pi^2)]\phi(\mathbf{x}) = \lambda_\mu = 1 \tag{60}$$

or $\phi = -\kappa^{-2}$ with $\kappa^2 = m_\pi^2 - m_\mu^2$. The full solution must contain the scattered wave $e^{-\kappa x}/x$ and in addition have the same behaviour for $x \to 0$ as Eq. (59). Consequently

$$\phi(\mathbf{x}) = -\kappa^{-2}\left(1 + \frac{a_0}{1+\kappa a_0}\frac{e^{-\kappa x}}{x}\right) . \tag{61}$$

We therefore see that $\text{Im}\, T_0(\omega_\pi = m_\mu) = \text{Im}\, a_0/|1 + \kappa a_0|^2$. The conclusion is therefore that the A_0-component of the μ-capture is given by

$$\sum_{\text{final states}} |A_0|^2 \propto \frac{\text{Im} a_0}{|1+\kappa a_0|^2} \frac{1}{(m_\pi^2 - m_\mu^2)^2} . \tag{62}$$

From Eq. (62) it is evident that there is a pole in the capture rate as $m_\mu \to m_\pi$. This is expected because this is the pole dominated, pseudoscalar μ-capture term. The capture strength depends on the absorption of the pion s-wave expressed in terms of the s-wave scattering length.

A Paradox in the μ-Capture Rate. Since the muon is a weakly interacting particle it absorbs by volume absorption even in a large nucleus. In a way it is therefore quite surprising that we find the $|A_0|^2$ absorption proportional to the π-absorption T-matrix in Eq. (58). For if we have physical pions interact with the nucleus their absorption in a heavy nucleus will certainly be a surface effect, since the wave cannot penetrate the nuclear surface. Therefore we must ask the question: Is it reasonable to have the μ-capture proportional to the π-capture amplitude? Is there an explanation to this puzzle?

The secret of the explanation is that the muon acts as a driving source for the pion wave function, which normally cannot develop spontaneously below threshold (but for bound states!): in this way a π wave function is generated everywhere in space. We will now investigate this phenomenon in more detail.

The Pion Wave Function Generated by the Muon. Although no spontaneous state develops for pions with $\omega_\pi < m_\pi$ there exists a pion wave function for any energy in this region provided there is a driving source. In our case the muon provides a source of this kind. The form of this source could vary according to circumstances; the source function will be a constant $\lambda_\mu = 1$ in our case, if we neglect the variation of the muon wave function over the nucleus. We should therefore determine the solution to the inhomogeneous wave equation for the pion

$$\underset{\sim}{\nabla}\cdot(1-\alpha(\underset{\sim}{x}))\underset{\sim}{\nabla}\phi_\pi(\underset{\sim}{x}) - Q(\underset{\sim}{x})\phi_\pi(\underset{\sim}{x}) + (\omega_\pi^2 - m_\pi^2)\phi_\pi(\underset{\sim}{x}) = \lambda_\mu \quad , \tag{63}$$

which describes the "muonic pions". As in the previous case of a point nucleus the spatial symmetry of the source makes the wave function have $\ell = 0$. There is, in principle, no particular difficulty in solving Eq. (63) numerically for any particular spatial distribution of $\alpha(x)$ and $Q(x)$ and to derive Im $T_{\ell=0}$ (ω_π) from this solution. This means that Eq. (63) contains the solution of the A_0 part of the muonic capture rate. We will only discuss the important special case of a very large nucleus.

Solution for Nuclear Matter. Inside a large uniform system the generated wave will be a constant ϕ_0. From Eq. (63) we find, since $\underset{\sim}{\nabla}\cdot(\text{constant}) = 0$:

$$\phi_0 = - \frac{\lambda_\mu}{Q + (m_\pi^2 - m_\mu^2)} \quad , \tag{64}$$

where $Q = Q(x) = $ constant. We therefore find the remarkable result that the generated wave is independent of the local p-wave interaction. The local absorption rate in the medium per unit volume is

$$2\ \mathrm{Im}V|\phi_0|^2 = m_\pi^{-1} \frac{\mathrm{Im}Q}{|Q + m_\pi^2 - m_\mu^2|^2} \quad . \tag{65}$$

It is clearly this rate which will determine the rate of muon absorption in a large nucleus. Since Eq. (65) is a rate per unit volume the corresponding absorption rate is a volume effect. We are now in a position to clarify the previous paradox.

Since we have nowhere made any use of the principle that the π-μ mass-difference has its physical value it will be instructive to treat the muon mass as a free parameter. There are then three different regimes.

a) $|m_\pi - m_\mu|$ small. The muon is surrounded by a Yukawa field with a wave number $\kappa = (m_\pi^2 - m_\mu^2)^{1/2}$. As long as the nuclear radius R is such, so that

$\kappa R << 1$

the nucleus will appear point-like to the generated pion. This is the condition for the point solution in terms of a scattering length. The pion behaves like an external pion interacting with the system, and the absorption of the pion will be as for physical pions, i.e. surface absorption on heavy nuclei. The capture rate has a pole denominator $(m_\pi^2 - m_\mu^2)^{-2}$.

b) $\Delta m = (m_\pi - m_\mu) < |V|$.

The mass difference is negligible compared to the strength of the local potential. The absorption rate is $\propto \frac{\mathrm{Im}V}{|Q|^2}$ according to eq. (65): there is now a volume absorption with no trace of a pole denominator. The muon wave is heavily influenced and distorted by the local potential.

c) $\Delta m > |V|$. According to eq. (65) the absorption rate is now $\propto \frac{\mathrm{Im}V}{|m_\pi^2 - m_\mu^2|^2}$. The muon generates a Yukawa field at any point, but this field is so bound that it damps experimentally before being distorted by the local potential. As a consequence we find that the absorption is described in the Born approximation, in our case the volume integral of the absorptive potential. There is a pole-like mass denominator $(m_\pi^2 - m_\mu^2)^2$, but this is a false pole unrelated to the physical amplitude.

We therefore see very clearly how a very small mass difference can give surface absorption, while actual mass differences naturally give volume absorption. This resolves the apparent paradox.

Axial Locality for μ-capture at $\omega_\nu = 0$: Point nucleus. We have previously discussed the A_0 contributions to this case. From Eq. (47) and the fact that $\underset{\sim}{A} = \int \underset{\sim}{A}(\underset{\sim}{x})d\underset{\sim}{x} = \int \underset{\sim}{A}^0(\underset{\sim}{x})d\underset{\sim}{x} = \underset{\sim}{A}^0$ (since $\int \underset{\sim}{\nabla}\phi(\underset{\sim}{x})d\underset{\sim}{x} = 0$)

$$\sum_{\text{final states}} |\underset{\sim}{A}|^2 = f_\pi^2 \sum_{\text{final states}} |S|^2 \quad . \tag{66}$$

We derived earlier how the sum over partial s-wave transitions gave the imaginary part of the s-wave amplitude also below threshold. For a point nucleus this amplitude was given in terms of the complex $\ell = 0$ scattering length by Eqs. (58) and (62). A very similar argument gives now the sum over the A terms proportional to

$$\sum_{\text{final states}} |\underset{\sim}{A}|^2 \propto \mathrm{Im}T_{\ell=1}(\omega_\pi = m_\mu) = \frac{\mathrm{Im}a_1}{|1 - \kappa^3 a_1|^2} \quad . \tag{67}$$

This means that we have unitarized the p-wave scattering volume a_1 defined by $k^3 \cot \delta_1 = a_1^{-1}$ for $k = i\kappa$ with $\kappa^2 = m_\pi^2 - m_\mu^2$. It is very important to note that even as $m_\mu \to m_\pi$ there is no pole term in Eqs. (66) and (67). One may note that in the usual Goldberger-Treiman relation it is frequent to let $m_\pi \to 0$ and to note that no pole term appears. In the present case such a procedure would be meaningless: the limit $m_\pi \to 0$ would completely modify the whole absorption process. The correct limit for a small nuclear system is $m_\mu \to m_\pi$, instead, if the purpose is to exhibit the absence of a pole. For most nuclei the preservation of the physics forces us to renounce even this extrapolation.

Capture Rate by $\underset{\sim}{A}$ terms in Nuclear Matter: $\omega_\nu = 0$. The optical theorem can be expressed locally in the nuclear medium for the transition source: the absorptive part of the potential is the sum of the local absorptive transitions to final states.

Such relations are valid also for physical pions, and they are implicit in the normal use of absorptive parameters in the optical potential. More exactly

$$\sum_{\text{final states}} s_{if}(\underset{\sim}{x})\; s^*_{fi}(x') \propto \mathrm{Im}Q(\underset{\sim}{x})\text{"}\delta(\underset{\sim}{x}-\underset{\sim}{x}')\text{"}$$

$$\sum_{\text{final states}} \underset{\sim}{S}_{if}(\underset{\sim}{x})\cdot \underset{\sim}{S}^*_{fi}(\underset{\sim}{x}') \propto \mathrm{Im}\alpha(\underset{\sim}{x})\text{"}\delta(\underset{\sim}{x}-\underset{\sim}{x}')\text{"} \quad . \tag{68}$$

We have deliberately surrounded the δ-functions by quotation marks, since they are not strict δ-functions. It is only when the pion has a "sufficiently" long wavelength that we can make the point approximation. Pionic absorption in nuclei is believed to be associated with distances of order 0.5 fm << 2 fm, the characteristic distance of muonic pions. Let us now consider the $\Sigma\, |\underset{\sim}{A}|^2$. Since the pion field will be constant in the medium, we have

$$\sum_{\text{final states}} |\underset{\sim}{A}|^2 \propto \sum_{\text{final states}} |\underset{\sim}{S}|^2 = \iint d\underset{\sim}{x}\,d\underset{\sim}{x}' \sum_{\text{final states}} \underset{\sim}{S}(\underset{\sim}{x})\cdot\underset{\sim}{S}^*(\underset{\sim}{x}')$$

$$\propto \iint d\underset{\sim}{x}\,d\underset{\sim}{x}'\; \mathrm{Im}\alpha(\underset{\sim}{x})\;\delta(\underset{\sim}{x}-\underset{\sim}{x}') = \int \mathrm{Im}\alpha(\underset{\sim}{x})\,d\underset{\sim}{x} \quad . \tag{69}$$

Consequently, the absorption rate per unit volume of the muon is $\propto \mathrm{Im}\,\alpha$. This means that it absorbs just as a p-wave pion in Born approximation. One notes that, as in the s-wave case, a heavy nucleus absorbs by volume absorption and not by surface absorption.

Numerical Application to ^{12}C. Collecting our results for the point system at $\omega_\nu = E = 0$ we have from Eqs. (53), (57), (62) and (67) that the axial capture is

$$\left.\frac{d\Gamma}{dE}\right|_{E\to 0} = \frac{2G^2E^2}{\pi}|\psi_{1s}(0)|^2 \{f_\pi^2[(\frac{m_\pi^2}{m_\mu})^2 \frac{\mathrm{Im}\, a_s(\omega_\pi = m_\mu)}{[m_\pi^2 - m_\mu^2]^2}$$

$$+ 9\, \mathrm{Im}\, a_p(\omega_\pi = m_\mu)]\}$$

To this should be added the vector-capture which can be estimated from total photo cross-sections. Using the experimental scattering lengths a_s and a_p for ^{12}C as determined from π-mesic atoms we find

$$\frac{d\Gamma}{d\varepsilon} = (0.8 + 3.9 + 5.9)\varepsilon^2 \times 10^4 \ \mathrm{sec}^{-1}$$

where the neutrino energy is measured in units of the muon mass: $\varepsilon = E/m_\mu$.

From this result we conclude the following:

a) the vector capture is quite small ($\lesssim$ 10%) with axial capture dominant.

b) the time component is nearly as important as the space component in axial capture in contrast to usual μ-capture for which $|\underset{\sim}{A}^2| >> |A_0|^2$.

c) there is surprisingly strong capture to the high excitation region with $\simeq$ 1% of the total capture into energy transfers $\gtrsim$ 80 MeV. This explains qualitatively the energetic neutrons and protons in μ-capture: if two fast neutrons are typically emitted with binding (15-20) MeV there will be about 1% capture into neutrons of energy > 30 MeV as observed.

Experimental Tests of Axial Locality. The central content of axial locality for μ-capture can be stated as follows:

> To every muon-capture reaction by the axial current, there is the corresponding pionic reaction. The matrix element for the pion determines the muonic one. Consequently, we expect a qualitative counterpart of pionic phenomena, like 2-nucleon absorption, in large energy transfer muonic processes, and these are quantitatively predicted, at least in principle.

There are thus in principle many reactions which could serve as qualitative tests. It will however, be important to have the kinematics well established which is not always so simple, since the neutrinos cannot be observed. This militates for 3-body final states (2-body states in the corresponding pion reaction).

In addition, since the pion amplitudes must be extrapolated below threshold, one should primarily chose cases for which the extrapolation can be made with confidence. Only in this way can we hope to have a stringent test of the axial locality hypothesis.

With these restrictions the following cases are of particular interest

a) $\mu^- + D \to n + n + \nu$ (high q_o region). An ideal case which can be directly related to $\pi^- + D \to n + n$.

b) $\mu^- + {}^3He \to D + n + \nu$ (high q_o) as above but form factor effects more important.

c) Total μ-capture rate in nuclei to high q_o region. Requires calorimeter experiments. Extrapolation of π-nuclear optical potential.

d) Emission of correlated np and nn pairs. High q. Correlations of emitted pairs partially consequence of kinematics.

e) Axial capture >> vector capture at large q_o. Direct test probably difficult.

f) Outside of μ-capture: Renormalization of g_A in nuc: :i. Weaker conditions but also perturbative effects.

Conclusion on Axial Locality

In the present lectures we have emphasized very strongly the advantages of looking at the nuclear pionic currents as generated by external perturbations or sources. In the specific illustrations we have chosen large energy transfer μ-capture; this permits to consider the problem of virtual pions nearly as Schroedinger problem. Still, this was not the whole story. In the muonic case, the sources in the Schröedinger problem depended not only on the external perturbation. In the axial current the muon field also generated contact sources in the matter associated with local p-wave pion interactions. In fact, it is exactly these contact terms which have the largest ressemblance to the usual pion exchange currents in nuclei. The reason is that these pions are generated somewhere in the medium, say at a nucleon, and then rescatter in the medium until they are absorbed, i.e. they are exchanged between initial and final point in the nucleus. In contrast to these we also found the contributions from the muon as a pion source by its virtual decay into ($\pi\nu$): these terms correspond to the pseudoscalar interaction of pions in weak interactions.

The entire discussion was concentrated to the region in which the ($\pi\nu$) mass difference is so small that the description of pions by the same optical potential as at threshold will be valid. We have not discussed the limitations of this picture. It should be emphasized however that this is not a very important point. The problems we have discussed are only formally dealing with muons as an illustration. The central aim has all the time been to obtain a clear physical picture of the nature of the axial current and of the physical phenomena that occur as the pion becomes increasingly virtual. It is only as a by-product of this aim that we have obtained a quantitative description of μ-capture at large energy transfer. Indeed, it is easy to extend our general arguments to cover all μ-capture, even with the normal kinematics. By a reasoning closely akin to the Goldberger-Treiman argument one can immediately conclude that all axial μ-capture is quantitatively strongly pion dominated.

The kinematics can vary radically, however, and thus also the physics. We may therefore conclude: All axial μ-capture is π-capture with varying kinematics. This statement unifies the view of nuclear μ-capture.

The description we have given of μ-capture seems to us very natural and appealing, and it makes quantitative predictions at large energy transfer. In the final analysis, of course, the real test is in experimental evidence, which at present is nearly totally missing.

The present picture has wider implications. Indeed, a closely similar viewpoint is valid for the renormalization of g_A in β-decay. We also believe that the present viewpoint should be useful for discussing electromagnetic pion exchange phenomena in nuclei. There is a good possibility that we can obtain a very clear physical picture of all pionic exchange effects under very varying physical conditions.

4. (π,2γ)-Reactions and the π Field[34]

Two major motivations for the interest in low-energy pion interactions in nuclei are:

a) the pion propagation inside the nucleus and its relation

to the pion-nucleon scattering amplitude; and

b) the pionic degrees of freedom of the nucleus and their coupling to nuclear excitations.

It is therefore important to find probes sensitive to the detailed behaviour of real and virtual pions inside the nucleus. In analogy with positron annihilation on the electrons of a solid, we point out here that pion annihilation into (e^+e^-) and (2γ) on the virtual pions in the nucleus can be used for this purpose. In addition, inside the nucleus it is possible to convert a low-energy charged pion into a π^0 for a short time interval, even when this is energetically forbidden in the asymptotic state. During this interval an almost physical $\pi^0 \rightarrow 2\gamma$ decay is possible and observable. Consequently, we will discuss here the magnitudes of the rare electromagnetic (2γ) and (e^+e^-) decay branches of pionic atoms with the aim of using them as sources of information on points (a) and (b) above.

Consider first the very low-energy pion-nucleus optical potential. In the standard multiple scattering picture , the virtual charge exchange process $\pi^- \rightarrow$ "π^0" $\rightarrow \pi^-$, the second order fluctuation scattering, is supposed to play an extremely important role because of the near-cancellation at low energies of the leading term proportional to the isoscalar πN amplitude. If one accepts this point of view, it follows that inside the nucleus of a π^- atom the pion is roughly 50% of the time in the neutral state, with the nucleus in a corresponding excited state. While this appears plausible, the conclusion is based upon purely circumstantial evidence, the observed magnitude of the pionic energy shifts and the structure of the multiple scattering equations. Now apart from a few of the lightest pionic atoms (^1H, ^3He) for which the charge exchange of stopped π^- into π^0 is observed[35], such real charge exchange is energetically forbidden in light nuclei and can only occur virtually. Since the phase space for the decay $\pi^0 \rightarrow 2\gamma$ is, to a good approximation, unchanged by a few MeV of nuclear binding, the lifetimes of these virtual π^0's are close to that of the free π^0. The summed photon energies will then reflect the momentary nuclear excitation energy E*, while the momentum balance is a measure of the instantaneous pion momentum.

Qualitative estimates of the branching ratio $R_{2\gamma}$ through the virtual π^0 decay may be obtained as follows. From the absorption widths of the 1s levels in pionic atoms[36], the effective imaginary part of the scattering length per nucleon is of the order of 10^{-3} m_π^{-1}, corresponding to a characteristic absorption time in nuclear matter of $\tau_{abs} \sim 10^{-22}$ sec. With a probability P_{π^0} for neutral pions inside the nucleus, the branching ratio is hence

$$R_{2\gamma}^{\pi^0} \stackrel{\sim}{\sim} P_{\pi^0} \times \frac{10^{-16}\ \mathrm{sec}}{10^{-22}\ \mathrm{sec}} = P_{\pi^0} \times 10^{-6} \quad .$$

The naive estimate of 50% for P_{π^0} neglects the influence of the nuclear excitation energy E* which permits only a limited propagation of the π^0. This would become the determining factor if it were less than the mean free path for charge exchange. Putting in numbers for the charge exchange potential, and also allowing for the π^- excitation of the analogue states, $P_{\pi^0} = 0.2$ is probably more realistic. This give a branching ratio $R_{2\gamma}^{\pi^0} \stackrel{\sim}{\sim} 0.2 \times 10^{-6}$. One may argue that this figure should be enhanced in the nuclear periphery due to the reduced absorption. While this is correct, there will also be some decrease in the amount of charge exchange so that the gains should not be large.

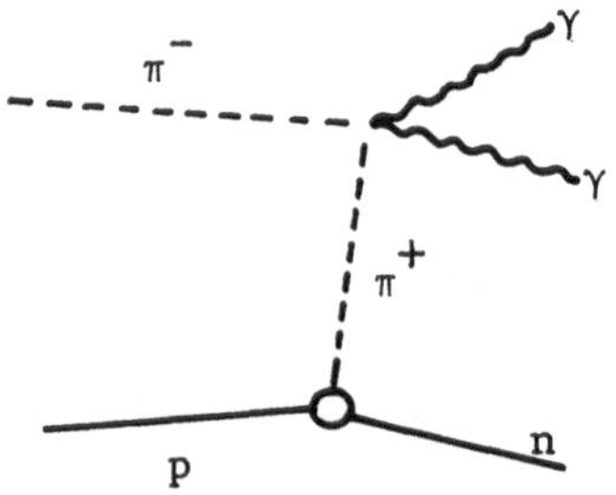

Fig. A. Dominant Born term in the Coulomb gauge for the reaction $\pi^- p \rightarrow \gamma\gamma n$ at low energies.

A lower limit for the branching ratio is obtained by considering the virtual charge exchange on an isolated nucleon with subsequent π^0 decay. Consider first the physical process $\pi^- p \rightarrow \pi^0 n$ with an energy release $E_A \stackrel{\sim}{\sim} 3$ MeV. Suppose now that because of nuclear binding the neutron energy is changed, and the energy release made negative -E*. Although the π^0 is virtual, from the overlap of the tail of the π^0 Breit-Wigner shape with the kinematically accessible phase space for decay, we find, neglecting all small energy variations of the matrix elements,

$$R_{2\gamma}^{\mathrm{Bound}}/R_{2\gamma}^{\mathrm{Free}} = \frac{1}{4\tau_{\pi^0}(E_A E^*)^{1/2}}$$

With a typical value $E^* \sim 10$ MeV, the above ratio is 3×10^{-7}. This can be related to the ordinary radiative capture by the Panofsky ratio, which for hydrogen is about 1.5[37]. The single photon process provides a convenient normalization also in nuclei, where it is usually less than 5% of the strong interaction mode[38]. Hence, the predicted single nucleon contribution to the (2γ) decay mode via virtual charge exchange is about 3×10^{-8}, an order of magnitude less than our nuclear estimate. In conclusion, π^0 decay should become an observable process in nuclei at the level of $10^{-6} - 10^{-7}$ simply due to the charge fluctuations of the pion field when a π^- propagates in nuclear matter.

In addition to the process of virtual π^0 decay, the propagating π^- can also directly annihilate on one of the virtual pions, already present in the nucleus, by the processes $\pi^- + "\pi^+" \rightarrow 2\gamma,\ e^+e^-$. These are the pionic analogies of the well-known positron process in solids $e^+ + "e^-" \rightarrow 2\gamma$. The (2γ) annihilation in hydrogen has been considered by Joseph[39] and various Soviet authors[40]. For a zero-energy pion, and neglecting terms of order m_π/M, the only Born term remaining in the Coulomb gauge is that of fig. A. This gives a branching ratio to the Kroll-Ruderman[41] estimate for the 1γ process on the proton at threshold, i.e. $\ell = 0$, of

$$R_{2\gamma}/R_{1\gamma} \approx \frac{1.12\alpha}{15} \approx 10^{-4} \quad .$$

As before we deduce a branching ratio in nuclei using the observation that the 1γ branch there is typically 5%. This yields $R_{2\gamma} \approx 5 \times 10^{-6}$, which is an order of magnitude larger than our estimate based on the virtual π^0 decay. Unlike the π^0 estimate though, this is a single nucleon mechanism with no enhancement of the pion density assumed because the nucleons are in a nucleus. It should, however, be noted that the annihilation mechanism is sensitive only to the relatively soft π^+ components in the nucleus with $p_\pi \ll m_\pi$ in order that the two photons might materialize freely.

Experimentally[42], an upper limit has been obtained for the (2γ) branch in ^{12}C of less than 4×10^{-5}. Even on the basis of our rough estimates this suggests that the pion content in ^{12}C is not anomalously high, as might be the case if a pion condensate had developed. Since the tests of a pion condensate proposed by the Migdal group[43] are rather indirect, a more detailed test along the lines suggested here would seem much more appropriate and clear cut. We can expect this old limit[42] should give a positive signal with only rather moderate improvements. We note in passing that π^0 decay and the annihilation processes are nearly incoherent. The reason is that the annihilation graph leads primarily to nuclear states excited by spin flip, whereas the charge exchange produces relatively little spin flip.

Apart from the isolated case of ^{208}Pb, true pion charge exchange becomes possible again for pionic atoms formed by nuclei heavier than about ^{40}Ca, and would be very interesting to measure for its own sake. Even in these cases the annihilation graphs might be competitive since only few nuclear levels are energetically accessible through charge exchange, and for these the nuclear overlap integrals tend to be rather small.

The first experimental determination of a $(\pi^-,2\gamma)$ decay reaction has recently been achieved. It is described elsewhere in this book by J. Deutsch. In order to determine whether it is influenced by a precursor to π-condensates or not it will be of importance to study its momentum-dependence.

REFERENCES

1. M. Ericson and T.E.O. Ericson, Ann. Phys. (N.Y.), 36 (1966), 323; M. Krell and T.E. O. Ericson, Nucl. Phys. B11 (1969), 521.

2. H.A. Bethe, Phys. Rev. Lett. 30 (1973), 105.

3. M. Johnson and H. Bethe, Nucl. Phys. A305 (1978), 418.

4. T.E.O. Ericson, The investigation of nuclear structure by scattering processes at high energy, ed. H. Schopper (North-Holland, Amsterdam, 1975), p. 165.

5. A.B. Migdal, Zh. Eksp. Teor. Fiz. 61 (1971), 2210; Engl. transl. Sov. Phys. JETP 34 (1972), 1184; G.E. Brown and W. Weise, Phys. Rep. 27C (1976), 1.

6. G. Baym and G.E. Brown, Nucl. Phys. A247 (1975), 395.

7. J.M. Eisenberg, J. Hüfner and E.J. Moniz, Phys. Lett. 47B (1973), 381.

8. L. Tauscher, Proc. 6th Intern. Conf. on High-Energy Physics and Nuclear Structure (Santa Fe, U.S.A., 1975), eds. D.E. Nagle, et al. (American Institute of Physics, New York, 1975), AIP Conf. Proc. No. 26, p. 541.

9. C.J. Batty, S.F. Biagi, E. Friedman, S.D. Hoath, J.D. Davies, G.J. Pyle, G.T.A. Squier and D.M. Asbury, Phys. Rev. Letters 40 (1978), 931.

10. R. Engfer et al., At. Data. Nucl. Data Tables 14 (1974), 509.

11. I. Sick, Proc. 6th Intern. Conf. on High-energy Physics and Nuclear Structure (Santa Fe, U.S.A., 1975), eds. D.E. Nagle et al. (American Institute of Physics, New York, 1975) AIP Conf. Proc. No. 26, p. 388.

12. J.W. Negele, in High Energy Physics and Nuclear Structure, p. 17, ed. M. Locher, Birkhauser, Verlag, Basel 1977.

13. M. Ericson and M. Krell, Phys. Lett. 38B (1972), 359.

14. M. Ericson and J. Delorme, Phys. Lett. 76B (1978), 182.

15. M. Ericson, "Structure of the Nuclear Pion Field" in Mesons and Nuclei, M. Rho and D.H. Wilkinson, Editors, in press.

16. J. Delorme, M. Ericson, A. Figureau and C. Thevenet, Ann. Phys. 102 (1976), 273.

17. M. Ericson, Thèse, Université de Paris (1958), Rapport CEA, 1189.

18. S.O. Bäckman and W. Weise, to appear in Mesons in Nuclei, M. Rho and D.H. Wilkinson, in press.

19. G.E. Brown, S.O. Bäckman, E. Oset and W. Weise, Nucl. Phys. A286 (1977), 191.

20. T.E.O. Ericson and F. Myhrer, Phys. Lett. 74B (1978), 163.

21. P.G. DeGennes, These, Université de Paris (1959), Presses Universitaires de France Edit.

22. S.A. Fayans, E.E. Saperstein, S.V. Tolokonnikov, J. Phys. G3 (1977), L51.

23. J. Meyer-ter-Vehn, preprint SIN (1978).

24. T.E.O. Ericson and J. Bernabéu, Phys. Lett. 68B (1977).

25. Y. Nambu, Phys. Rev. Letters 4 (1960); M. Gell-Mann and M. Lévy, Nuovo Cimento 16 (1960), 605.

26. M.L. Goldberger and S.B. Treiman, Phys. Rev. 111 (1958), 354.

27. M. Chemtob and M. Rho, Nuclear Phys. A163 (1971), 1.

28. M. Ericson, Particle and Nuclear Physics, vol. 1, p. 67, ed. D.H. Wilkinson, Pergamon 1978.

29. J. Le Tourneux, Nuclear Phys. 81 (1966), 665.

30. J. Bernabeu, T.E.O. Ericson and C. Jarlskog, Phys. Letters 69B (1977), 161.

31. T. Mitzutani and D.S. Koltun, University of Rochester Preprint UR-602, COO-2171-69 (1976); A.S. Rinat, Weizmann Institute Preprint WIS-77/9-Ph (1977).

32. T.E.O. Ericson, in Particle and Nuclear Physics, vol. 1, p. 173, ed. D.H. Wilkinson, Pergamon 1978.

33. H.F. Jones and M.D. Scadron, Phys. Rev. D11 (1975), 174.

34. T.E.O. Ericson and C. Wilkin, Phys. Lett. 57B (1975), 345.

35. P. Truöl, et al., Phys. Rev. Letters 32 (1974), 1268.

36. G. Backenstoss, et al., Ann. Rev. Nucl. Sci. 20 (1970), 467.

37. V.T. Cocconi, et al., Nuovo Cimento 22 (1961), 494.

38. H. Davies, H. Muirhead and J.N. Woulds, Nuclear Phys. 78 (1966), 673.

39. D.W. Joseph, Nuovo Cimento 16 (1960), 997.

40. A.M. Baldin, quoted by I.M. Vasilevsky, et al., Nuclear Phys. B9 (1969), 673; L.I. Lapidus and M.M. Musakhanov, Sov. J. Nuclear Phys. 15 (1972), 558.

41. N. Kroll and M. Ruderman, Phys. Rev. 93 (1954), 233.

42. V.I. Petrukhin and Yu.D. Prokoshkin, Nuclear Phys. 54 (1964), 414.

43. A.B. Migdal, Sov. Phys. JETP 34 (1972), 1184; A.B. Migdal, Phys. Letters 52B (1974), 264, and references contained therein.

THE MYSTERIOUS MUON

Pierre Depommier

Université de Montréal

Montréal, Québec, Canada

1 INTRODUCTION

It may seem somewhat peculiar in a summer school on nuclear physics to have a talk on the elementary-particle aspects of the muon. However there are several obvious justifications:

- at the new intermediate energy facilities (meson factories, proton or electron accelerators), particle physics and nuclear physics coexist. The similarities of the techniques used make it easy for the physicist to switch from one field to the other.
- in the field of intermediate energy the nuclear and particle aspects are strongly coupled and it becomes increasingly difficult for the physicist to ignore one or the other. Elementary particles are used as probes of the nucleus and therefore their fundamental interactions must be understood. On the other hand the atomic nucleus is often used as a laboratory for the study of fundamental interactions. A good example is given by the study of weak interactions. The fundamental questions in this field will be answered not only by high-energy physics experiments (done with neutrino beams, electron-positron colliding beams, hadron-hadron colliding beams) but also by careful experiments done in intermediate-energy physics (for instance, rare decay modes of pions and muons), in low-energy physics (for instance, parity violation in nuclear forces) and even in atomic physics (parity violation in atoms). In the past, the atomic nucleus has proven to be an efficient laboratory for the study of weak interactions: parity violation in β-decay, determination of the structure of charged weak currents, conserved-vector-current theory, current-current interaction, induced pseudoscalar interaction, second-class currents, and hopefully the existence of parity-violating neutral currents.

In the following I shall be mainly concerned with leptons, which are not permanent constituents of atomic nuclei but which can interact with them through electromagnetic or weak interactions. I shall review quickly the known leptons and their interactions. Then I will focus on the interesting problem of muon-number conservation and describe the present experimental situation.

2 REVIEW OF LEPTONS AND THEIR INTERACTIONS[1]

In the following it will be tacitly assumed that every particle has an antiparticle.

2.1 The Electron

It is certainly the most respectable member of the family of leptons. It carries all the prestige of the electromagnetic interaction. Electromagnetic theory is a beautiful theoretical framework which originated in the work of Maxwell, a unified theory of electricity and magnetism. It culminated in the modern quantum electrodynamics, a renormalizable theory. Electromagnetic theory has given us the concept of gauge invariance. It is in fact the simplest gauge theory, based on the group U(1). The electromagnetic interaction is transmitted by only one strictly massless boson and is characterized by a single coupling constant, the charge of the electron.

Since the beginning of the century we also know that the electron participates in the weak interactions since it is emitted in the β-decay of atomic nuclei.

2.2 The Neutrino

The first serious problem encountered in the history of weak interactions was the apparent violation of the conservation laws of energy-momentum and of angular momentum in β-decay. The neutrino was invented by Pauli in order to cure this problem. This neutrino hypothesis had its first success in the Fermi theory of β-decay, which could explain quantitatively the shapes of β-ray spectra. But one had to wait for many years before the neutrino became a "real" particle: a particle which can be produced, transported and detected.

The phenomenon of β-decay can be described as an interaction between a leptonic doublet and a nucleonic doublet:

$$\begin{pmatrix} \nu \\ e^- \end{pmatrix} \quad \begin{pmatrix} p \\ n \end{pmatrix}$$

Originally, Fermi postulated a vector coupling between these two doublets, in analogy with electromagnetism. Although we now know that this picture is not complete, it was a step in the right direction.

At that point, there is an obvious analogy between leptons and hadrons. Even the mass differences between doublet members are of the same order, a feature which will be upset as more particles are introduced.

2.3 The Muon

In the early days of nuclear physics one was confronted with the problem of understanding nuclear (strong) forces. Guided by an analogy with the electromagnetic interaction, which is mediated by the photon, Yukawa put forward the idea that nuclear forces are mediated by particles with masses of the order of 100 MeV (mesons). Such a particle was observed in the cosmic-ray radiation but it was soon realized that this was not the meson predicted by Yukawa. This particle is now known as the muon and it belongs to the family of leptons. Later on the Yukawa meson was discovered and also produced in the laboratory: this is the π-meson. It decays almost exclusively into a muon and a neutrino (99.99% branching ratio) through the weak interaction.

Shortly after the discovery of the muon it became apparent that there could exist a universal weak interaction between two leptonic doublets and a nucleonic doublet

$$\begin{pmatrix} \nu \\ e^- \end{pmatrix} \begin{pmatrix} \nu \\ \mu^- \end{pmatrix} \begin{pmatrix} p \\ n \end{pmatrix} .$$

This results in the unification of various processes:

- muon decay
- β-decay of nuclei and electron capture by nuclei
- muon capture by nuclei.

By universality is meant that all these processes can be accounted for by a common Hamiltonian and common coupling constants. Lorentz invariance restricts the Hamiltonian to a combination of five basic couplings: Scalar, Vector, Tensor, Axial-vector and Pseudoscalar. The coupling constants have to be determined from experiment.

The existence of two leptonic doublets poses a problem, especially in view of the very large mass difference between the electron and the muon.

2.4 The V-A Theory of Weak Interactions

Another serious problem in the history of weak interactions was the so-called "τ-θ puzzle". It could be solved only by the assumption that parity is violated in weak interactions. This parity violation was discovered and opened a wide field of new experiments. In a very short time numerous experiments established that the basic weak Hamiltonian is of the V-A type. This is strictly true for leptons. For weak interactions involving hadrons this simple picture is spoiled by the strong interactions: renormalization of the axial vector coupling constant, induced terms. Basically the weak interaction connects left-handed doublets:

$$\begin{pmatrix} \nu \\ e^- \end{pmatrix}_L \quad \begin{pmatrix} \nu \\ \mu^- \end{pmatrix}_L \quad \begin{pmatrix} p \\ n \end{pmatrix}_L .$$

These doublets are "charged", the electric charges between members differing by one. By analogy with electromagnetic and strong interactions, it is tempting to assume that weak interactions are transmitted by charged vector bosons $W^{\pm}$. Since weak interactions have a very short range, these bosons must be very heavy. At low energy one can forget about them and treat the interaction as point-like.

A further assumption is the "current-current" interaction. Each doublet can couple to itself. Consequences are:

- neutrino-electron scattering
- neutrino-muon scattering
- a weak-interaction contribution to the nucleon-nucleon interaction, giving rise to parity violation in nuclei. This effect has been observed experimentally but the experimental status is not very satisfactory.

2.5 The Cabibbo Theory

A theory of weak interactions would not be complete if it could not explain the decays of strange particles. A very important step was made by Cabibbo when he applied SU(3) symmetry to the weak interactions and showed that strange-particle decays can be unified with non-strange decays. Something new appears: we have a "weak" state which is a linear combination of two particles (different mass eigenstates), the neutron and the Λ^o. The nucleonic doublet considered so far must be replaced by a hadronic doublet

$$\begin{pmatrix} p \\ n \end{pmatrix}_L \longrightarrow \begin{pmatrix} p \\ n \cos\theta_c + \Lambda^o \sin\theta_c \end{pmatrix}_L$$

where θ_c is a new parameter, the Cabibbo angle, which has to be determined from experiment.

The mixing of a strange and a non-strange particle causes strangeness violation in weak interactions. Since $\theta_c \sim 0.2$ strangeness is badly broken: strange hadronic matter decays easily to non-strange hadronic matter. This is exemplified by the life-time of the Λ^0 which is 2.6×10^{-10}s.

2.6 The Two-neutrino Hypothesis

It was shown rather early that the muon decays into an electron and two different neutral fermions:

$$\mu^+ \rightarrow e^+ \nu \bar{\nu} .$$

In writing one of the neutral fermions as a neutrino and the other as an anti-neutrino, we have effectively postulated a conservation law for leptons (analogous to the conservation law of baryons). Furthermore we have assigned leptonic quantum numbers in the following way:

$L = +1$ for μ^-, e^- and ν
$L = -1$ for μ^+, e^+ and $\bar{\nu}$.

The muon decay can be depicted as the interaction between the basic doublets:

$$\begin{pmatrix} \nu \\ e^- \end{pmatrix}_L \quad \begin{pmatrix} \nu \\ \mu^- \end{pmatrix}_L .$$

In such a process there is obviously conservation of the leptonic quantum number L. But it is obvious that the process

$$\mu^+ \rightarrow e^+ \gamma$$

should also exist, since it satisfies lepton number conservation. In the modern theories assuming the existence of intermediate weak vector bosons it should go according to the diagram

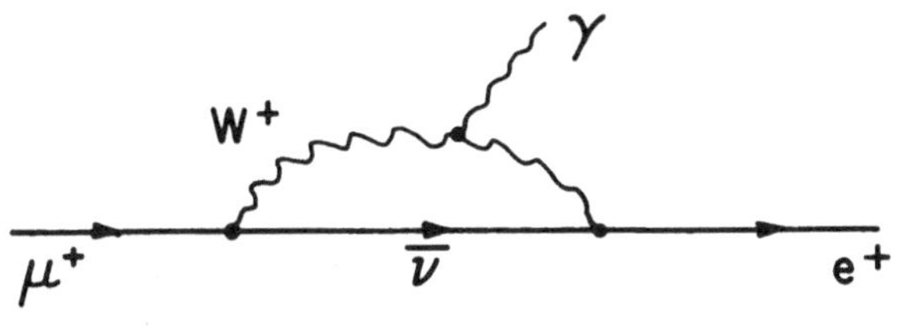

(note that in order to satisfy gauge invariance other diagrams should be considered where the photon is attached either to the μ^+ or the e^+). The existence of weak vector bosons $W^{\pm}$ is not a necessary condition for the decay $\mu^+ \rightarrow e^+\gamma$. In the old point-like weak interaction between four fermions it could occur through the second-order process

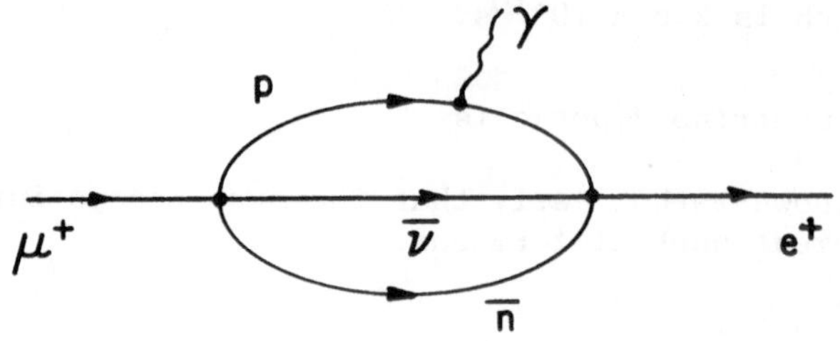

Until recently it was not possible to compute rigorously the preceding graphs because of the occurence of divergences. However, using reasonable cut-offs it was generally agreed that the branching ratio

$$R(\mu^+ \to e^+\gamma) = \frac{\mu^+ \to e^+\gamma}{\mu^+ \to e^+\nu\bar{\nu}}$$

should be of the order of 10^{-4}. But very rapidly the experimental upper limit on R was brought down to 10^{-7} (it reached 2.2×10^{-8} in the sixties).

One possibility for curing this new problem was to assume that there are two different neutrinos ν_e and ν_μ, associated with the electron and the muon respectively, and separate conservation laws for leptonic quantum numbers

$$L_\mu = 1 \quad \text{for } \mu^-, \nu_\mu$$
$$L_\mu = -1 \quad \text{for } \mu^+, \bar{\nu}_\mu$$
$$L_\mu = 0 \quad \text{for } e^-, e^+, \nu_e, \bar{\nu}_e$$

and

$$L_e = +1 \quad \text{for } e^-, \nu_e$$
$$L_e = -1 \quad \text{for } e^+, \bar{\nu}_e$$
$$L_e = 0 \quad \text{for } \mu^-, \mu^+, \nu_\mu, \bar{\nu}_\mu .$$

This assumption was verified experimentally by using muonic neutrino beams (from pion decay). These muonic neutrinos can convert into muons but not into electrons. The best experimental upper limit obtained so far is

$$\frac{\nu_\mu N \to e^- X}{\nu_\mu N \to \mu^- X} < 3 \times 10^{-3}$$

(90% confidence level).

We have now two leptonic doublets

$$\begin{pmatrix} \nu_e \\ e^- \end{pmatrix}_L \quad \begin{pmatrix} \nu_\mu \\ \mu^- \end{pmatrix}_L .$$

We have considered here the "additive" conservation law for muonic and electronic lepton numbers. Another possibility is the "multiplicative" law: one defines a leptonic parity:

+ 1 for electronic leptons e^-, e^+, ν_e, $\bar{\nu}_e$

- 1 for muonic leptons μ^-, μ^+, ν_μ, $\bar{\nu}_\mu$.

These different conservation laws make different predictions for various processes which otherwise conserve the total lepton number L (Table 1).

Table 1

Process	Allowed by additive law	multiplicative law
$\mu^+ \to e^+ \nu_e \bar{\nu}_\mu$	yes	yes
$\mu^+ \to e^+ \bar{\nu}_e \nu_\mu$	no	yes
$\mu^+ \to e^+ \nu_e \bar{\nu}_\mu \gamma$	yes	yes
$\mu^+ \to e^+ \gamma$	no	no
$\mu^+ \to e^+ \gamma\gamma$	no	no
$\mu^+ \to e^+ e^- e^+$	no	no
$\mu^+ e^- \to \mu^- e^+$	no	yes
$\mu^-(A,Z) \to e^-(A,Z)$	no	no
$\mu^-(A,Z) \to e^+(A,Z-2)$	no	no

We are now confronted with the existence of two kinds of leptonic matter: the electronic leptonic matter and the muonic leptonic matter. This is somewhat reminiscent of the situation which prevails in the hadronic world, where there is the strange hadronic matter and the non-strange hadronic matter. But already in the sixties the lifetime for the conversion of the muonic leptonic matter into the electronic leptonic matter was

$$\tau = \frac{\tau_\mu}{R(\mu^+ \to e^+ \gamma)} > \frac{2.2 \times 10^{-6} s.}{2.2 \times 10^{-8}} = 100s.$$

a lifetime which is very long compared to the lifetime of the strange hadronic matter.

For many years physicists have been puzzled by this problem. They have searched actively for differences in the behavior of the electron and the muon, hoping to detect an unknown interaction which would couple differently to those particles and would be responsible for the large mass difference between them. But the electron and the muon seem to have the same electromagnetic and weak interactions and no strong interaction. This leads to the idea of electron-muon universality, which means that these particles are interchangeable (if their corresponding neutrinos are also interchanged). The best test of electron-muon universality is provided by the agreement between theoretical and experimental values for the small branching ratio ($\sim 10^{-4}$)

$$R(\pi^+ \to e^+ \nu_e) = \frac{\pi^+ \to e^+ \nu_e}{\pi^+ \to \mu^+ \nu_\mu} .$$

Since the discovery of a new lepton (the τ), this problem has to be looked at more generally. Recent theories do not necessarily respect electron-muon universality and an accurate measurement of the branching ratio $R(\pi^+ \to e^+ \nu_e)$ is badly needed.

2.7 Quarks

Quarks are supposed to be the building blocks of hadrons. This idea has had so many successes that quarks are now familiar objects to the physicists in spite of the fact that free quarks have not been seen and that they might even be unobservable (quark confinement).

Since quarks are more "elementary" than the hadrons it is normal to depict the weak interaction in terms of quarks and leptons and to consider the basic doublets

$$\begin{pmatrix} \nu_e \\ e^- \end{pmatrix}_L \quad \begin{pmatrix} \nu_\mu \\ \mu^- \end{pmatrix}_L \quad \begin{pmatrix} u \\ d \cos\theta_c + s \sin\theta_c \end{pmatrix}_L$$

where u, d, s are the "up", "down" and "strange" quarks respectively in the original SU(3) theory; θ_C is again the Cabibbo angle,

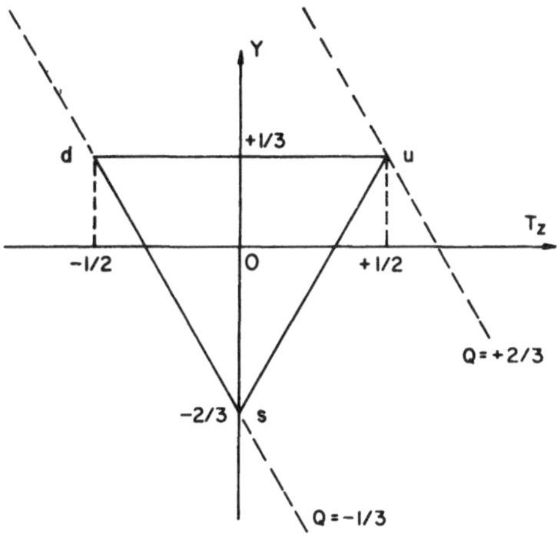

and Y = hypercharge = B + S
B = baryon number
S = strangeness
T_z = z-component of isospin
Q = electric charge = T_z + Y/2.

Again we note the strong mixing of the d and s quarks and the fact that leptons and hadrons behave differently.

2.8 Gauge Theories[2]

The V-A theory with charged currents and Cabibbo mixing has been a good description of all weak processes known at the end of the sixties. But it was not a satisfactory theory in the sense that it was not possible to make predictions beyond first-order of perturbation theory (the V-A theory was not renormalizable). For many years theorists have been aware of the difficulties which prevented them from building a respectable theory of weak interactions (like quantum electrodynamics). The main difficulty had to do with the heavy mass of the weak vector boson. But finally it was shown that it is possible to build a renormalizable theory provided one starts with massless particles (fermions and bosons). Then the concept of spontaneously broken symmetries provides us with a mechanism (the Higgs mechanism) which generates masses for all particles without breaking the gauge invariance and therefore leaves the theory renormalizable. Moreover it becomes evident that such a theory can unify electromagnetic and weak interactions, a very important step in unifying all interactions in nature. But there is a price to pay. In order to achieve renormalizability it is necessary to introduce new particles, either new intermediate bosons or new heavy leptons, or both. An attractive possibility is to base the theory on the gauge group U(2) or equivalently its decomposition $SU(2) \otimes U(1)$. It is the so-called "standard model" of Weinberg and Salam. This group has four generators and therefore predicts a SU(2) triplet of vector bosons W^+, W^o, W^- and a U(1) singlet neutral boson. The charged vector bosons $W^\pm$ can be identified with the mediators of the charged weak interaction whereas the two neutral bosons can be mixed to produce the photon (note that while all other bosons acquire finite masses the photon stays massless) and a heavy neutral boson Z^o associated with the weak interaction. Due to the group structure the theory has two coupling constants g and g' corresponding to the sub-groups SU(2) and U(1) respectively. The combination $gg'/(g^2 + g'^2)^{\frac{1}{2}}$ must be identified with the electric charge e and the ratio g'/g is equal to $\tan\theta_W$ where θ_W is a new parameter, the Weinberg angle. By postulating a doublet of Higgs scalar mesons, all particles (except the photon) acquire masses; the $W^\pm$ and Z_o masses can be predicted (they depend on the Weinberg angle).

This model predicts weak neutral currents mediated by the Z_o vector boson. These neutral currents had been rejected in the past on the basis of experimental evidence against the existence of strangeness-changing weak neutral currents (the decay $K^o \to \mu^+\mu^-$ is very rare, etc...). With the advent of the Weinberg-Salam model, weak neutral currents without change of strangeness were looked for experimentally, and found! (in high-energy neutrino experiments). This finding was indeed a big success for the new gauge theories. As for strangeness-changing weak neutral currents, which do not exist in nature, it became necessary to invent a suppression mechanism (Glashow-Iliopoulos-Maiani). The price to pay is the introduction of charm, a new hadronic flavor. There is a new quark c and a new kind of hadronic matter. The basic weak doublets are now:

$$\begin{pmatrix} \nu_e \\ e^- \end{pmatrix}_L \begin{pmatrix} \nu_\mu \\ \mu^- \end{pmatrix}_L \begin{pmatrix} u \\ d\cos\theta_c + s\sin\theta_c \end{pmatrix}_L \begin{pmatrix} c \\ -d\sin\theta_c + s\cos\theta_c \end{pmatrix}_L$$

while the right-handed projections of electron and muon $(e^-)_R$ and $(\mu^-)_R$ are two singlets. The two orthogonal combinations of the d and s quarks will cancel the weak neutral currents with change of strangeness. These neutral currents are associated with the $d\bar{s}$ and $s\bar{d}$ terms arising from the neutral doublets

$$\begin{pmatrix} d\cos\theta_c + s\sin\theta_c \\ d\cos\theta_c + s\sin\theta_c \end{pmatrix}_L \qquad \begin{pmatrix} -d\sin\theta_c + s\cos\theta_c \\ -d\sin\theta_c + s\cos\theta_c \end{pmatrix}_L .$$

With the discovery of charm, particle physics has gained a new dimension and this must be considered as another spectacular success of the new gauge theories. Another advantage of charm is that it makes the number of quarks equal to the number of leptons. In fact, charm was introduced the first time precisely for this reason. The concept of quark flavor being defined in order to distinguish between the various quarks u, d, s, c, it is possible to define "leptonic flavors" which distinguish between the various leptons. It is also possible to establish a correspondence between quarks and leptons.

For instance:

	u	d	s	c
Q =	2/3	-1/3	-1/3	2/3
	ν_e	e^-	μ^-	ν_μ
Q =	0	-1	-1	0

This picture is certainly not complete since a new heavy lepton has been discovered (the τ^-) and there is evidence for an associated neutrino ν_τ. Since the equality of the number of quarks and the number of leptons is not only an aesthetic assumption but is in fact required on theoretical grounds, there must exist other quarks and in fact there is strong experimental evidence for additional quark flavors. Two new quarks (b and t) would add to the existing ones (u, d, s, c).

3 MUON NUMBER VIOLATION[3]

The old problem of the electron and the muon (and their associated neutrinos) has now to be looked at in a more general manner and several questions arise:

- How many types of leptons really exist? Let us note here that astrophysical considerations provide us with an upper limit for the possible number of leptons.
- What are the laws which govern lepton spectroscopy? Is it possible to explain the mass spectrum of leptons? Are the neutrinos exactly massless, and if they are massless, what is the reason for it?
- How good is the lepton-quark analogy? How many quarks really exist? Here again we note that there is an upper limit coming from the concept of asymptotic freedom.

3.1 Neutrino Mixing[4]

At this point we can examine further the idea of the lepton-quark analogy. For instance, is there anything like Cabibbo mixing in the leptonic sector? If we believe that the neutrinos are stricly massless such a mixing is impossible (more generally if the neutrinos are degenerate in mass, mixing does not make sense since the weak states can always coincide with the mass eigenstates). But there is nothing in the new gauge theories which prevents us from giving finite masses to the neutrinos. Indeed the mechanism which gives masses to the fermions treats leptons and quarks on the same footing. Of course since there are experimental upper limits on the masses of neutrinos, we have some constraints. The possibility of giving small but finite masses to neutrinos allows us to introduce a mixing (à la Cabibbo) in the leptonic sector. This has been discussed by several authors. The "weak" states ν_e and ν_μ would be linear combinations of two mass eigenstates ν_1 and ν_2 according to

$$\begin{pmatrix} \nu_e \\ \nu_\mu \end{pmatrix} = \begin{pmatrix} \cos\phi & -\sin\phi \\ \sin\phi & \cos\phi \end{pmatrix} \begin{pmatrix} \nu_1 \\ \nu_2 \end{pmatrix}$$

where ϕ is a mixing angle. Such a mixing will give rise to an interesting phenomenon called "neutrino oscillations" which is reminiscent of the similar effect in the K_o-$\overline{K}_o$ system. The weak doublets will be

$$\begin{pmatrix} \nu_1 \cos\phi - \nu_2 \sin\phi \\ e^- \end{pmatrix}_L \quad \begin{pmatrix} \nu_1 \sin\phi + \nu_2 \cos\phi \\ \mu^- \end{pmatrix}_L .$$

Rewriting the Lagrangian one gets the following doublets

$$\begin{pmatrix} \nu_1 \\ e^- \cos\phi + \mu^- \sin\phi \end{pmatrix}_L \quad \begin{pmatrix} \nu_2 \\ -e^- \sin\phi + \mu^- \cos\phi \end{pmatrix}_L .$$

If we call "muonness" the lepton flavor which belongs to the muon, we see that muonness conservation is broken, in complete analogy with breakdown of strangeness. Processes like $\mu \to e\gamma$, $\mu \to 3e$, etc... become possible. The new gauge theories provide the appropriate theoretical framework which allows these processes to be calculated without ambiguity. For instance, the $\mu \to e\gamma$ decay can go via the graph

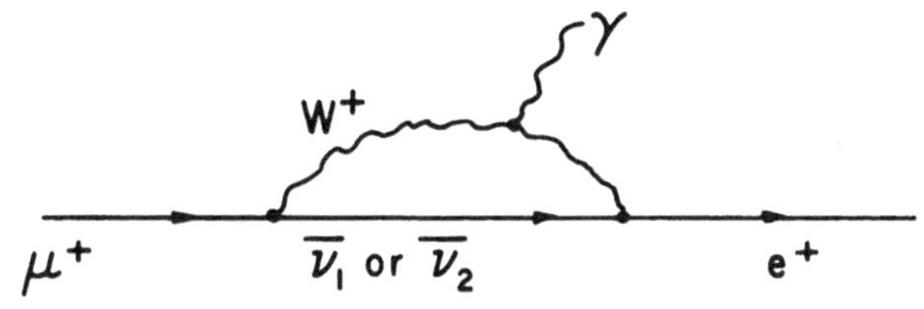

and the branching ratio reads

$$R(\mu \to e\gamma) = \frac{3\alpha}{32\pi} \left(\frac{m_{\nu_1}^2 - m_{\nu_2}^2}{M_W^2} \right)^2 \sin^2\phi \cos^2\phi$$

to a good approximation; α is the fine structure constant, M_W is the mass of the W vector boson, m_{ν_1} and m_{ν_2} are the masses of the ν_1 and ν_2 neutrinos respectively. Setting the neutrino masses m_{ν_1} and m_{ν_2} equal to their experimental upper limits, and taking for M_W the Weinberg-Salam prediction, the assumption of maximal mixing ($\phi = \pi/4$) gives

$$R(\mu \to e\gamma) \simeq 10^{-26}.$$

Such a small branching ratio is not measurable and will not be for a very long time.

3.2 Heavy Leptons, etc......[5]

It is impossible to affirm that we have discovered all the leptons which exist in nature. Also we cannot say that the standard model is the final story. This model has had spectacular successes in explaining various phenomena in a very economical way. For some time there were difficulties with the experiments on parity violation in heavy atoms (Bismuth) done in Oxford and in Seattle. These experiments did not find the parity violation effect predicted by the standard model. But a recent experiment, similar to the Oxford experiment, performed in Novosibirsk, showed a parity-violating effect consistent with the prediction. Another experiment done at SLAC with polarized electrons incident on hydrogen or deuterium has also given a result in good agreement with theory. Hopefully the experimental situation will soon be clarified. On the other hand there are still several questions at the conceptual level which have no answers:

- There is still some phenomenology in the Weinberg-Salam model in that the angle θ_W has to be taken from experiment. Of course the situation could change if the $W^{\pm}$ and Z^{0} bosons are discovered and their masses measured.
- The possibility of the existence of a right-handed weak current is not completely ruled out. Certainly we do not have these right-handed currents in low-energy phenomena where parity violation is maximal. But do we have to re-introduce right-handed currents at higher energies? This question is connected with the problem of the origin of parity-violation. Is this parity violation in the basic Lagrangian of nature or does it arise from spontaneous symmetry breaking?

The future of gauge theories might lie in a generalization of the standard model. There have been numerous attempts in this direction and the large number of papers which deal with the subject shows that there are many possibilities.

Following a rumor according to which the $\mu \to e\gamma$ decay had been observed at SIN (Zurich, Switzerland), a large number of papers (more than fifty) exploited various gauge models in order to predict $\mu \to e\gamma$ branching ratios in the range 10^{-9} to 10^{-8}. Almost all of them are based on hypothetical heavy leptons. Since it is impossible to review all these papers I will only give a few examples.

In addition to the usual neutrinos which have been considered above, one can assume the existence of a neutral heavy lepton L_0[6]. It is then possible to mix ν_1, ν_2 and the left-handed projection of L_0 to produce the "weak states". One gets the following doublets

$$\begin{pmatrix} a_{11}\nu_1 + a_{12}\nu_2 + a_{13}L_0 \\ e^- \end{pmatrix}_L \qquad \begin{pmatrix} a_{21}\nu_1 + a_{22}\nu_2 + a_{23}L_0 \\ \mu^- \end{pmatrix}_L$$

or equivalently

$$\begin{pmatrix} \nu_1 \\ a_{11}e^- + a_{21}\mu^- \end{pmatrix}_L \begin{pmatrix} \nu_2 \\ a_{12}e^- + a_{22}\mu^- \end{pmatrix}_L \begin{pmatrix} L_0 \\ a_{13}e^- + a_{23}\mu^- \end{pmatrix}_L$$

Due to the large mass of the L_0, one gets the branching ratio

$$R(\mu \to e\gamma) = \frac{3\alpha}{32\pi}\left(\frac{M_{L_0}}{M_W}\right)^4 (a_{13}\, a_{23})^2 .$$

With $(a_{13}\ a_{23}) < 3 \times 10^{-3}$ in order to be consistent with high-energy neutrino experiments, a neutral lepton with a mass around 30 GeV could explain a branching ratio of the order of 10^{-10}.

Another possibility is to assume the existence of two neutral heavy leptons N_1 and N_2 which mix [7] according to

$$\begin{pmatrix} N_e \\ N_\mu \end{pmatrix} = \begin{pmatrix} \cos\phi & -\sin\phi \\ \sin\phi & \cos\phi \end{pmatrix} \begin{pmatrix} N_1 \\ N_2 \end{pmatrix}.$$

The right-handed projections of the states N_e and N_μ are put in two doublets together with the electron and the muon. Therefore one has the basic doublets

$$\begin{pmatrix} \nu_e \\ e^- \end{pmatrix}_L \quad \begin{pmatrix} \nu_\mu \\ \mu^- \end{pmatrix}_L \quad \begin{pmatrix} N_e \\ e^- \end{pmatrix}_R \quad \begin{pmatrix} N_\mu \\ \mu^- \end{pmatrix}_R .$$

The $\mu \to e\gamma$ can proceed via the graph

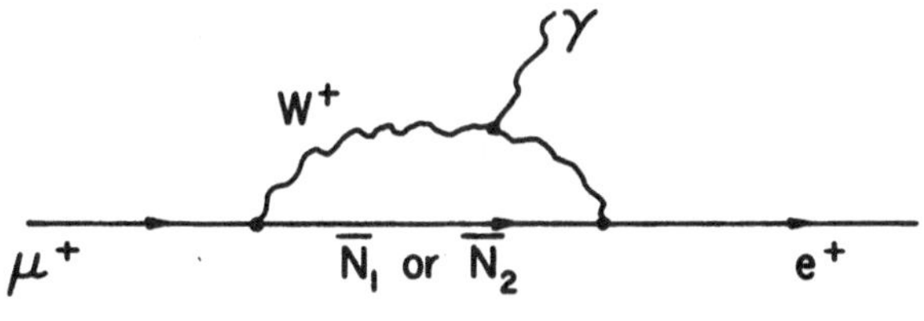

and the branching ratio is

$$R(\mu\to e\gamma) = \frac{3\alpha}{32\pi}\left(\frac{M_{N_1}^2 - M_{N_2}^2}{M_W^2}\right)^2 \sin^2\phi\, \cos^2\phi .$$

If N_1 and N_2 have masses in the GeV range, it is easy to predict branching ratios as large as 10^{-9}. In this model, the electronic weak neutral current is pure vector, so that it predicts no parity violation in heavy atoms.

All models can also make predictions for the other decays like $\mu^+ \to e^+e^-e^+$ and for the $\mu^- \to e^-$ conversion in a nucleus. In general, the $\mu^+\to e^+\gamma$ decay is faster than the $\mu^+\to e^+e^-e^+$ decay but the reverse can also exist. This is the case for a model which introduces lepton triplets including doubly charged leptons [8]

$$\begin{pmatrix}\nu_e\\ e^-\\ h^{=}_{\phi}\end{pmatrix}_L \quad \begin{pmatrix}\nu_\mu\\ \mu^-\\ k^{=}_{\phi}\end{pmatrix}_L$$

where

$$\begin{pmatrix}h^{=}_{\phi}\\ k^{=}_{\phi}\end{pmatrix} = \begin{pmatrix}\cos\phi & -\sin\phi\\ \sin\phi & \cos\phi\end{pmatrix}\begin{pmatrix}h^{=}\\ k^{=}\end{pmatrix}$$

one finds

$$\frac{R(\mu^+\to e^+e^-e^+)}{R(\mu^+\to e^+\gamma)} = \frac{32}{225}\left(\frac{M_W}{m}\right)^4$$

with

$$m^2 = (m_h^2 + m_k^2)/2$$

and

$$R(\mu^+\to e^+\gamma) = \frac{75\alpha}{32\pi}\left(\frac{m_h^2 - m_k^2}{m^2}\right)^2 \sin^2\phi\,\cos^2\phi\,.$$

Similar models differ by the gauge group, the number of postulated heavy leptons or their classification scheme.

Another possibility [9] is to invoke Higgs bosons ϕ according to the diagram

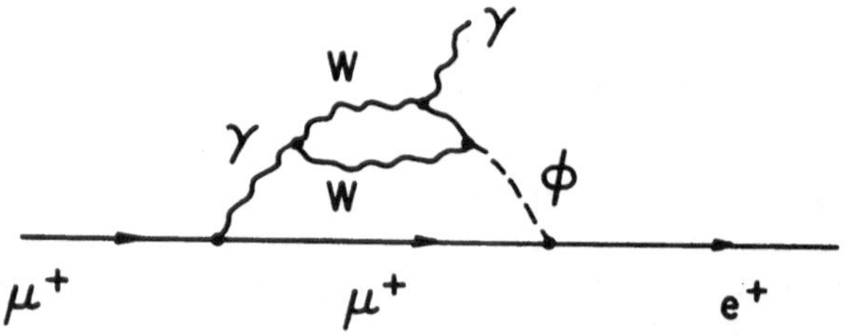

Finally, one could try to exploit the newly discovered τ^- lepton and the fact that it has an associated neutrino ν_τ. The basic doublets are

$$\begin{pmatrix}\nu_e\\ e^-\end{pmatrix}_L \quad \begin{pmatrix}\nu_\mu\\ \mu^-\end{pmatrix}_L \quad \begin{pmatrix}\nu_\tau\\ \tau^-\end{pmatrix}_L$$

where ν_e, ν_μ and ν_τ are mixtures of three mass eigenstates by means of a 3 x 3 unitary matrix. But we have already an upper limit for the ν_τ mass ($m_{\nu_\tau} < 250$ MeV). This would give at most a $\mu \rightarrow e\gamma$ branching ratio of $\sim 10^{-11}$.

We see that it will be necessary to provide experimental numbers for all these muon-number violating processes. They will put constraints on theoretical models and will help in building future versions of gauge theories.

4 THE EXPERIMENTAL SITUATION [3]

The revival of interest in the weak interactions following the birth of gauge theories coincided with the advent of the new meson factories (LAMPF, SIN, TRIUMF). These new facilities have created new opportunities for doing more precise experiments and experimental proposals were put forward with the aim of improving the upper limits on various muon-number violating processes. In the last two years considerable progress has been made and in some cases improvements of two orders of magnitude have been achieved.

4.1 The $\mu^+ \rightarrow e^+\gamma$ Decay

Until 1977 the best upper limit on this decay was

$R(\mu \rightarrow e\gamma) < 2.2 \times 10^{-8}$ (90% confidence level).

There is a natural limitation to the upper limit which can be achieved experimentally. It comes from the radiative muon decay

$$\mu^+ \rightarrow e^+ \nu_e \bar{\nu}_\mu \gamma .$$

The total branching ratio for this decay (integrated over all energies of the emitted particles) is 1.4×10^{-2}. This process can simulate a $\mu^+ \rightarrow e^+\gamma$ decay in the limit of vanishing energies for the neutrinos. The kinematics of the radiative muon decay are shown in Fig. 1 where x and y represent the photon and the positron energies normalized to their maximal values, i.e.:

$$x = 2E_\gamma/m_\mu \qquad\qquad y = 2E_{e^+}/m_\mu$$

(the mass of the positron being neglected); z is the invariant mass of the neutrino pair, also normalized to its maximal value

$$z = \left[(q_1 + q_2)^2 - (\vec{q}_1 + \vec{q}_2)^2\right]^{\frac{1}{2}} /m_\mu .$$

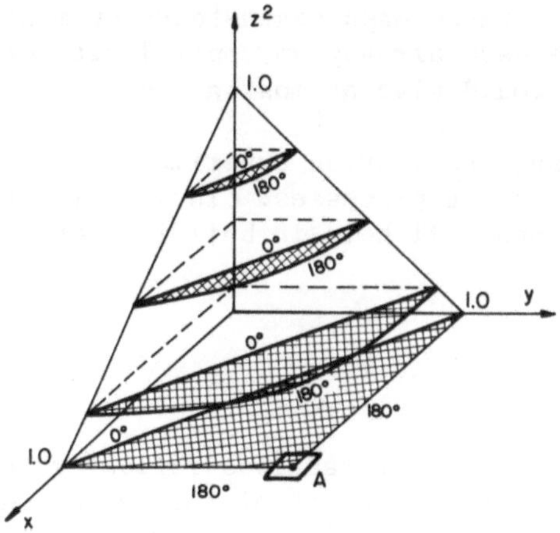

Fig. 1

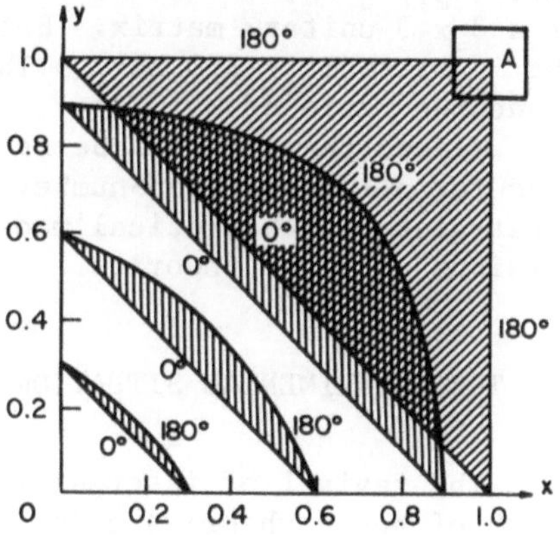

Fig. 2

The $\mu \to e\gamma$ events are located at the point A(1,1,0). The domain allowed to the radiative muon decay can be easily visualized from the various slices parallel to the (x,y) plane. If one projects this domain onto the (x,y) plane one sees that the radiative muon decay fills the square shown in Fig. 2(curves are drawn for z^2 = 0; 0.1; 0.4; 0.7). The density of events is very large for small x and y and goes down very rapidly to very small values as one approaches point A. Due to the finite energy resolution of the photon and positron detectors, $\mu \to e\gamma$ events have to be searched for in a small region surrounding A. This region overlaps with the region of the radiative muon decay. Using theoretical formulae based on the V-A theory it is possible to compute the contribution of the radiative muon decay in the region of interest for the $\mu \to e\gamma$ decay. One finds $R(\mu^+ \to e^+\nu\ \bar{\nu}_\mu\gamma$; x and y > $1-\varepsilon) =$ $1.0 \times 10^{-3}\varepsilon^6$ where ε is related to the energy resolution of the detectors. One immediately realizes the importance of having a good energy resolution. A 10% energy resolution for both the photon and positron detectors will allow a measurement at the 10^{-9} level whereas one would need $\sim$ 3% in order to get down to 10^{-12}. It is in principle sufficient to measure the two energies E_γ and E_{e^+} but another criterion can be used: the angle θ between photon and positron. It is exactly 180^0 for the $\mu \to e\gamma$ case whereas for the radiative muon decay all angles are possible. The volume which is allowed to the radiative decay by kinematics is bounded externally by a surface characterized by $\theta = 180^0$ and internally by a surface characterized by $\theta = 0^0$. Measuring the angle θ in addition to the energies E_γ and E_{e+} is not strictly necessary but it helps very much in rejecting various classes of background events.

Early in 1977 a TRIUMF group was working on the $\pi^+ \to e^+ \nu_e\gamma$ decay by using two large sodium iodide crystals (18" x 22" and 14" x 14") for γ-ray and positron detection. When we heard about the possibility that the $\mu \to e\gamma$ decay might have been seen at SIN we decided to modify our experimental set-up so that we could look for this process. This implied slight changes in the geometry and the electronics and a big improvement of the shielding against neutrons and cosmic-rays. We ran the experiment with a pion beam. Some of the runs were done in parallel with a measurement of the $\pi \rightarrow e\nu$ branching ratio. No attempt was made to measure the angle θ, the only limitation in angle being set by the geometry of the detectors and the collimators. Charged particles were identified by using thin plastic scintillators in front of the sodium iodide crystals. The energy resolution for positrons and γ-rays being approximately 10% the experiment was obviously limited to about 10^{-9} for the $\mu \rightarrow e\gamma$ branching ratio from the radiative muon decay alone. But it was found that the main source of background was due to accidental coincidences between a normal muon decay with E_{e^+} close to its maximal value and a radiative muon decay with E_γ close to its maximal value. In order to keep this background low enough we reduced the pion stopping rate to about 200,000 per second. Other sources of background were due to cosmic-rays and neutrons from the beam dump.

Using a pion beam has some advantages and some disadvantages. There are additional sources of background which can be eliminated electronically (like prompt events) at a small cost of detection efficiency. On the other hand, pions provide very nice energy-calibration points. In addition to the edge of the electron spectrum of μ-decay, one can use the $\pi \rightarrow e\nu$ line. By reversing the polarity of the beam channel, it is possible to stop negative pions and to observe γ-rays from π^0-decay which provide a good calibration point for γ-rays.

A first result based on 25% of our data has already been published. The number of stopped pions was 3.61×10^{10} and the overall efficiency 1.68% (including energy cuts in the x y plane). We observed one event in the region of interest whereas the background prediction was 1.74 event. From that we set an upper limit[10]

$R(\mu^+ \rightarrow e^+\gamma) < 3.6 \times 10^{-9}$ (90% confidence level).

We have now completed the analysis of the data and have obtained an improved limit

$R(\mu^+ \rightarrow e^+\gamma) < 2.0 \times 10^{-9}$ (90% confidence level).

Moreover, during our $\pi^+ \rightarrow e^+ \nu_e \gamma$ runs in 1978 we have accumulated more data on the $\mu^+ \rightarrow e^+\gamma$ decay and we expect to reach a final upper limit close to 10^{-9}.

At SIN they started with an experimental set-up very similar to ours although their sodium iodide crystals were smaller. Later on they added an active converter and a wire-chamber on the γ-ray side and a plastic scintillator hodoscope on the positron side, so that for a fraction of the events (for about 20% of the events the γ-ray was converted) they could measure the angle θ. They have published the upper limit[11]

$$R(\mu^+ \rightarrow e^+\gamma) < 1.1 \times 10^{-9}$$

and they are still working on the data, hoping to get below 10^{-9}.

The best experiment so far was performed at LAMPF. The positron was detected in a magnet and wire-chamber system. This gave a good resolution for the positron and a reconstruction of its trajectory. For the γ-rays they used an array of sodium iodide blocks which gave a reasonable energy resolution and a measurement of the γ-ray position at the detector. It was therefore possible to use the colinearity criterion for all the events. They have not completed their analysis yet but they have already reached an upper limit [12]

$$R(\mu^+ \to e^+\gamma) < 1.5 \times 10^{-10}.$$

This represents an improvement of more than a factor of hundred relative to the old upper limit. According to this result the lifetime of muonic leptonic matter is at least

$$\frac{\tau_\mu}{R(\mu\to e\gamma)} = \frac{2.2 \times 10^{-6} \text{ sec}}{1.5 \times 10^{-10}} \simeq 14{,}000 \text{ sec.}$$

This is already more than ten times the neutron lifetime (the neutron decay is a d → u transition).

4.2 The $\mu \to e\gamma\gamma$ decay

This decay is very difficult to measure directly: one would have to observe one positron and two γ-rays in coincidence. It is however possible to obtain an upper limit indirectly. One possibility is to look simultaneously for all neutrino-less events[13], essentially

$$\mu^+ \to e^+ \gamma$$
$$\mu^+ \to e^+ e^- e^+$$
$$\mu^+ \to e^+ \gamma\gamma .$$

One stops a muon (or pion) beam in the center of a large sodium iodide crystal. In all these decays the energy is carried by charged particles and γ-rays, and a total energy almost equal to the muon mass will be deposited in the detector. One therefore obtains an upper limit for the sum of the branching ratios. But since other experiments provide much better limits for the $\mu^+ \to e^+\gamma$ and the $\mu^+ \to e^+e^-e^+$ decays, the measured upper limit can be essentially attributed to the $\mu^+ \rightarrow e^+\gamma\gamma$ decay. An experiment done at the Berkeley 154" cyclotron has given $R(\mu^+ \rightarrow e^+\gamma\gamma) < 4 \times 10^{-6}$. Of course this number is model independent since it is the result of a measurement integrated over the whole 4π solid angle and over individual energies.

A better limit has been obtained by analysing the $\mu^+ \rightarrow e^+\gamma$ data collected at SIN and TRIUMF [14]. Since only a fraction of the energy range has been used the upper limit so obtained is model dependent: one has to make assumptions on the energy (or angular) distributions of the three emitted particles. The results are:

$R(\mu^+ \rightarrow e^+\gamma\gamma) < 6.1 \times 10^{-8}$ using the SIN data

$R(\mu^+ \rightarrow e^+\gamma\gamma) < 5.0 \times 10^{-8}$ using the TRIUMF data.

A particular model where one assumes a virtual axion and its subsequent decay into two γ-rays gives a very small upper limit:

$$R(\mu^+ \rightarrow e^+\gamma\gamma) < 10^{-9}.$$

An alternative possibility would be to look for the annihilation of muonium into two γ-rays

$$\mu^+ e^- \rightarrow \gamma\gamma.$$

This is an attractive possibility because the two γ-rays are emitted at an angle of 180° and share equally the available energy. One would however need a strong muonium source.

4.3 The $\mu^+ \rightarrow e^+ e^- e^+$ decay

Since this decay results in the emission of three charged particles it is experimentally easier than the preceding one. A recent experiment done in the USSR has given a very good upper limit[15]

$$R(\mu^+ \rightarrow e^+e^-e^+) < 1.9 \times 10^{-9}.$$

An alternative possibility would be to look for the annihilation of muonium into an electron-positron pair

$$\mu^+ e^- \rightarrow e^+ e^-.$$

Here again, the positron and electron are emitted at an angle of 180° and share equally the available energy.

4.4 The $\mu^- \to e^-$ conversion in a nucleus

This is the process

$$\mu^-(A,Z) \to e^-(A,Z)$$

and the quantity under study is the ratio

$$R(\mu^- \to e^-) = \frac{\mu^-(A,Z) \to e^-(A,Z)}{\mu^-(A,Z) \to \nu_\mu(A,Z-1)} .$$

Generally this branching ratio is supposed to be larger than the corresponding ones for the rare muon decays, because of the coherent effect of the protons in the nucleus. There are obvious sources of background:

- the bound-muon decay:

$$\mu^-(A,Z) \to e^- \nu_\mu \bar{\nu}_e(A,Z)$$

  where the nucleus participates in the energy-momentum balance, allowing the electron to be emitted in an energy range extending almost up to the muon mass. This background becomes negligibly small close to the electron maximal energy. Since this background increases rapidly with Z, one would favor light nuclei.
- the radiative muon capture:

$$\mu^-(A,Z) \to \nu_\mu(A,Z-1)\gamma$$

  if the γ-ray converts into a pair where the electron carries almost the full energy. One would favor nuclei with a large Q value for muon capture.

Because of these background problems the experiment consists in looking for $\mu^- \to e^-$ transitions where the nucleus is left in its ground state, the electron carrying an energy close to the muon mass. One has therefore to worry about the fraction of the $\mu^- \to e^-$ conversions which go to the ground-state. For medium-mass nuclei this fraction is expected to be of the order of 85%, hence the experiment becomes possible.

Before 1977, the best upper limit was observed in copper, the branching ratio being [16]

$$R(\mu^- \to e^-) < 1.6 \times 10^{-8} .$$

Recently, a SIN experiment on S^{32} has given a very good upper limit [17]

$$R(\mu^- \to e^-) < 4 \times 10^{-10}.$$

This limit will likely be improved when the SIN group has completed its data analysis. Here again, a factor of hundred has been gained in the last two years.

4.5 Future plans

Since most theorists still believe that muon-number conservation should be violated at some level not too far from the present experimental limits (this possibility looks so natural and there are so many mechanisms which can induce it) experimentalists are encouraged to try to extract the ultimate upper limits which can be achieved with the present meson factories. Big efforts are being made to build detectors with large solid angles, good energy resolution, good background rejection, and to improve beam intensity and beam quality. At TRIUMF we are building a Time-Projection-Chamber (TPC) [18] which will be first used for the study of $\mu^- \to e^-$ conversion in a nucleus. This detector will have a solid angle larger than 50% and an energy resolution which should be good enough to allow a measurement on the 10^{-12} scale. At other places there are proposals to look for muon-number violating processes as well.

5. CONCLUSION

We can hope that in a few years all these planned experiments will have been completed. If muon-number violation is established, one will be forced to perform additional experiments (for instance angular distribution of the $\mu^+ \to e^+\gamma$ decay with respect to the muon spin). We will certainly develop some understanding of the dynamics underlying the lepton sector. If however the muonic leptonic matter retains its identity all these experiments will only provide upper limits which will nevertheless be useful in putting very strong constraints on theoretical models. But then the leptonic world will keep its mystery.

ACKNOWLEDGEMENTS

I wish to thank S. Fallieros, B. Goulard, C. Leroy and J.M. Pearson for help in preparing the manuscript.

REFERENCES

1. For a detailed review of the historical development of weak interactions the reader is referred to L. Jauneau in "Weak interactions", a book based on lectures presented at the International School of Elementary-Particle Physics, Basko Polje (Yugoslavia), edited by M.K. Gaillard and M. Nikolic, available from the "Institut National de Physique Nucléaire et de Physique des Particules", 11 rue Pierre et Marie Curie, 75231 Paris CEDEX 05, France.

2. L. O'Raifertaigh in "Proceedings of the Spring School on Weak Interactions and Gauge Theories", Lyceum Alpinum, Zuoz, Switzerland, March 29-April 8, 1978, available from SIN, CH-5234 Villigen, Switzerland. O'Raifertaigh's article will also appear in "Reports on Progress in Physics". L. Maiani in "Proceedings of the 1976 CERN School of Physics", Wépion, Belgium, 6-19 June 1976, available from CERN, Geneva, Switzerland. J. Bernstein, Rev. Mod. Phys. 46 (1974), 7. H. Primakoff, lectures in this series.

3. For a review of early results on muon-number violation the reader is referred to S. Frankel in "Muon Physics", edited by C.S. Wu and V.W. Hughes (Academic Press, New York, 1975).

4. B. Pontecorvo, Soviet Physics Uspekhi 14 (1971), 235. A.K. Mann and H. Primakoff, Phys. Rev. D15 (1977), 655.

5. It would be difficult to give here an exhaustive list of papers dealing with the subject. Useful references can be found in the following papers: J.D. Bjorken, K. Lane and S. Weinberg, Phys. Rev. D16 (1977), 1474; B. Humpert, SLAC-PUB-1935, May 1977; B.W. Lee and R.E. Shrock, FERMI LAB-PUB-77/21-THY, February 1977.

6. J. Leite-Lopes and Ch. Ragiadakos, Lettere al Nuovo Cimento, 16 (1976), 261.

7. T.P. Cheng and L.F. Li, Phys. Rev. Letters, 38 (1977), 381.

8. F. Wilizek and A. Zee, Phys. Rev. Letters, 38 (1977), 531.

9. J.D. Bjorken and S. Weinberg, Phys. Rev. Letters, 38 (1977), 622.

10. P. Depommier, J.P. Martin, J.M. Poutissou, R. Poutisson, D. Berghofer, M.D. Hasinoff, D.F. Measday, M. Salomon, D. Bryman, M. Dixit, J.A. Macdonald and G.I. Opat, Phys. Rev. Letters, 39 (1977), 1113.

11. H.P. Povel, W. Dey, H.K. Walter, H.J. Pfeiffer, . Sennhauser, J. Egger, H.J. Gerber, M. Salzmann, A. van der Schaaf, W. Eichenberger, R. Eugfer, E. Hermes, F. Schlepütz, . Weidmann, C. Petitjean and W. Hesselink, Phys. Letters, B72 (1977), 183.

12. M. Cooper, private communication.

13. J.M. Poutissou, L. Felawka, C.H.Q. Ingram, R. MacDonald, D.F. Measday, M. Salomon and J. Spuller, Nucl. Phys. B80 (1974), 221.

14. J.D. Bowman, T.P. Cheng, L.F. Li and H.S. Matis, Phys. Rev. Letters, 41 (1978), 442.

15. S.M. Korenchenko, B.F. Kostin, G.V. Mitsel'makher, K.G. Nekrasov and V.S. Smirnov, Sov. Phys. JETP, 43 (1976), 1.

16. D.A. Bryman, M. Blecher, K. Gotow and R.J. Powers, Phys. Rev. Letters, 28 (1972), 1469.

17. A. Badertscher, K. Borer, G. Czapek, A. Flückiger, H. Hänni, B. Hahn, E. Hugentobler, A. Markees, U. Moser, R.P. Redwine, J. Schacher, H. Scheidiger, P. Schlatter and G. Viertel, Phys. Rev. Letters, 39 (1977), 1385.

18. J.M. Poutissou, in "Proceedings of the Spring School on Weak Interactions and Gauge Theories", Lyceum Alpinum, Zuoz, Switzerland, March 29-April 8, 1978, available from SIN, CH-5234, Villigen, Switzerland.

AXIAL CURRENTS AND PIONIC MODES IN NUCLEI

Mannque Rho

DPhT - CEN Saclay, Orme des Merisiers,

91190 Gif-sur-Yvette, France

In this lecture, I discuss how one can correlate and unify several aspects of medium-energy nuclear physics and learn about the mesonic and other degrees of freedom in nuclear structure. One of the main objectives of meson factories is precisely to probe such degrees of freedom and I shall show how this objective can actually be achieved.

First I shall use low-energy theorems to reduce the complicated amplitude relevant for axial currents to a model-independent and somewhat more tractable form. It will be shown from this that the axial charge distribution in nuclei is quite different from the vector charge distribution. Then the space part of the axial current will be related to a π-nuclear scattering amplitude and the "quenching" of the axial-vector coupling constant interpreted in terms of the Landau-Migdal g' or a generalized Lorentz-Lorenz effect. This last step resolves one of the outstanding problems in nuclear physics; namely the "quenching of g_A" observed in light nuclei can now be understood.

Most of the material used in this lecture are drawn from the works done in collaboration with Jean Delorme, Kuniharu Kubodera, Hans Pirner, Eulogio Oset and Koichi Yazaki. Since I present the material freely in my own fashion, none of them should be held responsible for any errors or incorrect statements that I may be making.

I. AXIAL CURRENTS

The first problem we address to is: What is the pion degrees of freedom in nuclei? This is essential for learning about possible phase transitions in nuclear matter involving pion field. To tackle this problem, we approach it from mesonic current point of view. We know that mesonic currents are important ingredients in nuclei when one goes deeper into the structure[1]. This is quite unambiguously demonstrated in the np capture[2] and the electrodisintegration of deuteron[3]. To learn something about excitations with pion quantum numbers, the axial current is a more natural probe since they are directly related and in fact its divergence extrapolates as a pion field (PCAC). Up until recently, there was little success in pinning down the mesonic effects in the axial current. The reasons will be seen later. I shall now discuss the recent developments which seem to provide a clue to the basic issue. This comes from chiral symmetry and low-energy theorems associated with it.

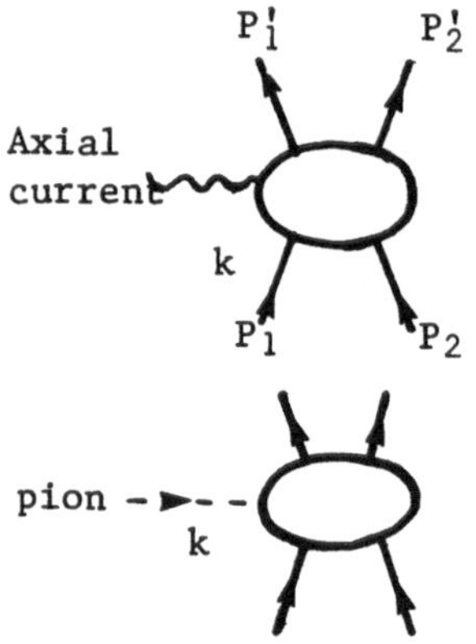

$\equiv M^a_\lambda(k)$ a=1,2,3

$\equiv M^a_\pi(k)$

Then PCAC states [$g_r(k^2)$ is the πNN form factor; $g_r(0) \simeq 13.6$]

$$k_\mu M^a_\mu = i \frac{Mg_A}{g_r(0)} \frac{m_\pi^2}{k^2 + m_\pi^2} M^a_\pi \tag{1}$$

from which one can derive the low-energy theorem[4]:

$$M^a_\mu = M^{a1}_\mu + \Delta M^a_\mu + O(k)$$

$$M_\mu^{a1} = \text{"impulse approximation"}$$

$$\Delta M_\mu^a = M_\mu^{a\,pair} + \Delta M_\mu^{a\,PCAC} + \Delta M_\mu^{a\pi} \qquad (2)$$

where $\Delta M_\mu^{a\,pair} =$

$$= \bar{u}(p_1')\bar{u}(p_2')\ \{T(p_1'p_2';Q_1p_2)S_F^{(-)}(Q_1)\ \Lambda_{1\mu}^a$$

$$+ \Lambda_{1\mu}^a S_F^{(-)}(Q_1')T(Q_1'p_2';p_1p_2) + (1 \leftrightarrow 2)\}\ u(p_1)u(p_2)$$

$$\Delta M_\mu^{a\,PCAC} = i\,\frac{g_A}{2g_r(0)}\ \bar{u}(p_1')\bar{u}(p_2')\ \{\frac{\partial}{\partial k_\mu}\ T(p_2'p_1';Q_1p_2)\Big|_{k=0}\ \Lambda_1^{5a}$$

$$+ \Lambda_1^{5a}\ \frac{\partial}{\partial k_\mu}\ T(Q_1'p_2';p_1p_2)\Big|_{k=0}\ \}\ u(p_1)u(p_2)$$

$$\Delta M_\mu^{a\pi} = -\,\frac{Mg_A}{g_r(0)}\ \bar{u}(p_1')\bar{u}(p_2')\ \frac{\partial}{\partial k_\mu}\ T_\pi^{NB\,a}(k)\Big|_{k=0}\ \ u(p_1)u(p_2)$$

where

$$\Lambda_\mu^a = ig_A(k^2)\gamma_\mu\gamma_5\,\frac{\tau^a}{2} \quad ,$$

$$\Lambda^{5a} = ig_r(k^2)\gamma_5\tau^a \quad ,$$

$$Q_1 = p_1+k \quad ,$$

$$Q_1' = p_1'-k \quad ,$$

and $T_\pi^{NB} = T_\pi - T_\pi^{Born}$ where T_π^{Born} includes the pair term as well as the positive energy term

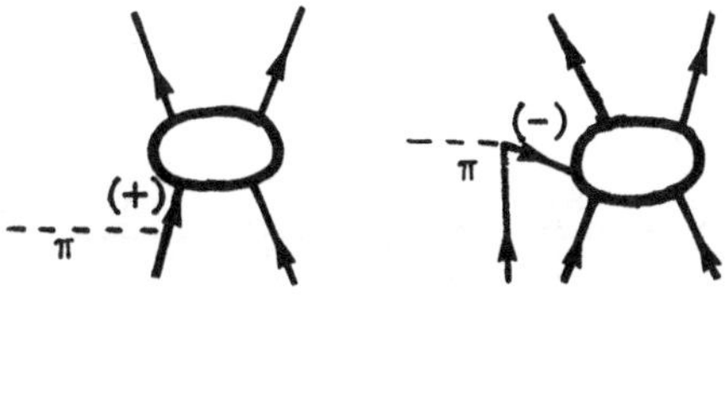

As they stand these expressions are not very transparent; things become better understandable in a model, particularly the OBE model. Let us see how they come out in this model.

Write

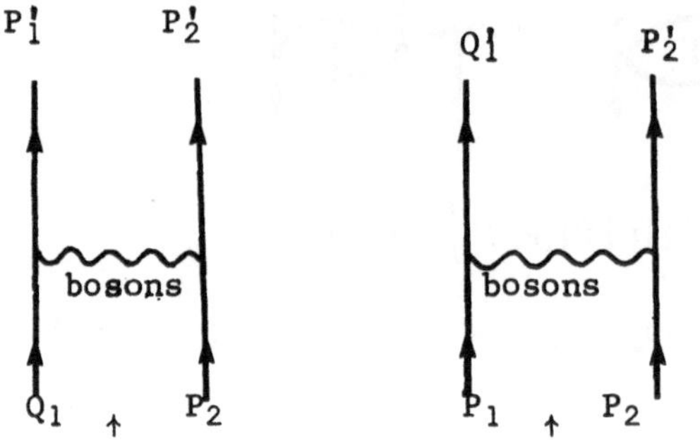

$T(P_1'P_2';Q_1P_2)$ $\qquad$ $T(Q_1'P_2';P_1P_2)$

$$= \sum_{\mu i} \Gamma^i_{1\mu}\left(-q - \frac{1}{2}k\right) \times \frac{1}{(q+\frac{1}{2}k)^2+m^2}\,\Gamma^i_{2\mu}\left(q+\frac{1}{2}k\right)$$

where Γ^i_μ are covariants of isospin index i and ensemble of Lorentz index μ, and m the mass of the exchanged particle. Similarly

$$T_\pi^{NBa} = \sum_{\mu i} T^{NBai}_{1\pi\mu} \frac{1}{(q+\frac{1}{2}k)^2+m^2}\,\Gamma^i_{2\mu}$$

$=$

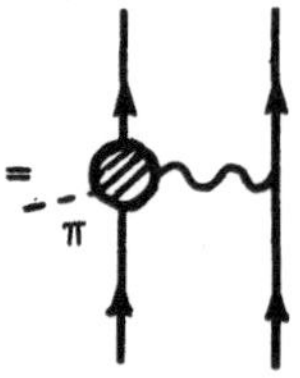

We then obtain

$$\Delta M^a_\mu = \Delta M^{aPair}_\mu + \Delta M^{aPCAC}_\mu + \overline{\Delta M}^{a\pi}_\mu \tag{3}$$

$\Delta M^{aPair}_\mu =$

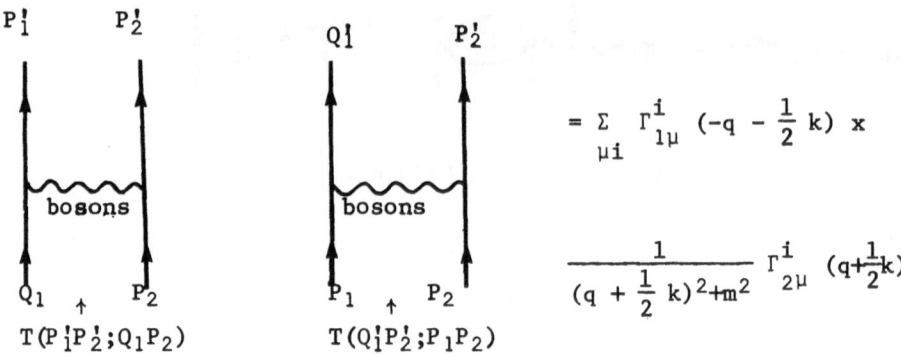

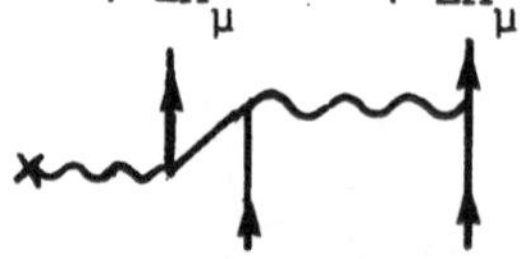

$+ (1 \leftrightarrow 2)$

$$\Delta M^{aPCAC}_\mu = i\sum_{\alpha i} \bar{u}(p_1')\,\frac{1}{2}\left\{\frac{g_A}{2}\,\frac{\partial}{\partial q_\mu}\,\Gamma^i_\alpha(-q),\ \frac{1}{g_r(0)}\,\Lambda^{5a}\right\}u(p_1)$$

$$\frac{1}{q^2+m^2}\, \bar{u}(p_2')\, \Gamma^i_\alpha\, u(p_2) + (1 \leftrightarrow 2): \quad \text{(PCAC constraint)}$$

$$\overline{\Delta M}^{a\pi}_\mu = -\frac{Mg_A}{g_r(0)} \sum_{\alpha i} \bar{u}(p_1') \frac{\partial}{\partial k_\mu} T^{NBai}_\alpha(k)\Big|_{k=0} u(p_1)$$

$$\frac{1}{q^2+m^2}\, \bar{u}(p_2')\Gamma^i_\alpha\, u(p_2) + (1 \leftrightarrow 2)$$

To make the matter even simpler, we look for a situation where one-pion exchange shows up predominantly. Considered in configuration space, heavy-meson exchanges are associated with short-ranged operators, so that their matrix elements are in general suppressed relative to the pion exchange unless the latter turns out to get accidentally suppressed due to symmetry or kinematical reasons. The thing to do is then to look for a case where one-pion exchange is unsuppressed while heavy-meson exchanges are cut down by other mechanisms in addition to short-range correlations. For this, we need to look at the vertex

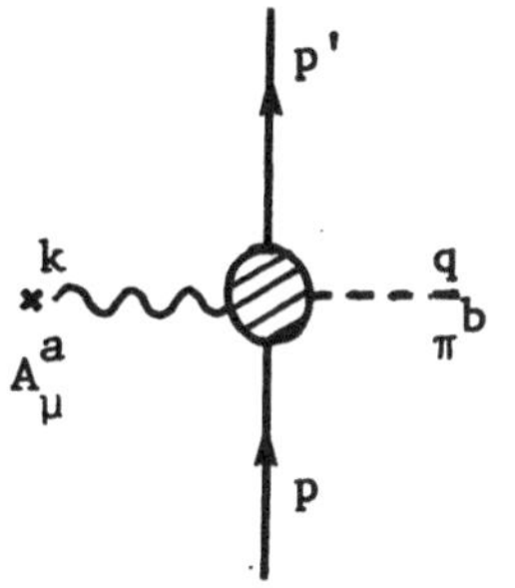

$\equiv m^{ab}_\mu$

The low-energy theorem for this, now looked at from the soft pion point of view, is of greater interest. To the extent that only the long-ranged piece is being looked at, the soft-pion regime is a highly relevant one. In this case, one can discuss a low-energy theorem for any current J^a_μ and the "master" equation for this is

$$m^{ab}_\mu(J^a_\mu) = \text{piece giving rise to impulse approximation}$$

$$+ \frac{1}{F_\pi}\, \bar{u}(p')\, [J^a_\mu(0), Q^b_5(0)] u(p) \dotplus O(q) \tag{4}$$

where $Q^b(0)$ is the axial charge

$$Q^b_5(0) = \int d^3x\, A^b_o(\underset{\sim}{x},0) \quad .$$

Referring back to Equation (3), we remark that the amplitude involving pion production essentially contributes to order O(q). So as long as the first term of Eq. (4), e.g. the commutator, is unsuppressed, the O(q) term should not matter much. With the axial current A_μ^a, we have

$$[A_\mu^a(0),\, Q_5^b(0)] = i\varepsilon_{abc}\, V_\mu^c(0) \quad .$$

Now the vector current goes like

$$V_\mu^c \sim \gamma_\mu \frac{\tau_c}{2}$$

so it has a large time component $O(\gamma_0) \sim O(1)$ whereas the space component is suppressed by a factor $|\vec{P}|/M$. Therefore to "see" this current algebra term, we have to be able to single out the process most sensitive to this component.

II. AXIAL CHARGE DENSITY

I now discuss two cases where the time component of the axial current can be exhibited. The first case we consider is the transition matrix element quite analogous to vacuum-to-pion processes, namely $0^+(T=0) \leftrightarrow 0^-(T=1)$, which reflects on the axial charge distribution in nuclei. This has been studied recently by Guichon, Giffon, Samour[6] in ^{16}O - ^{16}N transitions:

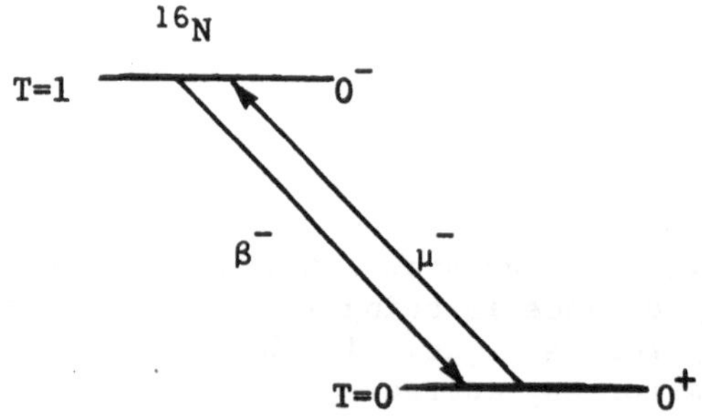

Separately, each process is quite sensitive to nuclear structure, i.e. nuclear models describing the wave functions. However the ratio

$$R = \frac{\Lambda(\mu)}{\Lambda(\beta)}$$

is expected to be practically independent of nuclear models. This turns out to be borne out by practical calculations. In terms of particle-hole configurations, the 0^- state may be given by

$$|0^-\rangle = |p_{1/2}^{-1} 2s_{1/2}\rangle + x|1p_{3/2}^{-1}\, 1d_{3/2}\rangle$$

Now using the canonical value of weak coupling constants including the Goldberger-Treiman result of pseudoscalar form factor

$$m_\pi F_P(q^2) = \frac{2m_\mu M F_A(q^2)}{q^2+m_\pi^2}$$

Guichon et al have found the following results

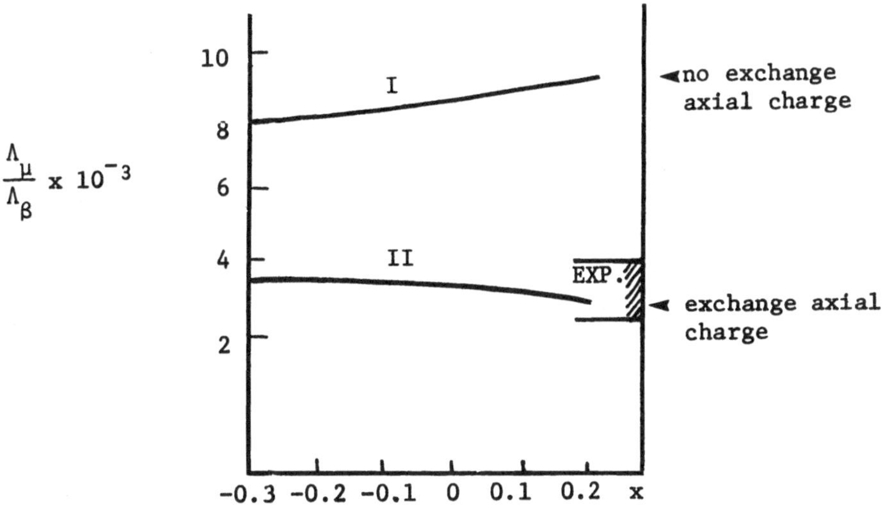

This indicates that the exchange axial charge is very important numerically

$$\frac{\Lambda^{Total}_{(\beta)}}{\Lambda_r(\beta)^I} \approx 4.3 - 4.9$$

$$\frac{\Lambda^{Tot}(\mu)}{\Lambda(\mu)} \approx 1.7 - 1.8$$

which accounts for the qualitative behaviour.

There are two caveats to this: one is that the experiments

are not yet corroborated and R_{exp} is yet to be agreed upon; the other is that the pseudoscalar form factor even for nucleon is still untested experimentally and there is a possibility that its value in nuclear medium be different from the Goldberger-Treiman prediction. Furthermore, Coulomb effects are a delicate matter in β-decay and need be carefully analyzed. Though the result is very nice, one should be cautious in the interpretation.

Let me now turn to the other case. This is the famous mass-12 β decay:

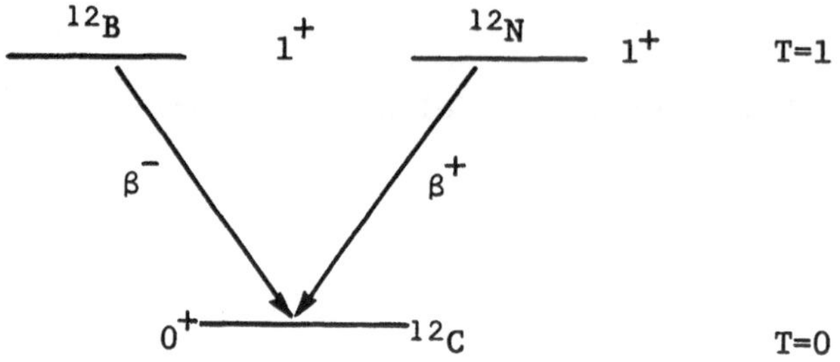

Let the initial spin polarization be $\hat{J}$, the electron or positron momentum be $\hat{e}$. Then in terms of $\hat{J}\cdot\hat{e} = \cos\theta$, the transition rate is given by

$$W(\theta,E)_{\pm} \sim 1 \pm P(1+\alpha_{\pm}E)P_1(\theta) + A\alpha_{\pm}EP_2(\theta) \tag{5}$$

where P is the polarization, and A the alignment of the parent nucleus. The coefficients $\alpha_{\pm}$ contain useful information on the desired matrix elements. To see this, it is convenient to write the transition matrix elements in terms of the global elementary-particle form factors:

$$\langle 0^+;P_1|V_\alpha^{\pm}(0)|1^+;P_2\rangle = \varepsilon_{\alpha\beta\gamma\delta}k_\beta\xi_\gamma P_\delta\ (F_M/4AM^2)$$

$$\langle 0^+;P_1|A_\alpha^{\pm}(0)|1^+;P_2\rangle = \xi_\alpha(F_A^{(1)} \pm F_A^{(2)}) + k_\alpha(\xi\cdot k)(F_P^{(1)} \pm F_P^{(2)})$$

$$+ \left(\frac{P_\alpha}{2M}\right)(\xi\cdot k)\ (F_T^{(1)} \pm F_T^{(2)})/2AM\ . \tag{6}$$

$$P = P_1+P_2\ ; \quad k = P_2 - P_1$$

Here ξ is a spin-1 polarization vector, F's are nuclear form factors, the superscripts 1 and 2 refer, respectively, to first and second class[7]. In terms of these form factors, we have

$$\alpha_{\pm} \simeq \frac{2}{3} [\mp(F_M - F_T^{(2)}) - F_T^{(1)}]/2MF_A^{(1)} \quad . \tag{7}$$

Now the combination $(\alpha_+ + \alpha_-)$ is free of F_M and $F_T^{(2)}$ and is directly proportional to $F_T^{(1)}$ to which the axial charge A_o contributes predominantly (there is a negligible contribution from the space component but this is unimportant for the present problem)

$$S \equiv \alpha_+ + \alpha_- \simeq -\frac{4}{3} \cdot F_T^{(1)}/2MF_A^{(1)} \quad . \tag{8}$$

Let S_1 and S_2 be contributions from one-body and two-body densities respectively. Then one finds theoretically

$$\frac{S_2}{S_1} \approx 0.4$$

quite consistent with the A = 16 case described above. This would imply (for S_1 available in the literature)

$$S_2 \approx 1.5/2M \quad . \tag{9}$$

If one takes the one-body density contribution S_1 calculated with the Cohen-Kurath wave function seriously, the recent experimental data do not agree with this prediction:

$$S_2^{exp} \approx 0.18/2M \qquad \text{(consistent with zero)} \quad . \tag{10}$$

There is thus a violent contradiction between the data and the theory. I cannot conceive of any theoretical means to cut the prediction by an order of magnitude or more and so I must conclude either of the following reasons is applicable:

1. Experimental data are wrong.

2. The one-body density contribution to $F_T^{(1)}$ is incorrectly calculated.

3. Or both.

III. AXIAL CHARGE DISTRIBUTION

One might think that the reason why axial charges are not given by the sum of single-particle charges (as with vector charges) is due to the non-conservation of the axial current. This is not so. In fact, one can show that the shared charges do not vanish in the limit $m_\pi \to 0$ and remain important. The reason for this difference from the conserved vector charge may have to do with the way chiral symmetry is realized. While the symmetry associated with the vector current is manifested in the Wigner mode, the symmetry associated with the axial current, chiral symmetry, is realized in the Goldstone mode. This suggests that the distribution of axial charges in nuclei as perhaps probed by neutrino must be vastly different from that of vector charges as probed by electron scattering. The study of axial charge distribution (when it becomes feasible) would provide us with those aspects of nuclei that have not been explored at all up to now.

IV. GAMOW-TELLER MATRIX ELEMENTS

We now consider the space component of the axial current, in particular the Gamow-Teller matrix element. As emphasized before, there is no reason to believe that a pion exchange can adequately describe the space component. In fact, a pion exchange with large momenta is involved, so the two nucleons feel more strongly the hard-core repulsion. Since the matrix element of such a process will be strongly suppressed, one expects that other short-range phenomena (exchange of heavier mesons such as ρ...) cannot be ignored. Thus the soft-pion theorem is expected to be powerless for this component. Without a model-independent theory, I cannot but resort to models so I shall describe a model which seems to give, surprisingly enough, a fairly good description of what's going on and which in addition provides an interesting connection to π-nucleus scattering and pion condensation.

IV.1 Triton Beta Decay[9]

As a first step toward gaining some insight into the problem and to reducing a complicated problem to a simpler one with a small number of parameters, let us look at the triton beta decay:

$$^3H \to {}^3He + e^- + \bar{\nu}_e$$

If one takes the strict impulse approximation, then the Gamow-Teller matrix element is given by

$$M_A \equiv \frac{|\langle {}^3He || \Sigma_i \tau_i^+ \vec{\sigma} || {}^3H \rangle|}{\sqrt{2}} = \sqrt{3}\,[1 - \frac{2}{3} P_D - \frac{4}{3} P_{S'} \ldots] \quad (11)$$

where $P_D, P_{S'}$ are the D-state and the mixed-S (S')-state probabilities and some negligible components are ignored here. $P_{S'}$ is small, of the order of 1.6 ∿ 2%, so the matrix element is quite strongly dependent upon the D-state probability. The standard calculation gives $P_D \approx 9\%$, so one expects

$$M_A = \sqrt{3}\,(0.927) \quad . \tag{12}$$

This is to be compared with the experiment

$$M_A^{exp} = \sqrt{3}\,(0.976 \pm 0.03) \tag{13}$$

Suppose now that a more accurate theoretical calculation includes exchange currents and other effects and denote it by E. Then M_A should read

$$M_A = \sqrt{3}\,[1 - \frac{2}{3} P_D - \frac{4}{3} P_{S'} + E] \quad . \tag{14}$$

Now if one can establish that

$$E \approx \frac{2}{3} P_D \tag{15}$$

then we get an agreement ($P_{S'} \approx 1.6 \sim 2\%$)

$$M_A \approx \sqrt{3}\,[1 - \frac{4}{3} P_{S'}] \approx \sqrt{3}\,[0.979 \sim 0.973] \tag{16}$$

The relation (15) has been discussed before[9]. I do not believe that one can make a very convincing demonstration of Eq. (15) since there are so many uncertainties, but it has been shown to hold roughly for the set of graphs

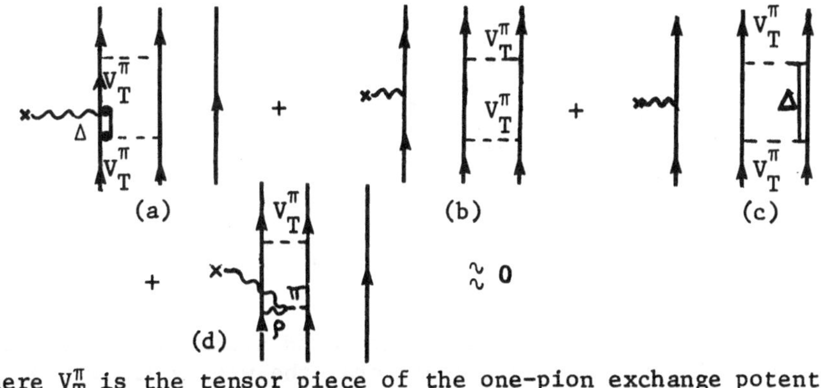

(17)

where V_T^π is the tensor piece of the one-pion exchange potential. The correct tensor force should contain both π and ρ exchanges[10], but even then this cancellation mechanism should hold still. Note that the figure (b) corresponds to $-\frac{2}{3} P_D$ in Eqs. (11) and (14), and the figure (c) to $-\frac{2}{3} P_\Delta$ where P_Δ is the probability of the Δ component in the 3-nucleon wave function. The figures (a) and (d) are conventionally calculated in terms of exchange currents (which I shall call D-terms).

I shall now assume that all the tensor-like terms as in (17) sum to a negligible result in the triton beta-decay. Since the intermediate states are of high energy involving excitation energies $\gtrsim$ 300 MeV, they are of short range and hence local. Therefore the cancellation is expected to persist in complex nuclei as well[11]. As with all local properties, it is very difficult to calculate them reliably; so I shall assume that we can completely ignore all the terms involving tensor force with large excitation energy. (This should be eventually verified by actual calculation.)

IV. COMPLEX NUCLEI

Having disposed of all the "D" terms, what turns out to be the most important is the Δ-hole excitations which propagate in the nucleus. This is the set of RPA-like graphs

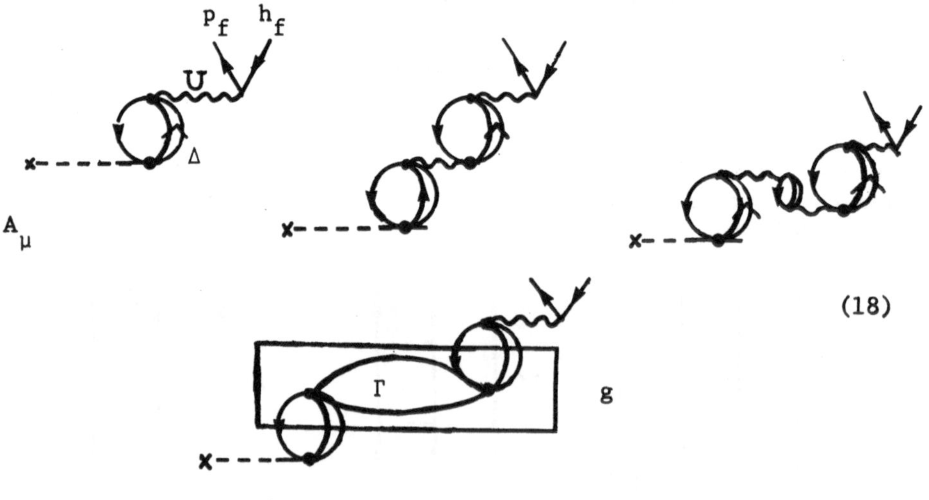

(18)

$$= \sum_{\Delta\Delta' hh'} \left\langle 0 \left| \frac{G_A}{2} T^{\alpha} \vec{S} \right| \Delta h \right\rangle g^{(\omega)}_{\Delta h, \Delta' h'} \left\langle \Delta' h' \right| \mathcal{U}(\omega) \left| p_f h_f \right\rangle \Big|_{\omega=0}$$

where G_A is the axial coupling constant to ΔN transition, $\vec{T}$ and $\vec{S}$ the transition isospin and spin operators, the wiggly lines the full Δh-ph and Δh-Δh interactions in direct channel and g is the exact Δh Green's function. One gets from a generalized Goldberger-Treiman relation

$$\frac{G_A}{g_A} = \frac{f^*_\pi}{f_\pi} \tag{19}$$

The problem of exchange terms will be discussed later. In order to evaluate Eq. (19), we note that the p-wave π-nucleus scattering can be described by the same Green's function g but at different kinematics,

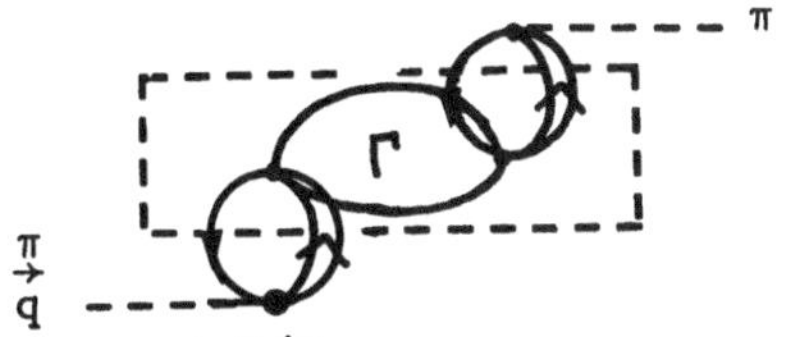

$$\omega = (q^2+m_\pi^2)^{1/2} \tag{20}$$

$$= \sum_{\Delta\Delta' hh'} \langle 0|\frac{f^*_\pi}{m_\pi} T^\alpha \vec{S}\cdot\vec{q}|\Delta h\rangle \, g_{\Delta h,\Delta' h'}(\omega)\Big|_{\omega=(g^2+m_\pi^2)^{1/2}} \langle \Delta' h'|\frac{f^*_\pi}{m_\pi} T^\beta \vec{S}\cdot\vec{q}|0\rangle$$

This suggests that we use the same model as the one which successfully describes π-nucleus scattering[12]; namely for the Δh-Δh interaction (and similarly for Δh-ph interactions);

$$U(\omega,\vec{q}) = \int \frac{d^3k}{(2\pi)^3}\, \Omega(\vec{q}-\vec{k})\, [V_\pi(\omega,\vec{k}) + V_\rho(\omega,\vec{k})]$$

$$V_\pi(\omega,\vec{k}) = \frac{f^{*2}_\pi(k^2)}{m_\pi^2}\, \frac{\vec{S}_1\cdot\vec{k}\; \vec{S}_2^{\,+}\cdot\vec{k}}{\omega^2-k^2-m_\pi^2}\, (\vec{T}_1\cdot\vec{T}_2^{\,+}) \tag{21}$$

$$V_\rho(\omega,\vec{k}) = \frac{f^{*2}_\rho(k^2)}{m_\rho^2}\, \frac{(\vec{S}_1\times\vec{k})\cdot(\vec{S}_2^{\,+}\times\vec{k})}{\omega^2-k^2-m_\rho^2}\, (\vec{T}_1\cdot\vec{T}_2^{\,+})$$

with

$$f^*_\pi(q^2) = f^*_\pi \left(\frac{\Lambda_\pi^2-m_\pi^2}{\Lambda_\pi^2+q^2}\right)$$

$$f^*_\pi(q^2) = f^*_\rho \left(\frac{\Lambda_\rho^2-m_\rho^2}{\Lambda_\rho^2+q^2}\right) \tag{22}$$

where Λ_π and Λ_ρ are cut-off masses fitted to experimental data[12] and $\Omega(q)$ the Fourier transform of the static baryon-baryon correlation function which seems rather reliably given by

$$\Omega(\vec q) = (2\pi)^3\delta^3(\vec q) - \frac{2\pi^2}{q^2}\,C(q), \tag{23}$$

with

$$C(q) \simeq \delta(q-q_c) \qquad q_c \simeq m_\omega = \omega\text{-meson mass}$$

One gets

$$U(\omega,\vec q) = V_\pi(\omega,\vec q) + V_\rho(\omega,\vec q) - [V^C_\pi(\omega,\vec q) + V^C_\rho(\omega,\vec q)]$$

with

$$V^C_{\pi,\rho}(\omega,\vec q) = \frac{2\pi^2}{q_c^2}\int \frac{d^3k}{(2\pi)^3}\,\delta(|\vec q-\vec k| - q_c)\,V_{\pi,\rho}(\omega,\vec k) \tag{24}$$

which in the limit $\omega \ll q_c$, $|\vec q| \ll q_c$, leads to

$$U(\omega,\vec q) \simeq V_\pi(\omega,\vec q) + \left(\frac{f^*}{m_\pi}\right)^2 g'\, \vec S_1\cdot\vec S_2\, \vec T_1\cdot\vec T_2 \tag{25}$$

to be used in direct matrix element; 

It is this g' which becomes crucially important for pion condensation and has, in the present model, the form

$$g' \simeq \frac{1}{3}\left\{\left(\frac{\Lambda_\pi^2-m_\pi^2}{\Lambda_\pi^2+q_c^2}\right)^2 \frac{q_c^2}{q_c^2+m_\pi^2} + \frac{2m_\pi^2}{m_\rho^2}\left(\frac{f^{*2}_\rho}{f^{*2}_\pi}\right)\left(\frac{\Lambda_\rho^2-m_\rho^2}{\Lambda_\rho^2+q_c^2}\right)\frac{q_c^{\,2}}{q_c^2+m_\rho^2}\right\}. \tag{26}$$

To complete the description of the model, let me give the numbers for the parameters[12]

$$\frac{f^*_\rho}{f_\rho} = \frac{f^*}{f} = 2 \qquad \text{(consistent with Chew-Low model)}$$

$$\Lambda_\pi = 8.6\ m_\pi$$
$$\Lambda_\pi = 17.92\ m_\pi \tag{27}$$

The ρNN tensor coupling is not well known and ranges widely

$$f_\rho = g_\rho (1 + K) \frac{m_\rho}{2M}$$
$$\frac{g_\rho^2}{4\pi} = 0.5 \tag{28}$$
$$3.7 \lesssim K \lesssim 6.6$$

The lower limit for K corresponds to the VDM prediction and the upper one to Höhler-Pietarinen value[13]. Since this is not precisely known, we shall take g' or equivalently

$$G'_\rho \equiv \frac{f_\rho^2}{m_\rho^2} \times \frac{m_\pi^2}{f_\pi^2} \tag{29}$$

to be a free parameter. Thus

G'_ρ	g'	K
1.47	0.5	5.6
1.90	0.6	6.5
2.33	0.7	7.3

Or

$$g' \simeq \frac{1}{3}(\underset{\text{"}\pi\text{"}}{0.47} + \underset{\text{"}\rho\text{"}}{G' \times 0.70})$$

Summing of the diagrams (19) can be done in the same way as was done for π-nuclear scattering except that here we are in the kinematical situation where

$$\vec{k} \to 0$$
$$\omega \to 0$$

Denoting the contribution from this sum δM_A and the single-particle Gamow-Teller matrix element M_A^o, the relevant quantity is

$$\delta_\Delta \equiv \frac{\delta M_A}{M_A^o} \quad .$$

The results for LS closed shell nuclei are given in Table 1[14].

There is one important "trivial" nuclear correction that should be added to δ_Δ before comparing with empirical results and that is the bona-fide core polarization of the type:

with the axial current attached to particle or hole line in all allowed combinations and the intermediate state with excitation energies much smaller than several hundred MeV involved with the D-state like terms. The intermediate state involves usually $2\hbar\omega$ excitation. The contribution has been studied extensively by many authors and there is a consensus on this[15]. This is given in the table as $\delta_{C.P.}$.

A non-trivial correction is the relativistic correction and here the situation is very tricky. It is very difficult to make a reliable estimate of the relativistic effect and in that matter even its meaning is obscure. In any event, estimates have been made and it seems that the results are not so different even when vastly different assumptions are made. I give the results averaged over those available in the literature in Table I. Each might have an error of $\pm(2 \sim 3)\%$. The total results are in good agreement with the empirical value within the uncertainties ascribed to δ_R and to possible contributions from other terms described below. The value of the g' somewhere around 0.6 and 0.7 is required by this analysis. Note that the attempt to explain the observed deviation in terms of core polarization induced through tensor forces[15] fails in two respects: it is incompatible with the triton β-decay and it does not explain the trend between A = 15 and A = 17. In our description, such a core-polarization is inoperative because of cancellation.

TABLE 1

	δ_Δ (%) [d]		$\delta_{C.P.}$ [c]	δ_R [a]	δ_{Th}		δ_{EXP}
	g'=0.6	g'=0.7			g'=0.6	g'=0.7	
$^{15}O \to {}^{15}N$	-5.7	-8.8	-2.3	-4.4	-12.4	-15.5	-11.8±0.9[a]
$^{17}F \to {}^{17}O$	-8.0	-9.7	-2.4	-2.7	-13.1	-14.8	-12.4±0.9[a] (-15.3±1.4[b])
$^{39}Ca \to {}^{39}K$	-12.2	-15.0	-5.8	-3.7	-21.7	-24.5	-31.4±0.9[a] (-25.5±2.9[b])
$^{41}Sc \to {}^{41}Ca$	-10.5	-12.4	-5.1	-3.4	-19.0	-20.9	-23.9±0.6[a]

a) Quoted from A. Barroso and R.J. Blin-Stoyle, Nucl. Phys. A251, 446 (1975).
b) Average value in the sense of B.A. Brown, W. Chung and B.H. Wildenthal, Phys. Rev. Lett. 40, 1631 (1978).
c) K. Shimizu et al., Nucl. Phys. A226, 282 (1974) and Ref. 15.
d) Unpublished results, Ref. 14.

Other Terms

The set of diagrams summed in (18) contains no exchange terms. To lowest order in particle-hole or Δ-hole interaction, we should include

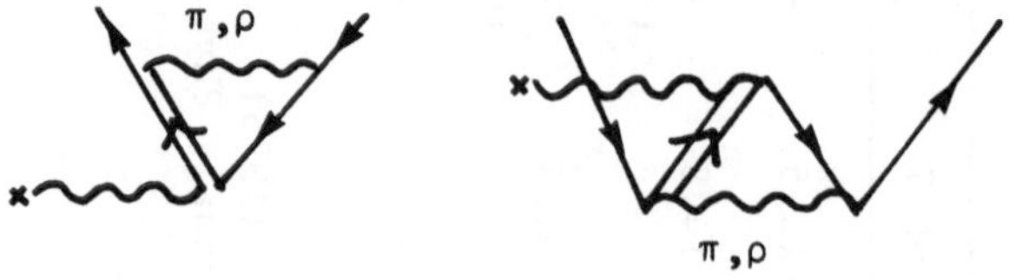

(30)

However the central piece of the ρ exchange, being of short range, can be re-written by means of $P_\sigma P_\tau P_x \stackrel{\sim}{\sim} P_\sigma P_\tau$ (since $P_x = 1$ for zero-range interaction) and incorporated into the direct term:

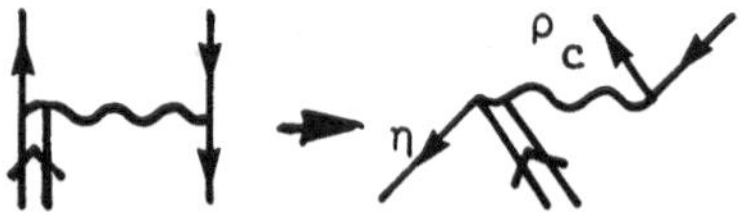

c ≡ central piece

To the extent that g' is a parameter in the theory, the contribution from the central part of the ρ exchange <u>need not</u> be calculated. The tensor piece of the ρ exchange cannot however be approximated by a zero-range form, so this should be explicitly included. There is a small contribution to g' from

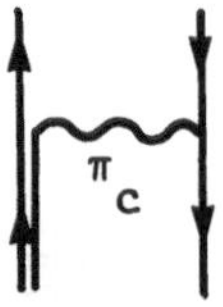

c ≡ central piece

but this is insignificant, so the π-exchange terms of (30) may be included totally without seriously double-counting.

There is an extra contribution in the exchange term and it is

(31)

which has both a "central" and a "tensor" component. Again the coupling constants involved are not known except in the soft-pion limit, so this cannot be calculated with great accuracy, but this term tends to be small because of the cancellation arising from

$$\frac{1}{q^2+m_\pi^2}\,\frac{1}{q^2+m_\rho^2} = \frac{1}{m_\rho^2-m_\pi^2}\left(\frac{1}{q^2+m_\pi^2} - \frac{1}{q^2+m_\rho^2}\right) . \tag{32}$$

We made a very rough calculation of the relevant central-piece contributions of (30) and (31), namely

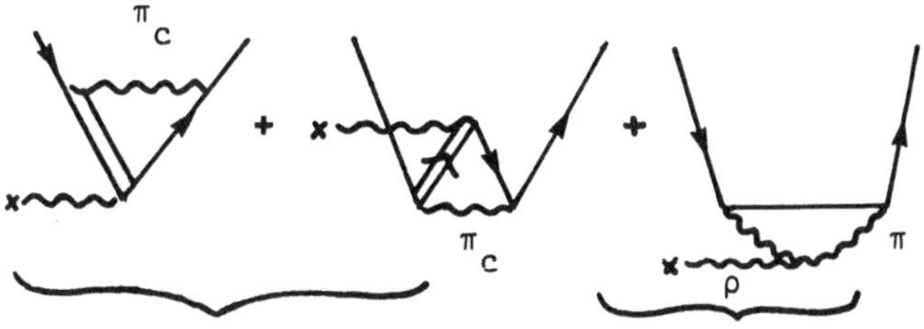

(33)

and found the sum to contribute something like

$$\delta \underset{\sim}{\sim} \underset{(a)}{(0.043} - \underset{(b)}{0.034)}\,\frac{\rho^T}{\rho_o} = 0.009\,\frac{\rho^T}{\rho_o} \tag{34}$$

where ρ^T is the relevant density (ρ_o is the nuclear matter density = $0.48\ m_\pi^3$). This shows the extent to which various cancellations occur and indicates that they can be indeed negligible. The tensor pieces are hard to evaluate, but cancellations are abundant; again the two graphs (30) and (31) are expected to sum to a small value. There are other factors which would make the tensor terms individually suppressed.

V. PIONIC MODE: PION CONDENSATION

The effective quenching of the Gamow-Teller matrix elements in nuclei has an important consequence on pion condensation phenomena. In fact I argue that the quenching of the G.T. matrix elements has the same physical origin as the Lorentz-Lorenz inhibition of pion condensation.

To motivate this argument, we recall in a few words how pion condensation is formulated in Chiral symmetric theories[17]. Consider the σ-model Lagrangian $\mathcal{L}$. The fact that chiral symmetry is an extremely good symmetry suggests that the pion-condensed phase of nuclear matter can be reached by a local chiral rotation on the Lagrangian $\mathcal{L}$, considered within the normal ground state:

$$\mathcal{L} \xrightarrow[\text{chiral transformation}]{\text{local}} \mathcal{L}$$

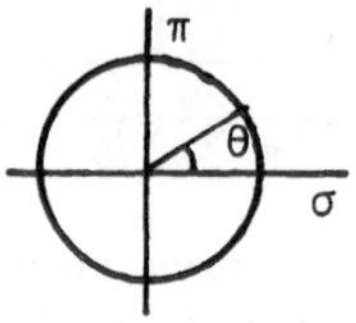

The transformation generates in a well-known way both vector (V^{α}_{μ}) and axial-vector (A^{α}_{μ}) currents. The vector current is conserved, but the axial vector current renormalizes by strong interaction to $g_A = 1.26$. The larger g_A, the larger the attraction, so that pion condensation sets in more readily. However in nuclear medium, the axial-vector coupling constant undergoes a medium correction and the way it gets influenced most significantly in the condensed phase is more or less the same as what happens to the Gamow-Teller matrix elements in normal nuclei: it is g' of Eq. (25) that plays a crucial role in both cases.

We now suggest that the Gamow-Teller matrix elements in normal nuclei should determine g' to be used in the pion-condensed phase. The question then arises: is it the same g' that is operative in both phases? The answer is provided by the chiral symmetry arguments: because of the approximate but excellent symmetry, as long as the condensed phase is generated by chiral rotation, the parameters of the theory must be the same in both phases. Therefore the g' reflected in the axial currents of normal nuclei is close to that relevant in the pion condensed phase[18]. Similar arguments have been used to extract g' from nuclear spectra with pionic quantum numbers: 0^-, 1^+, 2^- ... T = 1. The spectra, are, however, sensitive to the particle-hole channel (a);

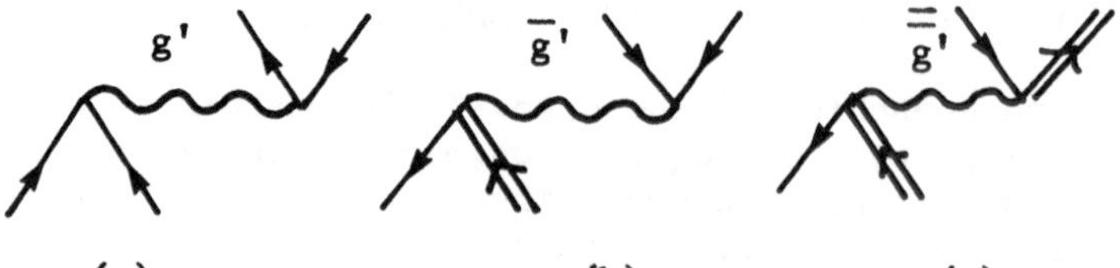

(35)

(a) (b) (c)

whereas pion condensation is affected by all three g', $\bar{g}'$, $\bar{\bar{g}}'$. The Gamow-Teller matrix elements provide information on (b) and to a certain extent on (c). One usually assumes that

$$g' = \bar{g} = \bar{\bar{g}}' \tag{36}$$

but of course there is no compelling reason why this should be so. In fact, Migdal[19] argues that $\bar{g}'$ and $\bar{\bar{g}}'$ might be much smaller than g'. The recent analyses[20] suggest that

$$g' = 0.5 \sim 0.7 \tag{37}$$

so the value $\bar{g}' = 0.6 \sim 0.7$ implied in the Gamow-Teller matrix elements is close enough to satisfy Eq. (36).

To complete this short excursion into pion condensation, let me just quote how the present picture of the density isomer caused by pion condensation looks like[21]

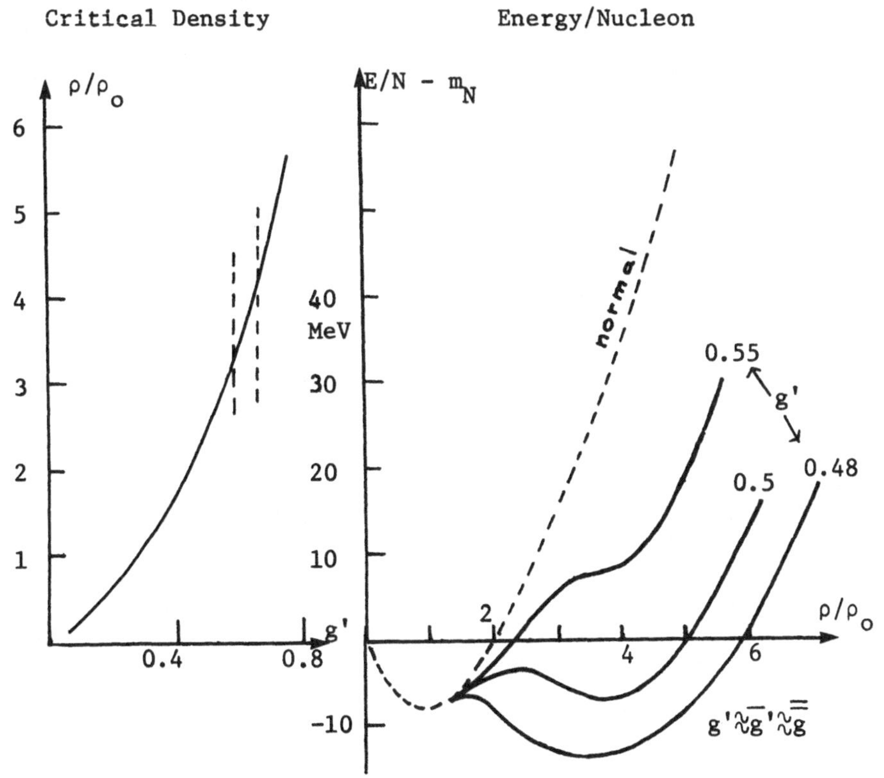

ACKNOWLEDGEMENTS

I am grateful for discussions with G.E. Brown, J. Delorme, K. Kubodera, E. Oset, H. Pirner and K. Yazaki.

REFERENCES

1. See Mesons in Nuclei, ed. by M. Rho and D.H. Wilkinson (North-Holland Publishing Co., in press).

2. D.O. Riska and G.E. Brown, Phys. Lett. 38B, 193 (1972).

3. J. Hockert, D.O. Riska, M. Gari and A. Huffman, Nucl. Phys. A217, 14 (1973); G.G. Simon et al., Phys. Rev. Lett. 37, 739 (1976).

4. H. Ohtsubo, J. Fujita and G. Takeda, Prog. Theor. Phys. 44, 1596 (1970); M. Rho, unpublished; K. Ohta, to be published. I. follow Ohta's notation.

5. K. Kubodera, J. Delorme and M. Rho, Phys. Rev. Lett. 40, 755 (1978).

6. P.A.M. Guichon, M. Giffon and C. Samour, Phys. Lett. 74B, 15 (1978).

7. There is a rather strong indication that the second-class form factor is negligibly small (see Ref. 8). Here we retain it for the sake of completeness.

8. K. Sugimoto, T. Minamisono, Y. Nojiri and Y. Masuda, to be published; K. Sugimoto, in Proceedings of the International Conference on Nuclear Structure, Tokyo, Japan, 1977 (International Academic Printing Co. Ltd., Tokyo, 1977); P. Lebrun, Ph. Deschepper, L. Grenacs, J. Lehmann, C. Leroy, L. Palffy, A. Possoz and A. Maio, Phys. Rev. Lett. 40, 302 (1978); H. Brandle, L. Grenacs, J. Lang, L.Ph. Roesch, V.L. Talegdi, P. Truttmann, A. Weis and Z. Zehnder, Phys. Rev. Lett. 40, 306 (1978); H. Brandle, G. Miklos, L.Ph. Roesch, V.L. Telegdi, P. Truttmann, A. Zehnder, L. Grenacs, P. Lebrun and J. Lehmann, Phys. Rev. Lett. 41, 299 (1978).

9. A.M. Green and T.H. Shucan, Nucl. Phys. A188, 289 (1972); M. Ichimura, H. Hyuga and G.E. Brown, Nucl. Phys. A196, 17 (1972); M. Rho, Erice Summer School Lectures, 1976.

10. G.E. Brown, NORDITA Lectures, 1978.

11. See ref. 9 and G.E. Brown, private communication.

12. E. Oset and W. Weise, Phys. Lett. 77B, 159 (1978) and to be published. I follow closely this paper.

13. G. Höhler and E. Pietarinen, B95, 210 (1975).

14. E. Oset and M. Rho, Phys. Rev. Lett. 42, 47 (1979).

15. A. Arima and H. Hyuga, in Mesons In Nuclei, ed. by M. Rho and D.H. Wilkinson (North-Holland Publishing Co., in press).

16. G.E. Brown, S.-O. Bäckman, E. Oset and W. Weise, Nucl. Phys. A286, 191 (1977).

17. G. Baym, Les Houches Summer School Lectures, 1977; G. Baym and D. Campbell, Mesons in Nuclei, ed. by M. Rho and D.H. Wilkinson (North-Holland Publishing Co., in press).

18. The close connection between g' operative in the quenching of Gamow-Teller m.e.'s and the generalized Lorentz-Lorenz was first pointed out in M. Ericson, A. Figureau and C. Thevenet, Phys. Lett. 45B, 19 (1973), and phrased in the form presented here by myself in Nucl. Phys. A231, 493 (1974). That the g' effect accounts indeed largely for the observed trends in light nuclei was re-emphasized in my Erice lectures, ref. 9. I believe that the recent calculations by Oset and myself confirm this assertion. See also F. Khanna and I. Towner in this proceeding, and also Phys. Rev. Lett. 42, 51 (1979).

19. A.B. Migdal, Rev. of Mod. Phys. 50, 107 (1978). It is easy to find an example where g' and $\bar{g}'$ can differ. Consider a scalar meson-exchange potential. In direct particle-hole channel, it contributes nothing to g' and $\bar{g}'$. But in the exchange term, it can contribute to g' but not to $\bar{g}'$.

20. J. Meyer-ter-Vehn, contribution to the Symposium on Relativistic Heavy-Ion Research, GSI Darmstadt, March 7-10, 1978; P. Bonche, H. Pirner, M. Rho and K. Yazaki, to be published.

MICROSCOPIC CALCULATIONS OF THE QUENCHING OF g_A AND OF THE GAMOW-TELLER MATRIX ELEMENTS IN A = 15, 17, 39 AND 41 SYSTEMS

I.S. Towner and F.C. Khanna

Atomic Energy of Canada Ltd.

Chalk River, Ontario, Canada K0J 1J0

ABSTRACT

A microscopic calculation of the Gamow-Teller matrix element is carried out to establish three points: a) cancellation of certain contributions in second-order of perturbation theory, b) quenching of g_A in finite nuclei, and c) comparison with the experimental β-decay matrix elements in A = 15, 17, 39 and 41 systems.

1. INTRODUCTION

The strong interactions are believed to play a significant role in changing the weak and electromagnetic interactions of nucleons inside nuclei. Theories[1] for modifying the electromagnetic currents were given many years ago while the modification of the weak currents has been considered only recently[2]. The partially-conserved-axial-vector-current (PCAC) hypothesis was exploited by Adler[3] and Weisberger[3] to investigate the renormalization of the axial-vector coupling constant in the decay of the neutron as compared to that in the decay of a muon. Most of the renormalization was attributed to the Δ-resonance (3/2,3/2) and the increase of g_A to a value of 1.23 (for n-decay) from its value of 1 (for μ-decay) was related to the π-nucleon scattering cross-sections off-the-energy-shell. In the case of nuclear systems first quantitative estimates[4,5] of the magnitude of g_A in nuclear matter were obtained by using PCAC. The effective value of g_A is related to the pionic current in nuclear matter. It is instructive to display this relationship of the effective value of g_A in nuclear matter and the polarizability of the medium. We begin with the PCAC equation:

The talk presented by F.C. Khanna

$$\partial_\lambda A_\lambda^{(\pm)} = f_\pi m_\pi^2 \phi_\pi^{(\pm)} \tag{1}$$

where f_π is the pion decay constant given by Goldberger-Treiman relation $\sqrt{2}\, M\, g_A = f_\pi g_{\pi NN}$, $A_\lambda^{(\pm)}$ is the axial-vector current and ϕ_π is the π-meson operator. The superscript + (-) denotes raising (lowering) the charge of the nucleus by one unit, corresponding to β^- (β^+) decay or π^+ (π^-) absorption.

Taking matrix elements of eq. (1) between initial ($|i\rangle$) and final ($|f\rangle$) nuclear states we get

$$iq_\lambda M_\lambda^\pm = \frac{f_\pi m_\pi^2}{q_\lambda^2 + m_\pi^2} M_\pi^\pm \tag{2}$$

where

$$M_\lambda^\pm = \langle f|A_\lambda^\pm|i\rangle$$

$$M_\pi^\pm = \langle f|J_\pi^\pm|i\rangle$$

$$q_\lambda = (P_i - P_f)_\lambda, \qquad q_\lambda^2 = \vec{q}^2 - q_0^2.$$

The equivalence of the matrix element of eq. (1) and eq. (2) can be proved by starting with $M_\pi^\pm$ and using the Klein-Gordon equation satisfied by $\phi_\pi^\pm$ i.e.

$$[-\nabla^2 + \frac{\partial^2}{\partial t^2} + m_\pi^2]\phi_\pi^\pm = J_\pi^\pm. \tag{3}$$

In the case of nuclear β-decay, equation (2) can be simplified by using the fact that q_0 and $\vec{q}$ are small. (This is consistent with the usual approximations made in the multipole analysis of the axial-current operator A_μ to obtain the Gamow-Teller operator $\vec{\sigma}\tau$). Then equation (2) reduces to the form

$$i\vec{q}\cdot\vec{M}^\pm = f_\pi M_\pi^\pm.$$

It is possible to expand $M_\pi^\pm$ in a Taylor series to obtain

$$i\vec{M}^\pm = f_\pi[\nabla_q M_\pi^\pm]_{\vec{q}=0}\,. \tag{4}$$

This indicates that if the pion-nucleus-nucleus vertex ($M_\pi^\pm$) is known completely it is possible to calculate the β-decay matrix element ($M^{\pm}$) precisely independent of the nuclear structure. This is an indication that the renormalization of the pion-nucleon-nucleon vertex to obtain $M_{\pi}^{\pm}$ has direct consequences on the renormalization of the β-decay vertex in a nucleus as compared to that for neutron decay.

The π-nucleus-nucleus vertex may be described by a set of graphs shown in Fig. 1. The non-static interaction between particle (isobar)-hole (nucleon) and the nucleus has one-pion contribution subtracted out. The sum over the momenta of the isobar particle-nucleon hole states in non-interacting nuclear matter is given by the Lindhard function $\chi_0(q,\omega)$. Except for the first diagram, the sum of the other diagrams reduces to

$$i \frac{g_{\pi NN}}{\sqrt{2}\,M} 2m_{\pi} \frac{V_{opt}(q,\omega)}{q^2} H(q,\omega)\tau_{+}\ \sigma\cdot q \tag{5}$$

where $V_{opt}(q,\omega)$ is the pion optical potential which can be investigated by considering the pion self-energy.

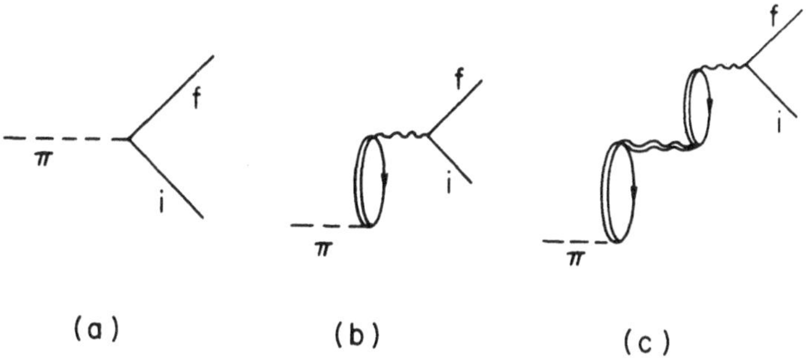

Fig. 1. Pion-nucleus-nucleus vertex $M_{\pi}^{\pm} = \langle f|J_{\pi}^{\pm}|i\rangle$. a) Lowest order; b) first order polarization correction due to particle (isobar)-hole (nucleon) states and c) higher order corrections.

The sum of the particle-hole bubbles (Fig 2) also expresses the pion-nucleus elastic scattering optical potential as

$$2m_{\pi} V_{opt}(q,\omega) = - \frac{4\pi\ \chi_0(q,\omega)q^2}{1 + 4\pi\ \chi_0(q,\omega)H(q,\omega)} \tag{6}$$

where

$$H(q,\omega) = g(q,\omega) - \frac{-q^2}{q^2+m_{\pi}^2-\omega^2} \tag{7}$$

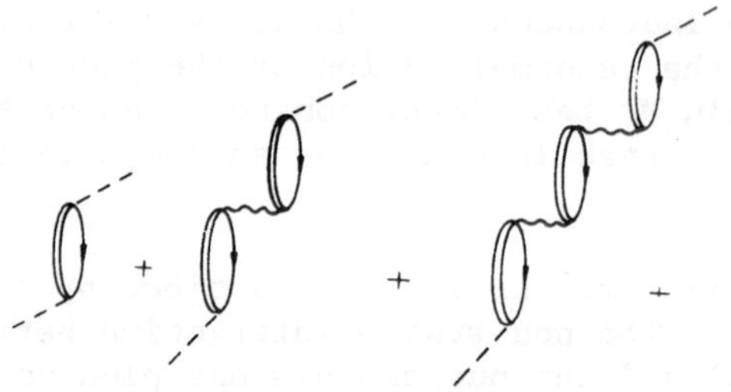

Fig. 2. Polarization corrections due to particle (isobar)-hole (nucleon) states to the pion self-energy or the pion optical potential.

and $g(q,\omega)$ is the non-static nucleon-Δ isobar interaction to all orders of perturbation theory. For simplicity using a one-pion potential with the δ-function removed for the function $g(q,\omega)$, we have

$$H(q,\omega) = \frac{1}{3}.$$

Now defining polarizability

$$\alpha(q,\omega) = \frac{4\pi}{3}\chi_0(q,\omega) \tag{8}$$

the pion optical potential takes the form

$$2m_\pi V_{opt}(q,\omega) = -\frac{3\alpha q^2}{1+\alpha}. \tag{9}$$

The presence of the factor α i.e. the polarizability in the denominator is called the Lorentz-Lorenz effect[6,7] which is the analog of light scattering from solid materials with due account of the discrete structure of the material.

Noting that π-nucleus-nucleus vertex in the impulse approximation (Fig 1a) may be written as

$$i\frac{g_{\pi NN}}{\sqrt{2}M}\tau_+ \sigma\cdot q, \tag{10}$$

the β-decay matrix element, $M^{\pm}$, becomes

$$M^{\pm} = \pm\langle i|\,[i\,g_A\,\tau_\pm\sigma - i\,g_A\,\frac{\alpha}{1+\alpha}\,\tau_\pm\sigma]\,|f\rangle \tag{11}$$

$$= \pm\,i\,\frac{g_A}{1+\alpha}\langle i|\tau_\pm\,\sigma|f\rangle.$$

Therefore in the case of nuclear matter, an effective axial-vector coupling constant is defined as

$$g_A^{eff} = \frac{g_A}{1+\alpha}. \tag{12}$$

This is a very interesting result (already obtained by Rho[5]) in that the quenching of the axial-vector coupling-constant is directly related to the Lorentz-Lorenz effect in π-nucleus scattering.

To get an estimate of the quenching, we evaluate the polarizability α as follows:

$$\alpha(\omega) = \frac{4\pi}{3} \chi_0(q,\omega).$$

For nuclear matter $\alpha(\omega)$ does not depend on ω; then

$$\alpha \equiv \frac{32\pi}{9} \frac{G^2}{\omega_R} \rho_0,$$

if $\chi_0(q,\omega)$ is assumed to have the form

$$\chi_0(q,\omega) = \frac{8}{3} \frac{G^2 \ \omega_R}{\omega_R^2 - \omega^2} \rho_0$$

where G is the $\pi N\Delta$ coupling constant, ρ_0 is the density of nuclear matter and ω_R is the centre of mass energy of the pion at the (3,3) resonance. Substituting the physical values $\rho_0 = 0.48\ m_\pi^3$, $G^2 = 0.137\ m_\pi^{-2}$ and $\omega_R \equiv m_\pi$ we get

$$\frac{g_A^{eff}}{g_A} \cong 0.74. \tag{13}$$

Thus the axial-vector coupling constant in nuclear matter (g_A^{eff}) is quenched - i.e. it is smaller than g_A, the value required to explain β-decay of the neutron.

So far we have only considered the role of the π-meson in the quenching of g_A and in the Lorentz-Lorenz effect in π-scattering from nuclei. Additional effects due to inclusion of ρ-meson etc. have been considered[8] and tend to decrease the polarizability of the system.

The questions we would like to ask here are two-fold: a) how can such a relationship between the quenching of g_A and the Lorentz-Lorenz effect in π-nucleus scattering be established in the case of finite nuclei, b) can the results of such calculations be compared with experiment? We shall attempt to answer these questions by considering the role of meson-exchange currents in finite nuclei. However, as will be established later, any comparison with experiments will necessarily involve detailed calculations of the core-polarization effects which actually dominate the corrections to the Gamow-Teller matrix elements as calculated in the shell model. In the course of this we shall examine the conjecture[10] about cancellation of some second-order core-polarization diagrams and second-order diagrams involving the isobar.

In order to calculate the quenching of g_A in finite nuclei, we choose several odd-A nuclei with a single valence particle or hole outside L-S closed shells. The contribution to the Gamow-Teller matrix element of the valence particle or the hole from exchange currents is calculated by summing over the core. Then the three matrix elements between the single particle states $J = \ell \pm \frac{1}{2}$ are re-expressed in terms of all tensors of rank-1 with positive parity i.e. the effective G-T operator is written[9] as

$$\mp \frac{1}{2} [\delta g_A \underset{\sim}{\sigma} + g_\ell \underset{\sim}{\ell} + \sqrt{8\pi}\, \Gamma_p (Y_2 \times \underset{\sim}{\sigma})^1] \tau_\pm \qquad (14)$$

where the upper sign refers to β^--decay and the lower sign to β^+ decay. The iso-spin raising and lowering operators are defined as $\tau_\pm = \tau_x \pm i\tau_y$. This effective G-T operator is to be compared with the free G-T operator $\mp \frac{1}{2} g_A \underset{\sim}{\sigma} \tau_\pm$. The coefficient δg_A directly gives the change in the axial-vector coupling constant g_A. The coefficient Γ_p is the analog of the pseudo-scalar coupling g_p in the decay of the neutron. It may be recalled that pseudo-scalar coupling arises from a physical process of the type shown in Fig 3. The non-relativistic reduction of the pseudo-scalar part of the

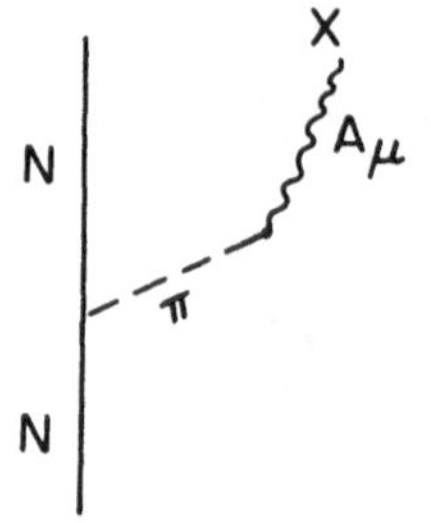

Fig. 3. A simple diagram with axial current (A) interacting with the nucleon via a pion intermediate state. This leads to the induced pseudo-scalar coupling in the weak decays.

current gives an operator of the form

$$\frac{g_3(k^2)}{2M}\,\vec{\sigma}\cdot\vec{k}\;\vec{k} \qquad (15)$$

which can be expressed in terms of irreducible tensors as

$$\frac{g_3}{2M}\,\frac{k^2}{3}\,[\underset{\sim}{\sigma} - \sqrt{8\pi}\,(Y_2 \times \underset{\sim}{\sigma})^1]. \qquad (16)$$

The induced pseudo-scalar coupling constant $g_P = m_\ell g_3$ where m_ℓ is the mass of the lepton is estimated to have a value of $\frac{g_P}{g_A} = 0.068$ (7.3) for e-capture (for μ-capture). The presence of $\frac{k^2}{2M}$ further reduces the magnitude of the pseudo-scalar term in all β-decay processes in finite nuclei. In the case of μ-capture with $\frac{k^2}{2M} \sim 0.1$, the pseudo-scalar term may be expected to be significant[10]. The coefficient Γ_p is to be added to g_P to give the renormalization of the pseudo-scalar coupling constant in finite nuclei. The coefficient g_ℓ in eq. [14] is reminiscent of a similar coefficient appearing in the M1-operator. As will be seen later the lowest order exchange currents do not give any contribution to g_ℓ but higher order core-polarization effects do give a non-zero magnitude to g_ℓ.

2. MODEL CALCULATIONS

Previously the exchange current contributions to β-decay have been calculated[9,11] using the expression for the two-body operator as given in the paper by Chemtob and Rho[2]. The main contribution was given by the diagram shown in Fig 4a with small corrections given by the pair term (Fig 4b). The vertex giving weak-production of pions was evaluated either by using PCAC or by using a phenomenological Lagrangian that describes the nucleon resonances in a quark model. The results (some are shown in Table I) thus obtained were encouraging and the following interesting conclusions were drawn.

a) δg_A increases in magnitude as the mass (A) of the L-S closed shell nucleus is increased while Γ_p decreases with A.
b) The magnitude of the quenching of g_A is larger for the decay of nucleons in the inner shells than that for the valence nucleons. If the results are extrapolated to infinite nuclear matter δg_A might saturate.
c) It was quite interesting to find that though the magnitude of the matrix elements for one particle or one hole in an L-S closed shell nucleus did not show any trend with mass number, the parametrization of the matrix elements (eq 14) showed that δg_A and Γ_p varied smoothly with mass number.

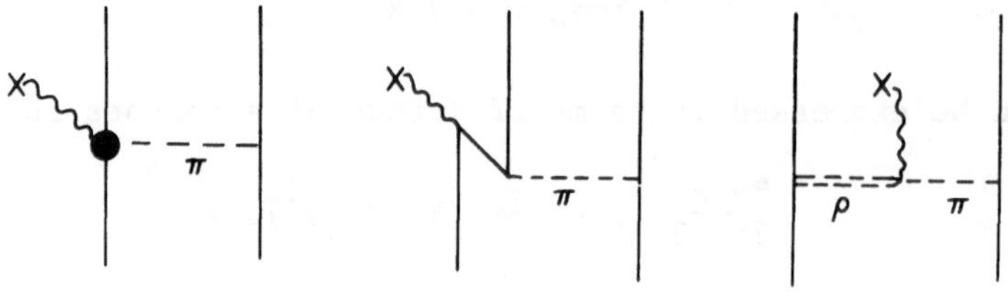

(a) (b) (c)

Fig. 4. Exchange current contribution to the weak decay rate in nuclei a) PCAC contribution with the weak pion-production vertex as given by Adler[3]; b) nucleon-pair term and c) ρ-π diagram.

TABLE 1

Variation of δg_A and Γ_p with single particle orbit

		0s	0p	0d	1s	0f	1p
A=16	δg_A	-.068	-.071	-.080	-.101		
		-.001	-.023	-.047	-.066		
	Γ_p	-	.116	.139	-		
		-	.068	.085	-		
A=40	δg_A	-.143	-.124	-.112	-.129	-.106	-.131
		-.054	-.053	-.058	-.072	-.065	-.087
	Γ_p	-	.074	.096	-	.110	.105
		-	.039	.054	-	.065	.057

First line: PCAC; second line: Phenomenological Lagrangian

The estimates for δg_A and Γ_p using the expressions obtained by using PCAC are believed to be somewhat more reliable than the phenomenological Lagrangian estimates. But the PCAC estimates include only the one pion exchange part of the two-body operator. Inclusion of ρ-, ω- and σ- exchange may change the results significantly. Then there is the contribution from the ρ-π diagram (Fig 4c). In addition there have been conjectures about the effect of tensor forces and cancellation of certain second-order diagrams with nucleons only and with nucleons and nucleon-isobars in the intermediate state. A consistent calculation to check this conjecture and to extend the usual PCAC results, we have resorted to a microscopic calculation of the Gamow-Teller matrix element in a nucleus assuming that the excitations in the nucleus may consist of nucleon particle-nucleon hole and of isobar particle-nucleon hole. The inclusion of isobar particle-nucleon hole is suggested by the earlier calculation of the Lorentz-Lorenz effect in nuclear matter (Fig 1). The inclusion of the nucleon particle-hole implies that core polarization effects are being included. (This follows from theories[12] of effective operators in nuclei). There will be physical processes that would imply a particle-hole chain that mixes with isobar particle-nucleon hole chain. Such effects are included also. The interactions among nucleons and among nucleons and isobars are mediated by exchange of π, ρ, ω and σ-mesons. The excitation of isobar states may be treated along the same lines as the conventional core-polarization calculations for magnetic moments[13], M1-decays[13] and E2-decays[14]. Since we are dealing with L-S closed-shell-plus or minus one nucleon, first order core-polarization contributions to the Gamow-Teller transition matrix element are zero. Some of the second-order diagrams belonging to the number-conserving set (Ellis and Siegel[14]) are shown in Fig 5. Additional diagrams obtained by replacing particle lines by an isobar line such that there is only one isobar in any intermediate state, are included in the number-conserving set. Dotted line with a cross at the end is the G-T transition operator. The potential between nucleons and isobars is a linear combination of one-boson (π,ρ,ω,σ) exchange potentials (Fig 6). Explicit expressions for the potentials as a function of the radial coordinate are not given here[7,15]. Similarly explicit expressions for the core-polarization diagrams shown in Fig 5 are not given here[14].

Furthermore the RPA series with an isobar particle-nucleon hole bubble at the end of the chain with G-T operator attached to it (Fig. 7) has been summed. As indicated in Fig. 7, nucleon p-h bubbles in the intermediate state starting with second order of perturbation theory are included. Various two-body interaction operators are picked as shown in Fig. 6.

There are additional corrections arising from ρ-π diagram (Fig 4c) and from pair-excitation (Fig 4b). Algebraic expressions for these two diagrams are given in the paper by Chemtob and Rho[2]. We have retained the proper q^2-dependence of the ρ-propagator in

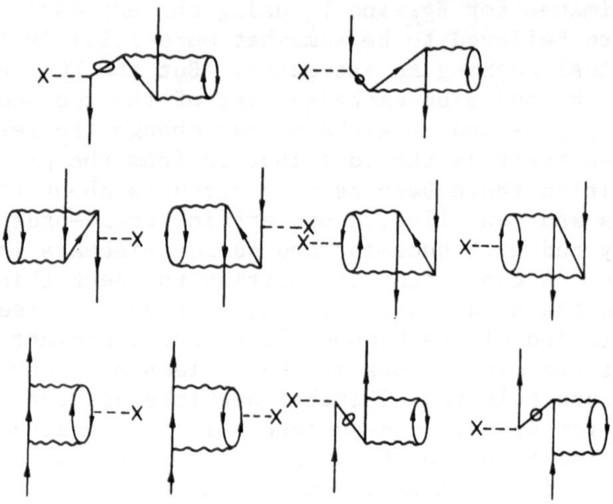

Fig. 5. Number conserving sets (Ellis and Siegel[14]) of second order diagrams. The first row is a set involving 2p-1h intermediate states, the second and third rows a set involving 3p-2h states. Additional diagrams, obtained by replacing nucleon lines by isobar lines such that no intermediate state has more than one isobar line, are also included in the number-conserving set.

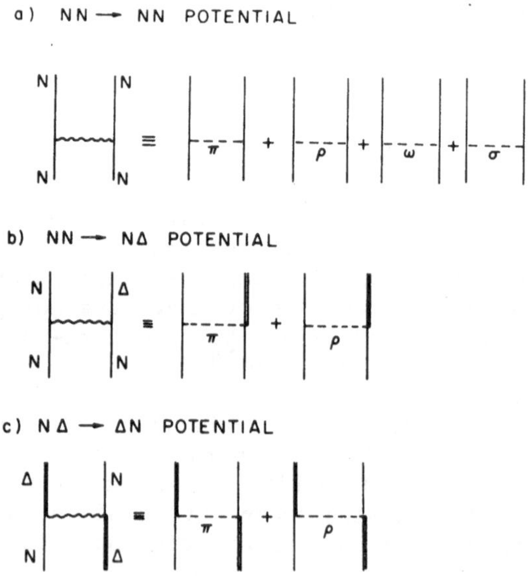

Fig. 6. Simple diagrammatic representation of the two-body potentials: a) nucleon-nucleon potential; b) NN → NΔ potential, and c) NΔ → NΔ potential. The coupling constants and the masses of the mesons π, ρ, ω and σ are given in Table II.

evaluating the two-body operator in r-space corresponding to the ρ-π diagram. As will be seen later, the pair excitation diagram with π-exchange contributes a small amount and so further contributions of ρ-, ω- and σ-exchange to this diagram are not included.

Single particle wave-functions are chosen to be that of a harmonic oscillator with $\hbar\omega = \frac{41}{A^{1/3}}$. Particle-hole excitations with energy up to 12 $\hbar\omega$ are included.

The coupling constants for meson-nucleon-nucleon and for meson-nucleon-isobar are given in Table II.

3. NUMERICAL RESULTS

The numerical results obtained are analysed so that quantitative statements about the following three topics can be made:

a) Cancellation within the number-conserving set (NCS);
b) Quenching of g_A in finite nuclei, and
c) Comparison with experiment.

It should be remarked that the results are quite sensitive to the cut-off parameter. Here we use a cut-off parameter $d = 0.5/m_\pi$ (0.71 fm). Let us consider the three aspects separately one by one.

a) The results for the number conserving set in second order of perturbation theory are shown in Table III. The first column gives results with the nucleon-nucleon potential given by one-pion-exchange potential. Columns two and three give results when successively one ρ- and one ρ- plus one ω-exchange potentials are added. The last column includes the effect of one isobar in the intermediate state. Contributions from the one π- and one ω-exchange have the same sign while one ρ-exchange and the isobar have the opposite sign and tend to decrease the total contribution to the Gamow-Teller matrix element. However with the parameters used, the isobar contribution is not large enough to cancel the nucleon contribution. Effects due to more than one isobar in the intermediate state are not included. The conjecture of Rho[10] is that the contributions from the NCS arising from high energy p-h excitations (say $\gtrsim 4\hbar\omega$) will be cancelled by the isobar contribution and the iterated ρ-π diagram. The contribution from these high energy p-h excitations is roughly half the totals listed in Table III. Thus if the iterated ρ-π diagram is to cancel this contribution it must necessarily be quite large. Indeed, estimates[10] do suggest that such a substantial cancellation may be possible.

TABLE II

Masses and Coupling Constants of Bosons

Coupling		a)	b)	c)
πNN	m_π	138 MeV	138 MeV	138 MeV
	$g_{\pi NN}$	13.49	13.49	13.29
	$f^2_{\pi NN}$ †	0.078	0.078	0.076
ρNN	m_ρ *	5.60	5.55	5.64
	$g_{\rho NN}$	2.84	2.63	3.54
	K_ρ	$=K_V=3.7$	6.6	$=K_V=3.7$
	$f^2_{\rho NN}$ †	0.109	0.092	0.172
ωNN	m_ω *	5.67	5.67	5.64
	$g_{\omega NN}$	7.60	7.60	10.63
	K_ω	$=K_s=-0.12$	-0.12	-0.12
	$f^2_{\omega NN}$ †	0.800	0.800	1.549
σNN	m_σ *			2.6
	$g_{\sigma NN}$			5.5
	$f^2_{\sigma NN}$ †			0.088

a) Chemtob[1]; b) Hohler and Pietarinan[16]; c) Gross[15]

† $f^2_{BNN} = \frac{1}{4\pi}\left(\frac{g_{BNN}m_B}{2M}\right)^2$; using quark model BNΔ coupling constants, $f^*_{BN\Delta}$, can be expressed in terms of f_{BNN} as $f^*_{BN\Delta} = \frac{6}{5}\sqrt{2}\, f_{BNN}$.

* in units of m_π.

TABLE III

Corrections from the NCS to the Gamow-Teller matrix element expressed as a percentage of the single-particle estimate

		Nucleons only			Nucleon (1π+1ρ+1ω) + isobars
		1π	1π+1ρ	1π+1ρ+1ω	
$0p_{1/2}^{-1}$	a)	-33.6	-23.9	-24.5	-19.9
	b)	-33.6	-17.2	-17.5	-13.4
	c)	-31.4	-18.4	-19.1	-13.4
$0d_{5/2}$	a)	-17.6	-14.8	-14.8	-12.8
	b)	-17.6	-12.9	-12.8	-11.0
	c)	-16.4	-12.9	-12.6	-10.8
$0d_{3/2}^{-1}$	a)	-34.5	-27.7	-27.6	-24.5†
	b)	-34.5	-22.7	-22.4	-19.7
	c)	-32.2	-23.2	-23.7	-20.6†
$0f_{7/2}$	a)	-19.6	-17.0	-16.8	-15.6†
	b)	-19.6	-15.3	-15.1	-13.3
	c)	-18.3	-15.0	-14.5	-12.8†

†The contribution of the isobar is estimated in these cases.

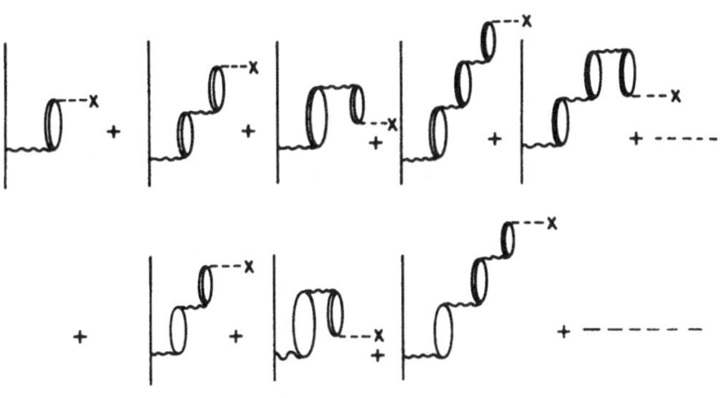

Fig. 7. Core-polarization diagrams. Double lines indicate isobar-particles and single lines in the bubble are nucleon-holes. Dotted line with the cross is the weak vertex.

b) Quenching of g_A in finite nuclei was calculated using the two-body exchange operator for β-decay as given by Chemtob and Rho[2]. In the present calculation we restrict ourselves to light nuclei. The change in G-T matrix element is decomposed in terms of irreducible tensors as given in eq. (14). Magnitude of δg_A, Γ_p and g_ℓ are tabulated in Table IV for the three sets of parameters (coupling constants, masses of bosons, etc.) listed in Table II. The first three rows give the results for each of the three parameter sets when the isobar particle-nucleon hole RPA series, π-pair diagram (Fig. 4b) and ρ-π diagram (Fig. 4c) are included. In this case g_ℓ is always zero. Signs of δg_A indicate quenching of g_A in the case of both a hole and a particle in A = 16 and A = 40 nuclei. The magnitude of both δg_A and Γ_p are smaller than the values given in Table I. The choice (b) of $g_{\rho NN}$ particularly reduces Γ_p a great deal. The choice (b) corresponds to a larger anamolous magnetic moment for the ρ-meson than is consistent with vector-dominance model.

The second set of three rows shows results when the effects of nucleons as given in Fig. 5 and Fig. 7 are included. The effect of core polarization, i.e. effects of diagrams with nucleon p-h is quite large. A finite value of g_ℓ emerges. In particular it should be noted that the magnitude of δg_A increases by a large amount. The total effect of all the diagrams calculated is to quench g_A from its free value of 1.23 to a value that is close to unity. In order to calculate G-T matrix elements with several particles in sd-shell or several holes in p-shell, an effective G-T operator with the parameters given in Table IV ought to be used.

It should be mentioned that the coefficient Γ_p given in Table IV implies quenching of the pseudo-scalar coupling constant. The overall quenching of g_p for μ-capture is ∿ 15-20% while in the case of β-decay g_p is negligibly small and the magnitude of Γ_p given in the table gives the overall value for the pseudo-scalar coupling constant. It should be recalled that the magnitude of Γ_p is to be compared with a value of $-\frac{1}{3} g_3 \frac{k^2}{2M}$ (see eq. (16)) of −0.147 (-6.7×10^{-6}) for μ-capture (e-capture).

The quenching of g_A and g_p in finite nuclei as obtained above can be used to estimate the total μ-capture rates in heavy nuclei. Assuming SU(4) symmetry total capture rates (Λ) are proportional to the following combination of coupling constants[16]

$$[1 + g_A^2 \{3 + (\frac{g_p}{g_A})^2 - 2 \frac{g_p}{g_A}\}] \quad .$$

Assuming the effective (free) value of g_A to be 1.0 (1.23) and

TABLE IV

Corrections to the Gamow-Teller matrix element decomposed in terms of the spherical tensors displayed in eq. 14

		$0p^{-1}$			$0d$			$0d^{-1}$			$0f$		
		δg_A	Γ_p	g_ℓ	δg_A	Γ_p	g_ℓ	δg_A	Γ_p	g_ℓ	δg_A	Γ_p	g_ℓ
	a)	-.079	.025	-	-.043	.056	-	-.050	.015	-	-.050	.020	-
(E)	b)	-.043	.011	-	-.055	.016	-	-.049	~ 0	-	-.062	.005	-
	c)	-.038	.024	-	-.047	.027	-	-.068	.012	-	-.058	.017	-
	a)	-.303	.058	.0111	-.250	.042	.0106	-.353	.010	.015	-.299	.003	.016
(T)	b)	-.244	.021	.0082	-.216	.018	.0071	-.318	-.006	.011	-.280	.010	.011
	c)	-.251	.037	.0093	-.218	.034	.0085	-.339	.007	.015	-.270	.001	.013

(E) = Exchange current contribution = ρ-π + π-pair + RPA with isobar particle-nucleon hole of $0\hbar\omega$ excitation.

(T) = total renormalisation = ρ-π + π-pair + Number-conserving set (Fig. 5) + RPA with p-h excitations up to $6\hbar\omega$ (both for isobar and nucleon particle states).

Rows a), b) and c) refer to the three sets of coupling constants listed in Table II.

the ratio of g_p to g_A to remain at the free value, i.e. $g_p/g_A \cong 7.3$, the total capture rates in heavy nuclei ($g_A = 1.0$) will be smaller than those in light nuclei ($g_A = 1.23$) by a factor of ~ 1.7. This is consistent with the trend in the experimental data as deduced by Duplain et al.[17].

c) Now we are in a position to compare our results with experiments[18] on the β-decay of A = 15, 17, 39 and 41 systems. The corrections to the G-T matrix element calculated from the diagrams shown in Figs. 4b, 4c, 5 and 7 are given in Table V. The results in the case of a p-hole ($0p_{1/2}^{-1}$) are very sensitive to the details of the calculation and the choice of parameters. There is a large cancellation between RPA and NCS results.

TABLE V

Corrections to the Gamow-Teller matrix element expressed as a percentage of the single-particle estimate, and compared with experiment.

		NCS	RPA	ρ-π	π-pair	SUM	Expt.[18]
	a)	-19.9	+17.5	-4.1	-1.0	-7.6	
$0p_{1/2}^{-1}$	b)	-13.4	+5.8	-6.7	-1.0	-15.4	-11.2
	c)	-13.4	+9.9	-4.1	-1.0	-11.2	
	a)	-12.8	-1.0	-0.7	0	-14.5	
$0d_{5/2}$	b)	-11.0	-2.5	-1.1	0	-14.6	-11.7
	c)	-10.8	-1.6	-0.7	0	-13.1	
	a)	-24.5†	-0.5	-2.2	-0.3	-27.5	
$0d_{3/2}^{-1}$	b)	-19.7	-3.0	-3.7	-0.3	-26.7	-32.1
	c)	-20.6†	-1.9	-2.2	-0.3	-25.0	
	a)	-15.6†	-2.7	-0.8	0	-19.1	
$0f_{7/2}$	b)	-13.3	-4.1	-1.4	0	-18.8	-24.4
	c)	-12.8†	-3.6	-0.8	0	-17.2	

Rows a), b) and c) refer to the three sets of coupling constants given in Table II.

†The contribution of the isobar is estimated in these cases.

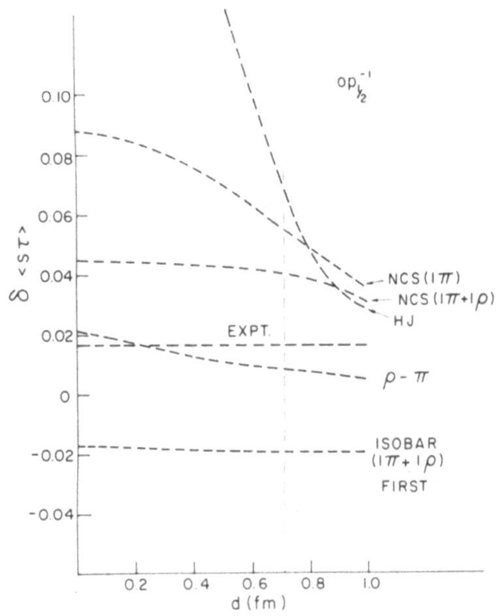

Fig. 8. Dependence of $\delta\langle s\tau\rangle$, the change in the matrix element $\langle s\tau\rangle$ (related to the Gamow-Teller matrix element by $2g_A \frac{J+1}{J} \langle s\tau\rangle$), on the cut-off parameter d. The magnitude of $\delta\langle s\tau\rangle$ required to explain the experimental results in weak-decay of $Op^{-1}_{1/2}$ is also shown. Results obtained with phenomonological Hamada-Johnston (HJ) potential are shown for comparison. Dependence of the contribution to $\langle s\tau\rangle$ from i) NCS (Fig. 5) with 1π exchange potential only, ii) NCS with 1π+1ρ exchange potential, iii) ρ-π diagram (Fig. 4c) and iv) first order isobar particle-nucleon hole contribution with 1π+1ρ exchange potential is illustrated.

With the choice of parameter set (a), most of the discrepancy in the Gamow-Teller matrix element can be explained by the contribution of ρ-π and π-pair terms. The parameter set (c) gives results closest to the experimental values.

In Fig. 8 we display the sensitivity of the results to the cut-off parameter d in the A = 15 case. The degree of sensitivity in the case of the crucial number conserving set (Fig. 5) is seen to depend on the choice of the nucleon-nucleon interaction, being particularly sensitive for the Hamada-Johnston potential and less so for the 1π- and 1π- plus 1ρ-exchange potentials. The contributions of the ρ-π diagram and the first-order isobar diagram (first term in the RPA series in fig. 7) are relatively insensitive to the cut-off parameter. The magnitude of the discrepancy of the experimental results from the shell model value is shown to provide a guide. Small adjustments of the cut-off parameter can bring the calculated results into agreement or into violent disagreement with experiment.

4. CONCLUSIONS

A microscopic study of the core polarization and exchange currents has been carried out primarily a) to calculate the quenching of g_A due to the presence of an isobar (J = 3/2, T = 3/2) among the excited particle states, and b) to attempt to resolve the problem of the G-T β-decay matrix element of A = 15, 17, 39 and 41 systems. The results appear very encouraging in that the experimental results can be quite adequately explained. Uncertainties in coupling constants, cut-off parameter and the single particle wave-functions are sufficient that the remaining discrepancy need not be worrisome.

The relationship of the quenching of g_A to Lorentz-Lorenz effect is not completely established in that the effect of the isobar particle-nucleon hole on the optical potential for π-scattering on finite nuclei has not been studied in the present framework.

ACKNOWLEDGEMENT

We would like to express our sincere thanks to Mannque Rho for correspondence and suggestions about the material of this talk.

REFERENCES

1. R.G. Sachs, Phys. Rev. 74 (1948), 433. M. Chemtob, "Meson theory of nuclear vector and axial vector exchange currents", to appear in Mesons in Nuclei, Ed. D.H. Wilkinson and M. Rho.

2. R.J. Blin-Stoyle, V. Gupta and H. Primakoff, Nucl. Phys. 11 (1959), 444. M. Chembot and M. Rho, Nucl. Phys. A163 (1971), 1.

3. S.L. Adler and R. Dashen, Current Algebras (Benjamin, N.Y. 1968); S.L. Adler, Phys. Rev. Lett. 14 (1965), 1051; W.I. Weisberger, Phys. Rev. Lett. 14 (1965), 1047.

4. M. Ericson, A. Figureau and C. Thevenet, Phys. Letters 45B (1973), 19; M. Ericson, Ann. Phys. 63 (1971), 562; J. Delorme, M. Ericson, A. Figureau and C. Thevenet, Ann. Phys. 102 (1976), 273.

5. M. Rho, Nucl. Phys. A231 (1974), 493.

6. M. Ericson and T.E.O. Ericson, Ann. Phys. 36 (1966), 323; S. Barshay, G.E. Brown and M. Rho, Phys. Rev. Lett. 32 (1974), 787; S.L. Adler, Phys. Rev. 126 (1962), 413.

7. G.E. Brown and W. Weise, Phys. Reports 22C (1975), 279.

8. M. Theis, Phys. Lett. 63B (1976), 43; G. Baym and G.E. Brown, Nucl. Phys. A247 (1975), 395; G.E. Brown, Meson Nuclear Physics - 1976, Ed. P.D. Barnes, R.A. Eisenstein and L.S. Kisslinger (A.I.P., New York, 1976).

9. I.S. Towner, F.C. Khanna and H.C. Lee, contribution to VII Int. Conf. on High Energy Physics and Nuclear Structure, 1977, Zurich, Switzerland. F.C. Khanna, I.S. Towner and H.C. Lee, Nucl. Phys. A305 (1978), 349.

10. M. Rho, Meson fields in nuclei, lectures, given at the Int. School of Nuclear Physics, Erice, Italy (1976).

11. A. Barroso and R.J. Blin-Stoyle, Nucl. Phys. A251 (1975), 446.

12. M. Harvey and F.C. Khanna, Nucl. Phys. A152 (1970), 588. B.H. Brandow, Rev. Mod. Phys. 39 (1967), 771.

13. H.A. Mavromatis and L. Zamick, Nucl. Phys. A104 (1967), 17. I.S. Towner, F.C. Khanna and O. Hausser, Nucl. Phys. A266 (1977), 285.

14. F.C. Khanna, H.C. Lee and M. Harvey, Nucl. Phys. A164 (1971), 612; P.J. Ellis and S. Siegel, Phys. Lett. 34B (1971), 177; K. Shimizu, M. Ichimura and A. Arima, Nucl. Phys. A226 (1974), 282.

15. F. Gross, Phys. Rev. D10 (1974), 223.

16. G. Höhler and E. Pietarinen, Nucl. Phys. B95 (1975), 210.

17. D. Duplain, B. Goulard and J. Joseph, Phys. Rev. C12 (1975), 28.

18. S. Raman, C.A. Houser, T.A. Walkiewicz and I.S. Towner, to Atomic Data and Nuclear Data Tables 21 (1978), 567; D.H. Wilkinson, Nucl. Phys. A225 (1974), 365: G. Azuelos and J.E. Kitching, Nucl. Phys. A285 (1977), 19.

CORRELATIONS AND TOTAL MUON CAPTURE RATES

Aram Mekjian

Rutgers University, Department of Physics

New Brunswick, New Jersey, U.S.A.

ABSTRACT

The total muon capture rate for s-wave muons can be accounted for by the Primakoff expression which gives the dependence of this rate on the mass number A and the proton number Z of the absorbing nucleus. The expression is a simple three parameter phenomenological formulae which accurately describes these rates from light weight nuclei to heavy nuclei. These parameters relate to the isospin structure of the squared isovector operator which appears in a sum rule approach to such rates. A microscopic analysis of the parameters appearing in the capture rate expression is presented in the light of recent developments concerning photonuclear reactions. A shell model analysis is given and it is found that the predictions of the unperturbed shell model and also Hartree-Fock theory are in complete disagreement with the data. Considerable improvement is obtained when long range correlations are included in the ground state wave function of the absorbing nucleus.

The capture rate of a muon in a 1S orbit is proportional to the probability of finding a muon at the nucleus and to the number of protons in the nucleus. Thus, the lowest order expression for the capture rate follows the Wheeler law $\Lambda \sim Z^4$ since the 1S probability density varies as Z^3 at the origin. Finite nuclear size corrections change Z to Z_{eff}. The effect of neutrons in the nucleus inhibit the transition $\mu^- + p \rightarrow n + \nu_\mu$ and reduce the capture rate. Specifically, the effect of the neutrons on the capture rate can be seen in the following semi-empirical formulae obtained by Primakoff[1] in the closure approximation

$$\Lambda_c(A,Z) = KZ^3_{eff}Z\left(1 - \frac{A-Z}{2A}\delta\right) \tag{1}$$

Here, the δ is a parameter and the value $\delta = 3.15$ fits the (A-Z)/2A dependence of the capture rate $\Lambda_c(A,Z)$ as shown in Fig. 1. The solid curve is the straight line result of Eq. (1).

More recently, a reinvestigation of the question of the A and Z dependence of these capture rates has been given [2,3] and the following dependence has been proposed.

$$\frac{\Lambda_c(A,Z)}{KZ^3_{eff}Z} = 1 + \frac{A}{2Z}(\tilde{\beta}_0 - 1) + \frac{A-2Z}{2Z}\tilde{\beta}_1 - \frac{A-Z}{2A}\tilde{\beta}_2 \tag{2}$$

The $\tilde{\beta}_0 = 0.97$, $\tilde{\beta}_1 = 0.25$ and $\tilde{\beta}_2 = 3.24$ are obtained by fitting Eq. (2) to the data.[1] The largest term in Eq. (2) is the (A-Z)/2A term since Eq. (1) accurately characterizes the data and contains only this term. Table 1 gives the values of the experimentally determined capture rate Λ_c^{exp} and the rate obtained from Eq. (2), Λ_c^{fit}, for a series of nuclei.[2]

Let us now try to understand Eq. (2) in more detail and then attempt a microscopic evaluation of the parameters $\tilde{\beta}_0$, $\tilde{\beta}_1$ and $\tilde{\beta}_2$.

To begin, a detailed analysis by Primakoff[1] based on the impulse approximation gives the following expression for the capture rate

$$\Lambda_c = \frac{m_\mu^2}{2\pi}|\phi_\mu|^2_{Av}\left[G_V^2 M_V^2 + 3G_A^2 M_A^2 + (G_P^2 - 2G_PG_A)M_P^2\right] . \tag{3}$$

The m_μ is the muon mass and the $|\phi_\mu|^2_{Av}$ is the square of the muon wave function averaged over the nucleus. The G_V, G_A and G_P are the effective vector, axial vector and induced pseudoscalar coupling constants of ref. (1). The vector matrix element M_V is

$$M_V^2 = \sum_b\left(\frac{\nu_{ab}}{m_\mu}\right)^2\int\frac{d\hat{\nu}}{4\pi}\left|(b|\sum t_i^+ e^{i\vec{\nu}_{ab}\cdot\vec{r}_i}|a)\right|^2 , \tag{4}$$

the axial vector matrix element is

$$M_A^2 = \sum_b\left(\frac{\nu_{ab}}{m_\mu}\right)^2\int\frac{d\hat{\nu}}{4\pi}\left|(b|\sum_i t_i^+\vec{\sigma}\, e^{i\vec{\nu}_{ab}\cdot\vec{r}_i}|a)\right|^2 \tag{5}$$

while the induced pseudoscalar matrix element is given by

TABLE 1

Nucleus	Λ_c^{exp} 10^6 sec^{-1}	Λ_c^{fit} 10^6 sec^{-1}	$\frac{\Lambda_c^{exp}-\Lambda_c^{fit}}{\Lambda_c^{exp}}$ %
$_{8}O$	0.0974±0.0031	0.1151	18.2
$_{16}S$	1.338 ±0.007	1.244	7.0
$_{20}Ca$	2.45 ±0.02	2.46	0.4
$_{24}Cr$	3.29 ±0.04	3.06	7.1
$_{30}Zn$	5.74 ±0.04	5.48	4.5
$_{42}Mo$	9.22 ±0.06	9.55	3.6
$_{48}Cd$	10.62 ±0.10	10.66	0.4
$_{54}Ba$	10.18 ±0.10	10.40	2.2
$_{64}Gd$	12.09 ±0.16	12.74	0.9
$_{82}Pb$	13.02 ±0.11	13.05	0.3
$_{92}U$	11.0 ±0.5	11.5	4.9

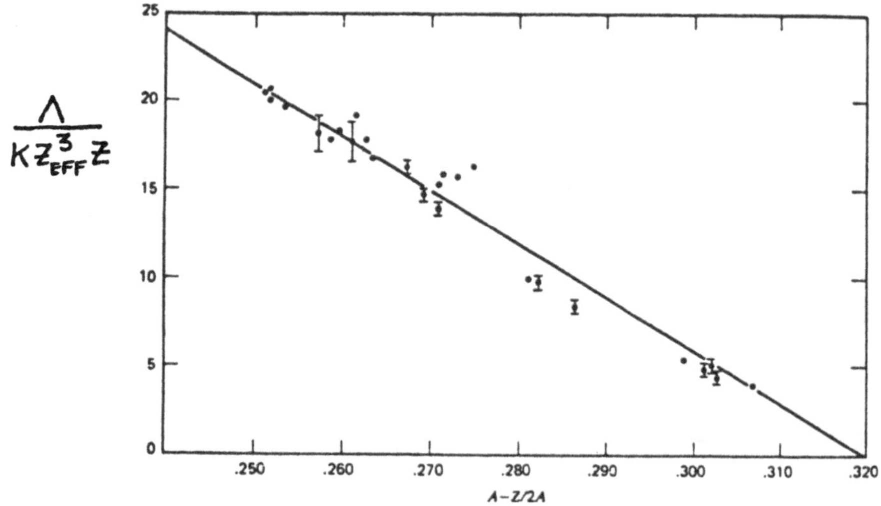

Fig. 1. Comparison between experimental muon (capture rate)/$Z_{eff}^3 Z$ and Primakoff theory.

$$M_p^2 = \Sigma \left(\frac{\nu_{ab}}{m_\mu}\right)^2 \int \frac{d\hat{\nu}}{4\pi} \left|(b| t_i^+(\vec{\nu}.\vec{\sigma}_i) e^{i\vec{\nu}_{ab}.\vec{r}_i} |a)\right|^2 \quad (6)$$

The ν_{ab} is the energy of the emitted neutrino accompanying the transition of the nucleus from its initial state $|a)$ to its final state $|b)$ in the nucleus of one less proton and one more neutron than $|a)$. Thus, $\nu_{ab} = m_\mu - (E_b - E_a)$, where E_a is the energy of $|a)$ and E_b is the energy of $|b)$. For a typical transition $E_b - E_a \simeq 25$ MeV, $\nu_{ab} = 80$ Mev and the momentum transfer $q = 2/5\ fm^{-1}$. The t^+ in Eqs. (4) - (6) is defined such that $t^+|p) = |n)$.

We will make the usual SU4 assumption

$$M_A^2 = M_v^2 = M_p^2 , \quad (7)$$

the validity of which is discussed in Ref. (4) and Ref. (5). We will therefore concentrate on M_v^2. The methods for evaluating M_v^2 are either: a) summing up partial transitions to individual states $|b)$, and b) using the closure approximation by introducing a mean neutrine energy $\bar{\nu}$. In the former case, the giant dipole resonance plays an important role for light nuclei and medium weight nuclei, while the giant quadrupole state contains most of the strength for the heavy nuclei. The contribution of a monopole state is due to retardation terms[6] since all simple t_i^+ transitions are blocked by neutron orbitals. These results are illustrated in Table 2 where the results of the calculation of Joseph, Ledoyen and Goulard[5] are quoted.

TABLE 2. Relative percentages of the vector matrix element $(M_v^2)_L$ in several nuclei using a shell model description.

	^{40}Ca	^{60}Ni	^{88}Sr	^{114}Sn	^{140}Ce	^{208}Pb
$0^+ \to 0^+$	7	6	15	13	19	28
$0^+ \to 1^-$	77	69	51	44	34	11
$0^+ \to 2^+$	14	22	28	35	39	51
$0^+ \to 3^-$	2	3	6	7	7	8
$0^+ \to 4^+$	0	0	0	1	1	2

Here, we will direct our attention to the second approach involving the closure approximation. This approximation is based on the result that if an operator θ acts on a state $|0\rangle$, its effect is, in general, to produce a multiple of the original state $|0\rangle$ and to admix other states $|\lambda\rangle$:

$$\theta|0) = a_o|0\rangle + \Sigma\, a_\lambda|\lambda\rangle \tag{8}$$

The amplitudes a_λ are just the matrix elements $a_\lambda = (0|\,\theta\,|\lambda)$. In the closure approximation, we use

$$\sum_\lambda |a_\lambda|^2 = \langle 0\,|\,\theta^{\,2}|0\rangle - \left|\langle 0|\,\theta\,|0\rangle\right|^2 \tag{9}$$

where the sum in Eq. (9) does not involve the initial state. Energy moments of the distribution of the a_λ's can also be defined by

$$N_m = \sum_\lambda (E_\lambda - E_o)^m\, |a_\lambda|^2 \tag{10}$$

with E_λ the energy of state $|\lambda)$. Our main concern will be the non-energy weighted sum rule approach for the evaluation of Eq. (4). To employ it, we have to introduce a mean neutrino energy through the defining relationship:

$$M_v^2 = \left(\frac{\bar{\nu}}{m_\mu}\right)^2 \sum_b \int \frac{d\hat{\nu}}{4\pi} \left|\langle b|\sum_i t_i^+\, e^{i\vec{\nu}.\vec{r}_i}\,|a\rangle\right|^2 \tag{11}$$

Then Eq. (11) reduces to

$$M_v^2 = \left(\frac{\bar{\nu}}{m_\mu}\right)^2 \langle a|J^{-+}|a\rangle\ , \tag{12}$$

where

$$J^{-+} \equiv \int \frac{d\hat{\nu}}{4\pi} \sum_{ij} t_j^-\, t_i^+\, e^{i\nu.(\vec{r}_i-\vec{r}_j)} = \int \frac{d\hat{\nu}}{4\pi}\ \theta^-\ \theta^+ \tag{13}$$

with $\theta^+ = \Sigma\, e^{i\vec{\nu}.\vec{r}_i}\, t_i^+$ and a similar definition for θ^-. Now, we note that the evaluation of M_V in the closure approximation involves the ground state expectation value of a two-body operator. Its value is therefore subject to ground state correlations. Our main purpose is to show that the slope of the muon capture rate as a function of (A-Z)/2A is very sensitive to such ground state correlations, changing by a factor of ten when shell model functions are improved on to include long-range correlations.

To proceed in this endeavor, we have to write Eq. (13) in the form of Eq. (2). This is accomplished by introducing the isoscalar, isovector and isotensor parts of the squared operator $\theta^- \theta^+$. In defining these isospin components we will remove that part which produces a multiple of the original state, the a_0 of Eq. (8), since we are only interested in transition matrix elements. Thus we define a two-body isoscalar part by

$$M_0 = \sum_{i,j} (e^{-i\vec{q}.\vec{r}_i} - \bar{\theta}_z)(e^{i\vec{q}.\vec{r}_j} - \bar{\theta}_z)\frac{1}{3}\vec{t}_i \cdot \vec{t}_j \qquad (14)$$

a one-body isovector part by

$$M_1 = \sum_{i} (e^{-i\vec{q}.\vec{r}_i} - \bar{\theta}_z)(e^{i\vec{q}.\vec{r}_i} - \bar{\theta}_z)\, 2t_{iz} \qquad (15)$$

and a two-body isotensor operator through

$$M_2 = \sum_{i,j} (e^{-i\vec{q}.\vec{r}_i} - \bar{\theta}_z)(e^{i\vec{q}.\vec{r}_j} - \bar{\theta}_z)\,\tfrac{1}{2}(\vec{t}_i\cdot\vec{t}_j - 3t_{iz}t_{jz}) \qquad (16)$$

The $\bar{\theta}_z$ serves the purpose of removing the multiple of the original state from the sum and it is given as

$$\bar{\theta}_z = \frac{2(0|\,\theta_z|0)}{(N - Z)} \qquad \text{with} \qquad \theta_z = \Sigma\, e^{i\vec{q}.\vec{r}_i}\, t_{iz}$$

Having established these definitions, it is a matter of straightforward isospin algebra to obtain

$$(\theta^- \theta^+) = Z\left[1 + \frac{A}{2Z}(\tilde{\beta}_0 - 1) + (\frac{A-2Z}{2Z})\tilde{\beta}_1 - (\frac{A-Z}{2A} + \frac{A-2Z}{8ZA})\tilde{\beta}_2\right] \qquad (17)$$

where

$$\tilde{\beta}_0 \equiv 2(||M_0||) + (||M_2||)$$
$$\tilde{\beta}_1 \equiv (||\,\theta_z||)$$
$$\tilde{\beta}_2 \equiv (||\,4M_2\,||) \qquad (18)$$

The reduced matrix elements are defined by

$$\frac{A}{2}\,(||M_o||) = (|M_o|)$$

$$T_z\,(||\theta_z||) = (|\theta_z|)$$

$$3(||M_2||)\,[3T_z^2 - T_z(T_z + 1)]/2A = (|M_2|) \tag{19}$$

The advantage of making an isospin decomposition is twofold. First, from the above results we learn that the isotensor sum rule determines the slope of the muon capture rate. Secondly, the various components have different dependencies on the model wave functions used. In particular, as shall be shown, the tensor part is strongly model-dependent, differing by a factor of ten when simple wave functions are improved upon. The scalar part is mildly model dependent and it differs by a factor of two in such improvements, while the vector part, given by

$$(0|M_1|0) = \tfrac{1}{4}(N - Z)\,[1 - |(||\theta_z||)|^2]$$

is nearly model independent since it involves the expectation value of a one-body operator $(||\theta_z||)$ which is simply the nuclear neutron-proton form factor difference:

$$(||\theta_z||) = \frac{\left(NF_n(q) - ZF_p(q)\right)}{(N - Z)} \tag{20}$$

Having established the connection of Eq. (2) with the isospin structure of our sum rules, let us next evaluate the coefficients $\tilde{\beta}_o$, $\tilde{\beta}_1$ and $\tilde{\beta}_2$ with a Slater determinant wave function

$$\Psi_A = \frac{1}{\sqrt{A!}} \begin{vmatrix} \phi_1(\vec{r}_1) & \cdots & \phi_1(\vec{r}_A) \\ \vdots & & \\ \phi_A(\vec{r}_1) & \cdots & \phi_A(\vec{r}_A) \end{vmatrix} \tag{21}$$

We will take the single particle neutron orbitals $\phi_{\nu_i}(n)$ to be the same as the corresponding proton orbitals $\phi_{\nu_i}(p)$. Now, note that the expectation value of isotensor operator M_2 vanishes in a T=0 state and in a T=½ state. From this observation we conclude that all the N=Z core orbitals do not contribute to (M_2), nor do the neutron-excess orbitals with the N=Z core. The M_2 has non-vanishing matrix elements only between two neutrons in neutron-excess orbitals. If we let ν_i and ν_i' be two such orbitals, then we have

$$(\Psi_A|M_2|\Psi_A) = \tfrac{1}{4}\left\{ \sum_{\nu_i} \left|(\nu_i|e^{iqz}|\nu_i)\right|^2 + \sum_{\nu_i \neq \nu_i'} \left|(\nu_i\, e^{iqz}|\nu_i')\right|^2 - (N - Z)\, \left|\, (\,||\theta_z||\,)\, \right|^2 \right\} \qquad (22)$$

Some further observations can also be made concerning (M_2). For one, odd multipoles vanish if all ν_i and ν_i' have the same parity. Secondly, $(\nu_i|e^{iqz}|\nu_i') = 0$ if ν_i and ν_i' belong to the same major shell since e^{iqz} excites oscillator quanta in the z-direction, leaving the x and y quanta unchanged. Thirdly, the monopole contribution vanishes through order q^6b^6 (b is the oscillator size parameter) since the average monopole interaction has been subtracted out. Thus, the lowest contribution in a q^2b^2 expansion is from the quadrupole term and it is of the order q^4b^4 and higher. The expectation value of $\theta^- \theta^+$ is also easily evaluated for a Slater determinant wave function and is

$$(\Psi_A|\theta^-\ \theta^+|\Psi_A) = Z - \sum_{\nu_i(p)\ \nu_j(n)} |(\nu_i(p)|e^{iqz}|\nu_j(n)\ )|^2 \qquad (23)$$

For ^{40}Ca

$$(\Psi_A|\ \theta^-\ \theta^+|\Psi_A) = 20 - 20e^{-(q^2/2\nu)} \left[1 + \frac{1}{2}\left(\frac{q^2}{2\nu}\right) - \frac{1}{10}\left(\frac{q^2}{2\nu}\right)^3 + \frac{1}{40}\left(\frac{q^2}{2\nu}\right)^4\right] \qquad (24)$$

Once $(\ \theta^-\ \theta^+)$, $(\ \theta_1)$ and $(\ \theta_2)$ are given, the $(\ \theta_0)$ follows from the isospin identity $\theta^-\ \theta^+ = 2[\ \theta_0\ -\ \theta_1\ +\ \theta_2/3]$. Using the above results, the unperturbed shell model results can be obtained and the results are given in Table 3. From the table we note that an order of magnitude discrepancy exists for the slope coefficient $\tilde{\beta}_2$, while a factor of two discrepancy exists in $(\ \theta^-\ \theta^+)/Z$. This latter disagreement between the evaluation using the shell model without correlations and the experimental result has previously been noted.[4,6]

Let us next turn to an improved wave function which includes ground state correlations and see what happens to the parameters β_0, $\tilde{\beta}_1$ and $\tilde{\beta}_2$. The effect of correlations on $(\ \theta^-\ \theta^+)$ using a linked cluster expension has already been discussed and a reduction of $\sim$ 30% was found for ^{16}O.[7] Here, we will consider the change in the ground state expectation value of a square operator θ^2 when the shell model wave function $|\Phi_0)$ is perturbed by a residual interaction H_1.

TABLE 3. Phenomenological and calculated values of the total capture rate $\langle \mathcal{O}^- \mathcal{O}^+ \rangle$ and the individual paramaters $\tilde{\beta}_0$, $\tilde{\beta}_1$, $\tilde{\beta}_2$.

	$\tilde{\beta}_0$	$\tilde{\beta}_1$	$\tilde{\beta}_2$	$\langle \mathcal{O}^+ \mathcal{O}^- \rangle_{N=Z/Z}$
Phenomenological values	0.97	0.25	3.24	0.16
Shell model without correlations	0.34	0.30	0.27	0.27
Shell model with correlations	0.57	0.30	1.6	0.17

The total Hamiltonian $H = H_0 + H_1$, where H_0 is the unperturbed part. For degenerate states we will take $H_0 = H$ and diagonalize H on these states. To first order in H_1, our new wave function is

$$|\Psi_0\rangle = |\Phi_0\rangle + \frac{Q}{E_0 - H_0} H_1 |\Phi_0\rangle \tag{25}$$

where $Q = 1 - |\Phi_0)(\Phi_0|$. Then, a first order evaluation in H_1 of the expectation value of θ^2 with the improved $|\Psi_0)$ gives

$$\langle\Psi_0| \theta^2 |\Psi_0\rangle = \langle\Phi_0| \theta^2 |\Phi_0\rangle + 2\left\langle \Phi_0 \right| \theta^2 \frac{Q}{E_0 - H_0} H_1 \left| \Phi_0 \right\rangle \tag{26}$$

To evaluate Eq. (26) we will use a multipole-multipole force to give an estimate of the effect of long-ranged correlations on θ^2. For example, a dipole-dipole interaction given by

$$H_1 = \tfrac{1}{2} \sum_{i,j} b_t \, \vec{r}_i \cdot \vec{r}_j \, \vec{t}_i \cdot \vec{t}_j \tag{27}$$

is used to evaluate the correlation corrections to the part of $e^{i\vec{q}\cdot\vec{r}_i}$ which excites the dipole state. The strength of the interaction is chosen to give the correct position of the dipole state. For the monopole and quadrupole parts of $e^{i\vec{q}\cdot\vec{r}_i}$, a monopole-monopole and quadrupole-quadrupole force will be used with the strength of each component, again adjusted to give the position of each isovector giant resonance, as given by the hydrodynamic model. These energies are $E_D = 2\hbar\omega$, $E_m = 4\hbar\omega$ and $E_Q = 3.2\hbar\omega$. It is straightforward to evaluate Eq. (26) for the isoscalar, isovector, and isotensor operators of Eqs. (14)-(16). Then, from these values, the coefficients $\tilde{\beta}_0$, $\tilde{\beta}_1$ and $\tilde{\beta}_2$ can easily be obtained. The results are summarized in Table 3 under the row "shell model with correlations."

Comparing these results with the phenomenological ones and the unperturbed shell model values, we see that the effect of long range correlations substantially improved on the unperturbed results. However, note that while the total transition matrix element agrees very well with the experimental one, the individual parameters still disagree by a factor of about 2 for $\tilde{\beta}_0$ and $\tilde{\beta}_2$. This discrepancy has of yet not been resolved.

Let us try to understand the large changes in the expectation value of an isotensor operator when wave functions are improved on. To do this, we will consider a problem closely related to muon capture, which is γ-absorption. Specifically, we will consider the excitation of the giant dipole resonance in γ-absorption. Recall that the integrated absorption cross section is related to the oscillator strength, f_{on}, through [8,9]

$$\int \sigma_{\gamma\text{-abs.}}(E)\,dE = \frac{2\pi^2 e^2\hbar^2}{mc} \sum_n f_{on} \tag{28}$$

with $f_{on} = |(n|D_z|0)|^2 (E_n - E_0)\, 2m/\hbar^2$ and where $D_z = \Sigma z_i\, t_{iz}$ is the dipole operator. Thus, the first moment of the dipole distribution is connected to the γ-absorption cross section. On the other hand, the zero'th moment of the dipole distribution is connected to the Bremsstrahlung cross section[8]

$$\int \sigma_{\gamma\text{-abs}}(E)\, dE/E = \frac{2\pi^2 e^2\hbar^2}{mc} \sum_n \frac{2m}{\hbar^2} |(n|D_z|0)|^2 \tag{29}$$

For $\gamma + {}^{208}Pb$ the integrated cross section is $\int\sigma_\gamma dE = 4.03$ b·MeV and $\int\sigma_\gamma\, dE/E = 260$ mb.[10] Using Eq. (29), and noting that

$$\sum_n |(n|D_z|0)|^2 = (0|D_z^2|0) \tag{30}$$

the expectation value $(0|D_z^2|0) = 93$ fm^2. The traditional shell model value is

$$(D_z^2) = \frac{Z}{8\nu_p} + \frac{N}{8\nu_n} \tag{31}$$

For ^{208}Pb, $(D_z^2) = 150$ fm^2 using the oscillator result of Eq. (31); Hartree-Fock theory with Skyrme III gives for (D_z^2) a value of about 130 fm^2.[12] Thus, the simplest approaches fail to account for the oscillator strength in ^{208}Pb. To see how dynamical correlations effect this evaluation, we introduce the center of mass coordinate in the z-direction, $Z_{ij} = \frac{1}{2}(z_i + z_j)$ and the corresponding relative coordinate $z_{ij} = z_i - z_j$. Then,

$$D_z^2 = \sum_{i=j} \tfrac{1}{4} z_i^2 + \sum_{i\neq j} Z_{ij}\, t_{iz} t_{jz} - \tfrac{1}{4}\sum z_{ij}^2\, t_{iz} t_{jz} \tag{32}$$

The last term in Eq. (32) is affected by correlations and we therefore consider the function

$$R = \frac{1}{16}\left\langle \sum_{2n} z_{ij}^2 \right\rangle + \frac{1}{16}\left\langle \sum_{2p} z_{ij}^2 \right\rangle - \frac{1}{16}\left\langle \sum_{np} z_{ij}^2 \right\rangle \qquad (33)$$

Now, for (D_z^2) to decrease, R must increase. Since the np force is more attractive than the nn or pp force, the $(\Sigma_{np} z_{ij}^2)$ is reduced with respect to the $(\Sigma_{nn} z_{ij}^2)$ and $(\Sigma_{pp} z_{ij}^2)$. Consequently, we can understand the qualitative effect of dynamical correlations on (D_z^2).

A more quantitative approach using the results of Eqs. (25) and (26) gives the following

$$(\Psi_o|D_z^2|\Psi_o) = \frac{(\Phi_o|\tfrac{1}{2}[D_z,[H,D_z]]|\Phi_o)}{E_{T.D.}} \qquad (34)$$

In obtaining Eq. (34), we have used the fact that $D_z|\Phi_o) = |\Phi_D)$, with $|\Phi_D)$ the dipole state, and that $H_oD_z|\Phi_o) = E_{T.D}|\Phi_o)$, where $E_{T.D}$ is the Tamm-Dancoff energy for this state. The numerator of Eq. (34) contains the exchange current enhancement κ to the integrated cross section. For ^{208}Pb the enhancement, $\kappa = ([D_z,[V,D_z]])/([D_z,[T,D_z]])$ is 40% of the model independent kinetic energy contribution in the vicinity of the dipole state; the $\tfrac{1}{2}[D_z,[T,D_z]] = (\hbar^2/2M)(NZ/A)$. The contribution of very distant states to the integrated γ-absorption cross section leads to an enhancement of 100%.[12] This is confirmed in the theoretical investigation of Ref. (13) and the effect is due to short-range tensor correlations. However, it should be realized that the distant state tensor correlation enhancement is essentially absent in the Bremsstrahlung weighted cross section because of the energy denominator of Eq. (29) and in Eq. (34) since we have used the dipole energy $E_{T.D.}$ Consequently we need to consider only the local enhancement. Using the Hamada-Johnston potential to evaluate $\tfrac{1}{2}[D_z,[F,D_z]]$ we find a κ of 0.4 with 0.2 of the enhancement coming from the one pion exchange part of this potential. Similarly, for the one boson exchange potential of Bryan and Scott,[14] the T=1 σ_1 and ρ mesons double the value of the enhancement from its π-meson value of 0.2. We also note that for a zero-range-velocity independent force, the double commutator $[D_z,[V,D_z]] = 0$, so that for such a force there is no exchange current enhancement. On the other hand, for a zero-range-velocity dependent force, as in Skyrme type interactions, this double commutator is not zero and an exchange current enhancement exists. For Skyrme III, κ = 0.4[11] for ^{208}Pb, while κ = 0.2 for Skyrme I[15] and κ = 0.8 for Skyrme II.[15] Here, we will use κ = 0.4.

The above results imply a 0.7 reduction in (D_z^2) from correlations which give a value of $\sim$100 fm^2, in much better agreement with the 93 fm^2 experimental result. This reduction is obtained by noting that the shell model value of Eq. (31) follows when the numerator is evaluated with only the kinetic contribution and the denominator is set equal to $\hbar\omega$. Since the dipole state is shifted to twice the energy $\hbar\omega$, we have $E_{T.D.} = 2\hbar\omega$ and for an enhancement of 0.4 we have $[D_z, [H,D_z]] = 1.4[D_z,[T,D_z]]$. These two results give the 0.7 reduction in (D_z^2) and we therefore conclude that the expectation value of the squared dipole operator has sizeable corrections from ground state correlations. Such effects can also be seen when the dipole-dipole force of Eq. (27) is used to evaluate Eq. (26).[16] When such a force is employed, the change in D_z^2 from its unperturbed value arises from 2p-2h states admixed into the ground state. Specifically, we obtain

$$(\Psi_o|D_z^2|\Psi_o) - (\Phi_o|D_z^2|\Phi_o) \sim - \frac{b_t}{4E_q} \left(\frac{1}{8\nu}\right)^2 [A^2] \tag{35}$$

where $E_q = 2(\hbar\omega + \Delta_{coll})$, with Δ_{coll} the collective shift of the dipole state. We observe that the change goes like A^2 since the square of a dipole excitation is linear in A from Eq. (31) when $\nu_p = \nu_n$ and we are considering two such independent excitations.

The above result can be simplified when use is made of the fact that in the schematic model[17] the collective shift is related to $(\Phi_o|D_z^2|\Phi_o)$:

$$\Delta_{coll} = b_t\ (\Phi_o|D_z^2|\Phi_o) \tag{36}$$

The result of this equation and that of Eq. (35) gives the change in the expectation value of (D_z^2) from 2p-2h excitations. For this change we have

$$\delta_{2p-2h}\ (D_z^2) = - \frac{\Delta_{coll}}{(\Delta_{coll} + \hbar\omega)} (\Phi_o|D_z^2|\Phi_o) \tag{37}$$

Next, let us look at the effect of 2p-2h excitations on the isotensor part of the squared isovector dipole operator, $D_2 = \frac{1}{2}(\Sigma\, z_i z_j\, \vec{t}_i \cdot \vec{t}_j - 3\Sigma\, z_i z_j\, t_{iz} t_{jz})$, and on $D_- D_+ = \Sigma\, z_i z_j\, t_{i-} t_{j+}$ for ^{208}Pb. The effect of D_+ on the ^{208}Pb ground state wave function is essentially zero since most of the transitions are dynamically blocked by the neutron-excess orbitals. In fact, only the $1h_{11/2}$ proton transition to a $2g_{9/2}$ or $1i_{11/2}$ neutron state contributes. We will therefore consider an idealized nucleus for which $D_+|\Phi_o) = 0$ or equivalently $(\Phi_o|D_- = 0$. Then to first order in H_1, the expectation value of $D_- D_+$ with $|\Psi_o)$ given by Eq. (25) is also zero since

$$(\Psi_o|D_-D_+|\Psi_o) = (\Phi_o|D_-D_+|\Phi_o) + (\Phi_o|H_1\ (E_o-H_o)^{-1} D_-D_+|\Phi_o)$$

$$+ (\Phi_o|D_-D_+Q(E_o-H_o)^{-1}\ H_1|\Phi_o) = 0$$

Thus $\delta_{2p-2h}(D_-D_+) = 0$, where δ is the change in (D) when $\Phi_0 \rightarrow \Psi_0$. Moreover, $\delta_{2p-2h}(D_1) = 0$ since the isovector operator $D_1 = \Sigma\frac{1}{2}z_i^2 t_{iz}$ is a one-body operator and is therefore not affected by 2p-2h states to first order in H_1. Having established the effect of 2p-2h correlations on D_-D_+, D_1 and D_z^2, we can use the isospin identity $D_-D_+ = 2[D_z^2 - D_1 + D_2]$ to evaluate its effect on the isotensor part D_2. We obtain

$$\delta_{2p-2h}(D_2) = -\delta_{2p-2h}(D_z^2) \tag{38}$$

Thus from this equation we conclude that the effects of correlations on (D_2) are large and make (D_2) positive since $\delta(D_z^2)$ is negative.

Large values of (D_2) have implications concerning the neutron radius as shall now be discussed. The total strength to T+1 is obtained from the expectation value of D_-D_+ since D_+ changes T_z by one unit and must therefore excite T+1 states when D_+ acts on a $T=T_z$ ground state. The cross section for γ-absorption to T+1 states is therefore proportional to (D_-D_+). But $D_-D_+ = 2[D_z^2 - D_1 + D_2]$. Since $(D_-D_+) \geq 0$, we have $(D_z^2) - (D_1) + (D_2) \geq 0$. The uncorrelated value of $(D_2) = 0$ for ^{208}Pb, since there are no E1 matrix elements between neutron-excess orbitals for ^{208}Pb. Since $(D_z^2) = 93$ fm^2, then it follows that 93 fm$^2 \geq (D_1)$. But $(D_1) = \frac{1}{12}(N(r^2)_n - Z(r_p^2))$. The precise value of $(r^2)_p^{\frac{1}{2}}$ for ^{208}Pb is determined, with some uncertainty, to be between 5.42 - 5.51 fm. If we use the value $(r_p^2)^{\frac{1}{2}} = 5.51$ fm, the $(r_n^2)^{\frac{1}{2}} \leq 5.31$ fm, and we conclude that neutrons are inside protons, contrary to all Hartree-Fock results. We can turn the problem around and ask what value of (D_2) is implied by taking the Hartree-Fock value of $(r_n^2)^{\frac{1}{2}}$. For this purpose we use the values $(r_p^2)^{\frac{1}{2}} = 5.45$ fm and $(r_n^2)^{\frac{1}{2}} = 5.68$ fm obtained by Negele and Vautherin[18] using a density dependent Skyrme interaction. Then $(D_1) = 135.788$ fm^2 which implies $(D_2) \geq 43$ fm^2, a large positive result; this result is also consistent with that obtained from Eq. (38).

Having established the effect of correlations on D_z, D_1, D_2, we can use these results to investigate the influence of correlations on the strength to specific isospin states. The excitation strength to a specific isospin state can be obtained from

$$M(T') = (\Phi_0(T,T_z)|D_z P_{T'} D_z|\Phi_0(T,T_z)) = C^2(T,T_z;10|T',T_z)M(||T'||)$$

The $P_{T'}$ is a proportion operator for T=T' states. The reduced matrix elements can be shown to obey

$$M(||T'||) = (D_z^2) - \frac{1}{2T}[T'(T'+1) - T(T+1) - 2](D_1 - D_2) + \delta_{T',T-1}\, 2\, \frac{2T+1}{2T-1} M_2 \tag{39}$$

From this result we see that 2p-2h correlations effect $M(||T||)$, reducing it compared to the $M(||T+1||)$ and $M(||T-1||)$ components.

Recently improvements in the sum rule approach to muon capture have been developed by Do Dang[19] and extended by Bernabeu[20] and Rosenfelder.[21] These improvements center around removing the dependence of the capture rate on the mean neutrino energy used in Eq.(11). This removal is accomplished by introducing a combined non-energy-weighted and energy-weighted expression for the capture rate. Specifically, the neutrino energy ν_{ab} which appears in Eq. (4) - Eq.(16) is written as

$$\frac{\nu_{ab}}{m_\mu} = 1 - \frac{E_b - E_a}{m_\mu} \tag{40}$$

The second term of this equation, which involves $(E_b - E_a)/m_\mu$, then introduces energy moments of the distribution of matrix elements into the expression for the capture rate. An evaluation of the total transition rate without including ground state correlations but using the moment expension and a Skyrme interaction gives results which are still in disagreement with the experimental data,[21] but less so than the simple shell model. Also, the combined non-energy-weighted and energy-weighted description for the parametric form of Eq. (2) as given in Ref. (2) has not be evaluated as of yet. It would be interesting to have such results.

REFERENCES

1. H. Primakoff, Rev. of Mod. Phys. 31, 802 (1959).

2. B. Goulard and H. Primakoff, Phys. Rev. C10, 2034 (1974).

3. A. Mekjian, Phys. Rev. Lett. 36, 1242 (1976).

4. J.R. Luyten, H.C. Rood and H.A. Tolhoek, Nucl. Phys. 41, 236 (1963).

5. J. Joseph, F. Ledoyen and B. Goulard, Phys. Rev. C16, 1742 (1972).

6. L.L. Foldy and J.D. Walecka, Nuovo Cimen. XXXIV, No. 4, 1026 (1964).

7. R.J. McCarthy and G.E. Walker, Phys. Rev. C11, 383 (1975).

8. J.S. Levinger, Nuclear Photodisintegration (Oxford University Press, Oxford, 1960).

9. A.M. Lane and A.Z. Mekjian, Phys. Rev. C8, 1981 (1973).

10. A. Veipsiere et al., Nucl. Phys. A159, 561 (1970).

11. D. Vautherin, private communication.

12. Ahrens et al., Proceedings Intern. Conf. on Nuclear Structure Studies, Sendoi, Japan, 1972.

13. W.T. Weng, T.T.S. Kuo and G.E. Brown, Phys. Lett. 46B, 329 (1973).

14. R.A. Bryan and B.L. Scott, Phys. Rev. 177, 1435 (1969).

15. D.M. Brink and R. Leonardi, Nucl. Phys. A258, 285 (1976).

16. A.Z. Mekjian and W.M. MacDonald, Phys. Rev. C15, 531 (1977).

17. G.E. Brown and M. Bolsterli, Phys. Rev. Lett. 3, 472 (1959).

18. J. Negele and D. Vautherin, Phys. Rev. C5, 1472 (1972).

19. G. Do Dang, Phys. Lett. 38B, 397 (1972).

20. J. Bernabeu , Nucl. Phys. A201, 41 (1973); A215, 411 (1973); J. Bernabeu and F. Cannata, Phys. Lett. 45B, 445 (1973); Nucl. Phys. A215, 424 (1973).

21. R. Rosenfelder, Nucl. Phys. A298, 397 (1978).

OBSERVATION OF TWO-PHOTON EMISSION IN NUCLEAR PION-CAPTURE:

PION-PION ANNIHILATION IN NUCLEI

J.P. Deutsch

Université de Louvain

B-1348 Louvain-la-Neuve, Belgium

This talk is based on an experiment performed by Dr. Favart, M. Lebrun, P. Lipnik, P. Macq, R. Prieels and myself at the CERN Synchrocyclotron. A description of the experiment and its result will be submitted shortly for publication to Physics Letters B.

The experiment consisted in stopping negative pions in a beryllium respectively carbon target and observing the emission of two energetic photons emerging from this reaction. The angle between the two photons and their energy was measured but the residual hadronic system remained unobserved. The aim of the experiment was a selective observation of the pion-content in nuclei.

The (hopefully) dominant graph (fig. 1.a) was estimated by Ericson and Wilkin[1] and independently by Leroy[2] to lead to a branching ratio of about 3×10^{-6} per stopped pion. The spurious graphs considered were: virtual charge-exchange (fig. 1.b) and "two-step" electromagnetic processes in which one of the gamma is emitted in normal radiative capture and the other gamma is emitted from any of the charged particles entering the reaction. The first graph was estimated in ref. 1 to be only about 3×10^{-7} per pion capture; for the "two-step" electromagnetic processes only crude estimates exist up til now by Beder (in hydrogen)[3] and Bernabeau (who considers mainly the gamma-desexcitation of virtual nuclear states formed in radiative pion capture). Both estimates range about one order magnitude lower than the estimates of the dominant graph.

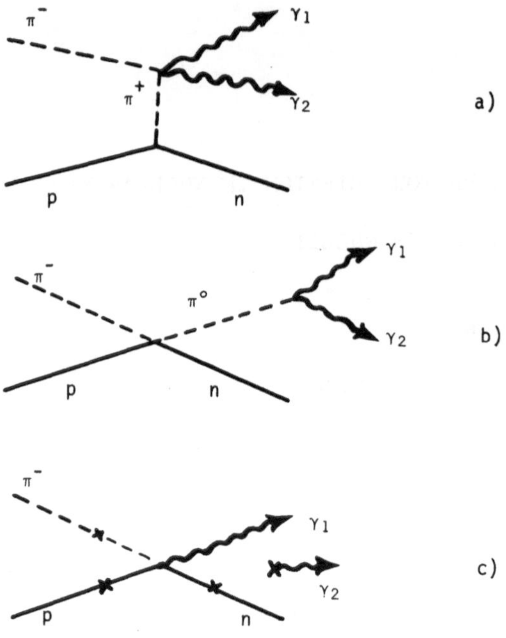

Fig. 1

As said above, the main <u>interest</u> of the experiment is to probe (in a hopefully selective way) the pion-content in nuclei. More specifically, it would indicate any strong deviation of the pion-nucleon coupling constant in nuclei from the value it has in free nucleons. As the momentum-transfer explored is rather low, the experiment is not expected to show up directly possible pion-condensates but may already indicate enhanced local oscillations of the pion field, i.e. a pre-condensate situation considered recently in ref. 5. This combined measurement of the pion-nucleon coupling in nuclei with that of the propagation of the virtual pion in nuclear matter, would be very useful also to understand the eventual renormalization of the induced pseudo-scalar weak coupling in nuclei[6].

The basic problem of the experiment is the observation of a very low decay branch (implying also an energy- and angular-measurement of photons) in the presence of an overwhelming hadronic ($\sim 10^6$) and single-photon ($\sim 10^4$) background. Consequently, the instrument has to feature not only a great solid angle but also a high selectivity against spurious "photon-photon" signals from accidental coincidences, inter-detector cross-talk and the decay of real π^0 formed in charge-exchange. For stopped π^-, this last process is energetically forbidden by about 9 MeV in the targets we choose, but remains allowed from high-energy pion-contaminants yielding charge-exchange either in the degrader or in the target itself.

The instrument constructed to meet these requirements is sketched in fig. 2. Its main element is a lead-glass hodoscope consisting in 20 elements placed around the target, perpendicularly to the beam axis. The total acceptance of this detector is about 20% of 4π solid angle and the energy resolution is typically 40 to 50%.

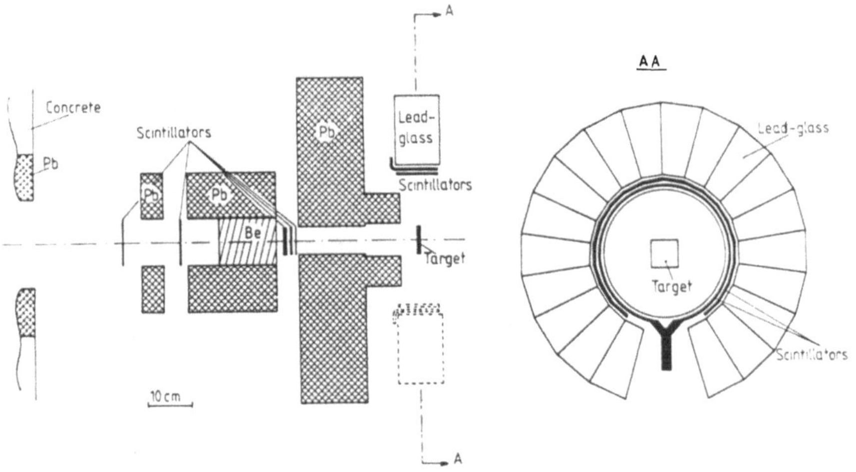

Fig. 2

The pions derived from an internal target of the CERN Synchro-cyclotron were degraded and brought to rest in the target as schematized in fig. 2. For an internal proton current of 4 μA and the beam transport set to 146 MeV/c, the incoming rate was about 1.6×10^6 s^{-1}, the pion muon electron ratio being .40 - .25 - .35. The total duty cycle was typically 60%.

In a typical 10 x 10 x 3 cm^3 graphite target we stopped about

20% of the incoming pions: so the stopped rate was typically $10^5\ s^{-1}$ leading to an expected count rate per angular bin of ~ 6/ (8 hour shift). In the configuration of fig. 2 the background (originating mainly from pions stopped in the absorber and the lead chimney) was about 30% of the total count-rate and was carefully measured.

A trigger is provided by a coincidence between the telescope and at least two photon detectors. For each trigger we registered on magnetic tape the pulse-height of each lead-glass detector and that of each scintillator placed in the beam line (fig. 2). The pulse height in the lead glass cells allows us to determine the energy of the two photons and their angle. It allowed us moreover to reject cross-talk, requiring that no lead glass detector but two should show signal. (Tests performed with π^0-decay photons and electrons of different energies showed us that this technic of cross-talk-rejection was sufficiently effective for inter-detector angles higher than 50°.) The pulse-height monitoring of the scintillators allowed us to reject accidentals due to multiple pion traversal followed by single radiative capture. The scintillators allowed us to reject also (by ionization-loss selection) incoming pions of energy high enough to yield π^0, which may have contributed to two-photon events especially at great inter-photon angles. The efficiency of this rejection criteria was tested by the stability of our results against modification of the incoming pion momentum and the cuts performed on the telescope pulse height.

Global tests performed changing the beam intensity or duty cycle showed that the accidental coincidences were negligible. Changing the target size and configurations we showed also that "photon to two photon" conversions in target were also negligible.

The results obtained with an energy threshold of 25 MeV for both photons is represented both for beryllium and carbon in fig. 3. It may be noted that gases of low charge-exchange threshold, possibly absorbed in graphite, may contribute to the corresponding data points for $\theta_{12} > 140^o$. As we shall explain later, this contribution is however of minor significance for our conclusions. We show also a theoretical expectation[4] computed assuming that 85% of the pions is captured from p-orbit. (In this computation no account was taken of the binding of the nucleons.) It may be noted that there is a rough qualitative agreement between the predicted angular distribution and the measured one.

Having calibrated our detector-efficiency with the single radiative capture in carbon[7] the weighted mean-value of our results yields the branching ratio $(\pi^- \to \gamma\ \gamma)/(\pi^- \to \text{all})$: $(0.97 \pm 0.13)\ 10^{-5}$ in beryllium and $(1.40 \pm 0.19)\ 10^{-5}$ in carbon. The branching ratio in graphite becomes $(1.23 \pm 0.17)\ 10^{-5}$ if we neglect the

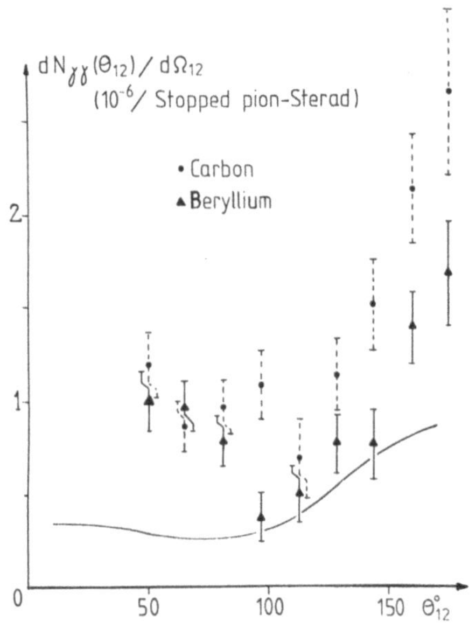

Fig. 3

data points above 140° in order to avoid possible contribution from the absorbed gases.

These results are about eight times greater than the first estimates of ref. 1. They are only twice the predictions of ref. 4. It should be noted, however, that the prediction of ref. 1 is obtained as the product of $[(\pi^- \rightarrow \gamma\gamma)/(\pi^- \rightarrow \gamma)] \times [(\pi^- \rightarrow \gamma\gamma)/(\pi^- \rightarrow \text{all})]$. The first factor is computed theoretically and the second taken from experiment. The higher prediction of ref. 4 comes from a direct computation of $(\pi^- \rightarrow \gamma\gamma)$ and an estimate of $(\pi^- \rightarrow \text{all})$ from phenomonological pion-nucleus optical potential parameters. It would seem that the first procedure takes into account more consistently some of the deviations from a free-nucleon approach. So there seems to be a serious disagreement between our results and the expectations obtained assuming that the two-photon branch we observed arises solely from the annihilation of the physical pions on virtual pions coupled to free nucleons. It is clear that before using this discrepancy to ascertain an anomalous pion-nucleon coupling in nuclei, a more realistic theoretical investigation of two-photon emission after pion capture in specific nuclei should be performed.

REFERENCES

1. T.E.O. Ericson and C. Wilkin, Phys. Lett. 57B (1975), 345.

2. Cl. Leroy, unpublished.

3. D. Beder, U.B.C. Report, January 1978, unpublished.

4. S. Barshay, Phys. Letters, 78B (1978), 384; according to an Erratum these predictions should be divided by two!

5. M. Ericson and J. Delorme, Phys. Letters 76B (1978), 132.

6. E.M. Nyman and M. Rho, Nucl. Phys. A287 (1977), 390.

7. J.A. Bistirlich et al., Phys. Rev. C6 (1972), 1867.

Part II

Quarks and Nuclear Structure

A QUASI-NUCLEAR COLOURED QUARK MODEL FOR HADRONS

H.J. Lipkin

The Weizmann Institute of Science

Rehovot, Israel

1. INTRODUCTION

The Implications of Two New Fermilab Experiments

Once upon a time physicists believed that nucleons and pions were elementary like electrons and photons, and that Yukawa's theory of nuclear forces was the analog of QED for strong interactions. Then the Δ (3-3 resonance) was discovered, and then the ρ and other pion resonances, and it became apparent that neither the pion nor the nucleon was elementary and that both had a composite structure. Today pions and nucleons seem to be very similar objects, instead of being very different like the electron and photon, and made of the same basic building blocks: spin 1/2 quarks bound by coloured gluons. But perhaps history will repeat itself. Maybe 25 years from now a lecture at the Banff Summer School will begin with the statement "Once upon a time physicists believed that quarks and gluons were elementary, and that Quantum Chromodynamics (QCD) was the analog of QED for strong interactions. Then ... ??????"

But we shall not enter into such speculations, and examine the situation as it appears today. We have the new QXD model for everything, where X = A, B, C, D, E, F, G, etc. So far there are only models for X = C, E, F and G, but no doubt the others will eventually be discovered as well. However, it is amusing that in the great excitement about non-Abelian gauge theory, the original non-Abelian gauge model for hadron dynamics has faded away. This was the gauge theory of strong interactions mediated by the octet of vector mesons ρ, ω, and K* coupled to conserved vector currents. The SU(3) group originally introduced by Gell-Mann and Ne'eman is now called flavour and dismissed as an irrelevant complication in the QCD description of strong interactions. Flavour is now discussed in QFD (Quantum Flavordynamics), the new fancy name for the kind of unified theory of weak and electromagnetic interactions discussed by Henry Primakoff in his lectures here. In this approach the mass differences between quarks of different flavours are assumed to come somehow from the weak and electromagnetic interactions and not from the strong interactions. The set of quarks with different flavours and masses is then given as input for the QCD description of strong interactions, and QCD does not attempt to explain flavour or mass differences. But the question of how many flavours there really are, and when and whether experimentalists will stop finding new and heavier bound states of new heavy quarks is still open. There is no theoretical clue to the answer yet.

I begin a review of the present status of the quark model by quoting some of the most recent results. These are motivated by two recent experiments at Fermilab. I take a "Galilean approach" which assumes that we learn about nature by making experimental observations like Galileo and trying to understand them, rather than by reading the words that great theorists like Aristotle have written. The quark model has grown out of such experimental observations, against the opposition of the Aristotelian establishment who have always found weighty theoretical reasons why quarks could not exist and why the regularities predicted from the quark model and found in experiment could not be significant.

The quark model is very ad hoc. It lacks a fundamental theoretical basis, yet it provides a very good description of much experimental data. Today theorists believe that the fundamental basis will eventually come from QCD, and there are many indications that this is indeed true. But there are many slips between cup and lip, and a measure of scepticism is always in place. We shall make free use of the approach and methods of QCD to guide our intuition in discussing the quark model, but we do not attempt any rigorous derivation, and are always looking at experimental data to see what nature is trying to tell us.

This discussion of quarks uses what might be called a "Newtonian approach". Newton was able to describe the motion of the earth around the sun with great precision without ever having heard of asymptotic freedom, while in fact, the gravitational field of the earth is indeed asymptotically free. At large distances the earth has a field given by Coulomb's law with a coupling constant proportional to the mass of the earth. But at short distances, less than the radius of the earth, the effective or "running" coupling constant decreases and goes to zero at the center of the earth. But this asymptotic freedom was irrelevant to Newton, who described the earth as a point mass with a pure Coulomb-like gravitational field. This is a drastic and unwarranted assumption for short distance phenomena. It ignores for example the fact that we exist on the earth. But it is certainly adequate for precise calculations of the earth's orbit.

In the same spirit we assume that the nucleon is made of three constituent quarks, which are treated as very simple objects. We know that they must in reality be much more complicated things including virtual gluons, quark-antiquark pairs, etc. But there seems to be a wide variety of phenomena successfully described by these simple constituent quarks. Somewhere in the fundamental theory there must be an explanation, just as there is an explanation justifying Newton's treatment of the earth as a simple point mass. But so far we do not have a good fundamental theory and do not have a satisfactory explanation. All we know is that the model works.

One interesting experimental result from Fermilab is the discovery[1] of the new fifth heavy quark in the bound states now called the Υ and Υ', with the surprising equality of the Υ-Υ' mass splitting to the mass splitting in the charmonium system, namely the ψ-ψ' splitting. Nature seems to be telling us that hadron mass splittings are simpler than expected: they do not depend very strongly on the flavour or mass of the quarks of which they are composed. This principle was incorporated formally in the logarithmic potential model of Quigg and Rosner[2] which has this mass scaling property. We can use the general philosophy of this model, without taking it too seriously in detail, and assume that flavour dependence of many mass splitting effects can be neglected. This might be called "Rolling off the Log".

A second interesting result from Fermilab is a new measurement[3] of the magnetic moment of the Λ to a precision of 1% by a Rutgers-Michigan-Wisconsin group. The number agrees with quark model predictions[4,5] of this moment to 1%. This agreement is completely unexpected, since the quark model is not expected to be that good. Let us review the simple-minded calculations of baryon magnetic moments in the quark model[6] to show what physics lies behind this surprising agreement.

The baryon octet is assumed to consist of three-quark states in a relative s-wave, with the total spin and magnetic moment given by simple vector addition of the quark spins and magnetic moments. The magnetic moment of a quark of flavour f is assumed to be the Dirac moment[7] which is

$$\mu_f = q_f(M_p/m_f) \quad \text{nuclear magnetons} \quad , \tag{1.1}$$

where q_f and m_f are the charge and mass of the quark of flavour f and M_p is the proton mass.

The magnetic moment of the baryon is then obtained by summing these quark moments in the SU(6) wave function which is totally symmetric in spin and flavour. For the Λ, the symmetry requirement means that the two nonstrange quarks which are coupled to isospin zero also have spin zero and do not contribute to the magnetic moment. The Λ magnetic moment is therefore given by the moment of the strange quark,

$$\mu_\Lambda = \mu_s = -(1/3)\ (M_p/m_s) \quad . \tag{1.2}$$

The remaining baryons can be described as consisting of two quarks of flavour a and one of flavour b, where a and b can be u, d or s. The two quarks of flavour a are symmetric in flavour and coupled to spin 1, and these are then coupled with the spin of the third quark to give total spin 1/2. The general result for the magnetic moment of such a baryon is

$$\mu(aab) = (4/3)\mu_a - (1/3)\mu_b \quad , \tag{1.3a}$$

where the coefficients come from the expectation values of S_z for the a and b quarks in this wave function. For the proton, where a = u, b = d, $q_a = +2/3$ and $q_b = -1/3$,

$$\mu_p = (8/9)(M_p/m_u) + (1/9)\ (M_p/m_d) = (M_p/m_u) \quad , \tag{1.3b}$$

where we neglect the difference between m_u and m_d. Similarly for the neutron, where a = d and b = u,

$$\mu_n = -(4/9)(M_p/m_d) - (2/9)\ (M_p/m_u) = -(2/3)(M_p/m_u) \quad . \tag{1.3c}$$

Combining Eqs. (3b) and (3c) gives the well known successful prediction for the neutron magnetic moment,

$$\mu_n = (-2/3)\mu_p = -1.86 \text{ nuclear magnetons} \quad , \tag{1.4}$$

in excellent agreement with the experimental value of -1.91.

The Λ magnetic moment cannot be related directly to the nucleon magnetic moments without some assumption regarding the difference between m_s and m_u. Let us make the very drastic assumption that this quark mass difference is exactly equal to the hadron mass difference,

$$m_s - m_u = M_\Lambda - M_p \quad . \tag{1.5}$$

This is a new ingredient leading to a very interesting prediction. A priori there is no reason to choose the Λ-N mass difference for the right hand side of (1.5) rather than Σ - N or Σ^* - Δ. The decuplet mass splitting has commonly been used because the equal mass spacing has been interpreted as indicating that decuplet mass splittings are simpler than octet splittings. However, arguments based on QCD show that the decuplet splitting involves a complicated interplay of both the quark mass differences (1.5) and the spin splittings while use of the Λ-N mass difference eliminates effects of spin splittings[8].

Combining Eqs. (2), (3b) and (5) gives the prediction

$$\mu_\Lambda = (-1/3)\ [(1/\mu_p) + (M_\Lambda - M_p)/M_p]^{-1} = -0.61 \quad \text{n.m.} \tag{1.6}$$

Another prediction is obtainable by assuming that the ratio of quark magnetic moments μ_u/μ_s is obtainable from hadron spin splittings like the ratio $(M_\Delta - M_N)/M_{\Sigma^*} - M_\Sigma)$. This ratio which is unity in the SU(3) limit is directly related to the ratio of quark magnetic moments under the assumption that the spin splittings come from a "colour magnetic" interaction[4] proportional to the colour magnetic moments of the quarks which are in turn proportional to electromagnetic moments. The result obtained is

$$\mu_\Lambda = -(\mu_p/3)\ (M_{\Sigma^{*+}} - M_{\Sigma^+})/(M_{\Delta^+} - M) = -0.61 \text{ n.m.} \tag{1.7}$$

Both predictions (1.6) and (1.7) are in remarkable agreement with the new experimental value $\mu_\Lambda = -0.6138 \pm 0.0047$ n.m. That they are also in remarkable agreement with one another suggests a new relation between hadron masses and the proton magnetic moment. Eliminating μ_Λ between (1.6) and (1.7) gives

$$[(M_{\Delta+} - M_p)/(M_{\Sigma*+} - M_{\Sigma+})] - 1 = \mu_p(M_\Lambda - M_N)/M_N . \qquad (1.8)$$

This peculiar relation is in excellent agreement with experiment. The left hand side is 0.523, the right hand side is 0.528. This unorthodox combination of hadron mass differences and the proton moment has a simple physical interpretation. The SU(3)-breaking quark mass parameter $(m_s - m_u)/m_u$ is computed in two ways. The LHS uses the quark mass ratio (m_s / m_u) obtained from hadron spin splittings. The RHS uses the quark mass difference $(m_s - m_u)$ obtained from hadron strangeness splittings, but needs the proton moment to provide a quark mass scale relating the mass difference to a mass ratio. Thus Eq. (1.8) says that the quark mass ratio and the quark mass difference determined in two different ways from hadron masses are consistent at the 1% level with the quark mass m_u determined from the proton mass and magnetic moment.

The success of these relations suggests a review of the underlying physics and its implications for hadron models. The prediction (1.7) is equivalent to a similar prediction obtained by DGG using explicit expressions involving quark mass ratios. Our derivation shows that explicit reference to quark masses is unnecessary and that all that is needed is proportionality between electromagnetic and colour magnetic moments. The prediction (1.6) and the relation (1.8) require the explicit assumption that quark magnetic moments depend upon masses like Dirac moments, and that the relevant quark mass difference is given by Eq. (1.5). This is a much more serious assumption which is generally not valid in conventional models. In the DGG model[4] Eq. (1.5) does not hold because the hadron masses include additional terms like kinetic energies which are inversely proportional to quark masses and do not cancel in the difference (1.5). The model of Ref. [8] avoids these terms by the use of scaling properties of the Quigg-Rosner[2] logarithmic potential model. In this model kinetic energies and mass splittings in the hadron specimen are independent of the quark mass and cancel out of mass differences like (1.5). This can be seen explicitly by the use of the virial theorem.

Consider a three-body system with non-relativistic Hamiltonian

$$H = \sum_{i=1} t_i + \sum_{i>j} v(r_{ij}) \quad , \qquad (1.9a)$$

where

$$t_i = p_i^2/2m_i \quad , \qquad (1.9b)$$

and

$$v(r_{ij}) = V \log(r_{ij}/r_o) \quad . \tag{1.9c}$$

The virial theorem then states that for any eigenfunction of H,

$$\langle \sum_i t_i \rangle = (1/2) \sum_{i>j} \langle r_{ij} dv(r_{ij})/dr_{ij} \rangle = (3/2)\, V \quad . \tag{1.10}$$

For the particular case of the log potential (1.9c) the right hand side of the virial theorem is a c-number, rather than the expectation value of an operator in the specific wave function, and is independent of both the wave function and the masses of the particles.

We now consider the flavour dependence of the spin splittings in more detail. These were first considered by Federman, Rubinstein and Talmi[9] in a nuclear shell model approach to baryon masses in 1966. In this model the low-lying baryon octet and decuplet were considered to all have the same "shell-model" wave functions and have their mass degeneracy split only by the strange quark mass difference $m_s - m_u$ and by a residual two-body interaction. If we denote the effective matrix elements for this interaction between two quarks of flavours a and b by V_0^{ab} and V_1^{ab} for the spin singlet and triplet states respectively, the following baryon mass differences are easily expressed in terms of these effective matrix elements.

$$M_\Delta - M_N = (3/2)(V_1^{ud} - V_0^{ud}) \quad , \tag{1.11a}$$

$$M_{\Sigma*} - M_\Lambda = (V_1^{ud} - V_0^{ud}) + (1/2)\,(V_1^{us} - V_0^{us}) \quad , \tag{1.11b}$$

$$M_{\Sigma*} - M_\Sigma = (3/2)\,(V_1^{us} - V_0^{us}) \quad . \tag{1.11c}$$

Combining these relations gives a relation between hadron masses,

$$M_\Delta - M_N = (1/2)\,(2M_{\Sigma*} + M_\Sigma - 3M_\Lambda) \quad . \tag{1.12}$$

This relation was found to be in good agreement with experiment, the LHS is 307 MeV; the RHS is 294 MeV, thus providing support for the assumption that baryon mass splittings are described by two-body forces.

The next development in the description of spin splittings was the assumption by DeRujula, Georgi and Glashow (DGG)[4] that these are due to a colour-magnetic hyperfine interaction having the form

$$V_{hf} = \mu_i^c \mu_j^c \vec{\sigma}_i \cdot \vec{\sigma}_j v(r_{ij}) \quad , \tag{1.13}$$

where μ_i^c denotes the colour magnetic moment of quark i. The particular form (1.13) immediately relates the triplet and singlet effective matrix elements by the eigenvalues of the operator $\vec{\sigma}_i \cdot \vec{\sigma}_j$.

$$V_0^{ab} = -3\, V_1^{ab} \quad . \tag{1.14}$$

This interaction (1.13) does not contribute to the difference $M_\Lambda - M_N$. If we assume that meson spin splittings are due to a similar interaction but with a different strength, we can construct linear combinations of meson masses to which an interaction satisfying the relation (1.14) does not contribute. The following relation[7] is obtained and is discussed in more detail in section 4.

$$M_\Lambda - M_N = m_s - m_u = (3/4)\,(M_{K*} - M_\rho) + (1/4)\,(M_K - M_\pi) \tag{1.15}$$

This relation is also in surprising agreement with experiment, the LHS is 177 MeV and the RHS is 180 MeV. Thus there seems to be experimental justification of the use of the hadron mass difference (1.15) as a quark mass difference in deriving the prediction (1.16).

We have now obtained three independent relations between hadron masses and moments which are experimentally confirmed at the 1% level, namely relations (1.6), (1.7) and (1.15). We can ask what we have put in to get these results.

1. We have identified hadron mass difference with quark mass differences in a very simple way, neglecting flavour dependence of binding effects and kinetic energies.

2. We have identified the quark masses in (1.15) with the masses appearing in the electromagnetic and colour magnetic moments of the quarks.

3. We have assumed a hyperfine interaction having the form (1.13).

However, it is still a big step further to use the quark mass difference of Eq. (1.5) as the mass parameter in the magnetic moments and to obtain results valid to a few per cent. The success of the relations (1.6) - (1.8) at this level indicates that the "quasinuclear coloured quark model" of Ref. [3] and the three basic assumptions above should be taken more seriously than indicated by their crude derivations. The underlying physics is that the <u>same quark mass parameter</u> appears in the simplest possible way in the electromagnetic moments, the colour magnetic moments and the hadron mass splittings. That electromagnetic and colour moments should depend upon the same mass parameter is not surprising. But the value of the magnetic moment is not expected to be determined to 1% by the mass parameter which enters hadron mass splittings and includes binding energies as well as quark masses.

The magnetic moment of a Dirac particle bound in an external potential depends upon a mass parameter which is a function of the Lorentz character of the potential [6]. For a Lorentz scalar potential this mass parameter is indeed the total energy of the bound state, including the binding energy. But for a Lorentz vector potential the magnetic moment is not affected by the binding (the magnetic moment of an electron strongly bound in the electrostatic field of a Van-de-Graaf accelerator is the same as that of a free electron). The results (1.6) and (1.7) suggest that the dominant binding potential for quarks in hadrons is Lorentz scalar rather than the Lorentz four-vector of a Coloumb or a one-gluon-exchange potential. But such an argument is not expected to hold to 1%. Note that Lorentz scalar confinement is implicit in bag models [11] which use Lorentz scalar bags as the principal confining mechanism and have only weak effects due to gluon exchange.

All the above leads to a deeper questioning of what indeed is the meaning of the quark mass. This mass appears as a parameter in many quark model calculations of observable hadron properties, but very different values are used in different calculations, varying from zero to infinity. There are "current quarks" which have nearly zero mass, bound "constitutent quarks" whose mass is of the order of hadron masses and free quarks, which have a very heavy mass or an infinite mass if quarks are permanently confined.

An intuitive picture of quark masses motivated by QCD shows that an isolated quark has a strong colour field at large distances and strong long range forces if there are no other quarks nearby to cut off the colour lines of force and confine colour. The mass of an isolated quark must include all the energy in the associated colour field at large distances, since this field must move with the quark and contribute to its inertial mass. In models with quark confinement, the energy in the field of an isolated quark is infinite and quarks have infinite mass and are unobservable.

Quarks bound in colour singlet hadrons do not have the large colour field at large distances and therefore do not have a large inertial mass. The mass parameter associated with the motion of these bound quarks inside hadrons and with their magnetic moments must be simply related to the energy in the colour field which moves with each quark. This may determine the value of the quark mass successfully used in constituent quark models and in the relations (1.1) - (1.5) of this paper. In scattering processes the mass to be used for the quark should depend upon how much of the associated colour field recoils with the quark. At very high momentum transfers the quark may have received a kick which moves it so fast that its colour field does not move with it. This would account for the small quark masses used for current quarks or quark partons, and the necessity to treat the colour field separately as a "gluon component" in the hadron wave function for deep inelastic processes.

Within this continuum of quark mass values from zero to infinity used for different processes there seems to be an intermediate region relevant to hadron spectroscopy where each valence quark has an inertia roughly given by its share of the hadron mass and only valence quarks need be considered [12]. These "constituent quark masses" determine the scales of mass splittings in the hadron spectrum and of hadron magnetic moments. There is no rigorous derivation as yet of these properties of constituent quarks from QCD, but the remarkable success and precision of nonrelativistic quark model predictions in describing the experimental spectrum suggest that a more fundamental derivation must exist.

In the remainder of these lectures we consider a quasinuclear constituent quark model in which constituent quarks are assumed to be made of constituent quarks interacting with a two-body colour-exchange logarithmic potential. In Section II we discuss the colour degree of freedom in detail. In Section III we consider some properties of the logarithmic potential. In Section IV we define the quasinuclear model and discuss its validity and compare some of its predictions with experiment.

To conclude this introduction we consider the validity of the nonrelativistic approximation used in constituent quark models and show that it cannot be valid for the nucleon. We consider this approximation in the spirit of similar approximations used elsewhere in physics. It is an expansion in powers of a "small" parameter, v/c, which however is manifestly not small. We cannot justify this expansion, but since it gives results that agree remarkably well with experiment, we continue to use it with the hope that some new idea like "relativistic freedom" or "asymptotic nonrelativity" will eventually come along to explain it.

To see that v/c is not small, we note that the velocity of a particle in an orbit is the product of the radius r and the angular frequency ω. Thus

$$v/c = \omega r/c = r/\lambda \quad . \tag{1.16}$$

where λ is the wave length of a wave with velocity c and angular frequency ω; i.e., the wave length of a photon with energy $\hbar\omega$.

The same result can be obtained more rigorously by use of the Heisenberg equation of motion for the quark co-ordinate x whose time derivative is the velocity v,

$$\frac{v^2}{c^2} = \langle \dot{x}^2 \rangle / c^2 = -\langle [H,x]^2 \rangle / \hbar^2 c^2 = \sum_i (E_i - E_o)^2 \langle o|x|i\rangle \langle i|x|o\rangle / \hbar^2 c^2, \tag{1.17a}$$

where E_i and E_o are the energies of the states i and o. Since the operator x changes parity, the minimum value of $E_i - E_o$ is the excitation energy ΔE_- of the lowest lying negative parity state of the three quark system. We thus obtain the inequality

$$v^2/c^2 \geq (\Delta E_-)^2 \langle x^2 \rangle / \hbar^2 c^2 \quad . \tag{1.17b}$$

The relation between the rigorous result (1.17) and the handwaving intuitive result (1.16) is now clear. The angular frequency ω to be used in the relation (1.16) is some average excitation energy for orbital excitation, and λ is just the wave length of a photon emitted in the transition from these excited states to the ground state. Thus Eq. (1.16) shows that the condition for the nonrelativistic approximation to be valid ($v/c \ll 1$) is the same as that for the validity of the multipole expansion for the radiative transitions from orbitally excited states to the ground state ($r \ll \lambda\!\!\!^{-}$).

For the case of the nucleon, where the measurement of the mean square radius by electron scattering shows a value of the order of one fermi, while the excitation energy of the first odd parity resonances is about 600 MeV and corresponds to a wave length of about 1/3 fermi, v/c is manifestly not small compared to unity.

Note that for a coulomb-like potential, the excitation energies are proportional to g^2/r, where g is the coupling constant, $\Delta E_- x$ is just g^2, and an expansion in v/c is equivalent to an expansion in powers of the coupling constant, or ordinary perturbation theory.

For charmonium, the upsilon system and heavier quarks, Eq. (1.17) show that the nonrelativistic approximation is probably all right. $\langle x^2 \rangle$ is presumably much smaller, of the order of the Compton wave length of heavy quarks having masses of 1.5 GeV, 4.5 GeV and higher, while ΔE_- for these systems seems to be around 300 MeV.

Note that the results (1.16) and (1.17) are model independent, as they use only the experimentally measured size of the system and excitation energies. Thus any model which fits the size of the proton and the excitation spectrum of low-lying negative parity states cannot have only nonrelativistic velocities.

1. COLOUR

We now consider in detail the colour degree of freedom and the properties of the colour exchange force which seems to be present in bound states of constitutent quarks. Much of this treatment is given in Refs. 13 and 14.

2.1 The Deuteron World

Some insight into the coloured quark models is given by the analogy of a world in which all low-lying nuclear states are made of deuterons and have isospin zero, free nucleons have not yet been seen and experiment has not yet attained energies higher than the deuteron binding energy or the symmetry energy required to excite the first I = 1 states. In this isoscalar world where all observed states have isospin zero the isovector component of the electromagnetic current would not be observed since it has vanishing matrix elements between isoscalar states. The deuteron energy level spectrum (something like that of a diatomic molecule) would indicate that the deuteron was a two-body system, but there would be no way to distinguish between the neutron and the proton. The deuteron would thus appear to be composed of two identical objects which might be called nucleons. Since the deuteron has electric charge +1, the nucleon would be assumed to have electric charge +1/2. Furthermore, the nucleon would be observed to have spin 1/2 and be expected to satisfy Fermi statistics. However, the ground state of the deuteron and all other observed states would be found to be symmetric in space and spin. Thus, the nucleon would appear to be a spin 1/2 particle with fractional electric charge and peculiar statistics.

Some daring theorists might propose the existence of a hidden degree of freedom expressed by having nucleons of two different colours. There would be a hidden SU(2) symmetry (which might be called isospin) to transform between the two nucleon states of different colours. All the observed low-lying states would be singlets in this new colour (or isospin) SU(2). Since the colour singlet state of the two-particle system is antisymmetric in the colour degree of freedom, the Pauli principle requires the wave function to be symmetric in space and spin, thus solving the statistics problem.

The direct analog of this deuteron problem in hadron quark models is the quark model for the Ω^-. In the conventional quark model, the Ω^- consists of three identical strange quarks (called λ-quarks by some people and s-quarks by others), with their spins of 1/2 coupled symmetrically to spin 3/2. Since the electric charge of the Ω^- is -1, the strange quark is required to have charge -1/3, and it is also required to have peculiar statistics because the system of three identical particles has a symmetric wave function in all known degrees of freedom. Some daring theorists have therefore proposed the existence of a hidden degree of freedom expressed by having strange quarks of three different colours[5], and a hidden SU(3) symmetry to transform between the three strange quark states of different colours. All the observed low-lying states are singlets in the $SU(3)_{colour}$ group. Since the colour-singlet state of the three-particle system is antisymmetric in the colour degree of freedom, the Pauli principle requires the wave function to be symmetric in the other degrees of freedom, in agreement with experiment and ordinary Fermi statistics. It is also possible to give these coloured strange quarks different integral electric charges, one with charge -1 and two neutrals, by analogy with the nucleons in the deuteron. However, as we are concerned primarily with strong interactions, we need not choose between models having different electric charges for coloured quarks.

We have chosen the example of the Ω^- for this discussion to simplify the treatment of the flavour degree of freedom by considering only strange quarks. When all flavours are considered, there are three colours for each flavour, and $3n_f$ quarks altogether. There are two SU(n) groups, the flavour $SU(n)_f$ and the colour SU(3), which are combined into the direct produce $SU(n)_f \times SU(3)_{colour}$.

2.2 The Puzzles of Quark Model Predictions of the Hadron Spectrum

Let us now consider some puzzles posed by one of the outstanding "successes" of the quark model, the prediction of the hadron spectrum. The empirical rule that all observed hadron bound states and resonances have the quantum numbers found in the three-quark and quark-antiquark systems is in remarkable agreement with experiment. Since no alternative explanation or description has been given for this striking regularity in the hadron spectrum, this rule may constitute evidence for taking quarks seriously. The quark model also predicts the energy level spectrum of the states constructed from the three-quark and quark-antiquark systems and observed experimentally as hadron resonances. These predictions also seem to be in reasonable agreement with experiment, but pose additional questions.

Why is the observed baryon spectrum fit only by the symmetric quark model[4] which restricts the allowed states of the three-quark system to those being totally symmetric under permutations in the known degrees of freedom rather than totally antisymmetric, as one expects for fermions? This can be explained by assuming that quarks obey peculiar statistics, or that there is a hidden degree of freedom sometimes called "colour". But this requires the additional ansatz that *all* observed hadrons are colour singlets. *Why* and *why* only 3q and $q\bar{q}$? *Why* not other configurations? *Why* does the low-lying meson spectrum show all the states "predicted by the quark model" without any supplementary conditions and with no allowed states conspicuously absent?

There is an inconsistency between the observation of bound states *in all channels* for $q\bar{q}$ scattering and the *absence of bound states* with quantum numbers of $2q\bar{q}$ and $3q\bar{q}$. If the quark-antiquark interaction is attractive in *all possible channels*, as indicated by the presence of bound states, an antiquark should be attracted by any composite state containing only quarks, like a diquark or a baryon, to make a bound state with peculiar quantum numbers that have not been observed.

In our discussion, we assume that free quarks are very heavy, and we consider only effects on the mass scale of the quark mass. All observed particles have zero mass on this scale. The observed hadron spectrum is a "fine structure" which we are unable to resolve in this approximation. This is a reasonable approach, since as long as we are not treating spin in detail, we are unable to distinguish between a pion and a ρ meson, and are neglecting mass splittings of the order of the $\rho - \pi$ mass difference. We therefore are only able to discuss whether a particle has "zero mass" and appears as an observed hadron, or whether it has a mass of the order of the quark mass and should not have been observed.

The question why only 3q and $q\bar{q}$ can be stated more precisely in terms of the following three whys:

1. *The triality why*. With attractive interactions between quarks and antiquarks, *why* are three quarks and an antiquark not bound more strongly than a baryon or two quarks and an antiquark bound more strongly than a meson? Note that we are not asking about four quarks vs. three quarks. Symmetry restrictions such as the Pauli principle with coloured quarks can prevent the construction of a four quark state which is totally symmetric in space, spin and flavour. But there is no Pauli principle which prevents an antiquark from being added to a system of three quarks in all possible states. Thus if each quark in the baryon attracts the antiquark, some additional mechanism must be found to prevent it from being bound to the quark system.

2. *The exotics why*. Even assuming some mysterious symmetry principle which prevents fractionally charged states from being seen, *why* are there no strongly bound states of zero triality, like

those of two quarks and two antiquarks or four quarks and one antiquark? The question of whether or not such bound four-quark states exist can be posed as follows: There are two analogs for the bound quark-antiquark meson state, the deuteron and positronium. If the meson is like the deuteron, then two mesons should form a bound four-quark system just as two deuterons bind together to form a much more strongly bound α particle. If the deuteron is like positronium, the forces saturate and the residual force between the two neutral systems is very small and does not produce a state more strongly bound than the original two particle states. From the experimental observation that there is no strongly bound doubly charged state of two positive pions, we conclude that the pion is more like positronium than like the deuteron.

However the positronium analogy is misleading because there is no bound state of three electrons while three quarks bind to make a baryon. The force between two positronium atoms is nearly zero because the repulsion between the electron pairs exactly cancels the attraction of the electron-positron pairs in the two positronium atoms. But in two positive pions the quark-quark force cannot be completely repulsive because the same quarks must have attractive forces to make baryons. If the quarks and antiquarks in two pions attract one another, why is there no net attraction between two positive pions to produce an $I = 2$ dipion resonance or bound state with a mass near the mass of two pions?

3. The diquark or meson-baryon why. Why is the quark-quark interaction just enough weaker than the quark-antiquark interaction so that diquarks near the meson mass are not observed, but three-quark systems have masses comparable to those of mesons? Vector gluons which are popular these days would bind the quark-antiquark system, but the force they provide between identical quarks is repulsive. Scalar or other gluons which are even under charge conjugation bind both the quark-antiquark and diquark systems equally. If the quark mass is very heavy, the single quark-antiquark interaction in a meson must cancel two quark masses, while the three quark-quark interactions in the baryon must cancel three quark masses. This suggests that the quark-quark interaction is exactly half the strength of the quark-antiquark interaction[17]. Such a result can be achieved by a suitable mixture of vector and scalar interactions, but it is not very satisfying to obtain such a simple fundamental property of hadrons by a model which fits it with an adjustable parameter.

In all of this discussion, we are considering one-particle states, with the assumption that multiparticle states exist which contain separated particles each having the properties we are trying to explain. Multiparticle states pose additional problems. The allowed spectrum for multiparticle states is not specified by a set of allowed quantum numbers, but by the condition that their constituent particles individually have allowed quantum numbers. Thus the whys cannot be answered by general symmetry principles which apply to all states. The triality why is not answered by a symmetry principle forbidding all states which do not have zero triality, because multiparticle states of zero triality must also be forbidden if they are made of particles which individually have nonzero triality. Similarly, the exotics why is not answered by a symmetry principle forbidding all states with exotic quantum numbers because multiparticle exotic states made from nonexotic particles are allowed. Thus any treatment which attempts to answer these whys must discuss both single-particle and multiparticle states, and must consider the space-time properties which distinguish between them. Algebraic arguments involving only internal symmetry groups cannot be sufficient.

Our three whys involve only the strong interactions which do not depend upon the couplings of quarks to the electromagnetic and weak currents. The following discussion thus applies to both fractionally charged and integrally charged models.

2.3 The Coloured Gluon Model

We now examine the three whys. In the coloured quark description of hadrons the restriction that only colour singlet states are observed immediately solves the triality why since only states of zero triality can be colour singlets. But requiring all low-lying states to be colour singlets is thus equivalent to requiring all low-lying states to have zero triality; it merely replaces one ad hoc assumption with another. What is needed is some dynamical description in which the colour singlets turn out to be the low-lying states in a natural way. To attack this problem we return to the fictitious deuteron world where all low-lying states are isoscalar and which is the analog of the coloured quark description of hadrons. We follow the treatment of ref. 44.

At first this isoscalar deuteron world seems very artificial. Why should all states with I = 0 be pushed down and all states with I ≠ 0 be pushed up out of sight? But there turns out to be a very natural nuclear interaction which creates exactly this isoscalar deuteron world; namely nuclear two-body forces dominated by a very strong Yukawa interaction provided by ρ exchange. This interaction is attractive for isoscalar states and repulsive for isovector states, in both nucleon-nucleon and nucleon-antinucleon systems. It thus binds only isoscalar states. The ρ-exchange interaction between particles i and j can be expressed in the form

$$v_{ij} = V \vec{t}_i \cdot \vec{t}_j \tag{2.1a}$$

where $\vec{t}_i$ is the isospin of particle i and V contains the dependence on all other degrees of freedom except isospin. If we neglect these other degrees of freedom we can write for any n-particle system containing antinucleons and nucleons,

$$V(n) = \frac{1}{2} \sum_{i \neq j} v_{ij} = \frac{V}{2} [\sum_{\substack{all \\ ij}} \vec{t}_i \cdot \vec{t}_j - \sum_i \vec{t}_i \cdot \vec{t}_i] = \frac{V}{2}[I(I+1)-nt(t+1)] \tag{2.1b}$$

where I is the total isospin of the system and t is the isospin of one particle; i.e., 1/2 for a nucleon.

The interaction (2.1b) is seen to be repulsive for the two-body system with I = 1 and attractive for all isoscalar states. A pair of particles bound in the I = 0 state is thus seen to behave like a neutral atom; it does not attract additional particles. Since the pair is "spherically symmetric" in isospace, a third particle brought near the pair sees each of the other particles with random isospin orientation, and its interaction with any member of the pair is described by the average of (2.1a) over a statistical mixture which is 3/4 isovector and 1/2 isoscalar. This average is exactly zero.

The neutral atom analogy is very appropriate for the description of the observed properties of hadrons. The forces between neutral atoms are not exactly zero, but are much weaker than the forces which bind the atom itself. These interatomic forces produce molecules which are much more weakly bound than atoms. Similarly the forces between hadrons do not vanish but are much weaker than the forces which bind the hadron itself. These interhadronic forces produce complex nuclei which are much more weakly bound than hadrons. In the approximation where we neglect energies much smaller than the quark mass these "molecular" effects are safely neglected.

We now generalize this picture for the coloured quark description of hadrons. If there are n colours, the interaction (2.1) must be generalized from SU(2) to SU(n). The quark-antiquark system then still saturates at one pair, but the multiquark system can be seen to saturate at n quarks. A quark-antiquark system which is a singlet in SU(n) exists for all values of n. However, the existence of a singlet in the two-quark system is an accident which occurs only in SU(2) and is not generalizable to SU(n). However the I = 0 two-quark state is also characterized as antisymmetric under permutation of the two particles. This antisymmetry is generalized easily to SU(n) where totally antisymmetric states exist for a maximum of n particles, and the n particle antisymmetric state is a singlet in SU(n).

We now construct the analog of the interaction (2.1b) for a model with three triplets of different colours. Then the Yukawa interaction produced by the exchange of an octet of "coloured gluons" has the form analogous to (2.1). For an n-particle system containing both quarks and antiquarks,

$$U(n) = \frac{1}{8} \sum_{i \neq j} u_{ij} \sum_{\sigma} \lambda_{i\sigma} \lambda_{j\sigma} \tag{2.2}$$

where u_{ij} depends on all the noncolour variables of particles i and j and $\lambda_{i\sigma}$ (σ = 1, ... , 8) denote the eight generators of $SU(3)_{colour}$ acting on a single quark or antiquark i.

If the dependence of u_{ij} on the individual particles i and j is neglected, the interaction energy of an n-particle system can be calculated by the same trick used in Eq. (2.1b) to give

$$V(n) = \frac{u}{2} (C - nc) \tag{2.3a}$$

where u is the expectation value of u_{ij}, integrated over the noncolour variables, C is the eigenvalue of the Casimir operator for $SU(3)_{colour}$ for the n-particle system and c = 4/3 is the eigenvalue for a single quark or antiquark. These eigenvalues are directly analogous to the SU(2) Casimir operator eigenvalues I(I + 1) and t(t + 1) in Eq. (2.1b).

In the approximation where all energies small compared to the quark mass M are neglected, the interaction (5.3a) gives the mass formula

$$M(n) = nM_q + V(n) = n(M_q - \frac{cu}{2}) + Cu/2 \tag{2.3b}$$

The interaction (2.2) and the mass formula (2.3b) were first proposed by Nambu[17], and the saturation properties of the interaction were considered by Greenberg and Zwanziger[18]. However, the remarkable properties of this interaction as demonstrated above in the simplified example of the analogous deuteron world have received little attention.

2.4 Answers to the Triality and Meson-Baryon Whys

The formula (2.3b) can test the triality why or the meson-baryon why by showing whether observable "zero mass" hadron states exist for a given number of quarks and antiquarks. However, it cannot test the exotics why, since it gives no information about the spatial properties of the states. It cannot distinguish between one-particle states and multiparticle scattering states and zero-triality exotic states are allowed as multiparticle states.

Since C is positive definite and has the eigenvalue zero only for a singlet[45] in $SU(3)_{colour}$, and $u \gtrsim 0$ as is evident from the two-body system, the state of the n-particle system with the strongest attractive interaction is a colour singlet. Since the interaction is a linear function of n all such singlet states have zero mass if $cu/2 = M_q$. For this case

$$M(n) = (C/c)M_q \quad \text{if } cu/2 = M_q \tag{2.3c}$$

The model thus gives observable hadron states for all quark and antiquark configurations for which C = 0 states exist. Since C = 0 states exist only for configurations of triality zero, this answers the triality why.

The meson-baryon why is also answered by this interaction, since zero mass is attained both in two-body and three-body systems. To obtain C = 0, the two-body system must be a quark-antiquark pair, while the three-body system must be a three quark state, totally antisymmetric in colour space. The approximation of neglecting the dependence of u_{ij} on i and j is justified in these two cases since there is only one pair in the two-body system, and a totally antisymmetric function has the same wave function for all pairs. The values[18] of the interaction parameter C-nc and the mass parameter C/c are listed in Table 2.1 for all states of the two-body system. These show that the quark-quark interaction in the baryon is exactly half of the quark-antiquark interaction in the meson, as required for the meson-baryon puzzle. The diquark mass is thus equal to one quark mass, since its interaction only cancels the mass of one of the two quarks.

Table 2.1 Values of the Interaction and Mass Parameters C-nc and C/c

System	$SU(3)_{color}$	Representation	C	C-nc	C/c
quark-quark	triplet	(antisymmetric)	4/3	-4/3	1
quark-quark	sextet	(symmetric)	10/3	+2/3	5/2
quark-antiquark	singlet		0	-8/3	0
quark-antiquark	octet		3	+1/3	9/4

The interaction averaged over all quark-quark states is seen to be zero and similarly for all quark-antiquark states. An antiquark or quark added to a meson or baryon thus has a zero net interaction, as there can be no colour correlations between particles in a singlet state and an external particle, and each pair feels the average interaction over all colour states. This suggests that the exotics puzzle is also answered, and that the states of zero mass obtained from the interaction (2.2) for exotic quantum numbers are multiparticle continuum states rather than bound states or resonances.

2.5 The Exotics Why--Spatial Properties of Wave Functions

To examine the exotics why in more detail we consider the spatial dependence of the interaction (2.2) for the specific case of the two-quark-two-antiquark system, with an interaction u_{ij} depending only on the positions of the particles and not on momenta, spin and flavour

We first note that the colour exchange force of the form (2.1) or (2.2) gives no bound α-particle-like states of two quarks and two antiquarks. We consider a wave function which is totally symmetric in space for the four particles and is a colour singlet,

$$\psi = \phi(r_1,r_2,r_3,r_4)\chi_o \quad , \tag{2.4a}$$

where χ_o depends upon the colour variables of the four particles and couples them to an overall colour singlet. Spin is disregarded. The expectation value of the interaction (2.2) with this wave function is given by Eq. (2.3a) with C = 0, n = 4 and

$$u = \langle\phi|u_{12}|\phi\rangle \quad . \tag{2.4b}$$

We can use u_{12} in Eq. (2.4b) since the wave function (2.4a) is symmetric in all pairs. Thus

$$\langle\psi|U(n)|\psi\rangle = -2uc \quad . \tag{2.4c}$$

Let us now consider a wave function

$$\psi' = \phi(r_1,r_2,r_3 + X, r_4 + X)\chi_o(12)\chi_o(34) \quad , \tag{2.5}$$

where X is a very large distance like 1 kilometer, and $\chi_o(12)$ and $\chi_o(34)$ couple the pairs (12) and (34) separately to a colour singlet. The expectation value of the interaction (2.2) with this wave function involves only two terms in the interaction, those for i,j = 1,2 and 3,4 since all other pairs are separated by the large distance X. For each pair the interaction is given by

Eq. (2.3a) with C = 0 and n = 2, and with u still given by Eq. (2.4b) because the dependence of the wave functions (2.4a) and (2.5) on r_{12} and r_{34} are identical. Thus

$$\langle\psi'|U(n)|\psi'\rangle = -uc + (-uc) = -2uc = \langle\psi|U(n)|\psi\rangle \quad . \qquad (2.6)$$

Thus any "α-particle-like" wave function can be broken up into two colour singlet quark-antiquark pairs separated by a large distance for which the interaction energy is the same. There is always a gain in kinetic energy by allowing such a breakup, namely the kinetic energy required by the uncertainty principle to keep the separation of the pairs very small. Thus any "α-particle-like" state will be unstable against immediate breakup into two pairs. Since the results (2.5) and (2.6) hold for any value of X, the breakup can occur continuously with no change in potential energy, and there can be no barrier hindering the breakup.

Let us now consider the case of two separated pairs and possible long range "Van-der-Waals" type forces. For simplicity we consider the deuteron world model (2.1a). The generalization to SU(3) is straight-forward. Let d be the distance between the centers of mass of pairs (12) and (34) and choose the x axis in the direction of d. It is then convenient to write the interaction operator (2.1) for the four particle system in the following form:

$$\begin{aligned} V(4) = \frac{1}{2}\sum_{i\neq j} v_{ij} &= (\vec{t}_1\cdot\vec{t}_2)v_{12} + (\vec{t}_3\cdot\vec{t}_4)v_{34} \\ &+ (\vec{t}_1+\vec{t}_2)\cdot(\vec{t}_3+\vec{t}_4)(v_{13}+v_{14}+v_{23}+v_{24})/4 \\ &+ (\vec{t}_1-\vec{t}_2)\cdot(\vec{t}_3+\vec{t}_4)(v_{13}+v_{14}-v_{23}-v_{24})/4 \\ &+ (\vec{t}_1+\vec{t}_2)\cdot(\vec{t}_3-\vec{t}_4)(v_{13}-v_{14}+v_{23}-v_{24})/4 \\ &+ (\vec{t}_1-\vec{t}_2)\cdot(\vec{t}_3-\vec{t}_4)(v_{13}-v_{14}-v_{23}+v_{24})/4. \qquad (2.7) \end{aligned}$$

We consider colour singlet wave functions for the four particle system. There are two independent couplings to an overall colour singlet. We choose the basis in which the colours of the pairs (12) and (34) are diagonal. Both pairs can either be singlets or triplets (octets in SU(3)). For the wave function in which both pairs are singlets, the first two terms on the right hand side of Eq. (2.7) give the binding energies of the two pairs, the next three terms give zero, since either $\vec{t}_1+\vec{t}_2$ or $\vec{t}_3+\vec{t}_4$ annihilate the wave function with 12 and 34 individually coupled to singlets. The last term gives an off-diagonal matrix element which can produce a "polarization force". We rewrite this term by expanding the potentials around r = d,

$$V_{pol} = (\vec{t}_1-\vec{t}_2)\cdot(\vec{t}_3-\vec{t}_4)x_{12}x_{34}(d^2v/dr^2)_{r=d} \quad . \tag{2.8}$$

This interaction is seen to connect the ground state configurations of the two bound pairs with excited states which are colour octets and are p-wave excitations in configuration space. This interaction can produce an energy shift in second order perturbation theory which depends upon the distance d and could give a long range Van-der Waals force. The magnitude of this energy shift is given by the square of the matrix element of the interaction (2.8) divided by an energy denominator. Since the intermediate state has pairs (12) and (34) in colour triplets, the excitation energy of this state and the corresponding energy denominator is dominated by the third term on the right hand side of Eq. (2.7) which does not vanish for colour triplet states and depends upon V(d). Thus the energy shift is given approximately by

$$\Delta E = [(d^2V/dr^2)_{r=d}]^2\cdot\langle x_{12}^2\rangle\langle x_{34}^2\rangle/8V(d) \quad . \tag{2.9a}$$

It is amusing that for a linear potential, the expression (2.9a) vanishes because the second derivative of the potential is zero. For a logarithmic potential, or for a harmonic oscillator potential,

$$E = V_o\langle x_{12}^2\rangle\langle x_{34}^2\rangle/8d^4\langle\log(d/r_{12})\rangle \quad \text{for } V = V_o\log(r) \tag{2.9b}$$

$$E = k\langle x_{12}^2\rangle\langle x_{34}^2\rangle/8d^2 \qquad \text{for } V = kr^2 \quad . \tag{2.9c}$$

Thus we see that even for confining potentials which increase to infinity at large distances, the residual force between two colour singlet hadrons decreases rapidly at large distances. Note that these estimates are large overestimates of the force because effects of retardation have been neglected and should add additional damping factors. The interaction (2.8) involves instantaneous colour correlations between the two pairs, with each jumping simultaneously from colour singlet to colour triplet. The colour-correlated triplet-triplet state involves instantaneous correlations which certainly cannot be maintained over large distances.

We have seen that exotic four quark states cannot be bound either in α-particle-like symmetric wave functions or in molecular type separated pairs. We now investigate a third possibility, a correlated four-particle state, which is neither an α particle nor a molecule. We consider in more detail the potential for any given spatial configuration which is a 2 x 2 matrix in colour space explicitly constructed in Ref. 19 by evaluation of the λ-matrices in Eq. (2.2) and then diagonalized. The colour degree of freedom was eliminated by use of a static approximation, analogous to the static Coulomb approximation in QED, which assumes that particle motion is slow in comparison with photon or coloured gluon exchanges. This approximation is implied in all charmonium potential calculations, where a static potential is assumed to hold for a colour singlet state of the two-body system, even though the colours of the individual constituents must be changing rapidly in time to make a colour singlet state. The condition for validity of this approximation; namely that the time scale of colour changes is much more rapid than the time scale of quark motion, is equivalent to the requirement that excitation of the colour degree of freedom requires a much greater energy than excitations in space-time[14,20].

Diagonalization of the potential matrix in colour space gives the following "eigenpotentials" in the static approximation for the two colour couplings[14,19].

$$U' = (7/16)(u_\alpha+u_\beta) + (1/8)u_q \pm (3/16)\sqrt{8(u_\alpha-u_\beta)^2+(u_\alpha+u_\beta-2u_q)^2}, \tag{2.10b}$$

where

$$u_\alpha = u_{13}+u_{24}; \quad u_\beta=u_{14}+u_{23}; \quad u_q = u_{12}+u_{34} \; . \tag{2.10b}$$

To test the exotics puzzle we look for coordinate configurations where four-particle correlations may give stronger binding than in two noninteracting clusters. Since u_α and u_β appear symmetrically in (2.10) we need only consider values of $u_\beta \leq u_\alpha$. For any value of u_α the value of $u_\beta \leq u_\alpha$ which minimizes the interaction (2.10) is $u_\beta = u_\alpha$ with the negative sign for the square root. This gives

$$U' = -(8/3)u_\alpha - (2/3)(u_\alpha - u_q) \; . \tag{2.11}$$

This expression is minimized by choosing the minimum values of u_q consistent with a given value of u_α. For monotonically decreasing potentials this is achieved by placing the four particles at the corners of a square with the like particles at opposite diagonals.

For a square well potential the particles can be arranged in a square with the diagonal greater than the range of the forces and the sides less than the range. This configuration has $u_q = 0$

and forms a stable four-particle state with a binding 25% greater than that of two quark-antiquark pairs. However, the sharp edge of the square well is essential for this binding and does not seem reasonable physically. For smooth potentials without sharp edges such as Coulomb, linear, Gaussian, Yukawa or harmonic oscillator potentials Eq. (2.11) shows that such a four-particle cluster is less strongly bound than two noninteracting quark-antiquark pairs, and the system simply breaks up into two clusters. This leads to a description in which all states having exotic quantum numbers are just scattering states of particles which individually have nonexotic quantum numbers, and answers the exotics puzzle.

The eigenpotentials (2.10) can also be used to examine configurations described to a good approximation as a diquark and an antidiquark separated by a distance large compared with the diquark size. The quark-antiquark interaction should then be the same for all four quark-antiquark pairs; i.e. we neglect correlations between the motion of one particular quark in the diquark and one particular antiquark in the antidiquark. Then $u_\alpha = u_\beta$ and Eq. (2.10a) simplifies to give the two solutions

$$U' = \frac{1}{2} u_\alpha + \frac{1}{2} u_q \quad , \tag{2.12a}$$

$$U' = \frac{5}{4} u_\alpha - \frac{1}{4} u_q \quad . \tag{2.12b}$$

The eigenfunction corresponding to Eq. (2.12a) has both diquarks in the colour triplet state. For Eq. (2.12b) both are in colour sextet states.

Another case of interest is that of two separated quark-antiquark pairs. Here the neglect of correlations between particles gives $u_\beta = u_q$, and Eq. (2.12a) simplifies to

$$U' = u_\alpha \quad , \tag{2.13a}$$

$$U' = -\frac{1}{8} u_\alpha + \frac{9}{8} u_q \quad .$$

The corresponding eigenfunctions have separated colour singlet pairs for Eq. (2.13a) and separated octet pairs for Eq. (2.13b).

Equation (2.13a) shows that the lowest state for two separated quark-antiquark pairs has an interaction which depends only on the spatial separations within each pair and is independent of the distance between pairs. There is no long range residual force between two separated colour singlet states, as expected from the saturation property of the colour charge force[14,19].

3. THE LOG POTENTIAL

We now consider some of the properties of the log potential and argue that it is a reasonable one to use for a quasinuclear model for hadrons. We first note the characteristic scaling property[5,21] which motivated its introduction by Quigg and Rosner. Consider the Hamiltonian for a particle in a log potential,

$$H = (p^2/2m) + V\log(r/r_o)$$
$$= (p^2/2m_o)(m_o/m) + V\log(r/r_o) \quad , \tag{3.1}$$

where m is the mass of the particle and m_o is some standard mass. Let us now introduce the scale transformation,

$$p' = p(m_o/m)^{1/2} \quad , \tag{3.2a}$$
$$r' = r(m/m_o)^{1/2} \quad . \tag{3.2b}$$

Then

$$H = (p'^2/2m_o) + V\log(r'/r_o) + \frac{1}{2}V\log(m_o/m) \quad . \tag{3.3}$$

The scale transformation thus reduces the Hamiltonian (3.1) for any mass m to the Hamiltonian (3.3) which has the standard mass m_o and the same potential and only an added constant term. Thus the energy spectrum of the Hamiltonian H depends upon the mass m only by the additive constant 1/2 V $\log(m_o/m)$; all energy splittings are independent of the mass m.

However, there is no theoretical justification from first principles or QCD for this log potential. It was only introduced because it is the potential which gives equality for the $\Upsilon - \Upsilon'$ and $\psi - \psi'$ splittings. The potential previously used with at least hand waving support from QCD was a combination of a Coulomb and a linear potential[21], giving Coulomb behaviour at short distances and linear confinement at large distances. The relation between this potential and the log potential is seen by considering the potential

$$V = (V_o/\chi)\sinh[\chi\log(r/r_o)] \quad . \tag{3.4a}$$

For $\chi = 0$, this is just the simple log potential. For $\chi = 1$ it is the Coulomb + linear potential. For intermediate values of χ, it is convenient to rewrite the potential in the form

$$V = (V_o/2\chi)\,[(r/r_o)^\chi - (r_o/r)^\chi] \quad . \tag{3.4b}$$

Equation (3.4) defines a family of potentials characterized by a parameter χ, which are all singular at the origin and confining at large distances. Both the singularity at the origin and the strength of the confinement at large distances become weaker when χ decreases, and at $\chi = 0$ these become the singularities at the origin and infinity of the log potential. In the vicinity of $r = r_0$, which defines the transition region between Coulomb and log for $\chi = 1$, the hyperbolic sine can be expanded to give the log potential. Thus the log potential is a good approximation to the potential (3.4) for all values of χ in the vicinity of $r = r_0$.

As long as the properties of the system being considered do not depend on the exact form of the potential at very small or at very large distances, the log potential may give a good approximation for any potential of the type (3.4) which is singular at short distances and confining at large distances. Thus the log may be very useful for calculations, even though it has no deep fundamental significance. We accept its use on this basis and do not attempt to justify it on any more serious grounds.

The log potential can also be placed in a hierarchy of power law potentials. Potentials which vary as a positive power of r, like the linear or harmonic oscillator potentials, are always confining at large distances and approach a constant at the origin which can be chosen as zero energy. The spectrum is discrete and there is no continuum. The splittings of the energy levels decreases with increasing mass of the particle. Potentials which vary as a negative power of r, like the Coulomb potential, are singular at the origin and go to zero at infinity. They are not confining and have a continuous spectrum, with the possibility of a discrete spectrum as well if the potential is attractive. The splittings of the energy levels increases with increasing particle mass.

The log potential, which is in some sense the limiting case of a power law potential with the power zero is intermediate between the two cases. It is singular both at the origin and at infinity, has a discrete spectrum and no continuum, and the splittings of energy levels remains constant with changing particle mass.

It is also instructive to note that the characteristics of the energy spectrum of the log potential are nearly midway between the Coulomb and harmonic oscillator cases. The harmonic oscillator has an equally spaced set of energy levels. The Coulomb potential has energy levels with a spacing that decreases very rapidly with increasing energy. The log potential has a spectrum of energy levels with a spacing that decreases with increasing energy but not as rapidly as the Coulomb case. A convenient quantitative measure of this feature of the spectrum is the ratio (D-P)/(P-S) of the spacing between the lowest D state and the lowest P state to the spacing between the lowest P state and the ground state . For the harmonic oscillator this is 1.0. For the Coulomb potential it is 0.2. For the log it is 0.6, just midway between the two.

We also recall the result from Eq. (1.10) that the expectation value of the kinetic energy in a multiquark system with two-body logarithmic potentials is given by a c-number rather than the expectation value of an operator. It depends only upon the strength parameter of the logarithmic potential and is independent of particle masses or the degree of excitation of the wave function. Here again it is in between the negative and positive power law potentials. In positive power potentials like the harmonic oscillator, the kinetic energy decreases with increasing mass of the particle and increases with the degree of radial excitation. In negative power law potentials, like the Coulomb potential the kinetic energy increases with increasing particle mass and decreases with higher radial excitation. In the log potential the kinetic energy is constant with both particle mass and radial excitation.

4. THE QUASINUCLEAR COLOURED QUARK MODEL

The discovery of the first pion-nucleon and pion-pion resonances as low-lying p-wave resonances with equal widths suggested that mesons and baryons were composite objects with very similar structures, rather than elementary objects as different from one another as photons and electrons. The non-relativistic quark model described these first resonances and their photoexcitation as magnetic dipole excitations of quark spin flip for both mesons and baryons. The quark model also succeeded in describing other similar properties of mesons and baryons, including high energy scattering and reaction processes and strong, electromagnetic and weak decay processes. The introduction of the colour degree of freedom[19] explained the difference between quark-quark and quark-antiquark interactions which made the low-lying states of the multiquark system appear as three-body states while the lowest states containing both quarks and antiquarks were two-body states.

Although meson and baryon spectra were seen to be qualitatively similar, quantitative relations between the mass splittings were difficult to obtain. Two basic physical assumptions are necessary to relate meson and baryon spectra: (1) a relation between the quark-antiquark forces binding mesons and the quark-quark forces binding baryons; (2) radial scaling properties of these interactions and of the baryon and meson wave functions. The assumption of colour exchange forces has successfully related the gross features of meson and baryon spectra, but has been inadequate to describe the finer details. Quantitative estimates of mass splittings are sensitive to the difference in sizes of meson and baryon wave functions, and to flavour-dependent size effects arising from mass differences between quarks of different flavours. Since these size effects are model-dependent, it has been difficult to obtain significant predictions without introducing too many free parameters.

The successful description of the charmonium spectrum using the nonrelativistic quark model[23,24] suggests that it is reasonable to describe heavy quark systems by a Schroedinger equation for coloured quarks interacting with a confining two-body colour-exchange potential and no additional bag. We extend this model to the three-quark system and assume that baryons are described by the same Schroedinger equation with the same two-body forces, and that the effective matrix elements of the two-body interaction in the three-body system are related to those for the two-body system by simple scaling laws obtained by analysis of heavy quarkonium spectroscopy. In this way we construct a "Quasinuclear Coloured Quark Model" for hadrons with the same approach as that of nuclear physics, to determine the properties of the n-body system from the known properties of the two-body system. In the hadron case, where free quark scattering is not observed, the only input comes from effective matrix elements of the two-body interaction in bound quark-antiquark states, and we must use this information for the quark-quark interactions in three-body systems. We also use a nonrelativistic formulation for light quarks which are certainly relativistic in light hadrons. However, the results for the magnetic moments shown in the introduction above indicate that this approach works, even though we do not yet understand why. We therefore continue to use it wherever we can, to see where it continues to work and where it might break down.

The recent discovery that the $\Upsilon - \Upsilon'$ and $\psi - \psi'$ mass splittings are equal has motivated the introduction of a model with a flavour-independent logarithmic potential[2] whose mass splittings depend only on the strength of the potential and are independent of reduced mass. We extend this model to the conventional hadron spectrum and compare its predictions for orbital, spin and strangeness mass splittings with experiment. This model makes possible quantitative predictions relating meson and baryon spectra in which baryon mass splittings are predicted with only meson mass splittings used as inputs. The log potential provides a unique prescription without free parameters. The success of this prescription should be considered as further evidence for the common structure and interactions of mesons and baryons, rather than as a test for the details of the potential. Any potential with similar scaling properties in the relevant radial domain would presumably make similar predictions.

We first consider orbital excitations. Since clearly defined radially excited states are not easily seen experimentally, in contrast with the heavy quarkonium systems, we choose for comparison of theory and experiment the S, P and D states with "stretched" angular momenta; namely those with the highest values of spin and J for a given configuration. These are the 1-, 2+ and 3- mesons and the 3/2+, 5/2- and 7/2+ baryons. The results are shown in Table 4.1. The masses of the isovector mesons ρ, A2 and g are taken as input along with the lowest states in the K*, ϕ and nonstrange baryon families. The masses of the remaining excited states are then predicted with no free parameters. Predictions from harmonic oscillator and Coulomb potentials are presented for comparison.

Three types of predictions were considered.

1. The ratio of the P-D and S-P splittings predicted to be 0.6 by the log potential model, 1.0 by a harmonic oscillator potential and 0.2 by a Coulomb potential. The values 0.68, 0.65 and 0.58 obtained for the ρ, K* and baryon systems respectively support the log potential model.

2. Flavour-independent mass splittings predicted for the ρ, K* and ϕ systems. The results in Table 4.1 show reasonable agreement.

3. Extension to baryons. This requires relating quark-quark and quark-antiquark interactions and treating the three-body problem. A straightforward analysis discussed in detail below leads to the prediction that baryon splittings are reduced by a factor 3/4 relative to meson splittings. The results in Table 4.1 show surprising agreement.

For the spin splittings we assume that the scaling property of the spin dependent part of the two-body interaction is that suggested by simple arguments based on the logarithmic potential. This leads immediately to predictions for baryon splittings with meson spin splittings used as input and no parameters, listed in Table 4.2.

For the strangeness splittings we assume two sources of SU(3) symmetry breaking: (1) a constant mass difference between strange and nonstrange quarks over the whole spectrum; (2) the dependence upon quark masses of the spin dependent interaction already used for spin splittings. This again gives predictions for baryon spin splittings with meson spin splittings used as input and no parameters, also listed in Table 4.2.

The details of the model which lead to these predictions are discussed below. The essential numerical factors which appear in relations between meson and baryon spectra are: (1) a factor 1/2 between strengths of quark-quark and quark-antiquark potentials which comes from colour couplings and (2) a factor 3/4 which comes from scaling of baryon and meson wave functions. It is the factor 3/4 which is new and which is responsible for the success of the predictions listed in Tables 4.1 and 4.2. It is clear from Table 4.2 that reducing this factor 3/4 would lead to better agreement for spin splittings, but this is parameter juggling, which obscures the physics if it has no strong theoretical motivations.

Table 4.1. Orbital Splittings from Logarithmic Potential Model Theoretical and Experimental Values of Masses in MeV

	I=1 Mesons	Kaon Family		Family		Baryon Family		Harmonic Oscillator
	Expt	Theory	Expt	Theory	Expt	Theory	Expt	Model
L=0	770*	*	892*	*	1020*	*	1232*	*
L=1	1310*	1432	1421	1560	1516	1637	1670	1700
L=2	1680*	1802	1765	1930		1914	1925	2020
	Ratio of P-D and S-P Splittings					(Coulomb Potential Gives 0.2)		
	0.68	0.6	0.65	0.6		0.6	0.58	1.0

*Denotes Input

Table 4.2. Spin and Strangeness Splittings from Logarithmic Potential Model Theoretical and Experimental Mass Differences in MeV

Spin Splittings			Strangeness Splittings		
Mass Difference	Theory	Expt	Mass Difference	Theory	Expt
M(ρ) −M(π)	*	630	M(K*)−M(ρ)	*	122
M(K*)−M(K)	*	396	M(Σ*)−M(Δ)	137	153
M(Δ) −M(N)	354	292	M(Ξ*)−M(Σ*)	149	148
M(Σ*)−M(Σ)	223	192	M(Ξ) −M(Σ)	149	125
M(Ξ*)−M(Ξ)	223	216	M(Ω) −M(Ξ*)	160	139
M(Σ) −M(Λ)	88	77	M(Λ) −M(N)	180	177

*Denotes Input

The overall agreement with predictions of the quasinuclear coloured model with the logarithmic potential is impressive and suggests further investigation, both theoretical and experimental.

We now consider multiquark systems in detail: The origin of the baryon factor 3/4 will be seen explicitly, and the formulation will be sufficiently general to be applicable to systems with more than three quarks. Our phenomenological "quasinuclear nonrelativistic coloured quark model" is motivated by ideas similar to QCD but not justified by any rigorous argument. The two basic assumptions of the model are: (1) the quasinuclear approximation of n constituents interacting with two-body forces without any additional parton-antiparton pairs, gluons or bag; (2) a colour-exchange force with a colour dependence given by Eq. (2.2); and no additional Wigner force.

The quasinuclear assumption is implicit in all conventional spectroscopy as well as in the successful charmonium calculations, where all the confinement effects are in the two-body potential and there is no bag. The pure colour exchange assumption is required to obtain a long range force which confines quarks within hadrons, but leaves no residual long range forces between physical hadrons. Any other long range force at the quark level gives long range hadron-hadron forces. Arguments from QCD which base this colour exchange force on the colour properties of the one gluon exchange contribution are misleading because the experimental evidence supporting colour exchange is in the long range (infra red) part of the force which is definitely not one gluon exchange. We accept colour exchange on phenomenological grounds as necessary to fit the experimental spectrum with no rigorous justification at this stage. We do not heed to choose between models with quark confinement and those with heavy liberated quarks, but can include both cases by appropriate choices of the radial dependence of the interaction.

The model hamiltonian for a system of n particles which can be any combination of quarks and antiquarks is (3,4):

$$H = \sum_{i=1}^{n} \frac{p_i^2}{2m} - \frac{3}{16} \sum_{\alpha=1}^{8} \sum_{i>j} \lambda_i^\alpha \lambda_j^\alpha u_{ij} \quad , \tag{4.1}$$

where λ_i^α and λ_j^α are the λ-matrices for the colour SU(3) group for particles i and j, u_{ij} depends on all the noncolour variables of particles i and j, and the normalization factor (-3/16) is chosen so that the potential is exactly equal to u_{ij} for the case of a quark-antiquark pair in the colour singlet state. Thus for the physical mesons

$$H_{mes} = \sum_{i=1}^{2} \frac{p_i^2}{2m} + u_{12} \quad . \tag{4.2}$$

The interaction (4.1) was first introduced by Nambu[17] who obtained it from one gluon exchange and considered only the colour degree of freedom to show that it gave a spectrum where all low-lying states were colour singlets. The dynamics of a similar phenomenological interaction were considered in a series of previous papers[25,19,14] which introduced the spatial dependence and analyzed the simplest non-trivial case where colour and space do not factorize. This treatment is given in detail in section 2.5. The results showed that the interaction (4.1) did not bind four-particle states and gave only two-meson scattering states if a reasonable radial dependence was assumed for the interaction and spin dependence was neglected. Spin dependence was later introduced by De Rujula et al[4], who showed that the sign of one gluon exchange predicted the right sign for the N-Δ and Σ-Λ mass splittings. Jaffe[26] has shown that the spin-dependent interaction responsible for these mass splittings can also lead to the binding of exotic multiquark configurations.

For a system of quarks which is totally antisymmetric in colour, as in the baryon case, the summation of the λ-matrices can be evaluated explicitly in Eq. (4.1) and space and colour factorize to give the result

$$H_{bar} = \sum_{i=1} \frac{p_i^2}{2m} + \frac{1}{2} \sum_{i>j} u_{ij} \quad . \tag{4.2b}$$

For the three particle system it is convenient to express the hamiltonian (2) in terms of the center-of-mass momentum P and two independent relative co-ordinates and their canonically conjugate momenta,

$$\vec{x} = \vec{r}_1 - \vec{r}_2 \quad , \tag{4.3a}$$

$$\vec{y} = (\vec{r}_1 + \vec{r}_2 - 2\vec{r}_3)/\sqrt{3} \quad , \tag{4.3b}$$

$$\vec{p}_x = (\vec{p}_1 - \vec{p}_2)/2 \quad , \tag{4.4a}$$

$$\vec{p}_y = (\vec{p}_1 + \vec{p}_2 - 2\vec{p}_3)/2\sqrt{3} \quad , \tag{4.4b}$$

$$H = \frac{p^2}{6m} + \frac{p_x^2}{m} + \frac{p_y^2}{m} + \frac{1}{2}\,[u(\vec{x})+u(\frac{\vec{x}}{2} + \frac{\sqrt{3}\vec{y}}{2}) + u(\frac{\vec{x}}{2} - \frac{\sqrt{3}\vec{y}}{2})] \;. \quad (4.5)$$

Two special cases of interest are the harmonic oscillator and logarithmic potentials,

$$u_{osc} = k\ r \quad , \qquad (4.6a)$$

$$u_{log} = V \log r. \qquad (4.6b)$$

For these potentials the Hamiltonian (4.5) becomes

$$H_{osc} = \frac{p^2}{6m} + \frac{p_x^2}{m} + \frac{p_y^2}{m} + \frac{3}{4}\,k[x^2 + y^2] \quad , \qquad (4.7a)$$

$$H_{log} = \frac{p^2}{6m} + \frac{p_x^2}{m} + \frac{p_y^2}{m} + \frac{1}{2}\,V\ [\log x + \log(\frac{\vec{x}}{2} + \frac{\sqrt{3}\vec{y}}{2}) + \log(\frac{\vec{x}}{2} - \frac{\sqrt{3}\vec{y}}{2})]. \qquad (4.7b)$$

Equation (4.7a) shows the factor (3/4) relating the strengths of the potentials in the baryon and meson cases. However, the harmonic oscillator model does not have the desired scaling property, as shown in Table 4.1. Its level splittings change with mass and are proportional to the square root of the potential strength for constant mass.

For the logarithmic potential (4.7b) the two internal degrees of freedom are not separable. The factor 3/4 does not appear explicitly in the Hamiltonian but can be seen from application of the virial theorem and simple sum rules. The "scale" of the energy spectrum can be defined by the quantity

$$\overline{E}(x) = \frac{\sum_i \langle 0|x|i\rangle\langle i|x|0\rangle\ (E_i - E_o)^2}{\sum_i \langle 0|x|i\rangle\langle i|x|0\rangle\ (E_i - E_o)} \;. \qquad (4.8)$$

This is the ratio of the mean square energy to the mean energy of the states excited from the ground state by the operator x, with a weighting factor proportional to the square of the transition matrix element. For the case where all the transition strength is dominated by a single excited state (as is exactly true for the harmonic oscillator) the quantity $\overline{E}$ is just the excitation energy of this state. For the case of the Hamiltonian (4.5) and the operators x and y, $\overline{E}$ can be evaluated exactly by using commutators with H and closure to give

$$\overline{E}(x) = \frac{\sum_i <0|[x,H]|i><i|[H,x]|0>}{\frac{1}{2} <0|[x,H]|i><i|x|0>-<0|x|i><i|[x,H]|0>} = \frac{\hbar^2<p_x>/m^2}{\hbar^2/2m}$$

$$= \frac{2<p_x^2>}{m} \quad , \qquad (4.8b)$$

and similarly for $\overline{E}(y)$.

The virial theorem gives for the general Hamiltonian (4.1)

$$<T> = (-3/32) \sum_{\alpha=1} \sum_{i>j} <r_{ij}(du_{ij}/dr_{ij})\lambda_i^\alpha\lambda_j^\alpha> \quad . \qquad (4.9a)$$

For the case where the space and colour factorize in this expression, the summation over colour can be evaluated explicitly using the trick of Eq. (2.3a) to give

$$<T> = (-3/16)(C-nc)<r_{ij}du_{ij}/dr_{ij}> = \frac{n}{4} <r_{ij}du_{ij}/dr_{ij}> \ , \qquad (4.9b)$$

where C = 0 for a colour singlet state and c = 4/3. Note that space and colour factorize trivially for the case of a logarithmic potential, where $r_{ij}du_{ij}/dr_{ij}$ is a c-number and is independent of i and j. For this case Eq. (4.9b) shows that the kinetic energy of any colour singlet state is proportional to the number of particles n and is the same for all colour singlet states with the same value of n. This tells us immediately that the mean kinetic energy of a baryon is 3/2 the mean kinetic energy of a meson. This can be seen explicitly for the case of the logarithmic baryon Hamiltonian (4.7b).

$$\frac{1}{2} \left[\frac{<p_x^2>}{m} + \frac{<p_y^2>}{m}\right] = \frac{3}{4} \left(\frac{1}{2} V\right) \quad . \qquad (4.9c)$$

Thus

$$\frac{1}{2} \left[\overline{E}(x) + \overline{E}(y)\right] = \frac{3}{4} V \quad . \qquad (4.9d)$$

For the meson case, with only one relative co-ordinate $\vec{r}$ and the log potential, E(r) = V. Thus Eqs. (4.9c) and (4.9d) show that the mean kinetic energy and the mean excitation energy for each degree of freedom is 3/4 of the value in the corresponding meson case.

Encouraged by the success of the scaling potential shown in Table 4.1, we attempt to extend this approach to the spin dependent part of the interaction u(r). We assume that u_{ij} contains a term u^s_{ij} proportional to $\vec{\sigma}_i \cdot \vec{\sigma}_j$ which can be treated as a perturbation. The spin splittings are therefore given by the expectation values of the spin dependent interaction in the eigenfunctions of the log potential. To determine the scaling properties of the interaction, we make the standard assumption[4,26] that it is a "colour-magnetic" interaction between two quarks and is therefore inversely proportional to the product of the quark masses.

$$u^s_{ij} = V^s \frac{\vec{\sigma}_i \cdot \vec{\sigma}_j}{m_i \cdot m_j} f(r) \quad , \tag{4.10a}$$

where V^s is the strength of the spin-dependent potential and includes the colour dependent factors in Eq. (4.1), and f(r) is some function of the radial distance between the quarks. If we are only interested in the scaling property, and assume that the spin-independent colour charge force is scale invariant; i.e. a log, then dimensional considerations force f(r) to have the form $(1/r^2)$ to compensate for the additional mass factors. For scaling properties this is equivalent to a factor p^2, where p is the relative momentum of the pair. The scaling properties of the spin-dependent interaction are therefore given by the expectation value of the modified interaction

$$w_{ij} = V^s \frac{\vec{\sigma}_i \cdot \vec{\sigma}_j}{(m\ + m_j)} \frac{p^2}{m_r} = V^s \frac{2\vec{\sigma}_i \cdot \vec{\sigma}_j}{m_i\ m_j} t_{ij} \quad , \tag{4.10b}$$

where $m_r = m_i \cdot m_j/(m_i + m_j)$ is the reduced mass of the pair and t_{ij} is the kinetic energy of the relative motion.

From the Hamiltonian (4.1), the ansatz (4.10b) for the spin-dependent part of the potential, and the assumption that hadrons containing strange quarks have an additional contribution to the mass proportional to the number of strange quarks, we obtain a simple unified formula for the description of meson and baryon masses,

$$M = A(n) + Bn_s + \sum_{i>j} K_{ij}\{[\langle u_{ij}\rangle/(n-1)] + [nK_{ij}\vec{\sigma}_i \cdot \vec{\sigma}_j \langle v_{ij}\rangle/2(m_i + m_j)]\} \quad , \tag{4.11a}$$

where the colour coupling factor

$$K_{ij} = (3/32) \sum_\alpha \lambda_i^\alpha \lambda_j^\alpha \quad , \tag{4.11b}$$

n is the total number of quarks and antiquarks in the system, n_s is the total number of strange quarks and antiquarks, λ_i^α are the λ-matrices for the colour SU(3) group for particle i, $<u_{ij}>$ and $<v_{ij}>$ are reduced matrix elements for the two-body spin independent and spin dependent interactions respectively, m_i is the mass of quark i, and A and B are parameters. The n-dependence of A is unknown. Thus we only relate mass splittings and not absolute mass values. The parameter B and the reduced matrix elements $<u_{ij}>$ and $<v_{ij}>$ are universal for all n. They may depend upon the angular momentum quantum numbers of particles i and j but not on spin, flavour or radial wave functions.

The spin dependent term (4.10b) is reduced to the form (4.11a) by using the virial theorem,

$$<t_{ij}> = <p_{ij}^2> (m_i+m_j)/2m_i m_j = nK_{ij}U/2 \quad , \tag{4.11c}$$

where t_{ij} and p_{ij} are the relative kinetic energy and the relative momentum of particles i and j and U is the strength of the log potential. The n-dependence of the overall scaling factor is needed to relate meson and baryon spin splittings, but specific flavour dependence of the quark-mass factors has no significant effect on our results.

For the meson and baryon cases, the colour factors can be evaluated explicitly to give the relations

$$M(\text{meson}) = A(\text{mes})+Bn_s+<u_{ij}>+\vec{\sigma}_i\cdot\vec{\sigma}_j<v_{ij}>/(m_i+m_j) \quad , \tag{4.12a}$$

$$M(\text{baryons}) = A(\text{bar})+Bn_s+ \sum_{i>j} [(1/4)<u_{ij}>+(3/8)\vec{\sigma}_i\cdot\vec{\sigma}_j<v_{ij}> /(m_i+m_j)] \tag{4.12b}$$

These differ from similar formulas in Ref. (4) by one essential new ingredient, the scaling factors (1/4) and (3/8) obtained from the log model which removes ambiguities of wave functions and matrix elements. These enable the quantitative predictions of baryon mass splittings from meson mass splittings (with no free parameters) displayed in Tables 4.1 and 4.2. The remarkable agreement shows that mesons and baryons do have a similar structure and should have a unified description, even though the simple nonrelativistic quark model may not be valid.

The most interesting prediction is eq. (1.15) discussed above which follows from very general grounds independent of the log potential and scaling factors. It holds in any model where all the flavour dependence appears in linear quark mass and spin-spin interaction terms proportional to n_s and $\vec{\sigma}_i \cdot \vec{\sigma}_j$ respectively. A $\vec{\sigma}_i \cdot \vec{\sigma}_j$ term does not contribute to the Λ-N mass difference because the interactions of the two strange-nonstrange pairs in the Λ cancel exactly,

$$\vec{\sigma}_u \cdot \vec{\sigma}_s + \vec{\sigma}_d \cdot \vec{\sigma}_s = (\vec{\sigma}_u + \vec{\sigma}_d) \cdot \vec{\sigma}_s = 0 \quad . \tag{4.13a}$$

We construct a linear combination of meson masses in which the contribution of the $\vec{\sigma}_i \cdot \vec{\sigma}_j$ term vanishes and obtain

$$M(\Lambda)-M(N) = (3/4)[M(K^*)-M(\rho)]+(1/4)[M(K)-M(\pi)] = 180 \text{ MeV} \tag{4.13b}$$

The result (4.13b) is in remarkable agreement with the experimental value of 177 MeV.

The remaining strangeness splitting predictions are dominated by this 180 MeV splitting due to the strange quark mass difference. Other model-dependent terms are down in the noise. Predictions for all baryons with strangeness 0 and -1 are given by combining Eq. (4.13) and spin splitting predictions listed in Table 4.2 and discussed below. The remaining three baryon masses must satisfy two general constraints from Ref. (9) and would be determined completely by the additional assumption of the equal-spacing rule for the decuplet. The specific model (4.1) used for Table 4.2 predicts insignificantly small deviations from the equal-spacing rule*.

The remaining predictions in Tables 4.1 and 4.2 depend upon the scaling factors (1/4) and (3/8) in Eq. (4.12b). The results for orbital excitations in Table 4.1 show reasonable agreement for the flavour-independent mass splittings predicted for the ρ, K* and ϕ systems. Surprising agreement is shown for the predicted baryon splittings which depend upon the scaling factor (1/4) in Eq. (4.12b) but not on the spin dependent term.

---

*Rubinstein[27] calculated strangeness splittings with the ad hoc assumption neglecting SU(3) breaking in the spin triplet state. This is now seen as a rough approximation for the $\vec{\sigma}_i \cdot \vec{\sigma}_j$ force whose triplet interaction is weaker than the singlet by a factor of three.

For the spin splittings the results in Table 4.2 arise from the scaling factor (3/8) in Eq. (4.12b) which gives the following new predictions:

$$[M(\Delta)-M(N)]/[M(\rho)-M(\pi)] = [M(\Sigma^*)-M(\Sigma)]/[M(K^*)-M(K)] = 9/16 \,, \tag{4.14a}$$

$$M(\Sigma)-M(\Lambda) = (3/8)\{[M(\rho)-M(\pi)] - [M(K^*)-M(K)]\} = 88 \text{ MeV} \,. \tag{4.14b}$$

The equality of the $\Sigma^* - \Sigma$ and $\Xi^* - \Xi$ splittings predicted in Ref. (9) holds for the most general two-body interaction. The 24 MeV discrepancy indicates the inherent error in the baryon sector in any model with only two-body interactions. Thus the 15-20% agreement of the new predictions (4.14) with experiment is as good as can be expected.

The same approach applied to charmed particles predicts the mass values 2305, 2435 and 2570 for the charmed baryons C_0, C_1 and C^* from the masses of the charmed mesons by the analogs of Eqs. (4.14a) and (4.14b). The spin splitting of the charmed-strange mesons $M(F^*) - M(F)$ is predicted by Eq. (4.12b) to be 122 MeV with the D^* and D masses used as input in addition to the ρ, π, K and K^*. This prediction is sensitive to the flavour dependent quark mass factor. If the sum of the quark masses is replaced by the product as in Ref. (4) the corresponding prediction is 87 MeV.

REFERENCES

1. S.W. Herb, et al., Phys. Rev. Lett. 39 (1977), 252; W.R. Innes, et al., Phys. Rev. Lett. 39 (1977), 1240.

2. C. Quigg and J.L. Rosner, Phys. Lett. 71B (1977), 153.

3. L. Schachinger, et al., Phys. Rev. Lett. 41 (1978), 1348.

4. A. De Rujula, H. Georgi and S.L. Glashow, Phys. Rev. D12 (1975), 147.

5. H.J. Lipkin, Phys. Rev. Lett. 41 (1978), 1629.

6. H.J. Lipkin, "Why Are Hyperons Interesting and Different from Nonstrange Baryons?", Particles and Fields 1975. Edited by H.J. Lubatti and P.M. Mockett, University of Washington (Seattle), p. 352.

7. O.W. Greenberg, Phys. Rev. Lett. 13 (1964), 598.

8. H.J. Lipkin, Phys. Lett. 74B (1978), 399.

9. P. Federman, H.R. Rubinstein and I. Talmi, Phys. Lett. 22 (1966), 208.

10. H.J. Lipkin and A. Tavkhelidze, Phys. Lett. 17 (1965), 331.

11. T. De Grand, R.L. Jaffe, K. Johnson and J. Kiskis, Phys. Rev. D12 (1975), 2060.

12. G. Altarelli, N. Cabibbo, L. Maiani and R. Petronzio, Nucl. Phys. B69 (1974), 531.

13. Harry J. Lipkin, Quarks, Partons, Triality, Exotics and Coloured Glue, in Proc. of Summer Inst. on Particle Physics, Stanford Linear Accelerator Center, 9-12 July 1973, Vol. I (National Technical Information Service, 1973), SLAC-167, pp. 239-253.

14. H.J. Lipkin, Erice Lectures 1977, to appear in the proceedings, presently available as Fermilab preprint Conf-77/93-THY.

15. O.W. Greenberg and C.A. Nelson, Physics Reports 3C (1977), 69.

16. H.J. Lipkin, Physics Reports 8C (1973), 173.

17. Y. Nambu, in Preludes in Theoretical Physics, eds. A. De-Shalit, H. Feshbach and L. Van Hove (North-Holland Publishing Co., Amsterdam, 1966), p. 133.

18. O.W. Greenberg and D. Zwanziger, Phys. Rev. 150 (1966), 1177.

19. H.J. Lipkin, Phys. Letters 45B (1973), 267.

20. H.J. Lipkin, Phys. Rev. Letters 28 (1972), 63 and Particle Physics (Irvine Conference, 1971), edited by M. Bander, G.L. Shaw, and D.Y. Wong (American Institute of Physics, New York, 1972), p. 30.

21. C. Quigg and J. Rosner, Comments on Nuclear and Particle Physics 8 (1978), 11; C. Quigg and Jonathan L. Rosner, Phys. Rev. D17 (1978), 2364.

22. T. Appelquist and H.D. Politzer, Phys. Rev. Lett. 34 (1975), 43; Phys. Rev. D12 (1975), 1404; E. Eichten, K. Gottfried, K. Lane and T.-M. Yan, Phys. Rev. D17 (1978), 3090.

23. E. Eichten and K. Gottfried, Phys. Lett. 66B (1977), 286.

24. For recent reviews see J.D. Jackson, in Proceedings of the 1977 European Conference on Particle Physics, edited by L. Jenik and I. Montvay (Central Research Institute for Physics, Budapest), p. 603; K. Gottfried, in Proceedings of the 1977 International Symposium on Lepton and Photon Interactions at High Energies, edited by F. Gutbrod (DESY, Hamburg), p. 667; T. Appelquist, R.M. Barnett, and K. Lane, SLAC-PUB-2100, to appear in Ann. Rev. Nucl. Sci., vol. 28.

25. H.J. Lipkin, in Physique Nucleaire, Les Houches 1968, eds. C. de Witt and V. Gillet (Gordon and Breach, New York, 1969).

26. R.J. Jaffe, Phys. Rev. D15, (1977), 267, 281.

27. H.R. Rubinstein, Phys. Lett. 22 (1966), 210.

PION FIELDS AND QUARKS IN NUCLEAR MATTER

Gordon Baym

Department of Physics, University of Illinois,

Urbana, Illinois 61801, U.S.A.

1. INTRODUCTION

While the basic description of nuclear matter is usually in terms of point nucleons interacting via two-body potentials*, other degrees of freedom can be important in determining the properties of the matter. Recently the effects of nucleon isobar, or Δ(1236), degrees of freedom on the effective nucleon-nucleon interaction in matter have been stressed (Green 1976, Holinde et al. 1977, 1978). In these lectures I would like to describe the roles of explicit pion and quark degrees of freedom in producing unusual states of nuclear matter, concentrating particularly on the possibilities of pion condensation, and formation of quark matter, in the density range from normal nuclear matter density, $n_0 \simeq 0.17$ fm^{-3}, to several orders of magnitude higher.

Briefly, the pion-condensed state of matter results when pion-like excitations become spontaneously created, and the pion field generated by the nucleons, which normally fluctuates about zero, develops a non-zero expectation value. Present calculations, described below, indicate the onset of pion condensation at a density $\sim$ 1-2 n_0. The quark matter phase occurs when the density of matter is sufficiently high that the cores of the nucleons become strongly overlapping, and the nucleons coalesce into a liquid formed of the quarks of which the nucleons were composed; in this phase the quarks are no longer confined to nucleons, but can run freely throughout the system.

*For recent reviews of nuclear matter theory see Day (1978) and Clark (1979).

Such unusual states of dense matter may occur in the dense cores of neutron stars, those highly compressed stellar objects of mass on the order of that of the sun and radius on the order of 10 km, which are observed in pulsars and compact X-ray sources. Effects of pion condensation or quark degrees of freedom may be observable in laboratory nuclei, in heavy ion reactions in which the nuclei are temporarily compressed as they collide, or possibly even in low-lying states.

2. PION CONDENSATION

Neutron star matter*, at densities $\gtrsim n_o$, is composed basically of degenerate neutron, proton and electron liquids filling Fermi seas, and in equilibrium under the beta reactions

$$n \to p + e^- + \bar{\nu} \quad , \qquad e^- + p \to n + \nu . \tag{2.1}$$

Neutrinos produced by these reactions escape from the matter, and consequently beta equilibrium requires that μ_n, μ_p and μ_e, the neutron, proton and electron chemical potentials, or energies of the particles at the top of the respective Fermi seas, obey

$$\mu_n = \mu_p + \mu_e \quad . \tag{2.2}$$

Charge neutrality implies that the proton and electron number densities obey

$$n_p = n_e \quad . \tag{2.3}$$

The nucleons are non-relativistic, while the electrons are relativistic. Thus (2.3) implies $\mu_p << \mu_e$, and to satisfy (2.2) only a few percent of the nucleons can be protons. By contrast, symmetric nuclear matter contains equally filled Fermi seas of neutrons and protons, and no electrons.

The question of pion condensation is whether pions, or more accurately pion-like excitations, spontaneously appear in the ground state of dense nuclear matter, through processes such as

$$n \to p + \pi^- \quad , \tag{2.4}$$

or

$$n \to n + \pi^0 \quad . \tag{2.5}$$

---

*See Baym and Pethick (1975, 1979) for general reviews of neutron star matter.

If we neglect the interactions of π^- with matter, then once $\mu_n - \mu_p = \mu_e > m_\pi$, process (2.4) would become favourable in neutron star matter, and the created pions would, because they are bosons, macroscopically occupy the lowest available mode, i.e., form a condensate, as in ordinary Bose-Einstein condensation.

One cannot ignore the interactions of the created pions with the background matter. The π^--n s-wave interaction is strongly repulsive, and this tends to inhibit formation of π^- in neutron rich matter. The pion-nucleon p-wave interactions, on the other hand, are strongly attractive, and as we shall see, can lower the self-energy of pions in matter sufficiently to enable the spontaneous appearance of both charged and neutral modes of the pion field.

2.1 The Pion Condensation Threshold

The actual mechanism of pion condensation turns out to be more subtle than a simple pion emission process such as $n \to p + \pi^-$. In order to understand the pion condensation thresholds, it is useful to study the spectrum of pionic excitations of nuclear matter*. These are determined from the pion Green's function, $D(k,\omega)$, whose inverse can be written as

$$D^{-1}(k,\omega) = \omega^2 - k^2 - m_\pi^2 - \Pi(k,\omega) \quad , \tag{2.6}$$

where ω and k are the pion frequency and wavenumber, m_π is the pion mass, and Π is the self-energy.

The modes of the pion field are solutions of $D^{-1}(k,\omega) = 0$, which for small Π are approximately of the form

$$\omega = \pm[\omega_k + \Pi(k,\omega_k)/2\omega_k] \quad , \tag{2.7}$$

where $\omega_k = (k^2+m_\pi^2)^{1/2}$. Thus we may identify $\Pi(k,\omega_k)/2\omega_k$ with the pionic optical potential

$$V_{opt}(k) = \Pi(k,\omega_k)/2\omega_k \tag{2.8}$$

in the medium.

The major contributions to the self-energy in nuclear matter are the s-wave, which for π^- has the form

---

*More detailed discussions of the pion condensation threshold may be found in Migdal (1978) and Baym (1977, 1978).

$$\Pi_s(k,\omega) \simeq \frac{\omega}{2f_\pi^2}(n_n - n_p) \quad , \tag{2.9}$$

where n_n and n_p are the neutron and proton densities, and f_π = 94.5 MeV is the pion decay constant. The s-wave interactions raise the energy of a π^- in neutron rich matter, while for π^0, and in symmetric nuclear matter, the s-wave interactions average to zero in a first approximation. In addition, as shown in Fig. 1, the p-wave coupling of pions to nucleon particle-hole states, Fig. 1a, and isobar particle (Δ)-nucleon hole states, Fig. 1b, are attractive and at higher density reduce the energy of a pion in the medium sufficiently to cause condensation. In Fig. 1 the wavy lines are pions, the solid lines nucleons, and the double line isobars. The shaded vertices in Figs. 1a and b represent renormalization of the πNN and πNΔ vertices by Landau Fermi-liquid effects; for low energy pions this renormalization is essentially the Ericson-Ericson Lorentz-Lorenz effect.

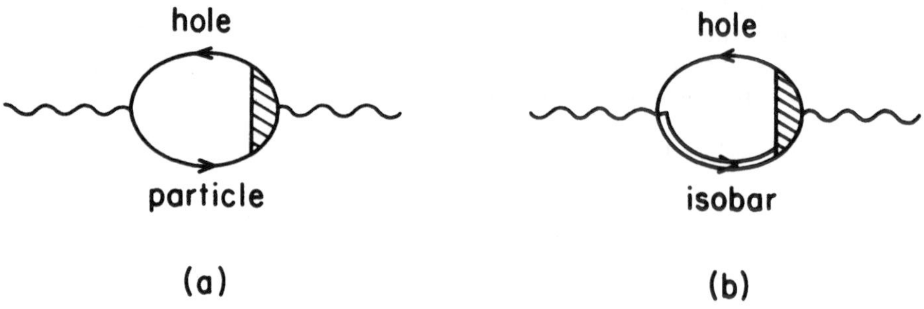

Fig. 1

In the absence of this effect the p-wave self-energy has the approximate form, for a π^- in matter:

$$\Pi_p^0 = \Pi_p + \Pi_{res} \tag{2.10}$$

where

$$\Pi_p \simeq \frac{2f^2k^2}{m_\pi^2}\frac{(n_p - n_n)}{\omega} \tag{2.11}$$

and

$$\Pi_{res} \simeq \frac{2g_\Delta^2 k^2}{9m_\pi^2}\left[\frac{3n_n+n_p}{\omega-\Delta} - \frac{3n_p+n_n}{\omega+\Delta}\right] \tag{2.12}$$

are the self-energies corresponding to Fig. 1a and 1b respectively with bare vertices; $f(\simeq\sqrt{4\pi(0.081)})$ is the πNN, and $g_\Delta \simeq 2f$, the $\pi N\Delta$ coupling constant; Δ is the isobar-nucleon mass difference, $\sim$ 300 MeV, and we have neglected nucleon and isobar kinetic energies in (2.11) and (2.12).

The renormalization of the vertices by the effective particle-hole interaction corresponds to repeated scattering of particle-hole excitations, and modifies the p-wave self-energy from (2.10) to

$$\Pi_P = \frac{\Pi_P^0}{1-g'(m_\pi/fk)^2\Pi_P^0} \quad . \tag{2.13}$$

The quantity g' is the spin and isospin anti-symmetric Landau parameter for the short range nuclear forces coupling particle-hole states. The repeated particle-hole scattering due to these interactions raises the energies of the particle-hole states, decreasing the effects of the p-wave coupling on the pion self-energy and thus tending to inhibit condensation. To a good approximation, g' can be taken to be a density and momentum independent constant. In (2.13) we have made the reasonable assumption that the coupling of isobar particle-nucleon hole states in the pion channel to other isobar particle-nucleon hole states, as well as to nucleon particle-hole states, is described by the same Landau parameter g' that couples nucleon particle-hole states.

The standard Ericson-Ericson contribution to g', from suppression of the attractive delta function part of the one pion exchange potential by short range nuclear correlations, corresponds to $g' = (1/3)(f/m_\pi)^2$. The finite range of the pion-nucleon vertex reduces this contribution. The most important contribution to g' most likely arises from exchange between correlated nucleons of ρ mesons in the one pion channel (Baym and Brown, 1975). The most complete estimates of g' are given by Brown et al. (1977); they are quite sensitive to the description of the short range correlations as well as to the empirically deduced $\pi\pi \to N\bar{N}$ helicity amplitudes assumed as input. These estimates lie in the range $g' \sim (0.6\text{-}1.0)(f/m_\pi)^2$. Migdal (1978) has estimated g' from an analysis of nuclear magnetic moments and finds it to be $\sim 0.5\ (f/m_\pi)^2$.

In nuclear matter the excited states with the quantum numbers of the charged pion relative to the ground state are found from the poles of the negative pion Green's function; those with positive residue have the quantum numbers of the π^-, while those with negative residue have the quantum numbers of the π^+ (and occur at minus the excitation energies). These states include the branch of "free" π^+ and π^- excitations, modified somewhat by the matter, and the continuum of proton particle-neutron hole ($p\bar{n}$) states with the quantum numbers of the π^+; there is in addition a continuum of neutron particle-proton hole ($n\bar{p}$) states with the quantum numbers of the π^-, and a continuum of isobar particle-nucleon hole states. Furthermore, as Migdal showed, due to p-wave interactions which couple the "free" pion states to the particle-hole states, neutron rich matter may develop a neutron hole-proton particle 0^-, T=1 collective mode or spin-isospin zero sound, denoted by π_s^+, with the quantum numbers of the π^+. This mode, analogous to a giant dipole mode, is an oscillation of the T_1 and T_2 components of the isospin density, together with a counteroscillation of the components of spin density parallel and antiparallel to the wave vector. In isospin space the mode corresponds to an oscillation of the directions of the nucleon state vectors (which for pure neutrons point along the negative T_3 axis) about the negative T_3 axis. Negative pion condensation corresponds to this mode becoming "soft". From Eqs. (2.6), (2.9) and (2.13) one finds that at the condensation threshold in neutron matter, $n \sim 2n_o$, the π_s^+ mode is $\sim$ 25 MeV below the neutron hole-proton particle continuum.

The states of the system with the quantum numbers of the π^o include the "free" π^o branch, the particle-hole ($n\bar{n}$ and $p\bar{p}$) and the isobar particle-nucleon hole continua; in addition for certain k the matter may develop a neutral 0^-, T=1 particle-hole collective mode, π_s^o, outside the particle-hole continuum, corresponding to $D_o^{-1}(k,\omega) = 0$, where D_o is the neutral pion Green's function. Where the branch overlaps the particle-hole continuum it is damped [Re $D_o^{-1}(k,\omega) = 0$ but Im $D_o^{-1}(k,\omega) \neq 0$].

The effects of the attractive p-wave pion-nucleon interaction on the "free" pion branch may lead to detectable pion bound states in nuclei, as pointed out by Ericson and Myhrer (1978) (and described in this school). These could occur if the "free" pion spectrum in matter falls in energy below m_π [as for small k in symmetric nuclear matter at n_o, when $g' < 0.15(f/m_\pi)^2$].

Charged pion-like excitations can appear in the ground state via several mechanisms. The first is for a neutron in the system to turn into a proton plus a "free" π^- excitation, $n \rightarrow p + \pi^-$. This process can be looked at as the spontaneous creation of a π^- and a neutron hole-proton particle excitation. A second possibility, which is the one that occurs in detailed calculations, is that the minimum of the π_s^+ spectrum falls below the $p\bar{n}$ continuum sufficiently that it becomes energetically favourable for the system spontaneously to create pairs of π^- and π_s^+:

$$(\ ) \to \pi^- + \pi_s^+ \quad .$$

In matter with protons intially present, condensation can also occur through the mechanism $p \to n + \pi_s^+$. This process becomes energetically possible when the minimum of the π_s^+ spectrum falls as low as the bottom of the $p\bar{n}$ continuum. The existence of this additional mechanism for condensation means that neutron matter in beta equilibrium can condense at a lower density than pure neutron matter. [However, realistic calculations with inclusion of short range correlations between nucleons (Brown and Weise, 1976) indicate that neutron matter in beta equilibrium condenses at a density not very different from that for pure neutron matter.] One finds that with $g' \simeq 0.5(f/m_\pi)^2$ and (2.13) that condensation occurs at $n \simeq 2n_0$, with critical frequency $\mu_n - \mu_p \simeq 150$ MeV and the critical wavevector $k_c \simeq 2.0\ \text{fm}^{-1}$. Migdal in his original work (1973) did not include the Fermi liquid corrections to the isobar particle-nucleon hole states--thus he takes

$$\Pi_p = \frac{\Pi_p}{1-g'(m_\pi/fk)^2\Pi_p} + \Pi_{res} \tag{2.14}$$

--and consequently found a somewhat lower π^- condensation threshold.

Neutral pion condensation occurs when a mode with the quantum numbers of the π^0 in the normal system falls to zero frequency for some wavenumber. Such a mode may possibly be one that is damped in the normal state, such as the π_s^0 branch within the particle-hole continuum, since the damping disappears as the frequency goes to zero. In the condensed state the mode spontaneously becomes statically excited. Theoretical estimates of the π^0 condensation threshold are in the range from somewhat below (Migdal, 1978) to at least twice nuclear matter density (see Brown and Weise, 1976 and Pirner et al., 1978), and thus it is still somewhat uncertain whether pion condensation is possible in ordinary nuclei. The theoretical estimates are uncertain not only because the Fermi liquid vertex corrections (g') are not well pinned down, but also because of uncertainty in the off-shell extrapolation of the s-wave pion-nucleon interactions (see Baym and Campbell, 1979, for further discussion of this point).

Experimental tests for pion condensation and associated precursor phenomena, such as "critical opalescence" (M. Ericson and Delorme, 1978, and discussed by T. Ericson in this school), in laboratory nuclei are reviewed in detail by Migdal (1978); see also Toki and Weise (1979). While not ruling out condensation in laboratory nuclei, such tests have so far not provided strong evidence for its existence; the search for possible effects of condensation remains an interesting problem.

Instabilities leading to pion condensation may play an important role in heavy ion collisions. As Ruck et al. (1976) pointed out, the softening of pion-like excitations leading to instabilities at threshold (at the increased densities in collisions) can enhance the dissipation, and hence increase the possibility of shock formation. The dynamics of this process has been discussed in further detail by Gyulassy (1977, 1978).

2.2 The Pion Condensed State

Because pions are bosons, if pion-like excitations appear in the ground state they will form a condensate. Such a state corresponds to a classical or coherent excitation of the pion field in the medium, or alternatively to an "off-diagonal long-range order" which can be described by a non-vanishing of the ground state expectation value of the charged pion field $\langle\pi(\mathbf{r})\rangle$. Similarly if neutral pion-like excitations appear in the ground state, and thus the system is in a neutral pion condensed phase, the neutral pion field $\pi^0(\mathbf{r})$ has a non-vanishing expectation value $\langle\pi^0(\mathbf{r})\rangle$. The expectation values $\langle\pi(\mathbf{r})\rangle$ and $\langle\pi^0(\mathbf{r})\rangle$ serve as the order parameters of the charged and neutral pion condensed phases. Such phases are states of broken symmetry. Because the pion field is pseudoscalar, its expectation value in uniform matter must vanish in a parity eigenstate; furthermore $\langle\pi(\mathbf{r})\rangle$ vanishes in the normal non-condensed state from charge conservation as well.

Since the attractive π-N p-wave interactions, which are proportional to $\nabla\pi$, make the appearance of pion-like excitations favourable in dense matter, in order that the condensed field have a p-wave interaction with the nucleons, $\nabla\langle\pi(\mathbf{r})\rangle$ must be non-zero. The simplest possibility is a running wave solution

$$\langle\pi(\mathbf{r})\rangle = e^{i\mathbf{k}\cdot\mathbf{r}}\langle\pi\rangle \quad , \tag{2.15}$$

corresponding to pion condensation in the single mode $\mathbf{k}$ (although it may well be possible that a more complicated spatial structure is more energetically favourable).

The charged pion condensed phase is superconducting. Since the p-wave source, $\sim \nabla\cdot\langle\psi_p^\dagger\boldsymbol{\sigma}\psi_n\rangle$, of the condensed charged pion field is non-zero, one has effective pairing of proton particles with neutron holes (and vice versa) in a T=1 state, analogous to ordinary BCS pairing.

The order parameter $\langle\pi^0\rangle$ of neutral pion condensed matter is real, and therefore this phase (in three dimensions) is not superfluid. Because the p-wave source of the π^0 field is $\sim \nabla \cdot \langle\psi_N^\dagger \tau_3 \sigma \psi_N\rangle$, the magnitude of $\langle\pi^0\rangle$ must vary in space, e.g., $\sim \cos \mathbf{k}\cdot\mathbf{r}$; thus this phase contains a spatially non-uniform spin ordering. [The charged pion-condensed phase with a running wave condensate is, by contrast, spatially uniform.] The violation of parity in a π^0 condensed phase is just that associated with the lack of translational invariance, since a spatially non-uniform state cannot be reflection invariant about all points. One can think of the π^0 condensed phase as having pairing of nucleon particles and holes. One interesting possibility is that the spatial ordering in the π^0 condensed phase leads to an actual solidification of the matter (Pandharipande and Smith, 1975, Takatsuka et al. 1978).

A complete description of the pion condensed state should take into account the basic s and p-wave pion nucleon interactions, π-π interactions, which limit eventually the strength of the condensed field, the effects of coupling to isobars, which favour condensation, and the important effects of nuclear correlations--the Ericson-Ericson effect--which tend to inhibit condensation. The relevant frequencies and wave vectors of condensed pion fields are relatively low, measured on a nucleon scale, and to a good approximation the low energy s and p wave interactions of the condensed field with nucleons and the π-π interactions of the condensed field with itself are determined in structure by the approximate SU(2) × SU(2) chiral invariance of low energy pion-nucleon physics. [The generators of the SU(2) × SU(2) group are constructed from the conserved vector current and the partially conserved axial current.]

One of the simplest realizations of the underlying chiral symmetry is the σ-model. By using it in a simple approximation one can get a rather good description of the charged pion condensed phase which includes the low energy pion interactions to all orders in the condensed field, and the Ericson-Ericson effect; the extension to include isobars is straightforward, though somewhat more algebraically complicated. We now summarize this calculation. [For fuller details see Baym 1977, 1978.]

The σ-model is specified by the Lagrangian

$$\mathcal{L}_\sigma = -\frac{1}{2}(\partial_\mu\sigma\,\partial^\mu\sigma + \partial_\mu\boldsymbol{\pi}\cdot\partial^\mu\boldsymbol{\pi}) - \frac{\lambda}{4}(\sigma^2+\boldsymbol{\pi}^2-\sigma_o^2)^2 +$$

$$+ \bar{\psi}_N i\gamma^\mu \partial_\mu \psi_N - g\bar{\psi}_N(\sigma + i\gamma_5 \underset{\sim}{\tau}\cdot\underset{\sim}{\pi})\psi_N + f_\pi m_\pi^2 \sigma \qquad (2.16)$$

where $\underset{\sim}{\pi}$ is the three component pion field, σ is a scalar field, and σ_0 is a constant $\simeq f_\pi$; ψ_N is the nucleon field and $g/2m_n = f/m_\pi$. We do not include isobars at this point.

For simplicity, let us assume that λ, and hence m_σ, tend to infinity. The σ and π fields are thus constrained to the 4-sphere $\sigma^2 + \underset{\sim}{\pi}^2 = f_\pi^2$. We treat the condensed pion field as well as the σ field as classical fields. In the normal ground state of a system of nucleons, $\langle\sigma\rangle \simeq \sigma_0$ and $\langle\underset{\sim}{\pi}\rangle = 0$. The charged pion condensed state has a non-vanishing expectation value of π as well as σ; we write these as

$$\langle\pi(\underset{\sim}{r})\rangle = \frac{1}{\sqrt{2}}\langle\pi_1 + i\pi_2\rangle = \frac{f_\pi}{\sqrt{2}} \sin\theta\, e^{i\underset{\sim}{k}\cdot\underset{\sim}{r}} \qquad (2.17)$$

$$\langle\sigma(\underset{\sim}{r})\rangle = f_\pi \cos\theta \quad ,$$

and $\langle\pi^0\rangle = 0$, so that $\langle\sigma\rangle^2 + \langle\underset{\sim}{\pi}\rangle^2 = f_\pi^2$. The angle θ measures the rotation of the expectation value of the vector $\langle(\sigma,\sqrt{2}\pi)\rangle$ away from the σ axis, as in Fig. 2. In a neutral pion condensed state $\langle\pi^0\rangle \neq 0$, $\langle\sigma\rangle \neq 0$, and $\langle\pi\rangle = 0$.

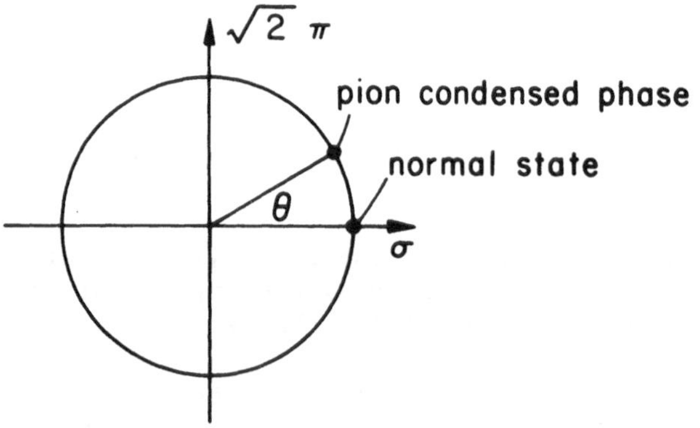

Fig. 2

After some manipulation one finds from (2.16) and (2.17) that the effective Hamiltonian $\hat{\mathcal{H}} = \mathcal{H} + \mu_\pi \rho_q$ (where ρ_q is the net charge density and $\mu_\pi = \mu_n - \mu_p$ is the pion chemical potential) describing the pion-condensed state becomes

$$\hat{\mathcal{H}} = \bar{\psi}[-i\underset{\sim}{\gamma}\cdot\underset{\sim}{\nabla} + m_n + \frac{1}{2}\mu_\pi\gamma^0$$

$$-\frac{1}{2}(\underset{\sim}{k}\cdot\underset{\sim}{\gamma}-\mu_\pi\gamma^0)(\cos\theta\tau_3 + g_A\sin\theta\ \gamma_5\tau_2)]\psi + \hat{E}_\pi \quad (2.18)$$

or non-relativistically

$$\mathcal{H} = \psi^\dagger[m_n + \frac{(i\underset{\sim}{\nabla} + \frac{1}{2}\underset{\sim}{k}\tau_3\cos\theta)^2}{2m_n} + \frac{1}{2}\mu_\pi(1+\tau_3\cos\theta + \frac{\mu_\pi}{4m_n}g_A^2\sin\theta)$$

$$-\frac{1}{2}g_A\sin\theta\ \underset{\sim}{\sigma}\cdot(\underset{\sim}{k} - \frac{\mu_\pi}{m_n}\underset{\sim}{p})\tau_2]\psi + \hat{E}_\pi \quad (2.19)$$

where the axial current renormalization constant g_A is taken from the Goldberger-Treiman relation to be $gf_\pi/m_n = 1.36$, and

$$\hat{E}_\pi = f_\pi^2[\frac{1}{2}(k^2-\mu_\pi^2)\sin^2\theta - m_\pi^2\cos\theta] \quad . \quad (2.20)$$

The terms in (2.18) and (2.19) $\propto \tau_3(\cos\theta-1)$ describe π-N s-wave interactions. To order θ^2, the s and p wave interactions included in (2.19) are equivalent to those we considered earlier in the normal state (for $g' = 0$). All condensed π-π interactions are automatically included as well in (2.19).

The ground state of $\hat{\mathcal{H}}$ is found by first diagonalizing the nucleon terms. The single nucleon eigenstates are not eigenstates of charge but are the linear combinations

$$|+\rangle = \cos\frac{\phi}{2}|p\rangle - i\sin\frac{\phi}{2}|n\rangle$$

$$|-\rangle = \cos\frac{\phi}{2}|n\rangle - i\sin\frac{\phi}{2}|p\rangle \quad (2.21)$$

of neutron states $|n\rangle$ of momentum $\underset{\sim}{p} + \underset{\sim}{k}/2$ and proton states $|p\rangle$ of momentum $\underset{\sim}{p} - \underset{\sim}{k}/2$, where, for $k,\mu_\pi \ll m_n$,

$$\tan\phi \simeq (g_A k\sigma/\mu_\pi)\tan\theta \quad , \quad (2.22)$$

and σ is the eigenvalue of $\underset{\sim}{\sigma}\cdot\underset{\sim}{\hat{k}}$.

Prior to condensation the mode π_s^+ corresponds to a small oscillation of the nucleon eigenstates about the T_3 axis in isospin space. At the condensation threshold, the frequency of this mode plus that of the π^- of momentum $\underset{\sim}{k}$ goes to zero; the mode becomes soft, and the T_3 axis is no longer a minimum energy direction, but rather a maximum. The minimum direction is shifted away from the axis, by the angle ϕ.

The ground state is constructed by filling all single nucleon states with energies $\epsilon_p \leq \mu_n$, the chemical potential for baryon number, which is determined by the requirement that the number of filled states per unit volume equals the prescribed baryon density n. For all densities of interest in pure neutron matter only the neutron-like branch $|->$ is occupied, and thus the ground state contains two (for spin) filled Fermi seas of $|->$ states. μ_π and k in the ground state are determined by extremizing the ground state eigenvalue of $\hat{\mathcal{H}}$ with respect to these quantities. After some algebra one finds that the ground state energy of neutron matter is given by

$$E(n,\theta) = E_o(n) - \frac{n^2}{8f_\pi^2}\frac{g_A^2(g_A^2-1)\sin^2\theta}{1+(g_A^2-1)\sin^2\theta} + f_\pi^2 m_\pi^2(1-\cos\theta) \quad , \qquad (2.23)$$

where $E_o(n)$ is the normal ground state energy. The terms involving θ are the condensation energy; for $n > n_c$, where

$$n_c = \frac{2f_\pi^2 m_\pi}{g_A\sqrt{g_A^2-1}} = \frac{0.32\ \mathrm{fm}^{-3}}{g_A\sqrt{g_A^2-1}} \qquad (2.24)$$

is the critical condensation density, the energy of the system is lowered by having a condensate present. Note that n_c is finite only if $g_A > 1$. The equilibrium θ, found by minimizing E with respect to θ, is a monotonically increasing function of density, approaching $\pi/2$ as $n \to \infty$. For $n \gtrsim n_c$, the energy of the pion condensed state has the form

$$E_{cond}(n) = E_o(n) - \frac{2g_A\sqrt{g_A^2-1}}{4g_A^2 - 1} n_c m_\pi (\frac{n}{n_c} - 1)^2 + \ldots \qquad (2.25)$$

The phase transition to the condensed state is second order. [However as Dyugaev (1975) has shown proper inclusion of field fluctuations always makes the transition weakly first order.] See also Migdal (1978) and Friman and Nyman (1978).

The Ericson-Ericson correction may be easily added to this calculation. The result is that the energy and critical density are again given by (2.23-2.25), only with the replacement $g_A \to g_A^* = g_A\sqrt{1-\gamma}$, where $\gamma \equiv (m_\pi/f)^2 g'$. Condensation is possible in this model only for $g_A^* > 1$, or $\gamma < 0.46$. Present estimates of γ are at least this value, and thus were all the relevant physics included in this model one would conclude that pion condensation could not take place in neutron stars. However we have so far not considered coupling to isobars, an effect favouring condensation, and essentially cancelling the repulsive effects of g'.

The effects of Δ's can be included by considering them as additional "elementary" particles which can be present in the pion condensed state. The basic argument to determine the fully renormalized coupling of the isobars is the PCAC result that, at low energies, the pion field is the divergence of the axial current of the matter. Note that in Eq. (2.18) the operators $\frac{1}{2}\bar{\psi}\gamma^\mu\tau_3\psi$ and $g_A(\frac{1}{2}\bar{\psi}\gamma^\mu\gamma_5\tau_2\psi)$ are essentially the nucleon vector current $V_{\mu,3}$ and axial current $A_{\mu,2}$. This expression of the p-wave coupling in terms of the axial current, and the s-wave in terms of the vector current density is actually a general consequence of the underlying chiral invariance. [See Campbell et al. (1975)].

The non-relativistic Hamiltonian, for $\mu_\pi, k \ll m_n$, can be written as $\hat{\mathcal{H}} = \hat{\mathcal{H}}_N + \hat{E}_\pi$ where

$$\hat{\mathcal{H}}_N = \psi^\dagger(m + \frac{p^2}{2m})\psi + \mu_\pi\cos\theta\, V_{0,3} - k\sin\theta\, A_{z,2} + \ldots \quad (2.26)$$

To include isobars we then construct the full 20 x 20 matrix formed from (2.26) evaluated between the 4 different spin and isospin states of the nucleon and the 16 different spin and isospin states of the isobar. The matrix elements of $A_{z,2}$ can be computed from the SU(4) quark model. The matrix elements of the vector current density are diagonal and simply equal to the eigenvalue of T_3. The free particle energy term is diagonal, and becomes $m_\Delta + p^2/2m_\Delta$ between isobar states, where m_Δ is the mass of the isobar.

The full 20 x 20 matrix can be block diagonalized into 4 x 4, 6 x 6, 6 x 6 , and 4 x 4 blocks, with each block having the same S_z value. At densities near threshold, the $S_z = \pm 3/2$ states have too high an energy, $\sim m_\Delta - m_n \equiv \Delta$, to be present in the ground state. To find the occupied eigenstates we need then to diagonalize only the 6 x 6 matrix $\mathcal{H}_N^*$ (in the basis Δ^{++}, Δ^+, p, n, Δ^0, Δ^-) corresponding to $S_z = 1/2$ or $-1/2$. The explicit matrix is too lengthy to write out here. The general scheme is to compute the eigenvalues of $\mathcal{H}_N^*$ and fill all states (with eigenvalues $\leq \mu_n$) until the requisite baryon density is present. $\mathcal{H}_N^*$ cannot be diagonalized analytically for general θ. However Weise and Brown (1975) have carried out a numerical diagonalization of $\mathcal{H}_N^*$, and conclude that the effects of isobars can be described approximately by letting $g_A \to g_A^{**}$ in expression (2.23) for the energy, where $g_A^{**2} = (1+S)(1-\gamma)g_A^2$, and the isobar enhancement factor S is ~ 0.8 over a wide range of densities. For $\gamma = 0.5$ in neutron matter, a first order phase transition occurs at a critical density $1.8\ n_0$, which is below the critical density $\simeq 2\ n_0$ for a second order transition. Further details, and calculations of the neutron matter equation of state can be found in Brown and Weise (1976), and Au (1976).

Calculations based on chiral symmetry similar to that outlined above have been carried out recently by Pirner et al. (1978) for the pion condensed state in symmetric nuclear matter, and by Dautry and Nyman (1979) for π^0 condensation in neutron matter. The calculations of Migdal and collaborators on the properties of the condensed state are reviewed by Migdal (1978).

3. QUARK MATTER

The picture that quarks are the basic constituents of strongly interacting elementary particles suggests that a more fundamental description of matter at very high densities is in terms of quarks. In particular, one expects that when matter is sufficiently compressed, the nucleons will merge together and undergo a phase transition to quark matter, a degenerate Fermi liquid, in which the basic constituents are the quarks that composed the original nucleons. In addition to its possible occurrence in neutron stars and production in heavy ion collisions, quark matter is of interest in the description of the early universe when the baryon density greatly exceeded that of nuclear matter (Chapline, 1976). One can very roughly estimate the density at which the transition to quark matter might occur by noting that nucleons begin to touch at a particle density $\sim (4\pi r_N^3/3)^{-1}$, where r_N is an effective nucleon radius; for r_N between 1 and 0.3 fm, this density is between $1.4\ n_0$ and $50\ n_0$.

In the basic quark model (see the lectures of Lipkin in this school for a fuller description), the quarks are spin 1/2 fermions, of baryon number 1/3, which come in at least four "flavours", u, d, s and c (up, down, strange, and charmed). The electrical charges of these four flavours are 2/3, -1/3, -1/3, and 2/3 respectively; all have strangeness zero, except s which has strangeness -1. For each quark q there exists a corresponding anti-quark $\bar{q}$ with opposite quantum numbers. Mesons are composed of a quark and anti-quark, and baryons (of unit baryon number) from three quarks. For example, a proton is a <u>uud</u> bound state, a neutron is <u>udd</u>, and a Λ is <u>uds</u>.

In addition quarks have an internal degree of freedom, colour, which we may take to have the three "values", red, blue and green.

Colour was originally introduced to enable quarks to obey the Pauli principle. In the fundamental model of quark-quark interactions--quantum chromodynamics (or nonabelian Yang-Mills SU(3) gauge theory), in which coloured quarks interact via exchange of eight massless vector gluons (analogues of photons in ordinary electrodynamics)--colour functions effectively as a charge for gluon interactions. Loosely speaking, two quarks of the same colour "repel", while two quarks of different colour "attract" with half the strength. Thus a combination of three quarks each of different colour (more correctly, a colour singlet) acts as a neutral object, producing no long range gluon "Coulomb" field.

The u and d quarks are believed, in order for chiral symmetry to be so nearly exact, to have a fairly small mass, m_u, m_d $\sim$ 10 MeV; the strange quark is heavier, with m_s perhaps 100-300 MeV, while the charmed quark is much heavier (m_c > 1 GeV). Because of its high mass the charmed quark is not expected to be present in quark matter that could occur in neutron stars, or be produced in normal heavy ion collisions.

The transition from nucleon matter to quark matter must occur via a sharp phase transition. One sees this as follows. In a gas of isolated nucleons the quarks are confined and thus such matter cannot have stationary colour currents, just as there can be no stationary electrical currents in a gas of un-ionized atoms. Thus the (dc) colour conductivity, the colour analogue of electrical conductivity, must be identically zero in a low density nucleon phase. However, the uniform quark phase can have stationary colour currents, and hence it has a non-vanishing colour conductivity. Since the colour conductivity changes non-analytically between the two phases, the transition cannot occur smoothly.

3.1 Free Quark Matter

The quark-gluon theory has the remarkable property of being asymptotically free, that is, quark interactions at sufficiently large momenta or short distances become arbitrarily weak. Furthermore in quark matter that is in a colour singlet (or colour neutral) state, the interactions between quarks at distances large compared to the interparticle spacing will be screened out, analogous to screening of long range Coulomb fields in a plasma in equilibrium. Thus, as Collins and Perry (1975) pointed out, at high densities the net quark interactions in quark matter in an overall colour singlet state should be sufficiently weak that the matter can to a first approximation be taken as a non-interacting relativistic Fermi gas. Let us discuss the properties of such matter. Imagine that we start with a gas of nucleons of proton fraction $x = n_p/(n_n+n_p)$, compress it into quark matter, and ignore the interactions between the quarks. Since the neutron is composed of two d and one u quark, and the proton two u and one d, the densities of up and down quarks in the matter obey

$$\frac{n_u}{n_d} = \frac{1+x}{2-x} \quad , \tag{3.1}$$

while the baryon density n_b equals $n_u/(1+x)$. Because quarks are fermions such matter will consist of <u>twelve</u> Fermi seas, two for u and d, times two for spin, times three for colour, e.g., a d-flavour, up-spin, blue sea, etc. Since hadrons are colour singlets, all three colours will be present in quark matter equally; thus in the absence of spin polarization, all six u Fermi seas will have the same Fermi momentum, p_u, as will all six d Fermi seas, p_d. In terms of p_u and p_d, $n_u = p_u^3/\pi^2$, $n_d = p_d^3/\pi^2$. If we assume p_u, $p_d >> m_u$, i.e., fully relativistic quarks [note that at $n_b = n_o$, $p_u = 234(1+x)^{1/3}$ MeV], then the energy density of the matter is

$$\begin{aligned} E_o &= \frac{3}{4}(n_u p_u + n_d p_d) \\ &= \frac{3}{4}\pi^{2/3}[(1+x)^{4/3} + (2-x)^{4/3}]n_b^{4/3} \quad . \end{aligned} \tag{3.2}$$

A second example of quark matter is that in complete beta equilibrium. Equilibrium under the weak interactions

$$d \to u + \ell + \bar{\nu} \quad , \quad u + \ell \to d + \nu \tag{3.3}$$

and

$$s \to u + \ell + \bar{\nu} \quad , \quad u + \ell \to s + \nu \quad , \tag{3.4}$$

where ℓ is an electron or (negative) muon, and ν the corresponding neutrino, implies that the chemical potentials of the species present in the matter obey

$$\mu_d = \mu_u + \mu_e = \mu_u + \mu_\mu = \mu_s \quad , \tag{3.5}$$

(where we have assumed the neutrinos not to be retained in the matter and have set $\mu_\nu = 0$). Furthermore electrical charge neutrality requires

$$\frac{2}{3}n_u = \frac{1}{3}(n_d + n_s) + n_e + n_\mu \quad . \tag{3.6}$$

If the Fermi momenta are sufficiently high, compared with the quark and lepton rest masses, that all the particles present are fully relativistic, and interactions are neglected, then no leptons are present in matter in beta equilibrium, and $n_u=n_d=n_s=n_b$, $n_e=n_\mu=0$. Such quark matter is equivalent to that resulting from compressing a gas of pure Λ particles. Its energy density is given by

$$E_o = \frac{3}{4}(n_u p_u + n_d p_d + n_s p_s) = \frac{9\pi^{2/3}}{4} n_b^{4/3} \quad . \tag{3.7}$$

One can see from Eq. (3.7) and from Fig. 3 (line labelled "free quarks") that in free quark matter resulting from symmetric nuclear matter, the quark kinetic energy per baryon is less than the nucleon rest mass for

$$n_b < (2/3)^7 m_n^3/\pi^2 = 0.64 \text{ fm}^{-3} \quad ;$$

for Λ-like quark matter this occurs for $n_b < 0.96$ fm^{-3}. Ordinary matter would thus be unstable under collapse into free quark matter; to have stability it is essential then to include the effects of interactions between the quarks. The reason for the instability is basically that the quarks have so many more internal degrees of freedom that the mean zero point energy is lower than in the nucleon phase.

3.2 Inclusion of Interactions

The fundamental model of the quark-quark interactions is the non-abelian Yang-Mills SU(3) gauge theory - or quantum chromodynamics - in which coloured quarks interact via exchange of eight massless vector gluons. This theory is described by the Lagrangian

$$\mathcal{L} = \bar{q}_{ai}[\gamma^\mu(i\delta_{ab}\partial_\mu + \frac{1}{2} g\lambda^\alpha_{ab}A^\alpha_\mu) - m_i\delta_{ab}]q_{bi} - \frac{1}{4} F^\alpha_{\mu\nu}F^{\mu\nu}_\alpha , \tag{3.8}$$

where q_{ai} is the quark field, i the flavour and a the colour index; A^α_μ is the gluon field, where $\alpha = 1, \ldots , 8$; g is the coupling constant, m_i the quark mass,

$$F^\alpha_{\mu\nu} = \partial_\mu A^\alpha_\nu - \partial_\nu A^\alpha_\mu + gf_{\alpha\beta\gamma}A^\beta_\mu A^\gamma_\nu \quad ;$$

the λ^α are the eight generators of the SU(3) colour gauge group, normalized so that tr $\lambda^\alpha\lambda^\beta = 2\delta_{\alpha\beta}$; and the f's are the structure constants of SU(3), defined by $[\lambda^\alpha,\lambda^\beta] = 2if_{\alpha\beta\gamma}\lambda^\gamma$. The gluons themselves carry colour and hence can emit, absorb and scatter gluons via the terms in FF of order A^3 and A^4.

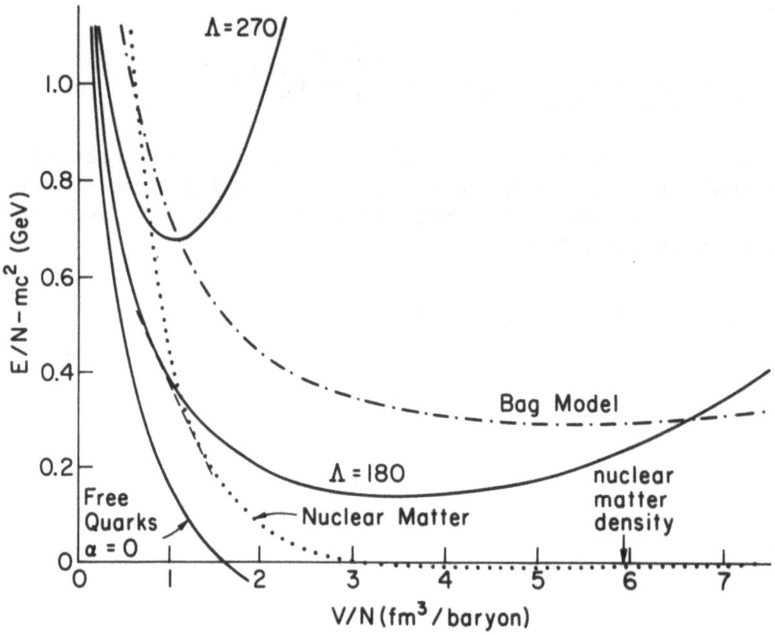

Fig. 3

At high densities one can look for corrections to the free quark picture by carrying out a perturbation theory calculation of the ground state energy as an expansion in terms of the fine structure constant $\alpha_c = g^2/4\pi$. Of course, as the density decreases the interactions become more and more important, leading eventually to confinement of the quarks in hadrons. The first correction to the free gas result is the ordinary exchange energy, corresponding to Fig. 4a, in which the wavy line is a gluon, and the solid lines quarks. To this order the ground state energy density is

$$E = \frac{3}{4\pi^2}\sum_i p_i^4\left(1 + \frac{2\alpha_c}{3\pi}\right) \quad , \tag{3.9}$$

where p_i is the Fermi momentum of flavour i. The next correction, coming from the sum of the ring diagrams, Fig. 4b, is computed exactly as in the relativistic electron gas, and is given by

$$E_{ring} = (\sum_i p_i^2)^2 \frac{\alpha_c^2}{4\pi^4} \ln \frac{\alpha_c N_f}{4\pi} \quad , \tag{3.10}$$

plus terms of order α_c and higher; N_f is the number of flavours present.

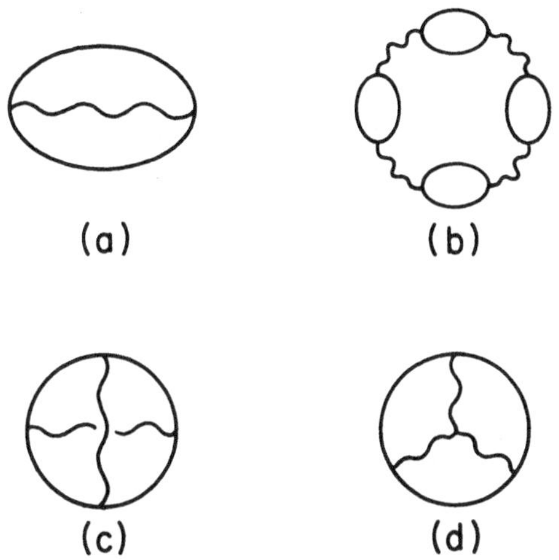

Fig. 4

An ordinary perturbation expansion of the energy, as begun above, fails to describe the confinement of quarks at lower densities and thus does not show why the nucleon phase is favoured at low density. One simple phenomenological picture of quark confinement, which provides a first description of the phase transition, is the MIT bag model (Chodos et al. 1974). In this model the quarks in a hadron are assumed confined to a finite region of space, the "bag", whose volume is limited by the introduction of a term in the hadron energy equal to the volume of the bag times a constant $B > 0$. With $m_u = m_d = 0$, the parameters $B \simeq 55$ MeV/fm^3 and $\alpha_c \simeq 2.2$ give a reasonable fit to hadron masses.

To describe quark matter in this model we simply add in a term B to the energy density (3.9). The resultant energy per baryon of symmetric nuclear matter as a function of n_b^{-1} is shown in Fig. 3, as the dash-dot curve labelled "bag model"; here the parameters $B = 55$ MeV/fm^3 and $\alpha_c = 2.2$ were used. The energy per baryon of normal symmetric nuclear matter, as calculated in the Chin and Walecka (1974) mean field model, a representative calculation of high density matter in the nucleon phase, is shown as the dotted curve. We see that the "bag" quark matter curve lies above the normal curve until a density $\sim 9\ n_o$, above which the quark phase is energetically favourable. [This phase transition from the normal to the quark phase is first order; the density discontinuity at the phase transition is determined by making the standard "double tangent" construction between the two energy curves; this construction is illustrated in Fig. 3 as the dashed line bitangent to the normal curve and that labelled $\Lambda = 180$.]

Similar calculations within the framework of the bag model of the phase transition from pure neutron matter to quark matter have been given by Baym and Chin (1976), Chapline and Nauenberg (1976), and Keister and Kisslinger (1976). The conclusion of these calculations is that, for all normal equations of state examined, the phase transition to quark matter takes place at too high a density for quark matter to be found in neutron stars. For example, with the Reid pure neutron equation of state, the density jumps from 3.7 to 7.1 fm^{-3}, or 14-40 x 10^{15} gm/cm^3, at the transition, while the maximum central density found in neutron stars described by the Reid equation of state is 4.1 x 10^{10} gm/cm^3.

The bag model calculations described above are quite phenomenological and ought not to be taken too seriously for densities far away from that inside nucleons, $n_n \sim 1.4\ n_o$. A more satisfactory approach is to attempt to compute the energy directly from the Lagrangian (3.8). From a renormalization group analysis (reviewed in Baym 1977, 1978), one can show that the ground state energy density of quark matter (for simplicity we assume all flavours present to have the same density and Fermi momentum p_f) must be of the form

$$E(n) = p_f^4 b(g(p_f/\nu)) \quad , \tag{3.11}$$

where b is a dimensionless function of the density dependent coupling constant $g(p_f/\nu)$, and ν is the point at which the vacuum theory is renormalized. The coupling constant obeys the renormalization group equation

$$\frac{\partial g(\kappa)}{\partial \ell n \kappa} \equiv \beta(g(\kappa)) = -\frac{g(\kappa)^3}{16\pi^2}\left(11 - \frac{2}{3}N_f\right) + \dots \quad ; \tag{3.12}$$

the expression on the right is valid for small $g(\kappa)$, and for momenta large compared with the rest masses of the quarks included in N_f. Integrating (3.12) one finds to leading order

$$\frac{1}{g^2(\kappa)} = \frac{1}{g^2(1)} + \frac{33-2N_f}{24\pi^2}\ \ell n\ \kappa \quad , \tag{3.13}$$

or that the "running coupling constant" is given by

$$\alpha(p_f) \equiv \frac{g^2(p_f/\nu)}{4\pi} = \frac{6\pi}{33-2N_f} \frac{1}{\ell n(p_f/\Lambda)} \quad , \qquad (3.14)$$

where $\ell n\Lambda \equiv \ell n\nu - [24\pi^2/(33-2N_f)]g^{-2}(1)$. This result is valid as long as $\alpha(p_f) << 1$. Note that $\alpha(p_f) \to 0$, i.e., the matter becomes asymptotically free, as the density $\to \infty$.

Estimates of Λ from analysis of e^+e^- annihilation into hadrons (Poggio et al., 1976) are in the range $\sim$ 150-600 MeV. However, because higher order corrections in α_c were not evaluated in this analysis, these estimates must be regarded as preliminary. [A fit to the bag model value $\alpha(p_f) = 2.2$, for density $n_h \sim 0.24\ fm^{-3}$, implies $\Lambda \sim 200$ MeV.]

To order $\alpha(p_f)^2 \ell n\alpha(p_f)$ the energy of a relativistic quark gas is given by (3.9) and (3.10). The calculation of terms up to order $\alpha(p_f)^2$ is rather complicated and requires computing diagrams of the types shown in Fig. 4c and d. The latter diagram, arising because gluons can emit and absorb gluons (the three-gluon vertex is of order g) does not occur in the relativistic electron gas. The energy of a system with equal numbers of quarks of each flavour present is, as found by Freedman and McLerran (1977, 1978) and Baluni (1978a, b):

$$E = \frac{3}{4} \frac{p_f^2}{\pi^2} N_f \left[1 + \frac{2\alpha}{3\pi} + \frac{\alpha^2}{3\pi^2} (N_f \ell n \frac{\alpha N_f}{\pi} + 0.02N_f + 6.75)\right] \qquad (3.15)$$

where $\alpha = \alpha(p_f)$. The terms of order α^2 are positive.

In Fig. 3 we show the energy per baryon of quark matter computed using (3.15) and (3.14), with Λ = 180 MeV and 270 MeV (as well as Λ = 0, the "free quark" curve). The range of results is very large. For small Λ, the transition density between nucleon and quark matter can be quite low; for Λ = 180 MeV the transition, indicated by the dashed line, occurs at a density 4 n_0 in the normal state.

3.3 Deconfinement as a Percolation Transition

The calculations summarized in Fig. 3 are only a first estimate of the deconfinement transition, and do not indicate whether the transition is first or second order. Another instructive way of studying the transition is from the point of view of percolation theory. The problem posed by percolation theory is the following:

suppose one makes a close-packed pile of children's blocks, in which individual blocks are randomly copper or wood. If the blocks are almost all wood the pile will be an electrical insulator, while if they are almost all copper, the pile will conduct. What fraction p of blocks must be copper before the pile conducts? Numerical studies (Shante and Kirkpatrick, 1971) indicate that for a cubic lattice the critical percolation fraction p_c is 0.31. For a fraction of copper blocks $p > p_c$, the system contains a spatially infinite conducting network, while for $p < p_c$, the conducting clusters are finite in extent. If one phrases the question alternatively in terms of spheres arranged in a cubic lattice, of a size that spheres on neighbouring sites just touch, then at the critical percolation density, the conducting spheres occupy a fraction $\rho_c = 0.16$ of space. For different 3-dimensional lattices one finds that p_c varies from 0.20 for fcc and hcp lattices to 0.43 for a diamond lattice, but what is remarkable is that the fraction of space occupied by the conducting spheres at the critical percolation density is always 0.15 to within 10% (the level of accuracy of the numerical calculations of p_c) (Scher and Zallen, 1970). That is, to a good approximation ρ_c is independent of the lattice structure [but not exactly independent, as exact calculations in two dimensions on the triangular lattice ($\rho_c = 0.4534$) and Kagomé lattice ($\rho_c = 0.4440$) show]. If one packs space with spheres at random locations, allowing the spheres to overlap, then numerical studies show that at the percolation threshold a fraction 0.29 of space (in three dimensions, and 0.68 in two dimensions) is covered, and that the sums of the volumes of the spheres is 0.34 of the total available space (and 1.13 in two dimensions).

To apply these results to nuclear matter, let us as a first approximation assume nucleons to be bags (in the sense of the MIT bag model) of radius r_o. For the relatively large value $r_o \simeq 1$ fm found by the MIT group the close packing density (for an hcp or fcc lattice),

$$n_{cp} = \frac{1}{4\sqrt{2}\, r_o^3} \simeq \frac{0.177}{r_o^3} \quad , \tag{3.16}$$

is $\simeq n_o$, that is, if the bags in a nucleus remain intact and non-overlapping, the nucleus is practically close packed. For a little bag of radius $r_o \simeq 0.3$ fm (corresponding to the hard core radius of nuclear forces) (Brown and Rho, 1979), n_{cp} is $\sim 40\, n_o$.

At what density n_c does one expect the bags to percolate and form an infinite connected cluster? For $r_o \simeq 1$ fm, we can estimate n_c from the calculation of percolation of spheres that can overlap, since nuclear forces, which act to make the filling of space non-random, are weak at distances ~ 2 fm, where only the tail of the one-pion exchange force remains. Thus for $r_o \simeq 1$ fm, one would conclude that $n_c \sim 0.34\ n_h \simeq 0.5\ n_o$ (where $n_h = 3/4\pi r_o^3$ is the density of matter inside hadrons) and hence that a system of nucleons at density n_o is already percolated. As the assumed bag radius decreases the percolation density n_c increases, and one does not expect bags with r_o as small as 0.3 fm to be percolated at nuclear matter density. If for small bags we estimate a lower limit to n_c from the lattice value $\rho_c = 0.15$, then $n_c \gtrsim 0.45/4\pi r_o^3$. Thus for $r_o \sim 0.3$ fm, the percolation transition would be at $n_c \gtrsim 8\ n_o$, which is in the same range as the transition density estimated from comparison of energies of the nucleon and quark phase (Fig. 3).

For a large ($r_o \sim 1$ fm) bag model, the picture we arrive at is thus the following. As the baryon density n of nuclear matter is raised from very low values, the bags begin to touch, and at a subnuclear density, the system undergoes a percolation transition. Beyond this density, space becomes filled with an infinite network of interconnected "bag", and the quarks are no longer confined to within individual nucleons. This does not mean however that the quarks are free to roam the nucleus; as the success of nuclear physics (based on the interacting nucleon picture) indicates, in the region of n_o the quarks remain essentially localized, or bound, in colour singlet trios within the nucleus, in a manner similar to the binding of electrons to atoms, or nucleons to nuclei, in low density matter. As the density is raised sufficiently above n_o the nucleon-like quark bound states will unbind in a Mott transition and the system will become uniform quark matter. It is this latter delocalization transition that one estimates from a comparison of the ground state energies of the uniform quark and nuclear matter phases. [The energy gain $\Delta\varepsilon$ due to quark localization in uniform symmetric nuclear matter density n_o is only $\sim$ 250-300 MeV per nucleon, rather than the full nucleon mass m; as one can estimate from the bag model description of uniform quark matter,

$$\Delta\varepsilon \simeq \frac{B}{n} + \frac{9}{4}\left(\frac{3\pi^2 n}{2}\right)^{1/3}\left(1 + \frac{2\alpha_c}{3\pi}\right) - m \quad , \tag{3.17}$$

where B is the bag constant, and the second term is the quark kinetic and exchange energies.]

The calculations of DeTar (1978) on the description of the deuteron in the (static) bag model provide a good illustration of the deconfined state of quarks in nuclei. Filling quark orbitals in a bag with three adjustable shape parameters, he finds that the optimal state of the deuteron, for "nuclear" separation < 2 fm, is a configuration of the six quarks in an elongated bag with a pronounced but not complete separation of the quarks into two colour singlet trios, a correlation strongly favoured by colour electrostatic attractions.

One can characterize the transitions from confined to bound to unbound quarks in terms of the behaviour of the gluon field. In the low density nucleon phase the gluon field has only short wavelength, high frequency cavity modes within individual nucleons. At the percolation transition, however, the field should develop a spectrum of long wavelength modes extending through the percolated network of bag over the entire space, with frequencies reaching, as long as the quarks are bound, down to zero frequency; the longitudinal static gluon Green's function should also become Coulomb-like, $\sim 1/r$, at large r. The calculation of the low-frequency long-wavelength spectrum of the gluon field in such a random infinite percolated network poses a challenging problem. [This situation is somewhat analogous to that of fourth sound in superfluid liquid ^{4}He filling porous Vycor glass, which contains a similar percolated network of channels through which the superfluid can flow (Kiewiet et al., 1975).] The change of spectrum of the gluon field at the percolation transition implies the possibility of long range van der Waals forces, due to two gluon exchange, between nucleons in nuclear matter. When the delocalization transition takes place the quarks can begin to screen long wavelength excitations of the gluon field, and the spectrum becomes analogous to that of the electromagnetic field in a plasma, bounded below by a colour plasma frequency ω_p, which approaches a value at high density (Baym and Chin, 1976a), $\omega_p = (g^2 \Sigma_i n_i/6\mu_i)^{1/2}$, where n_i and μ_i are the quark number density and chemical potential of flavour i present.

In normal uniform quark matter, the spectrum of quark-like excitations goes down to zero energy, while in the intermediate regime between the percolation and delocalization transitions, in which, for the MIT bag model one would expect normal nuclei to be, the system should have a set of finite energy quark excitations extending over the system. Detection of such excitations in nuclei would be an important test of this picture of the deconfinement transition. [See, e.g., Wong and Liu (1978) for a discussion of several possible experiments.]

Let me close by mentioning briefly the possibility of observing the transition to uniform quark matter in heavy ion collisions. In such a collision nuclear matter is both compressed and heated, and to a first approximation can be taken to be in thermal equilibrium at a finite temperature. Chin (1978) has estimated. in the spirit of the bag model calculation described earlier, the transition from nuclear to quark matter at finite temperature, and calculated, for an assumed nuclear compression, the lab energy required to produce sufficient heating to form quark matter.

Chapline and Kerman (1978), on the other hand, estimate the transition to occur when the energy per baryon in nuclear matter shocked in the collision equals that of zero temperature quark matter at the same density. Calculating the quark matter using (3.14), they find, in this way, that for $\Lambda \sim 300$ MeV, the transition could occur for center of mass kinetic energies in the experimentally accessible range of 300 to 700 MeV per nucleon, or 1.4 to 3.8 GeV per nucleon lab energy, estimates consistent with Chin's. Possible signatures of formation of quark matter discussed by these authors include formation of metastable quark nuclei, due to quark shell effects, and effects on the spectrum of nucleons and pions emitted in the collision.

At present, though, the strength of the quark-gluon interactions is not well enough pinned down experimentally, nor is the quark confinement problem and structure of the nucleon adequately understood for one to say more precisely whether quark matter can be made either in heavy ion collisions or in neutron stars.

This work has been supported in part by the U.S. National Science Foundation Grant DMR75-22241.

REFERENCES

C.-K. Au, Phys. Lett. 61B (1976), 300.

V. Baluni, Phys. Lett. 72B (1978a), 381.

V. Baluni, Phys. Rev. D17 (1978b), 2092.

G. Baym, Neutron Stars and the Properties of Matter at High Density (1977), Nordita, Copenhagen.

G. Baym, Proc. 1977 Les Houches Summer School, R. Balian, G. Ripka, eds. North-Holland, Amsterdam (1978).

G. Baym and G.E. Brown, Nucl. Phys. A247 (1975), 395.

G. Baym and D. Campbell, in Mesons and Fields in Nuclei, D. Wilkinson, M. Rho, eds. North-Holland, Amsterdam (1979).

G. Baym and S. Chin, Phys. Lett. 62B (1976), 241.

G. Baym and S. Chin, Nucl. Phys. A262 (1976a), 527.

G. Baym and C. Pethick, Ann. Rev. Nucl. Sci. 25 (1975), 27.

G. Baym and C. Pethick, Ann. Rev. Astron. Astrophys. 17 (1979), 1.

G.E. Brown, S.-O. Bäckman, E. Oset and W. Weise, Nucl. Phys. A286 (1977), 191.

G.E. Brown and W. Weise, Phys. Reports 27C (1976), 1.

G.E. Brown and M. Rho (1979), Phys. Lett. B (in press).

D. Campbell, R. Dashen and J. Mana sah, Phys. Rev. D12, (1975), 979, 1010.

G. Chapline, Nature 261 (1976), 550.

G. Chapline and M. Nauenberg, Nature 259 (1976), 377.

G. Chapline and A. Kerman (1978), preprint.

S.A. Chin, Phys. Lett. 78B (1978), 552.

S. Chin and J.D. Walecka, Phys. Lett. 52B (1974), 24.

A. Chodos, R.L. Jaffe, K. Johnson, C.B. Thorn and V.F. Weisskopf, Phys. Rev. D9 (1974), 3471.

J.W. Clark, Prog. in Particle and Nucl. Phys., Pergamon, Oxford (1979), in press.

J.C. Collins and M.J. Perry, Phys. Rev. Lett. 34 (1975), 1353.

F. Dautry and E. Nyman, Nucl. Phys. A. (1979), in press.

B. Day, Rev. Mod. Phys. 50 (1978), 495.

C. DeTar, Phys. Rev. D17, (1978), 302, 323.

A.M. Dyugaev, Pisma ZhETF 22 (1977), 181. (Engl. transl: JETP Lett. 22 (1975), 83).

M. Ericson and J. Delorme, Phys. Lett. 76B (1978), 182.

T.E.O. Ericson and F. Myhrer, Phys. Lett. 74B (1978), 163.

B. Freedman and L. McLerran, Phys. Rev. D16 (1977), 1130, 1147, 1169.

B. Freedman and L. McLerran, Phys. Rev. D17 (1978), 1109.

B. Friman and E. Nyman, Nucl. Phys. A302 (1978), 365.

A.M. Green, Rep. Prog. Phys. 39 (1976), 1109.

M. Gyulassy, LBL preprint 6525 (1977).

M. Gyulassy, LBL preprint 7704 (1978).

K. Holinde and R. Machleidt, Nucl. Phys. A280 (1977), 429.

K. Holinde, K. Machleidt, M.R. Anastasio, A. Faessler and H. Müther, Phys. Rev. C18 (1978), 870.

B. Keister and L. Kisslinger, Phys. Lett. 64B (1976), 117.

C.W. Kiewiet, H.E. Hall and J. Reppy, Phys. Rev. 35 (1975), 1287.

A.B. Migdal, Phys. Rev. Lett. 31 (1973), 247.

A.B. Migdal, Rev. Mod. Phys. 50 (1978), 107.

V.R. Pandharipande and R.A. Smith, Nucl. Phys. A237 (1975), 507.

H.-J. Pirner, M. Rho and K. Yazaki, Proc. 1977 Les Houches Summer School, op. cit., and to be published (1979).

E. Poggio, H. Quinn and S. Weinberg, Phys. Rev. D13 (1976), 1958.

V. Ruck, M. Gyulassy and W. Greiner, Z. Physik A277 (1976), 391.

H. Scher and R. Zallen, J. Chem. Phys. 53 (1970), 3759.

V.K.S. Shante and S. Kirkpatrick, Adv. Phys. 20 (1971), 325.

T. Takatsuka, K. Tamiya, T. Tatsumi and R. Tamagaki, Prog. Theor. Phys. 59 (1978), 1933.

H. Toki and W. Weise (1979), preprint.

W. Weise and G.E. Brown, Phys. Lett. 58B (1975), 300.

C.W. Wong and K.F. Liu (1978), preprint.

BARYON-ANTIBARYON SYSTEMS*

S.H. Kahana

Brookhaven National Laboratory

Upton, New York 11973, U.S.A.

I. INTRODUCTION

I would like to discuss a few simple situations involving the B-B and B-nucleus systems with the baryon label B = N for nucleon, B_c for a charmed nucleon, $N\bar{N}$, $B_c\bar{B}_c$, N-nucleus and B_c-nucleus systems. The unifying thread in this discussion will be the non-relativistic treatment of the dynamics, hopefully justified by the particular nature of the elementary two-body states considered. The successes of the quark model as the underlying theory for strong interactions are well known. Of particular relevance here is the spectroscopy of nucleons, isobars and mesons as 3-quark or 2-quark composites. Lipkin has discussed these structures elsewhere in these proceedings.

We want in nuclear physics to accept the composite structure and then to proceed to consider interactions between the composites (e.g., nucleons) as if they were themselves elementary. It is sometimes difficult to make the separation between composite and elementary objects. The spectrum of mesons near the NN threshold (Fig. 1) at ∿ 1800-1900 MeV most readily illustrates the possible usefulness of beginning with very simple composites. Light mesons with masses << $2M_N$ probably require a more complete dynamical theory and at the very least a relativistic potential theory. The light mesons, such as the π, ω, ρ, ε, are best described as quark-antiquark states, i.e., nuclear qq in structure. The MIT bag model[1] has been used to describe these with some success--although

*Research supported by the Dept. of Energy under Contract No. EY-76-C-02-0016.

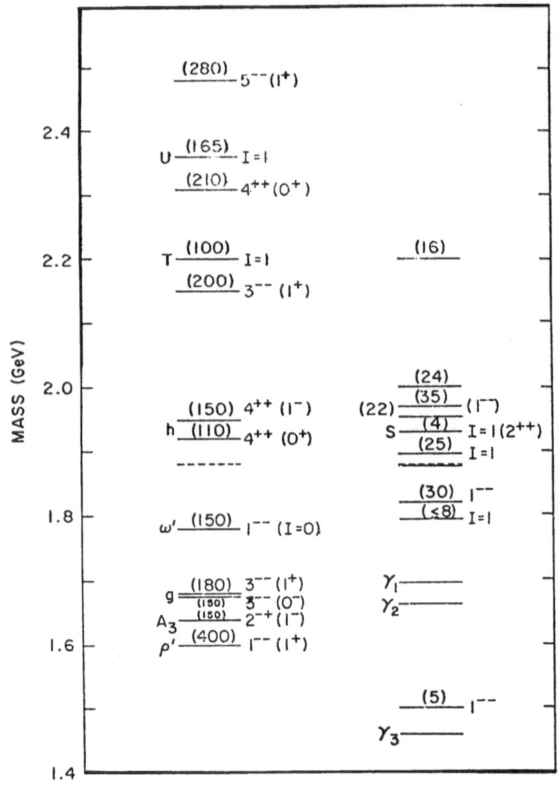

Fig. 1

failing notably for the π-meson. In contrast, meson states appearing near the $N\bar{N}$ threshold in mass might well be considered as consisting of a bound or resonant $N\bar{N}$ pair and could then be referred to as molecular. In fact the $N\bar{N}$ state also possesses structure (qqq) ($\bar{q}\bar{q}\bar{q}$) but hopefully with a definite spatial correlation among the quarks and separately among the antiquarks, i.e., two bags. A proper field theoretic description of any nuclear state would yield mixing between such components (and generally including qq $\bar{q}\bar{q}$ as well) but we will for the purposes of this talk assume the molecular states, with separation between N and $\bar{N}$ do on occasion exist.

If mesons with center of mass energies near $2M_N$ are truly $N\bar{N}$ in character, non-relativistic dynamics will be applicable. A spatial separation between the baryons implies the forces will not be large, while the binding energy E(meson) - $2M_N$ is small relative to the total mass. The relative kinetic energy may then reasonably be expected to also be small. In contrast treating pairs of light quarks u, $\bar{d}$ combining to form a π, ρ etc. meson non-relativistically is clearly non-rewarding. Even the nuclear or bag quark systems may be an occasion amenable to a Schrodinger equation analysis. A most striking example is the charmonium description of the J/ψ, ψ', χ family of mesons[2,3]. The assumed large mass (1.5 GeV) for the charmed quarks $c\bar{c}$ lends meaning to a non-relativistic dynamics. The mixing between nuclear quark states and molecular $N\bar{N}$ (or meson-meson) states mentioned above is particularly relevant for some of these charmonium states which are in some senses closer to the $\bar{B}_cB_c$ threshold than say the π, ρ are to the $N\bar{N}$ threshold. This mixing will be referred to further. One should re-emphasize that the molecular labeling for $B\bar{B}$ states will be most meaningful when the baryons are for the most part spatially separated. Excellent candidates then are high orbital angular momentum states in which the centrifugal barrier keeps the two 3-quark components apart.

Considerable experimental and theoretical[4] activity has recently concentrated on mesons near $2M_N$ in mass. Such structures have been seen earlier in $N\bar{N}$ total and elastic cross sections[5] (Fig. 2) and $\bar{p}d$ spectator experiments[6] (Fig. 3.). More recently the possible observation of high energy γ-rays from the $\bar{p}p$ system has been suggestive of states somewhat below threshold[7] (Fig. 4), (although an earlier experiment yielded a null result[8]), while

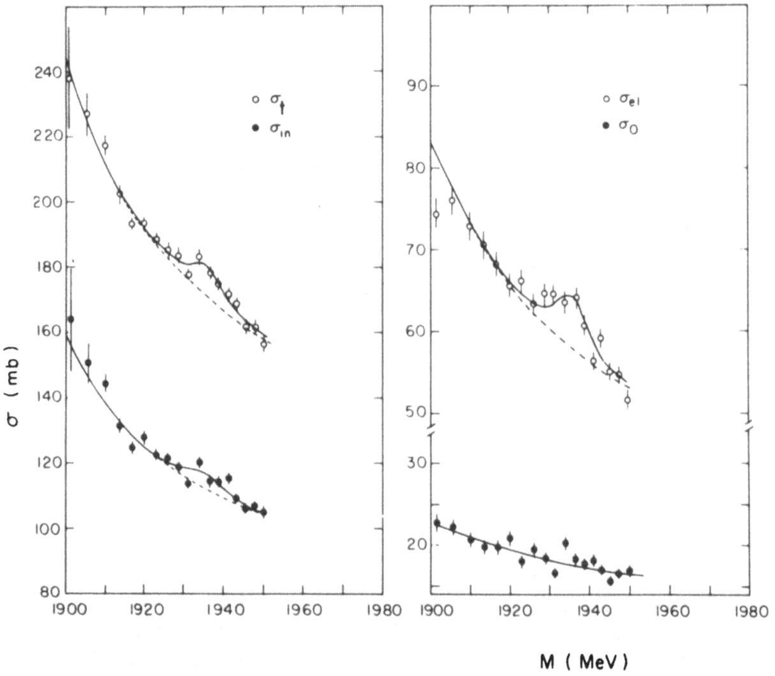

Fig. 2

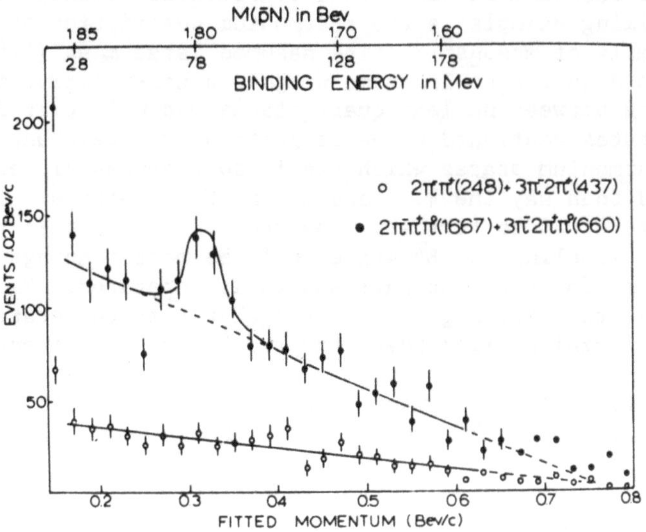

Fig. 3

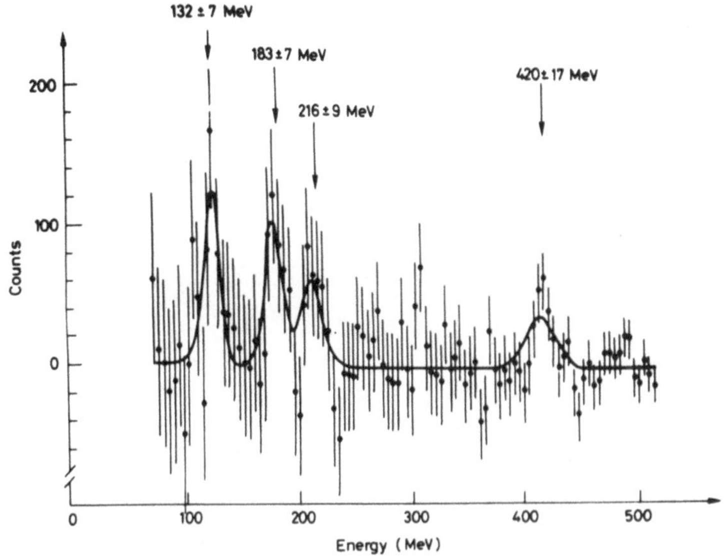

Fig. 4. Gamma-ray spectrum as obtained after subtraction of a third order polynomial. The solid line corresponds to a computer fit to the peaks in the original spectrum (see also Table 1).

rather narrow states above threshold were identified in πp production experiments. Attempts have been made to describe such states theoretically in the MIT bag model extended to the $\bar{Q}^2Q^2$ sector[9] as well as by $N\bar{N}$ molecular states[10].

TABLE I. Results

Energy (MeV)	Instrumental line width (MeV)	Confidence level (%)	Yield per annihilation
132 ± 6	16	99.3	$(5.1 \pm 2.7) \times 10^{-3}$
183 ± 7	19	99.0	$(7.2 \pm 1.7) \times 10^{-3}$
216 ± 9	21	97.5	$(6.0 \pm 1.9) \times 10^{-3}$
420 ± 17	34	98.2	$(8.5 \pm 2.0) \times 10^{-3}$

II. $\bar{B}B$ MOLECULAR STATES

A. Baryon-Baryon Meson Exchange Potentials

1. Nucleons. A key element for performing a non-relativistic investigation of the $B\bar{B}$ system is the inter-baryon force. Any field theoretic derivation of the BB force will also yield the $B\bar{B}$ force since these are related by G-parity. This relation is analogous to the simple sign change between electron-electron and electron-positron forces in quantum-electrodynamics. In QED the charge configuration symmetry C determines the $e^- - e^-$ to $e^+ - e^-$ relation while for strong interactions an additional isospin rotation about the y-axis in the formal charge-symmetry space is required. Thus

$$G = e^{i\pi I(y)}\ C.$$

One's knowledge of the N-N forces comes from observations of nucleon-nucleon scattering over a wide range of energies. Measurements of the deuteron, a BB "molecular" state at E = -2.23 MeV, adds to this knowledge. To represent these forces in a realistic fashion amenable to the G-parity transformation and extendable to two-body systems involving strangeness and charm, one introduces forces mediated by exchanged mesons of a pseudo-scalar (π,η,η'), scalar (δ,ε,S^*) and vector (ρ,ω,ϕ) character. The one-boson-exchange-potential (OBEP)[11] thus obtained is represented schematically in Fig. 5.

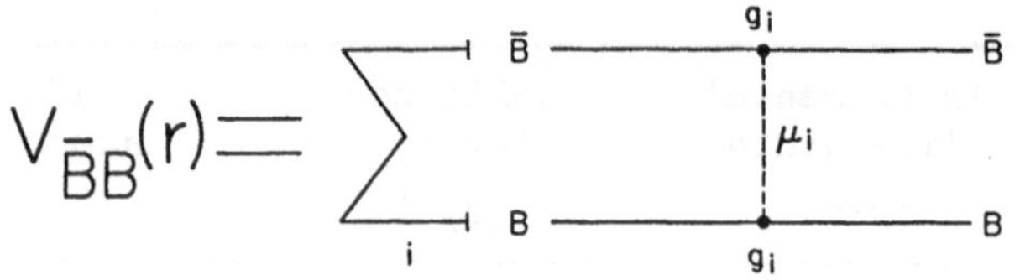

$$i = \begin{cases} \pi,\eta,\eta' \ (0^-) \\ \delta,\epsilon,S^* \ (0^+) \\ \rho,\omega,\phi \ (1^-) \end{cases}$$

Fig. 5

The best known of these exchange forces, that due to the π-meson is typical and described by the pseudo-scalar Hamiltonian

$$H = ig\,\bar{\psi}\,\gamma_5\,\tau_\alpha\,\psi\,\phi_\alpha \quad .$$

In a non-relativistic limit (static limit) for the baryons a Yukawa-like potential arises as an expansion in the ratio of meson (μ) to Baryon (m_B) masses:

$$V^i_{NN}(\gamma) = + (\vec{\tau}_1\cdot\vec{\tau}_2)\,(\vec{\sigma}_1\cdot\vec{\sigma}_2)\,V^i_\sigma(r) + S_{12}V^i_T(r)$$

with

$$V^i_\sigma = \mu_i(g_i^2/4\pi)\,\left(\frac{\mu_i}{2m_B}\right)^2 \frac{e^{-\mu_i r}}{\mu_i r}$$

and

$$V^i_T = \mu_i(g_i^2/4\pi)\,\left(\frac{\mu_i}{2m_B}\right)^2 \left(\frac{1}{3} + \frac{1}{\mu_i r} + \frac{1}{(\mu_i r)^2}\right) \frac{e^{-\mu_i r}}{\mu_i r} \quad .$$

The scalar and vector meson exchange potentials are obtained in a similar fashion and involve additional terms of a spin-orbit character. Because of the G-parity invariance referred to above the diagonal baryon-baryon and baryon-antibaryon potentials are related by

$$V^i_{B\bar{B}} = (-1)^{G_i} V^i_{BB}(r)$$

where

$$G_i = -1 \text{ for } \pi, \rho, \omega, \phi, \delta$$
$$= +1 \text{ for } \eta, \eta', \rho, \varepsilon, S^*.$$

General representations of the forces obtained for NN and $\overline{N}N$ are displayed in Figs. 6 and 7.

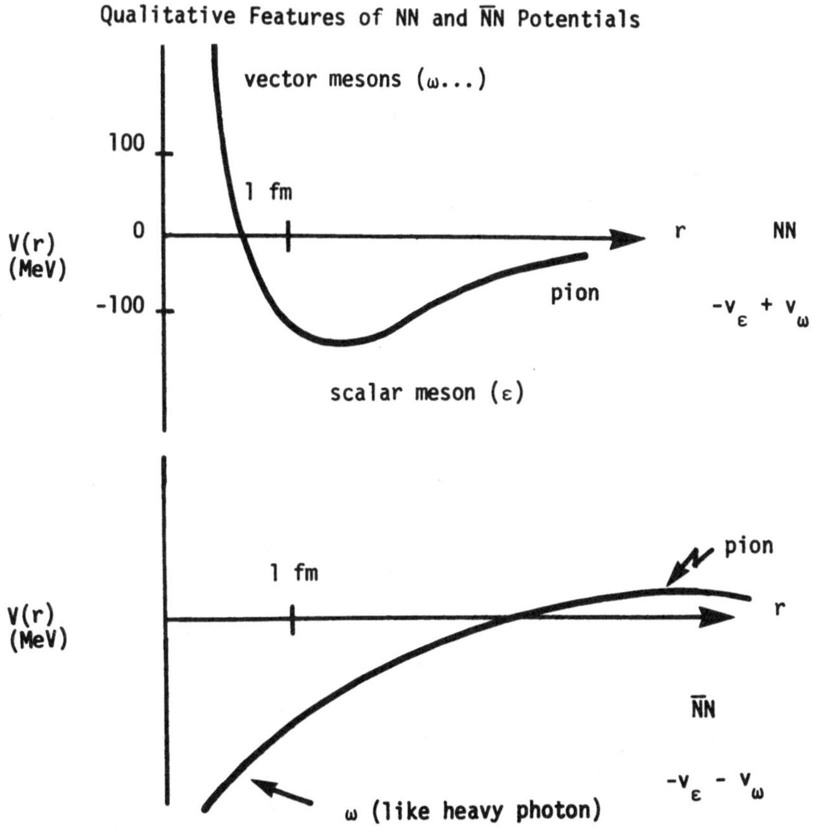

Fig. 6

TOTAL POTENTIAL FOR $\bar{p}p$ SYSTEM

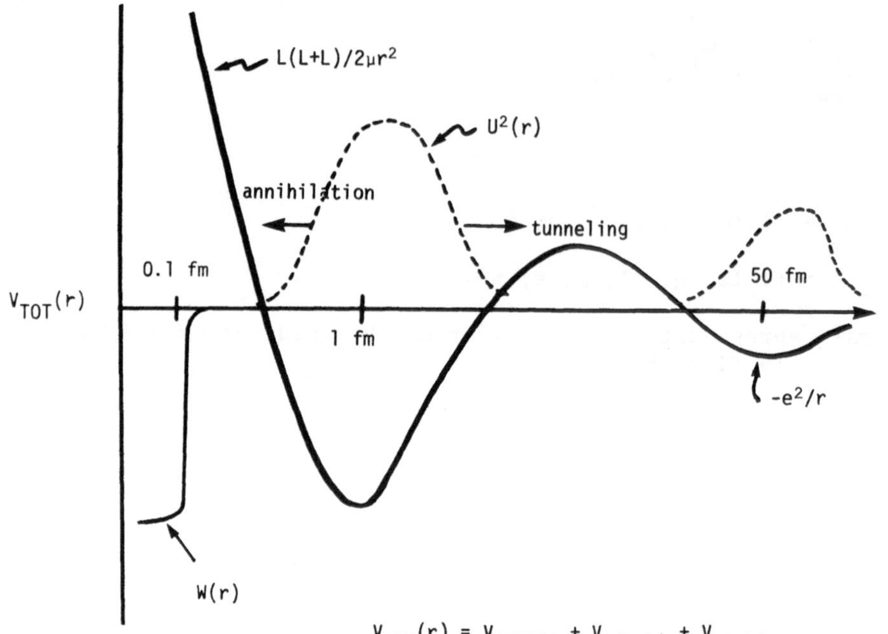

$$V_{TOT}(r) = V_{STRONG} + V_{L(L+1)} + V_{COULOMB}$$

Fig. 7

2. Potentials for Charmed and Strange Baryons

To extend the $N\bar{N}$ calculations to $B_c\bar{B}_c$ systems one can use an SU(4) framework[12]. The charmed baryons B_c together with the nucleons N, and strange baryons B_s form a (20) multiplet in SU(4). In addition to the (8) multiplet of SU(3) consisting of N, p, Λ, $\Sigma^{\pm}$, Σ^{o}, Ξ^{o} one adds the charmed baryons including isospin I = 0 members, $C_0(\Lambda_c^+)$, T and X_s, I = 1/2 members A, S, X and the I = 1 member $C_1(\Sigma_c^{++,+,0})$. The C_0, C_1 have been observed experimentally[13] with masses (2.26 and 2.42 GeV). Table 2 schematically indicates the expected quark structure of some of the baryons. The large mass differences between the N, Λ, Σ, Ξ masses on one hand and the charmed baryon masses on the other implies a large SU(4) breaking. In the calculations seeking $B_c\bar{B}_c$ states, described in detail in ref. 4, SU(4) symmetry is invoked only as a means of relating unknown $\bar{B}_c B_c$ coupling constants to those deduced from scattering data for the NN (and occasionally the ΣN) system. Indeed the large

SU(4) breaking is finally an advantage since the $\overline{B}_c B_c$ states can be considered in isolation from other $\overline{B}B$ states.

TABLE 2

Schematic quark structure of representative members of the SU(4) (20) multiplet of baryons. The quarks are labeled in the usual fashion u, d, s, c. A proper tensor decomposition may be found in Ref. 14.

Baryon	qqq	Isospin (I)	Strangeness	Charm (c)
p	[uud]	1/2	0	0
n	[udd]	1/2	0	0
Λ	$(\frac{ud + du}{\sqrt{2}})s$	0	-1	0
Σ^0	$(\frac{ud - du}{\sqrt{2}})s$	1	-1	0
Ξ^-	[dss]	1/2	-2	0
C_0^+	$(\frac{ud - du}{\sqrt{2}})c$	0	0	1
C_1^{++}	[uuc]	1	0	1
C_1^-	$(\frac{ud + du}{\sqrt{2}})c$	1	0	1
C_1^0	[ddc]	1	0	1

An SU(4) scalar interaction is constructed from the $\overline{B}B$ multiplets and mesons (15) + (1) multiplets. In determining the diagonal BB forces however, one needs only the non-charmed and non-strange meson exchanges described above for the NN two-body system. Some appreciable corrections and additional states might arise from K-meson exchange forces which enter through channel coupling for $\overline{B}_i B_i$ states and directly in $\overline{B}_i B_j$. Exchange of the charmed I = 1/2 D,$\overline{D}$ mesons is probably negligible in any case because of the large D-mass, ∿ 1.9 GeV[13].

III. BOUND AND RESONANT $B\bar{B}$ STATES

The $N\bar{N}$ system will, with the potentials described above, sustain both bound and resonant states near the $N\bar{N}$ threshold. If the non-relativistic calculations are taken literally, very deeply bound states with the quantum numbers of the π, ρ etc., are also present[15]. One cannot, however, trust the dynamics in the latter situation. It is useful in classifying the bound states to introduce the notation ${}^{IS}L_J$ for a given state and to then note for the $B\bar{B}$ system the parity and G-parity and charge configuration are given by

$$\pi = (-1)^{L+1}, \quad C = (-1)^{L+S}, \quad G = (-1)^{L+S+I}.$$

In the $B_c\bar{B}_c$ system the large charmed baryon mass $\gtrsim 2.5\ M_N$ aids in justifying the non-relativistic approximation but lowers the centrifugal barrier in the $L \neq 0$ states sufficiently to make positive energy resonant states unlikely. Only diagonal $B_c\bar{B}_c$ states are considered with the lowest lying $\bar{C}_0C_0$, $\bar{C}_1C_1$ ($\bar{S}S$, $\bar{A}A$) of most interest. Below a few brief examples of results of calculations are given.

A. The $N\bar{N}$ System

Perhaps the best established $N\bar{N}$ candidate is the S(1930) meson, which still presents something of an observational puzzle[16] (see Fig. 2). In Fig. 2 are displayed indications of the presence of this state in $\bar{p}p$ total, total elastic data and its absence in $\bar{p}p \rightarrow \bar{n}n$ data. The observed width of this rather narrow state is between 4 and 10 MeV. A possible interpretation[16] of this apparent paradox, consistent with charge independence, is obtained if two nearly degenerate resonances with identical J^π but different I = 0,1 exist at the S-mass and interfere constructively for elastic scattering but destructively in the charge exchange reaction $\bar{p}p \rightarrow \bar{n}n$. One notes in explanation the following isospin decompositions of the observable $\bar{N}N$ states.

$$(p\bar{p}) = \frac{1}{\sqrt{2}}\,[(I=0) + (I=1)]$$

while

$$(n\bar{n}) = \frac{1}{\sqrt{2}}\,[(I=0) - (I=1)] \quad .$$

It is also possible to achieve interference between a state of given isospin and a background of different isospin, but it is then necessary to have a strong energy dependence in the background.

A dynamical mechanism can be produced which would produce the interference naturally; NN resonance generated by purely isoscalar meson exchange (t-channel e.g., ε, ω) will appear as degenerate s-channel I = 0, 1 resonances of equal elastic width. In selected channels then the residual isovector exchange potential due to say π, ρ exchange, may be sufficiently weak to leave the resonances overlapping.

It is of interest to examine the $N\bar{N}$ potentials and Schrodinger calculation for the S-meson since it is, aside from the possible degeneracy, a quite representative situation. Figures 8 and 9 display the $\bar{N}N$ potential obtained from a particular OBEP parametrization of NN scattering, that of Bryan and Phillips (BP)[11]. The phenomenological hard core in the NN potentials for $r < r_c$ provides something of a problem in the $N\bar{N}$ system. One simply sets V = 0 for $r < r_c$ and notes that the philosophy of the molecular states would reject as possible candidates those states whose properties are too strongly dependent on the cutoff. States localized at a

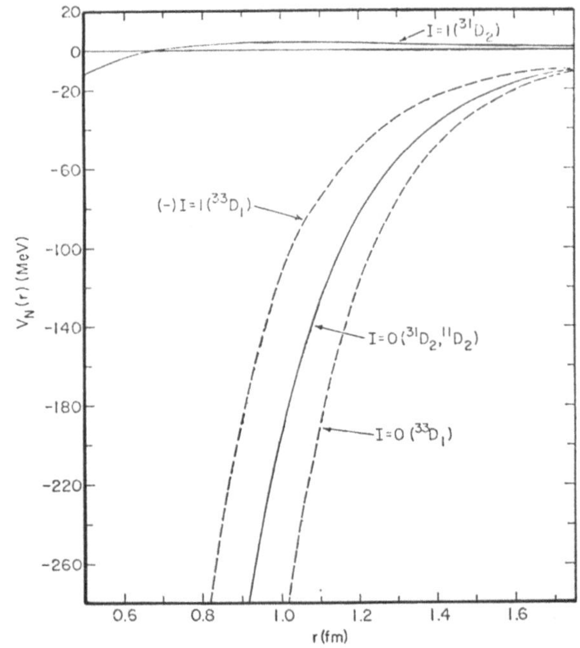

Fig. 8

large separation radius, e.g., near $r \approx 1$ fm ($r_c \approx 0.4$-0.5 fm), should present less of a problem. Fig. 8[14] displays the I = 0,1 forces for the L = 2 states $^{31}D_2$, $^{11}D_2$ and $^{33}D_1$. The strongest piece of the vector meson exchange force (π,ρ) is the tensor force which does not act in the singlet spin states. Thus $^{31}L_L$ and $^{11}L_L$ are suitable candidates for a degenerate I = 0,1 pair. Further the positions of the centrifugal barriers obtained from the BP potential (Fig. 9) suggests the P states are perhaps too broad

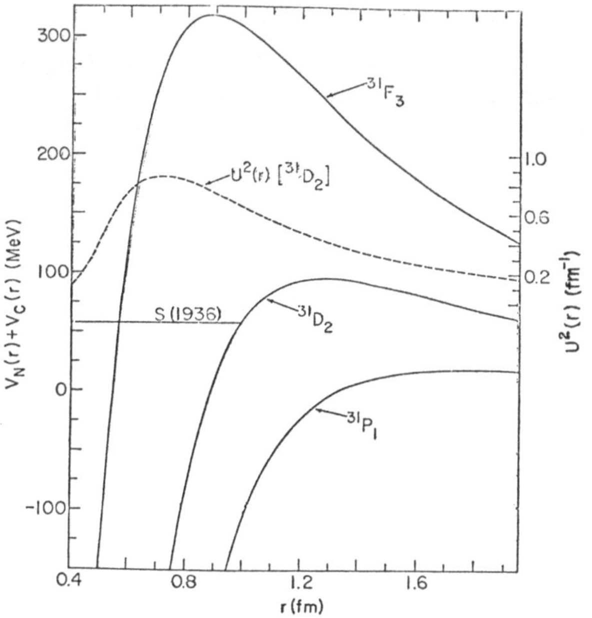

Fig. 9.

∴ only possibility for S(1930) doublet is $^{11}D_2$, $^{31}D_2$ $(J^{\pi C}(I^G)) = 2^{-+}(0^+,1^-)$)

and the F states too narrow. These simple considerations indicate that in this model the S(1930) is a $^1D_2$ I = 0,1 doublet, i.e., a pair of $J^{\pi C} = 2^{-+}$ resonances with $I^G = 0^+,1^-$. Unfortunately recent back angle data[5] may imply even parity for this state.

Finally in Fig. 10 we display a representation of typical wave functions obtained for some bound states in the general $N\bar{N}$ calculations[15].

One can usually make rather similar general arguments concerning the isospin and spin dependence of the $N\bar{N}$ force and determine the order and spacing of low-lying $N\bar{N}$ states if not their absolute positions[15].

Further experimental $\bar{N}N$ data is needed to place stronger

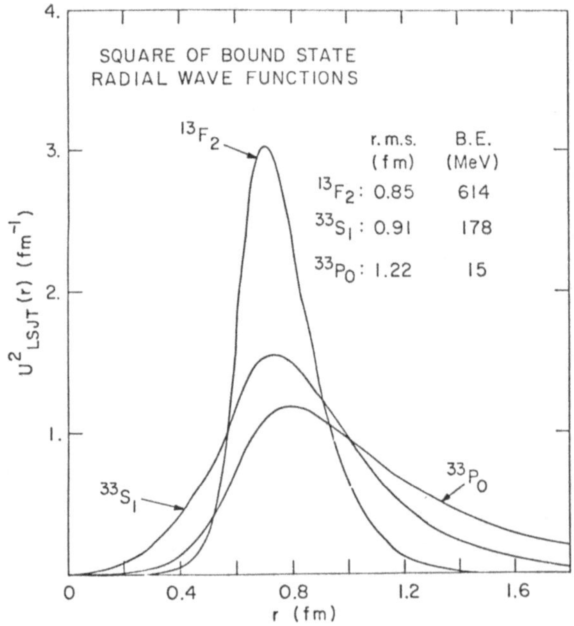

Fig. 10

constraints on the theoretical analyses. A particularly grey area is the annihilation widths of the states[15]. Some corrections to NN forces can be expected to come from diagrams like that in Fig. 11, but these alterations should occur at very small distances $\sim \frac{1}{M_B}$. If annihilation is sufficiently strong many bound states may be wiped out.

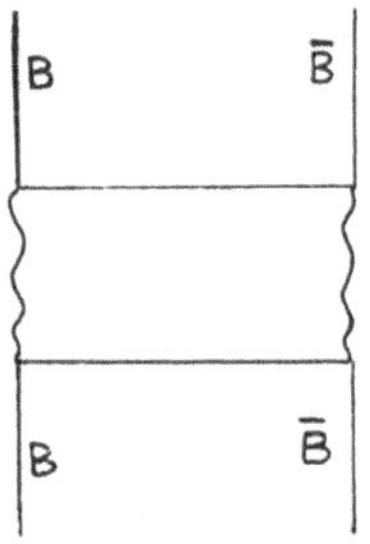

Fig. 11

B. $\bar{B}_c B_c$ Molecular States

Interest in a possible $\bar{B}_c B_c$ bound system arises from its relationship to charmonium as well as for its own sake. The calculations resemble those for $\bar{N}N$ with the heavier baryon mass $M(B_c) \lesssim 3M(N)$ altering the feature of the force somewhat but making the non-relativistic calculations perhaps more suitable. The spin dependent parts of the $B_c B_c$ and $\bar{B}_c B_c$ forces are greatly diminished, and the centrifugal barrier (permitting states for $E > 2M_B$) decreased in height. The two BB fitted OBEP parametrizations used as model potentials were those of Bryan-Phillips (BP)[11] and of Nagels, Rijken and de Swart (NRD)[17]. The coupling constants for the four lightest (expected) charmed baryons (C_o,C_1,A,S) are shown in ref. 14.

Quasi-molecular $\bar{B}_c B_c$ mesons are expected a priori in the mass region above 4 GeV, in view of the measured C_o,C_1 masses of 2.26 and 2.42 GeV. Structures between 3 and 4 GeV are strongly bound and presumably mix with corresponding $\bar{c}c$ states, when the latter exist. A principal difference between $\bar{B}_c B_c$ and $\bar{c}c$ states is the possibility of non-zero isotopic spin, i.e., I = 1,2 ($\bar{D}D$ can produce I = 1 as well). A principal mode of identification of the $\bar{B}_c B_c$ bound states would in fact be to observe π decay from I = 1,2 states to lower I = 0 states of $\bar{B}_c B_c$ or of charmonium. A model independent conclusion is that I = 2 $\bar{B}_c B_c$ states should exist just below threshold for the $\bar{C}_1 C_1$ ($\bar{\Sigma}_c \Sigma_c$) pair. From ref. 14 one notes only the C_1,S baryons couple strongly to the pion. Although the dominant mechanism for binding is the coherent central attraction from ω, ε exchange (coherent for central, spin and isospin dependent terms in some states with favourable quantum numbers[15]), the long range pion potential serves to increase the binding energy and numbers of bound states. Generally it also enables the $\bar{B}_c B_c$ wave function to be localized at comparatively large distances and hence is largely responsible for the sizable r.m.s. radius and molecular character of these states. It is expected the $\bar{C}_1 C_1$ and $\bar{S}S$ systems will support the greatest number of bound states.

Fig. 12 shows sample spectra obtained with the BP potential for nodeless (n = 0) states of $\bar{B}_c B_c$. For the most deeply bound $\bar{C}_1 C_1$ system states with n = 1 are also supported, perhaps 1 GeV higher in energy, (note n = 0,1 splitting in a deep square well $\Delta\varepsilon \sim \frac{3\pi^2}{M_B R^2}$). Fig. 13 shows the bound state L = 1 spectrum for the $\bar{C}_1 C_1$ system as a function of potential and of cutoff radius.

Thus the calculations described here indicate the existence of many quasi-molecular states $\bar{B}_c B_c$ below 4.8 GeV. Some of these states will have I = 1,2 in clear contrast with charmonium. Though

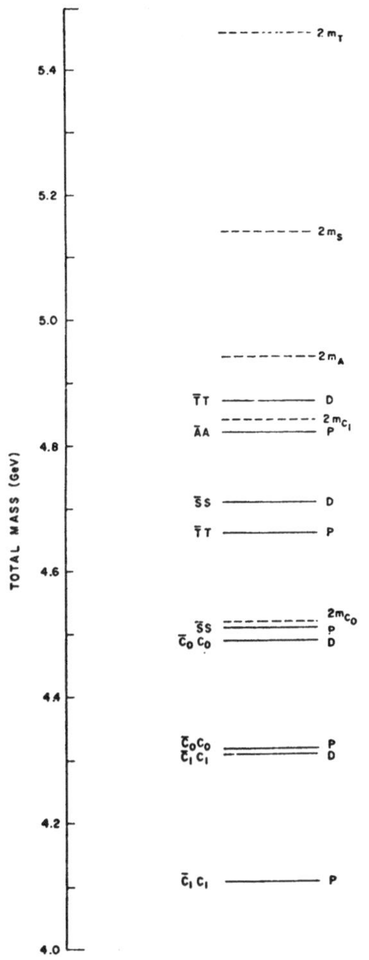

Fig. 12

discussion of L = 0 states rests on a weak foundation, some of the character of such states must be mixed, albeit weakly, into known charmonium states. The most likely result of the direct production of $\overline{D}D,\overline{B}B$ states is a cascade of π-emission to low lying S-admixtures. The rate for such decays is determined mainly by energy separations and given one's lack of knowledge of the spectra a reliable estimate of widths is impossible. Perturbation theory suggests $\Gamma \gtrsim 10$ MeV and often considerably higher.

The mixing between $\overline{B}_c B_c$ and $\overline{c}c$ clearly determines the rate at which π cascades terminate in J/ψ, e.g.,

$${}^3P_1 \rightarrow J/\psi + \pi$$

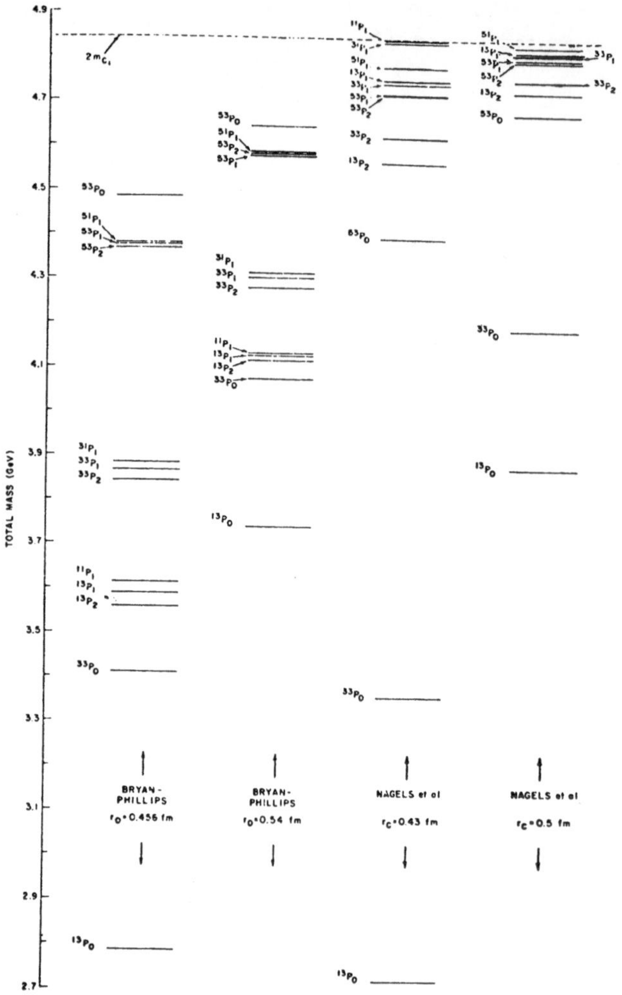

Fig. 13

or

$$^{33}P_J \to \chi + \pi \quad .$$

The first decay occurs because of S-state mixing and the second is due to the probably less likely P-state admixture. The mixing could have a significant effect on the properties of the charmonium states. The $\overline{B}_cB_c$ states are expected to be perhaps twice as large spatially as corresponding $\bar{c}c$. Calculations for the charmonium states[18] suggests radii $R(J/\psi) \approx .3-.4$ fm, confirmed by the measured $\psi' \to \chi$ electromagnetic decay widths. Assuming the ψ', χ and ψ/J states are described by oscillator functions with a common radius parameter yields $R(J/\psi) \approx 0.38 \pm 0.08$ fm. This size difference, which again emphasizes the molecular character of $\overline{B}_cB_c$, combined with the B_c unit electric charge (2/3 for c) leads to E1 transitions approximately an order of magnitude larger for $\bar{B}_c B_c$. Thus small admixtures of $\bar{B}_c B_c$ into $\bar{c}c$ could strongly effect calculated γ-ray widths and branching ratios.

What else can happen to the lowest lying $\bar{B}_c B_c$ states? Since one is starting with a state containing a baryon and an anti-baryon, it is tempting to think the decays will lead appreciably to final states containing $\bar{p}p$, $\bar{\Lambda}\Lambda$, etc. However, the large momentum transfer involved in annihilating $\bar{c}c$ quarks will tend to destroy any remnants of baryon-like wave functions of the residual light quarks. Thus S-states if they exist will likely decay by

$$^{33}S_1 \rightarrow J/\psi + \pi$$

and

$$^{33}S_0 \rightarrow J/\psi + \pi + \pi \text{ or } \eta_c + \pi \ .$$

Production of $\bar{B}_c B_c$ states will not be easy, e^+e^- production will be small to non-S-states. One might expect to see $\bar{B}_c B_c$ states wherever charmed baryon production is appreciable such as in photo-production, and despite the statements of the last paragraph one has a strong bias to look for these states in say $\bar{p} + p$.

IV. MANY-BODY SYSTEMS

A. Charmed Hypernuclei

One might expect in analogy with Λ-hypernuclei (and possibly Σ-hypernuclei) to find the B_c-nucleon interaction sufficiently strong to generate B_c-nucleus bound states, i.e., charmed hypernuclei. Indeed the increased mass of the charmed baryon should appreciably lower its kinetic energy either in the B_c-N or B_c-A systems and facilitate the generation of bound states. In the B_c-A system where potentials comparable to the nucleon-nucleus potential in depth are found, a rich spectroscopy results. In the two-body system the π exchange potential, important in spin and isospin unsaturated systems, is reduced in magnitude by the increase in mass. Nevertheless some bound states will appear. The estimated short lifetime of even the lowest mass charmed baryon, $\tau(C_0) \sim 10^{-11}$ to 10^{-14} sec may make it difficult to establish the existence of analogue bound states.

The B_c-N potential V(r), is given again by a sum of meson exchange potentials for $r \gtrsim r_c$ and is taken as an infinite hard core for $r \lesssim r_c$. Since only the C_0 is stable under strong interactions (M_{C_0} = 2.26 GeV, M_{C_1} = 2.42 GeV) it is likely the C_0-N and C_0-Nucleus systems are of greatest interest, although the C_1-nuclei are most deeply bound.

The $^1S_0$ states of B_c = N were examined with the thought that $^3S_1$-$^3D_1$, coupling by tensor forces might yield more deeply bound $^3S_1$ states, as in the deuteron. This tensor force due mainly to the π exchange is however, somewhat weakened by the increase in baryon mass. Bound states of the two-body system are found for $^1S_0$ C_1N (I = 3/2) and SN (I=1) states at 1.8 and 0.1 MeV respectively. Maximum I is favoured since the ρ, π isovector exchange potentials are then both attractive. The decrease in kinetic energy is here sufficient to compensate a similar $M(B_c)$ dependent reduction in potential, but in approximate calculations of channel coupling I = 1/2 $^3S_1$, C_1N and $^1S_0$, C_0N states are unbound. These conclusions about the two-body system are model-dependent; a slight increase in coupling constants leads to a bound $^1S_0$ (I=1/2) C_0N state, but it is clear no such states will ever be strongly bound.

In contrast the large mass of the charmed baryon should lead to many bound states in the many-body system. The spin and I-spin averaging which occurs in a finite nucleus leads to strong C_0, C_1-nucleus interactions, comparable to N-nucleus. Thus charmed nuclear states stable against strong decay will exist.

For J = 0, I = 0 nuclei the Hartree potential is

$$V_H(r) = \int \rho(\underset{\sim}{r}') \, V_0(\underset{\sim}{r}-\underset{\sim}{r}')d^3r'$$

neglecting spin-orbit and Coulomb contributions. Only Wigner forces are generated and these arise from exchange of ω, ϕ, ε mesons. The hard core may be removed by the use of the Moszkowski-Scott separation procedure[19], i.e., by finding a cutoff radius r_0 for which the $^1S_0$ B_c-N phase shift vanishes near zero kinetic energy. One takes

$$\begin{aligned} V'(r) &= 0 \qquad & r < r_0 \\ &= V_0(r) \qquad & r \geq r_o \end{aligned}$$

with

$$V_0(r) = \sum_i \mu_i g_i^2 \frac{e^{-x_i}}{x_i}$$

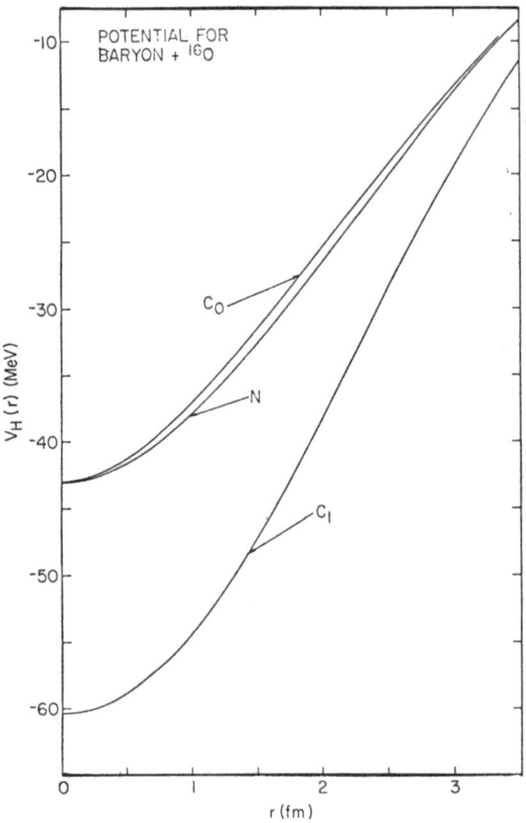
POTENTIAL FOR
BARYON + 16O
-10
-20
-30
-40
-50
-60
V_H (r) (MeV)
C0
N
C1
0
1
2
3
r (fm)

Fig. 14

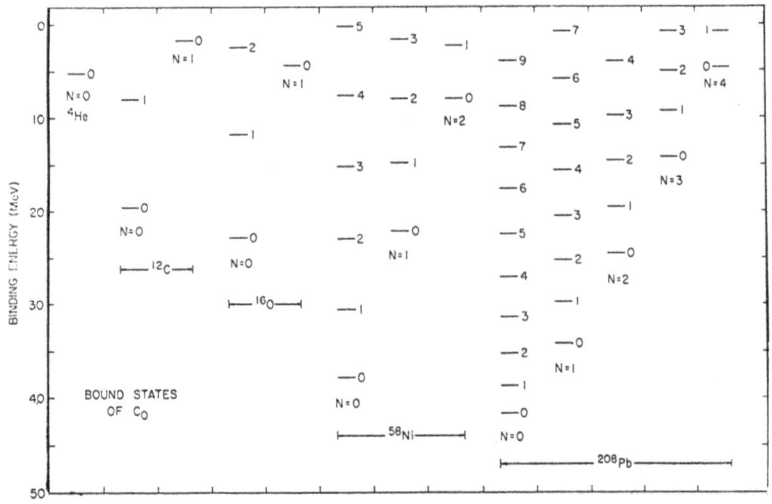
BINDING ENERGY (MeV)
0
10
20
30
40
50
4He
12C
16O
58Ni
208Pb
N=0
N=1
N=2
N=3
N=4
BOUND STATES
OF C0

Fig. 15

with specific values for $g_1^2/4\pi$ given in references 14 and 16.

A sample shell model charmed baryon-nucleus potential is shown in Fig. 14 for B_c-^{16}O, while in Fig. 15 calculated spectra for the C_0-nucleus shell model states are displayed. It is worth mentioning (a) the large reduced B_c mass leads to a compressed level spectrum, (b) states of rather high ℓ may be bound in heavy charmed nuclei, e.g., a C_1 $J = 25/2$ state coupled to a $i^{-1}_{13/2}$ neutron hole can lead to states with spins as high as 19 in $^{208}Pb_{\Sigma c}$.

B. A = 2 Bags

This is a subject which has been extensively considered by Jaffe and coworkers[20]. Rather than the molecular 3q + 3q states like the deuteron, or $3q + 3\bar{q}$ like $N\bar{N}$ considered above, Jaffe, et al have suggested the residual magnetic and electric colour forces may lead to rather deep binding of the $\Lambda\Lambda$ system, i.e., of a nuclear bag consisting of 4 ordinary and 2 strange quarks. Early emulsion work[21] on an event in which a doubly strange cascade Ξ^- particle is absorbed in a ^{12}C nucleus eventually producing a $^6He_{\Lambda\Lambda}$ fragment would seem to rule out the 6-quark bag, at least as a deeply bound system (binding energy $\sim$ tens of MeV), suggesting instead a Λ - Λ correlation energy in nuclear matter $\sim$ 4 MeV. Recent experiments at Brookhaven National Laboratory[22] on doubly strange system produced in p + p collision at few GeV also have seen no such structure, perhaps eliminating the possibility that some high repulsive barrier between the baryons prevents formation of the 6-quark bag at the low emulsion energy. Such systems, however, are of great interest and are relevant to the nuclear physicist if only because of exploration of the small distance behaviour in the A = 2 system.

C. $\bar{N}$ - Nucleus

A final topic which is under consideration (Dover, Zabek and Kahana) is the interaction between antibaryons and nuclei. Of prime interest is the possible existence of deeply bound $\bar{N}$ - nucleus states, presumably shielded from excessive annihilation by the localization of the antinucleon in the nucleon surface. Such states if not too wide could be detected, again, by the observation of high energy γ-rays.

REFERENCES

1. A. Chodos, R. Jaffe, K. Johnson, C. Thorn, and V. Weisskopf, Phys. Rev. D9 (1974), 1471.

2. J.J. Aubert et al, Phys. Rev. Lett. 33 (1974), 1404; J.E. Augustin et al, ibid 33 (1974), 1406.

3. T. Appelquist and B.A. Politzer, Phys. Rev. Lett. 34 (1975), 43; E. Eichten et al, Phys. Rev. Lett. 34 (1975), 369.

4. O.D. Dalkarov, V.B. Mandelzweig and I.S. Shapiro, Nucl. Phys. B21 (1970), 88; L.S. Shapiro, Usp. Fiz. Nauk 109 (1973), 431; C.B. Dover in Proceedings of the Fourth International Symposium on $\bar{N}N$ Interactions, Syracuse, New York, 1975; C.F. Chew and C. Rosenzweig, Phys. Lett. B58 (1975), 93.

5. A.S. Carroll et al, Phys. Rev. Lett. 32 (1974), 247; V. Chaloupka et al, Phys. Lett. 61B (1976), 487.

6. L. Gray, P. Hagerty, T. Kalogerpoulos, Phys. Rev. Lett. 26 (1971), 1491.

7. P. Pavlopoulos et al, Phys. Lett. 72B (1978), 415.

8. T.E. Kalogeropoulos et al, Phys. Rev. Lett. 35 (1975), 824.

9. R.L. Jaffe, Phys. Rev. D17 (1978), 1444.

10. C.B. Dover and M.C. Zabek, Phys. Rev. Lett. 41 (1978), 438.

11. R.A. Bryan and R.J.N. Phillips, Nucl. Phys. B5 (1968), 201; R.A. Bryan and B.L. Scott, Phys. Rev. 177 (1969), 1435.

12. J.L. Rosner, Rev. Mod. Phys. 47 (1975), 277.

13. E.G. Cazzoli et al, Phys. Rev. Lett. 34 (1975), 1125; B. Knapp et al, Phys. Rev. Lett. 37 (1976), 882; G. Goldhaber et al, Phys. Rev. Lett. 37 (1976), 255; I. Peruzzi et al, Phys. Rev. Lett. 37 (1976), 569.

14. C.B. Dover, S.H. Kahana, and T.L. Trueman, Phys. Rev. D16 (1977), 799.

15. W.W. Buck, C.B. Dover, and J.M. Richard, Annals of Physics (to be published).

16. C.B. Dover and S.H. Kahana, Phys. Lett. 62B (1976), 293.

17. M.M. Nagels, T.A. Rijen, and J.J. de Swart, Phys. Rev. D12 (1975), 744; 15 (1977), 2547.

18. J.M. Richard, private communication.

19. S.A. Moszkowski and B.L. Scott, Ann. Phys. 11 (1960), 65.

20. R.J. Jaffe, Phys. Rev. Lett. 38 (1977), 195; C. de Tar, Phys. Rev. D17 (1978), 302.

21. D. Prowse, Phys. Rev. Lett. 17 (1966), 782.

22. A.S. Carroll et al, Phys. Rev. Lett. (Sept. 1978).

Part III

Physics of Electrons, Protons, and Kaons

THEORETICAL INVESTIGATIONS OF THE OPTICAL MODEL FOR NUCLEONS AT NEGATIVE, LOW AND INTERMEDIATE ENERGIES

C. Mahaux

Institut de Physique, Université de Liège

Sart Tilman, B-4000 Liège 1, Belgium

TABLE OF CONTENTS

1. INTRODUCTION

Much effort has been devoted to the theoretical understanding of what Bethe[1] considers as the most striking feature of finite nuclei, namely that "it is a good approximation to consider each nucleon as moving in a smooth potential, possibly velocity-dependent". Our purpose is to discuss recent progress which has been achieved in the microscopic investigation of this average potential.

Since this Institute is concerned with common problems in low and medium energy nuclear physics, it appears fit to deal with bound as well as with scattered nucleons. More specifically, we shall cover an energy domain which extends approximately from -50 to +500 MeV. We intend to show that this whole range can be described in a unified way, the building stone being the nucleon-nucleon potential. At higher energies, one must abandon the latter concept, but one can then resort to the multiple scattering expansion[2-6] or to Glauber's theory[7,8], where one attempts to express the optical-model potential in terms of the nucleon-nucleon scattering amplitude.

Many of the numerical results presented below have been obtained from Reid's hard core interaction[9]. The reason is that if one wants to deal with intermediate as well as with low energies, it is necessary to adopt a realistic nucleon-nucleon potential with all its cumbersome ingredients: non-locality, short-range repulsion, tensor interaction. Furthermore, one must treat the Pauli principle with special care, since the incident particle is identical to those contained in the target. The basic tools for dealing with these two difficulties have been developed by Brueckner, Bethe and collaborators[1,10]. They were mainly interested in ground state properties. In collaboration with Hüfner, Jeukenne and Lejeune[11,12], we have combined this Bethe-Brueckner technique with nuclear reaction theory and with the Green function formulation of many-body physics. The resulting approach constitutes the leading thread of our lectures, which are organized as follows.

In sect. 2, we first briefly give an intuitive definition of the optical model and describe the main empirical properties of the optical-model potential. Then we proceed to a specific theoretical definition and explain our basic point of view: since the potential varies smoothly with mass number, it appears justified to study in detail the large target approximation, i.e. nuclear matter. The potential for finite nuclei can then be constructed by means of a local density approximation.

Section 3 is devoted to the scattering of nucleons with low energy, typically $0 < E < 170$ MeV. We successively consider the real and the imaginary parts of the potential, and its symmetry, Coulomb and spin-orbit components. We assess the successes and failures of the approach, and briefly discuss possible improvements.

In sect. 4, we deal with the intermediate energy domain, typically $150 < E < 500$ MeV. We devote particular attention to the change of sign of the real part of the optical-model potential, and to the relationship between our approach and the multiple scattering theory, and also with relativistic formulations.

Some rather general results on the relative importance of the non-locality and of the energy dependence of the optical-model potential can be derived from our theoretical approach, and also from a dispersion relation which connects the real and the imaginary parts of the optical-model potential. This constitutes the subject of sect. 5.

In sect. 6, we turn to the domain of negative energies, typically $-50 < E < 0$ MeV. We first discuss weakly bound single-particle states observed by pick-up reactions, and then the deeply bound states observed by (p,2p) or (e,e'p) knock-out relations.

2. THE OPTICAL MODEL

2.1. Quasi-definition of a Quasi-particle

For simplicity, we first consider the case of infinite nuclear matter with a uniform density ρ. The Fermi momentum k_F is defined by

$$\rho = \frac{2}{3\pi^2} k_F^{\ 3} \quad . \tag{2.1}$$

If we neglect ground state correlations (free Fermi gas), the momentum distribution is given by

$$n_<(k) = \langle \Phi_o | a^\dagger(k)\ a(k) | \Phi_o \rangle \tag{2.2}$$

$$= 1 \text{ for } k < k_F \quad , \quad 0 \text{ for } k > k_F \quad , \tag{2.3}$$

where Φ_o is a Slater determinant built with plane waves

$$|\underset{\sim}{k}\rangle = \exp(i\ \underset{\sim}{k}\cdot\underset{\sim}{r}) \quad .$$

Let Φ denote the true ground state wave function. The momentum distribution in this correlated state reads

$$\rho(k) = \langle \Phi | a^\dagger(k)\ a(k) | \Phi \rangle \quad . \tag{2.4}$$

A typical momentum distribution is plotted in fig. 1. We see that the depletion of the Fermi sea is fairly small. This suggests to

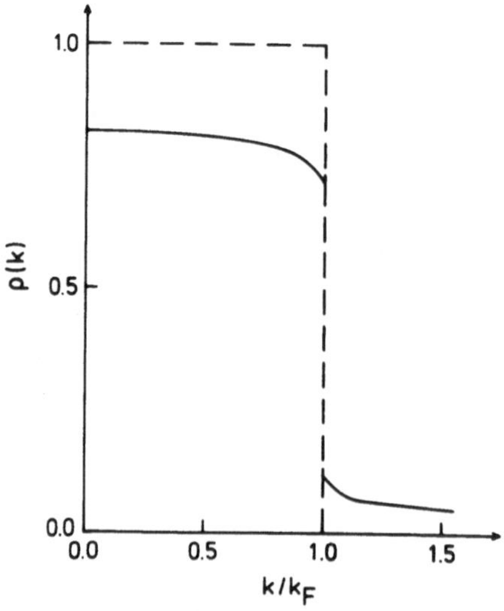

Fig. 1. Momentum distribution in the ground state of nuclear matter. The dashed line corresponds to the free Fermi gas (eq. (2.2)), the full curves to the correlated ground state. From ref. 12.

expand all calculated quantities in terms of the "smallness parameter"

$$\kappa = 1 - \rho(\bar{j}) \qquad , \qquad (2.5)$$

where $\bar{j}$ is the average momentum of a nucleon inside the Fermi sea:

$$\bar{j} \simeq 0.75\, k_F \qquad . \qquad (2.6)$$

The value of the smallness parameter is typically $\kappa \simeq 0.2$.

For definiteness, let us discuss the case of particle states; hole states can be handled in exactly the same way. Let us create at time $t = 0$ a nucleon with momentum $k > k_F$ on top of the correlated ground state. We obtain the initial state

$$\Psi(k,t=0) = a^{\dagger}(k,t=0)|\Phi\rangle \qquad . \qquad (2.7)$$

The probability amplitude of finding a particle with the same momentum k on top of the ground state at a later time $t > 0$ is given by the one-body Green function

$$G_p(k,t) = -i\,\langle\Phi|a(k,t)\;a^{\dagger}(k,t=0)|\Phi\rangle \qquad (t > 0) \qquad (2.8a)$$

$$= 0 \qquad (t < 0) \qquad . \qquad (2.8b)$$

For a hole state $k < k_F$, the Green function is defined by

$$G_h(k,t) = i\,\langle\Phi|a^{\dagger}(k,t)\;a(k,t=0)|\Phi\rangle \qquad (t > 0) \qquad (2.9a)$$

$$= 0 \qquad (t < 0) \qquad . \qquad (2.9b)$$

The one-body Green function proper is the sum of G_p and G_h[13].

In the case of a free Fermi gas where all nucleons move independently of one another in an average external potential U(k), eq. (2.8) yields ($\hbar = 1$)

$$G_p^{(0)}(k,t) = -i\;n_>(k)\;\exp(-i\;e(k)t) \qquad (t > 0) \qquad (2.10a)$$

$$= 0 \qquad (t < 0) \qquad , \qquad (2.10b)$$

where (compare with eq. (2.3))

$$n_>(k) = 1 \;\text{ for }\; k > k_F, \;\; = 0 \;\text{ for }\; k < k_F, \qquad (2.11)$$

and

$$e(k) = \frac{k^2}{2m} + U(k) \quad . \tag{2.12}$$

In the presence of correlations, the probability amplitude $G_p(k,t)$ must decrease in time since the collisions will modify the momentum of the nucleon. In the optical model as used in practice, it is <u>assumed</u> that this decrease is exponential:

$$G_p^{(OM)}(k,t) = -i\; n_>(k)\; e^{-t/2\tau(k)}\; \exp(-i\; e(k)t) \; . \tag{2.13}$$

As indicated, the lifetime

$$\tau(k) = (-2\; W(k))^{-1} \tag{2.14}$$

and the potential energy

$$V(k) = e(k) - \frac{k^2}{2m} \tag{2.15}$$

may depend on the nucleon momentum k.

By performing a Fourier transform over the time, the optical-model approximation (2.13) to the Green function reads ($k > k_F$)

$$G_p^{(OM)}(k,E) = \frac{n_>(k)}{E - e(k) + i\; W(k)} \tag{2.16}$$

This can be compared to the free Fermi gas value (see eq. (2.10))

$$G_p^{(0)}(k,E) = \frac{n_>(k)}{E - e(k) + i\delta} \tag{2.17}$$

This confirms that the imaginary part W is a consequence of the correlations. It exists for hole as well as for particle states. The comparison between eqs. (2.16) and (2.17) shows that in the optical model the nucleons are assumed to move independently of one another in a complex mean field

$$M(k) = V(k) + i\; W(k) \quad . \tag{2.18}$$

The momentum k of a nucleon with energy E is then complex:

$$k = \{2m(E - V - i\,W)\}^{1/2} = k^{(R)} + i\,k^{(I)} \quad . \tag{2.19}$$

and the single-particle wave function reads

$$\exp(ikz) = \exp(ik^{(R)}z)\,\exp(-z/2L) \quad . \tag{2.20}$$

The mean free path L is given by

$$L = (2\,k^{(I)})^{-1} = v^{(R)}\tau \quad , \tag{2.21}$$

where

$$v^{(R)} = k^{(R)}/m \simeq \{2m(E - V)\}^{1/2}$$

is the velocity of the nucleon inside nuclear matter.

Because of the uncertainty principle, the existence of a finite lifetime implies that the energy of the single-particle state is spread over a domain of width

$$\Gamma^{\downarrow} = 2\,W \tag{2.22}$$

centered on e(k). The quantity $\Gamma^{\downarrow}$ is called the single-particle spreading width (sect. 6).

All the quantities introduced above depend on the Fermi momentum k_F or equivalently on the density ρ of nuclear matter. Whenever useful, we shall recall this property by an index ρ; for instance, eq. (2.18) then becomes

$$M_\rho(k) = V_\rho(k) + i\,W_\rho(k) \quad . \tag{2.23a}$$

In a finite nucleus, the density is a function $\rho(r)$ of the distance r to the nuclear centre. The simplest version of the local density approximation (LDA) consists in assuming that at the location r the mean field is the same as in nuclear matter at the corresponding density $\rho(r)$:

$$M^{(LDA)}(r,k) = M_{\rho(r)}(k) = V_{\rho(r)}(k) + i\,W_{\rho(r)}(k) \quad , \tag{2.23b}$$

$$= M(r,k) = V(r,k) + i\,W(r,k) \quad . \tag{2.23c}$$

Equation (2.19) relates the energy and the momentum:

$$E = \frac{k^2}{2m} + V(k) \quad . \tag{2.24}$$

This equation yields k(E), or E(k), which enables one to transform the dependence on k of the mean field into a dependence on the energy:

$$M(r,k(E)) = M_L(r,E) \quad . \tag{2.25}$$

Expressed in the form (2.25), the mean field is a function of energy, and is local in coordinate space. We could also keep the original form M(r,k). By performing a Fourier transform over k, we would obtain a non-local but energy-dependent mean field

$$M_{\rho(r)}(k) \rightarrow M_{\rho(r)}(|\vec{r}-\vec{r}\,'|) = M(r,|\vec{r}-\vec{r}\,'|) \quad . \tag{2.26}$$

When used in a Schroedinger equation, the forms (2.25) and (2.26) yield, respectively,

$$-(2m)^{-1}\, \nabla^2\, \psi_E(\vec{r}) + M_L(r,E)\, \psi(\vec{r}) = E\, \psi(\vec{r}) \quad , \tag{2.27}$$

$$-(2m)^{-1}\, \nabla^2\, \psi_E(\vec{r}) + \int M(r,|\vec{r}-\vec{r}\,'|)\, \psi_E(\vec{r}\,')\, d^3r' = E\, \psi_E(\vec{r}) . \tag{2.28}$$

This method of constructing a non-local field "equivalent" to a local, energy-dependent potential is mainly due to Feshbach[14], Perey and Saxon[15], and Frahn[16]. We put the word "equivalent" between quotation marks because of the following reason. In a finite system, the "equivalent" potentials (2.25) and (2.26) yield practically the same elastic scattering phase shifts. However, we shall recall in chapter 5 that the amplitude of the elastic scattering wave function <u>inside</u> the nucleus is reduced by a factor $(\frac{\tilde{m}}{m})^{1/2}$ in the case of the non-local field (2.26) as compared with the local field (2.25)[17]. Here, $\tilde{m}$ is the effective mass which characterizes the non-locality (sect. 5). This effect is important when inelastic scattering is computed[18]. It may also introduce significant corrections to the calculated (p,2p) cross sections. These turn out to be larger than the experimental values by a factor two[19]. Since at the relevant energy $\tilde{m}/m \simeq 0.75$ (sect. 5), and since the Perey correction factor enters to the sixth power, the cross section calculated with an empirical local potential should be reduced by a factor close to 0.5. This may resolve the discrepancy between theoretical and experimental values.

2.2. Empirical Properties

The empirical density distribution in a nucleus with mass number A can be parametrized as follows[20]

$$\rho(r) = \rho_o \, f^{(\rho)}(r) \quad , \tag{2.29}$$

$$f^{(\rho)}(r) = \rho_o \, [1 + \exp(r - R_\rho)/a_\rho]^{-1} \quad , \tag{2.30}$$

$$R_\rho = (0.978 + 0.0206 \, A^{1/3}) \, A^{1/3} \text{ fm} \quad , \tag{2.31a}$$

$$\rho_o = 3A \, \{4\pi R_\rho^3 \, (1 + \pi^2 \, a_\rho^2 \, R_\rho^{-2})\}^{-1} \text{ fm}^{-3} \quad , \tag{2.31b}$$

$$a_\rho = 0.54 \text{ fm} \quad . \tag{2.31c}$$

The parametrization (2.30) is usually called a Woods-Saxon distribution by the practitioners of the optical model.

The optical-model potential is determined by the analysis of the experimental elastic scattering and total reaction cross sections. The real part of the potential is usually assumed to have a Woods-Saxon shape, at least for $E < 150$ MeV. A typical parametrization in the domain $10 < E < 70$ MeV is[21]

$$V(r,E) = V(E) \, f^{(V)}(r) \quad , \tag{2.32a}$$

$$R_V = r_V \, A^{1/3} \; , \; r_V = 1.17 \text{ fm} \; , \; a_V = 0.75 \text{ fm} \quad , \tag{2.32b}$$

$$V(E) = V_o(E) \mp V_1 \frac{N-Z}{A} + \Delta_C \quad , \tag{2.32c}$$

$$V_o(E) = -54 + 0.32 \, E \text{ (MeV)} \quad , \tag{2.32d}$$

$$V_1 = 24 \text{ MeV} \; , \; \Delta_C = 0.4 \, Z \, A^{-1/3} \quad . \tag{2.32e}$$

In eq. (2.32c), the upper sign in front of the symmetry potential V_1 refers to protons, the lower one to neutrons, and the Coulomb correction Δ_C vanishes for neutrons.

The parametrization (2.32) cannot be extrapolated to negative or to intermediate energies. It would for instance predict that the potential changes sign at 170 MeV, while empirically this change of sign is observed at an energy higher than 250 MeV (fig. 2). We shall discuss the intermediate energy range in sect. 4, and the negative energies in sect. 6.

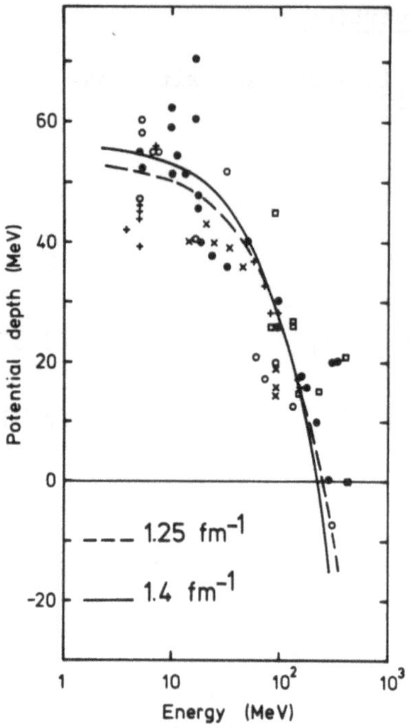

Fig. 2. The dots, crosses and squares show a compilation of empirical depths of the real part of the optical-model potential due to Passatore[22]. The full and the dashed curves represent the Brueckner-Hartree-Fock value at the densities $\rho = 0.185$ fm^{-3} ($k_F = 1.4$ fm^{-1}) and 0.132 fm^{-3} ($k_F = 1.25$ fm^{-1}). From ref. 23.

We have seen in sect. 2.1 that one can analyze the elastic scattering data with a non-local field instead of an energy-dependent potential. This was performed by Perey and Buck[24] who adopted the parametrization

$$V(r,|\vec{r}-\vec{r}\,'|) = V_o f^{(V)}(r) \exp(-s^2/\beta^2) \quad , \qquad (2.33a)$$

$$s = |\vec{r}-\vec{r}\,'| \quad . \qquad (2.33b)$$

For 10 < E < 25 MeV, they obtained good fits with the value

$$\beta \simeq 0.85 \quad \text{fm} \qquad (2.34)$$

for the non-locality range.

Experimental access to the mean free path L is very indirect and therefore subject to heated controversy[25,26]. It is related to the imaginary part W of the optical-model potential by eqs. (2.14), (2.21). At low energy, the radial dependence of W(r) is more complicated than that of V(r): it is peaked at the nuclear surface for E < 50 MeV, and takes a Woods-Saxon shape for higher energy. This leads to a variety of assumptions concerning the geometry of W(r) and is the main origin of the scatter of the empirical values shown in fig. 3. Besides, the value of W(r) is expected to be more sensitive to the detailed structure of the nuclear excited states.

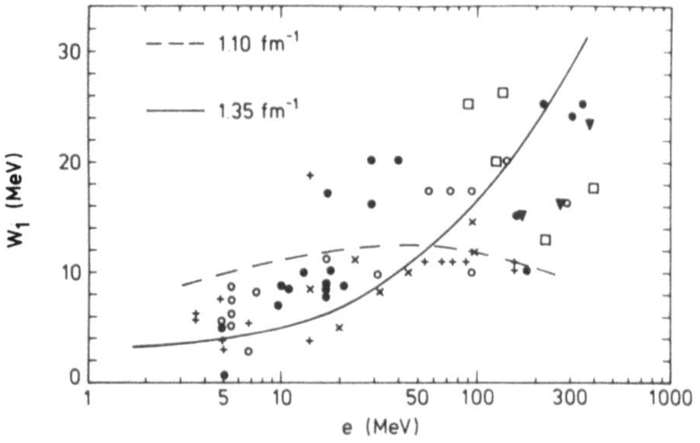

Fig. 3. Same as fig. 2, for the depth of the imaginary part of the optical-model potential; the curves correspond to the densities $\rho = 0.166$ fm^{-3} ($k_F = 1.35$ fm^{-1}) and 0.090 fm^{-3} ($k_F = 1.10$ fm^{-1}). From ref. 12.

While the shapes of V(r) and of W(r) are difficult to ascertain empirically, it appears[27,28] that the experimental data determine rather accurately their volume integrals per nucleon and their root mean square radii, i.e. the quantities

$$J_V/A = A^{-1} \int V(r)\, d^3r \quad , \quad J_W/A = A^{-1} \int W(r)\, d^3r \quad , \tag{2.35}$$

$$\langle R_V \rangle^{1/2} = (A\, J_V)^{-1/2} \left[\int V(r)\, r^2\, d^3r\right]^{1/2} \quad , \tag{2.36a}$$

$$\langle R_W \rangle^{1/2} = (A\, J_W)^{-1/2} \left[\int W(r)\, r^2\, d^3r\right]^{1/2} \quad . \tag{2.36b}$$

The dependence upon mass number for A > 40 of the volume integrals of V(r) and of W(r) for 35 MeV protons is plotted in figs. 4 and 5. There, we also show the contribution of the symmetry (V_1) and of the Coulomb (Δ_C) components. It is noticeable that the dependence of the volume integrals upon mass number is very weak. Since the main optical-model parameters are fairly independent of mass number and of the detailed structure of the target they can plausibly be derived from the study of nuclear matter. This is the basic point of view adopted in refs. 29-31 and below.

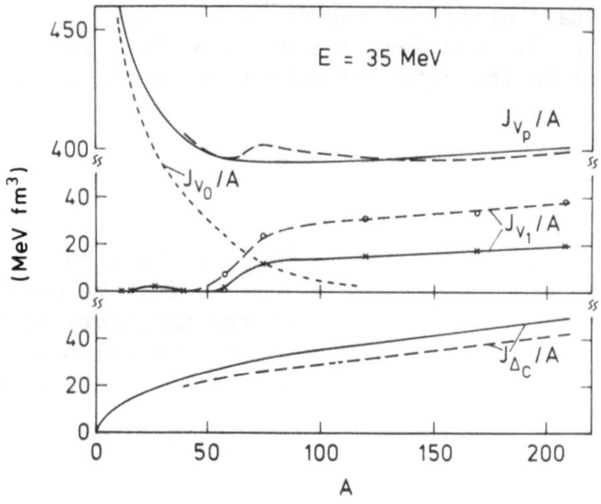

Fig. 4. Dependence upon mass number of the volume integral per nucleon of the real part of the optical-model potential, and of its symmetry and Coulomb components, for 35 MeV protons. The dashed curves correspond to the empirical potential of ref. 21 (eq. (2.32)). The dashed curves show the Brueckner-Hartree-Fock approximation. From ref. 29.

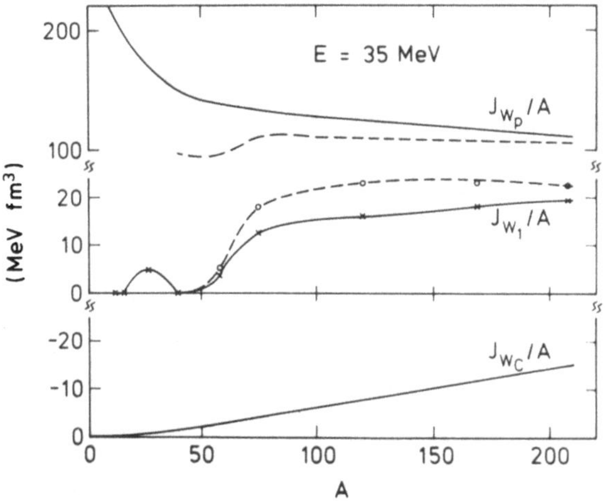

Fig. 5. Same as fig. 4, for the imaginary part of the optical-model potential. From ref. 29.

Despite numerous efforts, very little is known experimentally about the momentum distribution shown in fig. 1. Here the finite size of nuclei plays an essential role for momenta smaller than $\pi/R_0 \simeq 2\ \text{fm}^{-1}$. Larger momentum states are populated mainly because of the short-range correlations which are able to excite nucleons up to a relative momentum of order $\pi/2c \simeq 4\ \text{fm}^{-1}$, where c is the hard core radius[32]. It is probably a good approximation to assume that two-nucleon correlations dominate in the high-momentum range, and that this correlated pair behaves like a deuteron at short inter-pair distances[33].

Considerable interest is thus devoted to the analysis of physical processes which must involve at least two correlated nucleons. Pion absorption at rest is one example: the energy transferred to the nucleus is $\Delta E \simeq 140$ MeV, while the transferred momentum ΔQ is negligible. One has thus

$$\Delta E >> \frac{(\Delta Q)^2}{2m} \quad ; \tag{2.37}$$

one single nucleon can hardly "absorb" the energy[34] since one nucleon processes would correspond to $\Delta E \simeq (\Delta Q)^2/2m$. The opposite situation

$$\Delta E << \frac{(\Delta Q)^2}{2m} \tag{2.38}$$

is encountered in backward inelastic scattering of protons[35,36]: if the energy of the incident proton is 600 MeV, an outgoing 150 MeV proton corresponds to a momentum transfer such that $(\Delta Q)^2/2m \simeq 1.7$ GeV, while $\Delta E = 450$ MeV. Here again, at least $n \gtrsim 2$ nucleons probably take part in the process, with n determined by $\Delta E \simeq (\Delta Q)^2/2Mn$. However, it is not yet clear how to achieve a proper interpretation of these data[37-40].

2.3. Theoretical Definition

The one-body Green function is defined by eqs. (2.9) - (2.10). Equation (2.16) shows that in the practical optical model one assumes that this Green function can be approximated by a simple pole term. To what extent is this assumption valid? In order to discuss this point, it is useful to introduce two quantities which will play an important role in the following, namely the mass operator (or self-energy) M(k,E) and the spectral function S(k,E)[13,41].

Let $\phi_\omega^{(\pm)}$ denote the complete set of normalized eigenstates of the nuclear Hamiltonian for the system with (A±1) nucleons. The spectral function is equal to

$$S_p(k,\omega) = |\langle\phi_\omega^{(+)}|a^\dagger(k)|\Phi\rangle|^2 \tag{2.39a}$$

in the case of particle states, and to

$$S_h(k,\omega) = |\langle\phi_\omega^{(-)}|a(k)|\Phi\rangle|^2 \tag{2.39b}$$

in the case of hole states. One has

$$S(k,\omega) = \frac{i}{2\pi} [G(k,\omega+i\eta) - G(k,\omega-i\eta)] \quad , \tag{2.40a}$$

$$G(k,E) = \int d\omega \frac{S(k,\omega)}{E-\omega} \quad . \tag{2.40b}$$

The quantity $S_h(k,\omega)d\omega$ measures the joint probability of being able to extract one nucleon with momentum k from the correlated ground state, and of finding that the energy of the residual nucleus is contained in the energy interval $(\omega,\omega+d\omega)$. Hence, the spectral function is the quantity which is investigated in pick-up or in knock-out reactions. In the free Fermi gas model (2.10a), it reduces to a delta function

$$S^{(0)}(k,\omega) = \delta(\omega - e(k)) \quad . \tag{2.40c}$$

In the presence of interactions, we expect $S(k,\omega)$ to be spread over an energy range of size $\Gamma^{\downarrow}$, for fixed k (see eqs. (2.22), (2.45)).

The mass operator M(k,E) is related to the Green function by

$$G(k,E) = \frac{1}{E - k^2 - M(k,E)} \quad ; \tag{2.41}$$

it is a complex quantity

$$M(k,E + i\eta) = V(k,E) + i\, W(k,E) \quad . \tag{2.42}$$

The imaginary part W(k,E) is negative except at the Fermi energy ε_F defined by

$$\varepsilon_F = \frac{k_F^2}{2m} + M(k_F,\varepsilon_F) \quad , \tag{2.43}$$

where it vanishes quadratically:

$$W(k,E) \sim (E - \varepsilon_F)^2 \quad . \tag{2.44}$$

From eqs. (2.40), (2.41) and (2.42), one gets

$$S(k,E) = -\frac{1}{\pi} \frac{W(k,E)}{[E - \frac{k^2}{2m} - V(k,E)]^2 + \frac{1}{4}[2\, W(k,E)]^2} \quad . \tag{2.45}$$

The mass operator satisfies a dispersion relation (sect. 5):

$$V(k,E) = f(k) + \pi^{-1} \int_{-\infty}^{\infty} dE' \frac{W(k,E')}{E' - E} \quad , \tag{2.46}$$

where the energy-independent quantity f(k) is given by

$$\lim_{|E|\to\infty} M(k,E) = f(k) \quad . \tag{2.47}$$

We are now in a position to discuss the precise definition of the optical-model potential and the validity of the optical model. The comparison between the optical-model approximation (2.16) for the Green function and the exact expression (2.41) suggests to identify the mass operator M(k,E) with the mean potential. This is correct. Indeed, it has been proved by Bell and Squires[42] that, when introduced in the Schroedinger equation, the Fourier transform

$$M(|\vec{r}-\vec{r}\,'|,E) \tag{2.48}$$

of M(k,E) exactly reproduces the elastic part of the full scattering wave function[14,43,44]. Note that in a finite nucleus this mean field reads $M(\vec{r},\vec{r}\,';E)$: it is non-local and complex.

The practical optical model only deals with averaged quantities rather than with the exact elastic scattering wave function. This more practical viewpoint is attainable by expanding (2.41) in the vicinity of the "quasi-particle" pole. If e(k) is the solution of

$$e(k) = \frac{k^2}{2m} + V(k,e(k)) \quad , \tag{2.49}$$

one has[12,45]

$$G(k,E) = \frac{R(k)}{E - e(k) + i\, W(k)} + \text{background} \quad , \tag{2.50}$$

$$S(k,E) = \pi^{-1} \frac{\overline{W}(k)\ \mathrm{Re}\ R(k) - [E-e(k)]\ \mathrm{Im}\ R(k)}{[E - e(k)]^2 + \frac{1}{4}\,[2\,\overline{W}(k)]^2} + \text{background} \; ; \tag{2.51}$$

where

$$W(k) = W(k,e(k)) \quad , \tag{2.52a}$$

$$\overline{W}(k) = W(k)\,R(k) \quad , \tag{2.52b}$$

$$R(k) = \{1 - \frac{\partial V(k,E)}{\partial E}\}^{-1}_{E=e(k)} \quad . \tag{2.53}$$

Equations (2.16) and (2.50) show that the usual optical model deals with the pole approximation of the Green function or, equivalently, with that fraction ($\simeq |R(k)|$) of the incident wave packet whose amplitude decreases exponentially in time. This model is useful only inasmuch as the spectral function displays a well-defined "resonance peak" above the background. The spectral function calculated in the case of a hole and of a particle state are plotted in figs. 6 and 7, respectively. These show that the basic assumption of the optical model is fairly well fulfilled.

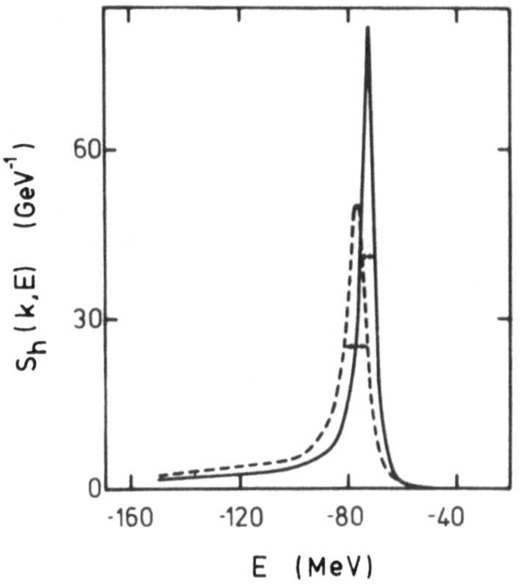

Fig. 6. Spectral function for a hole state with momentum $k = 0.55\ k_F$, $k_F = 1.36\ fm^{-1}$, in the case of the nucleon-nucleon interaction of Hamman and Ho-Kim[46,47]. The full line corresponds to eq. (2.51) and the dashed curve is obtained if one sets R(k) to unity. From ref. 48.

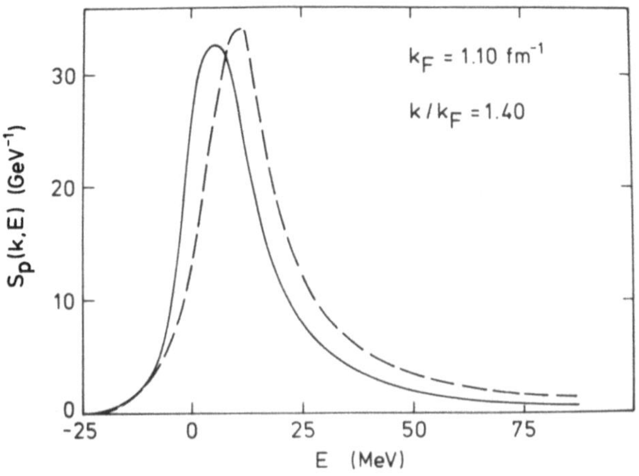

Fig. 7. Spectral function for a particle state with momentum $k = 1.40\ k_F$, $k_F = 1.10\ fm^{-1}$, in the case of Reid's hard core nucleon-nucleon interaction. The full curve is computed from eq. (2.47) and the dashes from the pole approximation (2.51). From ref. 12.

2.4. Calculational Procedure

If the nucleon-nucleon interaction v would be sufficiently soft, one could use straightforward perturbation theory. Each term of the resulting series can be represented by graphs. A few of these are shown in fig. 8.

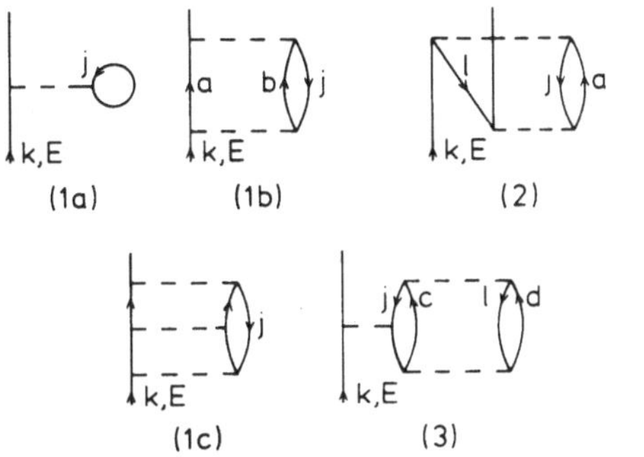

Fig. 8. Some graphs which appear in the expansion of the mass operator in powers of the strength of the nucleon-nucleon interaction. From ref. 12.

$$M(k,E) = \hat{M}_{1a}(k) + \hat{M}_{1b}(k,E) + \hat{M}_{1c}(k,E) + \hat{M}_2(k,E) + \hat{M}_3(k,E) + \ldots \tag{2.54}$$

where the hat refers to the perturbation expansion in powers of v. The diagrams (1a), (1b) and (1c) share the common feature of involving a summation over only one intermediate hole-line j. The first-order term

$$\hat{M}_{1a}(k) = \sum_j n_<(j) \langle k,j|v|k,j\rangle_A \tag{2.55}$$

is the Hartree-Fock approximation: it is real and independent of E. The index A refers to antisymmetrization:

$$|k,j\rangle_A = |k,j\rangle - |j,k\rangle \quad . \tag{2.56}$$

The algebraic expression of the other graphs reads

$$\hat{M}_{1b}(k,E) = \sum_{j,a,b} n_<(j)n_>(a)n_>(b) \frac{\langle k,j|v|a,b\rangle\langle a,b|v|k,j\rangle_A}{E + e(j)-e(a)-e(b)+i\delta} \tag{2.57}$$

$$\hat{M}_2(k,E) = \sum_{j,\ell,a} n_<(j)n_<(\ell)n_>(a) \frac{<k,a|v|j,\ell><j,\ell|v|k,a>_A}{E + e(a) - e(j) - e(\ell) - i\delta} \qquad (2.58)$$

The energy denominators in eqs. (2.57), (2.58) involve a potential U(k) (eq. (2.12)) which is in principle arbitrary but should be chosen in such a way as to optimize the rate of convergence of the expansion.

Because of the strength of v, the perturbation series (2.54) is not useful. However, one can rearrange it by summing up an infinite series of diagrams. For this purpose, one introduces Brueckner's reaction matrix g[w] which is the solution of the following (Bethe-Goldstone) integral equation

$$g[w] = v + v \sum_{a,b} n_>(a)n_>(b) \frac{|a,b><a,b|}{w - e(a) - e(b) + i\delta} g[w]. \qquad (2.59)$$

It can immediately be verified that the sum of the subseries

$$\hat{M}_{1a}(k,E) + \hat{M}_{1b}(k,E) + \hat{M}_{1c}(k,E) + \ldots \qquad (2.60)$$

is given by

$$M_1(k,E) = \sum_j n_<(j) <k,j|g[E + e(j)]|k,j>_A \quad . \qquad (2.61)$$

The expression (2.60) is called the Brueckner-Hartree-Fock approximation; it is represented by graph (1) of fig. 9, where a wiggly line corresponds to a g-matrix. In the same way, graph (2) in

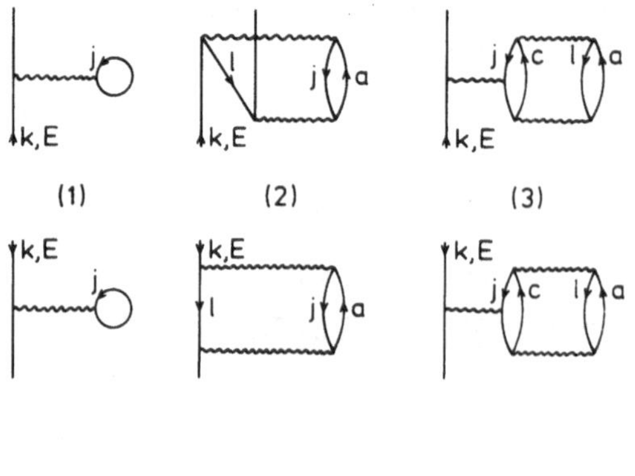

Fig. 9. Some graphs which appear in the low-density expansion of the mass operator. The upper part corresponds to particle states, the lower part to hole states. From ref. 12.

fig. 8 is the progenitor of a series of diagrams which can be summed up to yield graph (2) of fig. 9, etc.

The Bethe-Goldstone equation (2.59) is sometimes written in the abbreviated form

$$g[w] = v + v \frac{Q}{w - e} g[w] \quad , \tag{2.62}$$

where

$$Q = \sum_{a,b} n_>(a) n_>(b) \, |a,b\rangle\langle a,b| \tag{2.63}$$

is the "Pauli operator" which requires the intermediate states $|a\rangle$ and $|b\rangle$ to lie above the Fermi surface. If we would set

$$Q = 1, \tag{2.64}$$

and

$$U(k) = 0 \quad , \tag{2.65}$$

the reaction matrix would become identical to the transition matrix for free nucleon-nucleon scattering:

$$t[w] = v + v \sum_{a,b} \frac{|a,b\rangle\langle a,b|}{w - \frac{a^2}{2m} - \frac{b^2}{2m} + i\delta} t[w] \quad . \tag{2.66}$$

By comparing (2.55) and (2.61), we see that the reaction matrix plays the role of an effective interaction. Since it is similar to the transition matrix (2.66), it is rather soft and is in any case free of those singularities of v which are associated with the short range repulsion between nucleons. The rearranged series (fig. 9)

$$M(k,E) = M_1(k,E) + M_2(k,E) + M_3(k,E) + \dots \tag{2.67}$$

thus appears to be more satisfactory than the perturbation expression (2.54).

When is the series (2.67) likely to converge? One can characterize the terms on the right-hand side of (2.67) by their number of intermediate hole states, i.e. of integrations from 0 to k_F in the corresponding algebraic expression. Hence, we expect the nth term of this hole-line expansion to contain an additional factor of order $k_F^3 \propto \rho$ as compared to the (n-1)th term. If this is so, the expansion (2.67) is valid at low density, and can accordingly be called a low-density (or hole-line) expansion. A similar approach had been developed for π-nucleus scattering[49]. In the case of nucleons, it was introduced in ref. 11 by analogy with the Bethe-Brueckner hole-line expansion[1] for the binding energy of nuclear matter.

Is it possible to be more specific and define a smallness parameter? One can easily evaluate the ratio of the typical two-hole line graph M_3 to the one-hole line diagram M_1; one finds (see eq. (4.16))

$$M_3(k,E) \simeq - \kappa \, M_1(k,E) \quad , \tag{2.68}$$

where κ is defined in eq. (2.5). Hence, the rate of convergence of the series is measured by the amount of depletion of the Fermi sea in the correlated ground state. We recall that for realistic nucleon-nucleon interactions, one has (fig. 1)

$$\kappa \simeq 0.2 \quad . \tag{2.69}$$

We mentioned that the external potential $U(k)$ can be chosen in such a way as to improve the convergence of the expansion. The value quoted in (2.69) corresponds to the self-consistent Brueckner-Hartree-Fock choice

$$U(k) = \text{Re}\, M_1(k,e(k)) = \text{Re}\, M_1(k) \quad , \tag{2.70}$$

which is supported by arguments given in refs. 11, 12. In principle, one could adopt more complicated choices for $U(k)$, for instance by including the so-called "rearrangement" term M_2 in the choice of the external potential, i.e. by taking[50]

$$U(k) = \text{Re}\, [M_1(k,e(k)) + M_2(k,e(k))] \quad . \tag{2.71}$$

Finally, we note that once an approximation is known for the one-particle Green function $G(k,E)$, it is possible to compute the momentum distribution $\rho(k)$ (eq. (2.4)) and the average binding energy per nucleon[12].

3. THE LOW ENERGY DOMAIN ($0 < E < 170$ MeV)

3.1. Why Fly at Low Energy

The optical-model potential is best known experimentally for energies which range from 0 to about 80 MeV. Hence, this domain provides a good testing ground for the reliability of the various theoretical approaches. At low energy, there exist a variety of these[51], mainly because there one can tentatively use effective nucleon-nucleon interactions adjusted to the properties of ground and low excited nuclear states. One of the practical interests of the method described in sect. 2.4 is that it can be applied to a much wider energy range. The detailed investigation of the low energy domain will indicate to what extent the method is reliable, and what is its predictive power when applied to the foggier intermediate energy range (sect. 4). There, competitive approaches are very few, since the standard effective interactions are no longer meaningful, while the multiple scattering expansion is not yet reliable. Here, we enlarge the usual definition of the low energy range, mainly because the calculated optical-model parameters vary smoothly from 0 to 170 MeV.

The results presented below have been calculated from Reid's hard core nucleon-nucleon interaction (S, P and D partial waves with $J \leq 2$) and from the Brueckner-Hartree-Fock approximation (2.61). Some higher order terms of the low-density expansion have been considered in refs. 52 and 12, but have not yet been fully included in the calculations. We shall use the local density approximation given in eq. (2.23). This is admittedly a crude one, and improvements will be described in sect. 3.5. Unless otherwise specified, we assume that the neutron and proton distributions are the same and are given by eqs. (2.29)-(2.31). In sects. 3.2 to 3.4, we successively discuss the real, the imaginary and the small (isovector, Coulomb, spin-orbit) components of the optical-model potential.

3.2. Real Part of the Potential

In fig. 10, the momentum dependence (see eq. (2.70)) of the real part U(k) of the potential is represented by the full curve (left-hand scale), for the Fermi momentum $k_F = 1.36\ \mathrm{fm}^{-1}$ which is associated with the density at the center of a finite nucleus. To each value of k one can attach an energy e(k) by using eq. (2.12). The function e(k) is given by the dash-and-dots (right-hand scale).

In their analysis of elastic scattering data below 25 MeV, Perey and Buck[24] have assumed that the dependence of U(k) upon k can be approximated by a Gaussian (eq. (2.33))

$$U(k) = U_o \exp(-\frac{1}{4}\beta^2 k^2) \quad . \tag{3.1}$$

We first note that this cannot be exactly true since it is not

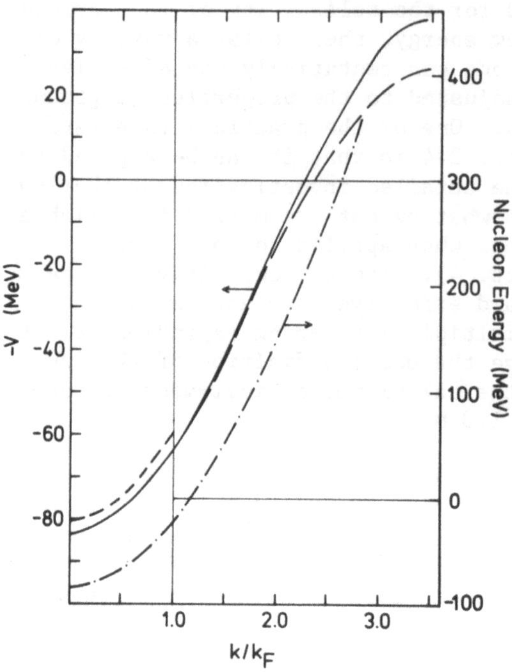

Fig. 10. Momentum dependence of the real part of the potential (eq. (2.70), left-hand scale, full curve) and of the nucleon energy (eq. (2.12), right-hand scale, dash-and-dots), for $k_F = 1.36\ fm^{-1}$. From ref. 53.

compatible with the fact that the real part of the potential changes sign (at $k \simeq 2.2\ k_F$). However, the Gaussian (3.1) is a very good approximation of U(k) in the momentum range ($k_F < k < 1.5\ k_F$) covered by the data. The corresponding non-locality range is $0.84\ fm^{-1}$ (eq. (8.5a) in ref. 12).

Most analyses use a local, energy-dependent potential. The energy dependence of the potential can directly be obtained from fig. 10 and is represented in fig. 11. It is approximately linear between 10 and 100 MeV:

$$U = -56 + 0.3\ E \quad (MeV) \quad , \tag{3.2}$$

which is very close to the depth of the empirical potential (2.32d) determined from the analysis of scattering cross sections of various targets, and with the depth obtained from the detailed analysis of proton scattering by ^{40}Ca and ^{58}Ni at several energies (fig. 11). This good agreement may be somewhat fortuitious, since the empirical depth is influenced by the geometry, e.g. by the value adopted for the potential radius.

For E > 100 MeV, the energy dependence of U is weaker than the one given by the low energy behaviour (3.2). Figure 2 shows

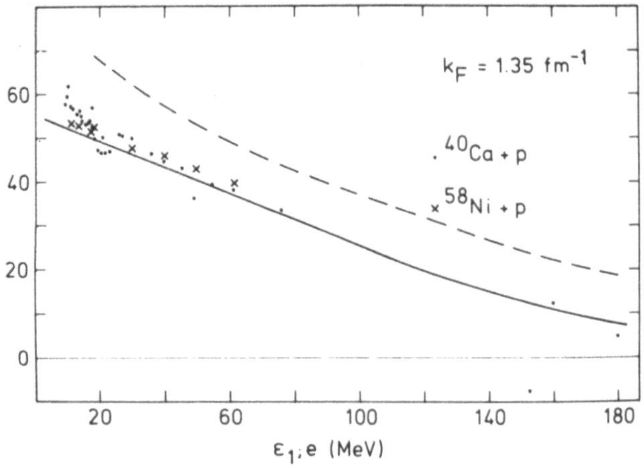

Fig. 11. The full curve represents the energy dependence of $|U|$, (eq. (2.70)), for k_F = 1.35 fm^{-1} (from ref. 12). The dots and the crosses are empirical depths (refs. 54, 55). From ref. 12.

that it becomes close to a logarithmic dependence. We note that the two curves in fig. 2 intersect one another. These features are confirmed in fig. 12 and will be discussed in sects. 4 and 5.

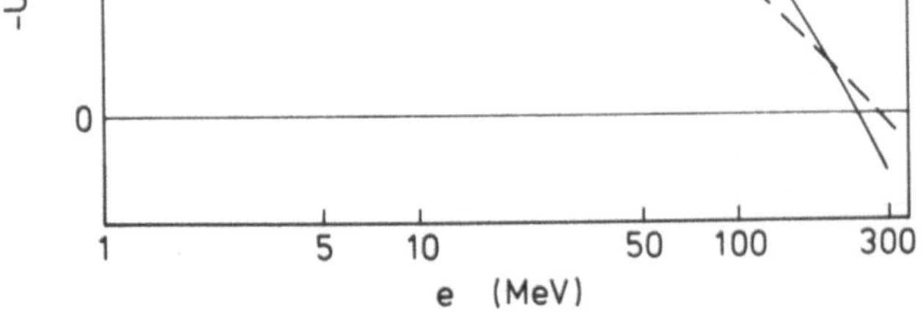

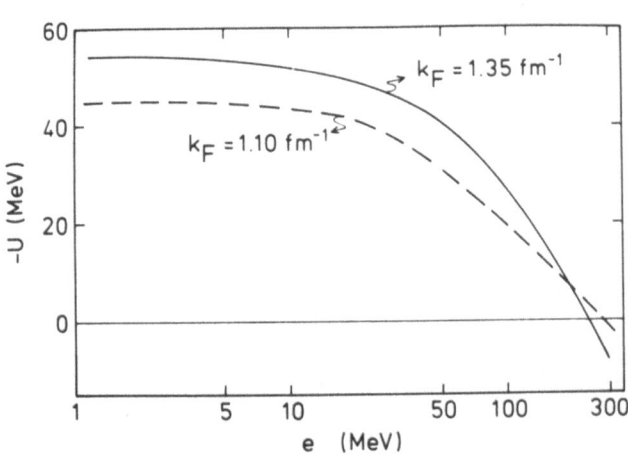

Fig. 12. Semi-logarithmic plot of the energy dependence of -U (eq. (2.70)), for k_F = 1.35 fm^{-1} (ρ = 0.166 fm^{-3}, full curve) and k_F = 1.10 fm^{-1} (ρ = 0.090 fm^{-3}, long dashes). From ref. 12.

We emphasized that the data determine better the volume integrals (2.35) than the value of each potential parameter. Figure 4 shows that the volume integrals of the real part of the calculated potential is in good agreement with that of the empirical potential (2.32), in the case of 35 MeV protons. In fig. 13, the computed volume integral of the real part is compared with a compilation of empirical values. Here again the agreement is impressive, in view of the fact that no parameter has been adjusted in the theoretical calculation.

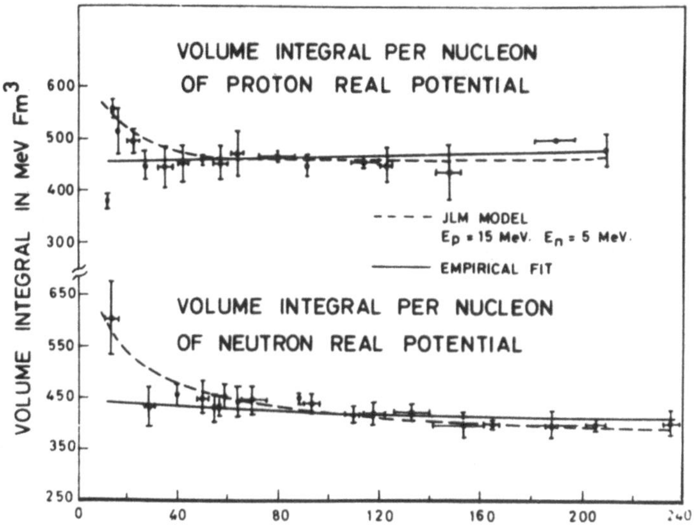

Fig. 13. Compilation of empirical volume integrals per nucleon for $10 < E < 20$ MeV protons (upper part) and for $1 < E_n < 8$ MeV neutrons (lower part). The dashed curves are the theoretical values, computed for $E_p = 15$ MeV and $E_n = 5$ MeV. From ref. 56.

The theory is less satisfactory when one turns to the root mean square radius: the calculated value is too small by about 0.4 fm. This is illustrated in fig. 14 in the case of ^{120}Sn; other nuclei lead to the same conclusion[12], which seems to be related to a too small diffuseness. We shall argue in sect. 3.5 that this is mainly due to a deficiency of the local density approximation.

The gap between E = 75 MeV and E = 150 MeV in the values shown in fig. 14 reflects the fact that there exist only very few reliable empirical analyses in this energy range. This domain deserves to be studied in more detail because the optical-model

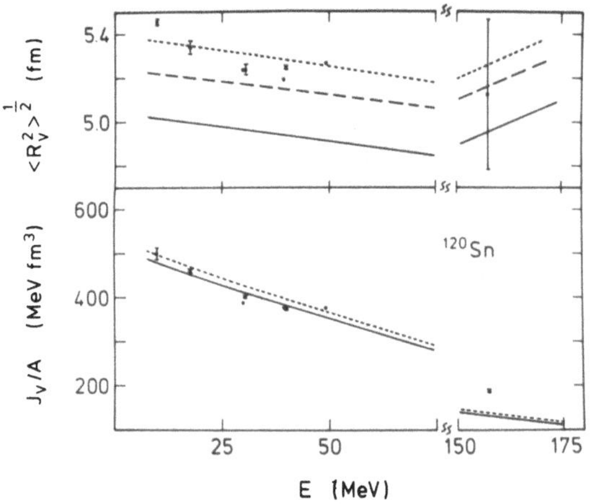

Fig. 14. Comparison between calculated volume integrals per nucleon and root mean square radii (full curves) and a compilation of empirical values, for the real part of the optical-model potential model for protons on ^{120}Sn. The long dashes correspond to the improved local density approximation described in sect. 3.5, and the short dashes to the improved local density approximation with, in addition, a hypothetical difference of 0.23 fm between the root mean square radii of the neutron and of the proton distributions. From ref. 29.

potential is needed not only for elastic scattering calculations, but also for studying the fate of a fast proton on its way out of a nucleus for instance after (γ,p), (π,p) or $(p,2p)$ collisions.

3.3. Imaginary Part of the Potential

The energy dependence of the calculated imaginary part

$$W(e) = W(k,e(\mathbf{k}))$$

of the optical-model potential (eqs. (2.52a), (2.61)) is represented in fig. 3, for two different densities. For $E < 60$ MeV, it is larger in absolute magnitude at the surface density than at the interior density. The calculated imaginary part will thus be surface peaked in a finite nucleus at low energy. This is illustrated in fig. 15 and is in keeping with experimental evidence (sect. 2.2).

The comparison with the empirical values should bear on the volume integral per nucleon and on the root mean square radius, eqs. (2.35), (2.36b). A global comparison is shown in fig. 5 and in fig. 16. The overall agreement is rather good.

However, one expects the imaginary part of the optical-model

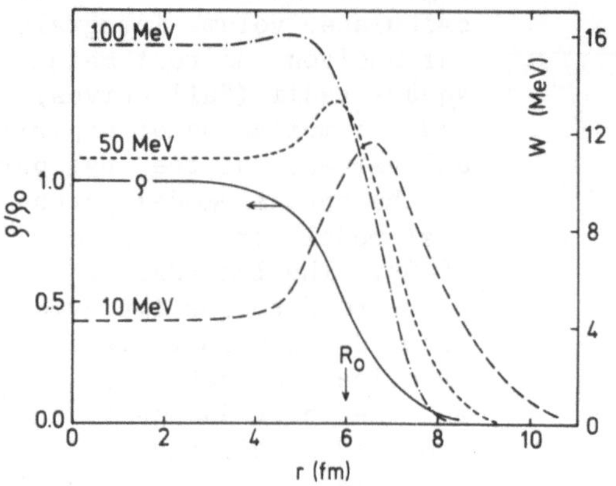

Fig. 15. Modulus of the imaginary part of the calculated optical-model potential at E = 10, 50 and 100 MeV (right-hand scale), for A = 170. The full curve shows the density distribution and the arrow labelled R_0 points to the half-density radius.

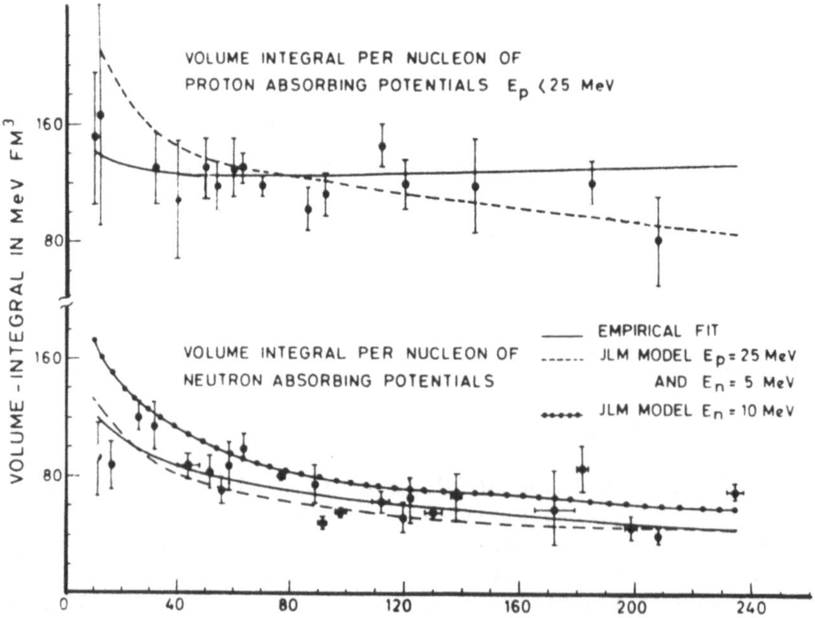

Fig. 16. Dependence upon A of the volume integral per nucleon of the imaginary part of the optical-model potential. The dots are obtained from a compilation of empirical values for 17 (± 8) MeV protons and for 5 (± 3) MeV neutrons. The theoretical curves refer to E_p = 25 MeV and E_n = 5 MeV (short dashes), E_n = 10 MeV (full curve with dots). From ref. 57.

potential to be more sensitive to finite size effects. A detailed comparison has been performed in ref. 29, mainly in the case of doubly closed shell nuclei. An example is shown in fig. 17. This confirms that the theory is more successful in rendering the global properties of the imaginary part than in individual cases.

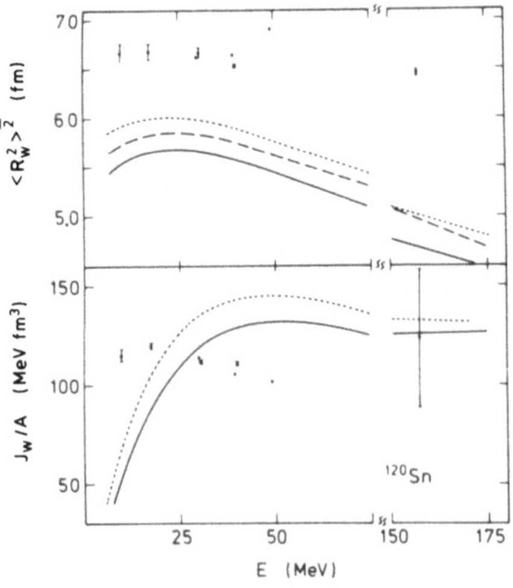

Fig. 17. Same as fig. 14, but in the case of the imaginary part of the optical-model potential. From ref. 29.

Another evaluation of the accuracy of the theory can be reached by computing elastic scattering cross sections. This will be discussed in sect. 3.5.

3.4. Small Components of the Potential

The empirical optical-model potential contains a symmetry (or isovector) component, which is proportional to neutron excess. The symmetry potential V_1 (eq. (2.32c)) can in principle be measured by studying the dependence upon (N-Z)/A of the depth of the potential. This yields $V_1 \simeq 14$ MeV in the case of low-energy neutrons and $V_1 \simeq 24$ MeV in the case of low-energy protons.

The theoretical value turns out to be $V_1 = 11$ MeV[30]. This is close to the empirical value for neutrons; the agreement becomes quite good if one notices that the radius of the calculated symmetry potential is slightly larger than that adopted in the phenomenological analyses: the calculation yields $r_{V_1} = 1.31$ fm, as compared to the value $r_{V_1} = 1.17$ fm, eq. (2.32b) assumed in the analyses. The volume integral of the calculated symmetry potential is almost equal to that determined from neutron scattering analyses.

The discrepancy between the empirical values of V_1 in the case of neutrons on the one hand and of protons on the other hand can also be understood on the basis of the theoretical calculations. These show that the geometry assumed in empirical analyses for the Coulomb correction Δ_C (eq. (2.32c)) amounts to an under-estimate of this effect. More specifically, the calculated radius of this correction is larger than that of the isoscalar part of the potential; it can therefore not be approximated by

$$0.4\ Z\ A^{-1/3}\ f^{(V)}(r) \quad . \tag{3.3}$$

If one insists for keeping the same geometry for all components of the potential, the parameter Δ_C should be increased by about 20% as compared to the assumed value (3.3). There exists recent experimental confirmation of this theoretical prediction[58]. This effect is illustrated in fig. 4, which shows that the calculated volume integral per nucleon of the Coulomb correction is larger than the one assumed (eq. (2.32)) in the global analysis, while the opppsite is true for the symmetry component. The theoretical value of the volume integral of the full proton potential is seen to be in good agreement with that of the full empirical potential.

The origin of the Coulomb correction is the following. For a given total energy the velocity of a proton inside the nucleus is less than that of a neutron because of the central Coulomb repulsion. This leads to an additional attractive field with strength

$$\Delta_C = V(E - V_C) - V(E) \quad . \tag{3.4}$$

A similar correction has to be included when comparing π^+ and π^- optical-model potentials.

It is very difficult to gain experimental information on the energy dependence of the symmetry potential. The good agreement between theory and experiment gives confidence in the calculated energy dependence, which shows that V_1 changes sign at about 130 MeV[30]. Thus, one expects that the (p,p') inelastic scattering will not excite the giant dipole resonance for $E_p > 100$ MeV. The bump in the outgoing p' spectrum which had been identified with this giant dipole state is now believed to correspond to the giant isoscalar monopole resonance[59].

The spin-orbit component of the optical-model potential has recently been computed by Brieva and Rook[60] in the framework of the approach outlined in sect. 2.4, with some improvements to the local density approximation (sect. 3.5). These authors use the Blin-Stoyle approximation[61] and a straightforward extension of the works of Scheerbaum[62] and Sprung[63]. The calculated strength of the spin-orbit potential appears to be in fair agreement with the empirical value. The authors claim that it is only about 10 per cent too small. This agreement may, however, be changed if the effect of the non-locality of the isoscalar part of the optical-model potential is taken into account. In the case of the symmetry potential, this non-locality leads to a reduction by a factor 0.7, but this correction becomes less important in the surface region where the spin-orbit component peaks.

3.5. Improved Local Density Approximations

We have seen that the Brueckner-Hartree-Fock approximation yields too small root mean square radii for the real and for the imaginary parts of the optical-model potential, when combined with the local density approximation. This was to be expected, because the latter is quite a crude approximation in the surface region. In order to trace its main drawback, let us consider the Hartree-Fock approximation for a central nucleon-nucleon interaction $\hat{v}(|\vec{r}-\vec{r}'|)$.

In nuclear matter, this Hartree-Fock approximation is given by eq. (2.55), which can be written in the equivalent form

$$M_\rho(k) = \rho \int \hat{v}(s) \, [1 - \tfrac{1}{4} j_0(ks) \, S_\rho(s)] \, d^3s \quad , \tag{3.5}$$

where $S_\rho(s)$ is the Slater function

$$S_\rho(s) = \frac{3}{(k_F s)^3} \, [\sin(k_F \, s) - k_F \, s \cos(k_F \, s)] \quad . \tag{3.6}$$

If the local density approximation (LDA) (2.23) is used to construct the mean potential in a finite nucleus, one obtains

$$M^{(LDA)}(r,k) = \rho(r) \int \hat{v}(s) \, [1 - \tfrac{1}{4} j_0(ks) \, S_{\rho(r)}(s)] \, d^3s \ . \tag{3.7}$$

Note that the local equivalent of this mean field is not proportional to $\rho(r)$ because the energy momentum relation

$$E = \frac{k^2}{2m} + M_{\rho(r)}(k) \tag{3.8}$$

introduces an r dependence of the local momentum k, and also because of the radial dependence of $S_{\rho(r)}(s)$.

In a finite nucleus, the Hartree-Fock potential reads ($\vec{s}=\vec{r}-\vec{r}'$)

$$M_{HF}(r,k) = \int \rho(r') \hat{v}(s)\, d^3r' - \frac{1}{4} \int \rho(r,r') \hat{v}(s)\, j_0(ks)\, d^3r \quad ,$$

provided that one uses the local energy approximation (3.8) to approximate the local equivalent of the non-local exchange (Fock) term[64]. If one further replaces the mixed density matrix by its Slater approximation one finds

$$M_{HF}(r,k) = \int \rho(r') \hat{v}(s) \left[1 - \frac{1}{4} j_0(ks)\, S_{\rho(r)}(s)\right] d^3r' \quad . \quad (3.9)$$

If we compare eqs. (3.7) and (3.9), we see that they are equivalent only when the effective interaction has zero range. Hence, the simple version (2.32) of the local density approximation does not take into account the effect in a finite nucleus of the range of the effective interaction.

Let us take the following example of a Gaussian-like interaction:

$$\left[1 - \frac{1}{4} j_0(ks)\, S_\rho(s)\right] \hat{v}(s) = \hat{v}_0 (t\sqrt{\pi})^{-3} \exp\left(-\frac{s^2}{t^2}\right) \quad . \quad (3.10a)$$

Then, eq. (3.5) gives

$$\hat{v}_0 = \frac{M_\rho(k)}{\rho} = \frac{M^{(LDA)}(r',k)}{\rho(r')} \quad . \quad (3.10b)$$

Equations (3.9), (3.10a) and (3.10b) yield the "improved local density approximation"[29]

$$M(r,k) = (t\sqrt{\pi})^{-3} \int M^{(LDA)}(r',k) \exp(-s^2/t^2)\, d^3r' \quad . \quad (3.11)$$

The long dashes in figs. 14 and 17 have been calculated from this expression, with the following adjusted value for the range parameter

$$t = 1.2 \text{ fm} \quad . \quad (3.12)$$

This improved local density approximation identifies the ratio $M_\rho(k)/\rho$ with the strength of an effective interaction, and attaches a finite range t to it. We emphasize that the density dependence of the ratio

$$\hat{v}_0 = M_\rho(k)/\rho \quad (3.13)$$

has a twofold origin. The first one is a possible density dependence of the effective interaction $\hat{v}(s)$ in eq. (3.9). The other one is the density dependence of the product

$$j_o(ks)\ S_\rho(s) \quad . \tag{3.14}$$

Therefore, the density dependence of (3.13) does not reflect that of an effective interaction which should be used in an optical-model folding calculation[64] where the exchange is included explicitly. Rather, it gives the energy dependence of an effective interaction which would be used in a folding calculation without antisymmetrization.

While representing a necessary improvement over the simple form (2.23) of the local density approximation, its improved version (3.11) is not quite satisfactory, since it involves an adjustable parameter. Moreover, and mainly, it would be of interest to compute the state dependent and finite-range effective interaction $\hat{v}(s)$ in order to investigate inelastic scattering processes[65] or the optical potential for composite particles[66,67]. This problem is being considered by Brieva, Rook and Von Geramb[31,68].

The comparison between the Brueckner-Hartree-Fock approximation (2.61) and the Hartree-Fock approximation (2.53) shows that Brueckner's reaction matrix $g[E+e(j)]$ (eq. (2.59)) plays the role of an effective interaction. Complications arise because g is a non-local and energy dependent operator which, moreover, depends on the target nucleon momentum j. The latter dependence can tentatively be averaged over, and one can try to construct a local (but energy dependent) operator $\hat{v}(E,s)$ which has the same matrix elements as g between plane wave states. This is not possible for all plane wave states, but appears feasible for the momenta relevant for elastic and inelastic scattering[20,31,66] and as far as the real part of g is concerned. Brieva and Rook[31] applied the same procedure to compute the imaginary part of $\hat{v}(E,s)$ as to calculate its real part, but the foundation of this prescription does not appear to have been justified.

As it stands, the procedure of Brieva and Rook[31] is more satisfactory than the improved local density approximation (3.11) because of two main reasons. Firstly, it does not involve the adjustable parameter t. Secondly, and mainly, it provides an interaction which can be used in the calculation of inelastic scattering[68]. The particular parametrization adopted in the latter work should be further improved in several respects: it appears plausible, and for many problems it is highly desirable[66] to require that $\hat{v}(E,s)$ reduces to the one-pion exchange potential for large values of s; the procedure used to calculate the imaginary part of $\hat{v}(E,s)$ should be justified or modified; large cancellations occur between direct and exchange terms which are inconvenient and could probably be avoided[20]; the use of the Slater approximation[69] and of the local momentum approximation could also be improved upon. A more detailed discussion can be found in ref. 70.

3.6. Discussion

The results presented in this chapter show that the Brueckner-Hartree-Fock approximation is able to account for the empirical properties of the optical-model potential in the low-energy domain with an accuracy of the order of ten per cent. One should not expect any better agreement in view of the size of the smallness parameter (2.69) which governs the convergence rate of the low-density expansion. The size of the accuracy is confirmed by the direct comparison between calculated and experimental elastic scattering cross sections. The full curves in fig. 18 are computed from the improved local density approximation (3.11). The dashed curves are obtained by multiplying the real and the imaginary parts of this theoretical potential by renormalization factors which range from 0.90 to 1.12.

The accuracy of the theory is sufficient to yield reliable information concerning properties of the optical-model potential which are otherwise difficult to determine experimentally, for instance the non-locality of the mean field, its energy dependence, its Coulomb correction term, its symmetry component, the energy dependence of its geometry, ...

4. THE INTERMEDIATE ENERGY DOMAIN (150 $<$ E $<$ 500 MeV)

4.1 Introduction

The intermediate energy domain to which this chapter is devoted is spanned by the last generation of cyclotrons. The number of accurate elastic scattering and total reaction cross section measurements in this energy range is still too limited for yielding a reliable empirical knowledge of the optical-model potential. As in the low energy domain, this will emerge only after an accumulation of systematics. Hence, it is of real practical interest to develop a calculable theory of the optical-model potential at intermediate energy. One expects the low-density expansion developed in ref. 11 and sketched in sect. 2.4 to be equally valid at intermediate energy as in the low energy domain, where as we have shown in sect. 3 it is sufficiently accurate to yield reliable information. This approach may thus provide a unique link between the low- and intermediate energy domains.

In sect. 4.2, we briefly discuss the relationship between the

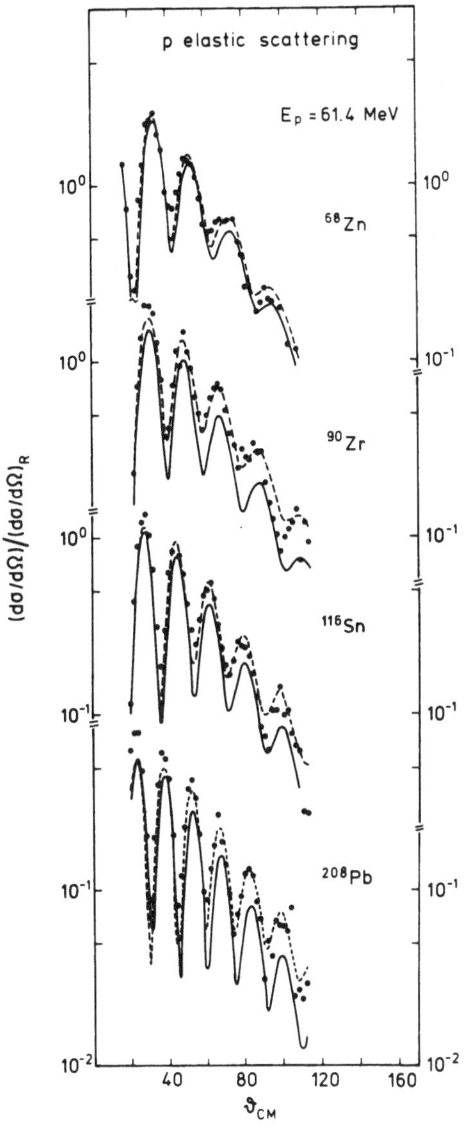

Fig. 18. Differential cross sections for the elastic scattering of 61.4 MeV protons. The full curves are calculated from the improved local density approximation (3.11), and the dashes from the same potentials after having renormalized their real and imaginary depths by less than 12 per cent. From ref. 71.

low-density and the multiple scattering expansions. The latter becomes convenient in the upper part of the intermediate energy range as defined here. At still higher energies, the concept of a non-relativistic nucleon-nucleon potential breaks down; one must then use the multiple scattering expansion whose basic input is the nucleon-nucleon transition matrix.

In sect. 4.3, we investigate the shape of the real part of

the optical-model potential in the vicinity of the energy where it changes sign. We shall argue that the Woods-Saxon form factor used in most empirical analyses is then not justified: the potential becomes repulsive in the interior of the nucleus at a lower energy than it does at the surface, thus acquiring a wine-bottle bottom shape.

In sect. 4.4, we turn to some relativistic theories. Relativistic effects may indeed become relevant at intermediate energy. Here, we don't have in mind the use of relativistic kinematics, but rather relativistic effects proper, which may be present in the nucleon-nucleon interaction, or require the use of a Dirac rather than of a Schroedinger equation to describe the elastic scattering process.

4.2. Relationship Between the Low-Density and the Multiple Scattering Expansions

The leading term of the low-density expansion for the optical model in nuclear matter is the Brueckner-Hartree-Fock approximation, eq. (2.61):

$$M_{BHF}(k,E) = \sum_j n_<(j) \langle k,j|g[E+e(j)]|k,j\rangle_A \quad . \tag{4.1}$$

The leading term of the multiple scattering expansion[2,3] is the impulse approximation:

$$M^{(IA)}(k,E) = \sum_j n_<(j) \langle k,j|t[E + \frac{j^2}{2m}]|k,j\rangle_A \quad , \tag{4.2}$$

where t is the transition matrix for free nucleon-nucleon collisions. In its practical form, the impulse approximation only uses on-shell values of the t-matrix; these are directly related to elastic scattering cross sections between free nucleons:

$$M_{IA}(k) = \sum_j n_<(j) \langle k,j|t[\frac{k^2}{2m} + \frac{j^2}{2m}]|k,j\rangle_A \quad . \tag{4.3}$$

If one neglects Fermi motion (k >> j), eq. (4.3) yields

$$M_{IA}(k) = \frac{\pi}{2m}\rho(3\ f_1 + f_0) \quad , \tag{4.4}$$

where the index T of the forward scattering amplitude f_T refers to the isospin. For later purposes, it is convenient to give the simplified form of eq. (4.4) which results from the neglect of the channel coupling associated with the tensor part of the nucleon-nucleon interaction:

$$f_T = (2\ i\ k)^{-1} \sum_{L,J} (2T + 1)\ [\exp(2\ i\ \delta_{LJT}) - 1] \quad , \qquad (4.5)$$

where δ_{LJT} is the nucleon-nucleon elastic scattering phase shift.

Comparing eqs. (4.1) and (4.3), we see that the difference between the Brueckner-Hartree-Fock and the impulse approximations is entirely contained in the difference between the operators g and t and between their arguments. The operators g[w'] and t[w] are related by the integral equation

$$g[w'] = t[w] + t[w]\ \{\frac{Q}{w'-e} - \frac{1}{w-\varepsilon}\}\ g[w'] \quad , \qquad (4.6)$$

where we used the notation (2.62), (2.63) and where ε is the kinetic energy operator:

$$\varepsilon|a,b\rangle = (\frac{a^2}{2m} + \frac{b^2}{2m})\,|a,b\rangle \quad . \qquad (4.7)$$

In second order, eq. (4.6) yields

$$g[w'] - t[w] = t[w]\ \frac{1-Q}{\varepsilon-w}\ t[w] + t[w]\ \{\frac{1}{w'-\varepsilon}\ (w-\varepsilon-w'+e)\ \frac{1}{w-\varepsilon}\}\ t[w]. \qquad (4.8)$$

The first term on the right-hand side of eq. (4.8) can appropriately be called a Pauli correction. It takes into account the fact that the intermediate states $|a,b\rangle$ (eqs. (2.59), (2.63)) must lie above the Fermi surface. Its size has been evaluated in ref. 11 in the case when k is sufficiently large to assume that

$$\langle\vec{k},\vec{j}|\,t[\frac{k^2}{2m} + \frac{j^2}{2m}]\,|\vec{k}+\vec{p},\vec{j}-\vec{p}\rangle \qquad (4.9)$$

strongly decreases with increasing momentum transfer p. It is found that the modulus of the Pauli correction term at 400 MeV amounts to about 15 per cent of the modulus of the impulse approximation (4.3). This Pauli correction can be expressed in terms of the Pauli correlation function[11].

The second term on the right-hand side of eq. (4.8) is a binding energy correction. It takes into account the fact that the incoming nucleon k, the target nucleon j and the excited nucleons a and b all feel an average potential. This binding energy correction can be evaluated at high energy by using the fact that p is small in (4.9). One can then replace the quantity $(w-\varepsilon-w'+e)$ by its average value

$$\Delta U \simeq U(a) + U(b) - U(k) - U(j)$$

$$\simeq U(k) - U(j) \simeq 40 \text{ MeV} \quad . \tag{4.10}$$

The second term on the right-hand side of eq. (4.8) then reads

$$- \Delta U \frac{\partial}{\partial w} t[w] \quad , \tag{4.11}$$

where we used the equation

$$\frac{\partial t[w]}{w} = - t[w] \frac{1}{(w-\varepsilon)^2} t[w] \quad . \tag{4.12}$$

This relation must be treated with some caution because $(w-\varepsilon)^{-2}$ is highly singular. At 400 MeV, the binding energy correction to the impulse approximation is a few per cent[11]; it is expected to become larger at small energy.

The Pauli and the binding energy corrections both make the real part of the optical-model potential less attractive in the Brueckner-Hartree-Fock than in the impulse approximation for a given momentum k:

$$V_{BHF}(k) > V_{IA}(k) \quad . \tag{4.13}$$

However, the momenta used in the two approximations are different for a given low energy E, since one has

$$E = \frac{1}{2m} k_{IA}^2 + V_{IA}(k_{IA}) = \frac{1}{2m} k_{BHF}^2 + V_{BHF}(k_{BHF}) \quad . \tag{4.14a}$$

From eqs. (4.13) and (4.14a), one concludes that

$$k_{IA} > k_{BHF} \quad . \tag{4.14b}$$

It turns out (see fig. 3 of ref. 51) that at low energy, the difference between the two momenta is sufficient to close to one another the potential depths calculated from the two approximations:

$$V_{BHF}(E) \simeq V_{IA}(E) \quad . \tag{4.15}$$

In contradistinction, the absorptive part of the optical-model potential at low energy is grossly overestimated by the impulse approximation, because the latter neglects the overwhelming Pauli effect. Approximate methods for estimating these Pauli corrections to the imaginary part of the impulse approximation have been proposed[72-75]. It would be of interest to test their accuracy by comparing their results to the Brueckner-Hartree-Fock approximation.

Let us now briefly discuss some higher order terms of the low-density expansion. It can be shown[11] that the dominant terms at high energy are those which are represented in fig. 19. Graph (A) is the Brueckner-Hartree-Fock approximation that we already discussed. Graph (B) accounts for the fact that the momentum state $j < k_F$ is partly unoccupied in the correlated ground state of nuclear matter (see fig. 1). Its contribution was called $M_3(k,E)$

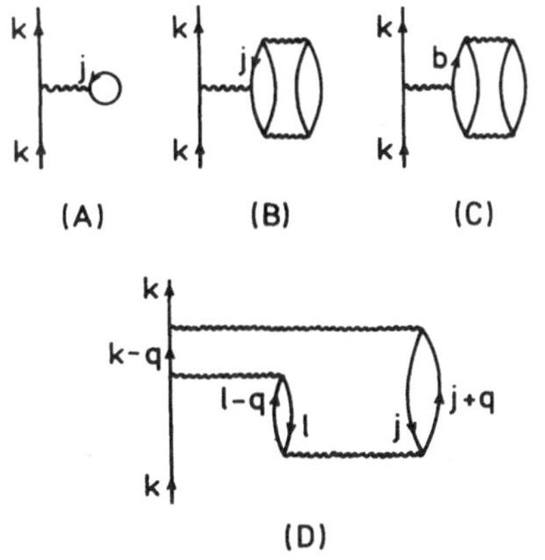

Fig. 19. Graphs which represent the leading terms of the low-density expansion at intermediate energy. Note that diagram (B) is identical to graph (3) of fig. 9.

in eq. (2.68); it is equal to (see eqs. (2.5) and (2.68))

$$M_B(k,E) = - \sum_j n_<(j)\, [1 - \rho(j)]\, \langle k,j|g[E+e(j)]|k,j\rangle_A \tag{4.16a}$$

$$\simeq - \kappa\, M_1(k,E) \quad . \tag{4.16b}$$

The sum $M_A + M_B$ is called the renormalized Brueckner-Hartree-Fock approximation. Since the number of nucleons is conserved, the nucleon j which has been excited outside the Fermi sea by its interaction with another nucleon ℓ (graph (B)) must be found in a particle state with momentum $b > k_F$. The effect is represented by graph (C) whose value for large k is given by

$$M_C(k,E) = \sum_b n_>(b)\, \rho(b)\, \langle k,b|g[E+e(b)]|k,b\rangle_A \quad . \tag{4.17}$$

Note that graphs (B) and (C) have a tendency to cancel each other. The sum of the graphs (A), (B) and (C) thus reads

$$M_{A+B+C}(k,E) = \sum_j \rho(j)\, \langle k,j|g[E+e(j)]|k,j\rangle_A \quad , \tag{4.18}$$

where $\rho(j)$ is the momentum distribution of the correlated ground state (fig. 1 and eq. (2.4)). Equation (4.18) is a natural modification of the Brueckner-Hartree-Fock approximation (2.61).

Graph (D) represents a double scattering correction. The incoming nucleon is scattered successively by nucleons ℓ and j; the latter had precedingly interacted and are thus correlated. This graph involves the dynamical two-body correlation function C(q,-q) for momentum transfer q; for large q, correlations arise predominantly from the short range repulsion between the nucleons. We recall that the long-range Pauli correlation are included in the Brueckner-Hartree-Fock term M_A. The ratio M_D/M_A has been evaluated in ref. 11:

$$M_D+M_A \simeq \frac{1}{2}\rho\, r_c\, \sigma_{tot} \quad , \tag{4.19}$$

where r_c is the correlation length (≃ hard core radius) and σ_{tot} is the total nucleon-nucleon cross section. The quantity (4.19) is nothing but the smallness parameter of the multiple scattering expansion which is thus, at intermediate energy, intimately related to the low-density expansion. The latter should thus be equally successful at intermediate as at low energy, as we claimed in sect. 4.1.

4.3. Change of Sign of the Real Part of the Potential

It is well known[76,77] that the real part of the optical-model potential changes sign at intermediate energy. This can be understood from the impulse approximation (4.4). However, the latter implies that the real part of the optical-model potential changes sign at the same energy for all densities, since the corresponding potential well is proportional to the density ρ. This is modified in the Brueckner-Hartree-Fock approximation. Indeed, figures 2 and 11 both show that the real part of the optical-model potential changes sign at a lower energy in the nuclear interior than at the nuclear surface. This is confirmed in fig. 20, where we show the theoretical optical-model potential for the scattering of 180 and 200 MeV protons by ^{56}Fe, respectively. These curves are strikingly similar to the empirical value (long dashes) determined by Elton[78] from the analysis of elastic scattering and total

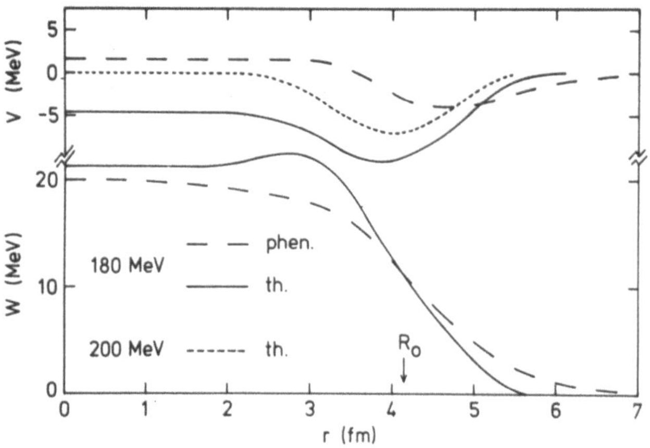

Fig. 20. Value of the real part V and of minus the imaginary part of the optical-model potential for protons on ^{56}Fe. The long dashes represent the empirical potential of Elton[78] for 180 MeV protons. The full curve (E = 180 MeV) and the small dashes (E = 200 MeV) are obtained from Reid's hard core interaction and the Brueckner-Hartree-Fock and local density approximations. The arrow points to the half-density radius. From ref. 79.

reaction data at 180 MeV. Elton's results at 180 MeV bombarding energy should be compared with the theoretical curves at 200 MeV, because of relativistic kinematics[80,81]. Elton emphasizes that "his fit has some claim to being unique" in the sense that "it does seem quite definite that V(r) must be peaked in the surface region, and is small and probably positive in the inner region of the nucleus; an interpretation in terms of a more orthodox optical potential would be difficult to achieve". As shown by fig. 20 and advertized in refs. 12, 51, 52, 79, 82, this empirical finding is substantiated by the theoretical calculation. Its existence is confirmed by fig. 4 of ref. 83, where the Hamada-Johnston nucleon-nucleon interaction is used, together with a technique for computing the g-matrix which is totally different from that described in ref. 53 and adopted in refs. 12, 51, 52, 79, 82. It thus appears that the wine-bottle bottom shape for the real part of the optical-model potential emerges naturally from the Brueckner-Hartree-Fock approximation.

In view of the practical importance of this result for intermediate energy reactions, it is highly desirable to analyze it in more detail, and to give a qualitative explanation. This is presently being investigated by A. Lejeune and myself[84]. In order to obtain more accurate results, we added to Reid's hard core interaction in the S,P and D ($J \leq 2$) partial waves (fig. 21) the Hamada-Johnston interaction[85] in the L = 3 and 4 states. We omitted the $^3F_4$ wave, which should be included only if one takes into account its coupling to the $^3H_4$ state. Preliminary results are

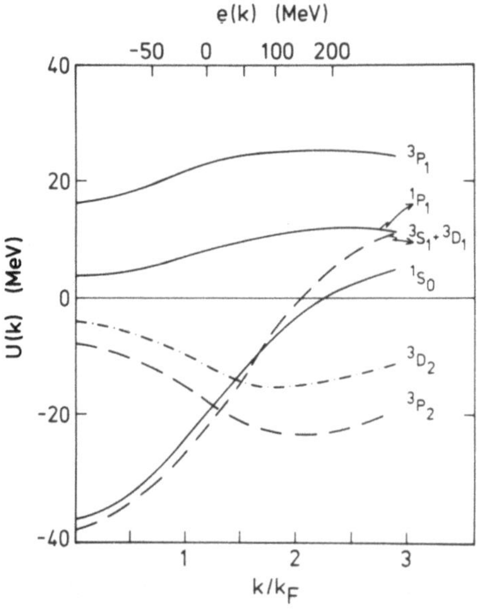

Fig. 21. Momentum and energy dependence of the contribution of various partial waves to the real part of the theoretical optical-model potential. From ref. 12.

shown in fig. 22. They confirm our previous findings, with small energy shifts: the calculated potential now changes sign in the tail at approximately 350 MeV, while it changes sign at 240 MeV in the inner part of the nucleus.

A qualitative interpretation of this phenomenon is the following. We argue in ref. 29 and in sect. 3.5 that the ratio V/ρ can be identified with the strength of the effective interaction. This quantity as computed from the Brueckner-Hartree-Fock approximation

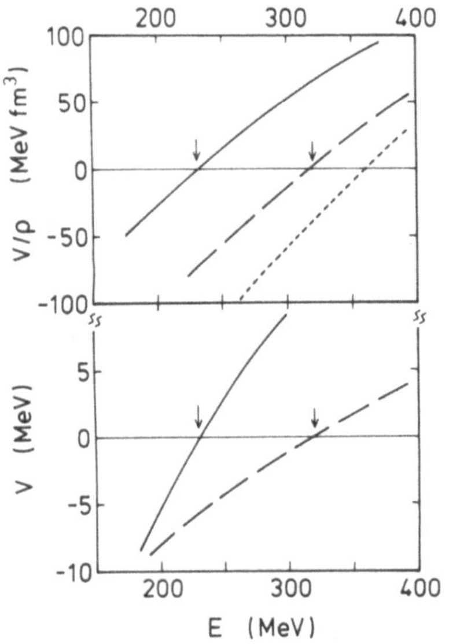

Fig. 22. The upper part of the figure shows the energy dependence of the strength V/ρ of the "effective interaction", for $k_F = 1.35\ fm^{-1}$ (full curve), $k_F = 1.00\ fm^{-1}$ (long dashes) and for the impulse approximation (4.3) (short dashes). The lower part of the figure shows the energy dependence of V for $k_F = 1.35\ fm^{-1}$ (full curve) and $k_F = 1.00\ fm^{-1}$ (long dashes).

is sketched in the upper part of fig. 22 for the inner and surface densities. We also show this ratio as computed from the impulse approximation (4.3); as expected from the discussion given in sect. 4.2, the corresponding effective interaction is more attractive than in the Brueckner-Hartree-Fock approximation, because of Pauli and binding energy effects. The difference increases with increasing density (and decreases with increasing energy). Hence, the quantity $\mathcal{V} = V/\rho$ changes sign at a higher energy for small ρ than for large ρ. When multiplied by ρ, this effective interaction yields a potential which also changes sign in the inner region of the nucleus at a lower energy than at the surface (lower part of fig. 22). This explains the results shown in fig. 20, and indicates that the wine-bottle bottom shape is a predictable outcome of the Brueckner-Hartree-Fock approximation. Further work on this phenomenon is in progress[84]. It should also emerge from the multiple scattering expansion once the Pauli and binding energy effects are included.

Recently, a "nuclear matter approach" to the energy dependence of the real part of the optical-model potential at intermediate incident energies has been published by Ray and Coker[86]. In effect, these authors simply calculate the impulse approximation (4.4) from Reid's soft core interaction. The latter is hardly realistic above 500 MeV, and we believe that it would be simpler and more justified to use directly the experimental value of the forward scattering amplitudes. Moreover, there exists no reason to use a nucleon-nucleon potential if off-shell effects are neglected. The results of Ray and Coker for the volume integral of the real part of the potential is represented by the dashes in fig. 23. It

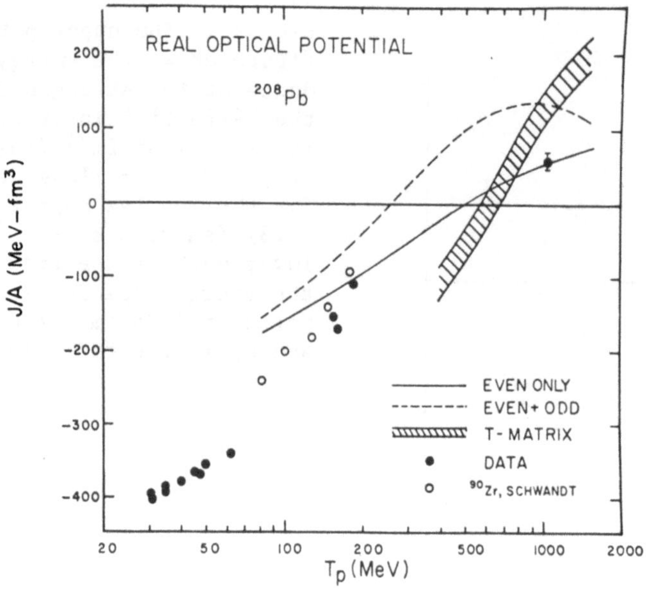

Fig. 23. Energy dependence of the volume integral per nucleon of the real part of the optical-model potential for protons on ^{208}Pb. The full dots are empirical values (see text for the dot at 1 GeV); the dashes correspond to the phase shift approximation (4.4) where the forward scattering amplitudes are computed from Reid's soft core interaction; the full curve is also obtained from eq. (4.4), but with omission of the odd partial waves and multiplication of the even partial waves by 4/3; the shaded region corresponds to a crude parametrization of the forward scattering amplitudes. From ref. 86.

changes sign at a surprisingly low energy, as compared with the empirical value, or with the result of Brueckner-Hartree-Fock calculations at low density based on Reid's hard core (see fig. 22) and on the Hamada-Johnston (see ref. 83) interactions, or with the impulse approximation (4.4) with experimental forward scattering amplitudes[87]. This may be due to the neglect of the spin-orbit component of Reid's interaction, or to the omission of exchange terms, or to the handling of the tensor interaction in ref. 86. The solid curve in fig. 23 is obtained by retaining only the contribution of the even orbital angular momentum partial waves, and multiplying these by 4/3. This procedure was advocated by Rajaraman[88] for the calculation of the binding energy of nuclear matter. There, it reflects the approximate cancellation of the contribution of odd partial waves to two three-hole line graphs. We believe that there exists no justification for adopting this prescription in the calculation of the optical-model potential. Furthermore, Rajaraman's prescription has no relationship with the difference between the operators t and g, contrarily to the claim made in ref. 86. Finally, we note that the empirical value of the volume integral per nucleon at 1 GeV given in refs. 76, 89 is considerably larger (116 MeV fm^3) than that shown in fig. 23. The shaded region in fig. 23 corresponds to the use of a simple and standard parametrization for the forward scattering amplitudes in eq. (4.4).

4.4. Relativistic Formulations

In the intermediate energy domain, it may turn out to be necessary or advantageous to take into account the fact that the nucleon-nucleon interaction is mediated by the exchange of mesons. A simplified one-boson exchange potential was constructed by A.E.S. Green and collaborators. Miller and Green[90] minimized the expectation value of the corresponding many-body relativistic Hamiltonian with respect to a Slater determinant. This relativistic Hartree-Fock method yields a relativistic expression for the single-particle potential. In the case of spherical nuclei, the average potential essentially reduces to the sum of a scalar field (U_s) and of the fourth component (U_v) of a vector field. The resulting Dirac equation for the single particle wave function reads

$$\{c\, \vec{\alpha}\cdot\vec{p} + \beta\, [m\, c^2 + U_s(r)] + U_v(r)\}\, \psi_D(\vec{r}) = E\, \psi_D(\vec{r}) \ , \qquad (4.20)$$

where we omit the Coulomb field for simplicity.

One can eliminate the lower components of ψ_D in the standard way. One then obtains an equation for the upper components ψ, with an effective momentum and energy dependent potential U_e

$$[\frac{p^2}{2m} + U_e + U_{s.o.}\, \vec{\sigma}\cdot\vec{L}]\, \psi = \frac{1}{2m}\, [E^2 - m^2]\, \psi \quad , \qquad (4.21)$$

where

$$U_e(E) = \frac{E}{m}\, U_v + U_s + \frac{1}{2m}\, [U_s^2 - U_v^2 + \frac{i}{rN}\, (\frac{\partial N}{\partial r})\, \vec{r}\cdot\vec{p}] \quad , \qquad (4.22)$$

$$U_{s.o.} = - \frac{1}{2m\, rN}\, \frac{\partial N}{\partial r} \quad , \qquad (4.23)$$

$$N = E + m + U_s - U_v \quad , \qquad (4.24.)$$

$$U_s = V_s + i\,W_s \quad , \quad U_v = V_v + i\,W_v \quad . \qquad (4.25)$$

One expects that V_s is attractive, since it is related to neutral scalar meson exchange, and that V_v is repulsive, since it is associated to the exchange of a neutral vector meson[90].

We have seen in chapter 2 that the real part of the optical-model potential is the extension to positive energy of the shell-model potential. Hence, it appears natural to use the wave equation (4.21) for scattering as well as for bound single-particle states. Good fits to the elastic scattering differential cross sections of 600-1000 MeV protons by ^{4}He can be obtained by using U_s or U_v alone[91]. However, the simultaneous fit of these differential cross sections and of the polarizations requires that U_s and U_v be both different from zero. This can be understood if one notices that U_s and U_v enter with different signs in the expression of U_e and $U_{s.o.}$, thus adding considerable flexibility to the empirical analysis. This is also better founded if one refers to the bound state calculations[90,92,93].

Noble[94] pointed out recently that the empirical value of the depth ($V_e \simeq -54$ MeV) and of the spin-orbit strength at low energy are reproduced if one takes $V_s = -295$ MeV, $V_v = 222$ MeV. Then, the energy dependence of the real part of V_e in the inner region of the nucleus is determined by eq. (4.22) and turns out to be (in MeV)

$$V_e = -53 + 0.24\,(E-M) \quad , \qquad (4.26)$$

in fair agreement with the empirical energy dependence at low energy (eq. (2.32d)).

Recently, fits to the differential and polarization elastic scattering cross sections of protons by ^{4}He ($350 \leq E \leq 1700$ MeV), ^{12}C (E = 800 and 1000 MeV), and ^{58}Ni, ^{90}Zr, ^{208}Pb (E = 800 MeV) have been obtained by Arnold, Clark and Mercer[95]. These authors adopted the parametrization

$$U_v = (V_v + i\,W_v)\, f^{(v)}(r) \quad , \qquad (4.27)$$

$$U_s = (V_s + i\,W_s)\, f^{(s)}(r) \quad , \qquad (4.28)$$

where the form factors $f^{(s)}(r)$ and $f^{(v)}(r)$ are identified with the density distribution $f^{(\rho)}(r)$ (eq. (2.29)). Several results of this empirical analysis are worth being reported in the present context.

(i) The ratio

$$R(E) = \frac{V_v}{V_s} \qquad (4.29)$$

is a linear function of the energy (fig. 24). It varies from -0.78 at E = 350 MeV to -0.66 at E = 1700 MeV. The value extrapolated at E = 0 is R(0) = -0.81, in striking agreement with the ratio R = 0.79 adopted by Walecka[92] in order to reproduce the binding energy and the saturation density of nuclear matter, and with the ratio R = -0.73 obtained by Noble[94] from the real part of the potential at low energy.

(ii) The fit to ^{4}He(p,p) is not good at 350 MeV, while being satisfactory to 400 MeV and above. Let us note that 350 MeV is close to the energy where the integral per nucleon of the real part of the effective potential changes sign (sect. 4.3).

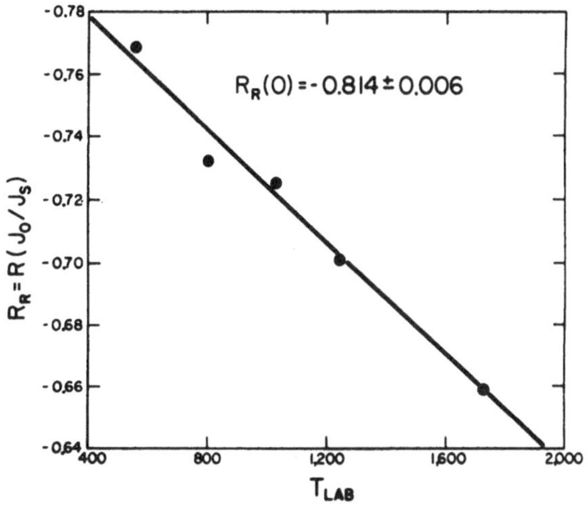

Fig. 24. Energy dependence of the ratio (4.29) between the strengths of the vector and of the scalar field. From ref. 95.

(iii) The fit to the analyzing powers of p + ^{90}Zr at 800 MeV is significantly improved if one allows the range of the scalar potential V_s to extend beyond that of the vector potential V_v by one per cent.

The discussion presented in sect. 4.3 suggests that items (ii) and (iii) are related. Indeed, a small difference in radius between $f^{(v)}(r)$ and $f^{(s)}(r)$ has the consequence that the real part of the effective optical-model potential changes sign at a lower energy in the nuclear interior than at the surface. In fact, this was the explanation put forward by Humphreys[96] for the empirical finding of Elton[78].

In fig. 25, we show the values of $V_v(r)$ and of $V_s(r)$ chosen by Humpreys[96] for E = 0 MeV, and in fig. 26 the energy dependence of his effective single-particle potential

$$V_{eff} = \frac{1}{\gamma} V_s + V_v \quad . \tag{4.30}$$

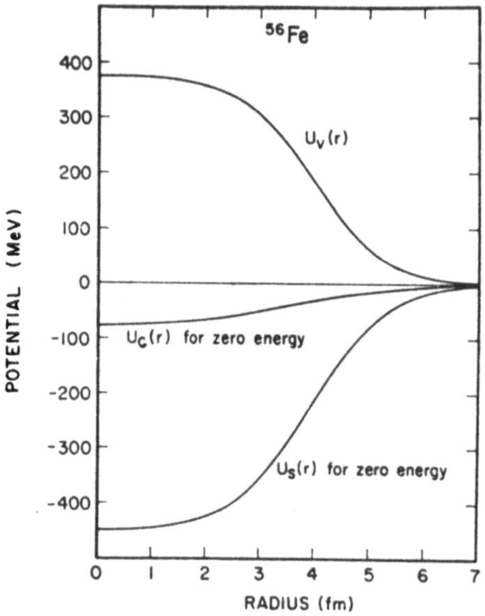

Fig. 25. Radial dependence at zero energy of the vector (U_v), scalar (U_s) components of the real part of the optical-model potential of ^{56}Fe, and of their sum U_c in Humphrey's relativistic model. From ref. 96.

Here, γ is the Lorentz contraction factor. Humphreys assumes that there exists a difference of 0.66 fm^2 between the mean square radii of V_v and V_s. We note that the energy dependence of V_{eff} arises from a decreasing contribution of the attractive scalar potential V_s, while that of V_e (eq. (4.26)) results from an increasing contribution of the repulsive vector potential V_v (eq. (4.22)). It seems that Humphrey's model is partly based on intuition. Figures 25 and 26 should therefore merely be regarded as illustrations of the fact that in the relativistic approaches a wine-bottle bottom potential shape emerges if one assumes that the vector potential extends slightly beyond the scalar potential.

Such a difference in range is not unexpected and might reflect the property that effective vector mesons have a larger mass than the effective scalar mesons[90,96].

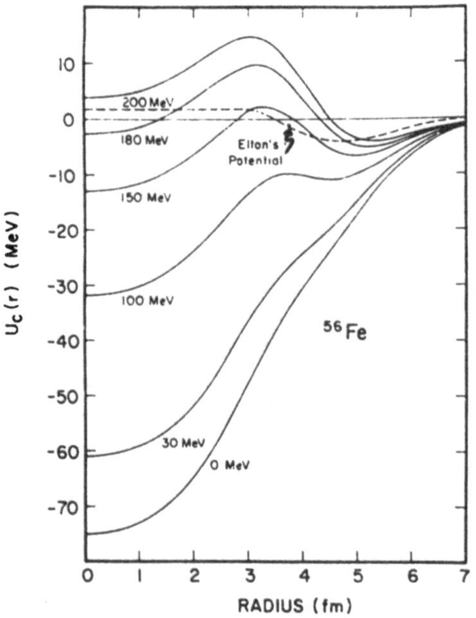

Fig. 26. Energy dependence of the real part U_c of the optical-model potential of ^{56}Fe in Humphrey's relativistic model. From ref. 96.

4.5. Discussion

The low-density expansion for the optical-model potential is of interest at intermediate as well as at low energies. Its leading term, the Brueckner-Hartree-Fock approximation, presents the advantage over the impulse approximation of including Pauli correlations and binding energy corrections. These are still important at intermediate energies. They give rise to a wine-bottle bottom shape for the real part of the optical-model potential in the 200-300 MeV range, i.e. near the energy where the volume integral per nucleon changes sign. Note that in a light nucleus the effective Fermi momentum in the nuclear interior is close to 1.2 fm^{-1}, so that the change of sign should occur at a somewhat higher energy. Moreover, the particular shape of the potential is such that its volume-integral changes sign at an energy significantly larger than that where the potential changes sign in the inner region of the nucleus. This wine-bottle bottom shape is also expected from relativistic treatments of the optical-model potential. In these approaches, it appears to be related to the relative masses of the mesons responsible for the repulsion and attraction between nucleons in free space. It would be of interest to investigate whether there exists a relationship between the two interpretations.

5. ENERGY-DEPENDENCE AND NON-LOCALITY OF THE POTENTIAL

5.1. Introduction

The optical-model potential is non-local and energy-dependent. It is not possible to disentangle its non-locality from its energy dependence from the analysis of experimental cross sections. Indeed, one can obtain good fits to the elastic scattering cross sections with a purely local and energy-dependent potential, as well as with a non-local and energy-independent potential. Therefore, it is necessary to resort to theoretical considerations. The problem is not only of conceptual interest: it also has practical fall-outs, as will be summarized in sect. 5.4. In particular, a local, energy-dependent potential and a non-local, energy-independent one can be equivalent only in the limited sense that they yield the same elastic scattering phase shifts, i.e. they are equal "on the energy shell". This is for instance apparent from the fact that we used the energy-momentum relation (2.24) to construct these "equivalent" fields:

$$M_{NL}(k) = M_{NL}(k,E(k)) \leftrightarrow M_L(k(E),E) = M_L(E) \quad ,$$

where the indices refer to "non-local" (NL) and to "local" (L), respectively. However, two phase-shift equivalent potentials yield different scattering matrices off the energy shell, which is a theorist's way of saying that they yield different wave functions inside the potential well. This difference may thus be important in analyses where distorted wave functions are needed, for instance in the calculation of inelastic scattering. We already alluded to this fact at the end of sect. 2.1. We shall see in chapter 6 that this difference is also relevant for bound single-particle states.

In sect. 5.2, we show that rather general conclusions on the relative importance of the non-locality and of the energy dependence of the potential can be derived from the existence of the dispersion relation (2.46) which connects its real and its imaginary parts.

In sect. 5.3, we discuss more detailed results which can be obtained from the Brueckner-Hartree-Fock approximation.

5.2. Dispersion Relation

5.2.a. Theoretical Formulae. The existence of a dispersion relation for the mean field is a general consequence of the causality principle[97]. Hence, it is not surprising that this relation emerges from any reasonable theoretical approach to the optical-model potential, based either on nuclear reaction theory[43,44,98] or on the nuclear matter approach[12,13,99-101].

In nuclear reaction theory, the dispersion relation takes the following form[22,44,98]

$$V(\vec{r},\vec{r}\,';E) = V_o(\vec{r},\vec{r}\,') + \sum_n \frac{A_n(\vec{r},\vec{r}\,')}{E - E_n} + \frac{1}{\pi} \int_{\varepsilon_o}^{\infty} \frac{W(\vec{r},\vec{r}\,';E')}{E' - E} dE' \;, \tag{5.1}$$

where all quantities are real and where ε_o is the energy of the first inelastic threshold. The energies E_n are essentially bound state energies, although they may lie above ε_o. Actually, the quantities V and W in eq. (5.1) pertain to the so-called generalized optical-model potential; the optical-model potential in the usual sense is obtained by replacing E by E + iI, where I is the averaging interval[102,103].

In most practical applications, the contribution to the right-hand side of eq. (5.1) of the states n with $E_n < \varepsilon_o$ has been dropped[22,104-108]. This is probably justified when E is not close to ε_o. In the case of nuclear matter, these terms correspond to the contribution of the energies $E' < \varepsilon_F$ to the dispersion relation (2.46), which we can rewrite in the form

$$V(k,E) = V_{HF}(k) + \pi^{-1} \int_{-\infty}^{\infty} dE' \frac{W(k,E')}{E' - E} \quad ; \tag{5.2}$$

We replaced the quantity f(k) in eq. (2.46) by the Hartree-Fock field (eq. (2.55))

$$V_{HF}(k) = \hat{M}_{1a}(k) \quad . \tag{5.3}$$

Equations (2.47) and (2.54) show that this is correct (except for a hard core interaction) since all terms other than (5.3) in the expansion (2.54) tend towards zero when $E\to\infty$. However, the Hartree-Fock field (5.3) should not be identified with the familiar mean field calculated from soft effective interactions. Indeed, it involves the actual nucleon-nucleon interaction and is thus a large quantity and wild function. Therefore, eq. (5.2) is usually not very useful as it stands and it is often preferable to use its subtracted form

$$V(k,E) = V_o(k,E_o) + \pi^{-1} (E-E_o) \int_{-\infty}^{\infty} dE' \frac{W(k,E')}{(E'-E_o)(E'-E)} , \quad (5.4)$$

where E_o is arbitrary.

5.2.b. Applications at Intermediate Energy. In the numerical applications of eqs. (5.2) and (5.4) to the optical-model potential at positive energy, the imaginary part W(k,E) has been set to zero for E < 0; this is also characteristic of the Brueckner-Hartree-Fock approximation. It becomes unjustified at negative energy, where one must include the spreading of hole states which is represented by $W(k.E < \varepsilon_F)$. At positive energy, we thus write

$$V(k,E) \simeq V_o(k,E_o) + \pi^{-1} (E-E_o) \int_{0}^{\infty} dE' \frac{W(k,E')}{(E'-E_o)(E'-E)} . \quad (5.5)$$

Equation (5.5) must be used with caution, since it relates off the energy shell quantities rather than their observable on the energy shell values. Furthermore, eq. (5.5) can at best give information on the slope of the real part of the potential since it contains a subtracted term. Various attempts to apply eq. (5.5) are discussed in refs. 109, 110. They are model-dependent, thereby altering somewhat the interest of using a model-independent dispersion relation. Usually, they amount to assume that one can neglect the non-locality of W(k,E) in comparison to its energy dependence. Then, one can write

$$W(k,E') \simeq W(k(E'),E') = W_L(E') \simeq W(k(E),E') , \quad (5.6)$$

where the function k(E) is defined by eq. (2.24), while $W_L(E')$ is the imaginary part of the local empirical optical-model potential at the energy E'. Equation (5.5) now reads

$$V(k,E) = V_o(k,E_o) + \mathcal{V}_L(E) , \quad (5.7)$$

where

$$\mathcal{V}_L(E) = \pi^{-1} (E - E_o) \int_{0}^{\infty} dE' \frac{W(E')}{(E'-E_o)(E'-E)} \quad (5.8)$$

is a local, energy-dependent potential which vanishes at $E = E_o$.

The energy dependence of W(E) is shown in fig. 3, and in a wider energy range in fig. 27. The curve (a) has been drawn through the empirical points and its expression has been used to calculate $\mathcal{V}(E)$[104]. The curve (a) of fig. 28 represents the energy dependence of

$$V(k(E_o),E) = V(k(E_o),E_o) + \mathcal{V}_L(E) \quad , \tag{5.9}$$

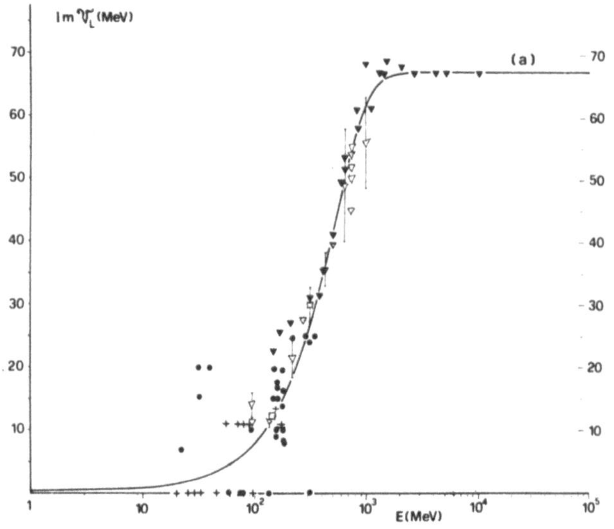

Fig. 27. The points show a compilation of empirical values of the imaginary part of the optical-model potential. The black triangles are obtained from nucleon-nucleon cross sections, with a Pauli correction. Curve (e) is fitted to the points. From ref. 104.

in the case E_o = 970 MeV, $V(k(E_o),E_o)$ = 22.3 MeV. The latter value is a fitted parameter. We see that curve (a) in fig. 28 is in good agreement with $V_L(E) = V(k(E),E)$ for E > 300 MeV. This indicates that in this energy range the assumption (5.6) is compatible with empirical evidence, and that furthermore the momentum dependence of V(k,E), which is contained in $V(k,E_o)$ can be omitted. This does not necessarily imply that the mean field is local for E > 300 MeV, but rather that the effects of a possible non-locality can be neglected. This would be expected when the wave length is sufficiently short for having k d << 1, where d is the non-locality range.

For E < 300 MeV, curve (a) in fig. 28 deviates from the empirical values. It even bends down and acquires a slope which differs in sign from that of V(E). In other words,

$$\frac{d\mathcal{V}_L(E)}{dE} < 0 \quad \text{for } E < 200 \text{ MeV} \quad , \tag{5.10}$$

$$\frac{dV_L(E)}{dE} \geq 0 \quad \text{for all energies.} \tag{5.11}$$

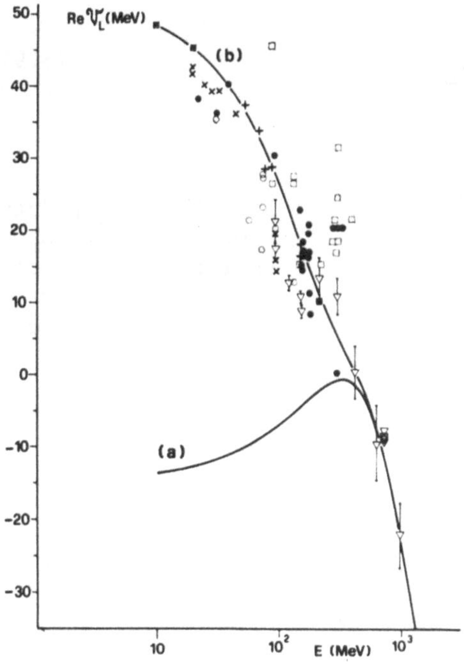

Fig. 28. The points show a compilation of empirical values of the real part of the optical model potential. Curve (a) represents the contribution of the local, energy-dependent part (5.9), curve (b) the sum of (5.9) and of a non-local, energy-independent Perey-Buck potential. From ref. 104.

This difference reflects the fact that the momentum dependence (or non-locality) of the mean field V(k,E), i.e. its non-locality, cannot be neglected for E < 200 MeV. This non-locality gives rise to an energy dependence of the local equivalent potential. Note that the limit E ≃ 200 MeV exhibited in fig. 28 probably depends on the choice of E_0; it would be of interest to repeat the analysis with a value of E_0 close to 300 MeV, as we further discuss below.

The difference between curve (a) in fig. 28 and the empirical points must reflect the momentum dependence of $V(k,E_0)$ if assumption (5.6) is justified. Hence, it is tempting to fit this difference with a non-local energy dependence field, for instance with a potential of the Perey-Buck type (eq. (3.1)):

$$V_0(k,E_0) \simeq N \exp(-\frac{1}{4} k^2 d^2) \quad . \tag{5.12}$$

Curve (b) in fig. 28, is obtained in this way, with a non-locality range d = 0.8 fm.

In conclusion, the dispersion relation is compatible with the assumption that the optical-model potential is a sum of a local, energy-dependent complex field and of a non-local, energy-independent real one.

We mentioned that it would be of interest to repeat this type of analysis with a whole range of values for the parameter E_o. This would give more detailed information (within the assumption (5.6) made on W(k,E)) on the relative importance of the non-locality and of the energy dependence of the mean field. For E close to E_o, one could indeed write the difference ΔV between the empirical depth and the calculated curve (a) (fig. 28) in the form

$$\Delta V = V(k(E),E) - V(k(E_o),E)$$

$$\simeq [k(E) - k(E_o)] \left[\frac{\partial V(k,E)}{\partial k}\right]_{E=E_o} \quad . \tag{5.13}$$

The resulting partial derivative is related to the true non-locality range of the full (non-local and energy-dependent) mean field (eq. (5.32b)). It could be compared with the results of the Brueckner-Hartree-Fock approximation given in sect. 5.3.

Since $W_L(E')$ approaches a constant W_m at large energy, eqs. (5.2) or (5.8) suggest that the energy dependence of $\mathcal{V}(E)$ (and thus of V(E) if non-locality effects are negligible for E large) is approximately logarithmic[104]

$$V_L(E) \sim \text{constant} + \pi^{-1} W_m \, \ell n(E) \quad . \tag{5.14}$$

This seems to be borne out by experimental evidence (see fig. 28), and also by the Brueckner-Hartree-Fock approximation (see figs. 2, 12). There exist, however, conceptual problems when applying (5.2) to high energy data. Some authors[105,106,108] take the point of view that the optical-model potential is defined within the framework of non-relativistic quantum mechanics. There, the imaginary part of a non singular potential approaches zero for E → 0. These authors thus introduce a cut-off in the energy dependence of the function W(E') used in eq. (5.8). This does not modify the qualitative behaviour below 200 or 300 MeV of V(E) as calculated from eq. (5.8).

<u>5.2.c. Applications at Low and Negative Energies</u>. When applying the dispersion relation (5.4) at low or negative energies, it is convenient[12,101,111,112] to choose for E_o the Fermi energy ε_F

(eq. (2.43)). Then eq. (5.4) reads

$$V(k,E) = V_o(k,\varepsilon_F) + \overline{V}_1(k,E) + \overline{V}_2(k,E) \quad . \tag{5.15}$$

with

$$\overline{V}_1(k,E) = \pi^{-1} (E - \varepsilon_F)\int_{\varepsilon_F}^{\infty} dE' \frac{W(k,E')}{(E-E')(\varepsilon_F-E')} \quad , \tag{5.16}$$

$$\overline{V}_2(k,E) = \pi^{-1} (E - \varepsilon_F)\int_{-\infty}^{\varepsilon_F} dE' \frac{W(k,E')}{(E-E')(\varepsilon_F-E')} \quad . \tag{5.17}$$

Equation (2.44) shows that the imaginary part W(k,E) vanishes at $E = \varepsilon_F$. More precisely, its threshold behaviour can be approximated by[112]

$$W(k,E) \simeq -\alpha^2(E-\varepsilon_F)^2 \left(1-2m\beta \frac{|E-\varepsilon_F|}{k_F^2} - 2\, m\, \beta' \frac{E-\varepsilon_F}{k_F^2}\right) \quad . \tag{5.18}$$

The coefficients α^2, β and β' are in general functions of k. Orland and Schaeffer[112] argue that their k-dependence can be neglected for k close to k_F, because the imaginary part is then mainly determined by phase space considerations. They suggest that

$$\beta \simeq \beta' \simeq \frac{1}{3} \quad .$$

In their work, however, these authors set β' to zero and thus take

$$\beta = \frac{1}{3} \quad , \quad \beta' = 0 \tag{5.19}$$

Then, W(k,E) becomes local and symmetric about $E = \varepsilon_F$:

$$W(k,E) \simeq W_L(E) \simeq -\alpha^2[(E - \varepsilon_F)^2 - \frac{2m}{3} \frac{|E-\varepsilon_F|^3}{k_F^2}] \quad .$$

The assumption $\beta' = 0$ is unfortunate because it eliminates from the model a remarkable property of the energy derivatives of $V_1(k,E)$ and of $V_2(k,E)$ at $E = \varepsilon_F$ (see ref. 12 and sect. 5.3). In order to deal with convergent integrals, Orland and Schaeffer[112] substitute in eq. (5.16) and (5.17) the following expression

$$W_L(E) = -\alpha^2(E-\varepsilon_F)^2 \exp[-\frac{2m}{3}\frac{|E-\varepsilon_F|}{k_f^2}] \quad (5.20a)$$

$$\simeq -\alpha^2(E-\varepsilon_F)^2 \exp(-\frac{|E-\varepsilon_F|}{114}) \quad (5.20b)$$

where E and ε_F are expressed in MeV in eq. (5.20b). This yields a maximum of W(E) at

$$E_m = (\varepsilon_F \pm 228) \quad \text{MeV} \quad .$$

It would be of interest to use the improved version

$$W_L(E) = -\alpha^2(E-\varepsilon_F)^2 \exp[-\frac{2m^*}{\beta}\frac{|E-\varepsilon_F|}{k_f^2}] \quad [1-\frac{2m^*}{\beta'}\frac{E-\varepsilon_F}{k_F^2}]^2, \quad (5.21)$$

and to choose β and β' in such a way as to reproduce the values of W(E) calculated from realistic approximations[12,48,50]. Note that the effective mass m* in eq. (5.21) depends on k_F[12]. Work in this direction is in progress and will be reported elsewhere.

The results obtained by Orland and Schaeffer[112] are the following. They computed the quantities

$$\overline{V}_1^{(L)}(E) \simeq \overline{V}_1(k,E) \quad , \quad \overline{V}_2^{(L)}(E) \simeq \overline{V}_2(k,E) \quad (5.22)$$

from their assumption (5.20b). The results are shown in fig. 29 for a value of α^2 adjusted to yield reasonable values for W(E). Figure 29 shows that irrespectively of the value chosen for α^2 one has

$$\frac{d\,\overline{V}_1^{(L)}}{dE} < 0 \quad , \quad \frac{d\,\overline{V}_2^{(L)}}{dE} < 0 \quad . \quad (5.23a)$$

In this model, the true energy dependence of the mean field is entirely contained in the local potential

$$\mathcal{V}_L(E) = \overline{V}_1^{(L)}(E) + \overline{V}_2^{(L)}(E) \quad .$$

Inequalities (5.23a) show that the slope of V(E) is opposite to that of the real part of the local empirical optical-model

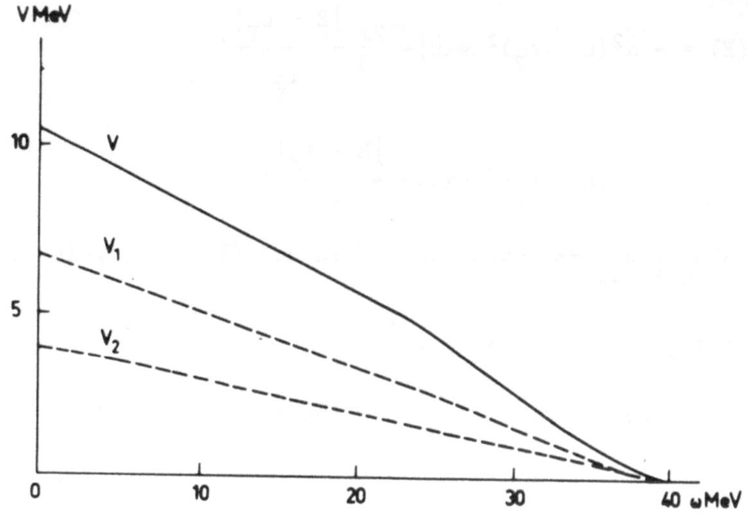

Fig. 29. Dependence upon $\omega = \varepsilon_F - E$ of the quantities $\overline{V}_1$ and $\overline{V}_2$ as computed from eqs. (5.16), (5.17) and (5.20). In this figure, extracted from ref. 112, $\overline{V}_1$ is labelled V_1; $\overline{V}_2$ is called V_2 and V denotes the sum $\mathcal{V} = \overline{V}_1 + \overline{V}_2$.

potential $V(k(E),E) = V(E)$:

$$\frac{d\,V(E)}{dE} > 0 \quad , \quad \frac{d\,\mathcal{V}(E)}{dE} < 0 \quad . \tag{5.23b}$$

This is in agreement with the conclusions drawn in sect. 5.2.b: the energy dependence of the real part of the empirical local optical-model potential at low energy mainly arises from the non-locality of the theoretical mean field, i.e. from the dependence of $V(k,\varepsilon_F)$ upon k.

This implies that the "true" non-locality of the mean field has to be larger than that of the empirical potential, e.g. than that obtained by Perey and Buck, eq. (2.34). In other words, the energy dependence of

$$V(k(E),\varepsilon_F) \tag{5.24a}$$

is larger than that of

$$V_L(E) = V_L(k(E),E) \quad . \tag{5.24b}$$

Using a notation which will be useful later on, we thus have

$$(1 - \frac{\tilde{m}}{m}) \simeq \frac{d\,V(k(E),\varepsilon_F)}{dE} > (1 - \frac{m^*}{m}) \simeq \frac{d\,V_L(k(E),E)}{dE} \quad . \tag{5.25a}$$

Orland and Schaeffer[112] identify the quantity (5.24a) with the standard Hartree-Fock field computed from an effective interaction. We disagree with this interpretation. The standard Hartree-Fock field is, like $V(k,\varepsilon_F)$, energy-independent and non-local. However, the parameters of the effective interaction are usually fitted to render the observed density occupied single-particle states near the Fermi surface which is given by (eq. (2.123) of ref. 12)

$$\frac{dn}{dE} = \frac{1}{\pi^2} k_F\, m^* \quad . \tag{5.26}$$

Hence, the familiar Hartree-Fock potential is a non-local potential which is equivalent to the true field (5.24b) at the Fermi surface, but whose energy dependence characterized by $1 - m^*/m$ is weaker than that $(1 - \tilde{m}/m)$ of the quantity (5.24a).

Let us make these comments more quantitative. We shall see in the following section that typical values of the energy derivatives in eq. (5.25) are

$$\left(\frac{m^*}{m}\right)_{E=\varepsilon_F} \simeq 0.75 \quad , \quad \left(\frac{\tilde{m}}{m}\right)_{E=\varepsilon_F} \simeq 0.55 \quad . \tag{5.27}$$

In nuclear matter, $\varepsilon_F = -16$ MeV at $k_F = 1.36$ fm^{-1}. Equation (2.43) then yields

$$V(k(\varepsilon_F),\varepsilon_F) = V_L(\varepsilon_F) = -54 \text{ MeV} \quad , \tag{5.28}$$

in excellent agreement with empirical evidence (eq. (2.32d)). For $1s_{1/2}$ hole states, one has typically (see chapter 6) $\varepsilon_F - E \simeq -40$ MeV; the corresponding depth of the potential (5.24a) is approximately equal to

$$-54 + (1 - \frac{\tilde{m}}{m})\,(E - \varepsilon_F) \simeq -72 \text{ MeV} \tag{5.29}$$

while that of the potential (5.24b), which we claim to be close to the standard Hartree-Fock field, is

$$-54 + (1 - \frac{m^*}{m})\,(E - \varepsilon_F) \simeq -64 \text{ MeV} \quad . \tag{5.30}$$

These results indicate that contrarily to the interpretation of Orland and Schaeffer the addition of V_1 and V_2 to (5.24a) will not worsen the discrepancy between the familiar Hartree-Fock single-particle energies (which are associated to (5.24b)) and the experimental value. Indeed, $(\overline{V}_1^{(L)} + \overline{V}_2^{(L)})$ should be added to the quantity (5.24a), which is deeper than the familiar Hartree-Fock field. We return in chapter 6 to the remaining discrepancy between the experimental removal energies and the single-particle energies obtained from a standard Hartree-Fock calculation.

Finally, we note that the slope of the local energy-dependent part of the field is approximately given by

$$\frac{d\,\overline{V}_1^{(L)}}{dE} \simeq -0.17 \quad , \quad \frac{d\,\overline{V}_2^{(L)}}{dE} \simeq -0.10 \quad , \quad \frac{d\mathcal{V}_L(E)}{dE} \simeq -0.27 \quad , \tag{5.31}$$

These numbers are sensitive to the choice of the constants β, β' and (mainly) α^2 in the model, but give a rough idea of the true energy dependence of the mean field. On detailed inspection of fig. 29 and of the corresponding analytical expressions, one can see that the slope $d\mathcal{V}/dE$ is somewhat larger in absolute magnitude in the domain $\varepsilon_F - E < 15$ MeV than in the energy range $\varepsilon_F - E > 15$ MeV. We shall see in sect. 5.3 that this feature is present in more detailed calculations.

5.2.d. Discussion. In the applications described in sects. 5.2.b and 5.2.c, one attempts to write the mean field as a sum of two terms, namely a non-local but energy-independent potential $(V(k,E_0))$ and a local but energy-dependent one $(\mathcal{V}_L(E))$. It is possible that the non-locality and the energy dependence of the field are more intimately entangled than assumed in these models. If the models are valid, they may lead to a useful parametrization of the mean field for scattering or bound state calculations. Hence, the basic assumption of the models is worth being investigated; this could be done in the framework of the Brueckner-Hartree-Fock approximation.

The dispersion relation enables one to draw the following qualitative conclusions; they will be confirmed and be made more quantitative in sect. 5.3:

(i) For energies larger than several hundreds MeV, the dispersion relation is compatible with the assumption that the energy dependence of the real part of the empirical optical-model potential M(E) is mainly due to the true energy dependence of the mean field M(k,E), i.e. to its dependence upon E.

(ii) In the energy range $-50 < E < 150$ MeV, the dispersion relation requires that the energy dependence of the real part of the empirical optical-model potential $M(E)$ mainly arises from the non-locality of the true mean field $M(k,E)$, i.e. from its dependence upon k.

In sect. 5.4, we shall discuss several practical consequences of the fact that the true mean field is non-local and energy-dependent. One of the practical interests of distinguishing between "true" non-locality and "true" energy-dependence may turn out to be the following. It was shown by Perey[17,24] that the amplitude of a wave function at a given energy E is decreased by a factor

$$\left(\frac{\tilde{m}}{m}\right)^{1/2} \tag{5.32}$$

inside a non-local potential well, as compared to its amplitude inside the local equivalent well. Here, $\tilde{m}$ is the effective mass that characterizes the non-locality of the field. We encountered it in (5.25):

$$\frac{\tilde{m}}{m} = 1 - \left[\frac{d\, V(k(\varepsilon),E)}{d\varepsilon}\right]_{\varepsilon=E} \tag{5.25b}$$

$$= \left\{1 + \frac{m}{k}\frac{\partial}{\partial k} V(k,E)\right\}^{-1}_{k=k(E)} \quad , \tag{5.25c}$$

where $k(E)$ is the on-shell momentum determined by

$$E = \frac{k^2}{2m} + V(k,E) \quad . \tag{5.33}$$

The standard effective mass is given by

$$\frac{m^*}{m} = 1 - \frac{d\, V_L(E)}{dE} = 1 - \frac{d}{dE} V(k(E),E) \tag{5.34a}$$

$$= \left\{1 + \frac{m}{k}\frac{d}{dk} V(k,e(k))\right\}^{-1} \quad . \tag{5.34b}$$

We have seen in (5.25) that

$$\frac{\tilde{m}}{m} < \frac{m^*}{m} \quad . \tag{5.35}$$

We emphasize that the effective mass m* is the one that can be determined from the analysis of the elastic scattering data or of the density of bound single-particle energies. If these are fitted with a non-local, energy-independent potential of the Perey-Buck type (see eqs. (2.33a), (3.1)), this effective mass is given by

$$\left(\frac{m^*}{m}\right)_{P.B.} = \{1 - \frac{1}{2} m\, \beta^2\, U(k)\}^{-1} \quad . \tag{5.36}$$

Note that the right-hand side of eq. (5.36) is a monotonically increasing function of k (or of energy) for a fixed non-locality range β. If for instance U(0) = -80 MeV and β = 0.85 fm, one has

$$\frac{m^*}{m} \simeq 0.60 \tag{5.37}$$

at the energy E = U(0) = -80 MeV, while

$$\frac{m^*}{m} \simeq 0.86 \tag{5.38}$$

at the energy for which U ≃ -20 MeV. Thus, the fact that the effective energy-dependence induced by the non-locality of a mean field becomes small does not imply that the non-locality range becomes small, but rather that its effect can be neglected.

Suppose now that one fits the experimental data (elastic scattering cross sections, density of bound single-particle states) with an empirical non-local and energy-independent potential (Perey-Buck potential, or Hartree-Fock potential computed from an effective interaction with adjusted parameters). Then the effective mass associated with this empirical non-local potential is equal to the value of m* in the relevant energy range. Inequality (5.35) shows that it is larger than the effective mass $\tilde{m}$ which characterizes the non-locality effects of the true mean field. In practice, this probably leads to an underestimate of the non-locality correction factor. Here, however, a caution signal must be raised: the value (5.32) of the correction factor has been obtained in the case of a non-local, energy-independent field[17,24]. Its value has not yet been derived for a non-local, energy-dependent mean field. I believe that it would still be given by (5.32), with $\tilde{m}$ defined by (5.25), but this should be checked.

5.3. Brueckner-Hartree-Fock Approximation

5.3.a. Effective Masses. We first recall a few results and concepts already encountered. The Brueckner-Hartree-Fock approximation (2.61) is a momentum and energy-dependent mean field:

$$M_1(k;E) = \sum_j n_<(j) \langle k,j|g[E+e(j)]|k,j\rangle_A \quad . \tag{5.39}$$

By Fourier transform over k, it yields the non-local and energy-dependent field $M(|\vec{r}-\vec{r}'|;E)$. Hence, we are able to calculate explicitly the energy and the non-locality properties of the mean field in the Brueckner-Hartree-Fock approximation.

The local potential equivalent to (5.39) is given by

$$M_1^{(L)}(E) = M_1(k(E),E) \quad , \tag{5.40}$$

where k(E) is the on-shell momentum determined by

$$E = \frac{k^2}{2m} + V_1(k,E) \quad . \tag{5.41}$$

The function E(k) defined by eq. (5.41) will indifferently be written E(k) or e(k), as previously.

The energy-independent (but non-local) field equivalent to (5.39) is given by

$$M_1^{(NL)}(k) = M_1(k,E(k)) \quad . \tag{5.42}$$

Unless otherwise stated, we deal henceforth with the real part V_1 of the optical-model potential.

The energy dependence of the local equivalent potential, i.e. in practice of the empirical optical-model potential is measured by the effective mass

$$\frac{m_1^*}{m} = 1 - \frac{d\, V_1^{(L)}(E)}{dE} \quad . \tag{5.43}$$

It arises from two sources, namely

(i) The energy dependence of the true mean field, measured by the E-mass $\bar{m}_1$:

$$\frac{\bar{m}_1}{m} = \left[1 - \frac{\partial\, V_1(k,E)}{\partial E}\right]_{E=e(k)} \quad . \tag{5.44}$$

(ii) The momentum dependence of the true mean field, measured by the k-mass $\tilde{m}_1$:

$$\frac{\tilde{m}_1}{m} = \{1 + \frac{m}{k}\frac{\partial}{\partial k} V_1(k,E)\}^{-1}_{k=k(E)} \quad . \tag{5.45}$$

The following relation can easily be derived

$$\frac{m^*}{m} = \frac{\tilde{m}}{m} \cdot \frac{\bar{m}}{m} \quad ; \tag{5.46}$$

it is of general validity and is in particular not limited to the Brueckner-Hartree-Fock approximation.

The Brueckner-Hartree-Fock field fulfills a subtracted dispersion relation:

$$V_1(k,E) = V_1(k,\varepsilon_F) + \pi^{-1} (E-\varepsilon_F) \int_{\varepsilon_F}^{\infty} dE' \frac{W_1(k,E')}{(E-E')(\varepsilon_F-E')} \quad . \tag{5.47}$$

Note the analogy with eq. (5.16):

$$V_1(k,E) \leftrightarrow \bar{V}_1(k,E) + V_o(k,\varepsilon_F) \quad . \tag{5.48}$$

The term corresponding to $\bar{V}_2(k,E)$ in eq. (5.15) does not appear in eq. (5.47) because in the Brueckner-Hartree-Fock approximation one has

$$W_1(k,E) = 0 \quad \text{for} \quad E < \varepsilon_F \quad . \tag{5.49}$$

We have seen in sect. 5.2 that the dispersion relation and the empirical values indicate that at low energy (see (5.31), (5.35))

$$\frac{\bar{m}}{m} > 1 \quad , \tag{5.50}$$

$$1 > \frac{m^*}{m} > \frac{\tilde{m}}{m} \quad . \tag{5.51}$$

This will be confirmed by the results of the detailed calculation given below. At high energy, the data are compatible with the inequalities

$$1 > \frac{m^*}{m} \quad ; \quad 1 > \frac{\bar{m}}{m} \quad . \tag{5.52}$$

5.3.b. Non-locality of the Brueckner-Hartree-Fock Field. At low energy, the energy dependence of the local empirical potential is mainly due to the non-locality of the true mean field. The latter contribution is measured by $(1 - \tilde{m}_1/m)$, where $\tilde{m}_1$ is given by eq. (5.45).

The k-mass is plotted in fig. 30, for the Fermi momentum k_F = 1.35 fm^{-1} which corresponds to the inner region of a nucleus. We see that it is a monotonically increasing function of the momentum. This behaviour is compatible with a Gaussian dependence of $V_1(k,E)$ upon k (for fixed E). Figure 30 can be fitted with a non-locality range equal to

$$d = 1.05 \quad \text{fm} \quad . \tag{5.53}$$

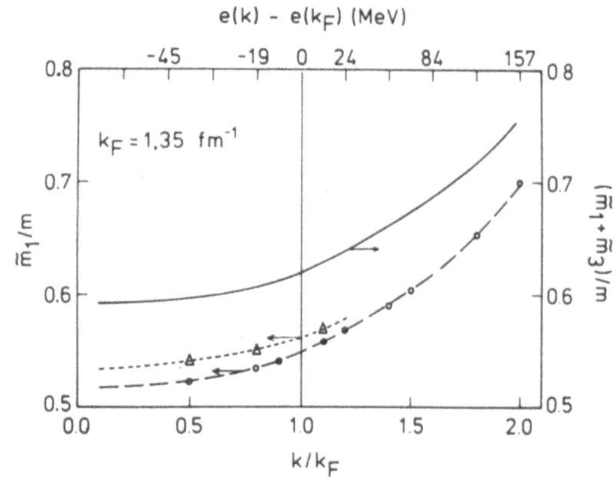

Fig. 30. The long dashes show the dependence upon momentum (lower scale) and upon $[e(k) - e(k_F)]$ (upper scale) of the k-mass $\tilde{m}_1$ (eq. (5.45)), for $k_F = 1.35\ fm^{-1}$. From ref. 12.

The smooth increase of $\tilde{m}_1/m$ is a consequence of the fact that d is almost independent of energy (see eqs. (5.36) - (5.38)). It reflects the fact that the non-locality range mainly arises from the exchange (Fock) term; its value is chiefly determined by the range of the effective interaction, in the present case by the range of the (density-dependent) g-matrix.

5.3.c. Energy Dependence of the Brueckner-Hartree-Fock Field. The energy dependence of the local empirical potential is influenced by the energy dependence of the true mean field. The latter contribution is measured by $1 - \bar{m}_1/m$, where the E-mass $\bar{m}_1$ is given by eq. (5.44).

The dependence of the E-mass upon energy is shown in fig. 31. We see that the inequality (5.50) is fulfilled; the true energy dependence of the mean field is opposite to that of the local equivalent potential, at negative and at low energy.

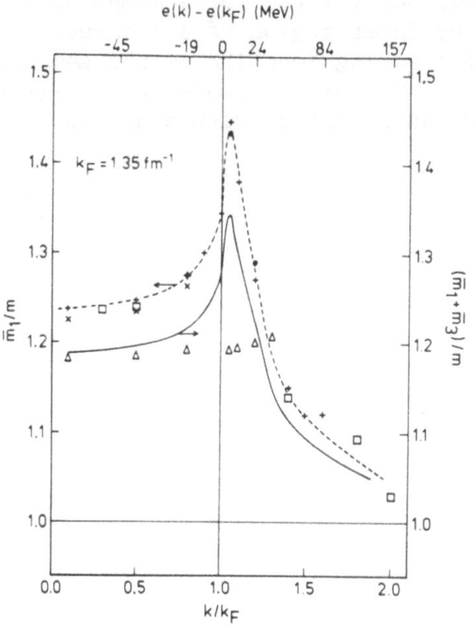

Fig. 31. The short dashes represent the dependence upon momentum (lower scale) and upon $[e(k) - e(k_F)]$ (upper scale) of the E-mass m_1 (eq. (5.44)), for $k_F = 1.35$ fm^{-1}. From ref. 12.

The E-mass displays a striking narrow enhancement just above the Fermi surface. This has been analyzed in detail in ref. 12. It has been shown to arise from the coupling of the single-particle state to two particle - one hole states with low excitation energy. In ref. 12, an analytical model has been constructed which reproduces this local enhancement. It would be of interest to analyze this phenomenon with an improved version of the model of Orland and Schaeffer[112] described in sect. 5.2.c. Work in this direction is in progress.

5.3.d. Energy Dependence of the Local Equivalent Potential. The energy dependence of the local equivalent potential is measured by the effective mass m_1^*, eq. (5.43). This quantity may be calculated directly, or can be obtained as the product of $\tilde{m}_1$ and of $\bar{m}_1/m$. This provides a useful check of the accuracy of the computational procedure[12].

The energy dependence of the effective mass m_1^* is shown in fig. 32. As expected from eq. (5.46) and from figs. 30 and 31, the effective mass has a narrow enhancement near the Fermi surface. The existence of a maximum in the energy dependence of the empirical effective mass near the Fermi surface had been pointed out by Brown, Gunn and Gould[113] from the analysis of the observed density

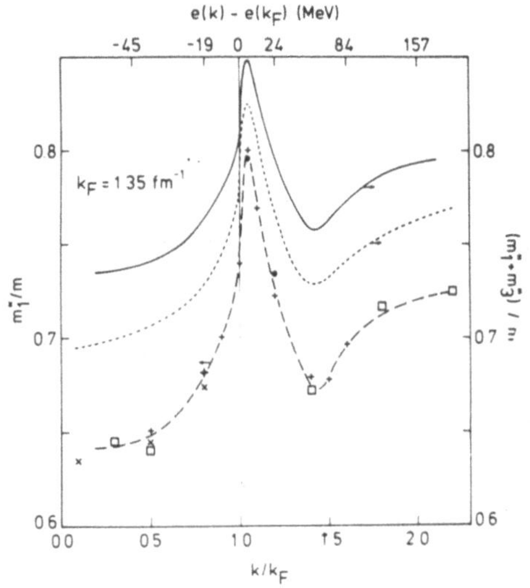

Fig. 32. The long dashes show the dependence upon momentum (lower scale) of the effective mass m_1^* (eq. (5.43)), for $k_F = 1.35\ fm^{-1}$. From ref. 12.

of bound single-particle states near the Fermi surface (see (eq. (5.26)). In finite nuclei, this phenomenon had been interpreted as arising from the coupling of the single-particle states to low-lying target excited states[114-116]. We have seen that it also exists in nuclear matter. It emerges from the Brueckner-Hartree-Fock approximation because the latter includes graphs (1b), (1c), ... of fig. 8. These correspond to the excitation of one particle-one hole target states. It had not been seen in previous nuclear matter calculations because of two main reasons: (a) these were limited to $k < k_F$; (b) they were based on the self-consistent choice (2.70) for the auxiliary potential U(k) for $k < k_F$ only, and adopted U(k) = 0 for $k > k_F$. This prescription suppresses the effect on the g-matrix of low-lying one particle-one hole target excited states[12,82,117]. On the basis of these earlier calculations, it was believed that the enhancement of the density of single-particle states at the Fermi surface is a finiteness effect.

The Brueckner-Hartree-Fock approximation does not include the correlation graph (2) of fig. 8, which accounts for the fact that the presence of the additional particle k blocks the admixture of the two particle-two hole state ($j^{-1}\ \ell^{-1}$, ak) in the target. This graph plays a significant role near the Fermi surface[12,114-116]: it increases the height of the peak of $\bar{m}/m$, and it is therefore of interest to investigate it in more detail.

A calculation of graph (2) of fig. 9 has been performed[48,50]. It confirms the existence of a narrow enhancement of the effective masses $\bar{m}$ and m* right above the Fermi surface, but is not sufficiently accurate to yield the detailed shape of the enhancement peak. This information could be gained from analytical models. For instance, an improved version of the model of Orland and Schaeffer[112] would yield the energy dependence of the contribution of graph (2) of fig. 9 to the E-mass, since this contribution is contained in $V_2(k,E)$ (eq. (5.17)). A model calculation of graph (2) of fig. 8 with a soft effective interaction would yield reliable information on the contribution of this graph (2) to the k-mass. Work in these directions is in progress.

5.4. Consequences and Conjectures

The idealized case of an infinite nucleus cannot pretend to yield an accurate picture of the situation that prevails in finite nuclei. However, it provides general trends and suggests useful parametrizations of the quantities of interest. The semi-quantitative conclusions drawn from this chapter are the following.

(I) It appears plausible to write the mean field as a sum of a non-local, energy-independent real potential and of a local, energy-dependent complex potential:

$$M_\rho(|\vec{r}-\vec{r}'|;E) = M_o(|\vec{r}-\vec{r}'|) + V_L(r;E) + i\ W_L(r;E) \quad , \qquad (5.54)$$

or equivalently

$$M_\rho(k,E) = M_o(k) + V_L(E) + i\ W_L(E) \quad . \qquad (5.55)$$

Even if this is not strictly justified, it provides a useful model.

(II) For energies larger than about 300 MeV, the effect of the non-locality can be neglected, i.e. M_o can be replaced by a constant.

(III) At negative and at low energy, the non-locality of M_o

is the principal responsible for the energy dependence of the real part of the empirical local equivalent average potential.

(IV) At negative and at low energy, the effect of the energy dependence of $V_L(E)$ has a sign opposite to that of the non-locality of M_0. This has the consequence that the non-locality of an empirical non-local energy-independent potential (Hartree-Fock potential or Perey-Buck potential) is an underestimate of the non-locality of the real part of the true mean field $M_\rho(|\vec{r}-\vec{r}'|;E)$.

(V) One consequence of item (IV) is that the non-locality correction factor (5.32) is then underestimated. This correction factor may have sizeable consequences in the analyses of inelastic scattering or of (p,2p) processes, for instance. It is often unduly neglected when a local empirical potential is used. The non-locality also affects the radial shape of bound single-particle states, as we now explain. The correction factor (5.32) decreases the amplitude of the single-particle wave function inside the potential well. Since it is normalized to unity, the single-particle wave function extends further outside in a non-local potential than in the local equivalent potential, for a given binding energy[142]. This affects the root mean square radius and the spectroscopic factors.

(VI) The energy dependence of $\mathcal{U}(E)$ becomes stronger just above the Fermi surface. This has the consequence that the energy dependence of the local equivalent field $M_L(E)$ is very weak for the unoccupied (valence) bound single-particle states. This is of importance for the discussion of their root mean square radius, or for the extrapolation to deeply bound or to positive states of an empirical potential which would have been fitted to the energies of the bound single-particle states near the Fermi surface.

In particular, it has recently been possible to measure the mean square radius of the valence particle wave function in ^{41}Ca and in ^{87}Sr by magnetic electron scattering[118-119]. This radius turns out to be smaller than that expected from a Hartree-Fock potential which fits deeply or semi-deeply bound single-particle states.

This can be qualitatively understood in terms of the enhancement of the effective mass at the Fermi surface[120]. Indeed, the plateau in the energy dependence of $M_L(E)$ above the Fermi surface implies that the actual average potential well is deeper than the extrapolated one. This shortens the mean wave length and is thus equivalent to a squeeze of the mean square radius.

(VII) The plateau in the energy dependence of $V_L(E)$ at the Fermi surface implies a plateau in the momentum dependence of the non-local (but energy-independent) field

$$V_{NL}(k) = V_1(k,e(k)) \quad . \tag{5.56}$$

This plateau is visible in fig. 33, where the dots show the value of the right-hand side of eq. (5.56). The curves represent the values of U(k) which have actually been used as input for the calculation of the Brueckner-Hartree-Fock field (see eqs. (2.12), (2.59), (2.70)). Including the plateau in the input potential would lead to a somewhat more pronounced enhancmenet of m* at the Fermi surface. This is indicated by the analytical model of ref. 12.

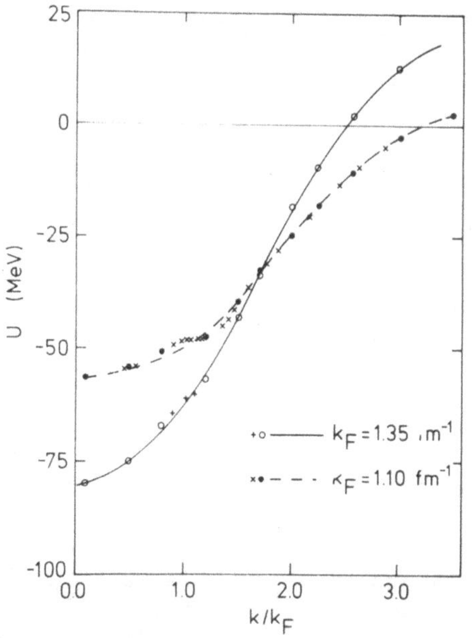

Fig. 33. The dots show the momentum dependence of the quantity $V_1(k,e(k))$ calculated from the Brueckner-Hartree-Fock approximation. The smooth curves represent the potential U(k) which has been used as input to calculate the reaction matrix (eqs. (2.12), (2.66)). From ref. 12.

(VII-I) One could take the Fourier transform of the momentum dependent field (5.56), including the plateau at the Fermi surface, and try to use this non-local energy-independent field for actual calculations. This raises two problems. The first one is that the accuracy of the local energy approximation to construct two phase-shift equivalent potentials should then be carefully reexamined. The second one is that the dependence of $V_{NL}(s)$ upon the non-locality parameter $s = |\vec{r}-\vec{r}'|$ would then be involved and hard to parametrize. These problems are worthwhile being investigated. Indeed, a non-local energy-independent average potential has the merit that its eigenstates form a complete orthogonal set. This is not true for an energy-dependent field, which may create problems if one attempts to use a non-local, energy-dependent potential to construct a nuclear wave function from a Slater determinant.

(IX) It has been argued that the dispersion relation (5.2) implies a plateau in the radial dependence of $V_L(E)$ in a finite nucleus, near the potential surface and at low energy[121]. The basis of the argument is that W(E) is largest at the surface (surface absorption). However, one must keep in mind that the energy dependence of W(E) is weaker at the surface (fig. 3), so that the argument requires careful examination. The prediction does not seem to be borne out by the calculation of ref. 108, but this may be due to the fact that a plateau in the radial dependence is barely visible unless one pays detailed attention to an effect of this type. We believe that the argument of Ahmad and Haider[121] may be qualitatively valid, in the sense that the diffuseness of the potential well may be larger at low energy. This would be a result of the fact that the plateau in the energy dependence of $V_L(E)$ near the Fermi surface is flatter at the surface density than at the inner density[12].

(X) If the empirical potential well used to fit the observed single-particle energies is assumed to be local and energy-dependent[122], the effect of the non-locality correction factor (5.32) is neglected altogether. It should be included a posteriori in an ad hoc fashion.

(XI) There exists a remote but exciting possibility that taking into account the fact that the non-locality corrections are underestimated in the empirical analysis (see IV, V, X) may help reducing the disagreement which exists in the inner region between the experimental charge density distribution and the calculated one. Indeed, these enlarged non-locality corrections will tend to smooth out the oscillations in the calculated distributions.

(XII) The Landau-Migdal theory establishes an intimate link between the residual interaction to be used in configuration mixing or in Hartree-Fock calculations on the one hand, and the properties of the mean field on the other hand[123]. Hence, the narrow enhancement of the effective mass may imply that this residual interaction should be state dependent whenever single-particle states close to the Fermi surface are involved.

6. OPTICAL-MODEL POTENTIAL AT NEGATIVE ENERGY

6.1. Introduction

It is intuitively clear that the familiar optical-model potential is the natural extrapolation to positive energy of the shell-model potential which is used to describe bound single-particle states. This was recently illustrated by Giannini and Ricco[124]. These authors succeeded in obtaining a decent fit of the energies of deeply and weakly bound single-particle states and of the elastic scattering cross sections up to 80 MeV with a single optical-model potential, for $N = Z$ nuclei with $A < 40$. The real part of the potential is non-local, with parameters which depend smoothly on mass number. Their fit is less good near the Fermi surface, as expected from the discussion in sects. 5.3 and 5.4.

It is customary to think of the shell-model potential as being real. However, the description given in sects. 2.1-2.3 applies to bound (hole) as well as to unbound (particle) states: a nucleon in a bound orbit has a finite mean free path; this can be taken into account by introducing an imaginary part in the mean field. Nevertheless, it is only occasionally that a complex potential has been used at negative energy, see e.g. refs. 125, 126. We believe that this unified description of bound and scattering single-particle states will become a familiar framework as the data on deeply bound single-particle states will continue accumulating.

In the present chapter, we show how the tools described above are useful in the interpretation of the experimental data on bound single-particle states. The basic formulae are gathered in sect. 6.2; then we proceed to the discussion of weakly (sect. 6.3) and of deeply (sect. 6.4) bound single-particle states.

6.2. Spectral Function

In pick-up (e.g. $(d,{}^3He)$), in knock-out (e.g. $(e,e'p)$) or in stripping (e.g. (d,p)) experiments, one can measure the distribution in energy of the spectroscopic factor of a bound single-particle state. In the idealized case of nuclear matter, this is given by the spectral function, eqs. (2.39) and (2.45):

$$S(k,E) = -\pi^{-1} \frac{W(k,E)}{[E - \frac{k^2}{2m} - V(k,E)]^2 + \frac{1}{4}[2\,W(k,E)]^2} \tag{6.1}$$

whose relationship with experimental knock-out cross sections has been discussed by Gross and Lipperheide[127] and others, see e.g. refs. 128, 129. We recall that S(k,E) gives the joint probability amplitude of finding a nucleon with momentum k and energy E in the correlated system.

We have seen in chapters 3-5 that W(k,E) and V(k,E) are smooth functions of E. Hence, the spectral function for fixed k will display a peak when plotted versus E. This peak is asymmetric, because the threshold behaviour (2.44) of W(k,E) entails

$$S(k,\varepsilon_F) = 0 \quad . \tag{6.2}$$

This asymmetry is visible on figs. 6 and 7. In the pole approximation (2.51), it is reflected in the term proportional to the imaginary part of the residue R(k) of the Green function[12].

The spectral function is directly related to other observables. For instance, the momentum distribution in the correlated ground state (eq. (2.4)) is given by

$$\rho(k) = \int_{-\infty}^{\varepsilon_F} dE\, S(k,E) \quad . \tag{6.3}$$

The binding energy B of nuclear matter can be calculated from the sum rule[130]

$$B = \frac{1}{2} \sum_k \rho(k) \frac{k^2}{2m} + \frac{1}{2} \sum_k \int_{-\infty}^{\varepsilon_F} E\, S(k,E)\, dE \quad , \tag{6.4}$$

whose practical interest has been stressed by Koltun[131]. The first term on the right-hand side of eq. (6.4) is obviously equal to one-half the kinetic energy of the correlated ground state. The centroid of the energy distribution is given by

$$E(k) = [\rho(k)]^{-1} \int_{-\infty}^{\varepsilon_F} E\, S(k,E)\, dE \quad . \tag{6.5}$$

Because of the asymmetry of the peak, it is smaller than the location of the maximum:

$$E(k) < e(k) \quad . \tag{6.6}$$

6.3. Weakly Bound States

The borderline between weakly and deeply bound single-particle states is somewhat arbitrary. An experimentalist's choice would be that weakly bound states are studied by stripping or pick-up reactions, while a deeply bound proton state is best investigated by knock-out reactions. However, the availability of intermediate energy p,d and ^{3}He beams enables one to observe deeply bound neutron states by pick-up reactions[132,133]. For our purpose, we shall call weakly bound states those whose binding energy is less than approximately 15 MeV.

The $g_{9/2}$ neutron single-particle state in the tin isotopes has been studied via (d,t) (ref. 134) and (^{3}He,^{4}He) (refs. 135, 136) reactions. The proton $d_{5/2}$ hole state in the Cl and K isotopes has been investigated via (d,^{3}He) (refs. 137, 138). In both cases, the dependence upon mass number of the width (2W) of the spectral function can be interpreted in terms of the coupling of the single-particle state to low-lying vibrational states. Therefore, a nuclear matter approach to the width of these states would not be realistic, and it is preferable to resort to configuration mixing calculations. This is because the density of nearby one particle-two hole states (see graph (d) in fig. 9) is too low for hole states with low excitation energy.

The systematics of the energies of the weakly bound states is of more interest in the present context. For instance, the dependence upon (N-Z)/A of the binding energy of the $g_{9/2}$ neutron single-particle state in the Sn isotopes is related to the symmetry potential (sect. 3.4).

As discussed previously, Brown, Gunn and Gould[113] had pointed out that the level density requires an effective mass $m^* \simeq m$, which is significantly larger than that ($m^* \simeq 0.6$ m) which corresponds to the level density of deeply bound states. This is the origin of the failure of standard Hartree-Fock calculations to reproduce the full spectrum of bound single-particle energies[12]. The early observation of Brown, Gunn and Gould is confirmed by a recent analysis: Bear and Hodgson[139] constructed a local energy-dependent potential well which fits the observed single-particle energies of the weakly bound states and, whenever known, of the deeply bound states, and which moreover provides a reasonable fit of the charge density distribution of ^{40}Ca, ^{48}Ca, ^{58}Ni, ^{90}Zr and ^{208}Pb. Figure 34 shows their result for the potential depth: it is compatible with a linear function of energy for $-50 < E < E_m \simeq -15$ MeV, with an effective mass

$$\frac{m^*}{m} = 0.50 \pm 0.15 \qquad ; \qquad (6.7)$$

the potential well then becomes practically independent of energy for $E_m < E < 0$ MeV, with

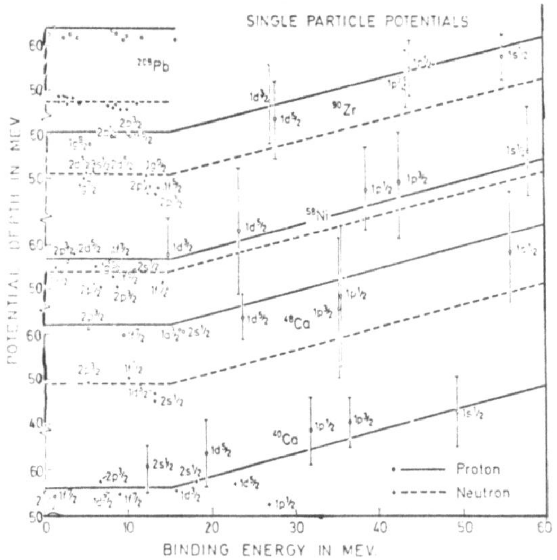

Fig. 34. Empirical depths for proton (dots) and for neutron (crosses) potential wells which reproduce the experimental binding energies. The full and the dashed straight lines are linear fits to these empirical values. From ref. 139.

$$E_m \simeq -15 \text{ MeV} \quad . \tag{6.8}$$

This is in semi-quantitative agreement with the theoretical results described in sect. 5.3. We note that the fit shown in ref. 139 also appears to be compatible with the values

$$\frac{m^*}{m} = 0.60 \pm 0.15 \quad , \quad E_m = -10 \text{ MeV} \quad , \tag{6.9}$$

which appear more realistic (see sect. 5.3).

It would be of interest to repeat this type of analysis with a potential which would be the sum of a non-local energy-independent field and of a local energy-dependent well, as described in sects. 5.2-5.4, and which would moreover include theoretical symmetry and Coulomb components (chapter 3).

We recall that the theory may help understanding the squeezed root mean square radius of the valence orbit (sect. 5.4).

6.4. Deeply Bound States

Deep hole states correspond to a high excitation energy of the residual nucleus. Therefore, the density of nearby one particle-two hole states is large and a nuclear matter approach is better justified than for weakly bound states.

The width of hole states is not given by the Brueckner-Hartree-Fock approximation. Its leading contribution in the low-density expansion is given by graph (2) of fig. 9, which is sometimes called the Pauli rearrangement graph or the correlation graph (sect. 5.3). This graph has been computed by Sartor[48,50], who used a semi-realistic nucleon-nucleon interaction. His result for W(k) is represented by that part of curve 2 in fig. 35 for which $k < k_F$.

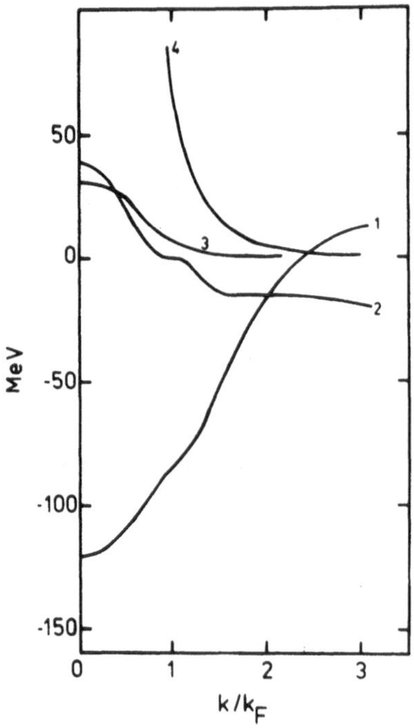

Fig. 35. Dependence upon momentum obtained by Sartor[48] for the real part of the Brueckner-Hartree-Fock potential (curve 1), and for its imaginary part (curve 2 for $k/k_F > 1$). Curve 2 for $k < k_F$ corresponds to minus the imaginary part of graph (2) of fig. 8, i.e. to the quantity $- W(k,e(k))$ for $k < f_F$. Curve 3 is the real part of graph (2) of fig. 8. The value of the Fermi momentum is $k_F = 1.36$ fm^{-1}. From ref. 48.

One can attempt to compare the computed widths with the experimental values. This raises the problem of finding a suitable local density approximation. Sartor[48,50] chose as a local variable the momentum, which is very model dependent. It would be preferable to adopt as variable the energy difference $E - \varepsilon_F$, since the width is mainly influenced by the density of two hole-one particle states.

Let us take for instance

$$E - \varepsilon_F \simeq -45 \text{ MeV} \quad , \qquad (6.10)$$

which approximately corresponds to the 1s shell in ^{40}Ca. The corresponding value of

$$\Gamma_{1s} = -2\ W(k(E),E) \tag{6.11}$$

can be obtained from fig. 35. One finds

$$\Gamma_{1s} \simeq 40\ \text{MeV} \quad , \tag{6.12}$$

which is approximately two times larger than the experimental value $\Gamma \simeq 20$ MeV[112]. This disagreement is probably due to the strength of the semi-realistic interaction used by Sartor. Indeed, he obtains -120 MeV for the depth of the real part of the Brueckner-Hartree-Fock field at k = 0 (fig. 33), while a more realistic interaction yields -80 MeV (fig. 10). Thus, his effective interaction used by Sartor appears to be too large by a factor 1.5. Since it enters to the square in the expression of the width, it is likely that a more realistic interaction would lead to much better agreement with the experimental data.

As they stand, Sartor's results are quite valuable. Indeed, we believe that he carried out a reliable calculation of the dependence upon $E - \varepsilon_F$ of the width of the hole states, up to a normalization factor. Sartor's results for $W_L < E - \varepsilon_F$ could thus be compared with the analytical form (5.20) adopted by Orland and Schaeffer[112], and which is represented in fig. 36. An equally promising procedure would consist in fitting the value of $W_L(E)$ for $E > \varepsilon_F$ with the analytic expression (5.21), which could then be used for $E < \varepsilon_F$. Note that the absolute value of Γ in fig. 36 is determined by the choice of the adjustable parameter α^2.

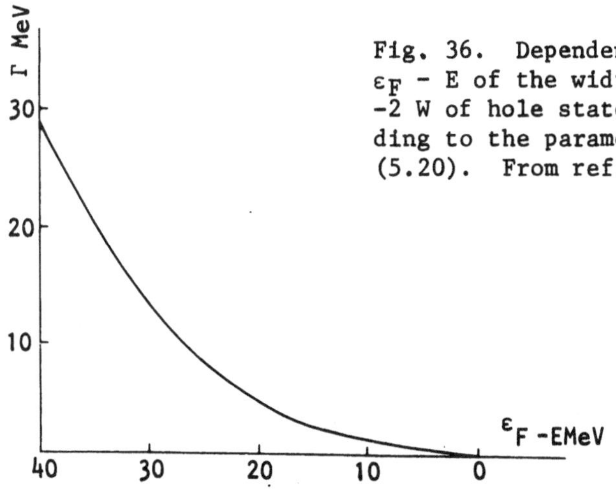

Fig. 36. Dependence upon $\varepsilon_F - E$ of the width $\Gamma = -2\ W$ of hole state, according to the parametrization (5.20). From ref. 112.

A calculation of the dependence upon mass number of the width Γ_{1s} in finite nuclei has been performed by Hughes, Fallieros and Goulard[140]. It would be of interest to compare their prediction with nuclear matter calculations.

We mentioned in sect. 6.1 that the peak of the spectral function has an asymmetric shape, with a steeper slope on the side of ε_F. A typical shape for the peak can be obtained by inserting the parametrization (5.20) in eq. (6.1). Figure 37 shows that this shape is in good agreement with experimental evidence.

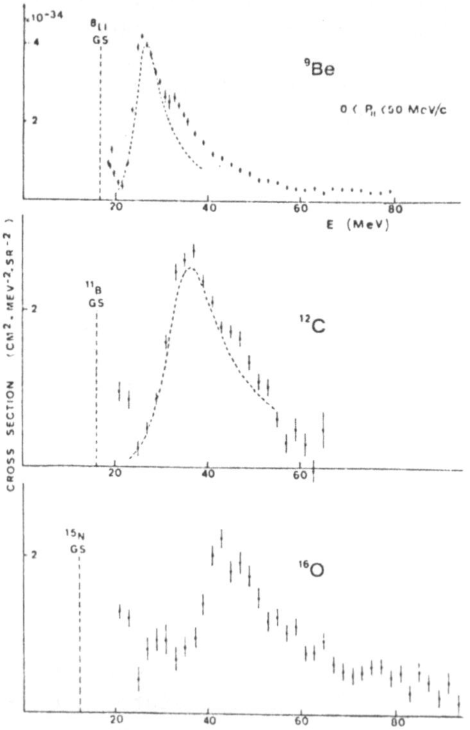

Fig. 37. The points show the measured strength of the 1s hole in ^{9}Be, ^{12}C and ^{16}O, plotted versus the proton separation energy. The dashed curves is the shape computed from eqs. (5.20) and (6.1). From ref. 141.

In principle, eq. (6.3) enables one to compute the occupation number $\rho(k)$ from the measured spectral functions. Since $S(k,E)$ becomes too small to be measurable for $E < -80$ MeV, it is difficult to ascertain the accuracy of this procedure, which yields typically[141]

$$\rho(0.75\ k_F) \simeq 0.6 \qquad . \tag{6.13}$$

This is somewhat smaller than the calculated value[12,112], see fig.

1, but is probably an underestimate because of the cut-off at -80 MeV: the short range correlations are responsible for a long tail of S(k,E) when $E \to -\infty$. The same kind of problems are encountered when one attempts to check the sum rule (6.4) experimentally[141].

Finally, we recall that the theoretical results described in sect. 5.4. may help understanding the disagreement between experimental single-particle energies for deeply bound states and theoretical values obtained from a Hartree-Fock field which would have been computed from an effective interaction adjusted to the properties of weakly bound single-particle states.

7. CONCLUSIONS

The low-density expansion enables one to compute the optical-model potential at negative, low and intermediate energies from a realistic nucleon-nucleon interaction. Its leading term is the Brueckner-Hartree-Fock approximation. The comparison with empirical values at low energy ($0 < E < 150$ MeV) indicates that its accuracy is of the order of ten per cent (chapter 3). This should be maintained at intermediate and at negative energies, and enables one to trust the following semi-quantitative predictions which in our opinion deserve particular attention.

The real part of the optical-model potential acquires a wine-bottle bottom shape in the vicinity of the energy ($\simeq$ 300 MeV) at which it changes sign (sect. 4.3). This is not contained in the simplest version of the impulse approximation because the latter does not include Pauli and binding energy corrections.

At high energy, the real part of the optical-model potential appears to be well represented by a local energy-dependent field (sect. 5.2). In contrast, it becomes predominantly non-local at low and negative energies. There, the energy dependence cannot, however, be neglected. As a result, the non-locality of most empirical potentials (e.g. a Hartree-Fock field computed from an effective interaction or a non-local energy-independent potential fitted to the elastic cross sections) is grossly underestimated. The theory also shows that the true energy dependence of the real part of the potential is enhanced right above the Fermi surface, while its non-locality remains unchanged. Several possible consequences of these properties have been enumerated in sect. 5.4. They concern, for instance, the mean square root radius of the valence orbit as compared to that of the core orbits, the energy of the deeply bound states, the non-locality correction factor by which one should multiply the amplitude of the single-particle wave function inside the potential well, etc.

The theory also yields information on the width of the deeply bound single-particle states. Possible improvements over existing calculations have been pointed out.

Useful discussions with B. Castel, A. Lejeune and P. Rochus are gratefully acknowledged.

REFERENCES

1. H.A. Bethe, Ann. Rev. Nucl. Sci. 21 (1971), 93.

2. M.L. Goldberger and K.M. Watson, Collision Theory (Wiley, New York, 1964).

3. A.K. Kerman, H. McManus and R.M. Thaler, Ann. Phys. (N.Y.) 8 (1959), 551.

4. H. Feshbach, A. Gal and J. Hüfner, Ann. Phys. (N.Y.) 66 (1971), 20.

5. E. Boridy and H. Feshbach, Phys. Lett. 50B (1974), 433.

6. N. Austern, F. Tabakin and M. Silver, Am. J. Phys. 45 (1977), 361.

7. J. Saudinos and C. Wilkin, Ann. Rev. Nucl. Sci. 24 (1974), 341.

8. M.M. Sternheim and R.R. Silbar, Ann. Rev. Nucl. Sci. 24 (1974), 249.

9. R.V. Reid, Ann. Phys. (N.Y.) 50 (1968), 411.

10. H.S. Köhler, Phys. Reports 18 (1975), 217.

11. J. Hüfner and C. Mahaux, Ann. Phys. (N.Y.) 73 (1972), 525.

12. P.-P. Jeukenne, A. Lejeune and C. Mahaux, Phys. Reports 25C (1976), 83.

13. A.A. Abrikosov, L.P. Gorkov and I.E. Dzyaloshinsky, Methods of Quantum Field Theory in Statistical Physics (Prentice Hall, Inc., 1963).

14. H. Feshbach, Ann. Rev. Nucl. Sci., 8 (1958), 49.

15. F.G. Perey and D.S. Saxon, Phys. Rev. 10 (1964), 107.

16. W.E. Frahn, Nucl. Phys. 66 (1965), 358.

17. F.G. Perey, Phys. Rev. 131 (1963), 745.

18. F.G. Perey and A.M. Saruis, Nucl. Phys. 70 (1965), 225.

19. A. Miller, Contribution to this Institute.

20. J.W. Negele, Phys. Rev. C1 (1970), 1260.

21. F.D. Becchetti and G.W. Greenlees, Phys. Rev. 182 (1969), 1190.

22. G. Passatore, Nucl. Phys. A95 (1967), 694.

23. J.-P. Jeukenne, A. Lejeune and C. Mahaux, Proceedings of the Twelfth International Winter Meeting on Nuclear Physics, Villars (Switzerland), 1974.

24. F. Perey and B. Buck, Nucl. Phys. 32 (1962), 353.

25. M. Blann, Phys. Rev. C17 (1978), 1871.

26. E. Gadioli, E. Gadioli-Erba and G. Tagliafermi, Phys. Rev. C17 (1978), 2238.

27. G.W. Greenlees, G.J. Pyle and Y.C. Tang, Phys. Rev. 171 (1968), 1115.

28. F.G. Perey, in Nuclear Spectroscopy and Reactions, edited by J. Cerny (Academic Press, New York, 1974) part B, page 137.

29. J.-P. Jeukenne, A. Lejeune and C. Mahaux, Phys. Rev. C16 (1977), 80.

30. J.-P. Jeukenne, A. Lejeune and C. Mahaux, Phys. Rev. C15 (1977), 10.

31. F.A. Brieva and J.R. Rook, Nucl. Phys. A291 (1977), 317.

32. B.D. Day, Rev. Mod. Phys. 39 (1967), 719.

33. E. Hadjimichael, S.N. Yang and G.E. Brown, Phys. Lett. 39B (1972), 594.

34. D.S. Koltun, Adv. Nucl. Phys. 3 (1969), 71.

35. S. Frankel, W. Frati, O. Van Dyck, R. Wesbeck and V. Highland, Phys. Rev. Lett. 36 (1976), 642.

36. V.I. Komarov, G.E. Kosarev, H. Muller, D. Netzband and T. Stiehler, Phys. Lett. 69B (1977), 37.

37. T. Fujita, Phys. Rev. Lett. 39 (1977), 174.

38. T. Fujita and J. Hüfner, Nucl. Phys. A (submitted, 1978).

39. R.D. Amado and R.M. Woloshyn, Phys. Lett. 69B (1977), 400.

40. Y. Alexander, E.F. Redish and N.S. Wall, Phys. Rev. C16 (1977), 526.

41. P. Nozieres, Theory of Interacting Fermi Systems (W.A. Benjamin, Inc., 1964).

42. J.S. Bell and E.J. Squires, Phys. Rev. Lett. 3 (1959), 96.

43. H. Feshbach, Ann. Phys. (N.Y.) 5 (1958), 357.

44. H. Feshbach, Ann. Phys. (N.Y.) 19 (1962), 287.

45. A.L. Fetter and K.M. Watson, in Advances in Theoretical Physics, ediited by K.A. Brueckner (Academic Press, New York, 1965), p. 115.

46. T.F. Hamman and Q. Ho-Kim, Nuovo Cim. 64B (1969), 356.

47. Q. Ho-Kim and R. Provencher, Nuovo Cim. 14A (1973), 633.

48. R. Sartor, Nucl. Phys. A267 (1976), 29.

49. C. Dover, J. Hüfner and R.H. Lemmer, Ann. Phys. (N.Y.) 66 (1971), 248.

50. R. Sartor, Nucl. Phys. A289 (1977), 329.

51. J.-P. Jeukenne, A. Lejeune and C. Mahaux, Proceedings of the International Conference on the Interactions of Neutrons with Nuclei, Lowell (edited by E. Sheldon), ERDA (1976), 451.

52. J.-P. Jeukenne, A. Lejeune and C. Mahaux, Nukleonika 20, n. 2 (1975), 181.

53. J.-P. Jeukenne, A. Lejeune and C. Mahaux, Phys. Rev. C10 (1974), 1391.

54. W.T.H. Van Oers, Phys. Rev. C3 (1971), 1550.

55. G.L. Thomas and E.J. Burge, Nucl. Phys. A128 (1969), 545.

56. S. Kailas and S.K. Gupta, Phys. Rev. C17 (1978), 2236.

57. S. Kailas and S.K. Gupta, Phys. Lett. 71B (1977), 271.

58. J. Rapaport, Phys. Lett. 70B (1977), 141.

59. F. Bertrand, Contribution to this Institute.

60. F.A. Brieva and J.R. Rook, Nucl. Phys. A297 (1978), 206.

61. R.J. Blin-Stoyle, Phil. Mag. 46 (1955), 973.

62. R.R. Scheerbaum, Nucl. Phys. A257 (1976), 77.

63. D.W.L. Sprung, Nucl. Phys. A182 (1972), 97.

64. B. Sinha, Phys. Reports 20C (1975), 1.

65. G. Bertsch, J. Borysowicz, H. McManus and W.G. Love, Nucl. Phys. A284 (1977), 399.

66. Y. Eisen, B. Day and E. Friedman, Phys. Lett. 56B (1975), 313.

67. Y. Eisen and B. Day, Phys. Lett. 63B (1976), 253.

68. H.V. Von Geramb, F.A. Brieva and J.R. Rook, preprint (1978).

69. X. Campi and A. Bouyssy, Phys. Lett. 73B (1978), 263.

70. C. Mahaux, Proceedings of the 1978 Workshop on Microscopic Optical Potentials, Hamburg, 1978 (edited by H.V. Von Geramb).

71. A. Lejeune and P.E. Hodgson, Nucl. Phys. A295 (1978), 301.

72. E. Clementel and C. Villi, Nuovo Cim. 2 (1955), 176.

73. G.L. Shaw, Ann. Phys. (N.Y.) 3 (1959), 509.

74. I.R. Afnan and Y.C. Tang, Nucl. Phys. A141 (1970), 653.

75. J. Cugnon, Nucl. Phys. A165 (1971), 393.

76. W.T.H. Van Oers and H. Haw, Phys. Lett. 45B (1973), 227.

77. R.C. Barrett and D.F. Jackson, Nuclear Sizes and Structure (Clarendon Press, Oxford, 1977).

78. L.R.B. Elton, Nucl. Phys. 89 (1966), 69.

79. J.-P. Jeukenne, A. Lejeune and C. Mahaux, Journal de Physique, 35, suppl. 11 (1974), C5-7.

80. P.G. Roos and N.S. Wall, Phys. Rev. 140 (1965), B12, 37.

81. A. Ingemarsson, Physica Scripta 9 (1974), 156.

82. J.-P. Jeukenne, A. Lejeune and C. Mahaux, in Proceedings of the International Conference on Nuclear Self-Consistent Fields (Trieste, February 1975), edited by G. Ripka and M. Porneuf, p. 155 (North-Holland Publ. Comp., Amsterdam, 1975).

83. F.A. Brieva and J.R. Rook, Nucl. Phys. A291 (1977), 299.

84. A. Lejeune and C. Mahaux, to be published.

85. T. Hamada and D. Johnston, Nucl. Phys. 34 (1962), 382.

86. L. Ray and W.R. Coker, Phys. Rev. C16 (1977), 340.

87. G. Passatore, Nucl. Phys. A248 (1975), 509.

88. R. Rajaraman, Phys. Rev. 129 (1963), 265.

89. W.T.H. Van Oers, Huang Haw, N.E. Davison, A. Ingemarsson, B. Fagerström and G. Tibell, Phys. Rev. C10 (1974), 307.

90. L.D. Miller and A.E.S. Green, Phys. Rev. C5 (1972), 241. A.E.S. Green, F. Riewe, M.L. Nack and L.D. Miller, in the Nuclear Many-Body Problem, edited by F. Calogero and C. Ciofi degli Atti, p. 415 (Editrice Compositori, Bologna, 1973).

91. L.G. Arnold, B.C. Clark, R.L. Mercer, D.G. Ravenhall and A.M. Saperstein, Phys. Rev. C14 (1976), 1878.

92. J.D. Walecka, Ann. Phys. (N.Y.) 83 (1974), 491.

93. K.P. Lohs and J. Hüfner, Nucl. Phys. A296 (1978), 349.

94. J.V. Noble, Phys. Rev. C17 (1978), 2151.

95. B.C. Clark, L.G. Arnold and R.L. Mercer, Bull. Am. Phys. Soc., 23, nr. 4 (1978), 571.

96. R. Humphreys, Nucl. Phys. A182 (1972), 580.

97. J.M. Cornwall and M. Rudermann, Phys. Rev. 128 (1962), 1474.

98. R. Lipperheide, Nucl. Phys. 89 (1966), 97.

99. J.M. Luttinger, Phys. Rev. 121 (1961), 942.

100. A.S. Reiner (Rinat), Phys. Rev. 133 (1964), B1105.

101. A.B. Migdal, Nucl. Phys. 30 (1962), 239.

102. R. Lipperheide, Z. für Physik 202 (1967), 58.

103. C. Mahaux and H.A. Weidenmüller, Shell-Model Approach to Nuclear Reactions (North-Holland Publ. Comp., Amsterdam, 1969).

104. G. Passatore, Nucl. Phys. A110 (1968), 91.

105. R. Lipperheide and A.K. Schmidt, Nucl. Phys. A112 (1968), 65.

106. H. Fiedeldey and C.A. Engelbrecht, Nucl. Phys. A128 (1969), 673.

107. I. Ahmad and M.Z. Rahman Khan, Nucl. Phys. A132 (1969), 213.

108. G. Eckart and M.K. Weigel, J. of Phys. G2 (1976), 487.

109. M. Bertero and G. Passatore, Z. Naturforsch. 28a (1973), 519.

110. G. Passatore, in Nuclear Optical-Model Potential, edited by S. Boffi and G. Passatore (Springer-Verlag, Berlin, 1976).

111. A.B. Migdal, Theory of Finite Fermi Systems (John Wiley, New York, 1967).

112. H. Orland and R. Schaeffer, Nucl. Phys. A299 (1978), 442.

113. G.E. Brown, J.H. Gunn and P. Gould, Nucl. Phys. 46 (1963), 598.

114. G.E. Brown, J.A. Evans and D.J. Thouless, Nucl. Phys. 45 (1963), 164.

115. G.F. Bertsch and T.T.S. Kuo, Nucl. Phys. A112 (1968), 204.

116. I. Hamamoto and P. Siemens, Nucl. Phys. A269 (1977), 199.

117. J.-P. Jeukenne, A. Lejeune and C. Mahaux, Phys. Lett. 59B (1975), 208.

118. I. Sick, J.B. Bellicard, J.M. Cavedin, B. Frois, M. Huet, P. Leconte, A. Nahada, Phan Xuan Ho, S. Platchnov, P.K.A. de Witt Huberts and L. Lapidas, Phys. Rev. Lett. 38 (1977), 1259.

119. M.V. Hynes, H. Miska, B. Noram, W. Bertozzi, S. Kowalski, F.M. Rad, C.P. Sargent, T. Sasanuma, W. Turchinetz and B.L. Berman, preprint (1978).

120. L. Zamick, preprint (1978).

121. I. Ahmad and W. Haider, J. Phys. G2 (1976), L157.

122. F. Malaguti, A. Uguzzoni, E. Verondini and P.E. Hodgson, Nucl. Phys. A297 (1978), 287.

123. G.E. Brown, Rev. Mod. Phys. 43 (1971), 1.

124. M.M. Giannini and G. Ricco, Ann. Phys. (N.Y.) 102 (1976), 458.

125. R. Shanta, Nucl. Phys. A199 (1973), 624.

126. G.M. Vagradov, F.A. Gareev and J. Bang, Nucl. Phys. A278 (1977), 319.

127. D.H.E. Gross and R. Lipperheide, Nucl. Phys. A150 (1970), 449.

128. C.A. Engelbrecht and H.A. Weidenmüller, Nucl. Phys. A184 (1972), 385.

129. S. Boffi, C. Giusti and F.D. Pacati, Lett. Nuovo Cim. 20 (1977), 168.

130. S. Boffi, Lett. Nuovo Cim. 1 (1971), 931.

131. D.S. Koltun, Phys. Rev. Lett. 28 (1972), 182.

132. J. Källne and B. Eagerström, Proceedings of the Fifth International Conference on High Energy Physics and Nuclear Structure, edited by G. Tibell, Uppsala (1973).

133. E. Gerlie, H. Langevin-Joliot, P. Roos, J. Van de Wiele, J.P. Didelez and G. Duhamel, Phys. Rev. C12 (1975), 2106.

134. S.Y. Van der Werf, B.R. Kooistra, W.H.A. Hesselink, F. Iachello, L.W. Put and R.H. Siemssen, Phys. Rev. Lett. 33 (1974), 712.

135. E. Gerlie, J. Källne, H. Langevin-Joliot, J. Van de Wiele and G. Duhamel, Phys. Lett. 57B (1975), 338.

136. M. Sekinguchi, Y. Shida, F. Soga, T. Hattori, Y. Hiras and M. Sakai, Phys. Rev. Lett. 38 (1977), 1015.

137. G.J. Wagner, P. Doll, K.T. Knöpfle and G. Mairle, Phys. Lett. 57B (1975), 413.

138. P. Doll, G.J. Wagner, K.T. Knöpfle and G. Mairle, Nucl. Phys. A263 (1976), 210.

139. K. Bear and P.E. Hodgson, preprint (1978).

140. T.A. Hughes, S. Fallieros and B. Goulard, Nuclei & Particles, 1 (1971), 93.

141. J. Mougey, Proceedings of 1977 Tokyo Conference, J. Phys. Soc. Japan 44 (1978), suppl. p. 420.

142. J.W. Negele, Phys. Rev. C9 (1974), 1054.

NUCLEAR STUDIES INVOLVING INTERMEDIATE ENERGY PROJECTILES

G.E. Walker

Physics Department, Indiana University

Bloomington, Indiana 47401, U.S.A.

I. INTRODUCTION

Recently the nuclear physics community has begun to receive the first dividends on a major investment in the construction of intermediate energy (100 MeV - 1 GeV) accelerators. High precision data from such new accelerators in North America as Bates (MIT), IUCF (Indiana), LAMPF (Los Alamos), and TRIUMF (Vancouver) involving medium energy electron, photon, pion and proton projectiles is currently available for comparison with present theory and may stimulate new insights into the multi-faceted aspects of nuclear structure. Of course the flow of new data involving these probes is just beginning. In addition, high quality kaon-nucleus data is anticipated shortly, for example, from the AGS at Brookhaven. Since each probe has associated with it characteristic limitations and complications it is especially important to study the same nuclear excitations in similar regions of momentum transfer with different probes to obtain maximum benefit from the new accelerators. The purpose of these lectures is to give selected examples of the kinds of new data available and anticipated and to compare theory and experiment (or discuss predictions) for the various probes. We stress the complementarity of the various projectiles in elucidating nuclear structure.

The nucleus is a finite many-body system held together by incompletely understood strong forces. The primary constituents are "dressed" neutrons and protons, however meson and, or, isobar degrees of freedom must enter into studies of nuclear structure (especially in the domain of large energy loss or high momentum transfer). Since the basic two body interaction is incompletely known and one cannot in general solve the many-body problem it is necessary to propose "models" of the nucleus and then to make predictions concerning static nuclear properties, the nuclear excitation spectrum, and nuclear reactions to be compared with experiment. Unfortunately the reaction mechanism is often also incompletely known and so the interpretation of experimental data, when compared with the results of predictions based on proposed nuclear models, is clouded.

A brief summary of the situation might include the following in a list of the desired nuclear information

1. Nuclear size, shape, excitation spectrum and single nucleon binding energies,
2. Nuclear symmetries or near symmetries (approximately good quantum numbers),
3. Relevant degrees of freedom (microscopic and macroscopic),
4. Momentum distributions, hole lifetimes and orbital occupation probabilities.

The experimental results from nuclear reactions often do not yield clean nuclear structure information because of such difficulties as

1. Uncertainties in the projectile-nuclear constituent elementary interaction,
2. Validity of the "Distorted Wave Impluse Approximation, DWIA"
   a. Distortion treated correctly?
   b. Importance of multistep processes?
3. Signature of long and short range correlation effects, and the effects of nuclear Δ's and N*'s and their prediction,
4. Importance of relativistic effects,
5. Technical problems
   a. Inability to apply existing theories because of the lack of appropriate computer codes
   b. Validity of the experimental information coming from the new machines.

Of course the obvious way to attempt to overcome some of these difficulties is to vary the probe, energy, momentum transfer, nuclear final state, target, etc. Theoretical guidance to permit some limitations on such a scattergun approach is required.

The new generation of intermediate energy accelerators each have some of the following characteristics

1. Variable energy - this allows one to study the effects of multistep processes and work in regions where one

expects the DWIA to be more nearly valid.

2. Intense beam current and/or high duty cycle - this allows one to contemplate experiments where the counting rate is relatively low (such as $(p,\pi-)$ or coincidence experiments) and still obtain high quality data.
3. Excellent energy resolution - this allows one to separate nuclear final states which may be $\lesssim$ 100 keV apart when the projectile energy is in excess of 100 MeV.

Just the fact that one is working at higher energy or a certain rare event can be studied often allows one to work in the high momentum transfer region which is a relatively poor understood region of nuclear research.

In concluding these introductory remarks, we should note that the "utility" of a given nuclear probe depends on the quality of data available, the "cleanliness" of the theoretical interpretation, what properties of the nucleus the probe "sees" and the availability of alternate probes that emphasize some of the same nuclear properties. For reaction mechanism studies each probe has something to offer and this can be especially interesting if the relevant nuclear structure is already understood from other reactions.

In sec. II we will discuss selected electromagnetic reactions concentrating on (e,e') and (γ,p). We shall see that there remain both structure and reaction mechanism questions raised by the "well-understood" electromagnetic probes.

In sec. III pion and kaon nucleus reactions will be reviewed. In that section it will be argued that, although there remain considerable uncertainties in the π and K^- nucleus reaction, useful nuclear structure information can be obtained simply because of the kind of nuclear excitations that are predicted to dominate the nuclear response to these probes. The K^+ projectile holds great promise as a "clean" probe of nuclear structure.

Proton-nucleus inelastic scattering is discussed in sec. IV. Comparison between theory and recent experiments at Indiana is presented. We also will discuss some recent predictions for (p,p') and (p,n) using a few medium energy pseudo-potential fitted to angular distribution and polarization data in the energy region (50 - 450 MeV).

Although these lectures do not emphasize weak nuclear reactions it should be clear that they will play a major role in the future of intermediate energy nuclear physics. In this area we have particularly in mind neutrino-nucleus reactions.

II. ELECTROMAGNETIC PROBES

A large portion of our definitive information regarding nuclear structure has come from electromagnetic interactions (particularly involving electrons and real photon absorption or emission) with the nucleus. This results from the fact that the elementary electromagnetic interaction is well understood and is weak enough that it can be treated in first order (single photon exchange for (e,e')).

We consider as our first example, inelastic electron scattering. We shall be interested in the nuclear response ($d^2\sigma/dqd\omega$) to a given probe as a function of momentum transfer q, and projectile energy loss ω.

In the one photon exchange approximation and ignoring, for simplicity, the energy of the electron compared to the mass of the target one can write the inelastic electron scattering cross section as[1]

$$\frac{d^2\sigma}{d\Omega d\omega} = 4\pi\,\sigma_M\,\{(\frac{q_\mu^4}{q^4})\sum_{J=1}^{\infty}\frac{|\langle J_f||\hat{M}_J^{coul}(q)||J_i\rangle|^2}{2J_i+1}$$

$$+(\frac{q_\mu^2}{2q^2}+\tan^2\theta/2)\sum_{J=1}^{\infty}(\frac{|\langle J_f||\hat{T}_J^{el}(q)||J_i\rangle|^2}{2J_i+1}$$

$$+\frac{|\langle J_f||\hat{T}_J^{mag}(q)||J_i\rangle|^2}{2J_i+1})\} \qquad (1)$$

where σ_M is the Mott scattering cross section

$$\sigma_M \equiv \frac{\alpha^2\cos^2(\theta/2)}{4\varepsilon_i^2\sin^4\theta/2} \quad . \qquad (2)$$

The quantity $q_\mu = (\vec{q},i\omega)$ is the four momentum transfer, $q = |\vec{q}|$ is the three momentum transfer (for $q^2 >> \omega^2$, $q \approx 2\,E_0\sin(\theta/2)$) and J_i and J_f denote the initial and final nuclear states, respectively. The term involving $\hat{M}_J^{coul}$ is called the longitudinal form factor, involves only the protons, and results from the interaction of the electron with the nuclear charge density. The terms involving T_j^{el} and T_j^{mag} are, together, called the transverse form factor and they each result from the interaction of the electron with the nuclear convection current density, $J_n(x)$, and magnetization density, $\mu_n(x)$. For a nucleus composed of point nucleons the current and magnetization densities are given by

$$J_n(x) = \sum_i [\delta(r_i - x) \frac{(1 + \tau_3(i))}{2} \frac{1}{iM_N} \nabla]_{sym}. \tag{3a}$$

$$\mu_n(x) = \sum_i \delta(r_i - x) \{(\frac{\mu_p + \mu_n}{2}) + \frac{\mu_p - \mu_n}{2} \tau_3(i)\} \frac{1}{2M} \quad . \tag{3b}$$

Because of the $\tan^2\theta/2$ factor one can enhance the effect of the transverse form factor over the longitudinal form factor by an order of magnitude or more by working at large angles ($\theta \gtrsim 120°$). For medium energy electrons this implies that one is working at reasonably large momentum transfer, $q \gtrsim 300$ MeV/c. For $q^2 >> \omega^2$, (which is satisfied, for example, for $|q| \gtrsim 100$ MeV/c and $\omega < 30$ MeV, in inelastic electron scattering at moderate momentum transfer leading to states up through the giant resonance region) $q_\mu^2 \approx |q|^2$ and the term multiplying the transverse form factor can be written $B = (1/2 + \tan^2\theta/2)$. A plot, of the differential cross section vs. B, can be used to separate the longitudinal and transverse contributions, the slope yielding the transverse term. A deviation of the plot from a straight line would be an indication that the single photon exchange, impulse approximation, was not a valid description of the reaction mechanism.

The parity change associated with the Jth multipole of the charge, convection current, and magnetization current densities is given by[1]

$$M_J^{coul}, \; T_J^{el} \quad \Delta\pi = (-1)^J \tag{4a}$$

$$T_J^{mag} \quad \Delta\pi = (-1)^{J+1} \quad . \tag{4b}$$

For elastic scattering, time reversal invariance and parity conservation eliminate all electric multipoles and allow only even coulomb multipoles and odd magnetic multipoles[1]. It turns out that, generally, the magnetization current dominates over the convection current at medium to large momentum transfer. Since $\mu_p - \mu_n/\mu_p + \mu_n \approx 6$, the isovector magnetic moment is much larger than the isoscalar moment in eq. (3b). Combining the two statements above means that $\Delta T = 1$, $\Delta S = 1$ states are predicted to dominate the response spectrum at large angles.

It is a general feature of weakly absorbed probes that high spin states tend to dominate at large momentum transfer in inelastic scattering. This can be understood by noting that (1) transition densities tend to peak near the nuclear surface R, (2) for probes that can qualitatively be treated in the plane wave limit of the distorted wave approximation, $X^{+}_{\mu f}(r)\, X_{\mu i}(r) \approx e^{iq\cdot r}$, (3) if one does a partial wave expansion of $e^{iq\cdot r}$ conservation of angular momentum will require, for example, that only $J_L(qr)$ enters in the radial matrix elements for a longitudinal type transition from a 0^+ ground state to a state $J_f = L$, $\pi = (-1)^L$, (4) $J_L(qR)$ has its peak near $q = L/R$ and finally (5) q increases with scattering angle. The general feature of high spin states dominating at large momentum transfer will be present, in the appropriate incident laboratory kinetic energy region, for all projectiles we discuss in these lectures.

If the distorted wave impuluse approximation (DWIA) is valid then the inelastic nuclear transition operator can be written as a sum of single (target) nucleon operators. In this case that part of the final nuclear excited state "linked" to the ground state by the transition operator can be obtained by a particle-hole creation or annihilation operator on the ground state. An often used approximation for obtaining nuclear wavefunctions (a linearization of the equations of motion) that differ from the g.s. in this way is the random phase approximation (RPA)[2]. If one considers only p-h creation operators (as would be appropriate if the g.s. were a closed shell) then the RPA reduces to the Tamm-Dancoff Approximation (TDA). If one wishes to include correlations in the g.s. before applying the RPA then the open-shell RPA, OSRPA[3] is aopted. We shall discuss results using all three of the approximation techniques. Generally the largest difference between RPA and TDA results occur for low-lying normal parity excitations[4].

An example of inelastic electron scattering showing the difference between the simple j-j coupling shell model, the TDA, and the RPA is shown in fig. 1 for the 2^+, T = 0 state of ^{12}C. The transition involves M^2_{coul} predominantly. The enhancement of the predicted cross section in the RPA is characteristic of the results for low-lying normal parity T = 0 excited states.

At higher excitation energies the giant dipole, T = 1, state dominates at low momentum transfer. The excitation energy and large concentration of dipole strength in the 20 - 25 MeV region for light nuclei can be understood qualitatively in terms of the schematic model of Brown and Bolsterli[7].

Results from Stanford about a decade ago indicated a few sharp states dominating the spectrum at large momentum transfer[8-10]. Theoretical calculations on ^{12}C, ^{16}O and ^{28}Si predict that these states should be "stretched" p-h high spin states, a 4^-, T = 1 $(1d^5/_2)(1p^3/_2)^{-1}$ state for ^{12}C and ^{16}O (M4 transition)

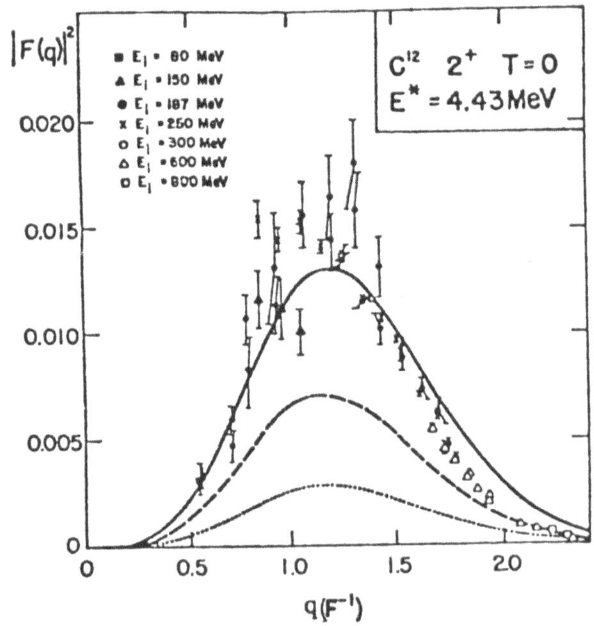

Figure 1. Comparison of the calculations of Gillet and Melkanoff[4] with experimental results of Fregeau[5] and Ehrenberg et al[6]. Figure taken from ref. 1. The dotted line is the pure p-h result while the dashed (solid) line is the TDA (RPA) result.

and a 6^-, T = 1 $(1f^7/_2)(1d^5/_2)^{-1}$ state in ^{28}Si (M6). Fig. 2 presents a typical comparison between theory and experiment for the case of ^{28}Si.

Typically, the data available at that time was averaged over one MeV energy intervals. We shall see that the recent data from Bates has considerably better energy resolution. For these high spin states one typically needs a factor of two reduction in the TDA theory to give agreement between theory and experiment (the shape of the angular distribution is generally satisfactorily reproduced)[8-10]. Generally the RPA and OSRPA or ground state correlations have yielded a reduction factor but not enough to bring theory in agreement with experiment.

Recently Rowe et al[11] have used the OSRPA to obtain wavefunctions to look at the M6 transition in ^{24}Mg. The results indicate that a factor of two reduction is still required to bring theory into agreement with experiment. It would be very useful to excite these high spin T = 1 states with other probes. One suspects

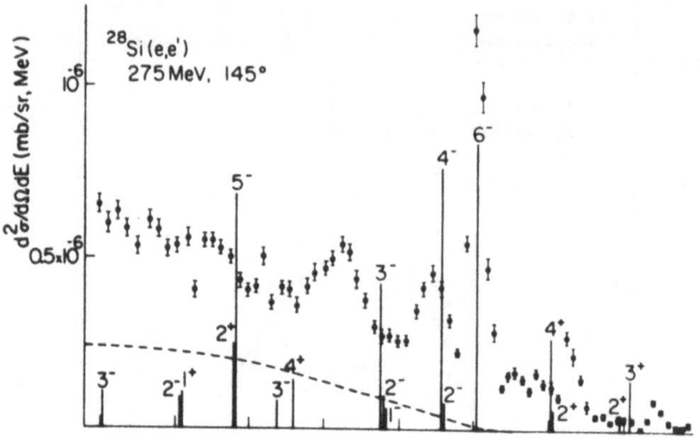

Figure 2a. The cross section $d^2\theta/d\Omega dE$ for inelastic electron scattering from ^{28}Si at $E_e = 275$ MeV, $\theta = 145^o$ unfolded for radiative processes. The non-negligible cross sections predicted by the p-h model (T = 1 states) for q = 525 MeV/c are shown as spikes (arbitrary overall scale). The dashed line is the computed quasi-elastic spectrum.

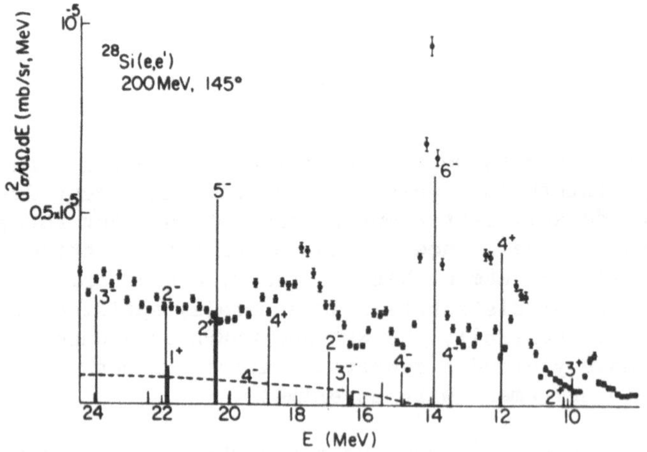

Figure 2b. Same as (a), except $E_e = 200$ MeV and theoretical cross sections computed for q = 275 MeV/c. (Figure taken from ref. 10.)

the major problem is with the nuclear wavefunctions although at high momentum transfer the effect of exchange currents (electromagnetic reaction mechanism uncertainty) may be non-negligible. Theoretical calculations predict that the 6^- T = 1 state should also dominate the (e,e') spectrum from ^{32}S and ^{40}Ca at large momentum transfer[12]. It would be very useful to have these experiments performed, especially in light of recent (p,p') results, which will be discussed more fully later - where these states were predicted but not seen in ^{32}S and ^{40}Ca.

More recently high spin non-normal parity states have been reported, using the BATES accelerator, in ^{58}Ni[13] (8^-) and ^{208}Pb[14] (12^-).

One of the interesting questions raised by the ^{58}Ni results is where the rest of the strength has gone since the summed strength is only 17% of that predicted in the simple shell-model[13]. Here again (p,p') and (p,n) results would be very useful.

Finally, in our brief review of selected recent results, elastic electron scattering data[15] from ^{17}O have indicated a suppression in the M3 moment. This may be understood in terms of core polarization in the Hartree-Fock theory[16]. Recently the oxygen isotopes$^{18,17,16}O$ have been studied at IUCF via proton elastic and inelastic scattering. The results are currently being analyzed and compared with theoretical predictions. The theoretical and experimental results will be presented at the Fall Asilomar Conference. We will see if the same wavefunctions can fit the electron and proton results and study the sensitivity of the polarization to the wavefunctions used and form of the spin dependent nucleon-nucleon transition operator.

As another example of a medium energy electromagnetic process we consider the (γ,p) reaction. Recently at the Bates Linac, the reaction $^{16}O(\gamma,p)^{15}N$ has been studied for photon energies in the interval 100 - 350 MeV[17,18]. High quality data in the region 40 - 100 MeV has been obtained at Glasgow[19]. The experimental results for medium energy photons is shown for three scattering angles in figures 3, 4 and 5. In the limit of the PWIA, averaging over initial photon polarizations and summing over nucleon and nucleus polarizations allows the differential cross section for photo-induced nucleon knockout to be written[20]

$$\frac{d\sigma}{d\Omega}(\gamma,N) = \frac{e^2}{16\pi^2}\frac{k_N^2}{k_\gamma}\frac{dk_N}{dE_F}\frac{(2j+1)}{(2\ell+1)}\sum_m |\phi_{n\ell jm}(\vec{k}_N-\vec{k}_\gamma)|^2$$

$$\times \left\{\delta_{\tau_{z,1/2}}\left[\frac{k_N^2}{M^2}\sin^2\theta + \frac{k_\gamma^2}{2M^2}\mu_p^2\right] + \delta_{\tau_{z,-1/2}}\frac{k_\gamma^2}{2M^2}\mu_n^2\right\} \qquad (5)$$

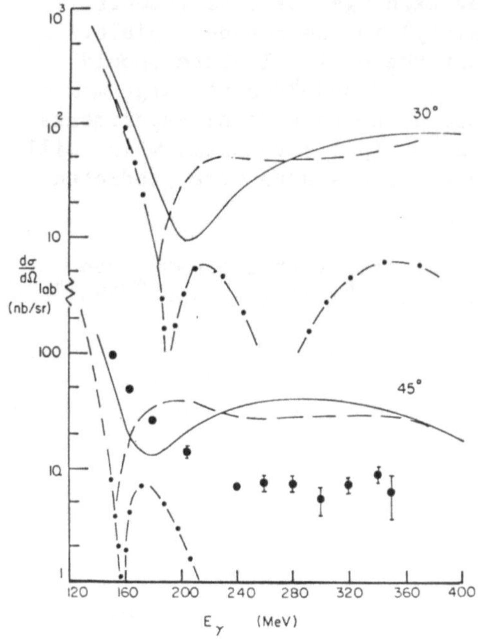

Fig. 3. Laboratory differential cross section $\frac{d\sigma}{d\Omega}$, in nanobars/steradian, vs. photon energy E in MeV for the reaction $^{16}O(\gamma,p_o)^{15}N$. The results are shown for laboratory proton scattering angles of 30° and 45° relative to the photon direction. In all curves shown here the final state proton-nucleus interaction has been neglected. Solid curve: theoretical calculation including direct PWIA amplitude plus isobar amplitude using SHO single-particle wave functions; dashed curve: PWIA plus isobar amplitudes using oscillator expansion of Negele wave functions; dot-dashed curve: PWIA amplitude only with Negele single particle wave functions. Data (at 45°): results of Matthews et al. Figs. 3-6 taken from Ref. 20.

where k_N (k_γ) is the ejected nucleon (inital photon) momentum, $n\ell jm$ are the quantum numbers of the initial bound state wavefunction of the ejected nucleon. The first term results from the photon interaction with the (ejected) proton current and magnetic moment while the second term contains only the interaction with the neutron magnetic moment (for (γ,n) processes). The bound state momentum wavefunction ϕ is evaluated at the momentum transfer $\vec{k}_N-\vec{k}_\gamma$ which can be several hundred MeV/c for medium energy incident photons since the ejected nucleon absorbs most of the photon energy. At these energies ($100 \lesssim k_\gamma \lesssim 350$ MeV) one cannot ignore the

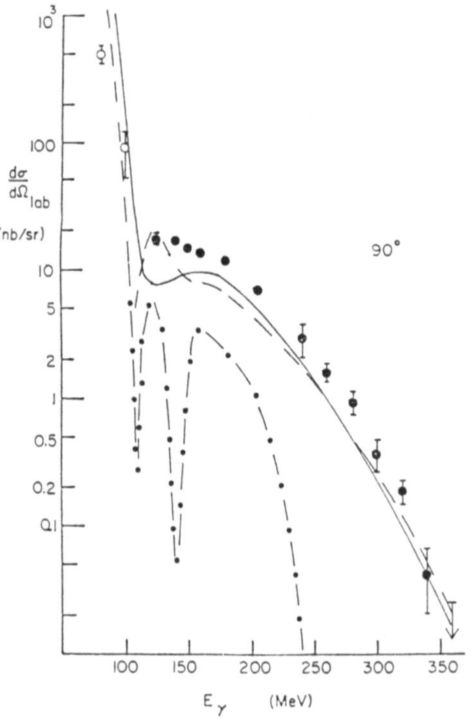

Fig. 4. Lab differential cross sections for $^{16}O(\gamma,p_0)^{15}N$ vs. photon energy for scattering angle of 90°. Notation is that of Fig. (3). Data: open circles: results of Findlay and Owens (Ref. 19); closed circles: results of Matthews et al.

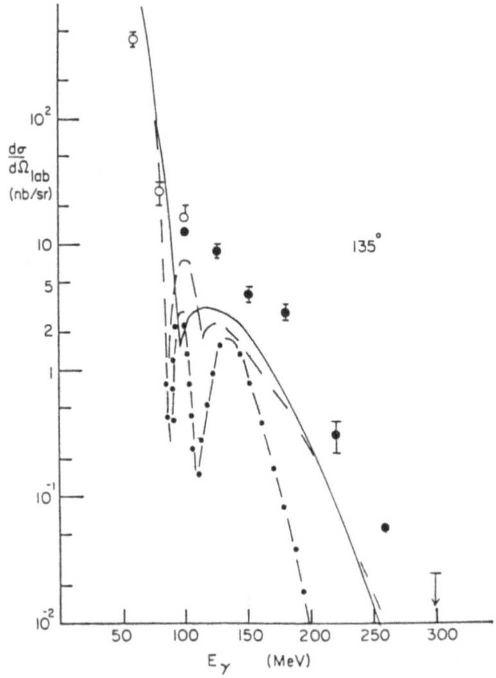

Fig. 5. Lab differential cross sections for $^{16}O(\gamma,p_0)^{15}N$ vs. photon energy for scattering angle of 135°. Notation is that of Fig. (3).

magnetic term in comparison with the convection current contribution[20]. Londergan and Nixon[20] have used a "modified plane wave", $X_p^+(r) = \sqrt{A}\, e^{ip'\cdot r}$, and have calculated the one step knockout term using single particle wavefunctions determined by Negele. The results of their PWIA calculation with Negele single particle wavefunctions are shown as dot-dashed curves in figs. 3-5. Clearly the single step process does not fit the shape of the energy dependence of the (γ,p) process and significantly underestimates the magnitude of the cross section above E_γ = 100 MeV. An additional mechanism that suggests itself at these energies is intermediate Δ production[21]. As the momentum transfer increases two step processes where the momentum transfer can be absorbed in two pieces become relatively more important. In addition as the photon energy increases it becomes relatively more easy to virtually photo-excite the nucleon to the isobar state. The process considered is shown in fig. 6a.

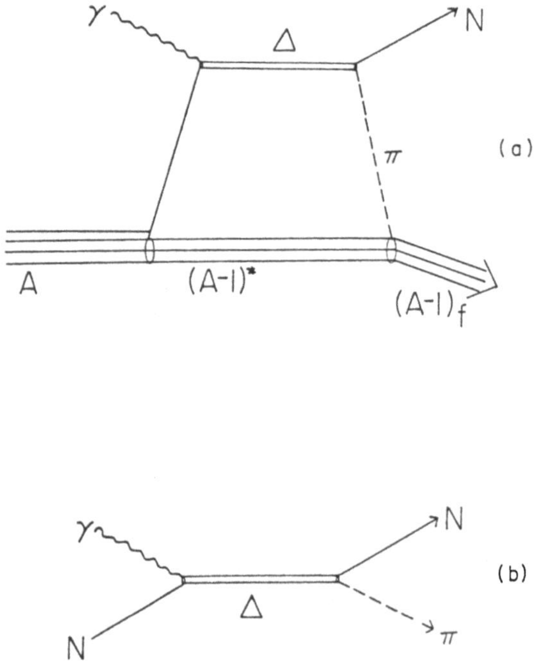

Fig. 6. Diagrammatic representation of the isobar contribution to the nuclear photoeffect. (a) The Δ(1232) is created by photoexcitation of a nucleon, propagates and then decays to a nucleon plus pion, the pion being absorbed by the residual nucleus. (b) The isobar contribution to the pion photoproduction amplitude.

The effect of including the intermediate Δ process is quite dramatic[20,21] and is shown as the solid (dashed) curve in figs. 3-5 using harmonic oscillator (Negele) single particle wavefunctions in the PWIA. Nixon and Londergan[20] have recently completed an exhaustive study of the sensitivity of their results to proton distortion, center-of-mass effects, ρ exchange, and varying the Δ width in the medium. They conclude that inclusion of these effects does not change the result that isobar production plays an important role in the medium energy (γ,p) reaction.

As we conclude this section on medium energy electromagnetic probes we can summarize as follows

1) At high momentum transfer, high spin, non-normal parity (spin-flip, isospin flip) excitations dominate the (e,e') cross section for medium energy electrons.

2) Present calculations significantly overestimate the cross section for these states. One expects the discrepancy arises because of inaccurate nuclear wavefunctions but exchange current contributions should be investigated. Experiments to locate these same states with other probes are strongly urged (see later discussion - some of these states have been studied via (p,p') at IUCF).

3) Isobar production (an exchange contribution) apparently plays an important role in (γ,p) reactions at medium energies.

III. MESON PROBES

a. Pion-Nucleus Reactions

There is, at present, an incomplete knowledge of the pion-nucleus reaction mechanism appropriate for describing even inelastic scattering and charge exchange. If one assumes that multi-step inelastic processes are relatively unimportant then there are the usual problems associated with (a) obtaining reliable distorted waves, (b) adoption of a reliable off-shell extrapolation of the pion-nucleon transition operator and (c) the adequacy of a non-field theoretic approach. Therefore it seems most likely that, initially, our present knowledge of nuclear structure obtained from other probes will be needed to clarify some aspects of the pion-nucleus inelastic scattering reaction mechanism. We note that to some extent for pion kinetic energies above 100 MeV this has already occurred. For example Edwards and Rost[22] have carried out calculations, using collective degrees of freedom in the rotational model, for pion inelastic scattering in the region 120-180 MeV leading to the lowest 2^+ and 3^-, T = 0 states in ^{12}C. They find that, except at large angles (using parameters motivated by multiple scattering theory), their theoretical results give good agreement with experiment using a value of $\beta_L \simeq .55$ which is consistent with β_L obtained from other probes.

It turns out that we predict that pions preferentially excite spin-flip non-isospin flip states which are most easily studied using microscopic models. Thus in the following we briefly review the techniques basically followed by several researchers for studying pion-nucleus inelastic scattering using single nucleon degrees of freedom and a separable form for the pion-nucleon transition operator. The details of the procedure depends on whether one uses the fixed scatterer approximation and "laboratory" phase shifts[23,24] or two-body center-of-mass phase shifts and an angle transformation for going from the two-body to the many body center-of-mass[25]. The results of the calculations do not differ significantly, except at very large angles, and for the purposes of identifying the spin-isospin structure of the kinds of states expected to dominate in medium energy (π,π') either should be sufficient. In what follows we use the notation adopted in ref. 25 for pion scattering and in ref. 27 for kaon inelastic scattering and charge exchange (we shall later refer to these same equations with minor modifications for kaon nucleus interactions).

The appropriate two body (πN or KN) on-shell t-matrix associated with the channel $\alpha = \{\ell jt\}$ is written

$$T_\alpha(k,k;E(k)) = -\frac{1}{2K}\frac{E_1+E_2}{E_1E_2}e^{i\gamma_\alpha(k)}\sin\gamma_\alpha(k) \tag{6}$$

where

$$E_{1,2} = (M^2_{N,\pi} + k^2)^{1/2}, \quad E(k) = E_1 + E_2 \tag{7}$$

$$\gamma_\alpha(k) = \delta_\alpha(k) - \frac{i}{2}\ln\eta_\alpha(k) \quad . \tag{8}$$

The fully off-shell t-matrix required for calculating inelastic scattering is related to the on shell t-matrix (assuming a separable form) by

$$T_\alpha(k_1,k_2;E(k)) = \frac{g_\alpha(k_1)g_\alpha(k_2)}{g^2_\alpha(k)}T_\alpha(k,k;E(k)) \quad . \tag{9}$$

The form factors, $g_\alpha(\kappa)$, can be obtained from a knowledge of the complex phase shift $\gamma_\alpha(\kappa)$, at all energies by solving the inverse scattering problem[25] and thus obtaining the expression

$$g_\alpha^2(\kappa) = \pm\frac{1}{k}\,\frac{E_1(\kappa)+E_2(\kappa)}{2E_1(\kappa)E_2(\kappa)}\,\sigma_\alpha\, e^{-\Delta_\alpha(\kappa)}\quad \sin\gamma_\alpha(\kappa) \tag{10}$$

where $\sigma_\alpha = \pm 1$ and $\Delta_\alpha(\kappa)$ is the principal value integral

$$\Delta_\alpha(\kappa) = \frac{1}{\pi}\, P \int_0^\infty dp\, \frac{\gamma_\alpha(p)\; p\; E(p)/E_1(p)\; E_2(p)}{E(p) - E(\kappa)} \quad . \tag{11}$$

Eqs. (10) and (11) are appropriate if there are no bound states in the two-body system and if $\delta_L \to 0$, $\eta_L \to 1$ for $E \to \infty$. Since the phase shifts are known only up to some finite energy, $E(p_0)$, it is necessary to join the existing phase shifts onto some smooth high energy cutoff form for $p \geq p_0$.

The procedure used to obtain the differential cross sections for inelastic scattering from an initial state J_0 to a final nuclear excited state J is discussed in detail in references 24 and 27. The resulting expression is given by

$$\frac{d\sigma_{JJ_0}}{d\Omega} = \frac{E(p_f)p_f}{E(p_i)p_i}\; |F_{JJ_0}(p_f,p_i)|^2 \tag{12}$$

where $E(p_{i,f}) = (M^2+P_{i,f}^2)^{1/2}$ and p_i and p_f are the initial and final meson momenta, respectively. Using the DWIA, assuming the transition operator can be written as a sum of single meson-nucleon amplitudes f_{E_i}, normalized so that $(d\sigma/d\Omega)_{\text{two-body}} = |f_{E_i}|^2$, treating the nuclear wavefunctions in the TDA approximation, assuming a separable form for the elementary transition operator, and making liberal use of standard techniques for angular momentum recoupling allows one to write $|F_{JJ_0}(p_f,p_i)|^2$ as follows;

$$|F_{JJ_0}(p_f,p_i)|^2 = (2\pi)^{-4}\; 8/9 \sum_{J_z} \Big| \sum_{\ell_p\ell_h j_p j_h} \alpha_{j_p j_h}^{\ell_p\ell_h} M_{\ell_p\ell_h j_p j_h}^{JJ_z} \Big|^2 \tag{13}$$

where

$$M_{\ell_p\ell_h J_p J_h}^{JJ_z} = \sum_i a^*_{n\ell_a}\, a_{n'\ell_b}\; \alpha\;\beta\;\gamma\;(N+\hat{N}) \tag{14a}$$

$$\alpha = \begin{pmatrix} \ell & \ell_3 & \ell_b \\ 0 & 0 & 0 \end{pmatrix} \begin{pmatrix} \ell & \ell_4 & \ell_a \\ 0 & 0 & 0 \end{pmatrix} \hat{\ell}_3\hat{\ell}_4 (\hat{\ell}_a\hat{\ell}_b\hat{\ell}_p\hat{\ell}_h\hat{J})^{1/2}\; i^{\ell_3-\ell_4}\, (-1)^{\ell} \tag{14b}$$

$$\beta = (-1)^{m_b} \begin{pmatrix} J & \ell_a & \ell_b \\ -J_z & -m_a & m_b \end{pmatrix} Y^{+}_{\ell_b m_b}(r_{p_f}) Y_{\ell_a m_a}(r_{p_i}) \tag{14c}$$

$$\gamma = \int_0^\infty r^2 dr\, R_{\ell_p}(r) R_{\ell_h}(r)\, j_{\ell_3}(k_{n'} r)\, j_{\ell_4}(k_n r) \tag{14d}$$

$$N = (-1)^{j_h + 1/2} \begin{pmatrix} J & \ell_3 & \ell_4 \\ o & o & o \end{pmatrix} \begin{pmatrix} \ell & \ell_p & \ell_h & J \\ o & o & o & o \end{pmatrix} \begin{Bmatrix} \ell & \ell_p & 1/2 \\ j_h & & j_h \end{Bmatrix} \begin{Bmatrix} J & \ell & \ell_b^{a} \\ \ell & \ell_3 & \ell_4 \end{Bmatrix}$$

$$\times \left(A^{\Delta S=0}_{\Delta T=0} + B^{\Delta S=0}_{\Delta T=1}\right) \tag{14e}$$

$$\hat{N} = \sqrt{6} \sum_{\bar{J}} \begin{pmatrix} J & \ell_3 & \ell_4 \\ o & o & o \end{pmatrix} \begin{pmatrix} \ell_p & \ell_h & \bar{J} \\ o & o & o \end{pmatrix} \begin{Bmatrix} \ell_p & 1/2 & j_p \\ \ell_h & 1/2 & j_h \\ \bar{J} & 1 & J \end{Bmatrix} \begin{Bmatrix} \ell & \ell_4 & \ell_a \\ \ell & \ell_3 & \ell_b \\ 1 & \bar{J} & J \end{Bmatrix}$$

$$\times \hat{\bar{J}}\, \hat{\ell}^{1/2} (-1)^{\ell_p} \left(C^{\Delta S=1}_{\Delta T=0} + D^{\Delta S=1}_{\Delta T=1}\right) \tag{14f}$$

where $i = \ell, \ell_3, \ell_4, \ell_a, \ell_b, n, n', m_a, m_b$ and $\hat{\ell} \equiv 2\ell + 1$. In eq. (14a-f), the factor β contains the angular dependence and γ contains the radial overlap integrals. The single nucleon bound state wavefunctions, $R(r)$, are normalized by $\int_0^\infty r^2 R^2(r) dr = 1$. The factors $a_{n\ell_a}$, a_{n',ℓ_b} arise because the configuration space meson distorted waves, $X^{\ell}_{E}(p_i r_k)$, have been expanded in terms of spherical Bessel functions $j_o^{\ell}(k_n r_k)$ where the k_n are chosen so that $j_\ell(k_N r_k)\big|_{r_k = R} = 0$

After taking the Fourier Transform, $X^{\ell}_{E}(p_i, k)$ can be written

$$X^{\ell}_{E}(p_i k) = \sum_n a_{n\ell}(p_i) \frac{\pi}{2k^2} \delta(k_n - k) \,. \tag{15}$$

The $\alpha^{\ell_p\ \ell_p}_{J_p\ J_h}$ are the admixture amplitudes of the pure p-h states obtained for the configuration mixed p-h state.

For pions the transition amplitudes A, B, C, D are given by

$$A^{\Delta S=0}_{\Delta T=0} = \ell\ (f^{1/2}_{\ell-} + 2\ f^{3/2}_{\ell-}) + (\ell + 1)\ (f^{1/2}_{\ell+} + 2\ f^{3/2}_{\ell+}) \tag{16a}$$

$$B^{\Delta S=0}_{\Delta T=1} = \ell\ (f^{1/2}_{\ell-} - f^{3/2}_{\ell-}) + (\ell + 1)(f^{1/2}_{\ell+} - f^{3/2}_{\ell+}) \tag{16b}$$

$$C^{\Delta S=1}_{\Delta T=0} = \sqrt{\ell(\ell+1)}\ (f^{1/2}_{\ell+} - f^{1/2}_{\ell-} + 2\ f^{3/2}_{\ell+} - 2\ f^{3/2}_{\ell-}) \tag{16c}$$

$$D^{\Delta S=1}_{\Delta T=1} = \sqrt{\ell(\ell+1)}\ (f^{1/2}_{\ell+} - f^{1/2}_{\ell-} + f^{3/2}_{\ell-} - f^{3/2}_{\ell+}) \tag{16d}$$

The quantities $f^{I}_{\ell\pm}$ are the appropriate off-shell meson-nucleon amplitudes for isospin I and $J = \ell \pm {}^1/_2$. If we wish the off-shell "lab amplitudes", these can be obtained from the corresponding on-shell c.m. amplitudes via

$$f^{I}_{\ell\pm(\mathrm{lab})} \equiv \frac{P_{\mathrm{lab}}}{k} \frac{g^{I}_{\ell\pm}(k'_n)g^{i}_{\ell\pm}(k_n)}{[g^{I}_{\ell\pm}(k)]^2} f^{I}_{\ell\pm} \quad \mathrm{c.m.} \tag{17}$$

where p_{lab} and k are the two-body lab and c.m. incident momenta, respectively, and the form factors are obtained from eq. 10. We note

$$f^{\mathrm{c.m.}}_{\ell\pm} \equiv \frac{1}{k}\, e^{i\gamma_{\ell\pm}} \sin \gamma_{\ell\pm} \qquad \text{(in the center of mass)} . \tag{18}$$

In order to use eqs. (13), (14a-f) for meson-charge exchange one simply sets A = C = 0, since $\Delta T = 1$ only, and multiplies the remaining <u>cross section</u> by a factor of two (one) for kaons (pions). The constants appearing in the equations above are correct for pions. The minor changes required for treating kaons will be given in a later section.

The ΔS and ΔT labels on A-D (eqs. 16a-d) indicate the final spin and isospin (spin-flip, iso-spin flip character) of the final nuclear excited state reached via that part of the transition operator if the ground state has S = T = 0. The terms A-D will be found to provide a simple and accurate means for predicting the spin and isospin structure of the dominant states excited by meson probes. For example, the $J = {}^3/_2$, $t = {}^3/_2$, p wave ($\ell = 1$) pion-nucleon amplitude is the dominant component in the pion nucleon interaction in the region below 250 MeV, thus, for the purposes of obtaining qualitative estimates of the most strongly excited states we ignore the other amplitudes. This means we retain only the $f_{\ell+}^{3/2}$ term in eqs. 16a-d. We immediately obtain the result for the amplitudes $A = 4\, f_{\ell+}^{3/2}$, $B = -2\, f_{\ell+}^{3/2}$, $C = 2\sqrt{2}\, f_{\ell+}^{3/2}$, $D = \sqrt{2}\, f_{\ell+}^{3/2}$. This means that (as usual) normal parity, T = 0, states are predicted to be most strongly excited, but the next most strongly excited states would come from $C(\Delta T=0, \Delta S=1)$ and would be responsible for the excitation of high spin, non-normal parity $\underline{\Delta T = 0}$ states at large momentum transfer (in contrast to the high spin T = 1 states excited via electrons). A knowledge of the location of these "stretched states" for both T = 0 and T = 1 as a function of the number of nucleons, as we shall discuss later, would be valuable information for the study of nuclear structure.

Before examining the results of DWIA (π,π') calculations we note one further characteristic of pion and kaon induced excitations that can be seen from examining eqs. (14a-f) - these mesons cannot excite 0^- states via a transition from a 0^+ target (since the π and K have zero intrinsic spin). Thus, if one could detect characteristic γ rays from the decay of low lying 0^- states reached via (π,π') or (K,K') this should yield an estimate of the relative importance of parity non-conserving processes.

We have used the DWIA, TDA, separable fixed scatterer form factors, and optical potentials whose parameters are obtained by using two body input as suggested by multiple scattering theory to calculate (π,π') on ^{16}O at 69.5 and 180 MeV[24]. The results are shown in figs. 7 and 8. We represent T = 1 states by solid lines and T = 0 states by dashed lines. For both pion laboratory kinetic energies one predicts that at forward angles (low momentum transfer) the familiar T = 1, $J^\pi = 1^-$, giant dipole state will dominate the nuclear response. At higher values of the momenum transfer high spin 3^- and 4^- states are predicted to dominate the spectrum. The predicted 4:1 ratio for the 4^-, T = 0 to 4^-, T = 1 cross section, obtained from eqs. (16a-d) is, in fact, essentially reproduced in the detailed calculations.

Because there is uncertainty in the quantitative reliability of a given pion calculation due to a lack of knowledge of the reactions mechanism, we regard tests of the sensitivity of the predictions, to the various inputs, as crucial. In figure 9 typical results[28], for the sensitivity of the prediction to the optical potential used to generate the distorted waves, are shown and compared with experimental data for (π,π') leading to the T = 0, 2^+ (4.43 MeV) state of ^{12}C. The solid curve was generated using distorted waves derived from the momentum space optical potential obtained by Landau and Tabakin. The dashed curve was derived using distorted waves from a Laplacian optical potential.

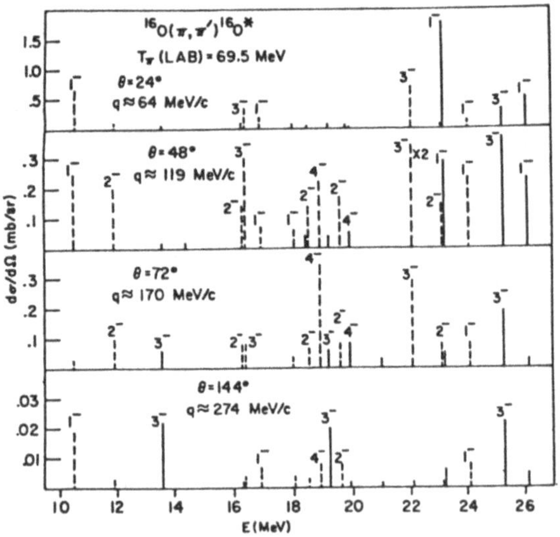

Fig. 7. Pion-^{16}O inelastic scattering differential cross sections as a function of the final nuclear excitation energy E. The initial lab kinetic energy of the pion T_π(lab) is 69.5 MeV. The differential cross sections are shown for four different scattering angles (momentum transfers). Solid lines correspond to T = 1 final nuclear excited states while T = 0 states are represented by dotted lines. The spin and parity J^π of the more predominantly excited states is indicated. Only states with appreciable cross sections are included.

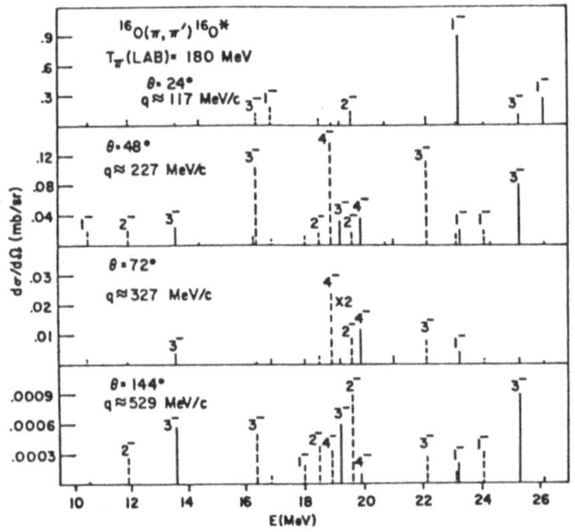

Fig. 8. Inelastic scattering cross sections as in Fig. (1) except T_π(lab) is 180 MeV.

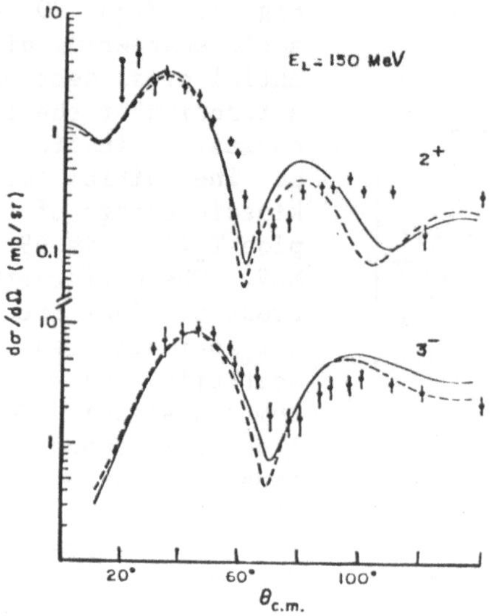

Fig. 9. Pion-^{12}C inelastic scattering cross sections using the same pion-nucleus transition potential but different distorted waves. The solid curve (dashed curve) uses distorted waves generated from the Landau-Tabakin momentum space optical potential (Laplacian optical potential). (Figure taken from ref. 28).

Our conclusion is that while the predictions vary at large angle, as a function of the optical potential adopted, the variations are not large enough to qualitatively change the shape of the angular distribution or diminish selectively states predicted to be relatively strongly excited. It is necessary to adopt some optical potential, primarily because of the important absorptive part of the optical potential, as can be seen from fig. 10 where the DWIA calculation (solid line) is smaller by approximately an order of magnitude compared to the PWIA (dashed line) for the 4^-, $T = 0$ state in ^{16}O[24].

Another major uncertainty in (π,π') calculations results from questions regarding the appropriate form for the required off-shell pion-nucleon to matrix. In ref. 28 the inelastic scattering predictions for the 2^+ (4.43 MeV) state in ^{12}C are compared for the three off-shell t matrices given by

$$\langle\kappa'|t(w_o)|\kappa\rangle = a(\tilde{w}_o) + b(\tilde{w}_o)\kappa'\cdot\kappa \qquad \text{(dashed)} \qquad (19a)$$

(Kisslinger Model)

$$= a(\tilde{w}_o) + b'(\tilde{w}_o)\ (\kappa'-\kappa)^2 \quad \text{(dashed-dot)} \qquad (19b)$$

(Laplacian Model)

$$= \sum_{\alpha} \langle k_o | t(w_o) | k_o \rangle \; \alpha \; \frac{g_\alpha(\kappa') g_\alpha(\kappa)}{g_\alpha^2(k_o)} \quad \text{(solid)} \tag{19c}$$

(separable model)

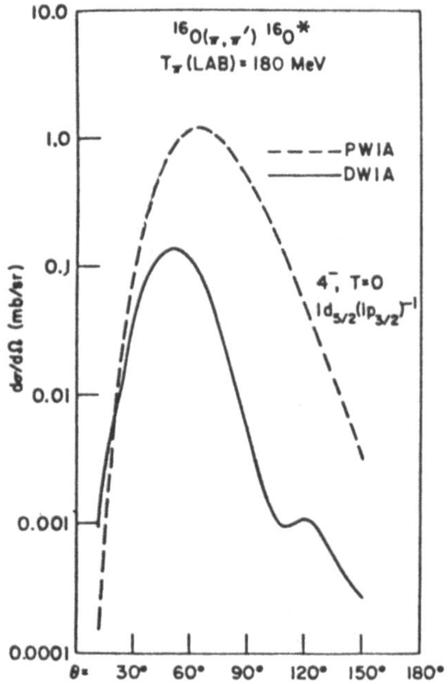

Fig. 10. A comparison of the pion-^{16}O inelastic scattering angular distribution for the 4^-, $T = 0$, $1d_{5/2}(1p_{3/2})^{-1}$ final nuclear state as calculated in the plane wave impulse approximation, PWIA (dotted line), and in the distorted wave impulse approximation, DWIA (solid line).

The results are shown in fig. 11. Although at large angles there can be a 50% difference in the predictions, up to the first diffraction minimum the predictions are relatively insensitive to the off-shell extrapolation adopted. Thus we argue that the sensitivity of the calculations to the ambiguities (except for multi-step processes where we do not as yet have enough information) is not severe and one should, for example, at least be able to use the pion as a probe for locating high angular momentum, spin-flip, $T = 0$, strength in nuclei.

Finally we note that charge exchange (π^-,π^o) experiments may be useful for locating $T_>$ giant dipole strength in $T \neq 0$ nuclei. Because $|T_z|$ increases by one in the process only $T_>$ states are present in the spectrum. A model calculation in ref. 24, demonstrates that $T_>$-dipole states should dominate the spectrum at low momentum transfer. At large momentum transfer high spin states are once again dominate. Of course pion charge exchange on T = 0 nuclei results in the excitation of T = 1 states only and is useful for testing the conjecture that T = 0 is the proper isospin identification for selected strong excitation seen in the inelastic spectrum.

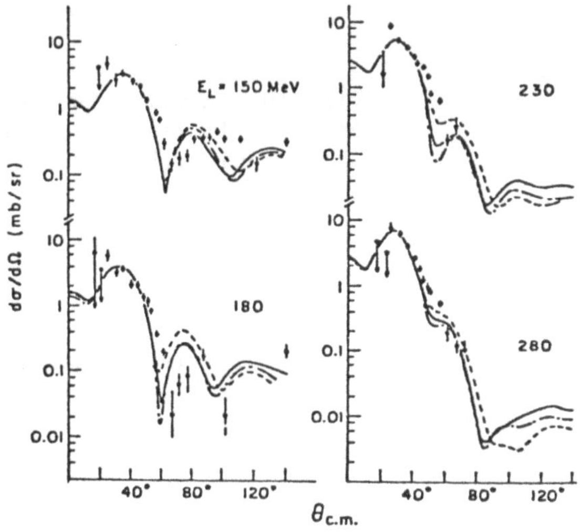

Fig. 11. The sensitivity of inelastic scattering to different forms for the pion-nucleon off-shell transition matrix. The three transition potentials are constructed using the same 2^+ (4.43 MeV) collective rotational form factor but three different off-shell models of the pion-nucleon collision matrix. The pion-nucleon off-shell models used are the Landau-Tabakin (solid curve), the Kisslinger (dashed curve), and the Laplacian (dashed-dot curve) extrapolations. (Figure taken from ref. 28).

Since one cannot identify, definitively, $T_>$ states in inelastic scattering, charge exchange reactions of the type (n,p), $\mu^- + p_{nuc} \rightarrow \nu + n_{nuc}$, (π^-,π^0) and $(K^-,\overline{K}^0)$ are motivated. The (π^-,π^0) reaction has the advantage over muon capture of allowing one to vary the momentum transfer delivered to the nucleus. The energy resolution however, is unlikely to be less than a couple of MeV because of difficulties associated with detection of the two gamma rays from the π^0 decay. Because of this $(K^-,\overline{K}^0)$, which we discuss in detail later may be more favoured.

b. Positive Kaon-Nucleus Reactions

The energy dependence of the importance of multistep processes and the ability of various hadronic projectiles to probe the nuclear interior can be estimated using the various hadronic nuclear mean free paths shown in fig. 12. We find the different energy dependence of the mean free path for protons, pions and positive kaons to be potentially very useful. The relative

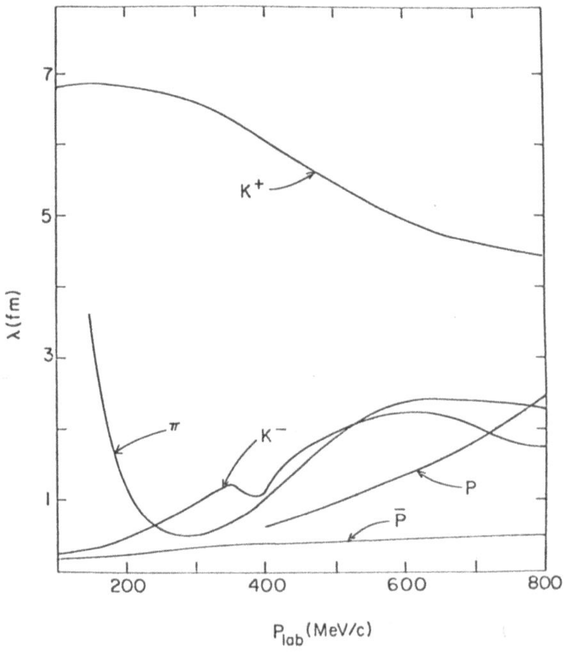

Fig. 12. Mean free path λ of various hadrons in nuclei as a function of lab momentum p_{lab}. We use the nuclear matter estimate $\lambda = (\rho\sigma_{AV})^{-1}$, where $\rho = 0.16\ fm^{-3}$ is the density and $\sigma_{AV} = (\sigma_p+\sigma_n)/2$ is the average of elementary hadron-proton and hadron-neutron cross sections σ_p and σ_n. For incident protons (P), we use $\sigma_{AV}(b) = \sigma_{np} \approx 8.5/T_{lab}$ (MeV), where T_{lab} is the lab kinetic energy. For antiprotons $(\bar{P})$, we use $\sigma_{\bar{p}n} \approx \sigma_{\bar{p}d} - \sigma_{\bar{p}p}$ and $\sigma^{\bar{P}}_{AV}(mb) \approx 64.5 + 39/P_{lab}$ (GeV). For K^+, the amplitudes of B. Martin[30] were used to evaluate σ_{AV}, while for K^-, the amplitudes of Gopal et al[31] were adopted. The elementary cross sections were not Fermi-averaged, which would tend to smooth out the rapid energy dependence of λ for pions (π) and for K^- near 400 MeV/c (the Y* (1520) resonance).

constancy and large mean free path of the K^+ (5.7 fm) in the energy region $100 \lesssim E_{lab}^{K+} \lesssim 500$ MeV make it a weakly absorbed and therefore an exciting new reliable probe for elucidating nuclear structure.

Because kaons possess non-zero strangeness, S ($S = \pm$ for $K^{\pm}$), which is conserved in strong interactions, the K^+ and K^- differ significantly in their strong interaction with the nucleus. This situation is in strong contrast to that for the π^+ and π^-. The basic difference in the kaons arises because the only "light" strange baryons stable with respect to strong interactions have $S = -1$ (Λ and Σ) and therefore the two-body channel, K^+ + nucleon, has no open inelastic channels below pion production threshold. The K^- + nucleon channel does have open channels (even at zero energy!). We postpone further discussion of K^- - nucleus interactions until later.

In this lecture we report on our recent studies of K^+ inelastic scattering on ^{12}C for $P_K^{lab} = 300$ MeV/c ($T_K^{lab} = 84$ MeV) and 800 MeV/c ($T_K^{lab} = 447$ MeV)[27]. As in the case of the (π,π') predictions presented, the calculations utilize the DWIA, TDA and assume a separable form for the off-shell kaon-nucleon transition t-matrix. Once again, s, p and d waves are included in the elementary transition amplitude. We find that the kaon-nucleon interaction is dominated by s waves in the region of interest (the interaction has, apparently a short effective range) and is relatively weak and energy independent. Thus, one may hope that the K^+ - nucleus interaction should be relatively easily understood and quantitatively reliable calculations should be possible (if the appropriate two-body data is available).

Only minor modifications of eqs. 13-16 are necessary for applying them to kaon-nucleus interactions. All the modifications result because the isospin Clebsch-Gordan sums have been evaluated in these equations and the kaon has isospin $^1/_2$ while the pion had unit isospin. The changes to be made are

a) replace the constant $^8/_9$ by $^1/_2$ in eq. (13),
b) redefine the terms A-D given by eqs. (16a-d) so that for kaons we have:

$$A_{\Delta T=0}^{\Delta S=0} \equiv \ell(f_{\ell-}^0 + 3f_{\ell-}^1) + (\ell+1)(f_{\ell+}^0 + 3f_{\ell+}^1) \quad (16a)$$

$$B_{\Delta T=1}^{\Delta S=0} \equiv \ell(f_{\ell-}^0 - f_{\ell-}^1) + (\ell+1)(f_{\ell+}^0 - f_{\ell+}^1) \quad (16b)$$

$$C_{\Delta T=0}^{\Delta S=1} = \sqrt{\ell(\ell+1)}\,(f_{\ell+}^0 - f_{\ell-}^0 + 3f_{\ell+}^1 - 3f_{\ell-}^1) \quad (16c)$$

$$D^{\Delta S=1}_{\Delta T=1} = \sqrt{\ell(\ell+1)}\ (f^{0}_{\ell+} - f^{0}_{\ell-} + f^{1}_{\ell-} - f^{1}_{\ell+}) \tag{16d}$$

where now the quantities $f^{I}_{\ell+}$ are the "lab" off-shell kaon-nucleon amplitudes for isospin I = 0, 1 and $J = \ell \pm {}^{1}/_{2}$. Once again $0^{+} \rightarrow 0^{-}$ transitions are not allowed if parity is conserved. As in the case of pions, kaon charge exchange on T = 0 ground state targets can be calculated using eqs. 13-16 by setting A = C = 0 and multiplying the resulting cross section by a factor of two.

Again we concentrate on the terms A-D as a guide to the spin and isospin structure of the dominant nuclear excitations. In figs. 13 and 14 the functions $|A|^2$ -- $|D|^2$ are compared for $\ell = 0$ and 1 for three different sets of the elementary two body amplitudes. In the region near 300 MeV/c the S_{wave}, T = 0 term, $|A|^2$, completely dominates. Thus T = 0 normal parity states are predicted to dominate K^{+} + closed shell nucleus inelastic scattering at all momentum transfers for $T^{K}_{lab} \simeq 84$ MeV. This should be contrasted with electron and pion probes for which spin flip states dominate at large q. We note that different two body solutions predict qualitative differences in the ratio of iso-spin flip spin flip and non-spin-flip excitations, $|B|^2$ and $|D|^2$. One way to allow such states to be seen in the nuclear response would be to consider positive kaon charge exchange (K^{+},K^{0}) where the terms $|A|^2$ and $|C|^2$ are not present.

In figures 15 and 16 we show the results of $(K^{+},K^{+\prime})$ on ^{12}C for T^{K}_{lab} = 300 MeV/c using the DWIA, TDA and the Martin and BGRT (iD) elementary amplitudes respectively. The results are as predicted by looking at the terms $|A|^2$ -- $|D|^2$ except that the nuclear collectivity causes the giant dipole state to dominate the spectrum. We find that the form factor g(p) is slowly varying with momentum transfer and therefore setting the form factor to 1 in the transition operator changes the predicted results by less than 50% at 300 MeV/c. So at this lower energy the main uncertainties result from incomplete knowledge of the two body input.

The first experiments, which are currently being performed at Brookhaven, will utilize an 800 MeV/c K^{+} beam and so we now turn our attention to this energy region. An examination of figures 13 and 14 for $|A|^2$ -- $|D|^2$ again suggests that T = 0, normal parity states will dominate the nuclear response at all momentum transfers as at 300 MeV/c. Spin-flip states are predicted to be relatively more important than at 300 MeV/c. The elementary amplitudes obtained by different authors do not vary significantly in this region so that there is less ambiguity associated with the results for the spin-slip states (both 4^{-} T = 0

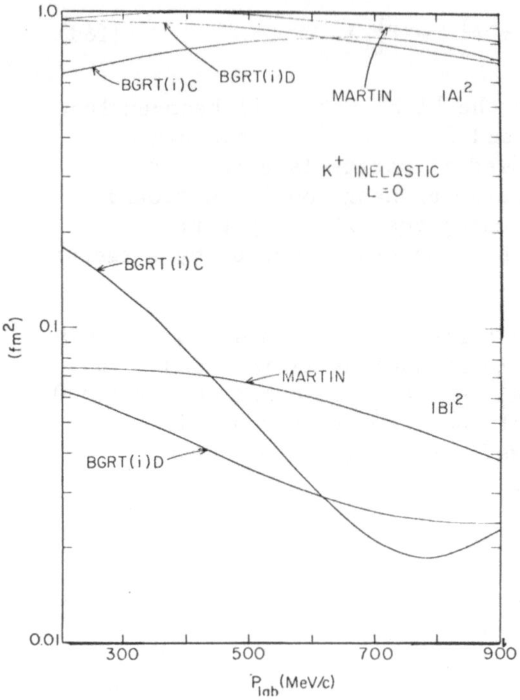

Fig. 13. The inelastic scattering functions $\left|A^{\Delta S=0}_{\Delta T=0}\right|^2$ and $\left|B^{\Delta S=0}_{\Delta T=1}\right|^2$ corresponding to the S-wave part of the K^+N amplitude. We use the definitions of Eqs. (16a) and (16b), with the two body c.m. amplitudes $f^{J,c.m.}_{\ell\pm}$ replacing the lab amplitudes for convenience. Three models for the elementary interaction, labeled MARTIN (Ref. 30), BGRT(i)C (soln. C of Ref. (32)) for isospin zero, soln. (i) of Ref. (32) for isospin one) and BGRT(i)D (soln. D of Ref. (32) for I = 0) are shown.

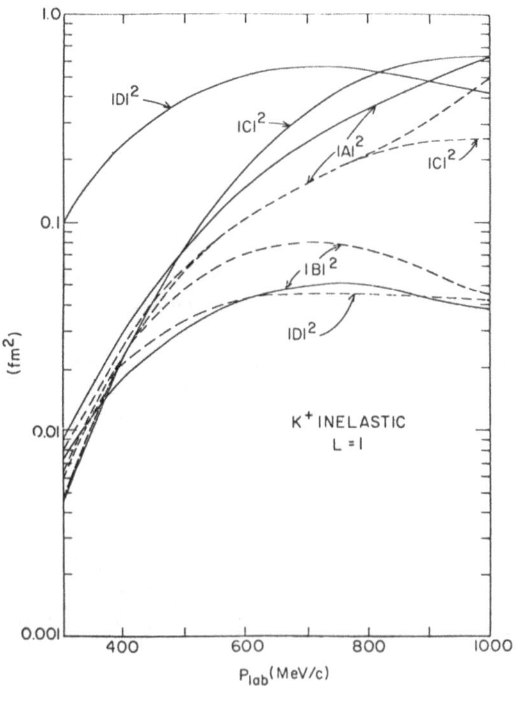

Fig. 14. The inelastic scattering functions $\left|A^{\Delta S=0}_{\Delta T=0}\right|^2$, $\left|B^{\Delta S=0}_{\Delta T=1}\right|^2$, $\left|C^{\Delta S=1}_{\Delta T=0}\right|^2$, and $\left|D^{\Delta S=1}_{\Delta T=1}\right|^2$ corresponding to the p-wave part of the K^+N amplitude. The solid curves correspond to model BGRT(i)D of Ref. (32) while the dashed curves follow from the amplitudes of B.R. Martin (Ref. (30)).

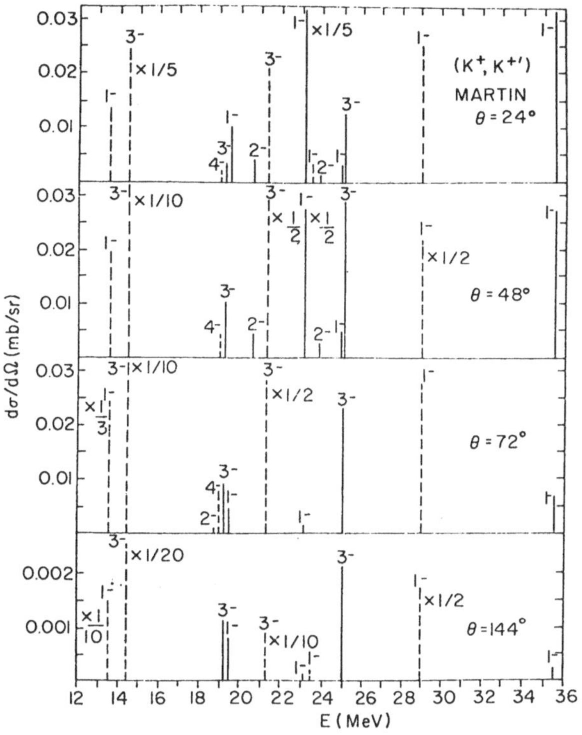

Fig. 15. Differential lab cross sections for K^+ inelastic scattering on ^{12}C at P_{lab} = 300 MeV/c, for various choices of lab angle θ. The excitation energy E is measured with respect to the ground state of ^{12}C. The dashed lines correspond to isospin zero (T=0) final states, while the solid lines are for T = 1 states. The final states are configuration-mixed particle-hole states of negative parity, as described in the text. Final states with very small cross sections are not plotted. The cross sections for several strong states have been multiplied by numerical factors of $^1/_2$ to $^1/_{20}$ as indicated. We have used the K^+N amplitude BGRT(i)D of Ref. (32). The bar graphs plotted here represent the cross sections to diagonalize particle-hole states; we have omitted the widths and energy shifts of these states which arise from nucleon emission or from mixing with more complicated states (2p2h).

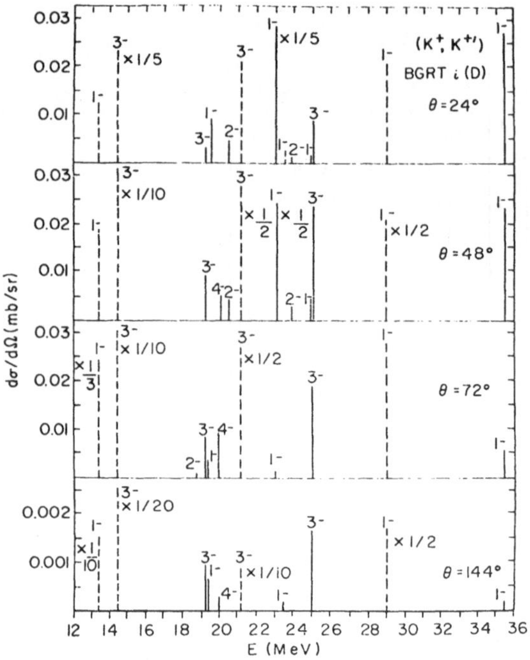

Fig. 16. Differential lab cross sections for K^+ inelastic scattering on C at P_{lab} = 300 MeV/c. The kinematics and final states are the same as Fig. 15 except that we use the K^+N amplitudes of B. Martin (30). No off-shell form factor was included here (g(p)=1), or in any of the other calculations in this paper which employ the Martin amplitude for K^+N.

and 4^- T = 1 states should be seen). The results of a DWIA calculation is shown in fig. 17. Once again the giant dipole T = 1 state dominates the spectrum at low q but except for this collective state the predictions obtained by looking at $|A|^2$ -- $|D|^2$ are re-enforced by the more detailed calculations. In this energy region the angular distributions obtained by using the Martin and BGRT (iD) amplitudes are quite similar (as is shown in Fig. 18). As at 300 MeV/c the form factor g(p) is slowly varying in momentum space (a short range interaction) and so setting g(p) = 1 in the transition operator causes only small differences in the predicted angular distributions.

We conclude our brief discussion of K^+ inelastic scattering by trying to offer some suggestions on how to use the K^+ in connection with some other probes. Since the K^+ predictions should be reliable one should compare the K^+ results with those obtained using the other "reliable" probe, the electron, for the low excitation, 2^+ and 3^-, T = 0, states in ^{12}C. The same wavefunctions which give agreement with the electroexcitation experiments (remember the RPA was required) should also yield satisfactory agreement for K^+ inelastic scattering. If they don't, something may be wrong with our understanding of the reaction mechanism for one of these two probes. Note, from our earlier discussion that, especially at larger momentum transfers, exchange currents may play a role in the electromagnetic experiments.

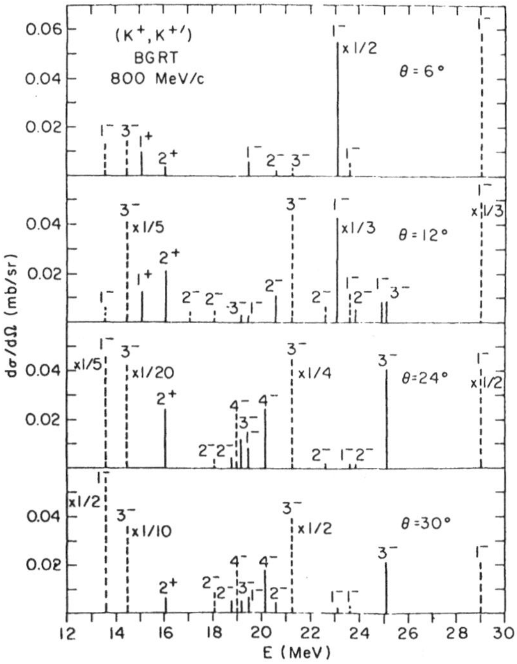

Fig. 17. Differential lab cross sections at selected angles for K^+ inelastic scattering on ^{12}C at 800 MeV/c, using the BGRT(i)D amplitudes of Ref. (32). The notation is the same as in Fig. 15.

At higher excitation energies the characteristic T = 0, normal parity states excited by the K^+ should be compared with the $\Delta T = 0$, $\Delta S = 0$ states excited by (α,α') where it is not clear whether one is dealing with single nucleon or cluster excitations.

Another example of the utility of the K^+ may be its use in investigating the concentration of strength associated with the isoscalar monopole and quadrupole resonances (T = 0, normal parity). Again by doing charge exchange, (K^+K^o) on T = 0 nuclei all isoscalar excitations should disappear from the spectrum.

c. Negative Kaon-Nucleus Reactions

As mentioned earlier, the K^- meson has a radically different nucleon and nuclear interaction as compared to the K^+. As shown in fig. 12 the K^--nucleus interaction is strong (the probe is strongly absorbed) and has an appreciable energy variation. The K^--N interaction is further complicated by the fact that there exist several reaction channels, the $\Lambda + \pi$ and $\Sigma + \pi$ channels, which have negative energy thresholds. Therefore it may well be necessary, especially at low energies, to treat the elementary K^- + nucleon amplitude as a coupled channels problem, complicated by the many body environment. However we find that the elementary form factors are relatively smoothly varying in the momentum region of interest (300 MeV/c and 800 MeV/c). This may seem somewhat surprising because in the πN problem, when the $\Delta(1236)$ channel was open, a modified treatment of the problem was necessary to obtain smoothly varying form factors[33]. But this was an elastic resonance coupled only to the N system whereas for the K^- the Y* resonances are generally very inelastic and often weakly coupled to the KN channel. In any event, we find that the choice of the K^-N form factor, g(p), has little qualitative influence on the inelastic scattering results.

For the understanding of K^- + nucleus inelastic scattering and charge exchange it is, once again, quite useful to study the relative magnitudes of the functions A-D given by eqs. 16a-d for this strongly absorbed probe.

The functions $|A|^2$ -- $|D|^2$ are shown for ℓ = 0, 1, 2 in fig. 18, adopting the K^-N amplitudes of Gopal et al[31]. In the region 300 MeV/c-1 GeV/c the ℓ = 0, $\Delta S = \Delta T = 0$ term, $|A|^2$, dominates. Therefore one predicts that for $p_K^{lab} \sim$ 300 MeV/c and 800 MeV/c the states strongly excited by the K^+ and K^- (for example on ^{12}C) inelastic scattering should be identical and would be normal parity T = 0 states for all values of momentum transfer.

K^- charge exchange, $(K^-,\bar{K}^o)$, which would be an excellent way to study $T_>$ states in $N \neq Z$ nuclei, can be studied by setting $|A|^2 = |C|^2 = 0$. From fig. 18 one can investigate the relative importance of the $|B|^2$ and $|D|^2$ terms. The ℓ = 0, $|B|^2$ term dominates with the ℓ = 1, $|B|^2$ term being the next most important. Thus the prediction would be that normal parity (ΔS = 0) excited states dominate the spectrum for charge exchange at both 300 and 800 MeV/c.

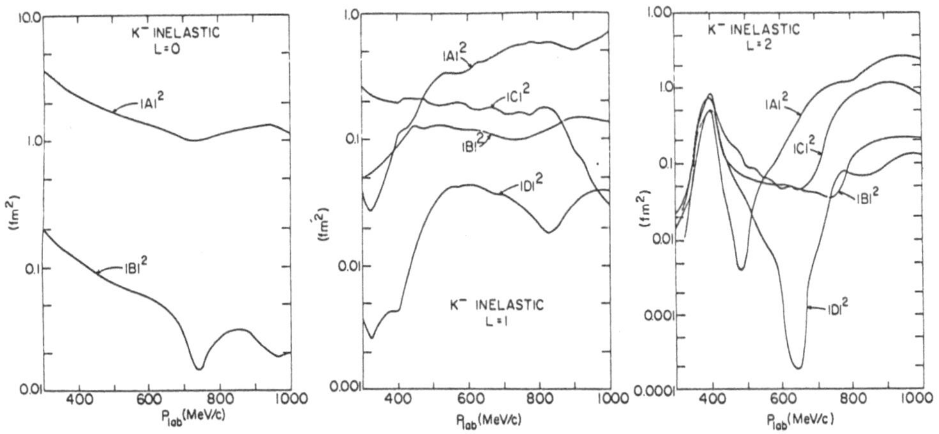

Fig. 18. The inelastic scattering functions $|A|^2$, $|B|^2$, $|C|^2$, and $|D|^2$ of Eqs. (16a-d) for the S, P and D waves of the K^-N system, using the amplitudes of Ref. 31 which have not been Fermi-averaged; this would tend to wash out some of the rapid energy dependences, for instance the peak near the Y* (1520) resonance at 400 MeV/c in L = 2.

The results of the 300 MeV/c and 800 MeV/c $^{12}C(K^-,K^{-\prime})^{12}C$ calculations are shown in figs. 19 and 20 respectively. At both lab momenta the excitation spectrum is very similar to the results obtained earlier, also using the DWIA, for $(K^+,K^{+\prime})$ with normal parity 1^- and 3^- states being dominant. This is important because, as discussed earlier, the spectroscopy of such states may be expected to be well understood from electron and K^+ inelastic scattering and thus one will be learning about the K^- reaction mechanism by studying these states via $(K^-,K^{-\prime})$. Once one has a better understanding of the K^- reaction mechanism one can use K^- charge exchange (as discussed below) to excite and study states not easily seen with other probes.

For the 800 MeV/c DWIA calculations it is necessary to include the $\ell = 2$ K-N amplitudes in the optical potential. Since some of the $\ell = 2$ amplitudes are not slowly varying it was necessary to Fermi-average the amplitudes and then use the "ρt" approximation to obtain the higher energy optical potential. The substantial difference between including the $\ell = 2$ term and using only the $\ell = 0$ and 1 terms in the optical potential is shown in fig. 21.

Finally, for K^- inelastic scattering, we have less confidence in the low energy ($\lesssim$ 500 MeV/c) K^- predictions than for the K^+ or K^- 800 MeV/c results because of the rapid change, as a

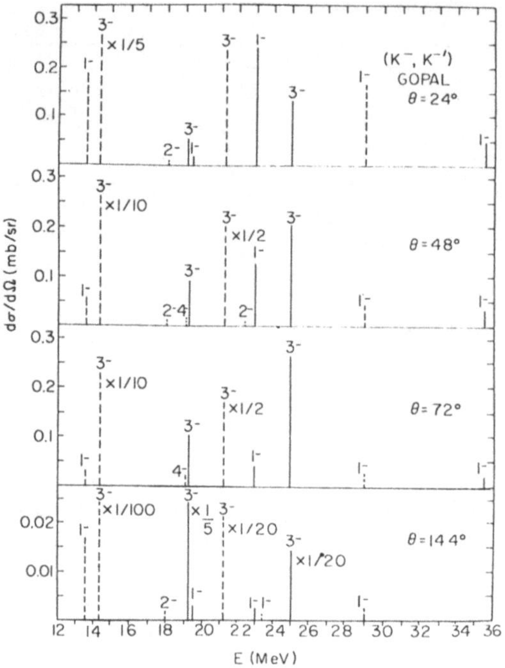

Fig. 19. Differential lab cross sections for K^- inelastic scattering on ^{12}C at p_{lab} = 300 MeV/c. The K^-N amplitudes are taken from Gopal et al[31]. The notation is the same as in Fig. 15.

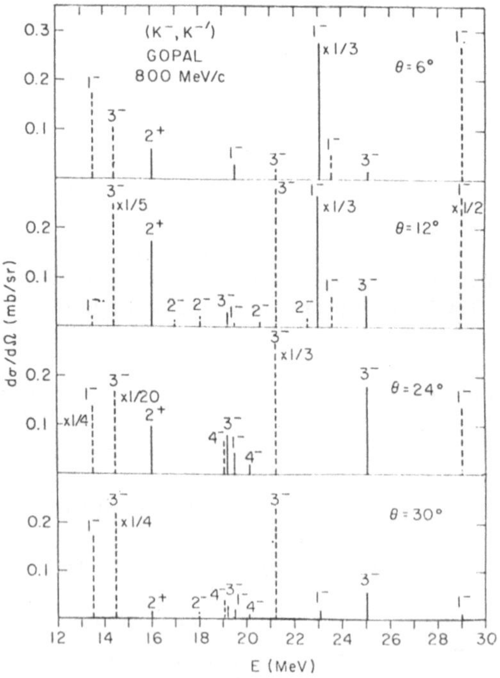

Fig. 20. Differential lab cross sections for K^- inelastic scattering on ^{12}C at 800 MeV/c. We use the K^-N amplitudes of Ref. (31). The notation follows that of Fig. 15.

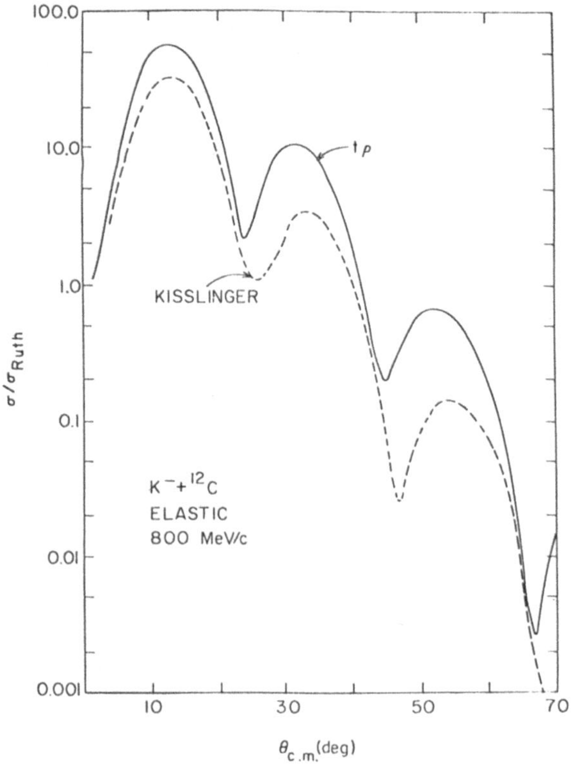

Fig. 21. Elastic scattering differential cross section (divided by Rutherford cross section (σ_{Ruth}) for $K^- + {}^{12}C$ at 800 MeV/c. The solid curve represents the "tρ" approximation to the K^- optical potential V(r), including S, P and D-wave K^-N amplitudes according to Gopal et al[31]. The dashed curve results from using the Kisslinger form for V(r), with only S and P waves.

function of energy, of the $\ell = 2$, K^- amplitudes near 400 MeV/c and the coupled channels complications associated with the open channels below threshold.

We now briefly introduce and discuss the utility of the process $(K^-, \bar{K}^0)$. Starting from a $T = -T_z$ neutron excess ground state and using this change exchange reaction, then the final states reached would have $T_f \geq T_z + 1$. Two reasons for studying these isospin stretched configurations are to investigate the energy and transition strength splitting of the $T_>$ and $T_<$ giant dipole resonances and also to obtain the same information regarding stretched angular momentum configurations. As an example we consider ${}^{30}Si(K^-,\bar{K}^0){}^{30}A\ell$. The results of the calculation for $p_{lab}^{K^-} = 300$ MeV/c are shown in fig. 22. (From an investigation of the properties of B^2 and D^2 the same qualitative features are predicted for this reaction at 800 MeV/c.) From fig. 22 one sees that normal parity states completely dominate the spectrum with high spin states dominating at large momentum transfer. The 1^-

state near 16.3 MeV excitation is predicted to contain ∿ 70% of the $T_>$ dipole strength. It is important in understanding the systematics of collective excitations in nuclei to be able to study the isospin splitting of, for example, dipole, quadrupole and octupole excitations as a function of neutron and proton numbers. Because of the energy resolution expected (by time of flight techniques) and the kind of states excited the K^- charge exchange reaction shows considerable promise for carrying out such studies.

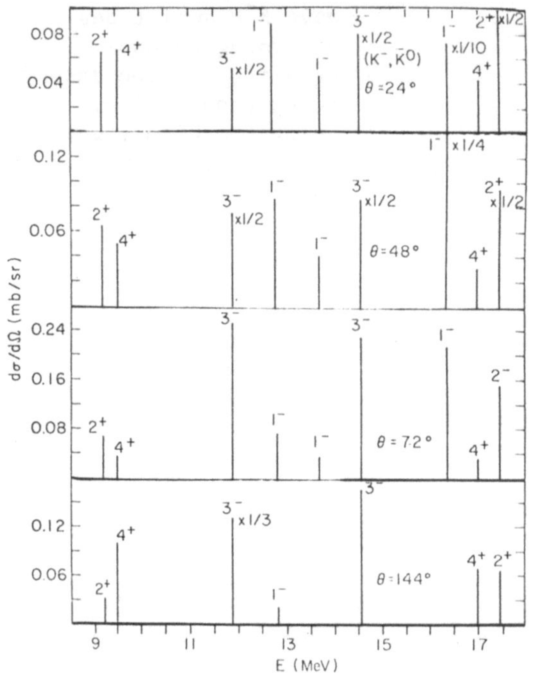

Fig. 22. Differential lab cross sections at selected angles for the charge exchange reaction $^{30}Si(K^-,\bar{K}^0)^{30}A\ell$ at p_{lab} = 300 MeV/c, leading to various diagonalized particle-hole states of $^{30}A\ell$. The notation is the same as in Fig. 15.

The $K^\pm$ meson should also be a useful probe for studying non-stretched isospin final states from $T \neq 0$ targets. Suppose one considers a $T = 1$, $T_z = -1$ g.s. Then the $T = 2$, $T_z = 2$ p-h state with a given spin and single nucleon orbitals (with parity different from the g.s.) is unique from the isospin coupling perspective. However there are two ways of forming a $T = 1$, T_z + -1 p-h state - starting with the $T = 1$, $T_z = -1$ g.s. and carrying out non-isospin-flip inelastic scattering or to consider an iso-vector transition with $\Delta T_z = 0$. If these states were pure the relative cross sections (in a simple model) for forming them should be given by the ratio $|\frac{A}{B}|^2$ discussed above for $K^{\pm}$ inelastic scattering.

In addition to the reactions discussed above the reaction (K^+,K^+N) and quasi-elastic scattering show promise for studying "conventional" nuclear structure. Of course the double strangeness exchange reaction (K^-,K^+) and the (K^-,π^-) reaction offer exciting possibilities for studying hypernuclei.

IV. NUCLEON-NUCLEUS INELASTIC SCATTERING AND CHARGE EXCHANGE

Nucleon-nucleus interactions are complicated by the well-known problems associated with a strong complex spin and velocity dependent elementary interaction that is not completely known, anti-symmetrization, and the general adequacy of the distorted waves used in the DWIA (especially for polarization studies associated with inelastic transitions where the spin-orbit term in the optical potential is expected to play an important role). Despite these complications there is strong motivation for studying nucleon inelastic scattering and charge exchange (including polarization) above 100 MeV. The N-N total cross section decreases rapidly with energy and has a minimum around $T_p^{lab} \approx 200$ MeV. Thus the mean free path of the nucleon is a few fermis in this energy region. This situation should be contrasted with that observed for the π (decrease in mean free path around 200 MeV) and the K^+ (constant long mean free path in this energy region). Thus the importance of multistep processes as a function of energy from 100-300 MeV should be considerably different for the proton, pion and K^+ probe. Above 100 MeV the DWIA should be a reasonable approach for analyzing (p,p') and (p,n) reactions and thus one has some hope of using the reaction as a probe of nuclear structure. Actually the fact that one has a very complicated interaction also means the variety of nuclear excitations one can study is also quite large. Of course comparison with electron and K^+ probes will be useful to clarify reaction mechanism uncertainties.

One of the main reasons for our discussion of 100-400 MeV proton-nucleus inelastic scattering is that there has recently been a lot of new high quality, excellent energy resolution (p,p') and (p,n) results from, for example IUCF, that show great potential for studying nuclear structure - we shall discuss some of these results below. In addition, at IUCF, we anticipate, in the next few months, similar experiments using high quality polarized beams. Excellent energy resolution is crucial in these experiments to resolve the nuclear states.

Examples of typical theoretical and experimental results previously available are given in ref. 34. For several cases, the fits to the angular distributions are satisfactory while the meager polarization data is not well represented by theory. It is our goal to obtain a nucleon-nucleon transition operator that should be adequate, if the SWIA is valid, for studying polarization and anti-symmetrization effects at medium energies and finally appropriate to study nuclear structure using (p,p') and (p,n).

We also would like to use the interaction to study the excitation of high spin states at large momentum transfer. Earlier work by Moffa and Walker[35] had predicted that high spin non-normal parity, T = 1, states would dominate the nuclear response due to a strong iso-vector tensor term in the effective nucleon-nucleon transition operator. However in these earlier calculations a realistic transition operator, fitted to the free nucleon-nucleon data in the energy region under consideration, was not available. In addition the effects of exchange terms were largely ignored. Although work on the new pseudo-potential and the relevant experiments to look for high spin states in ^{28}Si and ^{24}Mg were carried out simultaneously, we anticipate the theoretical results a bit by showing, in fig. 23, the results obtained by Adams et al[36] for these nuclei. Of particular importance is the strong excitation of the 6^-, T = 0 states as well as the strong 6^-, T = 1 state at high momentum transfer in ^{28}Si (only the latter was predicted to be strongly excited in ref. 35). Preliminary data[37] does not seem to show evidence of the strong excitation of these stretched, 6^-, states in ^{32}S and ^{40}Ca. This is in disagreement with our expectations and is currently quite puzzling.

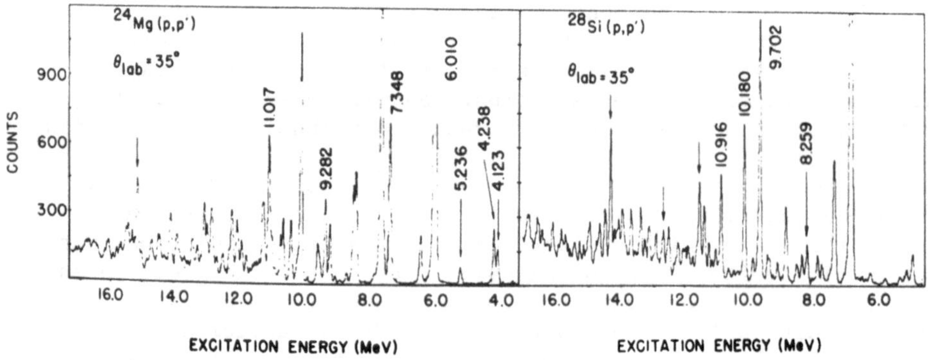

Fig. 23. Representative spectra for the reactions ^{28}Si(p,p') wave and ^{24}Mg(p,p') taken with 135-MeV protons. Some of the peaks used for energy calibration are identified by their excitation energies. Analysis of the unlabeled peaks by arrows is discussed in the text. Energy resolution was about 70-keV FWHM. Figure taken from Ref. (36).

We wish a transition operator fitted to the nucleon-nucleon differential cross section and polarization data in the energy region 50-400 MeV. It is necessary that the transition operator include realistic spin dependent terms (Bartlett, Heisenberg, tensor, and two-particle spin-orbit), possess real and imaginary components, and be constructed so that exchange effects can be explicitly calculated.

At the time this work was actually performed we took the available analysis of MacGregor, Arndt and Wright[38] (MAW) to obtain the free N-N phase shift and mixing parameters. In our subsequent fits we also compared directly with the appropriate data. The transition operator obtained is complicated (but can be used in its full power-treating exchange exactly and including the two particle spin-orbit force in a modified version of a code due to Ken Amos and co-workers which is now running at Indiana University). It is complex, explicitly energy-independent but momentum-dependent, with Yukawa-type form factors. The details of the interaction and the fitting procedures are given in detail by Picklesimer and Walker[39]. The quality of the fit to the free data obtained using the transition operator is shown in figs. 24-27. Overall the reproduction of the data is good. In addition to its utility in studying inelastic scattering and charge exchange, the interaction obtained should also be useful input for calculating the lowest order optical potential and studying higher order corrections.

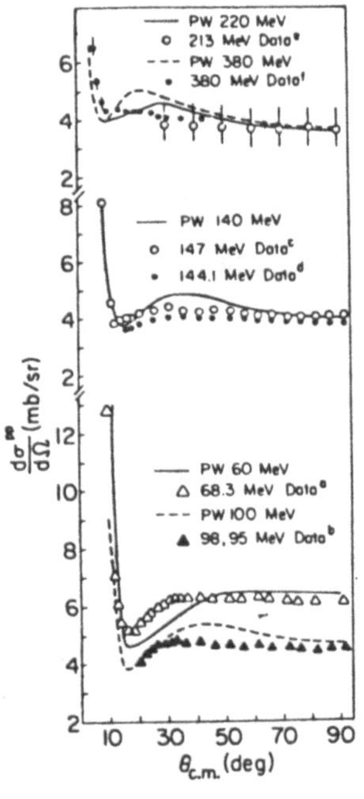

Fig. 24. Comparison of proton-proton angular distribution data with theoretical results obtained using the interaction discussed in the text. The theoretical results are denoted PW on the figure. (a) Reference 40; (b) Ref. 41 for MeV data, Ref. 42 for 95 MeV data; (c) Ref. 42; (d) Ref. 43; (e) Ref. 44; (f) Ref. 45.

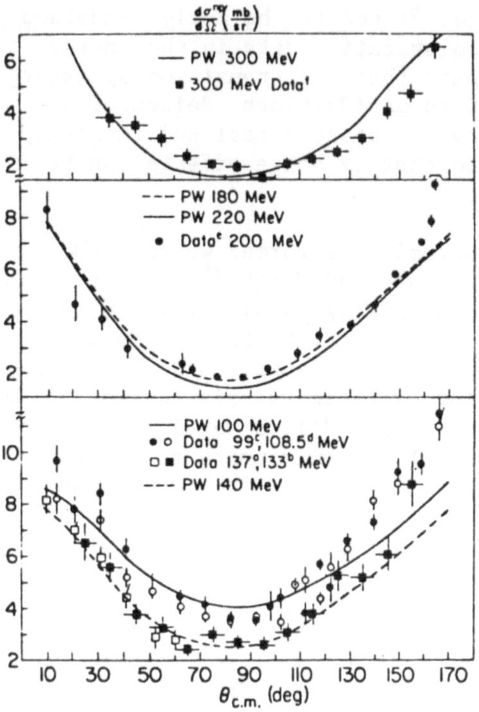

Fig. 25. Same as for fig. 24 except the comparison is for neutron-proton angular distributions. (a) Reference 46; (b) Ref. 47; (c) Ref. 48; (d) Ref. 48; (e) Ref. 49; (f) Ref. 50.

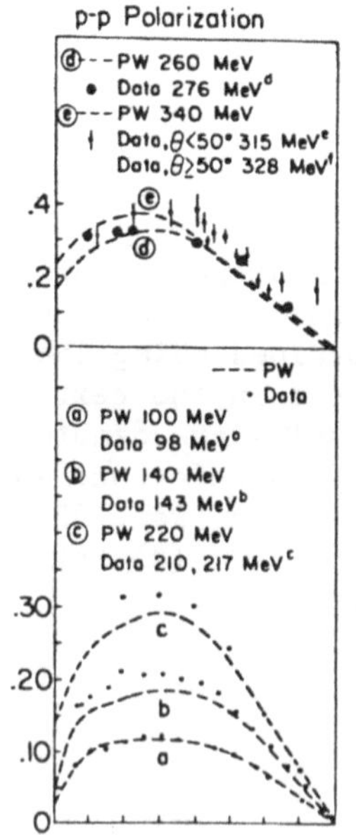

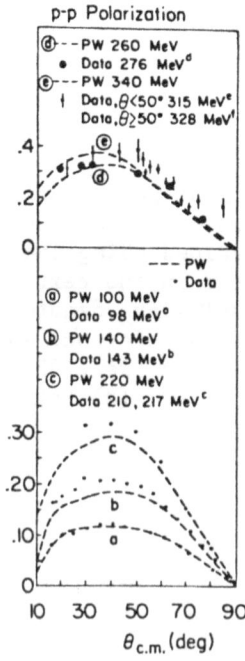

Fig. 26. Comparison of proton-proton polarization data with theoretical results obtained using the interaction discussed in the text. The theoretical results are denoted PW on the figure. (a) Reference 41; (b) Ref. 43; (c) Ref. 44; (d) Ref. 51; (e) Ref. 51; (f) Ref. 52.

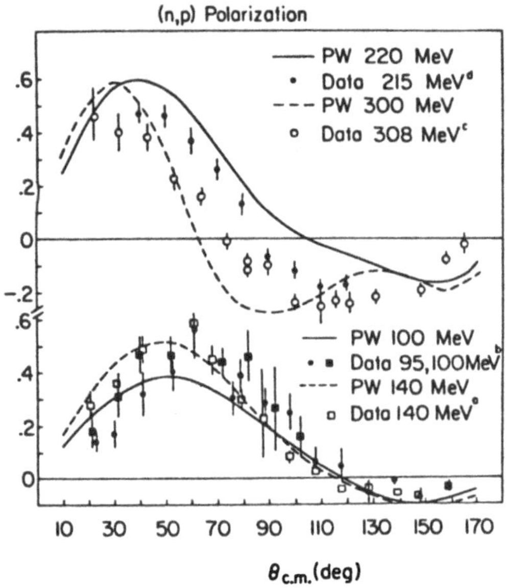

Fig. 27. Same as for fig. 26 except the comparison is for neutron-proton polarization. The 215 and 308 MeV data are uncorrected for binding effects of the deuteron. (a) Reference 53; (b) Ref. 54 (95 MeV), Ref. 55 (100 MeV); (c) Ref. 51; (d) Ref. 44.

We do not yet have available results using the interaction obtained above in the DWIA. However the main effect of distortion in the region 120-160 MeV has been found to be, for the direct terms, an overall reduction of a factor of two with the angular distribution remaining essentially unaffected[35]. Therefore, we show in figs. 28 and 29 the results obtained using the new transition operator in the plane wave impulse approximation (PWIA) for (156 MeV) ^{16}O (p,p') and (134 MeV) ^{28}Si (p,p'). The details of the calculations are given in ref. 39. In each case the (p,n) cross section can be obtained by setting all T = 0 excitations equal to zero and multiplying the T = 1 cross sections by a factor of two. As usual high spin states dominate the spectrum at large momentum transfer. Most of the high spin states strongly excited between 12 and 18 MeV have even spin. Note that (due to exchange effects) high spin T = 0 states are also predicted to be strongly excited as in ref. 35.

As a rough guide we find that for T = 1 p-h states the exchange term reduces the predicted angular distribution by $\lesssim$ 20% for strongly excited states. For T = 0 states the exchange term is extremely important with non-normal parity states (normal parity states) being substantially increased (decreased) by the inclusion of the exchange term. We are currently pursuing studies on (p,p') and (p,n) for light nuclei using the new transition operator. Comparison with new angular distribution and polarization data from IUCF will be a major objective of this research.

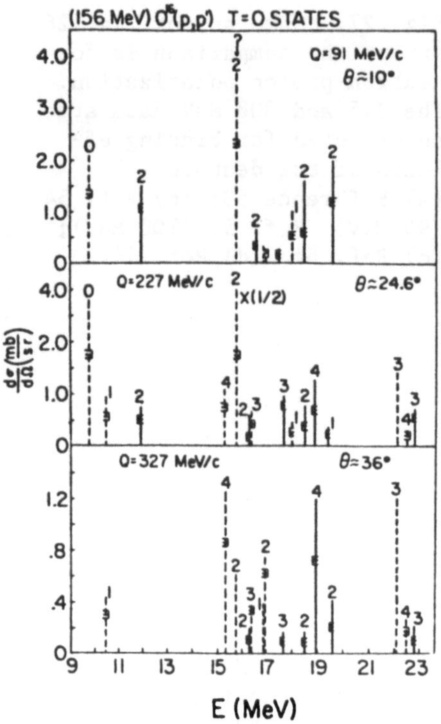

Fig. 28a. Inelastic proton scattering on ^{16}O for T_p (lab) = 156 MeV. Only T = 0 final excited states with appreciable cross section are shown. The spin of prominent peaks is indicated. Normal parity (non-normal parity) states are denoted by dashed (solid) lines. The predictions are based on the PWIA, the particle-hole model, and the transition operator discussed in the text. The effects of distortion should reduce the predicted cross section by approximately a factor of 2. The symbol E(3) denotes states whose differential cross section is enhanced (reduced) by at least a factor of 2 by inclusion of the exchange terms. The symbol (?) designates a state containing an appreciable spurious component.

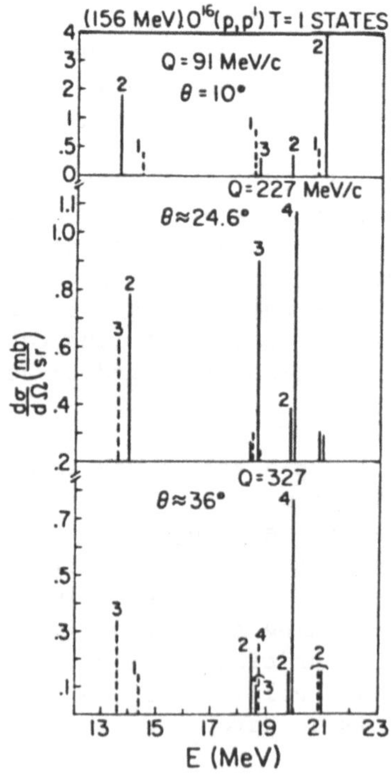

Fig. 28b. Same as fig. 28a except only T = 1 final states with appreciable cross section are shown.

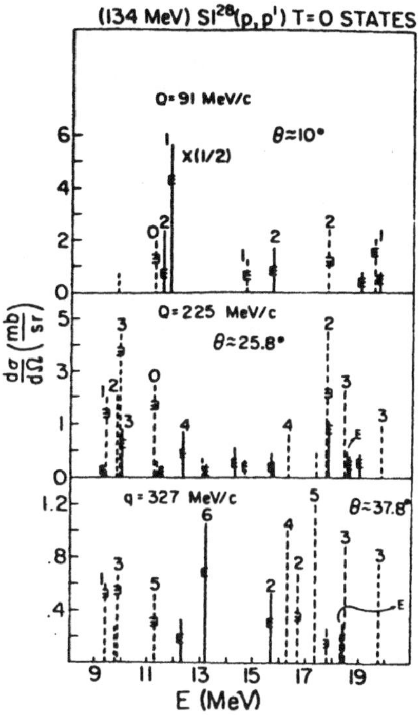

Fig. 29a. Same as fig. 28 except for inelastic proton scattering on ^{28}Si for T_p(lab) = 134 MeV. Only T = 0 final excited states with appreciable cross section are shown.

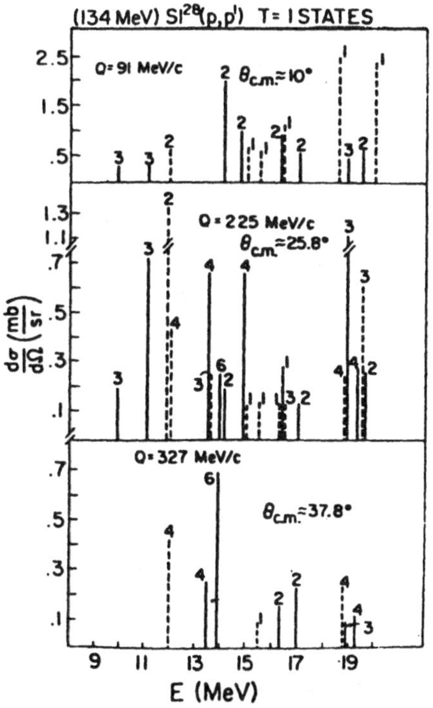

Fig. 29b. Same as fig. 29a except only T = 1 final states with appreciable cross section are shown.

Recently Amos et al[56] have used another nucleon-nucleon potential[57] fitted to lower energy reactions (including antisymmetrization effects explicitly) to compare with the ^{28}Si and ^{24}Mg high spin states and have found that substantial reduction factors (∿ 70% in the potential) are required to bring theory into qualitative agreement with experiment for the angular distribution. Whether the disagreement results from nuclear structure effects or the transition potential remains to be clarified.

Before concluding these lectures I should point out that, for example, at IUCF, there are many exciting new experiments planned or underway that will significantly broaden and deepen our understanding of the nucleus. A selection of those I find most interesting (and for brevity including only those involving proton projectiles) include (p,γ), $(p,2p)$, (p,d), $(p,\pi^{\pm})$, and finally $(p,\mu^{\pm})$ below pion production threshold[58].

REFERENCES

1. T. DeForest, Jr. and J.D. Walecka, Advan. Phys. 15 (1966), 1.

2. K. Sawada, Phys. Rev. 106 (1957), 372.

3. D.J. Rowe and S.S.M. Wong, Nucl. Phys. A153 (1970), 561.

4. V. Gillet and M.A. Malkanoff, Phys. Rev. B133 (1964), 1190.

5. J.H. Fegeau, Phys. Rev. 104 (1956), 225.

6. H.F. Ehrenberg et al., Phys. Rev. 113 (1959), 666.

7. G.E. Brown and M. Bolsterli, Phys. Rev. Lett. 3 (1959), 472.

8. T.W. Donnelly, J.D. Walecka, I. Sick and E.B. Hughes, Phys. Rev. Lett. 21 (1968), 1196.

9. I. Sick et al., Phys. Rev. Lett. 23 (1969), 1117.

10. T.W. Donnelly, J.D. Walecka, G.E. Walker and I. Sick, Phys. Lett. 32B (1970), 545.

11. D.W. Rowe, S.S.M. Wong and H. Chow, Nucl. Phys. A298 (1978), 31.

12. T.W. Donnelly and G.E. Walker, Ann. Phys. (N.Y.) 60 (1970), 209.

13. R.A. Lindgren, C.F. Williamson and S. Kowalski, Phys. Rev. Lett. 40 (1978), 504.

14. J. Lichtenstadt et al., Phys. Rev. Lett. 40 (1978), 1126.

15. W. Bertozzi, in Proceedings of the Conference on Electron and Photoexcitation, Sendai, Japan, 12-13 September 1977 (unpublished).

16. L. Zamick, Phys. Rev. Lett. 40 (1978), 381.

17. J.L. Matthews et al., Phys. Rev. Lett. 38 (1977), 8.

18. D.J.S. Findlay et al., Phys. Lett. 74B (1978), 305.

19. D.J.S. Findlay and R.O. Owens, Nucl. Phys. A279 (1977), 385.

20. J.T. Londergan and G.D. Nixon (submitted for publication).

21. J.T. Londergan, G.D. Nixon and G.E. Walker, Phys. Lett. 65B (1976), 427.

22. G.W. Edwards and E. Rost, Phys. Rev. Lett. 26 (1971), 785.

23. M.G. Piepho and G.E. Walker, Phys. Rev. C9 (1974), 1352.

24. M.K. Gupta and G.E. Walker, Nucl. Phys. A256 (1976), 444.

25. R.H. Landau and F. Tabakin, Phys. Rev. D5 (1972), 2746.

26. T.S.H. Lee and F. Tabakin, Nucl. Phys. A226 (1974), 253.

27. C.B. Dover and G.E. Walker (accepted for publication in Phys. Rev. C).

28. T.S.H. Lee, University of Pittsburgh Ph.D. Thesis (1974).

29. F. Binon et al., Nucl. Phys. B17 (1970), 168.

30. B.R. Martin, Nucl. Phys. B94 (1975), 413.

31. G.P. Gopal et al., Nucl. Phys. B119 (1977), 362.

32. G. Giacomelli et al., Nucl. Phys. B71 (1974), 138 (Solution D is used for the isospin zero amplitudes); G. Giacomelli et al., Nucl. Phys. B20 (1970), 301 (Solution (i) is used for the isospin one amplitudes).

33. J.T. Londergan, K.W. McVoy and E.J. Moniz, Ann. Phys. 86 (1974), 147.

34. H.K. Lee and H. McManus, Phys. Rev. 161 (1967), 1087.

35. P.J. Moffa and G.E. Walker, Nucl. Phys. A222 (1974), 140.

36. G.S. Adams et al., Phys. Rev. Lett. 38 (1977), 1387.

37. A.D. Bacher and G.T. Emery (private communication).

38. M.H. MacGregor, R.A. Arndt and R.M. Wright, Phys. Rev. 182 (1969), 1714.

39. A. Picklesimer and G.E. Walker, Phys. Rev. C17 (1978), 237.

40. D.E. Young and L.H. Johnston, Phys. Rev. 119 (1960), 313.

41. M.R. Wigan et al., Nucl. Phys. A114 (1968), 377.

42. J.N. Palmieri, A.M. Cormack, N.F. Ramsey, and R. Wilson, Ann. Phys. (N.Y.) 5 (1958), 299.

43. G.F. Cox et al., Nucl. Phys. B4 (1967), 353.

44. J. Tinlot and R.E. Warner, Phys. Rev. 124 (1961), 890.

45. J.R. Holt, J.C. Kluger and J.A. Moore, Proc. Phys. Soc. (London) 71 (1958), 781.

46. R.G.P. Voss, J.J. Thresher and R. Wilson, Proc. Roy. Soc. (London) A229 (1958), 493.

47. T.C. Randle, listed as private communication in R. Wilson, The Nucleon-Nucleon Interaction (Wiley, New York, 1963), p. 217.

48. J.P. Scanlon et al., Nucl. Phys. 41 (1963), 401.

49. Y.M. Kazarinov, V.S. Kiselev, I.N. Silin and S.N. Sokolov, Zh. Eksp. Teor. Fiz. 41 (1961) 197 [Sov. Phys. - JETP 14 (1962)], 143.

50. J. dePangher, Phys. Rev. 99 (1955), 1447.

51. O. Chamberlain et al., Phys. Rev. 105 (1957), 288.

52. F. Betz et al., Phys. Rev. 148 (1966), 1289.

53. G.H. Stafford, J.M. Dickson, D.C. Salter and M.K. Craddock, Nucl. Instrum. 15 (1962), 146.

54. G.H. Stafford, C. Whitehead and P. Hillman, Nuovo Cimento 5 (1957), 1589.

55. A. Langsford et al., Nucl. Phys. 74 (1965), 241.

56. K. Amos, J. Morton, I. Morrison and R. Smith, Aust. J. Phys. 31 (1978), 1.

57. H. Eikemeier and H.H. Hackenbroich, Nucl. Phys. A169 (1971), 407.

58. G.E. Walker and D.L. Weiss, Bull. Am. Phys. Soc. 22 (1977), 1006; and to be submitted for publication.

EXAMPLES WHERE WE ACTUALLY LEARN SOMETHING ABOUT NUCLEAR STRUCTURE FROM MEDIUM ENERGY EXPERIMENTS

Larry Zamick

Serin Physics Laboratory, Rutgers University

Piscataway, New Jersey, 08854, U.S.A.

I shall comment on experiments that have been done at Bates, Sendai, Lampf, Gatchina and Saclay and probably several other places.

The Bates and Sendai experiments are concerned with magnetic electron-nucleus scattering. I will try to convince you that to interpret the results of these experiments we must invoke spin-quadrupole correlations in the nucleus. In other words the quadrupole moment of spin up particles (in an odd nucleus) is different from the quadrupole moment of spin down particles.

The possibility of measuring the radius of a valence orbit by magnetic electron nucleus scattering has been proposed by Sick. One must of course worry about exchange current effects and core polarization before one can be sure one has extracted such a radius. But here I will be more concerned with structure aspects, and discuss a variable effective mass model, in which the radius of a valence orbit is smaller then what one gets in usual Hartree-Fock calculations. Whether this model turns out to be relevent or not, at least these experiments force us structure theorists to think more carefully about what a radius really means.

The results of muonic X ray work in the Calcium isotopes leads to a formula for the r.m.s. charge radius of Ca(40+n) which has the same form as the famous expression for the binding energy of a single j shell, as given by Talmi and de Shalit. The common bond is that both the energy and radius operators are scalars.

Lastly, I will discuss high energy ($\sim$ 1 GeV) proton nucleus elastic scattering. I hope to convince you that at least at the present time these experiments give us the best determination of neutron radii in nuclei, with error bars of about ±0.05 fm. The results are in good agreement with the predictions of the calculations of Negele (D.M.E. and D.D.H.F.).

Example 1

Magnetic Multipole Moments as Probes of Hartree-Fock

It has been found by Bertozzi et al.[1] that the M3 part of magnetic scattering in ^{17}O is strongly suppressed in the region of momentum transfer where it is expected to be strong when a single particle picture, of a closed ^{16}O core plus a valence $d_{5/2}$ neutron, is used. A similar suppression in ^{51}V has been reported by A. Enomoto.[2]

It is not the intent of this work to reproduce the momentum transfer dependence of M3 scattering. Rather, only the zero momentum transfer limit will be considered. The intention is to show that by keeping in mind a Hartree-Fock picture to describe a valence particle and a polarized core, the physical reason for M3 spin supression becomes transparent.

The spin part of the M3 operator is basically a product of an E2 operator and an M1 operator. This ties in nicely with a correlation between quadrupole defomation and spin which had been noted by Zamick, Golin and Moszkowski (ZGM).[3]

Let us recall the argument which was concerned with the quadrupole deformation of the core due to the presence of a valence nucleon e.g. ^{16}O and $d_{5/2}$ nucleon. A simple model of the deformed ^{16}O core is one in which all orbits have the same deformation. For example, if one uses harmonic oscillator wave functions, then all orbits in the core would have the same oscillator length parameters $b_x = b_y \neq b_z$.

Such a model had indeed been considered by Mottelson.[4] By also assuming that the potential followed the density he was able to show that the quadrupole moment of the core protons is $\frac{Z}{A} Q_{VALENCE}$. One could also express this result in terms of an effective charge.

$$e = Qcore/Q_v = Z/A$$

But ZGM pointed out[3] that if the interaction between the valence particle and the core is a zero range interaction, then the above 'trial solution' must be modified. Let the valence particle be a spin up neutron. With a zero range interaction this valence neutron cannot interact with spin up neutrons in the core. Hence the spin up neutrons in the core should not be deformed. On the other hand, the valence neutron can interact with spin down neutrons and so we expect the spin down neutrons to be deformed.

The magnetic multipole moment is

$$M(ML,N) = \frac{e\hbar}{2mc} [L(2L+1)]^{\frac{1}{2}} \sum_i r_i^{L-1} [\mu_i (Y_{L-1}\sigma]_N^L + \frac{2g_\ell}{L+1}(Y_{L-1}\ell)_\mu^L]$$

where

μ_i = 2.793 for a proton and -1.913 for a neutron

g_ℓ = 1 for a proton and 0 for a neutron

In the Hartree-Fock approximation the expectation value of the one body operator M(ML,0) is

$$\sum_i (i|M(ML,0)|i)$$

summed over occupied states i.

We note that the M3 operator is basically a quadrupole operator coupled with the magnetic moment to L= 3, plus a term with the quadrupole operator coupled with the orbital angular momentum to L = 3.

In this work we will consider mainly the suppression due to spin, not to orbital angular momentum. We therefore consider the $(Y_{L-1}\ \sigma)^L_\mu$ term. If we use ℓ-s wave functions then the only part of the operator which contributes is $Y_{z,0}\sigma_Z$. Therefore we will work with the operator

$$M_s(3) = \left(\frac{16\pi}{5}\right)^{\frac{1}{2}} \sum_i r_i^2 Y_{2,0}(i)\sigma_z(i)\ \mu_i$$

The value of this operator for the $d_{5/2}$ valence particle in the M = 5/2 state (this is a pure spin up state) is -1.91 $Q_{VALENCE}$

$$(Q_v = -[(2j-1)/(2j+2)]\ \langle r^2\rangle)$$

We now consider the core. Since the spin up neutrons are not deformed the expectation value of the $M_s(3)$ operator is zero. We introduce the notation

$$Q_{\pi\uparrow},\ Q_{\pi\downarrow} \text{ and } Q_{\nu\downarrow}$$

for the quadrupole moments of the core particles-protons spin up, protons spin down and neutrons spin down. The value of the $M_s(3)$ operator in the core is then

$$(Q_{\pi\uparrow} - Q_{\pi\downarrow})2.79 + 1.91Q_{\nu\downarrow}$$

We now assume the interaction of the valence spin up neutron with the core in a delta interaction $-A(1 + (-1)^T x)\delta(\vec{r}_1 - \vec{r}_2)$. A not uncommon choice is $x = 1/3$, for which the strength in a $T = 0$, $S = 1$ state is twice that for a $T = 1$ $S = 0$ state.

We assume the quadrupole moment of the spin up and spin down protons in the core is proportional to the strength of the interaction. We introduce Q_o such that

$$Q_{\pi\uparrow} = \frac{1}{2} Q_o (1+x)$$

$$Q_{\pi\downarrow} = \frac{1}{2} Q_o$$

We arrived at these results by noting that the $\pi\uparrow$-valence neutron interaction is in a pure $s = 1$ state, while the $\pi\downarrow$-valence interaction is half singlet and half triplet. The proton contribution to the $M_s(3)$ expectation value is then

$$\frac{1}{2} Q_o (2.79) x$$

Rather than calculate Q_o we can make an association with an E2 effective charge. The effective E2 charge for a valence $(d_{5/2})$ neutron is defined as

$$e_\nu = \text{Qcore proton}/Q_v = (Q_{\pi\uparrow} + Q_{\pi\downarrow})/Q_v$$

$$= Q_o (1+x/2)/Q_v$$

Thus the proton contribution to $M_s(3)$ is

$$[x/2(1+\tfrac{1}{2}x)] e_\nu \ (2.79) Q_v$$

This is clearly the opposite sign of the value of $M_s(3)$ for the $d_{5/2}$ neutron $(-1.91\ Q_v)$. We recall that x is about 1/3 and e_ν is often taken to be 1/2. Since the proton contribution is proportional to x, the result is due to the spin dependence of the interaction.

We now come to the neutron core contribution $1.91\ Q_{\nu\downarrow}$. The interaction with the valence neutron is in a pure $s = 0$ state so if we assume that the deformation is proportional to the interaction (which is equivalent to first order perturbation theory) we get

$$Q_{\nu\downarrow} = \frac{Q_o}{2} (1-x)(1.91) = \frac{(1-x)e_\nu}{2(1+x/2)} (1.91) Q_v$$

This is also the opposite sign of the value for the $d_{5/2}$ neutron. The reason is clear, only spin down neutrons contribute plus the fact that a quadrupole distortion of the core has the same sign as that of the particle.

Up to now, the result for the value of $M_s(3)$ is

$$Q_v\left(-1.91 + \frac{xe_\nu}{2(1+x/2)}\, 2.79 + \frac{(1-x)e_\nu}{2(1+x/2)} \times 1.91\right)$$

We next consider a modification of the above result due to R.P.A. correlations. We will do this in a simplified way. We first note that the neutron core contribution can be written in terms of the E2 effective charge correction for a valence proton. The point is that if charge symmetry holds the valence proton-core proton interaction is the same as the valence neutron-core neutron interaction.

We define the effective charge for a valence proton (^{17}F) as

$$e_\pi = \frac{Q_o}{2}\,\frac{(1-x)}{Q_v}$$

Hence the result for the value of $M_s(3)$ can be written as

$$Q_v(-1.91) + \frac{x\ e_\nu}{2(1+x/2)} \times 2.79 + \frac{e_\pi\ (1.91)}{(1+x/2)}$$

[Remember that the term with e_ν is due to the protons and the term with e_π is due to the neutrons.]

When one calculates e_ν and e_π in first order perturbation theory e_π comes out to be much less than e_ν. However when one does an R.P.A. calculation they come out to be much closer to each other. Hence the last term, due to the core neutrons might be very sensitive to R.P.A. correlations.

We now illustrate the effect of the R.P.A. by doing a simplified calculation. We call the first order charges $e_{F\nu}$ and $e_{F\pi}$ and the R.P.A. values $e_{R\nu}$ and $e_{R\pi}$. We introduce isoscalar and isovector charges as follows:

$$e_{F\nu} = (|e_F^0| + |e_F^1|)/2$$

$$e_{F\pi} = (|e_F^0| - |e_F^1|)/2$$

The R.P.A. results can be obtained by changing the energy denominators. If one uses zero range effective interactions e.g. of the Skyrme type, then the effective mass is one which means that the unperturbed single particle-single hole splitting for the quadrupole state is very close to $2\hbar\omega$. An approximate way of simulating the R.P.A. is to change the energy denominators from $2\hbar\omega$ to the energies of the isoscalar and isovector giant quadrupole states. These are approximately $\sqrt{2}\hbar\omega$[5] and $4\hbar\omega$ respectively. We then obtain

$$e_{R\nu} = (\sqrt{2}\ |e_F^o| + \frac{1}{2}\ |e_F^1|)/2$$

$$e_{R\pi} = (\sqrt{2}\ |e_F^o| - \frac{1}{2}\ |e_F^1|)/2$$

or

$$e_{R\nu} = \frac{1}{2}(\sqrt{2} + \frac{1}{2})e_{F\nu} + \frac{1}{2}(\sqrt{2} - \frac{1}{2})e_{F\pi}$$

$$e_{R\pi} = \frac{1}{2}(\sqrt{2} - \frac{1}{2})e_{F\nu} + \frac{1}{2}(\sqrt{2} + \frac{1}{2})e_{F\pi}$$

We shall see that it is not so easy to get the effective charges, especially e_π unambiguously, either from theory or experiment. We quote two recent analyses. B. A. Brown et al.[6] give for $d_{5/2} - d_{5/2}$ in $e_\nu = 0.33 \pm 0.01$, and $e_\pi = 0.14 \pm 0.23$ according to one analysis and 0.24 ± 0.27 according to another. In analyzing mass 18 they get $e_\pi = -0.07 \pm 0.03$. J. L. Durrell et al.[7] determine the radial integrals $\langle d_{5/2}|r^2|d_{5/2}\rangle$ from sub-Coulomb heavy ion transfer and deduce for $d_{5/2} = d_{5/2}$ $e_\nu = 0.43 \pm 0.02$, $e_\pi = 0.48 \pm 0.33$.

The range in variation of e_π is too wide for our purposes. Let us keep in mind that by e_π we here really mean the <u>neutron</u>-core deformation due to a valence <u>neutron</u>, not the proton-core deformation due to a valence proton. One can have large breakdown of charge symmetry due to loose binding effects. Basically what happens is that if a proton is loosely bound it is far away form the core and therefore cannot polarize the core very well. In that case e_π would be very small. But if the corresponding neutron is more tightly bound it may well be able to polarize the core neutrons more strongly.

Despite the difficulties mentioned above, we feel that meaningful limits on e_ν and e_π can be obtained from theory. Let us consider several cases which are in order of increasing believability (in the author's opinion).

1) Take $e_{F\nu} = 1/2$ and $x = 1/3$. Use first order perturbation theory. Then $e_{F\pi} = 2/7\ e_{F\nu} = 1/7$. The value of $M_s(3)$ is Q_V (-1.91 + 0.2 + 0.23) = -1.48 Q_V (corresponding to valence, core proton and core neutron contribution). The suppression factor is 1.48/1.91 = 0.77.

2) Use the R.P.A. Assume $e_{F\pi} = 2/7\, e_{F\nu}$ as above and arrange for $e_{R\nu}$ to be 1/2. We then find $e_{R\pi} = 0.34$. Using the R.P.A. result we find $M_s(3) = Q_V\,(-1.91 + 0.20 + 0.56) = -1.15\, Q_V$. The suppression factor is 1.15/1.91 = 0.6.

3) We now change x from 1/3 to 1. This looks like a drastic step, but it can be justified. The value x = 1/3 is chosen for spectroscopy of a few valence nucleons. But with such an interaction the symmetry energy is much too low. If we change to x = 1 we get a much better symmetry energy. Since core polarization is a one body field effect one should choose an interaction which yields the best one body potential, including the symmetry energy.

In this case $e_{F\pi} = 0$. However, the proton charge becomes finite in the R.P.A. Again we take $e_{R\nu} = 0.5$. We find

$$e_{R\pi} = \frac{\sqrt{2}-\frac{1}{2}}{\sqrt{2}+\frac{1}{2}}\, e_{R\nu} = 0.24$$

Now

$$M_s(3) = Q_V\,(-1.91+0.46+0.31) = -1.14 Q_V$$

The suppression factor is 1.14/1.91 = 0.6, the same as in case 2.

It is our feeling that the R.P.A. calculation is more sound than the first order one. Theoretical support for this argument comes from the work of G. E. Brown in "Facets of Physics"[8] who notes that when one linearizes the Hartree-Fock equations for a core plus one particle, the effective charges satisfy the R.P.A. equations of motion.

We now consider the $[Y_2 \ell]^3_0$ term. This can be written as a $r^2 Y_{2,0}\, L_z + b\,[r^2 Y_{2,1}\, L_{-1} + r^2 Y_{2,-1}\, L_1]$ where a and b are constants. We can further reduce this to the form $C_1 z^2\, L_z + C2\,(x^2 + y^2)\, L_z$. We now can easily see that this will vanish if we choose the usual form for the deformation of the core. It is most convenient to discuss this in terms of the asymtotic Nilsson quantum numbers N, M_3 and Λ.

For an axially symmetric deformation the values of z^2 and $(x^2 + y^2)$ are the same in the states $+\Lambda$ and $-\Lambda$. Hence $z^2 L_z$ and $(x^2 + y^2)\, L_z$ will be equal and opposite, and will cancel. Also the matter distribution is the same for $+\Lambda$ and $-\Lambda$, so it is hard to see why these two states should have different deformations.

In conclusion, we hope we have convinced the reader that there is indeed a spin-quadrupole correlation present in a nucleus and that it has observable consequences. When a valence neutron is added to the core, it is the Pauli principle which causes this correlation in the core neutrons, and it is the spin dependence of the interaction which causes this correlation in the core protons. These two effects add coherently in causing a suppression of the M3 moment.

A list of M3 moments by Migdal[10] indicates that there is a suppression, relative to the single particle model in a large number of nuclei.

We now give an alternate explanation. If we use deformed oscillators $\psi(x/bx, y/by, z/bz)$ with $b_x = b_y \neq b_z$ then we can make the deformed core "spherical" by the transformation $\hat{x} = x \frac{b}{b_x}$, $\hat{y} = y \frac{b}{b_y}$, $\hat{z} = z \frac{b}{b_z}$. But this will distort the operator. However under this transformation $L_z = x \frac{\partial}{\partial y} - y \frac{\partial}{\partial x}$ is equal to $\hat{x} \frac{\partial}{\partial \hat{y}} - \hat{y} \frac{\partial}{\partial \hat{x}} = L_{\hat{z}}$ since $\frac{b_x}{b_y} = \frac{b_y}{b_x} = 1$. Thus L_z, which is a vector goes into a vector. The other quantities z^2 and $(x^2 + y^2)$ will go into linear combinations of scalars and quadrupoles.

But only a scalar in the coordinates $\hat{x}$, $\hat{y}$ and $\hat{z}$, will be non-vanishing in the "spherical core". But a combination of a scalar and vector on a quadrupole and vector cannot yield a scalar. Hence, by this second argument, we see that the convection term $[y^2 \vec{\ell}]$ will vanish in the core.

REFERENCES

1. W. Bertozzi, in Proceedings of the Conference on Electron and Photoexcitation, Sendai, Japan, 12-13 September 1977 (unpublished).
2. A. Enomato, in Proceedings of the Conference on Electron and Photoexcitation, Sendai, Japan, 1977, edited by Y. Kawazoe (Laboratory of Nuclear Science, Tohoku University, Sendai, Japan, 1977), p. 173.
3. L. Zamick, M. Golin and S. Moszkowski, Phys. Lett. 66B (1977) 116.
4. B. R. Mottelson, The Many Body Problem (Wiley, New York, 1958).
5. M. Golin and L. Zamick, Nucl. Phys. A249 (1975) 320.
6. B. A. Brown, A. Arima and J. B. McGrory, Nucl. Phys. A277 (1977) 77.
7. J. L. Durell, C. Harter and W. R. Phillips, Phys. Lett. 70B (1977) 405.
8. G. E. Brown, In Facets of Physics, edited by D. A. Bromley and V. W. Hughes (Academic, New York, 1970), p. 141.
9. A. Arima, private communication.
10. A. B. Migdal, Nuclear Theory: The Quasi Particle Method (Benjamin, New York, 1968), p. 101.

Example 2

Comment on the problem of determining the radius of a valence orbit from magnetic electron nucleus scattering.

Recently, several authors, Sick et al.,[1] deWitt Huberts et al.[2] and Hynes et al.[3] have noted an apparent anomaly in the results of magnetic electron-nucleus scattering experiments. The cross section as a function of momentum transfer is larger at high q than is given by a single particle model using harmonic oscillator wave functions. If the scattering is from a nucleus consisting of a closed shell plus one particle, then in the single particle model all the magnetization is carried by the valence particle. Thus one expects to learn something about the orbit of the valence particle - in particular its r.m.s. radius. This is in contrast to Coulomb scattering in which the charge radius of the entire nucleus is obtained.

In the oscillator model the form factor consists of a polynomial times the exponential factor $\exp(-q^2b^2/4)$ where b is the oscillator length parameter. Sick et al.[1] noted that one way of slowing down the theoretical falloff was to make b smaller. That is instead of using only one oscillator parameter for all the orbits, presumably the one that fits the charge radius of the nucleus, allow the odd nucleon to have a different, in this case, smaller radius. However, several of the currently employed Hartree-Fock calculations do not have this feature.

Several authors, especially Arita,[4,5,6] have noted that the single particle model is too severe and that core-polarization is important. Indeed, several features of the cross section, in particular the M3 suppression are easily explained in terms of core polarization. Next, Arima et al.[6] and also Duback[7] have considered the importance of exchange currents. These are of comparable importance to core polarization, and indeed they help to keep the cross section up at high momentum transfer. According to the work of Hynes et al.,[3] however, the combination of the core polarization and exchange currents while improving the results considerably, are still inadequate to quantitatively explain the cross sections in ^{17}O. Also, changing from harmonic oscillator to Saxon-Wood potentials does little to resolve this problem.

In this work I would like to return to the original idea of Sick et al.[1] and reconsider the possibility that the valence orbit might indeed be less than that given by an oscillator length parameter b which is fitted to electron scattering. I will be referring to a previous work,[8] in which I did indeed obtain such an effect in a restricted Hartree-Fock calculation with density dependent interactions.

In this previous work, "Variational Study of a Closed Shell Plus One Particle Using Zero-Range Interaction"[8] a ^{40}Ca core plus an f nucleon was considered. Three interactions were considered all of which are special cases of the Skyrme interaction.[9]

The Skyrme interaction is

$$-t_o\delta(\vec{r}_{12}) + \frac{t_1}{2}[k^2\delta(\vec{r}_{12}) + \delta(\vec{r}_{12})\vec{k}^2]$$

$$+ t_2\vec{k}\delta(r_{12})\vec{k} + t_3\delta(\vec{r}_{12})\delta(\vec{r}_{13})$$

$$\vec{r}_{12} = \vec{r}_1-\vec{r}_2 \qquad \vec{k} = \frac{\vec{\nabla}_1-\vec{\nabla}_2}{2}$$

+ spin orbit terms.

The three cases considered were:

	t_o	t_1	t_2	t_3
Zero Range	996.895	0	0	16259.2
VB11'	1208.346	585.6	0	10011.116
Finite Range	1314.24	1199.1	0	0

In the zero range case, the velocity dependent terms are set equal to zero. All the repulsion in the interaction comes from the three body delta term, which is equivalent to a linear density dependent interaction $\rho(\vec{r}_1)\ \delta(\vec{r}_1 - \vec{r}_2)$.

Finite range is the other extreme. Here we set $t_3 = 0$ so the entire repulsion comes from the velocity dependent term. The term VB11' is an intermediate case,[10] close to the Vautherin-Brink II interaction[9] (except that VB11 has t_2 small and negative and we set it equal to zero).

The parameters were chosen so that when a trial Slate determinant with Harmonic Oscillators was used, the energy of ^{40}Ca came out to be 346.1086 MeV and the value of b which resulted was 1.95 fm.

As we go from zero range to finite range the effective mass $\frac{m^*}{m}$ gets smaller. Indeed in infinite nuclear matter we have

$$\left(\frac{m^*}{m}\right)^{-1} = [1 + \text{FINITE RANGE ENERGY/KINETIC ENERGY}]$$

In the oscillator model the radius of an orbit n,ℓj is $\langle r^2\rangle^{1/2}_{n,\ell} = (2n + \ell + \frac{3}{2})^{1/2}b$. In the previous work we had considered a situation in which the f orbit had a different value of b than did the ^{40}Ca core.

In the following we write down the effective mass and the value b_f for the f orbit

	m^*/m	b_f fm	
Zero Range	1	1.825	
VB11'	0.6	1.998	$b_{core} = 1.95$ fm
Finite Range	0.4	2.105	

Why do we do this? First of all, we see that indeed for the ZERO RANGE Force, with $m^*/m = 1$, the equivalent radius of the f orbit is indeed pushed in. The percentage change is $\left(\frac{1.95 - 1.825}{1.9}\right) \times 100 = 6\%$. This is a bit less but comparable to what Sick et al.[1] want.

However, most of the current Hartree-Fock calculations favor an interaction closer to VB11'[9,10] for which the effective mass is 0.6. For VB11' the value of b_f is 1.998, somewhat greater than b_{core}.

Would it therefore not be cheating to use the ZERO RANGE interaction for the f orbit and claim that the anomaly has been resolved?

Perhaps not. One outstanding feature of the Skyrme interaction[9,10] is that the effective mass m^*/m is nearly state independent. Yet there has been considerable literature on the fact that the effective mass should vary. We cite the work of G. E. Brown. J. H. Gunn and P. Gould[11], of G. F. Bertsch and T. T. S. Kuo[12] and Jeukenne and Mahaux[13]. The point of these works is that whereas the effective mass is 0.6 - 0.7 well below the Fermi surface and well above the Fermi surface, it should peak to a value of one just at their Fermi surface. This is shown schematically in figure 1.

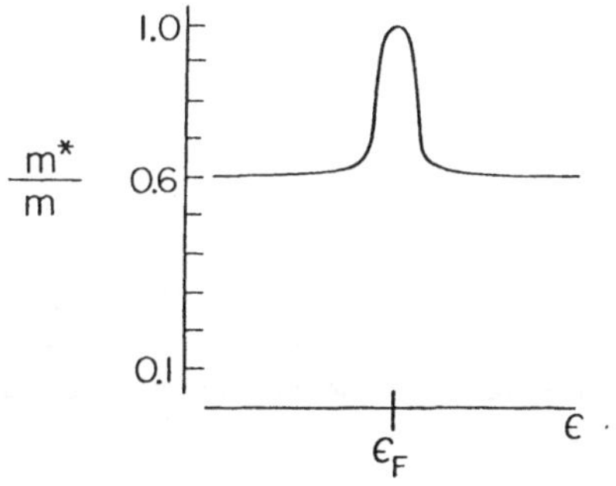

Fig. 1. Schematic diagram of the effective mass as a function of energy.

This peaking is due to the admixture of 2 particle-one hole and three particle - two hole components in the wave function, whose main component is the one particle state. The relevant diagrams are shown in figure 2.

Now it is true that the Skyrme interaction[9] simulates part of the diagrams in Fig. 2. This manifests itself in the rearrangement

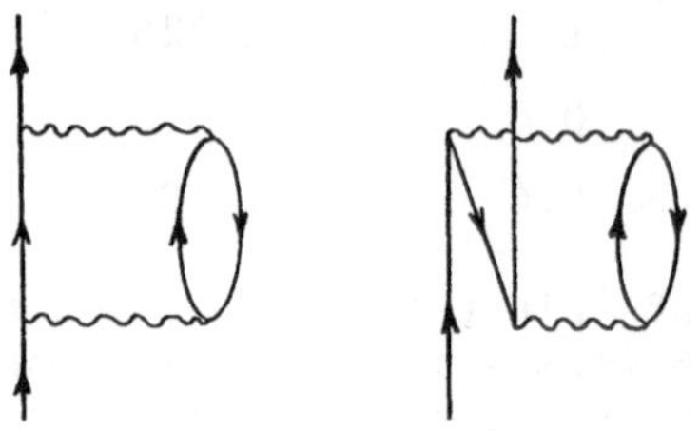

Fig. 2. Second order corrections to the single particle energy, which contribute to the variable effective mass.

effects which have been discussed by several authors.[9,10] But the rearrangement effects due to rather low lying configurations are not properly taken into account, only those due to high lying intermediate states arising from the hard core and tensor part of the interaction. Thus the Skyrme interactions,[9] and probably most of the others in common use, do not have the feature of a variable effective mass.

We are therefore justified in modifying the existing Hartree-Fock calculations so that the feature of a variable effective mass is present. But how should we do it? One way is to add the low lying configurations of Fig. 2 in explicitly, as was done by Brown, Gunn and Gould[11]. Another way is to add terms to the Skyrme interaction[9] of higher power than k^2 (or combinations of k^2 and densities) so that the effective mass is a more complicated behavior of energy, as in figure 1.

In this work, however, we will do something more simple. Let us just speculate that the parameters t_0, t_1, t_2, t_3 are state dependent. For states well below the Fermi sea and well above the parameters are close to those of VB11'[10] with an effective mass $m^*/m \sim 0.6$. However, for states near the Fermi sea we shoose them to be close to those of the zero range for which $m^*/m = 1$.

This will certainly bring out the feature of a variable effective mass in just the way that Brown, Gunn and Gould predicted (ref 11). It also makes it plausible that we should use the zero range interaction (with $m^*/m = 1$) for the f orbit even though we use something like VB11' everywhere else. And we get immediately the result that $b_f = 1.825$ fm, which is 6% smaller than b_c.

Obviously, the above considerations need further investigation, using a wider variety of variable effective mass models, and less restrictive Hartree-Fock calculations. Only then can we be more certain of the quantitative results. It is premature to say that we have definitely solved the anomaly that Sick et al.[1] have pointed out. On the other hand, I think a good case has been made for why we should not trust the usual Hartree-Fock results, even though they give excellent fits to many things, in particular, form factors for electron scattering. The usual H. F. calculations do miss an important feature at the Fermi surface, namely the peaking of the effective mass, so it should not be surprising that they might miss some other features as well. It seems physically reasonable that raising the effective mass at the Fermi surface will cause the radius to come in.

Added Note: It has been suggested by G.E. Brown and by G.F. Bertsch that the result of perturbation theory is to make m* peak only at the nuclear surface, being less than unity in the interior and greater than unity near the surface. This is in contrast to the above example where m* was unity for all r. With a surface peaked m* the valence orbit does not necessarily shrink. Whether it shrinks, expands or remains unchanged depends on the details.

REFERENCES

1. I. Sick, J. B. Bellicard, J. M. Cavedin, B. Frois, M. Huet, P. Leconte, A. Nahada, Phan Xuan Ho, S. Platchknow, P. K. A. de Witt Huberts and L. Lapidas, Phys. Rev. Letts. 38 (1977) 1259.
2. P. K. A. de Witt Huberts et al., Physics Letters 71B (1977) 317.
3. M. V. Hynes, H. Miska, B. Noram, W. Bertozzi, S. Kowalski, F. M. Rad, C. P. Sargent, T. Sasanuma, W. Turchinetz and B. L. Berman, preprint.
4. K. Arita, Proceedings of the Meeting on Giant Resonances and Related Topics, University of Tokyo Tanashi-shi (1977).
5. L. Zamick, Phys. Rev. Lett. 40 (1978) 381.
6. A. Arima, Y. Harikawa, H. Hyuga and T. Suzuki, Phys. Rev. Letters 40 (1978) 1001.
7. J. Dubach, private communication.
8. L. Zamick, Nuclear Physics A260 (1976) 241.
9. D. Vautherin and D. M. Brink, Phys. Rev. 65 (1972) 626.
10. P. W. Sharp and L. Zamick, Nucl. Phys. A208 (1973) 130.
11. G.E. Brown, J.H. Gunn and P. Gould, Nucl. Phys., 46 (1963), 598.
12. G. F. Bertsch, T. T. S. Kuo, Nucl. Phys. A112 (1968) 204.
13. J. P. Jeukenne, A. Lejeune and C. Mahaux, Physics Reports, Physics Lett. 25C (1976) 85.

Example 3

Comment on the Isotopic Dependence of Charge Radii

It was previously noted by this author[1] that for a closed shell of say protons, plus n neutrons in a single j shell, the mean square radius might obey a formula similar to the one for the binding energies as given by Talmi and de Shalit.[2]

$$r^2(A+n) - r^2(A) = nC + \frac{n(n-1)}{2}\alpha + [\frac{n}{2}]\beta$$

where A is the mass number of the double closed shell.

It was argued that the effective radius operator in the model space of a single j shell had at least an effective two body part. Then the link between energy and square radius is that both are scalars. It was noted that although the above form could be obtained in perturbation theory, quantitative calculations were an order of magnitude off, and that undoubtedly, in the Calcium isotopes, the presence of highly deformed many particle-many hole states would have to be taken into account, as well as other higher order effects.

At the time of the writing of this work there were no measurements of r^2 for an odd Calcium isotope. Hence one could only obtain two parameters, α and $D = 2C + B$.

However, there has been a recent publication of muonic isotope shifts in the stable calcium isotopes by H. D. Wohlfahrt et al.[3] In this work the calcium isotopes with A = 40, 42, 43, 44, 46, 48 are considered.

The authors make a distinction between the charge radius and the proton radius. The former contains the entire charge distribution, due both to protons and to the neutrons.

We will give fits to both of these using the subscript c for charge and p for proton.

Furthermore, we have

$$\frac{r^2(A+n) - r^2(A)}{\sqrt{r^2(A+n)} + \sqrt{r^2(A)}} = \sqrt{r^2(A+n)} - \sqrt{r^2(A)}$$

we take $\sqrt{r^2(A+n)} + \sqrt{r^2(A)}$ to be constant = 7.0 fm. Then we can work directly with the differences in r.m.s. radii

$$\sqrt{r^2(A+n)} - \sqrt{r^2(A)} = n\hat{C} + \frac{n(n-1)}{2}\hat{\alpha} + [\frac{n}{2}]\hat{\beta}$$

For the charge radii we make a fit to table 1 of the work of Wohlfahrt et al.[3] We find, using units of $\underline{10^{-3}\ \text{fm}}$, that

$$\hat{\alpha}_c = -4.867$$

$$\hat{\beta}_c = 38.056$$

$$\hat{C}_c = 2.285$$

Note the large odd-even or pairing term $\hat{\beta}_c$. The comparison is given in our table A.

Table A

$\sqrt{r^2(A+n)}_{CH} - \sqrt{r^2(A)}_{CH}$ in units of $\underline{10^{-3}\ \text{fm}}$

n	EXPT.[a]	OUR FORMULA
1		-2.3
2	30.1	28.6
3	16.6	16.6
4	40.1	37.7
5		16.0
6	21.1	27.4
7		-4.0
8	0.9	-2.3

[a]Data taken from table 1 of ref. 3. Note that they also have data in table 2 which in some cases differs from table 1.

Our formula makes some predictions for which there is not yet any empirical results. For example, we predict that the charge radius of ^{41}Ca will be slightly less than that of ^{40}Ca, and that it certainly will not be half way between the values for ^{40}Ca and ^{42}Ca. In general we predict a strong odd even staggering. Such effects are certainly present in other parts of the periodic table e.g. the mercury isotopes, and have been considered theoretically by Sorensen.[5]

For the proton radii, we find, again by fitting to the results of Wohlfahrt et al. the following (again in units of $\underline{10^{-3}\ \text{fm}}$).

$$\hat{\alpha}_p = -4.992$$

$$\hat{\beta}_p = 38.611$$

$$\hat{C}_p = 0.455$$

This leads to table B.

$$\sqrt{r^2(A+n)}_p - \sqrt{r^2(A)}_p \text{ fm}$$

n	EXPT.	OUR FORMULA
1		0.45
2	36	34.5
3	25	25.0
4	53	49.1
5		29.6
6	37	43.7
7		14.2
8	21	18.3

The fits are not as impressive as the ones Talmi[2] obtained for the binding energies, but they seem to this author to be good enough to indicate the trend. Clearly, the measurement of the missing radii, ^{41}Ca, ^{45}Ca and ^{47}Ca would be a good test.

Finally, as has been emphasized by Kirson,[4] the validity of this formula for binding energies or radii, does not mean that there are not significant three body effects present. For example, referring to binding energies, the α, β and C. obtained from the spectrum of ^{42}Ca alone are different from the ones obtained from a least squares fit to all the Calcium isotopes. This may well be true for the radii.

REFERENCES

1. L. Zamick, Annals of Physics 66 (1971) 784.
2. I. Talmi, Rev. Mod. Phys. 34 (1962) 704.
3. H. D. Wohlfahrt, E. B. Shera, M. V. Hoehn, Y. Yamazaki, G. Fricke and R. M. Steffen, Phys. Lett. 73B (1978) 131.
4. M. W. Kirson, private communication.
5. R. A. Sorensen, Phys. Lett. 21 (1966) 333.

Example 4

Proton - Nucleus Scattering as a Probe of Hartree-Fock Densities

It should be made clear at the outset that electron nucleus scattering does not test the Hartree-Fock theory. Rather it tests the one particle charge density that results from a Hartree-Fock calculation. In the Born approximation the electron nucleus cross section is

$$\frac{d\sigma}{d\Omega} = \left(\frac{d\sigma}{d\Omega}\right)_{MOTT} |S(q)|^2$$

where

$$S(q) = \frac{1}{Ze} \int \rho_\pi(\vec{r}) d^3\vec{r}\, e^{i\vec{q}.\vec{r}}$$

is the Form Factor, or Fourier transform of the charge density. Here, ρ_π is normalized so that $\int \rho_\pi(\vec{r})\, d^3\vec{r} = Ze$.

There may be all sorts of correlations present in the nuclear wave function, but, to the extent that the Born approximation holds, the electron - nucleus elastic scattering process will be insensitive to them.

Since proton nucleus scattering has been discussed so much in the past, I will not talk about this. Rather I will consider briefly high energy proton nucleus scattering, in which one might hope to learn about the matter distribution. Together with the charge distribution obtained from electron scattering, one might hope to learn about the distribution of neutrons in the nucleus.

The Glauber theory is used. This gives excellent agreement with experiment. We follow a discussion of Harrington and Varma.

The proton nucleus scattering amplitude is given by

$$F(q) = \frac{-k}{2\pi i} \int d^2\vec{b} e^{-i\vec{q}.\vec{b}}\ \Gamma(b)$$

$$\Gamma(b) = 1 - \langle\psi \int \prod_{i=1}^{A} [1 - Y(\vec{b}-\vec{b}_i)]\psi\rangle$$

where

$$Y(b) = \frac{1}{2\pi ik} \int d^2\vec{q} e^{i\vec{q}.\vec{b}}\ f(q)$$

where $f(\vec{q})$ is the nucleon-nucleon scattering amplitude.

Let $\langle Y(b)\rangle = \int d^3\vec{r}_1\ \hat{\rho}(r_1)\ Y(\vec{b} - \vec{b}_1)$ where $\rho(r)$ is the density normalized so that $\int \hat{\rho}(\vec{r})d^3\vec{r} = 1$ (in contrast to the density ρ which is normalized to A).

If we neglect correlations in the nucleus then

$$\Gamma(b) = 1 - [1-\langle Y(b)\rangle]^A$$

$$= A\langle Y(b)\rangle - \frac{A(A-1)}{2}\langle Y(b)\rangle^2 + \ldots + (-1)^A \langle Y(b)\rangle^A$$

This is the multiple scattering series. It is finite, with A terms. It is necessary to keep many terms in the series, e. g., in ^{208}Pb up to about 36 fold scattering must be included. Note that at this level (no correlations) it is just as easy to do the multiple scattering as the single scattering. The only quantity that enters is the matter density.

One can get excellent fits in this way, without assuming any correlations. This means that we will probably never learn about nuclear correlations by examining high energy elastic proton nucleus scattering.

On the other hand one has to put in the correlation terms to be sure that they are not so large as to obscure the extraction of the matter density distribution. This has been done by Harrington and Varma. They find that Pauli correlations are significantly more important than short range correlations, and that the correlations in general are more important as the angles become larger.

In practice the neutron radii obtained with and without correlations do not differ too much.

We present a list of proton and neutron radii obtained by the above described method. They are in excellent agreement with the Hartree-Fock calculations of Negele.

The results for $\Delta = r_n - r_p$ extracted from proton scattering data at 1.05 GeV (Saclay), 1.00 GeV (Leningrad), and 0.8 GeV (LAMPF), together with the predictions of density dependent Hartree-Fock calculations (DDHF) of Negele,[22] Density Matrix Expansion (DME) of Negele and Vautherin,[23] Skyrme II interactions (Sk II) of Vautherin and Brink[24] and calculations[25] due to Campi and Sprung (CS). As discussed in text, the analyses of refs. 18 and 19 neglect all correlation effects. Ref. 16 includes c.m. correlations whereas ref. 15 includes both c.m. and Pauli correlations. The typical error bars in Δ from proton scattering are $\sim \pm 0.05$ fm.

Nucleus	$\Delta = r_n - r_p$ (fm) from proton scattering (GeV)			DDHF	DME, SkII	CS
	1.05	1.0	0.8			
^{16}O	-0.02[18] -0.04[19]	~0[16]	-	-0.02	-0.03	-0.02
^{40}Ca	-0.04[15] -0.03[18] -0.03[19]	-0.07[16]	-	-0.04	-0.04	-0.05
^{48}Ca	0.19[15] 0.16[18] 0.17[19]	0.21[16]	-	0.23	0.18	0.18
^{90}Zr	-	0.13[16]	0.09[19]	0.12	0.08	0.07
^{208}Pb	0.21[15] 0.21[19]	0.08[16]	0.21[15] 0.19[19]	0.20	0.20	0.21

REFERENCES

G. Varma and L. Zamick, Nucl. Phys., A306 (1978) 343.
D. R. Harrington and G. K. Varma, Phys. Lett., 74B (1978) 316.

References from previous table

15. G. K. Varma, Bull. Am. Phys. Soc. 22 (1977) 1009.
16. G. K. Varma and L. Zamick, Phys. Rev. C16 (1977) 1308.
18. A. Chaumeaux, V. Layly and R. Schaeffer, Phys. Lett. 72B (1977) 33.
19. L. Ray and W. Coker, preprint (1977), L. Ray et al., preprint (1978) and private communication.
22. J. W. Negele, Phys. Rev. C1 (1970) 1260.
23. J. W. Negele and D. Vautherin, Phys. Rev. C5 (1972) 1472.
24. D. Vautherin and D. M. Brink, Phys. Rev. C5 (1972) 626.
25. X. Campi and D. W. Sprung, Nucl. Phys., A194 (1972) 401.

THE NEW GIANT RESONANCES - AN EXPERIMENTAL REVIEW

Fred E. Bertrand

Physics Division, Oak Ridge National Laboratory

Oak Ridge, Tennessee 37830 U.S.A.

An interesting and important development in nuclear physics during the past 7 years has been the experimental observation of non-dipole giant resonances, the so called "new" giant resonances. Unlike the giant dipole resonance (GDR) which has been established mainly through the photo-nuclear reaction, the new giant resonances were first observed by and have been studied almost exclusively through inelastic scattering of medium-energy hadrons and electrons. (Reference 1 provides a review of this subject as of early 1976. I will refer to this review and the references therein in order to shorten the present reference list. More recent work will be explicitly referenced here.)

In this talk I will provide an experimental review with emphasis on data and interpretation rather than on experimental methods. This discussion will be limited to the non-dipole[2] giant resonances and further I will only deal with the electric multipoles[3]. Finally, I shall describe only direct observations of giant resonances arising through inelastic scattering.

In nearly all respects, excitation of giant resonance states via inelastic scattering is experimentally the same as inelastic scattering to low-lying nuclear levels. As you will see, the giant resonances are located at rather high excitation energies (10-20 MeV), are quite broad (3-10 MeV) and appear on top of a large nuclear continuum, facts that necessitate that considerable care be taken to insure a background free-experiment.

Similarly, the analysis of angular distributions for the giant resonances is performed using the same techniques as are used for low-lying states, namely by use of the Distorted Wave Born approximation (DWBA). In the case of electron scattering the transition rate [B(EL)] for the state in question can be directly obtained. In inelastic hadron scattering however, the situation is not nearly so direct. Comparison of the calculated cross sections with those measured yields a quantity called the deformation parameter, β_L, as:

$$\beta_L^2 = \frac{d\sigma(L)}{d\Omega} \text{ measured } / \frac{d\sigma(L)}{d\Omega} \text{ calculated} \tag{1}$$

If β_L is assumed to be proportional to the mass multipole moment for a uniform distribution then

$$\beta_L^2 = B(EL) \left(\frac{4\pi}{3ZR^L} \right)^2 \quad . \tag{2}$$

Whether or not the transition rate as directly determined in inelastic electron scattering or deduced through use of equations (1) and (2) yield the same result has been an often debated topic. However, it is clear that hadron inelastic scattering is more model dependent than electron scattering. The application of the DWBA theory to giant resonance excitation by hadron inelastic scattering has been detailed by Satchler[4,5] and the experimental results of such studies have usually been analyzed following Satchler's formalisms. As is traditional in giant resonance work, the cross section observed in a resonance state is described in terms of a sum rule, a theoretical limit for the strength. The linearly energy-weighted sum-rule limit (EWSR) is especially useful because nearly model-independent estimates of its value can be made. The EWSR is related to the transition rate as:

$$S_L = B(EL)\,(E_f - E_i) = \frac{L(2L+1)^2}{4\pi}\frac{\hbar^2}{2m} Ze^2 < r^{2L-2} > , \tag{3}$$

where E_f and E_i denote the energies of the final and initial states in the transition and $<r^L>$ is the RMS charge radius of the ground state. For T=S=0 transitions and $<r^L> = (3/L+3)R^L$ for a uniform mass distribution, the EWSR for a transition of multipole L can be written as:

$$S_L = \frac{3A\hbar^2}{8\pi m} LR^{2L-2}$$

where m is the nucleon mass and A is the nuclear mass number.

Using equation (2) relating β_L^2 and B(EL) the following expression for β_L^2 in terms of the EWSR is obtained (for T=S=0):

$$\beta_L^2 = \frac{2\pi\hbar^2}{3m} \frac{L(2L+1)}{AR^2} \frac{1}{E} \sim \frac{60L(2L+1)}{A^{5/3}E} , \quad (4)$$

if $R = 1.2\ A^{1/3}$ fm and E = excitation energy in MeV.

Equation (4) provides the limit of the EWSR for a transition of multipole L and energy E. Throughout this talk the measured giant resonance cross section will be presented as a fraction of the limit from equation (4).

Although I said I would not discuss the GDR, it is helpful to remind ourselves of the properties of what was for so many years the only giant resonance established experimentally. Figure 1 shows the spectrum of ^{208}Pb as seen in the (γ,n) reaction[2]. The only structure observed in the spectrum is the peak from the GDR centered at about 13.5 MeV of excitation energy. The fact that photoabsorption proceeds overwhelmingly by dipole absorption leads to such beautiful GDR spectra which are largely uncomplicated by competing reactions. As shall be seen later such is not the case for excitation of giant resonances via inelastic scattering. As is well known[2], for $A \gtrsim 100$ the GDR is located at a systematic excitation energy of $\sim 77 \times A^{-1/3}$ MeV and $\sim 100\%$ of the sum rule strength is exhausted in the resonance.

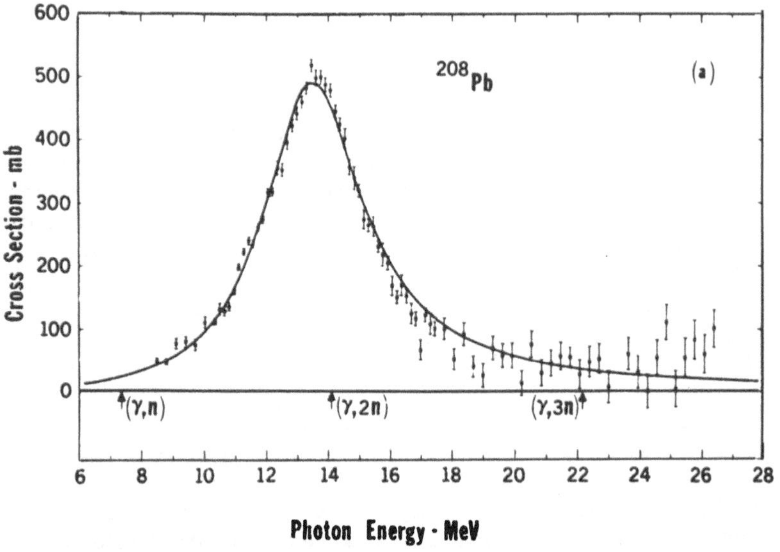

Fig. 1. Giant dipole resonance in ^{208}Pb as observed in the (γ,n) reaction (reference 2).

The many systematic studies of the GDR allow us to list a few features that seem characteristic of giant resonances and may be useful for experimental searches for new resonances.

1) Giant resonances are a general property of nuclei and as such should be observable in all nuclear mass regions.
2) The excitation energy of giant resonances varies smoothly with nuclear mass (at least over most of the mass range).
3) Resonance strength is generally localized in excitation energy (more of an experimental than theoretical necessity).
4) The resonance exhausts an appreciable fraction of an appropriate sum rule.

Giant resonances are often described as highly collective modes of excitation in which an appreciable fraction of the nucleons in a nucleus move together--so much so that it is appropriate to describe these modes in hydrodynamic terms like the oscillation of a liquid drop.

One of these modes, the GDR is often viewed as an oscillation in which the neutrons oscillate in bulk against the protons in the nucleus. There are, of course, other nuclear shape oscillations. For example, the quadrupole oscillation where the nucleus is characterized as oscillating from prolate to oblate shape. The low-lying collective 2^+ levels are formed by oscillations of this type. Of particular interest is the monopole (E0) oscillation in which the nucleus compresses or "breathes". There are higher order forms of oscillation such as octopole and hexadecapole.

Since the nuclear fluid has four components, neutron, proton, spin up and spin down, for each type of oscillation there are four possible combinations of these components. Modes in which the neutrons and protons move together (in phase) are characterized by isospin zero (T=0), i.e. isoscalar modes; but when the neutrons and protons move against each other the vibration has isospin one (T=1), i.e. isovector. The GDR is an example of an isovector mode while the first excited 2^+ state in nuclei is of the isoscalar type. Similarly, spin up and spin down nucleons moving in phase yield S=0 modes while if moving out of phase yield S=1 modes. In this talk I will almost exclusively discuss giant resonances of the T=S=0 variety.

Schematically represented in figure 2 are single-particle transitions between shell-model states of a hypothetical nucleus. Collective transitions result from coherent superpositions of such single-particle transitions. Major shells are denoted as N, N+1 and N+2 and within each major shell a few subshells are shown. The major shells are separated by 1 $\hbar\omega$ or $\sim$ $41/A^{-1/3}$ MeV. The transitions shown represent some of the variety of vibrational (collective)modes that may occur by exciting a nucleon from one orbit to another via inelastic scattering, for example. Within a shell-model framework, giant resonances can be considered to result from transitions of nucleons from one major shell to another, under the influence of an interaction that orders these transitions into a coherent motion. The interactive operator for inelastic scattering can excite a nucleon by at most $L\hbar\omega$, or, to state it differently, the nucleon can be promoted by at most L major shells. The number of shells is either odd or even in order to conserve parity.

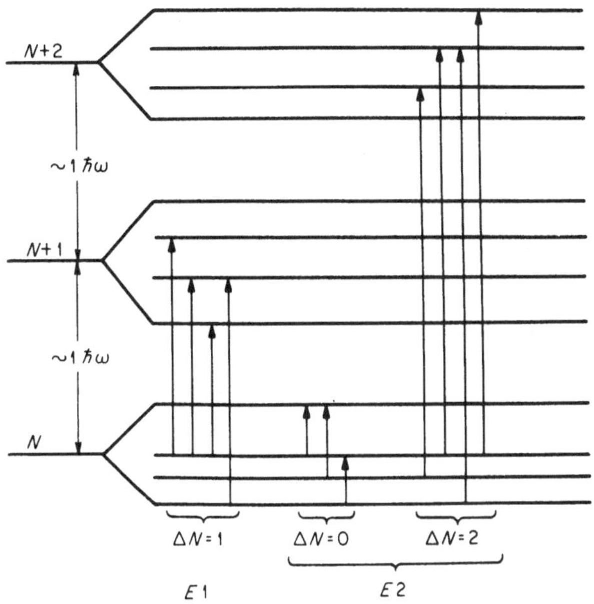

Fig. 2. Schematic representation of E1 and E2 single-particle transitions between shell-model states of a hypothetical nucleus. Major shells are denoted as N, N+1 and N+2 and lie $\sim 1\hbar\omega$ or 41 x $A^{1/3}$ MeV apart.

Thus, the GDR (E1) is pictured as built up of transitions spanning $1\hbar\omega$ ($\Delta N=1$). The GDR might then be expected to be located at an excitation energy of $\sim 41\ A^{-1/3}$ MeV; however, it is located at $\sim 77\ A^{-1/3}$ MeV. This difference arises from the fact that the interaction between the nucleons in the nucleus is repulsive for the isovector mode, so that the excitation energy is pushed up from that expected. Conversely, the interaction is positive for isoscalar modes, thus pushing the excitation energy down from what might be expected.

Two different sets of E2 transitions are allowed. The first

of these, with lowest energy, is comprised of transitions within a major shell, the so-called $0\hbar\omega$ ($\Delta N=0$) transitions. A second set is comprised of transitions between shells N and N+2 ($\Delta N=2$) and would have energy of $2\hbar\omega$, pushed down or up for isoscalar or isovector modes respectively. While the first class ($0\hbar\omega$) of E2 excitations is identified with the familiar low-lying 2^+ levels, the $2\hbar\omega$ class of E2 transitions had not been identified until a few years ago. By similar arguments E3 excitations of $1\hbar\omega$ and $3\hbar\omega$ and E4 excitations of $0\hbar\omega$, $2\hbar\omega$ and $4\hbar\omega$ are expected.

For each class of transitions (E1, E2, etc.) the sum rule should be exhausted by the sum of the strength in all the transitions. For example, for E2 transitions the sum rule should be exhausted by the sum of the strength in the $0\hbar\omega$ and $2\hbar\omega$ transitions.

For the GDR we know 100% of the sum rule is accounted for. What about other multipoles? Figure 3 shows the percent of the EWSR depleted in the first 2^+ level of even-even nuclei. Except for the lightest nuclei, only $\sim$ 10-15% of the EWSR is accounted for in the first 2^+ level. Table 1 shows a few cases from inelastic scattering where the entire bound state excitation region ($\lesssim$ neutron separation energy) was studied. Only for ^{24}Mg is any

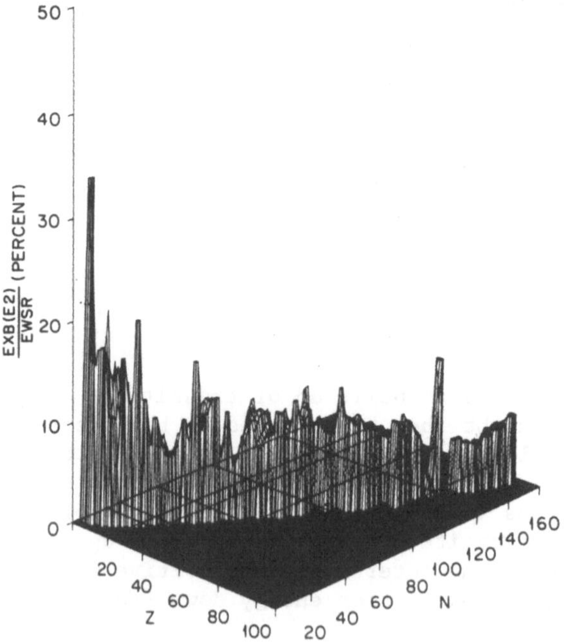

Fig. 3. Percent energy weighted sum rule depleted in the first excited 2^+ state plotted versus Z and N.

appreciable fraction of the L=2 EWSR strength found in the bound-state region. Thus, there is good reason to expect considerable quadrupole strength to be in the $\Delta N=2$ excitations. Indeed, less than one half of the possible sum-rule strength for any multipole is located in the bound states for the nuclei studied. In fact, there were predictions[8] that the $\Delta N=2$ quadrupole strength would be appreciable and localized at an excitation energy of $\sim 60 \times A^{-1/3}$ MeV.

TABLE 1. Percentage of isoscalar EWSR multipole strength depleted in bound states of ^{24}Mg, ^{40}Ca, and ^{208}Pb (Reference 1).

Nucleus	Multipole								
	0	1	2	3	4	5	6	7	8
^{24}Mg			40	10	3				
^{40}Ca	0	0	14	38	7	11	1	0.2	0
^{208}Pb	0	0	20	47	14	3	3	2	1

How should one search for these high excitation energy, non-dipole giant resonances? Since the photonuclear reaction proceeds so dominately by dipole absorption, these reactions are not good choices for observing other, higher order, multipoles. (Excitation of the GDR is 10-100 times stronger than E2 excitation via the photonuclear reaction.) On the other hand, it has long been known that the inelastic scattering reaction provides strong excitation of collective T=S=0 states. In addition, as shown on Table 2, variation of the particle type used for inelastic scattering can provide different strengths to isoscalar and isovector states. For example, the (e,e') reaction excites isoscalar and isovector states with equal probability (for same EWSR, J^{π} and E_x) while the (p,p') and (^{3}He, ^{3}He') reactions preferentially excite isovector states by the ratios of $\sim 1/9$ and $\sim 1/30$ respectively. On the other hand, the (α,α') and (d,d') reactions do not excite isovector states by an observable amount while the charge exchange reaction preferentially excite the isovector states. Comparison of the various reactions leading to giant resonance states can be of great help in unraveling the isospin makeup of the resonances.

Since the inelastic scattering reaction has been utilized for low-lying level studies for so many years and in addition we experimentalists were told[8] by our theoretical colleagues where to find the giant multipole resonances, why have they not been seen much earlier? At least part of the reason lies in what studies are fashionable and what are not at a given time. The advent of high resolution semi-conductor detectors and magnetic spectrometers led to nearly a complete preoccupation by nuclear experimentalists with the first few MeV of nuclear excitation energy. The excitation region above ∿ 10 MeV was simply thought to be uninteresting. We now know this is not the case.

TABLE 2. Relative cross sections[a] of isoscalar and isovector excitations for various reactions.

	Isoscalar	Isovector
(e,e')	1	1
(p,p')	1	∿ 1/9
(^{3}He,^{3}He')	1	∿ 1/30
(α,α')(d,d')	1	∿ 0
(n,p)(t,^{3}He)	0	1
(p,n)(^{3}He,t)N = Z nuclei	0	1

[a]Relative cross section normalized to 1 for the stronger excitation.

Figure 4 shows a plot of a complete inelastic proton spectrum[9]. The spectrum is complete in that nearly all protons emitted at 27 degrees by the ^{54}Fe target were observed (some protons are undetected for $E_p \lesssim 2$ MeV). The region of this spectrum above $\gtrsim$ 10 MeV, the region which has been so often studied, contains only $\gtrsim$ 5% of the integrated inelastic cross section. At the highest excitation energy end of the spectrum (lowest emitted proton energy) a large peak is observed which is produced by the nuclear evaporation process. The rather flat region between the evaporation peak and the neutron separation energy is called the nuclear continuum. This region is often described as arising from a pre-equilibrium particle emission process. The topic of this talk and for that matter all the giant multipole resonance work during the past years is the small bump located at ∿ 16 MeV of excitation. Systematic observation of this peak in many nuclei showed that it was not produced by kinematical effects or by GDR excitation. Rather the peak has been shown to arise from excitation of a giant quadrupole resonance (GQR).

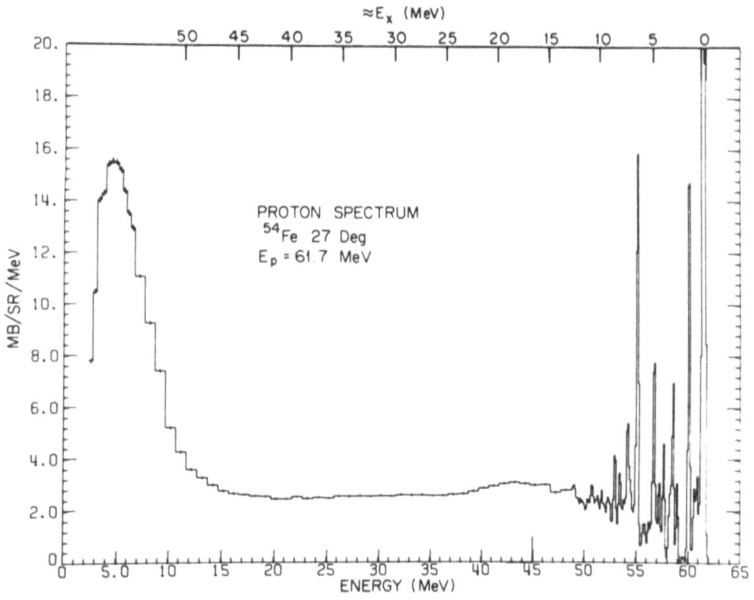

Fig. 4. Proton spectrum at 27° from 62 MeV protons in ^{54}Fe. The energy of the outgoing proton is plotted at the bottom of the figure, while the approximate excitation energy is plotted at the top. Data have been plotted in ∿ 1 MeV-wide bins up to ∿ 49 MeV, then plotted in 50 keV-wide bins. Protons below ∿ 1.5 MeV were not detected in the experiment. The small, broad peak near E_x ∿ 16 MeV is identified as arising from excitation of the giant quadrupole and dipole resonances.

Figure 4 serves to illustrate what is perhaps the most serious problem in inelastic scattering studies of giant resonances--the giant resonance cross section is only a small part of the total continuum cross section. Some assumption must be made about both the magnitude and shape of the continuum which lies under the resonance peak. It is assumed that the peak cross section does not mix with the underlying continuum and the peak is "stripped off" of the continuum by extrapolating the continuum magnitude and shape from higher excitation energies. I feel that cross sections for the giant resonances cannot be obtained from inelastic scattering with less than a 15-20% absolute uncertainty.

Figure 5 shows that proper selection of the reaction and

incident projectile energy can enhance the resonance peak. This spectrum of ^{120}Sn was obtained using 152 MeV alpha particles[10] and shows a resonance cross section nearly equal to the cross section of the underlying continuum.

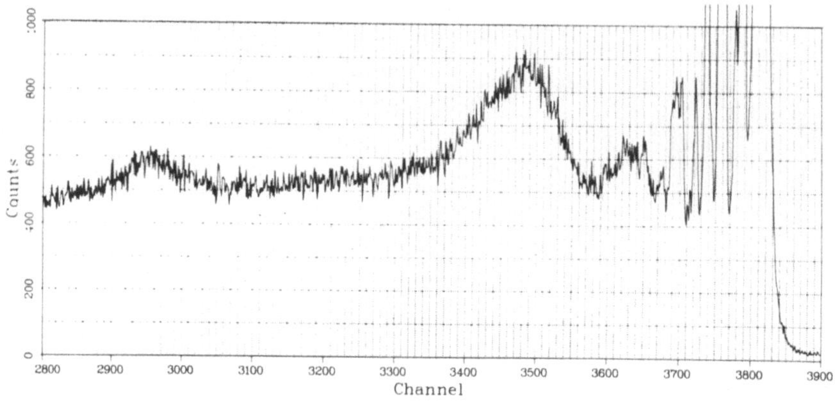

Fig. 5. Spectrum at 12 degrees from the $^{120}Sn(\alpha,\alpha')$ reaction using 152 MeV incident alpha particle (ref. 10). The broad peak observed near channel 3485 arises from excitation of giant resonances, mostly the GQR located at 13.3 MeV of excitation. The elastic scattering peak is in channel 3805.

Only a portion of the total inelastic proton spectrum from several nuclei is shown in figure 6, where the spectra are plotted to emphasize the resonance structure. In each case, a smooth curve has been drawn through the data to guide the eye. The peak energy, indicated by the arrow, is found to change with nuclear mass as $\sim 63 \times A^{-1/3}$. The inelastic proton reaction is expected to excite the isovector GDR, and in each of the cases shown, the GDR energy is known to be 2-3 MeV below the centroid energy of the broad peak.

Giant resonance spectra obtained on many nuclei using the (α,α') reaction are shown in figure 7. These data[11] are from Texas A and M. The (α,α') reaction should excite the GDR to only a negligible extent. A peak is again found in each of the nuclei at an energy of $\sim 63 \times A^{-1/3}$ MeV. Data for nuclei with mass less than 40 are not shown here. Those nuclei have presented a special problem that I will discuss later.

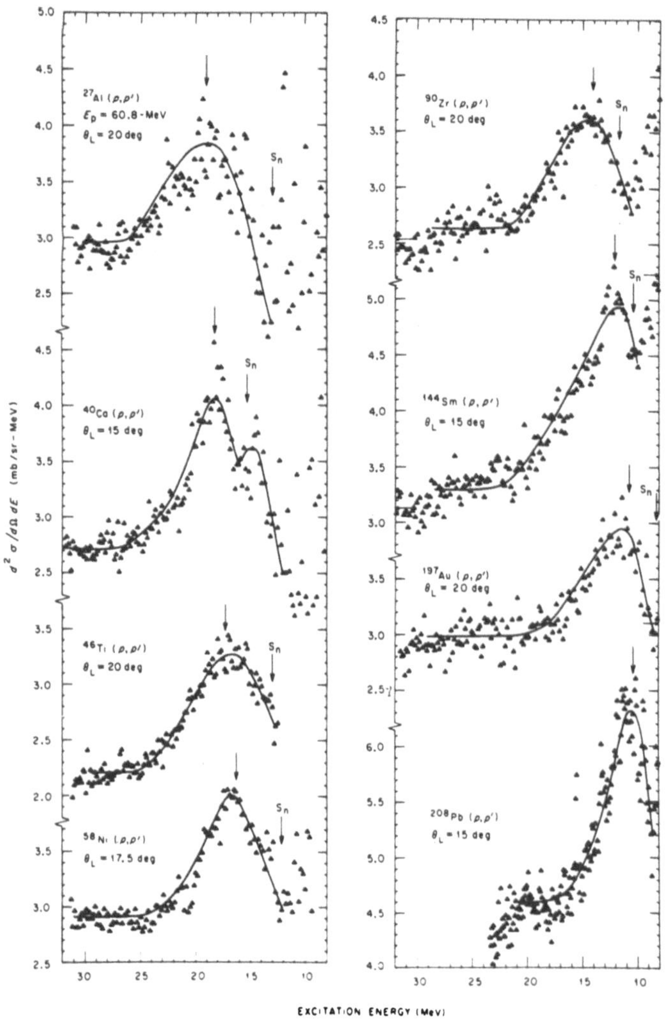

Fig. 6. Inelastic proton spectra from 61 MeV protons incident on several targets (ref. 1). The neutron separation energy and the energy 63 x $A^{-1/3}$ MeV are marked with arrows.

Figure 8 shows the angular distributions for the resonance peak from the (α,α') reaction on several nuclei. In each case, the L=2 DWBA calculation provides good agreement with the measured cross sections. The percentage of the L=2, T=0 EWSR depleted in each nucleus as derived from the measured cross sections, are shown on the figure. These numbers should be considered to have an uncertainty of 15-20%. As is seen, the EWSR strength depleted in the resonance peak varies from ∿ 45% in ^{40}Ca to essentially 100% for heavy nuclei. I will return to systematic trends later.

While the experimental evidence for a GQR in nuclei having A $\gtrsim$ 40 was rather quickly established, evidence for the GQR in lighter nuclei was more difficult to obtain. In fact, for a period of 2 or 3 years there existed several conflicting reports of observation or non-observation of the GQR in light nuclei.

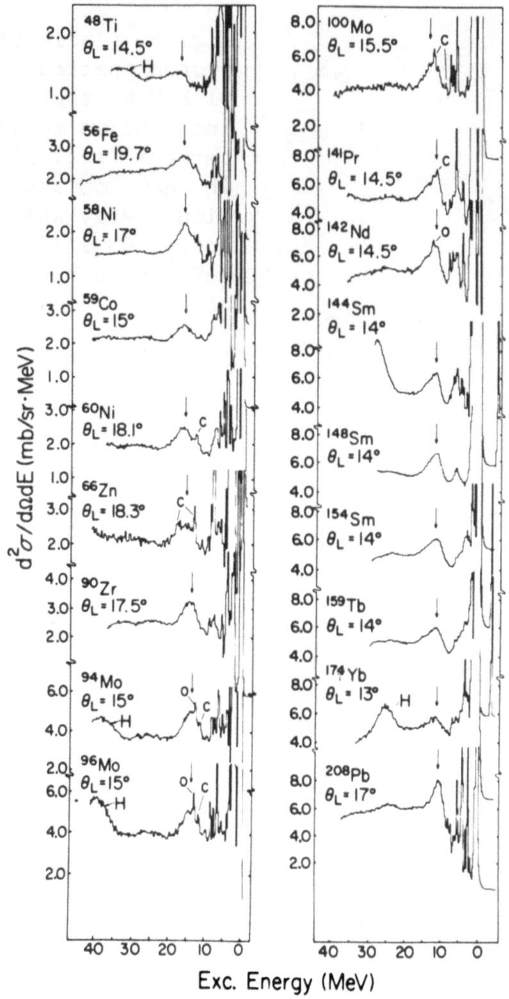

Fig. 7. Inelastic spectra from 96 and 115 MeV alphas incident on several targets (ref. 11). The arrow indicates the excitation energy 63 x $A^{-1/3}$ MeV. Hydrogen, carbon and oxygen contaminants are labelled.

Figure 9 snows several spectra from the reaction $^{28}Si(\alpha,\alpha')$ with E_α = 120 MeV[12]. With an energy resolution of $\sim$ 90 keV one finds a large number of peaks in the excitation region where the GQR is expected. Angular distributions for the observed peaks show that most of the structure arises from L=2 excitation. However, a few of the peaks are identified as L=3 or 4. It was clear from this measurement and high resolution measurements[13] in ^{16}O that the GQR strength in light nuclei is fragmented into many peaks having rather narrow widths. This contrasts sharply with the single broad GQR peak consistently observed for heavier nuclei.

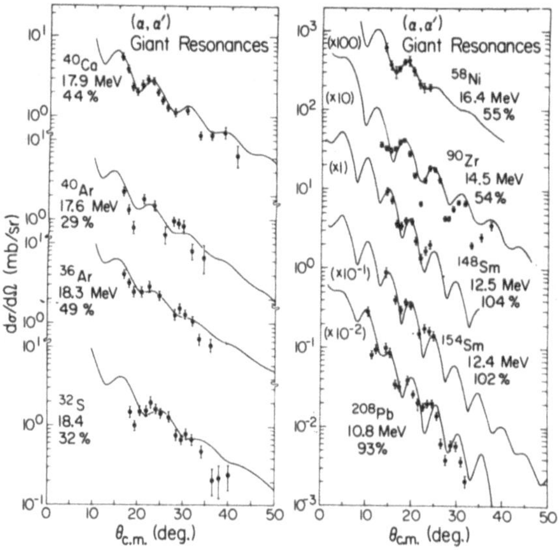

Fig. 8. Angular distributions for the giant resonance peak from the (α,α') reaction at 96 and 115 MeV (ref. 11). The percentage of the T=0, L=2 EWSR strength depleted in the resonance is shown for each nucleus.

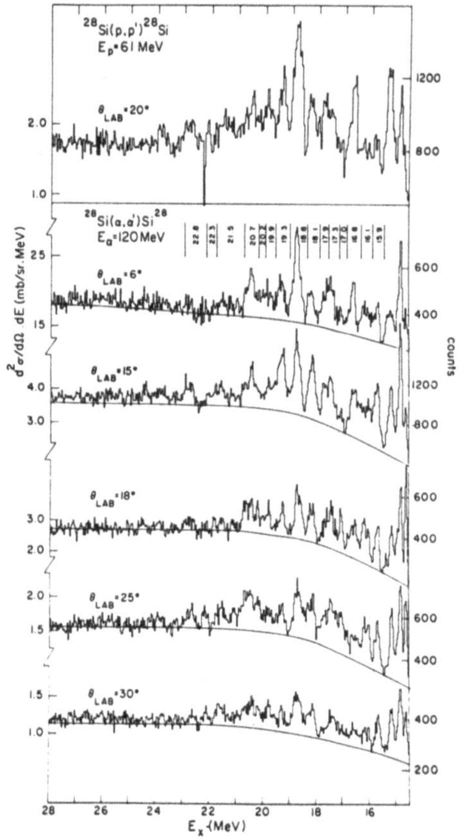

Fig. 9. Spectra of inelastic proton (top) and alpha particle scattering on ^{28}Si (ref. 12).

Figure 10 shows that a similar situation exists for ^{24}Mg. The top two spectra are from inelastic proton and alpha scattering[14]. As is the case for ^{28}Si, these spectra show considerable structure, most of which is L=2, in the GQR region of excitation. It is interesting to note the almost perfect correspondence between the proton and alpha spectra. This is unexpected since it had been calculated[4] that ∿ 50% of the cross section in the proton spectrum should arise from excitation of the GDR which is not excited in alpha scattering. This apparent contradiction led to recent re-evaluation of the (p,p') calculations for GDR excitation showing the early estimates to be too large by about a factor of three. The lower part of figure 10 shows the E2 strength distribution obtained[15] from the ^{20}Ne(α,γ) reaction. All of the structure observed in the capture measurement is also seen in the inelastic scattering measurements.

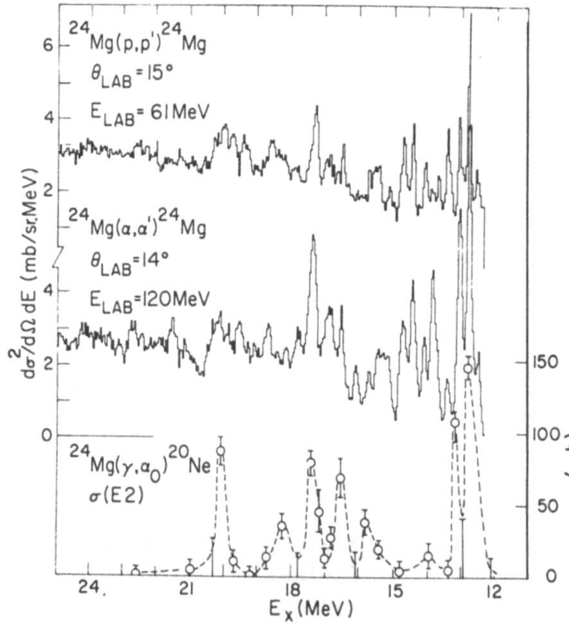

Fig. 10. Spectra from the (p,p') (top) and (α,α') (middle) reactions on ^{24}Mg at 15 deg and 14 deg respectively (ref. 14). The bottom spectrum shows the E2 strength in ^{24}Mg obtained from the ^{20}Ne(α,γ_0) reaction (ref. 15). Most of the peaks seen in the (α,α') spectrum are L=2, however, a few arise from excitation of other multipolarities.

Further insight into the structure of the GQR in light nuclei is being obtained through measurements of the decay properties of the resonance. Figure 11 shows results[16] from recently published work on the decay of ^{16}O by the Heidelberg-Jülich group. The top figure shows the singles (α,α') spectrum from ^{16}O obtained using 155 MeV alphas. The structure centered at ∿ 21 MeV of excitation is the GQR. (Higher resolution measurements[13] on ^{16}O show that part of the resonance structure observed in this region is not L=2). Part b) of the figure shows the inelastic alphas that are in coincidence with all Z=1 and Z=2 decay particles and those inelastic alphas (dark shading) in coincidence with protons only. Figures 11c and 11d show the inelastic alphas in coincidence with decay alphas to the residual nucleus ground state and first excited state respectively. From these results it is clear that the GQR in ^{16}O decays predominantly by alpha emission. In fact, alpha decay provides $\sim$ 80% of the T=0, L=2, EWSR depletion deduced in the ^{16}O GQR singles measurements. Studies such as these offer a possible technique by which some detailed structure information of the GQR states may be obtained.

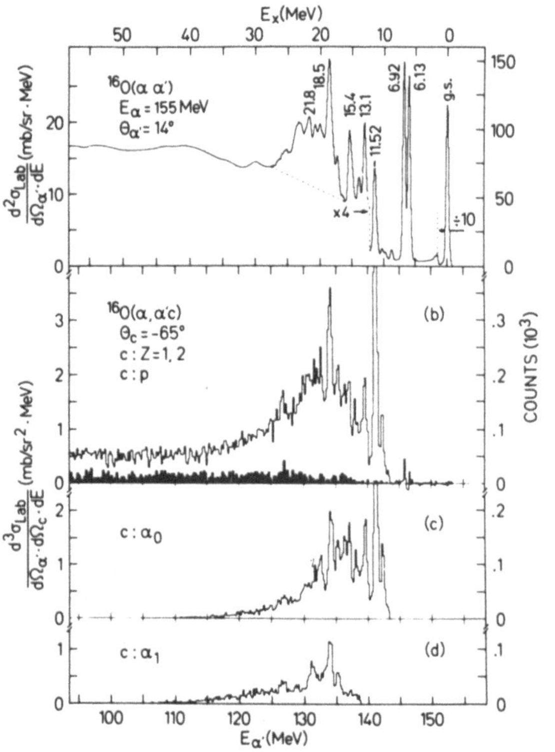

Fig. 11. Spectrum of inelastically scattered alpha particles at 14 deg: a) singles, b) in coincidence with all Z=1 and Z=2 decay particles and with protons only (dark curve), c) in coincidence with alpha-particles decaying to the ground state, d) in coincidence with alpha-particles decaying to the first excited state of ^{12}C (ref. 16).

The next few figures show systematics of the giant quadrupole resonance. The values shown represent averages of the many available measurements or that measurement which I feel provides the best available result.

Figure 12 shows the energy of the T=0, GQR as a function of nuclear mass number. The solid curve is the energy, 63 x $A^{-1/3}$ MeV.

In general, the data agree with this energy rather well. However, there is a definite trend at higher mass for the energy to be above the $63 \times A^{-1/3}$ MeV systematic and for the energy to be below $63 \times A^{-1/3}$ MeV for A $\lesssim$ 40. You will recall that the GDR follows the systematic energy of $\sim 77 \times A^{-1/3}$ MeV for A $\gtrsim$ 50.

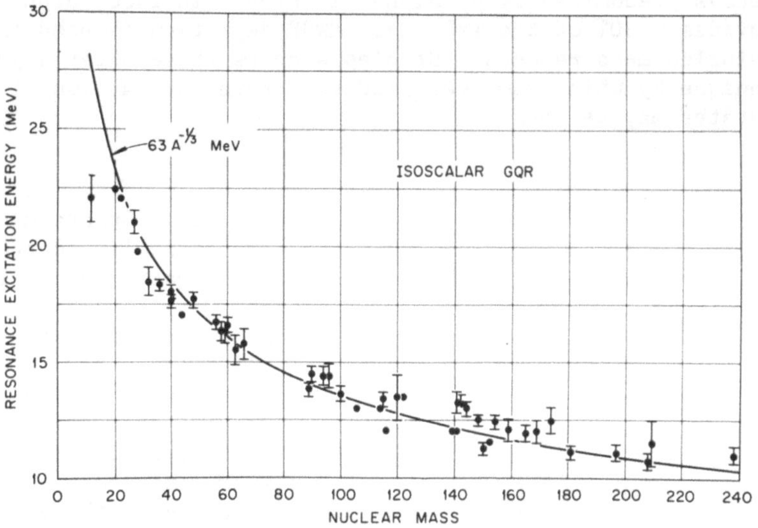

Fig. 12. Excitation energy of the isoscalar GQR plotted against nuclear mass number. The solid curve represents the energy $63 \times A^{-1/3}$ MeV.

The width of the GQR as a function of nuclear mass number is shown in figure 13. It is found that the resonance width is narrowest for closed-shell nuclei and that the width generally decreases with increasing nuclear mass. This trend is similar to that established for the GDR.

The percentage of the T=0, E2 EWSR strength depleted in the GQR is plotted on figure 14. These values do not include contributions to the L=2 EWSR strength depleted in bound state excitations. A trend to larger sum-rule depletion with increasing mass is clearly evident. For nuclei having A $\gtrsim$ 100 nearly 100% of the EWSR is found in the GQR peak (the $2\hbar\omega$ transitions), while for the s-d shell nuclei 30-40% EWSR is typically found in the GQR states. However considerable EWSR strength is located in the low-lying 2^+ states of light nuclei, so that the sum of the GQR strength and low-lying quadrupole strength exhausts 100% of the T=0, L=2 EWSR in light as well as heavy nuclei.

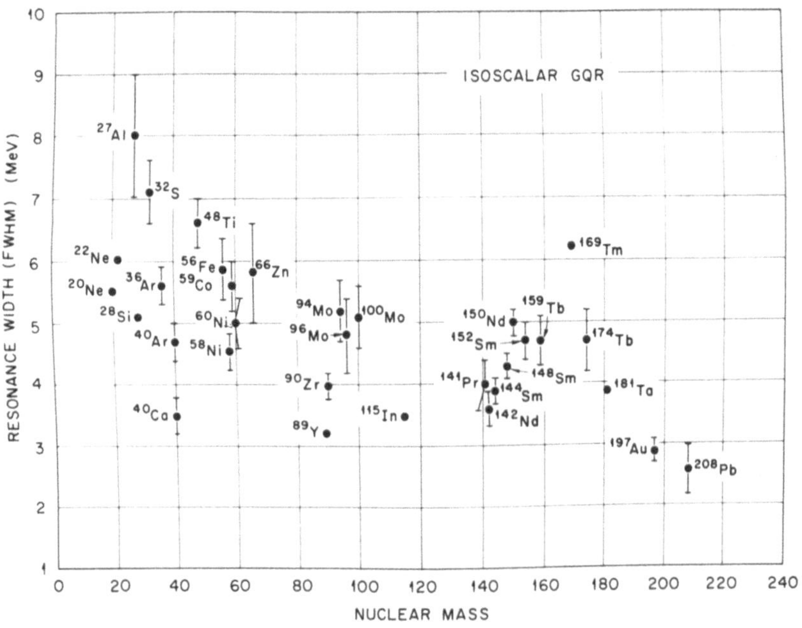

Fig. 13. Widths of the T=0 GQR plotted against nuclear mass number.

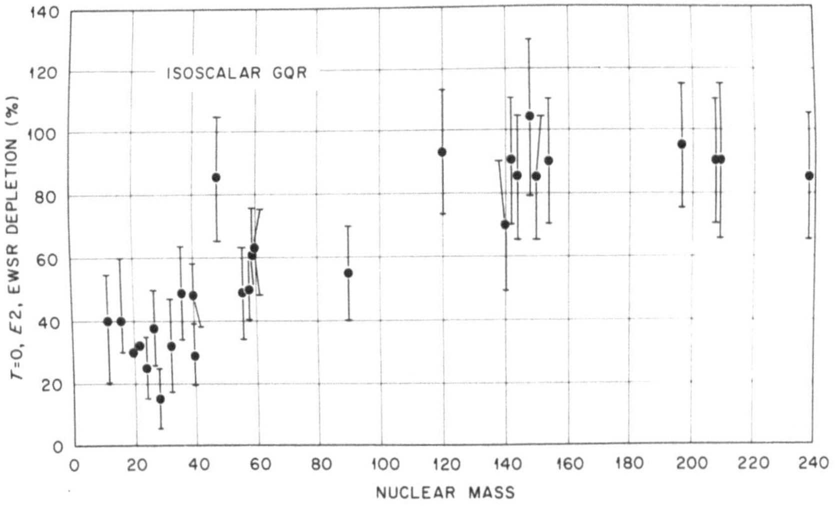

Fig. 14. Percentage of the T=0, E2 EWSR strength in the GQR plotted against nuclear mass number. The values do not include the EWSR depletion in low-lying 2^+ states.

It is in the measurement of the strength in the giant resonance cross section that use of inelastic hadron scattering is most uncertain. A shape and magnitude must be assumed for the nuclear continuum under the resonance peak, there is the additional uncertainty in the transformation of the deduced deformation parameter β to a transition rate, and the DWBA calculated cross sections are model and parameter dependent. It is significant to note that most of the hadron measurements agree with each other and generally agree with the available electron scattering data (within the rather large uncertainties inherent to both measurements).

Although the most thoroughly studied of the new giant resonances is the GQR (in fact, some would argue the GQR is the only new giant resonance for which there is convincing evidence), the resonance which has generated the most interest is the monopole. Observation of the monopole, "breathing", or compressional mode of nuclear excitation is of special significance because knowledge of its excitation energy could provide direct information on the nuclear compressibility. During the past few years several candidates for the E0 resonance have appeared but most have not withstood the test of further measurements. These early measurements are discussed in reference 1. It is important to note that some early indirect evidence for an E0 resonance[17,18] placed the resonance at an excitation energy of $\sim 80 \times A^{-1/3}$ MeV, a value in agreement with the recent more direct observations I will describe.

Figure 15 shows spectra[19] at two angles from the ^{208}Pb(α,α') reaction for 120 MeV incident alphas. The 14-degree spectrum shows the presence of what appears to be two broad peaks (indicated by the dashed lines) in the giant resonance region of the spectrum. The larger peak located at 11 MeV ($63 \times A^{-1/3}$ MeV) is the now familiar GQR peak. The smaller peak is located at 13.9 MeV or $\sim 80 \times A^{-1/3}$ MeV. This peak is near the energy of the GDR in ^{208}Pb (13.6 MeV), however the isovector GDR will not be excited by the (α,α') reaction with nearly enough cross section to account for this new peak. A similar peak was also found[19] in ^{206}Pb, ^{209}Bi, and ^{197}Au.

It was thus established that a here-to-fore not directly observed resonance peak was located at $\sim$ 14 MeV for nuclei in the lead region. The obvious question was what is the nature of the peak? Comparison of the 12 deg and 14 deg spectra shown on figure 15 shows that the ratio of GQR peak cross section to "new" peak cross section changes considerably in only two degrees. This observation suggests that the two peaks have different multipolarities. In addition, if the new peak is 2^+ then the L=2, T=0, EWSR would be significantly overdepleted; an unhappy situation theoretically. Figure 16 shows the angular distribution[19] of the 13.9

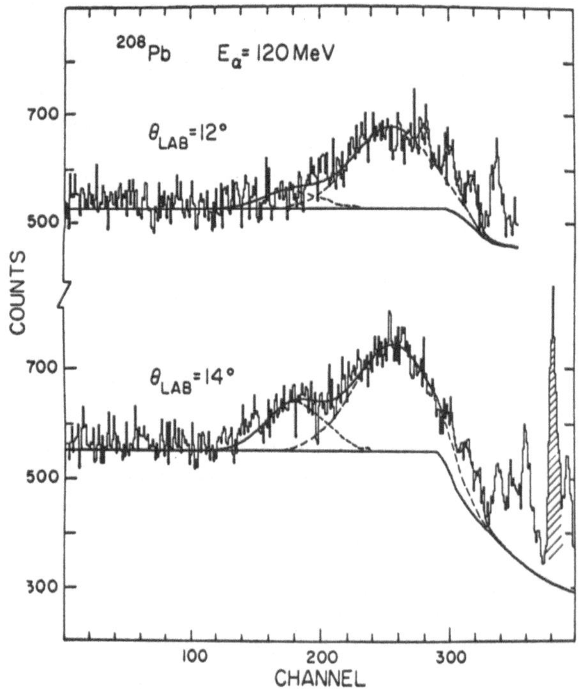

Fig. 15. Giant resonance spectra at 14 deg and 12 deg from the $^{208}Pb(\alpha,\alpha')$ reaction for 120 MeV incident alpha particles (ref. 19). The broad peak observed at both angles in channel 240 occurs at $\sim$ 11 MeV and is the GQR. In the 14 degree spectrum a second broad peak is observed at channel 180, or $\sim$ 13.9 MeV. This second peak is now identified as the E0 giant resonance.

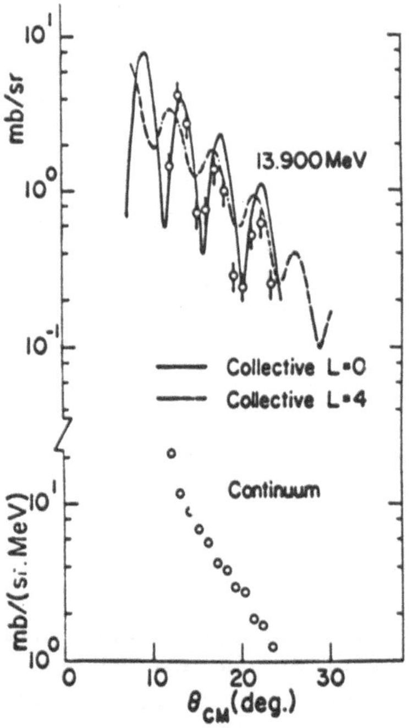

Fig. 16. (Upper) Angular distribution for the 13.9 MeV peak observed in the $^{208}Pb(\alpha,\alpha')$ reaction for E_α = 120 MeV (see figure 15). The dashed curve is the DWBA calculation for L=4 (nearly identical for L=2) and the solid curve is for L=0. (Lower) Angular distribution for the nuclear continuum near the giant resonance peaks.

MeV peak in ^{108}Pb compared with DWBA calculations for L=0 and L=4 (nearly identical with L=2). It is an unfortunate fact, as can be seen in figure 16, that in the angular region studied in reference 19 the L=0 and L=4 (and L=2) angular distributions are in phase. The only difference is in the larger peak to valley ratio for the L=0 calculation than for the higher multipoles. In fact, the L=0 calculation fits the data points better than the L=4 (or L=2) calculation, but the differences depend on only a few points and therefore the evidence is not very strong. If the peak is in fact E0 then comparison with the L=0 calculation indicates that $\sim$ 90% of the T=0, L=0, EWSR would be accounted for[19].

While the L=0 and L=2 angular distributions are nearly identical at angles greater than $\sim$ 10 degrees (for $\sim$ 120 MeV alphas) the two angular distributions are considerably different (nearly a factor of ten at some angles) at smaller angles. Thus, measurements, all be they very difficult, of the giant resonances at very small angles could more convincingly provide an L=0 assignment to the $\sim$ 14 MeV resonance. Figure 17 shows results from just such measurements[20] made at Texas A and M using 96 MeV incident alpha particles on ^{208}Pb and ^{144}Sm. The small angle spectra on ^{208}Pb show the same two peaks, 11.0 and $\sim$ 14 MeV that were seen in the previous work at 120 MeV. However, as demonstrated in figure 17 the assignment of the 13.9 MeV peak as an E0 excitation can be made with far more confidence on the basis of the small angle measurements. For ^{144}Sm the small angle measurements showed the GQR peak at 12.4 MeV and a second peak with L=0 angular distribution located at 15.1 MeV.

Thus, the location of the E0 resonance has been rather well established in several nuclei in the lead region and in ^{144}Sm. However, for lighter nuclei inelastic alpha scattering measurements for energies near 100 MeV encounter serious spectral contamination[21] which interferes with the observation of the monopole and quadrupole resonances. This contamination arises from the fact that the alpha particle may pick up a neutron (similar arguments hold for proton pickup) form ^{5}He and subsequently decay back into a neutron and alpha particle. The alphas from the ^{5}He decay form a broad spectral distribution which may obscure or interfere with (depending on the incident alpha energy) observation of the giant resonance peaks. Since no similar problem occurs for incident protons, (p,p') results may be useful for observations of the monopole resonance in lighter nuclei.

As pointed out earlier, two giant resonance peaks have always been observed in the inelastic proton spectra, the GQR and a second peak located at the energy of the GDR. However, the alpha particle results show the monopole resonance in heavy nuclei to be also located at an energy near the GDR energy. The monopole resonance must be excited by protons as well as alphas so that the (p,p') spectra must now accommodate three (at least) resonances, the GQR, GDR and the giant monopole resonance (GMR).

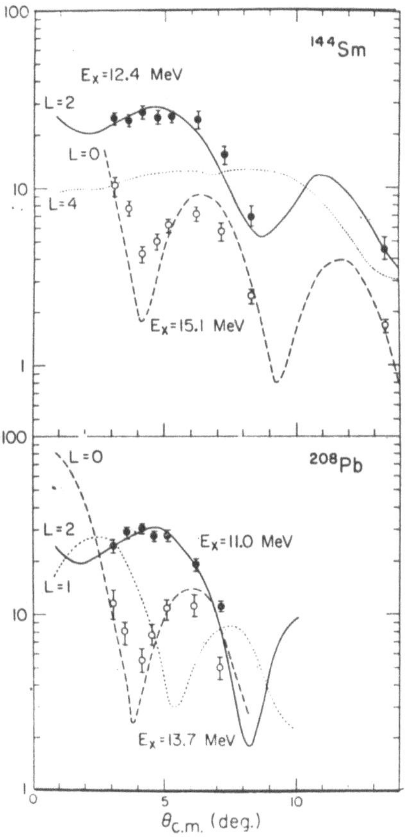

Fig. 17. Angular distribution for two peaks observed in the giant resonance spectra of ^{208}Pb and ^{144}Sm. The results are from the (α,α') reaction for E_α = 96 MeV. For both nuclei the lower excitation peak is the GQR while the angular distribution of the higher excitation peak is well described by the L=0 calculations and is thus, identified as the E0 giant resonance (ref. 20).

Although most calculations of the (p,p') cross section for the GDR and GQR agreed well with the data there were significant exceptions. For example, it was shown[22] in 1974 that (p,p') spectra on ^{144}Sm and ^{154}Sm which showed two peaks, a GQR and a peak at the GDR of ^{144}Sm, were not consistent with photonuclear measurements of the GDR spectra of the same nuclei. In addition, the cross section for the assumed GDR peak in ^{144}Sm was shown to be considerably larger than the DWBA calculations could account for. Further, as I mentioned earlier, recent (α,α') measurements on sd-shell nuclei[13,14] when compared to (p,p') measurements on the same nuclei indicated that the cross section for proton excitation of the GDR must be considerably smaller than had been predicted[4].

These observations have led to a re-evaluation[5] of the DWBA calculations for proton excitation of the isovector GDR. The largest change in the new calculations is in the T=1 interaction which has been a major source of uncertainty in previous calculations. The new calculations use energy-dependent global optical potentials derived from a recent unified analysis[23] of (p,p), (n,n) and (p,n) data. The real and imaginary strengths now used are smaller than those used previously and lead to considerably lower T=1, L=1 cross sections.

Figure 18 shows an example of the recent calculations for 61 MeV protons on ^{208}Pb for L=0, 1 and 2 excitations. For angles between $\sim$ 5 deg and 22 deg the L=0 and L=2 excitations normalized to sum rule depletions as indicated in the figure, have nearly the same cross section while at larger angles the L=0 cross section

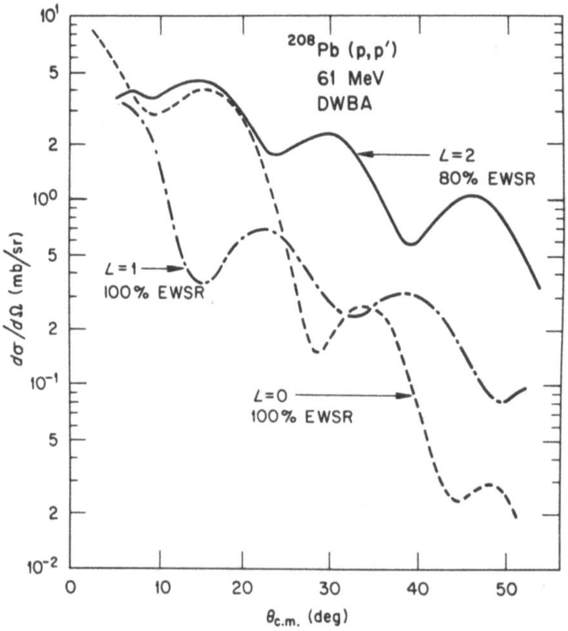

Fig. 18. Calculated cross sections for L=0, 1(T=1), and 2 excitations in the reaction ^{208}Pb(p,p) for 61 MeV incident protons.

decreases rapidly with respect to the L=2 cross section. Of particular note is the L=1 cross section which is calculated to be much smaller (up to an order of magnitude) than the L=0 and L=2 cross section for $10^{\circ} \lesssim \theta \lesssim 20^{\circ}$. At these angles, accepting the correctness of the calculations, any strong peak at the GDR energy would arise predominantly from the GMR not the GDR. At angles near 20 degrees the L=0 and L=1 cross sections are more nearly the same, a fact we have used to indicate the correctness of these calculations.

A spectrum at 20 degrees from the reaction ^{144}Sm(p,p') is shown on figure 19. The broad resonance structure appears rather clearly to be composed[5] of two peaks as indicated on the figure. The lowest excitation peak falls at ∿ 12.8 MeV and is the GQR while the second peak is located at ∿ 15.5 MeV, the GDR energy. Cross sections[5] for the two peaks are shown in figure 20 along with DWBA calculations for L=2, 1 and 0 excitations. The 12.8 MeV peak is well described by the L=2 calculation normalized to 80% T=0, L=2, EWSR depletion in agreement with previous measurements . For the 15.5 MeV peak, the cross sections at 15 and 20

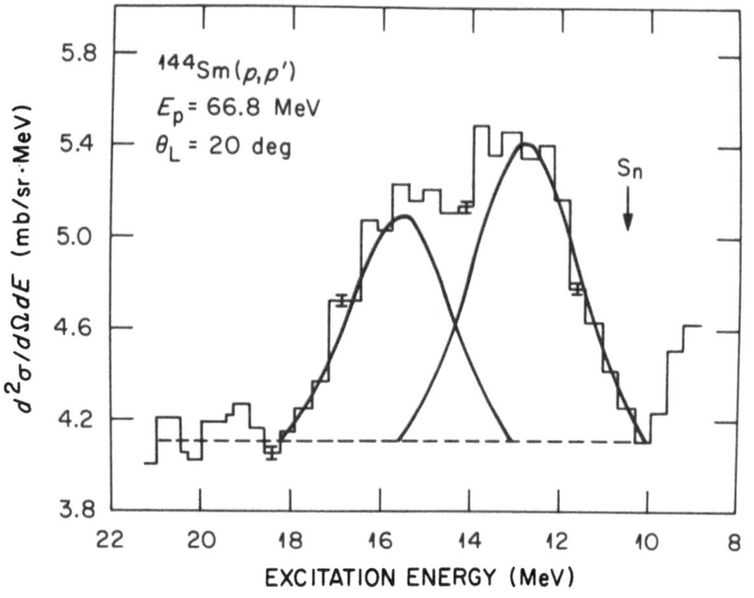

Fig. 19. Giant resonance spectrum observed in the ^{144}Sm(p,p') reaction at 20 deg; E_p = 66.8 MeV. The data are shown as a histogram. Error bars represent statistical uncertainties only. The resonance has been decomposed into two separate peaks as indicated.

degrees are much larger than can be accounted for by the L=1 (T=1) calculation normalized to 100% of the EWSR. At those angles the L=0 calculation agrees very well with the data. However, the cross sections at 25 and 30 degrees are too large to be explained by the L=0 calculations alone, but a sum of the L=1 and L=0 calculations provides good agreement with the measurement. These results indicate that the 15.5 MeV peak is composed of both the GMR, which dominates the smaller angles and the GDR which is dominant beyond 30 degrees. The peak energy, width and EWSR are in excellent agreement with the results from the small angle (α,α') measurements[20]. Further indication that the 15.5 MeV peak cross section is dominated by the monopole at forward angles but shared nearly equally by the GMR and GDR at 25 degrees comes from comparison[5] of the ^{144}Sm and ^{154}Sm (p,p') data.

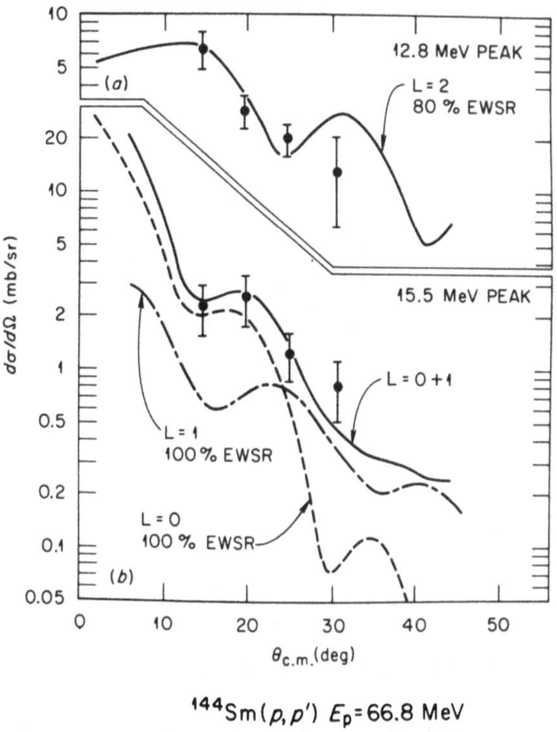

Fig. 20. Cross sections for the two peaks in ^{144}Sm compared with DWBA calculations (ref. 5).

We have applied[24] the same analysis technique to (p,p') data on ^{208}Pb, ^{197}Au, ^{120}Sn, ^{90}Zr, ^{58}Ni and ^{40}Ca in order to establish systematics for the GMR. Figure 21 shows spectra for three of these targets for angles where the L=0 cross section is expected to be very dominate. As was the case for the samarium nuclei we have assumed the broad, asymmetric, resonance peak to be composed of two peaks, a symmetrically shaped GQR and a second higher excitation peak. In the case of ^{58}Ni three peaks are included in the analysis to account for the presence of a now well studied but not so well understood 13.6 MeV resonance[25].

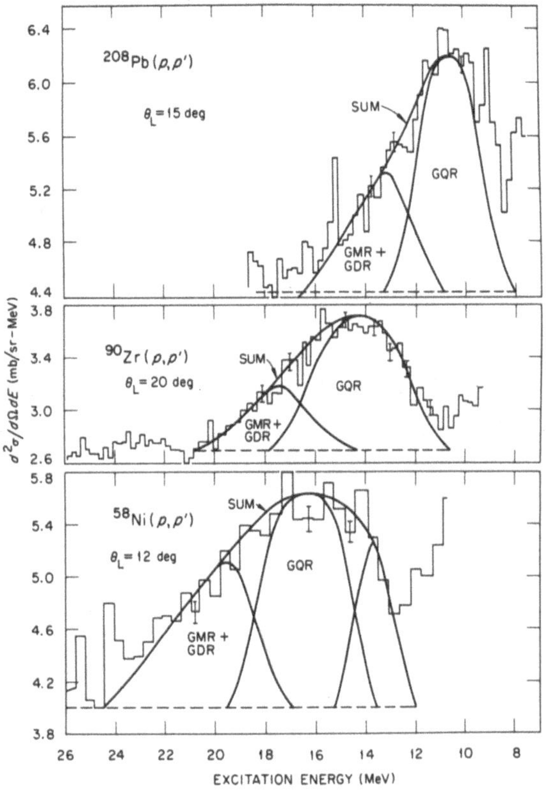

Fig. 21. Giant resonance spectra from the 60 MeV (p,p') reaction in ^{208}Pb, ^{90}Zr and ^{58}Ni. The resonance structure is assumed to be composed of two separate peaks (three for ^{58}Ni) as shown by the solid curves and described in ref. 24.

The results of our analysis are shown on figure 22 and provided us with an unexpected conclusion. For nuclei with atomic number greater than $\sim$ 90 we find a rather compact E0 giant resonance located at $\sim 80 \times A^{-1/3}$ MeV and accounting for nearly 100% of the T=0, L=0, FWSR. However, for the two lighter nuclei we have studied the situation is quite different. In ^{58}Ni although we find evidence for a peak at $\sim 80 \times A^{-1/3}$ MeV the peak cross section contains only 30 ± 10% of the GMR, EWSR. For ^{40}Ca, the magnitude of the peak observed at $\sim 80 \times A^{1/3}$ MeV can be completely accounted for by excitation of <u>only the GDR</u>.

What happens to the E0 strength? Assuming the DWBA model for the breathing mode is correct for light as well as heavy nuclei, the most likely explanation is that the E0 strength becomes highly fragmented and spread out in light nuclei. If this is the

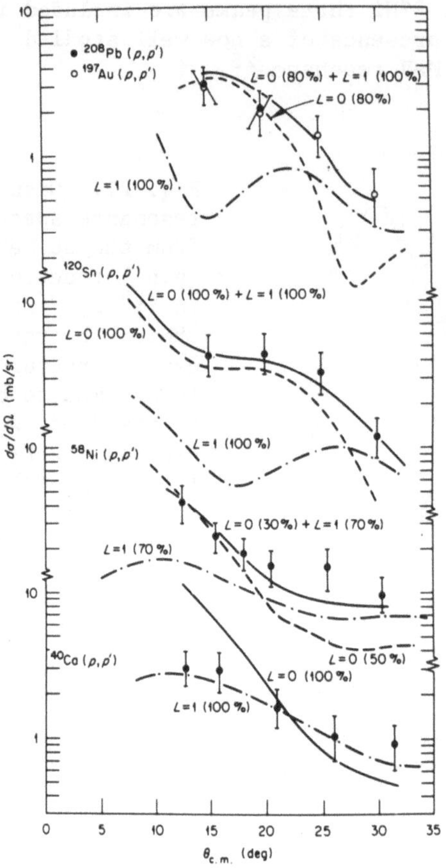

Fig. 22. Angular distributions for the GMR-GDR peak from the 60 MeV (p,p') reaction on several nuclei compared with DWBA calculations. The fraction of the EWSR strength used for normalization of the calculated cross section is shown on the curves (ref. 24).

case then observation of this strength would be very difficult in the inelastic scattering measurements because of the large nuclear continuum. It remains to be seen whether calculations support this possible explanation.

Figure 23 provides a summary of the GMR observations available at this time. The peak energies agree quite well with the systematic energy 80 x $A^{1/3}$ MeV. Table 3 lists the parameters of the monopole resonances that have been observed.

As I pointed out earlier, an important question is what value of the nuclear compressibility do the GMR energies yield? For an excitation energy of 80 x $A^{-1/3}$ MeV the liquid drop model yields a value of $\sim$ 200 MeV for the compressibility. However, no allowance for Coulomb, surface and neutron-excess effects have been made. Although theoretical estimates of these effects are

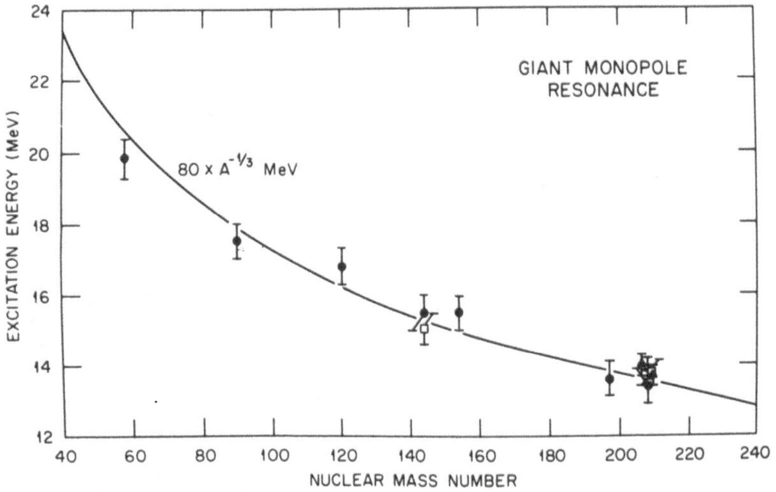

Fig. 23. Excitation energy (MeV) of the GMR for nuclei in which observations have been made. Circles - 60 MeV (p,p'), ref. 24; squares - 96 MeV (α,α'), ref. 20; triangles - 120 MeV (α,α'), ref. 19. The solid curve represents the energy 80 x $A^{-1/3}$ MeV.

very uncertain, at least one approach[26] yields compressibilities as small as 150 MeV, based on a GMR energy of 80 x $A^{1/3}$ MeV.

Although I spent the opening minutes of this discussion describing reasons to expect a large variety of high excitation energy giant resonances, I have spent the bulk of the lecture describing only two "new" giant resonances, the isoscalar quadrupole and monopole. This is justifiable since most of the data and perhaps most of the interest has been generated over these resonances. It is however, important to establish the existence of other giant resonances. Where are the isovector modes, the spin-flip modes, the high-L resonances? There are many resonances yet to observe. The results of some of our theoretical colleagues lead us to think we might obtain spectra as shown in figure 24 for ^{208}Pb. This figure shows the spectrum of states predicted in RPA calculations[27] published by Ring and Speth a few years ago. Their calculations indicate 2^+ states located in the bound state region and at ∿ 11 MeV and 19 MeV. The predicted 11 MeV state agrees well with the measured energy of the isoscalar GQR in ^{208}Pb. The calculations also predict L=3 continuum strength at ∿ 21 MeV and L=4 strength at ∿ 11 MeV and at ∿ 24 MeV.

Do we have any evidence for these higher L resonances and isovector resonances? The answer is - only a little. Figure 25 shows spectra from the ^{208}Pb(e,e') reaction[28] for several momentum

TABLE 3. GMR Parameters

Nucleus	Excitation Energy (MeV)	Width (MeV)	% EWSR	Reaction	Energy	Reference
^{209}Bi	13.7 ± 0.3	2.5 ± 0.4	100 ± 25	(α,α')	120 MeV	19
^{208}Pb	13.9 ± 0.3	2.5 ± 0.6	110 ± 25	(α,α')	120 MeV	19
	13.7 ± 0.4	3.0 ± 0.5	105 ± 20	(α,α')	96 MeV	20
	13.4 ± 0.5	3.0 ± 0.5	90 ± 20	(p,p')	61 MeV	24
^{206}Pb	14.0 ± 0.3	2.5 ± 0.4	100 ± 25	(α,α')	120 MeV	19
^{197}Au	13.6 ± 0.5	3.0 ± 0.5	100 ± 25	(p,p')	61 MeV	24
^{154}Sm	15.5 ± 0.5	2.5 ± 0.5	100 ± 25	(p,p')	67 MeV	5
^{144}Sm	15.1 ± 0.5	2.9 ± 0.5	100 ± 20	(α,α')	96 MeV	20
	15.5 ± 0.5	2.5 ± 0.5	100 ± 25	(p,p')	67 MeV	5
^{120}Sn	16.8 ± 0.5	3.5 ± 0.5	100 ± 25	(p,p')	61 MeV	24
^{90}Zr	17.5 ± 0.5	3.0 ± 0.5	60 ± 25	(p,p')	61 MeV	24
^{58}Ni	19.8 ± 0.5	3.5 ± 0.5	30 ± 10	(p,p')	61 MeV	24

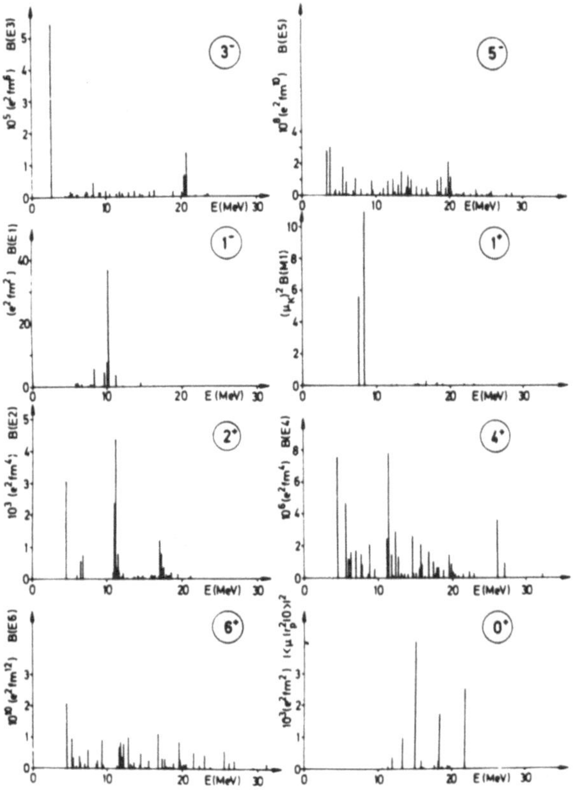

Fig. 24. Spectrum of various J^{π} states in ^{208}Pb predicted from the RPA calculations of ref. 27.

transfers. The broad downward sloped band is the assumed continuum underlying the giant resonances and the shaded peak is from GDR excitation. The structure below the GDR, at ∿ 11 MeV, is from GQR excitation. The authors report the broad peak at ∿ 22 MeV to have an L=2 angular distribution and attribute the peak to excitation of the isovector GQR. The excitation energy agrees with the predicted 2^+ state in the RPA calculations[27] shown on figure 24. The angular distribution for the 19 MeV peak is found by the authors to be L=3, thus this peak is assumed to arise from excitation of an isoscalar E3 giant resonance. Note that the energy of the E3 resonance also agrees with the RPA calculations.

A few other observations of T=1 GQR strength have been reported[1] in other (e,e') measurements. The resonance energies reported[1] in these measurements cluster around the value $120 \times A^{-1/3}$ MeV. EWSR depletion of 70%-100% are reported[1] for the T=1, GQR in the heavier nuclei.

Even fewer L=3 resonance observations have been made, so few that no systematics can yet be established. In fact, it is fair to say that the existence of a giant octopole resonance (3ħω) has not yet been established.

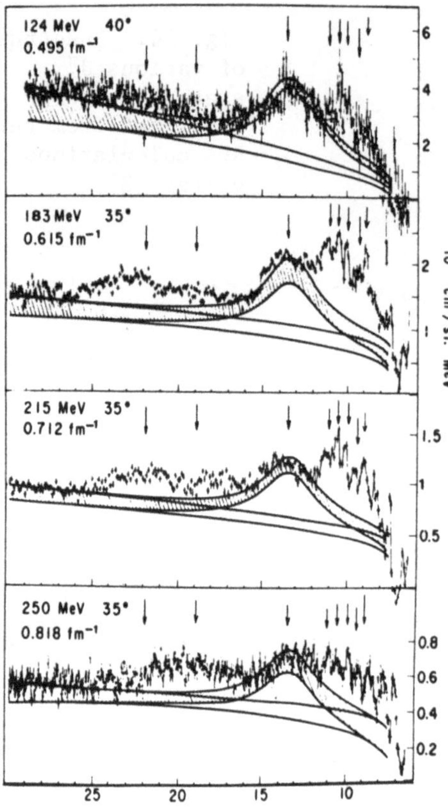

Fig. 25. Giant resonance spectra from the reaction ^{208}Pb(e,e') for several different momentum transfers (ref. 28).

Finally, Halbert et al.[29] have shown by comparing microscopic DWBA model calculations to (p,p') and (α,α') results on ^{208}Pb that the measured giant resonance cross section could be consistent with depletion of 80% of the GQR strength plus ∿ 20% of the L=4 EWSR. Unfortunately, the large uncertainties in the data allow the measurements to also agree with only the L=2 calculation. Nevertheless the possibility still remains that the GQR peak may not be purely L=2 but might be a sum of L=2+4.

Let me summarize the existing experimental situation, as I see it. Certainly, the GQR is now firmly established. I believe there is now good evidence for the existence of a GMR. The present status of the GMR is similar to that of the GQR in 1973 - a few measurements have been made on a few nuclei. What is needed now are systematics generated by a variety of inelastic reactions.

There is some evidence for the existence of the isovector GQR at a systematic energy of $\sim 120 \times A^{-1/3}$ MeV. Very little evidence exists, all from inelastic electron scattering, for a giant octopole resonance, and only "calculational evidence" and speculation exist for an L=4 resonance.

I would like to close with a look at two future efforts in the giant multipole resonance field.

Figure 26 shows a plot of calculated (DWBA) angular distributions for the reaction $^{208}Pb(p,p')$ at E_p = 200 MeV. Unlike the case with 60 MeV protons large differences exist between the

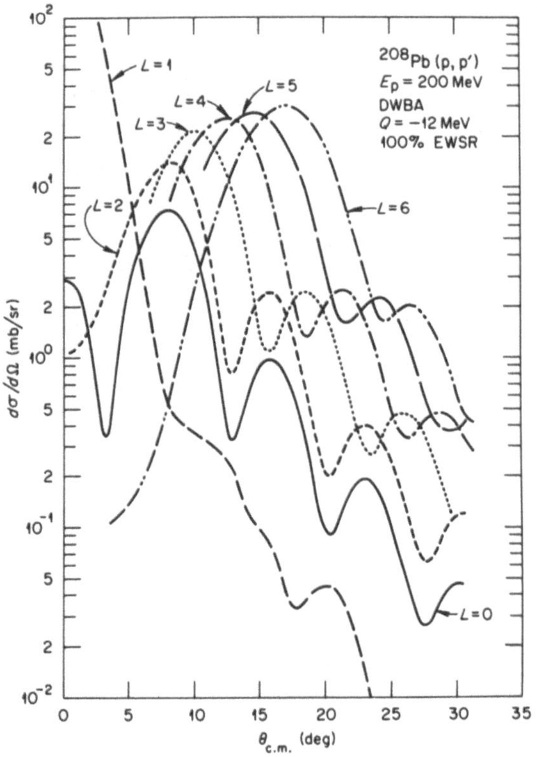

Fig. 26. DWBA calculations for excitation of various multipoles (all isoscalar except L=1 is isovector) by the reaction $^{208}Pb(p,p')$ for E_p = 200 MeV.

angular distributions of neighbouring L-transfer values. For example, the difference between the angular distribution for a pure L=2 giant resonance and a mixed L=2+4 (even for only 15-20% L=4 EWSR) is very large). In addition, the cross sections are considerably larger in the angular distribution maximum than for the same

L-transfer using 60 MeV protons. An additional advantage lies in the fact that the L=1 (isovector GDR) excitation proceeds overwhelmingly by Coulomb excitation which decreases rapidly with increasing angle. For angles greater than ∿ 5 degrees the GDR cross section is predicted to be as much as thirty times smaller than the L=0 cross section. Thus, the monopole should be strongly and clearly excited in the (p,p') reaction at intermediate energies.

Figure 27 shows another possibility for observation of higher L giant resonances. This figure plots cross section versus incident energy for L=2 and L=4 excitation (Q=-12 MeV) in the reaction $^{208}Pb(^{16}O,^{16}O')$. For comparison the cross sections obtainable using 60 MeV protons are also plotted. The exciting aspect to this

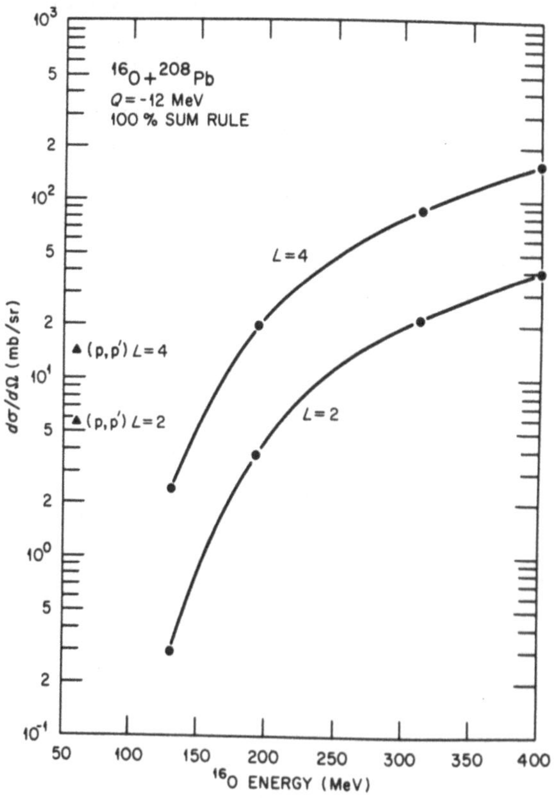

Fig. 27. Maximum cross section calculated for L=2 and L=4 excitations by the reaction $^{208}Pb(^{16}O,^{16}O')$ plotted as a function of incident ^{16}O energy. Also shown for comparison are maximum cross sections for the same excitations induced by 60 MeV proton inelastic scattering.

reaction is the large cross section for exciting high L (L ≳ 3) resonances with the higher energy heavy ions. A factor of ten increase in cross section is realized between the use of 60 MeV protons and 400 MeV ^{16}O beams. In addition, the heavy ions preferentially transfer higher angular momentum. Thus, for 400 MeV ^{16}O while the L=2 cross section is no larger than could be obtained with 150 MeV alpha particles, the L=4 cross section is several times larger. Such heavy ion beams will be available from the Phase I Heavy Ion Accelerator at ORNL and we eagerly await their use.

In my opening remarks I noted that giant resonances are fundamental properties of nuclei. We experimentalists should not limit our observations of such general nuclear properties by limiting our investigative scope through artificial categories such as photonuclear reaction or light-ion inelastic scattering reaction. Rather we must seek out whatever tools become available to elucidate the properties we wish to study. Much has been learned and much will still be learned about giant multipole resonances by the use of many different techniques.

The author would like to acknowledge the contribution of other ORNL Physics Division staff who have contributed to the giant resonance data and calculations shown here: E.E. Gross, D.C. Kocher, D.J. Horen, G.R. Satchler and A. van der Woude and his coworkers at the Kernfysisch Versnelle Institut, Groningen, The Netherlands.

Research is sponsored by the U.S. Department of Energy, under contract W-7405-eng-26 with the Union Carbide Corporation.

REFERENCES

1. F.E. Bertrand, Ann. Rev. Nucl. Sci. 26 (1976), 457.

2. For a recent review of the giant dipole resonance see: B.L. Berman and S.C. Fulty, Rev. Mod. Phys. 47 (1975), 713.

3. For a recent review of results on the M1 giant resonance see: L.W. Fagg, Rev. Mod. Phys. 47 (1975), 683.

4. G.R. Satchler, Nucl. Phys. A195 (1972), 1. G.R. Satchler, Part Nuclei 5 (1973), 105.

5. F.E. Bertrand, G.R. Satchler, D.J. Horen and A. van der Woude, to be published in Phys. Rev.

6. O. Nathan and S.G. Nilsson in "Alpha-, Beta- and Gamma-Ray Spectroscopy", Vol. 1, p. 601, ed. K. Siegbahn, Amsterdam, North Holland (1965).

7. S. Raman and P.H. Stelson, to be published.

8. B.R. Mottelson, Proc. Solvay Conf. Physics, 15th, Brussels (1970), ed. I. Prigogine (Gordon and Breach, New York).

9. F.E. Bertrand and R.W. Peelle, Phys. Rev. C 8 (1973), 1045.

10. Unpublished results obtained at the Indiana University Cyclotron by F.E. Bertrand, D.J. Horen, G.R. Satchler, A. van der Woude, G.T. Emery, D.E. Miller, W. Jones and A.D. Bacher.

11. D.H. Youngblood, J.M. Moss, C.M. Rozsa, J.D. Bronson, A.D. Bacher and D.R. Brown, Phys. Rev. C 13 (1976), 994.

12. K. van der Borg, M.N. Harakeh, S.Y. van der Werf, A. van der Woude and F.E. Bertrand, Phys. Lett. 67B (1977), 405.

13. M.N. Harakeh, A.R. Arends, M.J.A. de Voigt, A.G. Drentje, S.Y. van der Werf, A. van der Woude, Nucl. Phys. A265 (1976), 189.

14. F.E. Bertrand, K. van der Borg, A.G. Drentje, M.N. Harakeh, J. van der Plicht and A. van der Woude, Phys. Rev. Lett. 40 (1978), 635. Proton data: F.E. Bertrand, D.C. Kocher, E.E. Gross and E. Newman, ORNL Report No. 5137, 1975 (unpublished).

15. E. Kuhlman, E. Ventura, J.R. Calareo, D.G. Mavis and S.S. Hanna, Phys. Rev. C 11 (1975), 1525.

16. K.T. Knopfle, G.T. Wagner, P. Paul, H. Breuer, C. Mayer-Boricke, M. Rogge and P. Turek, Phys. Lett. 74B (1978), 191.

17. N. Marty, M. Morlet, H. Willis, V. Comparat, R. Frascaria and J. Kallne, Institute of Nuclear Physics, Orsay, Report No. IPNO-Ph-N-75-11 (1975), unpublished.

18. S. Fukuda and Y. Tarizuka, Phys. Lett. 62B (1976), 146; M. Sasao and Y. Torizuka, Phys. Rev. C 15 (1977), 217.

19. M.N. Harakhe, K. van der Borg, T. Ishimatsu, H.P. Morsch, A. van der Woude and F.E. Bertrand, Phys. Rev. Lett. 38 (1977), 676.

20. D.H. Youngblood, C.M. Rozsa, J.M. Moss, D.R. Brown and J.D. Bronson, Phys. Rev. Lett. 39 (1977), 1188.

21. A. Kiss, C. Mayer-Boricke, M. Rogge, P. Turek and S. Wiktor, Phys. Rev. Lett. 37 (1976), 1188.

22. D.J. Horen, F.E. Bertrand and M.B. Lewis, Phys. Rev. C 9 (1974), 1607.

23. D.M. Patterson, R.R. Doering and A. Galonsky, Nucl. Phys. A263 (1976), 261.

24. F.E. Bertrand, G.R. Satchler, D.J. Horen and A. van der Woude, to appear in Physics Letters.

25. D.C. Kocher, F.E. Bertrand, E.E. Gross, R.L. Lord and E. Newman, Phys. Rev. Lett. 31 (1973), 1070; C.C. Chang, F.E. Bertrand and D.C. Kocher, Phys. Rev. Lett. 34 (1973), 221.

26. C.Y. Wong, Phys. Rev. C 17 (1978), 1832.

27. P. Ring and J. Speth, Phys. Lett. B44 (1973), 477. P. Ring and J. Speth, Nucl. Phys. A235 (1974), 315.

28. M. Nagao and Y. Tarizuka, Phys. Rev. Lett. 30 (1973), 1068.

29. E.C. Halbert, J.B. McGrory, G.R. Satchler and J. Speth, Nucl. Phys. A245 (1975), 189.

MICROSCOPIC THEORIES OF GIANT RESONANCES

K. Goeke

Institut fur Kernphysik

KFA Jülich, D-5170 Jülich, West Germany

CONTENTS

1. INTRODUCTION

A very exciting development in experimental and theoretical nuclear physics has been the discovery and the description of new giant resonances. The most thoroughly studied one is located at an excitation energy of $\sim 63\ A^{-1/3}$ MeV and has been identified as an isoscalar (T=0) giant quadrupole resonance (GQR). Its observation in many nuclei reveals that the GQR represents a general behaviour of nuclei like the giant dipole resonance (GDR) known for many years[1]. Recently there have also been reports of further

---

*Some of the work presented here has been performed in collaboration with B. Castel, A.M. Lane, J. Martorell and P.G. Reinhard.

giant resonances like the isovector GQR and giant E3 resonances. In particular the isoscalar giant monopole resonance (GMR) has been attracting the interest of many physicists. Although the experimental evidence for GMR has probably only very recently been established[2], it has been for many years a playground for theorists predominantly due to its symmetry and to its relation to the nuclear compression modulus, a quantity being intimately related to the properties of nuclear interaction, and linked to experimentally easily accessible properties like nuclear isotope shifts.

During the past four or five years many theoretical microscopic calculations of the energies and the excitation probabilities of various GR have been performed. Although many of these approaches have the same objective, namely to calculate, e.g., the position of a GR with a certain multi-polarity in a certain nucleus, the results sometimes differ tremendously. This is particularly odd for the isoscalar GMR in ^{16}O which has not been fully established experimentally. The theoretical predictions range from 20 to 35 MeV excitation energy. No matter at which position the GMR in ^{16}O will once be observed, the experimentalist will always find its value already predicted (although it is not clear why one cannot compare it with another prediction).

Recently there have been several attempts[3-6] to classify the various theoretical approaches, and to extract clearly the physical meaning of their predictions and their relationship to experimental data. The main emphasis of this work will be placed on reviewing these approaches and to present a unifying view in terms of TDHF which allows to relate virtually all existing microscopic theories to each other in a precise way. Furthermore we will discuss some of the difficulties connected with the presently used interactions, and finally, a short review will be given about the possibilities to describe large amplitude corrections to GR.

2. MICROSCOPIC APPROACHES: A BRIEF REVIEW

All the microscopic approaches to GR make use of a dominating mean field in the nucleus and the corresponding single particle structure and are thus related to theories like TDHF or constrained Hartree-Fock (CHF).

One type of approach assumes GR to be high lying eigenmodes of the system describable in terms of small harmonic vibrations around the stable Hartree-Fock minimum. Theories of this sort include the Random-Phase Approximation, the Tamm-Dancoff approach and various particle-hole theories. This concept is theoretically very appealing and allows clearly to define the approximations involved. In the following the RPA will be considered as a representation of these approaches since it possesses the soundest theoretical base[7-10] (at least in doubly closed shell nuclei) and has been widely applied in recent years.

Another kind of approach uses sum rules, mostly energy weighted. Sum rules are a reliable tool to describe average properties of the strength function with respect to a given transition operator. For certain transition operators they can be evaluated exactly if the nuclear interaction fulfills some assumptions. For such a case they are very useful since the GR by definition are supposed to exhaust the respective sums to a large extent. If the strength distribution is narrow, the position and the transition strength of the GR can often be evaluated[11].

There are two further approaches, both making use of the supposedly simple geometric structure of GR. This is assumed to manifest itself in the existence of collective operators, usually multipole operators, whose expectation value with respect to the time dependent wave function during the vibration characterizes this wave function and is intimately connected with its time evolution. The collective operator is used by means of constrained HF techniques for the construction of a collective path[12]. This one is a set of quasistatic Slater determinants which are believed to dominate the collective vibration in a way such that coupling to noncollective excitations can be neglected.

The two approaches are conceptually different and include on one hand semi-classical techniques for potential energy surfaces combined with dynamical approaches like Inglis cranking[13], self-consistent cranking[14], time dependent coordinate cranking[15] and momentum CHF[12]. In particular in recent years a systematic unification of these models has been achieved by the formulation of the adiabatic time dependent Hartree-Fock theory (ATDHF) which, in this talk, will be treated as a representative of the semi-classical approaches[16-20].

On the other hand there is the generator coordinate method[21-23] (GCM) which also makes use of a collective path. It is a fully quantal approach and consists essentially in diagonalizing the total microscopic Hamiltonian in the space spanned by the various members of the collective path. This theory is very much related to the cluster approach of the resonating group method[24].

3. SEMI CLASSICAL APPROACHES

In the classical approaches one starts with the construction of a collective path with respect to a collective operator Q related to the supposed geometric features of the GR. The collective paths are usually constructed by means of constrained Hartree-Fock (CHF)

$$\delta\langle\lambda|H - \lambda Q|\lambda\rangle = 0 \tag{3.1}$$

or by means of scaling the HF ground state wave function

$$|\lambda\rangle = \exp\{\lambda[H,Q]\}|HF\rangle \tag{3.2}$$

There are various other ways to construct the collective paths. For isoscalar monopole vibrations, e.g., the Q is usually taken to be $Q = \Sigma r_i^2$. The next step in the semi-classical approach consists in the construction of a dynamic generalization of the collective path, i.e., in the construction of a p-h operator S as generator for a momentum coordinate p_λ conjugate to the coordinate

$$|\lambda\rangle \to |\lambda,p_\lambda\rangle = (1 + ip_\lambda S)|\lambda\rangle \tag{3.3}$$

The expectation value

$$H_c(\lambda,p_\lambda) = \langle\lambda,p_\lambda|H|\lambda,p_\lambda\rangle \tag{3.4}$$

can be interpreted as a classical collective Hamiltonian and has a structure in terms of a classical collective mass M (λ) and a classical collection potential V(λ) (potential energy surface):

$$H_c(\lambda,p_\lambda) = \frac{p_\lambda^2}{2M_{cl}(\lambda)} V(\lambda) \tag{3.5}$$

These are many suggestions in the literature on how to construct the operator S in eq. (3.3) and the collective mass in eq. (3.5). In recent papers[5,6] it has been shown that these theories require only some minor modifications in order to be made identical to ATDHF. These modifications mostly affect the formalisms off stationary points $|\lambda\rangle$ and are even often zero for HF minima, which we are mostly concerned with in dealing with GR.

Assuming small amplitude vibrations around the HF minimum one expands

$$V(\lambda) = E_{HF} + \frac{1}{2}\lambda^2 \left(\frac{\partial^2 V}{\partial\lambda^2}\right)_{HF} \tag{3.6}$$

and obtains immediately for the excitation energy of the GR

$$\hbar\omega_{cl} = \left(\frac{1}{M_{cl}} \frac{\partial^2 V}{\partial\lambda^2}\right)^{1/2} \tag{3.7}$$

where all values have to be taken at the HF point $\lambda = 0$.

The semi-classical approaches, not initially quantized require a a posteriori quantization of H_c which is by no means trivial and has only recently been studied thoroughly[19]. In addition without a quantization one has no wave function at disposal and the extraction of other properties but energy and transition moment is difficult. However nonharmonic corrections are easily calculated since the ATDHF-like approaches are basically large amplitude theories. A disadvantage consists in the fact that one has to chose a priori a collective path by means of a collective operator Q, which means that one has a basic ingredient which is not determined by the system itself but rather by an educated guess. The classical approaches have frequently been applied to monopole and quadrupole GR by Engel et al[16], Blaizot et al[25], Vautherin et al[26] and recently by Castel and Goeke[27].

4. QUANTUM MECHANICAL APPROACHES

The prominent and frequently used representative of the quantum mechanical approaches involving a collective path is the generator coordinate method (GCM).

One uses the collective path $|\lambda\rangle$ in order to construct a stationary microscopic wave function of the GR existed state

$$|\psi\rangle = \int d\lambda\, f(\lambda)\, |\lambda\rangle \tag{4.1}$$

The variation $\delta\langle\psi|H - E|\psi\rangle$ with respect to $f(\lambda)$ leads to the Griffin, Hill-Wheeler equation (GHW) for $f(\lambda)$

$$\int d\lambda\, \langle\lambda'|H - E|\lambda\rangle\, f(\lambda) = 0 \tag{4.2}$$

In the Gaussian overlap approximation[28,29] (GOA) one can transform the GHW integral equation to a Schrödinger equation

$$\left(-\frac{d}{d\lambda}\frac{1}{2M_{GC}(\lambda)}\frac{d}{d\lambda} + V(\lambda) - \frac{\beta}{4}\frac{\partial^2 V}{\partial\lambda^2} - \frac{1}{4\beta}\frac{1}{M_{GC}}\right) g(\lambda) = Eg(\lambda) \tag{4.3}$$

which in the harmonic case reduces to the differential equation of a harmonic oscillator with the excitation energy:

$$\hbar\omega_{GC} = \left(\frac{1}{M_{GC}} \frac{\partial^2 V}{\partial \lambda^2} \right)^{1/2} \tag{4.4}$$

where all values are to be taken at $\lambda = 0$.

The result (4.4) is very similar to the result of the classical approaches (3.7). The problem[6], however, is that $M_{GC} \neq \mathcal{M}_{cl}$ and that therefore also $\hbar\omega_{GC} \neq \hbar\omega_{cl}$. Non-harmonic corrections are easily performed by solving Eq. (4.3) or even (4.2) directly. Obviously the results depend on the choice of the collective path. One obtains exactly one energy for the GR with no width. The GCM has been applied to GR by Abgrall et al[30], Caurier et al[31], Vautherin et al[32], Krewald et al[33], Galonska et al[34] and Giraud et al[35].

5. THE RANDOM PHASE APPROXIMATION

The RPA has been used frequently in the last years to describe GR. It is the small amplitude approximation to TDHF. If one writes the TDHF equation as

$$\delta \langle \phi(t) | H - i \frac{\partial}{\partial t} | \phi(t) \rangle = 0 \tag{5.1}$$

and expands the time dependent Slater determinant $\phi(t)$ of the system around the HF minimum $|HF\rangle$ by means of Thouless theorem one obtains

$$|\phi(t)\rangle = \{1 + \sum_{ph} C_{ph}(t)\, a_p^+ a_h\} |HF\rangle \tag{5.2}$$

Assuming further harmonic motion

$$C_{ph}(t) = X_{ph}\, e^{i\omega t} + Y_{ph}\, e^{-i\omega t} \tag{5.3}$$

and inserting the expansion (5.2) into the TDHF equation gives two equations, where one is the static HF equation

$$\delta \langle HF | H - E | HF \rangle = 0 \tag{5.4}$$

and the other is the RPA equation

$$\begin{pmatrix} A & B \\ -B^* & -A^* \end{pmatrix} \begin{pmatrix} X \\ Y \end{pmatrix} = \hbar\omega_{RPA} \begin{pmatrix} X \\ Y \end{pmatrix} \tag{5.5}$$

with the PRA matrices given by

$$A_{php'h'} = \langle HF|a_h^+ a_p \;(H-\langle H\rangle) a_{h'}^+ a_{p'}|HF\rangle \tag{5.6}$$

$$B_{php'h'} = \langle HF|a_h^+ a_p a_{h'}^+ a_{p'} \; H|HF\rangle$$

The above sketched TDHF derivation of RPA shows that this theory is to be considered together with static HF. Only if the single particle static are determined by HF, the RPA is a fully consistent theory fulfilling all stability conditions. This will always be assumed in this lecture although many practical applications use, e.g., Saxon-Woods wave function and experimental single particle energies. The essential difference to the semi-classical theories and GCM lies in the fact that RPA does not assume anything about a collective path. Furthermore, if the continuum is taken into account properly one can also calculate the width of the GR. There are some disadvantages in the RPA since for density dependent interaction one has explicitly to construct a p-h interaction by means of functional derivatives of the energy density. Furthermore any step beyond the harmonic approach is rather tedious.

The RPA has been used with non-schematic interactions by Bertsch and Tsai[38], Liu and Brown[37], Speth et al[38], Krewald et al[39], Blaizot et al[40], Knupfer et al[41] and others.

The essential point in comparing RPA with ATDHF and GCM consists in the fact that ATDHF and GCM calculate just one energy for GR with respect to a collective path associated with a collective operator Q, whereas RPA essentially calculates a strength function with respect to a collective operator Q, or in other words, the linear response function to an external perturbing field generated by Q. Hence we have the situation sketched in Fig. 1. For the following considerations it does not matter if one uses a finite or infinite space spanned by some basis functions as long as one is consistent in the sense that one performs all calculations in the same space. If the space is infinite (e.g., coordinate space) certain sums have to be replaced by integrals.

The key quantity in the comparison between the theories are the energy weighted moments of the RPA strength distribution with respect to some one-particle operator Q

$$m_k(Q) = \sum_{\alpha>0} (\hbar\omega_\alpha)^k |\langle\alpha|Q|0\rangle|^2 \tag{5.7}$$

where $|\alpha\rangle$ indicates the one-phonon RPA excited states.

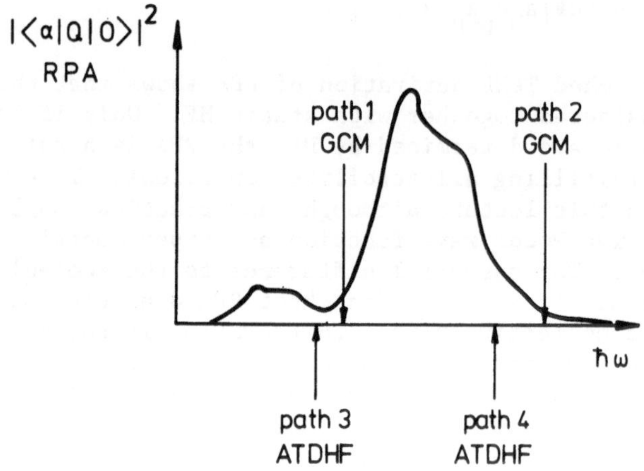

Fig 1

The moments can actually be expressed explicitly in terms of the matrices A and B of eq. (5.6) as shown by Goeke et al[6].

6. CONSTRAINED HARTREE-FOCK

For a one body operator Q, the CHF solution $|\lambda\rangle$ is obtained by the variational principle

$$\delta\langle\lambda|H - \lambda Q|\lambda\rangle = 0 \tag{6.1}$$

If we introduce the generator R of the path $|\lambda\rangle$ by

$$|\lambda + d\lambda\rangle = (1 - i\delta\lambda R)|\lambda\rangle \tag{6.2}$$

then one obtains by deriving (6.1) w.r. to λ at $\lambda = 0$ the equation

$$R = i(A + B)^{-1} Q \tag{6.3}$$

This has the immediate consequence that the polarisibility and the spring constant are given by

$$\frac{\partial^2}{\partial\lambda^2} <\lambda|H|\lambda> = 2m_{-1}(Q) = \frac{\partial}{\partial\lambda}< \lambda|Q|\lambda> \tag{6.4}$$

Hence the spring constant used in the collection theories eq. (3.7, 4,4) is in a clear way related to the RPA strength distribution if one uses the constraining operator Q also in the RPA distribution.

There is an easy way to generalize[6] the result (6.4) to all odd moments of RPA. To this end we define the operator Q_o on the p-h part of Q and construct a hierarchy of operators Q_n in the following way

$$\begin{aligned} Q_{-n} &= i[H, Q_{-n-1}] \ldots Q_{-1} = i[H, Q_{-2}]^1; \; Q_o = i[H, Q_{-1}]^1 \\ Q_1 &= i[H, Q_o]^1; \; Q_2 = i[H, Q_1]^1 \ldots Q_n = i[H, Q_{n-1}]^1 \end{aligned} \tag{6.5}$$

where the prime indicates the p-h part of the commutator. Obviously one has

$$i(A + (-)^{n+1}B)Q_n = Q_{n+1} \tag{6.6}$$

This allows to equate certain moments of the RPA distribution with respect to certain operation of the hierarchy by using eqs. (5.8):

$$m_{2n+1}(Q_m) = m_{2(n-k)+1}(Q_{m+k}) \tag{6.7}$$

Particularly one obtains

$$m_{-1}(Q_n) = m_{2n-1}(Q_o) \tag{6.8}$$

Hence if one uses the Q_n in a CHF procedure

$$\delta<\lambda_n|H-\lambda_n Q_n|\lambda_n> = 0 \tag{6.9}$$

then the spring constant with respect to Q_n is

$$\frac{\partial^2}{\partial\lambda_n^2} <\lambda_n|H|\lambda_n>= m_{-1}(Q_n) = m_{2n-1}(Q_o) \tag{6.10}$$

and is identical to the (2n-1)-moment with respect to Q_o. The generator of the path $|\lambda_n>$ is defined as

$$|\lambda_n+\delta\lambda_n> = (1 - i\delta\lambda_n R_n)|\lambda_n> \tag{6.11}$$

Another remarkable feature of the hierarchy is that it contains special cases which allow an interpretation of CHF in terms of scaling: From (6.11) follows that for small λ_n the $|\lambda_n\rangle = \exp(-i\lambda_n R_n)|HF\rangle = \exp(\lambda_n[H,Q_{n-2}])|HF\rangle$.

In the special case, when Q_{n-2} corresponds to a multipole operator satisfying $\nabla^2 Q_{n-2} = 0$ the effect of the exponential just consists in replacing every coordinate r in $|HF\rangle$ by $r - (\lambda\hbar^2/m)\nabla Q_{n-2}$ and one says that $|HF\rangle$ has been scaled. Hence CHF in Q_n is identical to scaling in Q_{n-2}. In case of $Q_{n-2} = \Sigma\, r_i^2$ one obtains the frequently used case for isoscalar monopole vibrations. The corresponding CHF operator is $Q_n = [iH,[iH,Q_{n-2}]']'$. Obviously with eq. (6.8) the spring constant of a scaled path is $m_{-1}(Q_n) = m_3(Q_{n-2})$.

7. ADIABATIC TIME DEPENDENT HARTREE-FOCK

As shown in Refs. 5 and 6 the classical theories can all be identified with ATDHF if minor corrections regarding the p-p and h-h - elements of Q are performed. Thus only the ATDHF is considered here. The ATDHF to first order has been considered by Engel et al[16], Brink et al[17], Villars[18], Goeke and Reinhard[19] and Baranger and Veneroni[20]. If one assumes the collective path given by Eq. (3.1), then one expands the TDHF wave function $|t\rangle$ in terms of small momenta around $|\lambda\rangle$

$$|t\rangle \simeq |\lambda,p_\lambda\rangle = |\lambda\rangle + p_\lambda|\lambda\rangle_{(1)} + p_\lambda^2|\lambda\rangle_{(2)} + \cdots \tag{7.1}$$

The $|\lambda,p_\lambda\rangle$ is to first order in p_λ given by

$$|\lambda,p_\lambda\rangle = (1 + ip_\lambda S)|\lambda\rangle \tag{7.2}$$

where S obeys the equation[19]

$$([H,S] + iR/M_{ATDHF})_{ph} = 0 \tag{7.3}$$

$$M^{-1}_{ATDHF} = \langle\lambda|[S,[H,S]|\lambda\rangle \tag{7.4}$$

The evaluation of (7.4) by means of eq. (7.3) and (6.3) gives

$$M_{ATDHF}(\lambda) = 2m_{-3}(Q) \tag{7.5}$$

and hence for the excitation energy

$$\hbar\omega_{ATDHF} = (m_{-1}(Q)/m_{-3}(Q))^{1/2} \tag{7.6}$$

This is a remarkable result which shows that ATDHF and together with it all other classical approaches yield a collective mass to be identified with the third inverse moment of RPA. The excitation energy is a ratio of two moments m_{-1} and m_{-3}. The generalization[6] of this result to arbitrary n is straightforward: If the CHF path is constructed by Q_n, one obtains

$$\hbar\omega_{ATDHF}^{(n)} = \left(\frac{m_{-1}(Q_n)}{m_{-3}(Q_n)}\right)^{1/2} = \left(\frac{m_{2n-1}(Q_o)}{m_{2n-3}(Q_o)}\right)^{1/2} \tag{7.7}$$

and $$M_{ATDHF}(\lambda_n) = 2m_{2n-3}(Q_o) \tag{7.8}$$

$$\frac{\partial^2 V}{\partial\lambda_n} = 2m_{2n-1}(Q_o) \tag{7.9}$$

Eqs. (7.7-7.9) show the intimate link between the classical theories for GR and the RPA. Apparently each odd energy weighted RPA moment can equally well be calculated by ATDHF using the proper collective page. This can be done either by evaluating the collective mass $M(\lambda)$ or by calculating the spring constant. Furthermore one obtains in ATDHF a multitude of energies $E_n = (m_k/m_{k-2})^{1/2}$ which altogether cover the range of the RPA strength distribution.

8. THE GENERATOR COORDINATE METHOD

As has been mentioned already in sect. 4 the GCM can be transformed to a collective Schrödinger equation, if one assumes the GOA. The mass in eq. (4.3) is then given by

$$\frac{1}{M^{GCM}(\lambda_n)} = \frac{\langle\lambda_n|\{R_n,\{H,R_n\}\}|\lambda_n\rangle}{4\langle R_n^2\rangle^2} \tag{8.1}$$

where R_n is the generator of the CHF path constructed by Q_n. If one evaluates Eq. (8.1) by means of Eqs. (6.6, 6.7) one obtains[6]

$$M^{GCM}(\lambda_n) = \frac{2\{\tilde{Q}_n(A+B)^2Q_n\}^2}{\tilde{Q}_n(A+B)^{-1}(A-B)(A+B)^{-1}Q_n} \tag{8.2}$$

This expression is in general not identical to an RPA moment. The inequality holds

$$M^{GCM}(\lambda_n) \leq 2m_{-3}(Q_n) = 2m_{2n-3}(Q_o) \tag{8.3}$$

Hence the excitation energy is

$$\hbar\omega_{GC}^{(n)} \geq \left(\frac{m_{2n-1}(Q_o)}{m_{2n-3}(Q_o)} \right)^{1/2} = \hbar\omega_{ATDHF}^{(n)} \tag{8.4}$$

As a matter of fact the classical masses and energies are uniquely related to the odd moments of the RPA strength distribution. The GCM however does not fit into this scheme and it has therefore been argued that ATDHF also would not agree with RPA-moments if the classical Hamiltonian (3.5) was properly quantized. However, there have been developed recently reliable techniques[19] for an unambiguous quantization of H_c leading to a quantized Hamiltonian $\overline{H}_c$. If one assumes the existence of a canonical transformation $x_A \ldots x_A \rightarrow \xi_A, \ldots, \xi_{A-1}, \Lambda$ of the particle coordinates to intrinsic coordinates and a collective coordinate Λ, then the H_c can be written as

$$H_c(\lambda,p_\lambda) = \int d\xi_i \, d\Lambda \, \langle\lambda p_\lambda|\xi_i\Lambda\rangle \, \overline{H}(\xi_i\Lambda) \, \langle\xi_i\Lambda|\lambda,p_\lambda\rangle \tag{8.5}$$

and the collective $\overline{H}_c$ is defined as that quantity which remains in eq. (8.5) after averaging over the intrinsic ξ_i:

$$H_c(\lambda,p_\lambda) = \int d\Lambda \, \langle\lambda,p_\lambda|\Lambda\rangle \, \overline{H}_c(\Lambda, \frac{d}{d\Lambda}) \, \langle\Lambda|\lambda,p_\lambda\rangle \tag{8.6}$$

In the GOA approximation it turns out that

$$H_c = -\frac{d}{d\Lambda} \frac{1}{M_{ATDHF}(\Lambda)} \frac{d}{d\Lambda} + V(\Lambda) + \text{zero point corrections} \tag{8.7}$$

The mass is not changed by the quantization and if one neglects in the Taylor expansion the zero point corrections one obtains again Eq. (7.7) for the excitation energy.

Actually one has to generalize the GCM to obtain agreement with ATDHF and the RPA moments. This is indicated already by the fact that the GCM gives in case of translations only the correct mass, if one integrates also over the velocities and not only over the displacements[42]. This feature can be generalized to arbitrary collective paths[43]. If one compares Eq. (7.4) with Eq. (8.1) one realizes that the GCM mass agrees with the ATDHF mass if the collective path $|\lambda\rangle$ fulfills the equation ($\sigma=2\langle R^2\rangle$)

$$(S + iR/\sigma)|\lambda\rangle = 0 \tag{8.8}$$

Unfortunately for an ordinary collective path this is not fulfilled. However, if S and R are given in ATDHF, Eq. (8.8) can be used to define a new, correlated path $|\lambda\rangle_c$ associated with $|\lambda\rangle$ which fulfills by construction

$$(S + iR/\sigma)|\lambda\rangle_c = 0 \tag{8.9}$$

The $|\lambda\rangle_c$ is obviously the local boson vacuum[44,45]. A GCM along $|\lambda\rangle_c$ must give a collective mass identical to ATDHF:

$$|\psi\rangle = \int d\lambda\ f(\lambda)|\lambda\rangle_c \tag{8.10}$$

As recently shown by Reinhard et al[43] this GCM along $|\lambda\rangle_c$ is identical to a GCM with two conjugate parameters (dynamical GCM):

$$|\psi\rangle = \int d\lambda\ dp_\lambda\ f(\lambda,p_\lambda)|\lambda,p_\lambda\rangle \tag{8.11}$$

In the GOA limit of DGCM one obtains a collective Schrödinger equation which has a mass

$$M^{DGCM}(\lambda_n) = M^{ATDHF}(\lambda_n) = 2m_{-3}(Q_n) = 2m_{2n-3}(Q_0) \tag{8.12}$$

and hence

$$\hbar\omega^{(n)}_{DGCM} = \hbar\omega^{(n)}_{ATDHF} = \left(\frac{m_{-1}(Q_n)}{m_{-3}(Q_n)}\right)^{1/2} = \left(\frac{m_{2n-1}(Q_0)}{m_{2n-3}(Q_0)}\right)^{1/2} \tag{8.13}$$

9. INTERPRETATION OF THE THEORETICAL RESULTS

The above identification of results of ATDHF and (D)GCM certain moments of the RPA strength distribution allows a clear interpretation of the ATDHF and GCM data (see Fig. 2). Suppose the GR has a rather wide strength distribution. Then the values $E_k = (m_k/m_{k-2})^{1/2}$ are all different and they are distributed over the range of the RPA strength function. There is no obvious link between E_k and the peak energy E_p. Furthermore none of the values E_k can be considered as "the excitation energy of the GR". One rather has the situation that the linear response to the perburbing field associated with the collective operator Q consists in the excitation of many eigenmodes rather than one and this is reflected in the variance of the values of E_k obtained by ATDHF and DGCM. Actually the variation of E_k can then be used to estimate the width of the GR. An appropriate measure is $E = (m_3/m_1)^{1/2} - (m_{-1}/m_{-3})^{1/2}$.

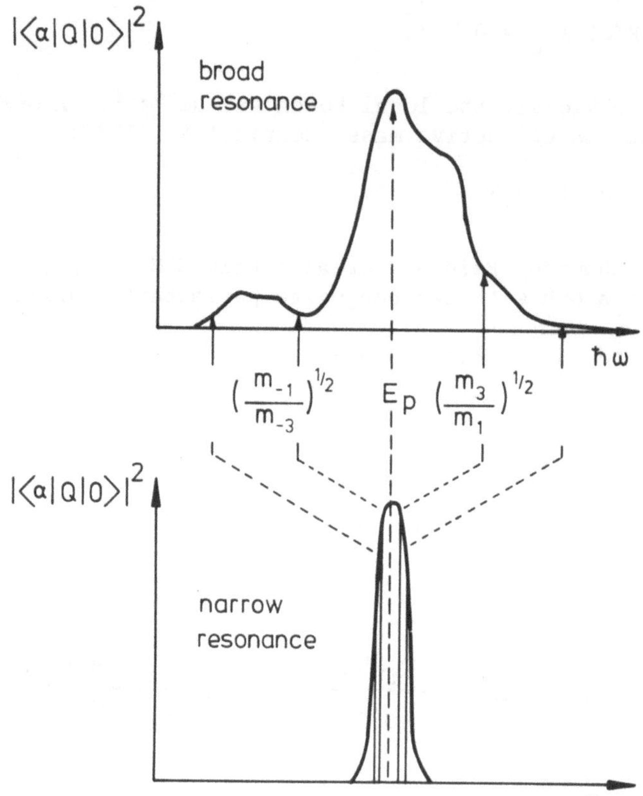

Fig 2

If on the other hand the GR is narrow then all values of E_k are roughly identical and agree with the peak energy. In such a case it obviously does not matter which collective operator of the hierarchy of Q_n is taken to construct the path, since they all describe the GR equally well and are in agreement with RPA. There might be problems connected with small but long ranging tails of the RPA distributions if one calculates E_k for large positive or negative k.

Hence one can conclude, that the classical methods (ATDHF, Cranking, ...) and the quantum mechanical methods (DGCM, GCM, ...) in their small amplitude approximation cannot reveal anything which could not be calculated by RPA. If one, however, is only interested in certain moments like, e.g., the polarizability of the nucleus or if we deal with narrow resonances, then the transparent structure of ATDHF, GCM, etc. and the simple numerical handling is really a virtue. It is, e.g., no problem to solve ATDHF in coordinate space[46] taking into account the full continuum whereas RPA in most of the practical application suffers under the use of a finite basis whose convergence properties are moderate. Furthermore the classical theories help in understanding recent linearized Hartree-Fock techniques[47,48].

10. ISOSCALAR AND ISOVECTOR GIANT MONOPOLE RESONANCES

The above considerations can easily be applied to calculate properties of the isoscalar giant monopole. In table 1 there are given various moments of the RPA strength distribution with respect to the operator $Q = \Sigma\, r_i^2$. They are calculated[27] by means of ATDHF and Skyrme-3 interaction in a basis consisting of $N_0 \geq 20$ harmonic oscillator shells. This large number of N_0 is actually required to obtain convergence. Smaller values of N_0 are used in the literature but have been shown to cause too high excitation energies[46]. The comparison with the experimental data of Bertrand et al[2] shows that the width ΔE is properly reproduced and ranges from ∿ 12 MeV in ^{16}O to ∿ 2 MeV in ^{208}Pb. The excitation energies are slightly overestimated probably due to the rather high compression modulum of the Skyrme-interactions (∿ 340 MeV) compared to the generally accepted value of ∿ 200 MeV.

Table 2 reports similar results for the isovector giant monopole resonance, where the values of $(m_3/m_1)^{1/2}$ are taken from Stringari et al[49]. The isovector GMR is obviously extremely broad with a width of at least 16 MeV in the doubly closed shell nuclei and it is probably very difficult to observe it experimentally.

TABLE 1: Isoscalar Monopole Giant Resonance Properties with Skyrme-3 Interaction

	^{16}O	^{40}Ca	^{90}Zr	^{208}Pb
$M(R_0)$ (mass units)	23.3	50.9	98.1	228.0
m_{-3} [fm^4 MeV^{-3}]	0.02	0.12	0.64	7.25
m_{-1} [fm^4 MeV^{-1}]	11.6	59.5	282.9	1893.3
m_1 [fm^4 MeV^1]	9182.	38083.	137481.	541316.
m_3 [fm^4 MeV^3]	$1.09 * 10^7$	$3.09 * 10^7$	$7.52 * 10^7$	$17.3 * 10^7$
$(m_3/m_1)^{1/2}$	34.5	28.5	23.4	17.9
$(m_1/m_{-1})^{1/2}$				
$(m_{-1}/m_{-3})^{1/2}$				
ΔE	11.4	6.1	2.4	1.8

TABLE 2: Isovector Monopole Giant Resonance Properties with Skyrme-3 Interaction

	^{16}O	^{40}Ca	^{90}Zr	^{208}Pb
$(m_3/m_1)^{1/2}$	45	42	39	35
$(m_{-1}/m_{-3})^{1/2}$	25	24	25	19
ΔE	20	18	14	16

11. LARGE AMPLITUDE CORRECTIONS

All above considerations were restricted to small amplitudes, i.e., to first order corrections of the static HF wave function. This had the consequence that a p-h algebra at the HF point was sufficient to describe the system whichever method was used. As mentioned already in sect. 3 and 4 the ATDHF and GCM allow in an easy way to incorporate large amplitude corrections if the constraining operator Q is known. Usually Q is obtained by an educated guess and is taken to be the same for small as well as large amplitudes. However a Q appropriate near the HF minimum is by no means guaranteed to be the proper one, i.e., at the second barrier of a fissioning nucleus. Hence one must allow Q to vary along the collective path. This means one needs a self-consistency condition which couples Q to the path, e.g., of the form $\delta\langle\lambda|H-\lambda Q(\lambda)|\lambda\rangle = 0$. Such a self-consistency condition is actually suggested by ATDHF in order to achieve a decoupling of collective and non-collective degrees of freedom along the collective path $|\lambda\rangle$. As shown in refs. (19) and (50), the decoupling is achieved by identifying the operator S of eq. (7.2, 7.3) with the constraining operator yielding the set of ATDHF eqs:

$$\delta\langle\lambda|H-\lambda S|\lambda\rangle = 0 \quad ; \quad \delta\langle\lambda|[H,S]+iR/M|\lambda\rangle = 0$$

with

$$(1-i\delta\lambda R)|\lambda\rangle = |\lambda+\delta\lambda) \qquad \text{and} \qquad M^{-1}(\lambda) = \langle\lambda|[S,[H,S]]|\lambda\rangle$$

The resulting collective path is due to its decoupling particularly appropriate for the quantization procedure[19] outlined in sect. 7. If one generalizes $|\lambda\rangle$ to $|\lambda,p_\lambda\rangle$ given by eq. (7.2) then one also has a well defined dynamic collective path to be used as input into a GCM with conjugate parameters. As recently shown[43] this is also in the case of large amplitudes identical to the quantized ATDHF.

There are various open questions in this sort of large amplitude collective theory. First, there may be several collective paths which are solutions of the set of eqs. (12.1-12.4). It is not known if one needs a further selection by any means or if each of them reflects certain properties of the system and allows then a physical interpretation of the associated results. Second, it is generally assumed that a description of collective motion in terms of TDHF makes sense. Collective motion is interesting since it allows to calculate certain eigenstates of the system in terms of one (or few) collective coordinate rather than treating all 3A coordinates. It is not known if TDHF is flexible enough a theory for such a reduction. Another challenging problem is the comparison of ATDHF with, e.g., boson expansion techniques, which within certain limits also allow for large amplitude corrections.

12. SUMMARY

Besides sum rule approaches there exist three different sorts of microscopic theories for giant resonances: Classical theories like ATDHF, Cranking, etc., quantum mechanical theories like GCM and particle-hole theories like RPA. The first two need as input a collective path constructed by CHF, they calculate the excitation energy of the GR without width. The RPA, on the other hand, needs essentially no input and calculates the strength distribution of the GR with respect to some collective operator Q. Although the philosophy behind all these theories is rather different they are stricIty related to each other in terms of energy weighted moments of the RPA strength distribution. If the operator, with respect to which the moment is calculated, is used as constraining operator, then the masses of the classical theories are $m_{-3}(Q)$ and the spring constants are $m_{-1}(Q)$. One is able to generalize this concept by systematically constructing collective paths which eventually give all odd RPA moments by means of classical collection masses and spring constants. The GCM does not fit into this scheme. However, it needs a rather simple generalisation of GCM to achieve this. It actually requires to use not only λ but also its conjugate momentum p_λ as generator coordinate. With this generalization virtually all existing microscopic theories turn out to have a clear correspondence to the RPA. This actually allows to interpret their results and to extract the physical relevance: For a broad resonance the variation of $(m_k/m_{k-2})^{1/2}$ is a measure of the position and the width of the resonance without being related to the peak energy, for a narrow resonance all those values agree more or less with each other and with the peak energy. In the latter case or if one is interested merely in a single moment, like, e.g., the polarization, the classical theories and the GCM have technical advantages to the RPA. If one likes to compare the full strength distribution with experiment the RPA is probably the best tool.

The investigation of the isoscalar and isovector monopole resonance in doubly closed shell nuclei by means of ATDHF with various collective paths provides an interesting application of this unifying view of GR in terms of energy weighted moments. It turns out that the isoscalar GMR has a width of $\sim$ 12 MeV in ^{16}O which narrows to $\sim$ 2 MeV in ^{208}Pb. The isovector GMR appears to be very broad with at least $\sim$ 16 MeV width which make it probably very hard to observe it experimentally.

Many calculations for GR have been performed using density dependent interactions of Skyrme type. Actually these forces show deficiencies which affect certain ground state properties as well as particle-hole properties. For instance, a comparison of properties of HF valence particles with experiment shows that the radius of the valence orbit and the density of states at the Fermi surface are not properly described. There have been suggestions to attribute this to the fact that the effective mass m^*/m of these interactions is constant, whereas it should show a local enhancement for single particle states at the Fermi surface. A calculation of static HF properties in the Ca-isotopes using a Sk2-interaction for deeply bound orbitals and a Sk6-interaction for the $f_{7/2}$-valence orbit showed indeed an improvement of the HF results concerning proton and neutron radii of the valence orbit[51]. Obviously there are also other ways to obtain a state dependent m^*/m, however, these calculations show that one can expect a noticeable improvement of the HF calculations using this type of interaction. This should of course affect also the position and transition probability of giant resonances and may lead out of the dilemma that Skyrme interactions either describe the position correctly or the transition strength, but rarely both simultaneously. This problem certainly deserves more investigations.

There occur various problems if one tries in the description of GR (or any sort of collective motion) to go beyond the harmonic limit and to describe large amplitude collective motion. In such a case the choice of the constraining operator Q in the CHF principle is crucial and one cannot rely on guesses but one has to formulate a theory which determines Q. The most recent approaches in this direction are more or less related to ATDHF and indeed achieve to construct a collective path along which collective and non-collective excitation are decoupled to a certain extent. However, these theories remain to be investigated numerically.

REFERENCES

1. F. Bertrand, Ann. Rev. Nucl. Sci. 26 (1976), 457.

2. F. Bertrand, see these proceedings.

3. O. Bohigas, J. Martorell and A.M. Lane, Phys. Lett. 64B (1976), 1.

4. J. Martorell, O. Bohigas, S. Fallieros and A.M. Lane, Phys. Lett. 60B (1976), 31.

5. K. Goeke, Nucl. Phys. A265 (1976), 315.

6. K. Goeke, A.M. Lane, J. Martorell, Nucl. Phys. A296 (1978), 328.

7. D. Rowe, Collective Motion, Methuen, 1967.

8. A.M. Lane, Nuclear Theory, Benjamin, New York, 1974.

9. A.B. Migdal, Theory of Finite Fermi Systems, New York, 1967.

10. J. Speth, E. Werner, W. Wild, Phys. Rep. 33C (1977), 128.

11. A.M. Lane, A. Mekjian, Adv. in Nucl. Phys. Vol. 7, Ed. M. Baranger, E. Vogt.

12. F. Villars, Proceedings of the Mount Tremblant Summer School, 1971, ed. D. Rowe et al., p. 3.

13. D.R. Inglis, Phys. Rev. 96 (1954), 1059.

14. D.J. Thouless, J.G. Valatin, Nucl. Phys. 31 (1962), 211.

15. M.K. Pal, Trieste Lectures 1973, vol. 2, IAEA 1975, p. 59.

16. Y.M. Engel, D. Brink, K. Goeke, S.J. Krieger and D. Vautherin, Nucl. Phys. A249 (1975), 215.

17. D.M. Brink, M. Giannoni and M. Veneroni, Nucl. Phys. A258 (1976), 237.

18. F. Villars, Nucl. Phys. A285 (1977), 269.

19. K. Goeke and P.-G. Reinhard, Ann. Phys. 112 (1978), 328.

20. M. Baranger and M. Veneroni, Ann. Phys. 114 (1978), 123.

21. D.L. Hill, J.A. Wheeler, Phys. Rev. 89 (1953), 1102.

22. J.J. Griffin, J.A. Wheeler, Phys. Rev. 108 (1957), 311.

23. D.M. Brink, A. Weiguny, Nucl. Phys. A120 (1968), 59.

24. K. Wildermuth, Y.C. Tang, A Unified Theory of the Nucleus, Vieweg, 1977.

25. J.P. Blaizot and B. Grammeticos, Phys. Lett. 53B (1974), 231.

26. D. Vautherin, Phys. Lett. 57B (1975), 425.

27. K. Goeke and B. Castel, Phys. Rev. 19C (1979), 201.

28. J. DaProvidencia, J.V. Urbano, L.S. Ferreira, Nucl. Phys. A170 (1971), 129.

29. P.-G. Reinhard, Nucl. Phys. A261 (1976), 291.

30. Y. Abgrall, E. Caurier, Phys. Lett. 56B (1975), 229.

31. E. Caurier, B. Bourotte-Bilwes, Y. Abgrall, Phys. Lett. 44B (1973), 411.

32. H. Flocard, D. Vautherin, Phys. Lett. 55B (1975), 259.

33. S. Krewald, J.E. Galonska, A. Faessler, Phys. Lett. 55B (1975), 267.

34. S. Krewald, R. Rosenfelder, J.E. Galonska and A. Faessler, Nucl. Phys. A269 (1976), 112.

35. B. Giraud, B. Grammaticos, Nucl. Phys. A255 (1975), 141.

36. G.F. Bertsch, S.F. Tsai, Phys. Rep. 18 (1975), 125.

37. K.F. Liu, G.E. Brown, Nucl. Phys. A265 (1976), 385.

38. P. Ring and J. Speth, Nucl. Phys. A235 (1974), 315.

39. S. Krewald, V. Klemt, J. Speth and A. Faessler, Nucl. Phys. A281 (1977), 166.

40. J. Blaizot, D. Gogny and B. Grammaticos, Nucl. Phys. A265 (1976), 315.

41. S. Krewald, J. Birkholz, A. Faessler and J. Speth, Phys. Rev. Lett. 33 (1974), 1386.

42. R.E. Peierls, D.J. Thouless, Nucl. Phys. 38 (1962), 154.

43. P.-G. Reinhard and K. Goeke, Journ. Phys. G4 (1978), L245.

44. A. Alves et al., Nucl. Phys. A284 (1977), 420.

45. D.J. Rowe, R. Bassermann, Can. Journ. Phys. 54 (1976), 1941.

46. K. Goeke, Phys. Rev. Lett. 38 (1977), 212.

47. B. Castel and K. Goeke, Phys. Rev. C16 (1977), 2092.

48. B. Castel, K. Goeke, G.R. Satchler and I.S. Towner, to be published.

49. S. Stringari, E. Lipparini, G. Orlandini, M. Traini and R. Leonardi, Nucl. Phys. A305 (1978), 189.

50. P.G. Reinhard and K. Goeke, Nucl. Phys.

51. B. Castel and K. Goeke, Phys. Letters, in print.

GIANT MONOPOLE RESONANCES IN NUCLEI AND COMPRESSIBILITY OF NUCLEAR MATTER

J.P. Blaizot

Service de Physique Théorique, CEN Saclay

91190 Gif-sur-Yvette, France

Within the last three years one has obtained more and more evidence on the excitation of a giant monopole resonance in the inelastic scattering of various particles on nuclei. (For references see the review presented by F.E. Bertrand at this Summer School.) The importance of this mode of excitation of the nucleus lies in the fact that its frequency is related to the compressibility of nuclear matter. We briefly discuss here the problems encountered when one attempts to extract the value of the compressibility of nuclear matter from the observed frequency of the giant monopole resonance (the so-called breathing mode)[1].

First of all let us recall a few definitions. The compressibility of nuclear matter is defined as usual by

$$\chi = -\frac{1}{\Omega}\frac{\partial\Omega}{\partial P} \tag{1}$$

where Ω and P are respectively the volume and the pressure of the system. It is however more usual to use a compression modulus K_∞

$$K_\infty = k_F^2 \frac{d^2E/A}{dk_F^2} \tag{2}$$

A single calculation shows that at the saturation

$$K_\infty = \frac{9}{n\chi} \tag{3}$$

(n being the density of nucleons). K_∞ is thus inversely proportional to the compressibility χ. For this reason it is sometimes called incompressibility. It is convenient for our forthcoming discussion to separate in K_∞ two contributions. We write:

$$K_\infty = 6\varepsilon_F(1 + F_o) \quad (4)$$

where ε_F is the Fermi energy:

$$\varepsilon_F = \frac{\hbar^2 k_F^2}{2m^*}$$

m* being the effective mass of a nucleon having a momentum k_F. The first contribution to K_∞, $6\varepsilon_F$, is just the compression modulus of a gas of free particles (with an effective mass). The second term $6\varepsilon_F F_o$ is the correction (F_o turns out to be small) due to the interactions; the parameter F_o is related to the variation with the density of the average potential.

Macroscopic arguments can be used to show quickly how the compressibility of nuclear matter determines the frequency of the breathing mode. Indeed, using conservation laws and assuming that the pressure is a function of the local density, one gets the following wave equation:

$$\frac{\partial^2(\delta n)}{\partial t^2} - \vec{\nabla}\,(c^2\vec{\nabla}\delta n) = 0 \quad (5)$$

where δn is the density fluctuation and $c^2 = \frac{\partial P}{\partial n}$ is the sound velocity, in general a function of $\vec{r}$. If one assumes, as in the liquid drop model, that the density is constant, equal to n_o, inside a sphere of radius R_o and zero outside it can easily be shown[2] that (5) has the following eigenfrequencies:

$$\omega_n = \frac{n\pi}{R_o}\, c_1 \quad (6)$$

where $c_1 = \sqrt{\frac{K_\infty}{m}}$ is the velocity of ordinary sound. In this model, the lowest mode has the frequency:

$$\omega = \left(\frac{A}{m}\,\frac{K_\infty}{\langle r^2\rangle}\,\frac{\pi^2}{15}\right)^{1/2} . \quad (7)$$

This model has two important drawbacks. First, the assumption of a sharp surface is a very poor one in the present case, as we shall see. Secondly, the model cannot predict whether or not a collective monopole mode can exist in nuclei. This is because it does not take into account the shell structure of the single particle spectrum which plays a crucial role in the properties of the collective modes of finite systems. It therefore appears that a microscopic approach is required for the calculation of the breathing mode. It is such an approach that we now describe.

A convenient starting point to the theory of collective excitations, both in nuclear matter and in finite nuclei is to consider the total energy of the system as a functional of the one body density matrix ρ. All the information which is needed to calculate the collective modes is contained in an expansion of this functional around ρ_o, the density matrix which corresponds to the ground state:

$$E[\rho] = E[\rho_o] + \sum_{\alpha\beta} \frac{\delta E}{\delta\rho_{\alpha\beta}} \delta\rho_{\alpha\beta} + \frac{1}{2} \sum_{\alpha\beta\gamma\delta} \frac{\delta^2 E}{\delta\rho_{\alpha\beta}\delta\rho_{\gamma\delta}} \delta\rho_{\alpha\beta}\delta\rho_{\gamma\delta} \ . \tag{8}$$

The quantity $\frac{\delta E}{\delta\rho_{\alpha\beta}}$ can be identified with a matrix element of the single particle Hamiltonian, while the second functional derivative is the particle-hole interaction. In practice, $E[\rho]$ will be constructed by taking the expectation value in a Slater determinant of an effective (density dependent) Hamiltonian $H(\rho) = T + \nu[\rho]$, in which case:

$$E[\rho] = \sum_{\alpha\beta} T_{\alpha\beta}\, \rho_{\beta\alpha} + \frac{1}{2} \sum_{\alpha\beta\gamma\delta} \nu[\rho]_{\alpha\beta,\gamma\delta}\, \rho_{\gamma\alpha}\, \rho_{\delta\gamma} \ . \tag{9}$$

To discuss nuclear matter, it is convenient to use the Wigner representation of the density matrix:

$$n_{\vec{k}}(\vec{k}) = \int d^3r\, e^{-i\vec{k}\cdot\vec{r}}\, \rho(\vec{R} + \frac{\vec{r}}{2}, \vec{R} - \frac{\vec{r}}{2}) \ . \tag{10}$$

The equation (8) can then be rewritten as follows:

$$E=E_o+\int d^3R\ \{\sum_{\vec{k}}(\varepsilon^o_{\vec{k}}-\psi)\delta n_{\vec{k}}(\vec{R})+\frac{1}{2}\sum_{\vec{k}\vec{k}'} f(\vec{k},\vec{k}',\vec{R})\delta n_k(\vec{R})\delta n_{\vec{k}'}(\vec{R})\} \tag{11}$$

In an infinite system, the equation (11) reduces to the well known expression used by Landau in his theory of Fermi liquid[4].

When the system is excited, the density matrix fluxtuations $\delta\rho_{\alpha\beta}$ are functions of time and can be used as complex (classical) coordinates for describing the departure from equilibrium. The energy (8) is the Hamiltonian for these classical coordinates; it is a Hamiltonian of coupled harmonic oscillators which, upon diagonalization, leads to the well-known RPA equations.

Thus in this approach the calculation of the collective modes is carried out in two steps[5]. First the ground state density matrix ρ_0 is determined by solving the Hartree-Fock equations. This provides the single particle energies and wave functions. In the second step, the RPA equations are solved, which gives the energies of the collective modes. One should emphasize that the only parameters which enter the calculation are those of the effective interaction. Various effective interactions, or various sets of parameters, yield in general different predictions for the properties of nuclear matter and finite nuclei. It is possible in particular to choose a set of effective interactions which give the same saturation properties of nuclear matter (binding energy and saturation density) but very different values for the compression modulus K_∞ and the effective mass m*. By calculating with these effective interactions the giant monopole resonances in nuclei, one can therefore study the relation between the excitation energy of the breathing mode and the compression modulus or the effective mass. Such calculations have been already performed and the results published in the reference[6]. We summarize here the essential features of these results.

A typical strength function is displayed in the figure 1. The bare particle-hole strength corresponds to those excitations which can be reached by the action of a monopole operator, such

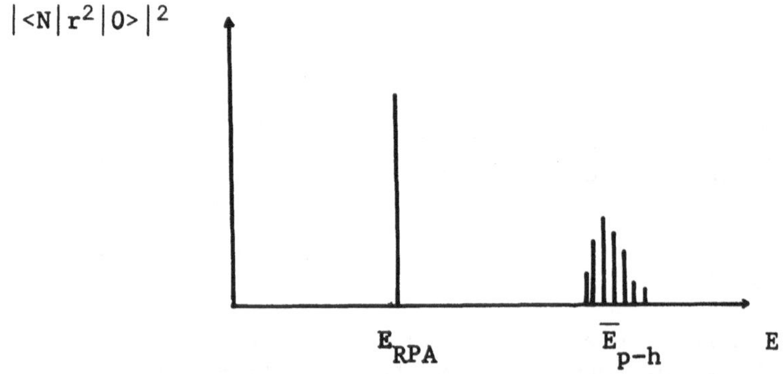

Fig. 1. Typical distribution of monopole strength in a heavy nucleus. The small peaks around $\bar{E}_{ph}$ correspond to the bare particle-hole configurations excited by the operator r^2. E_{RPA} is the energy of the collective mode which exhausts all the strength.

as r^2, on the Hartree-Fock ground state. When the particle-hole interaction is turned on, all the strength concentrates in a single state shifted by several MeV below the average particle-hole excitation energy. This situation is typical of medium and heavy nuclei where the collective mode exhausts most of the sum rule. In light nuclei the RPA strength is fragmented over several MeV. These features of the strength function as a function of the mass of the nucleus are in agreement with the recent analysis of (p,p') data by F.E. Bertrand et al.[7].

Concentrating on heavy nuclei, one can make use too, of the fact that the collective mode depletes the entire sum rule to write its frequency as the ratio of two moments of the strength function, for example as:

$$\omega = \frac{1}{\hbar} \left\{ \frac{\sum_n (E_n - E_o) |\langle n|r^2|0\rangle|^2}{\sum_n (E_n - E_o)^{-1} |\langle n|r^2|0\rangle|^2} \right\}^{1/2} . \tag{12}$$

The interest of formula (12) is that the numerator and the denominator of the ratio under the square root can be interpreted as the mass parameter and the restoring force constant for the collective mode. More precisely one has:

$$\omega = (C/B)^{1/2} \tag{13}$$

with

$$C = A\,K_A = \frac{2\langle r^2\rangle}{\sum_n (E_n - E_o)^{-1} |\langle n|r^2|0\rangle|^2} \tag{14}$$

$$B = m\langle r^2\rangle = \frac{1}{2}\frac{m^2}{\hbar^2} \sum_n (E_n - E_o) |\langle n|r^2|0\rangle|^2 .$$

As it appears from the formulae (14) the mass parameter B is independent of the interactions in so far as these interactions are precisely fitted in order to reproduce the r.m.s. radii. The restoring force constant C is strongly dependent on the interactions and from the formula (13) one expects that the frequency will vary with the force as the square root of an effective compression modulus K_A. Our numerical calculations show that K_A follows regularly the increase of the compression modulus of nuclear matter K_∞. In fact one can use the following parametrization in order to describe the variations of K_A as a function of A and K_∞:

$$K_A = K_\infty + K_{surf} A^{-1/3} + K_{sym} \left(\frac{N-Z}{A}\right)^2 + \ldots \qquad (15)$$

Typical values of the various coefficients are given in the table 1. The parametrization (15) suggests that K_A goes to K_∞ when $A \to \infty$ (hence the symbols). This is far from being obvious a priori and will be discussed later on. The regular relation between K_A

TABLE 1

Numerical values of the different coefficients introduced in eq. 15 in order to explain the variation of the compression modulus K_A as a function of K_∞ and A.

	B1	D1	Ska	SIV	SIII
K	190	228	263	325	356
K_{surf}	-300	-315	-394	-513	-568
K_{sym}	-500	-500	-610	-580	-630

and K_∞ induces a regular relation between ω and K_∞ which can be used to extract from the experimental data a value for the compression modulus:

$$K = 210 \pm 30 \text{ MeV} \qquad (16)$$

(see figure 2).

This is an important result which has motivated further analysis of our numerical calculations in order to understand better whether the relation observed between ω and K_∞ was accidental (due, for example to the necessarily limited choice of effective interactions used) or could be expected to hold even in a more sophisticated theory. There are two features of the results of our numerical calculations which deserve further explanation. The first one is the link observed between K_A and K_∞. The second one is connected with the role of the effective mass. Remembering that the effective mass is proportional to the density of single particle levels at the Fermi surface, one might expect, and this is what occurs in traditional RPA calculations that the position

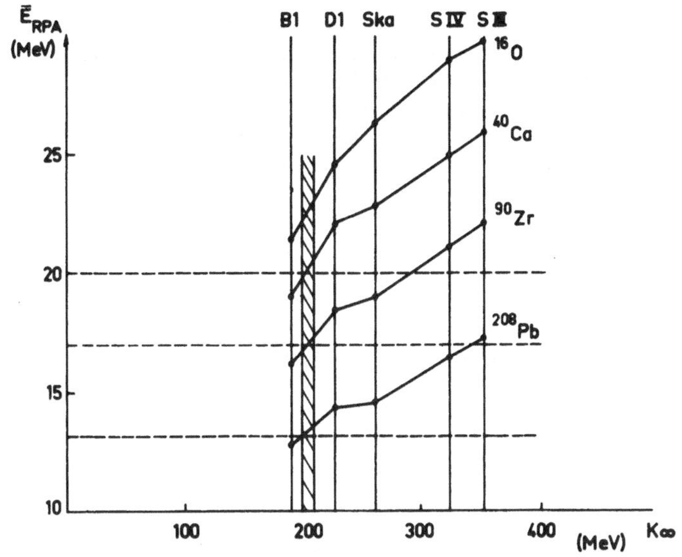

Fig. 2. The energy of the monopole resonance (E_{RPA}) is plotted as a function of K_∞. The horizontal dotted line corresponds to the experimental data.

of the collective mode will be higher the smaller the effective mass. This is indeed true for the average particle-hole excitation energy but not for the energy obtained in solving the RPA equations, as can be seen in table 2.

TABLE 2

Variations with the compression modulus K and the effective mass m* of the energy of the collective mode E_{RPA} and the average particle-hole energy $\bar{E}_{ph}$, in ^{208}Pb.

K_∞	190	228	263	325	356
$\frac{m^*}{m}$	0.43	0.66	0.61	0.47	0.76
$\bar{E}_{ph}$	28.2	23.2	24.1	30.2	21.9
E_{RPA}	12.5	14.4	14.7	16.5	17.2

More insight into these questions can be obtained by doing a local density approximation in order to express the restoring force as an integral over local quantities[3]. This approach makes use of the fact that in heavy nuclei the transition density of the collective mode can be obtained by making a single scaling transformation on the ground state density. In ref. (3) one derives the following expression for the restoring force:

$$C = \sigma \int d^3R \, \varepsilon_F(\vec{R}) \, (1+F_o(\vec{R})) \, n(\vec{R}) \qquad (17)$$

Comparing the expressions (17) and (4) one sees that in an infinite system K_A ($K_A = \frac{1}{A} C$) is equal to K_∞. The same is true if one assumes that the nucleus has a sharp surface. Thus most of the difference between K_∞ and K_A in a finite nucleus finds its origin in the existence of a diffuse surface. Furthmore, formula (17) makes obvious the fact that no effect can be correlated with the variations of the effective mass alone; indeed the effective mass combines, through ε_F, with an independent force parameter F_o to make the compression modulus. By distinction, in the case of the giant quadrupole resonance, one finds that the restoring force is proportional to ε_F and therefore to the inverse of the effective mass.

Finally it is worth pointing out that (17) results from an approximation on the RPA calculations. It is useful only in so far as it helps to understand the main features of the numerical calculations of ref. (6) which have been used to extract the value (16) of the compression modulus of nuclear matter. The main result of these calculations is to predict the existence in medium and heavy nuclei of a collective monopole mode of excitation with a single structure. The transition density of this mode can be obtained with a good accuracy from the ground state by a scale transformation. It is this simple property of the collective mode which explains the relation we have found between its frequency and the compressibility of nuclear matter. Let us emphasize again that the microscopic calculation clearly show that this simple property holds only for heavy nuclei; in light nuclei the monopole strength remains fragmented over the few particle-hole configurations which can be excited by a monopole operator.

REFERENCES

1. The material presented here will appear in a more complete form in Physics Reports (to be published in 1979).

2. A. Bohr and B. Mottelson, Nuclear Structure II. (W.A. Benjamin, Inc. 1975), p. 666.

3. J.P. Blaizot, Phys. Lett., 78B (1978), 367.

4. L.D. Landau, Sov. Phys. JETP 3 (1956), 920; 5 (1957), 101.

5. J.P. Blaizot and D. Gogny, Nucl. Phys. A284 (1977), 429.

6. J.P. Blaizot, D. Gogny and B. Grammaticos, Nucl. Phys. A265 (1976), 315. This paper contains a list of references of previous theoretical works on the breathing mode.

7. F.E. Bertrand, G.R. Satchler, D.J. Horen and A. Van der Woude. To be published in Phys. Rev. Lett.

STRANGENESS-EXCHANGE AND STRANGE NUCLEI

A. Gal

Racah Institute of Physics, The Hebrew University

Jerusalem, Israel

ABSTRACT

In these lectures, the nuclear strangeness exchange reaction (K^-,π^-) is discussed. Possible reaction mechanisms are reviewed in Section 2, while the spectroscopic Λ-hypernuclear information derived from observation of strangeness exchange is discussed in Section 3. The more general question of multiply-strange nuclei is introduced in Section 1.

1. STRANGE NUCLEI

Atomic nuclei are made of protons and neutrons which, ignoring weak interaction processes, are stable baryons. These, of course, are not the only stable baryons known to date. In Fig. 1 the non-charmed stable baryons are shown, together with their mass values and dominant decay modes. These baryons are grouped, vertically, according to their strangeness S. The latter is violated in weak interaction processes, a fact which is exemplified by the $\Delta S = 1$ weak mesonic decays sketched in the figure. All lifetimes associated with these mesonic decays are of the order of 10^{-10} sec, the appropriate leptonic decays proceeding with considerably lower rates (by about three orders of magnitude). The only exception to this rule is provided by the relatively fast electromagnetic M1 transition $\Sigma^0 \to \Lambda\gamma$ of rate about 10^{19} sec^{-1}.

Supported in part by US Department of Energy under Contract No. EY-76-C-02-0016.

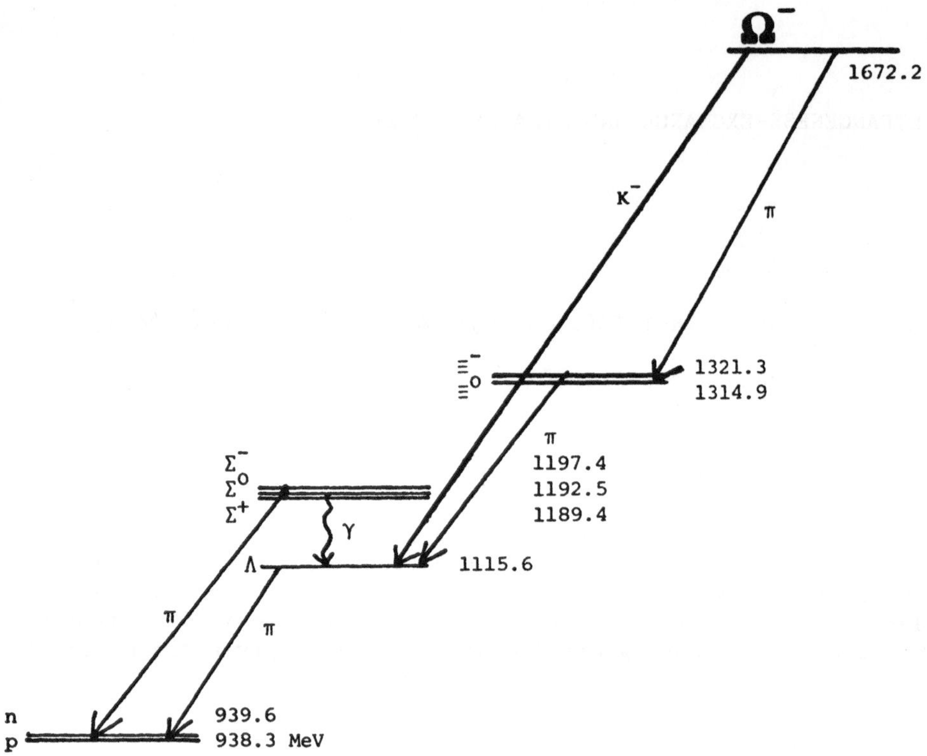

Fig. 1 - Dominant decay modes of non-charmed strange baryons

In deciding which strange nuclei are stable against strong decays, in consequence of trapping stable strange baryons, two immediate questions arise. The first one concerns the usefulness of various reaction mechanisms. For two-body reactions initiated by K^- beams, (K^-,π) and (K^-,K), the resultant strange nuclei are limited to $S = -1$ and $S = -2$ systems, respectively. More complex K^- initiated reactions, or associated production with non-strange projectiles, are not expected to practically change these considerations. The second question to be asked is which of the stable baryon states depicted in Fig. 1 survive strong conversion in the presence of nucleons. The situation is schematically illustrated in Fig. 2, where the lowest two-baryon configurations with S = -1, -2, -3 are shown.

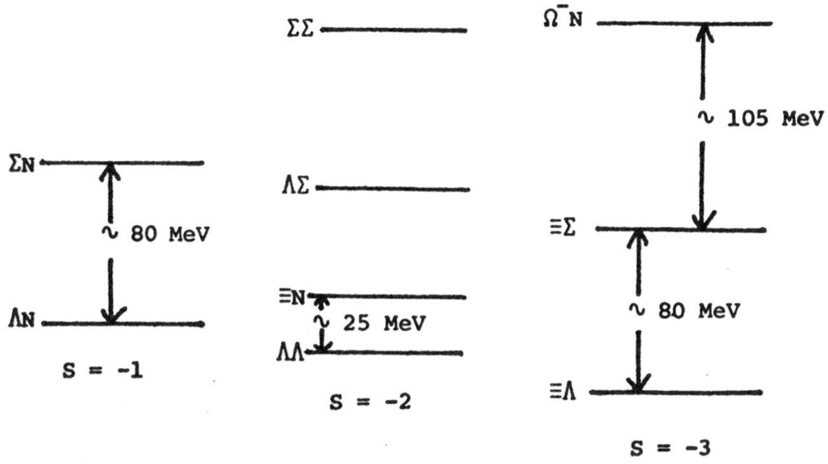

Fig. 2. Two-baryon configurations classified according to their strangeness, S = -1, -2, -3.

It is clear from this figure that, due to the large energy release, the baryon most unlikely to undergo nuclear binding is the Ω^-. We shall briefly analyse the S = -1, -2 cases in order to show that Σ's and Ξ's are also not expected to participate in stable nuclear systems. Thus, only Λ's are expected to form such bound systems with nucleons, and this is in accord with the observation of many S = -1 Λ-hypernuclei, with baryon-number greater than two, and two species of S = -2 Λ-hypernuclei ($^{6}_{\Lambda\Lambda}He$ and $^{10}_{\Lambda\Lambda}Be$).

1.1. S = -1 Systems

Most of hypernuclear information to date is traced back to emulsion work, where the formation sequence is hard to analyse and identification is necessarily linked to the dominant charged pionic decay mode $\Lambda \to p\,\pi^-$. The result is that mostly binding energy values (B_Λ) for ground states of light hypernuclei have been determined[1]. To determine a full hypernuclear spectrum, the two-body strangeness exchange reaction

$$K^- N \to \pi Y \qquad (Y = \Lambda, \Sigma) \tag{1a}$$

is applied to nuclei, e.g.

$$K^- \; {}^{A}Z \to \pi^- \; {}^{A}_{\Lambda}Z^* \quad . \tag{1b}$$

The main advantage of reaction (1b) over other possible reactions, e.g. associated production, is that the momentum transfer in this reaction can be made very small by appropriately choosing the kinematics . This will be discussed in Section 2.

As indicated in Fig. 2, the formation of Σ hyperons in nuclei will generally be accompanied by strong conversion $\Sigma N \to \Lambda N$, with energy release of about 80 MeV. However, since the ΛN system has isospin value of $^1/_2$, a bound ΣN system with $I = {}^3/_2$ cannot be ruled out. The most obvious examples are the unique charge symmetric states $\Sigma^- n$ and $\Sigma^+ p$. For the latter, scattering experiments in s-wave energies indicate $\Sigma^+ p$ nuclear interaction which is not strong enough to lead to a bound state. This is expected to hold for the singlet state, in spite of the strong attraction in this channel, whereas Coulomb-nuclear interference points to a repulsive nuclear interaction in the triplet state[3]. We note that one-step Σ^- formation in nuclei can be uniquely achieved by observing (K^-,π^+) which occurs only on protons. Thus, the $\Sigma^- n$ system may be formed in

$$K^- d \to \pi^+ \, (\Sigma^- n) \quad , \tag{2}$$

but this would overwhelmingly favour the triplet $\Sigma^- n$ component for π^+ detected at small angles. Since Σ^- is converted into Λ with any available proton, the only chance of binding Σ^- in heavier nuclear systems would be provided by $\Sigma^- nn$. The two-body formation

$$K^- \; {}^3H \to \pi^+ \, (\Sigma^- nn) \tag{3a}$$

is not considered practical and other formation sequences have been mentioned in the literature, such as

$$K^- \; {}^4He \to \pi^+ \, p \, (\Sigma^- nn) \tag{3b}$$

$$K^- \; {}^6Li \to \pi^+ \; {}^3He \, (\Sigma^- nn) \tag{3c}$$

It is unlikely, however, that the Σ^-nn system will turn out to be bound. Even if the Σ^-n s-wave interaction were attractive in both spin states, comparison of the observed Σ^+p[4] and Λp[5] cross sections in the intervals $160 \lesssim p_Y \lesssim 180$ MeV/c and $120 \lesssim p_Y \lesssim 200$ MeV/c, respectively, indicates that the statistical combination $\overline{V}_{YP} = (^3/_4)V^t_{YP} + (^1/_4)V^s_{YP}$ is still stronger for $Y = \Lambda$ than for $Y = \Sigma^+$. We may thus compare the system Σ^-nn with the Λnn system, which is known to be unbound. Since it is this same statistical combination of the YN s-wave interaction which enters both, we conclude that the Σ^-nn system is also unbound.

Strangeness minus-one is, hence, realized in nuclei only through the formation of Λ hypernuclei. Their formation (Eq. (1b)) via the one-step strangeness exchange reaction (1a) with $Y = \Lambda$ is the most promising experimental procedure to date. For kaons of several hundreds of MeV the cross sections involved are of the order of mb/sr in the forward direction and are thus readily measured. Hypernuclear spectra of ${}^9_\Lambda Be$, ${}^{12}_\Lambda C$, ${}^{16}_\Lambda O$, ${}^{32}_\Lambda S$ and ${}^{40}_\Lambda Ca$ have already been measured[6] in this way, in addition to the ground states of stable hypernuclei, from ${}^3_\Lambda H$ to ${}^{15}_\Lambda N$, determined in emulsion work[1]. These data yield new types of information, which are either hard or impossible to deduce from two-body data. Such information may be, broadly speaking, divided into two groups, (i) properties of ΛN interaction, e.g. exchange mixture, spin-orbit and (ii) weak non-mesic decays $\Lambda N \to NN$. In Section 3 we shall briefly discuss the strong-interaction aspect of this information.

1.2. S = -2 Systems

As mentioned earlier, two examples of double Λ hypernuclei have been established in the Sixties, ${}^6_{\Lambda\Lambda}He$[7] and ${}^{10}_{\Lambda\Lambda}Be$[8]. The $\Lambda\Lambda$ interaction in the $(s)^2\ {}^1S_0$ state in both these systems is estimated to be given by $\overline{V}_{\Lambda\Lambda} = B_{\Lambda\Lambda} - 2B_\Lambda$, where $B_{\Lambda\Lambda}$ is the separation energy of the strange pair and B_Λ is the separation energy of a single Λ in ${}^{(A-1)}_{\Lambda}Z$. This quantity is evaluated then to be $\overline{V}_{\Lambda\Lambda} \sim (4.5 \pm 0.5)$ MeV. Although the $\Lambda\Lambda$ interaction strength, as determined above, would be insufficient to bind two Λ's, the onset of stability for $\Lambda\Lambda$ hypernuclei is estimated[9] to occur for about A = 4, namely ${}^4_{\Lambda\Lambda}H$.

Formation of double hypernuclei is possible through double strangeness exchange[10,11] (K^-,K), e.g.

$$K^-\ {}^4He \to K^0\ {}^4_{\Lambda\Lambda}H \tag{4a}$$

$$K^- \, {}^6Li \rightarrow K^0 \, {}^{6}_{\Lambda\Lambda}He \qquad (4b)$$

$$K^- \, {}^{12}C \rightarrow K^+ \, {}^{12}_{\Lambda\Lambda}Be \qquad (4c)$$

In passing we remark that it would be useful to determine in reaction (4b) whether or not the "Fermi forbidden" low-lying $^3S_1$ state of ${}^{6}_{\Lambda\Lambda}He$ exists, as there lacks to date an experimental demonstration of the Λ statistics (except through the spin-statistics theorem). The only other low-lying state, presumably the ^1S ground state, cannot be formed in the forward direction. We note that reactions (4) proceed necessarily by two steps, for example:

$$K^-N_1 \rightarrow \pi\Lambda \quad , \quad \pi N_2 \rightarrow K^+\Lambda \qquad (5)$$

$$K^-N_1 \rightarrow K^+\Xi \quad , \quad \Xi N_2 \rightarrow \Lambda\Lambda \qquad (6)$$

Preliminary estimates call for dominance of mode (6) over mode (5) for K^+ observed in the forward direction. The overall hypernuclear formation cross section is estimated as several $\mu b/sr$[12].

Since the Ξ baryon may be formed, Eq. (6), in one step, the question arises whether or not it is likely to be bound in some nuclei. Looking back at Fig. 2 we see that a $\Xi N \rightarrow \Lambda\Lambda$ conversion will generally occur with the release of about 25 MeV, unless (i) charge conservation is effective, as for the unknown systems Ξ^-n, Ξ^-nn and their charge symmetric images, or (ii) Ξ binding is so large, saturating to more than 60 MeV in heavy nuclei, that the above conversion would become kinematically forbidden in nuclei. Both these exceptions depend, for their realization, on the strength of ΞN interaction. The ΞN $I = 1$ interaction may be studied in

$$K^-d \rightarrow K^+(\Xi^-n) \quad . \qquad (7)$$

We note that for a missing-mass determination, the $I = 1$ $\Sigma\Lambda$ and $\Sigma\Sigma$ channels enter only at higher excitation than is here relevant. On the theoretical side, there exist some indications[13] that the ΞN interaction is quite weak. These are based on OBE models[14] and the observation that, with the accepted range of F/D value for the PS octet, one pion exchange is of negligible importance for ΞN interaction. With a Ξ^-n interaction which is thus expected to be weaker than Λn, one readily concludes by comparing the unbound Λnn system to Ξ^-nn that the latter is unlikely to be bound.

Another $S = -2$ system, which due to charge conservation is decoupled from the rest, is $\Sigma^-\Sigma^-$ (and its charge symmetric image

$\Sigma^+\Sigma^+$). In a 8 $\otimes$ 8 SU(3) decomposition of two-baryon states this system is purely classified into the 27 representation, for which the baryon-baryon interaction is believed to be very strong (other members of this representation include the virtual $^1S_0$ nucleon-nucleon and the $^1S_0$ Σ^-n state). The simplest production reaction would be

$$K^- d \rightarrow K^+ \pi^+ (\Sigma^- \Sigma^-) \quad . \tag{8a}$$

with a generalization to proton-free final baryonic system:

$$K^- \, {}^4He \rightarrow K^+ \pi^+ \pi^+ (\Sigma^- \Sigma^- nn) \quad . \tag{8b}$$

It is clear that with spectator protons a strong conversion $\Sigma^- p \rightarrow \Lambda n$ will occur.

In conclusion of this Section we emphasize that strangeness in bound nuclear states is likely to be realized only through trapping of Λ particles. This, however, does not mean that observation of other strange baryons in nuclei is impossible, notably in quasi-free situations for the strange baryon.

2. REACTION MECHANISMS FOR (K^-,π^-)

The significance of the strangeness-exchange reaction (K^-,π^-) to the study of hypernuclei stems from the low momentum-transfer, $\underset{\sim}{q} = \underset{\sim}{p}_{K^-} - \underset{\sim}{p}_{\pi^-}$, which can be realized in the forward direction[2]. This ensures that hypernuclear states for which a neutron in the nuclear target is coherently substituted by a lambda in the same space-spin configuration are favourably formed[10,15]. Such states, except for the very light hypernuclei with $A \lesssim 4$, do not belong to the low-lying hypernuclear ground state configuration, and their study offers the possible determination of ΛN interaction parameters in Λ-states other than 1s states. The general kinematical plot[16] of momentum transfer q as function of the incident (laboratory) momentum p_{aL} is schematically shown in Fig. 3 for the exoergic two-body reaction

$$a + A \rightarrow b + B \quad , \tag{9}$$

with initial and final baryon masses, A and B respectively, satisfying $B > A$. For any scattering angle for the meson b in the cm system, the momentum transfer q varies between the values $|q_-|$ and q_+ appropriate to 0° and 180° respectively, the minimal value of q for given incident momentum being achieved at the forward direction, $\theta_b = 0$. This minimal value can be made equal to zero by solving for $(p_{aL})_0$:

$$(p_{aL})_0 = \sqrt{(B-A+a+b)(B-A+a-b)(B-A-a+b)(B-A-a-b)}/2(B-A) \quad (10)$$

with the result that for $p_{aL} < (p_{aL})_0$ the baryon B recoils backwards in the lab system, whereas for $p_{aL} > (p_{aL})_0$ it recoils forwards.

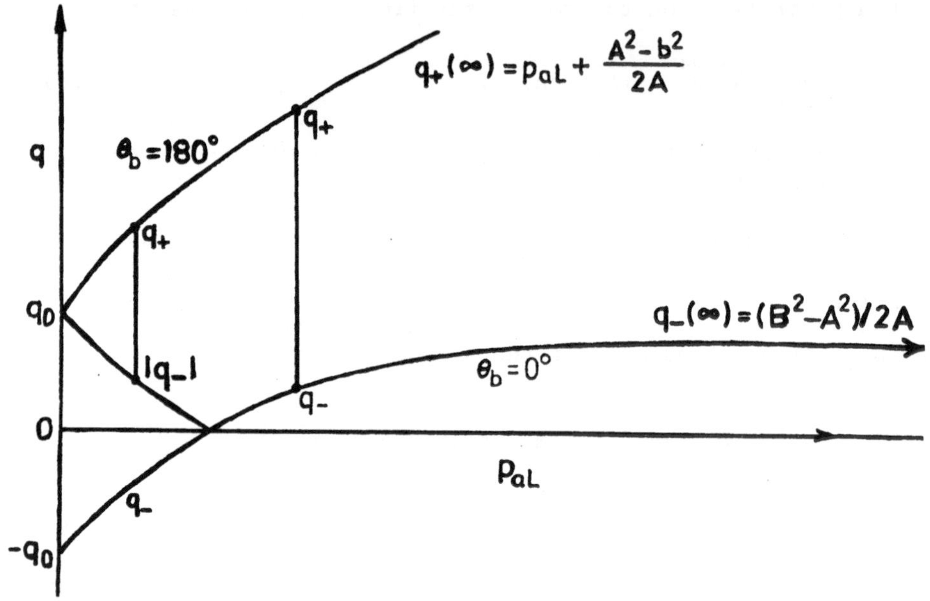

Fig. 3. The curves for q_+ and q_- for the reaction (9) on a stationary target A as function of incident laboratory momentum p_{aL}. The curve $|q_-|$ is the reflection of the curve q_- in the axis p_{aL}.

The actual situation[16] for $K^- n \to \pi^- \Lambda$ is sketched in Fig. 4, where the function $|q_-(p_K)|$ is given by the dash line. One observes that the value q_0 (253 MeV/c) appropriate to K^- capture at rest is considerably larger than the values of q in the region $0.5 \lesssim p_K \lesssim 2$ GeV/c [note that $(p_K)_0$ = 531 MeV/c], the asymptotic value q_∞ being given by 192 MeV/c. In this latter region the forward two-body cross-section, plotted as function of p_K given by a continuous line, varies between about 0.5 to 5 mb/sr, so that the choice of particular values of incident momentum, specifically about 750 MeV/c, and 1700 MeV/c, is expected to maximize hypernuclear formation with relatively small momentum-transfer. The situation may, however, prove more complicated than argued above due to the meson-nuclear initial and final state interactions which are somewhat stronger at the beginning of the region mentioned above.

We shall now deal with two extreme mechanisms for the (K^-,π^-) reaction, both of which rely heavily on the one-step nature of this reaction and implicitly assume that the effects of meson-nuclear interactions are unimportant for the shape of the obtained hypernuclear spectrum.

2.1. Quasi-Free Formation

This type of formation[17] means that the strangeness-exchange nuclear reaction (1b), with energy and momentum transfers

$$\omega = E_{K^-} - E_{\pi^-} \;, \quad \underset{\sim}{q} = \underset{\sim}{p}_{K^-} - \underset{\sim}{p}_{\pi^-} \;, \tag{11}$$

occurs on a single neutron, the remainder of the target nucleus acting as a "spectator". Thus, for a Fermi gas description of the target nucleus, for which neutron momenta $\underset{\sim}{k}$ are uniformly distributed below the Fermi momentum k_F, the quasi-free relationship between ω and $\underset{\sim}{q}$ is given by:

$$\omega = [M_\Lambda - U_\Lambda + (\underset{\sim}{k}+\underset{\sim}{q})^2/2M_\Lambda] - [M_N - U_N + k^2/2M_N]$$

$$= (M_\Lambda - M_N) + (U_N - U_\Lambda) - (1 - M_N/M_\Lambda)k^2/2M_N + q^2/2M_\Lambda + \underset{\sim}{k}\cdot\underset{\sim}{q}/M_\Lambda \;, \tag{12}$$

where U_N and U_Λ are the neutron and lambda nuclear-potential well-depths, respectively. Choosing the z axis along $\underset{\sim}{q}$ and, in view of the smallness of the term depending on $k_\perp^2$ in (12), averaging over $\underset{\sim}{k}_\perp$, the following expression obtains

$$\omega \simeq (M_\Lambda - M_N) + (U_N - U_\Lambda) - (M_N^{-1} - M_\Lambda^{-1})k_F^2/4 + q^2/2M_\Lambda + qk_z/M_\Lambda \;. \tag{13}$$

Of course, the magnitude as well as the direction of q generally depend on $\underset{\sim}{k}$, the momentum of the struck neutron. For first orientation this dependence will be ignored. For $\theta = 0^\circ$, where $\underset{\sim}{q}$ necessarily points in the forward direction, the magnitude q will be taken constant, with a value $\overline{q}$ appropriate to $k_z = 0$. Hence, ω is essentially linear in k_z and, since the Fermi k_z-distribution

$$\frac{dN}{dk_z} = \frac{3}{4k_F}(1 - k_z^2/k_F^2) \;, \quad |k_z| \le k_F \tag{14}$$

is a parabola in k_z, the resultant ω-distribution is also a parabola:

$$\frac{dN}{d\omega} = \frac{dk_z}{d\omega}\frac{dN}{dk_z} \simeq \frac{3M_\Lambda}{4\bar{q}k_F}\left(1 - \frac{M_\Lambda^2}{\bar{q}^2k_F^2}(\omega-\bar{\omega})^2\right) \quad (15)$$

for $|\omega-\bar{\omega}| \lesssim \bar{q}k_F/M_\Lambda$, and zero elsewhere. In this expression, $\bar{\omega}$ is the value $\omega(k_z = 0)$, corresponding to the energy transfer at the peak of the distribution:

$$\bar{\omega} = (M_\Lambda-M_N) + (U_N-U_\Lambda) - (M_N^{-1}-M_\Lambda^{-1})k_F^2/4 + \bar{q}^2/2M \quad . \quad (16)$$

In Fig. 5 quasi-free (QF) fits for the shapes observed[6] in ^{16}O, ^{32}S and ^{40}Ca for p_{K^-} = 900 MeV/c are shown. The position of the peak, $\bar{\omega}$, was fitted by eye and is seen to assume almost a constant value for all species considered. According to Eq. (16), this then allows the determination of $(U_N-U_\Lambda) \sim 31$ MeV, a value which is compatible with U_Λ being between 25 and 30 MeV, as also implied by the indirect observation of heavy hyperfragments in emulsion[18].

The crucial test of the QF hypothesis would experimentally be given by varying q with variation of the π^- detection-angle. According to Eq. (16), the QF peak will move to the region of higher excitation energy in the final hypernucleus, with a typical $q^2/2M_\Lambda$ behaviour. An effective Λ-mass may then be determined. The width of the QF peak will increase, roughly linearly with q.

The QF assumption for (K^-,π^-) becomes fully justified for $qR \gg 1$, where R is the nuclear radius. This allows values of q which are considerably smaller than k_F, and which are therefore completely inappropriate for a QF description of usual nuclear reactions in which the final (produced) baryon obeys the Pauli principle. For the species shown in Fig. 5, the product $\bar{q}R$ assumes values roughly between 1 and 2, so that the QF assumption is not that wild. The failure of the Fermi gas model to reproduce also the coherent excitations, clearly seen in the data, is due to $R \to \infty$ in this model. The proper consideration of such excitations requires the introduction of a finite size nuclear potential and residual interactions. The finite nuclear size is generally responsible for the dominance of coherent excitations for $qR \lesssim 1$, whereas the residual interactions are essential for splittings, mixings and shifts of any given group of coherent excitations. The QF shape calculated above does not, therefore, realistically account for the somewhat narrow excitations (coherent by expectation) observed[6] in the ^{16}O and ^{32}S (and similarly for ^{12}C and ^{9}Be) spectra.

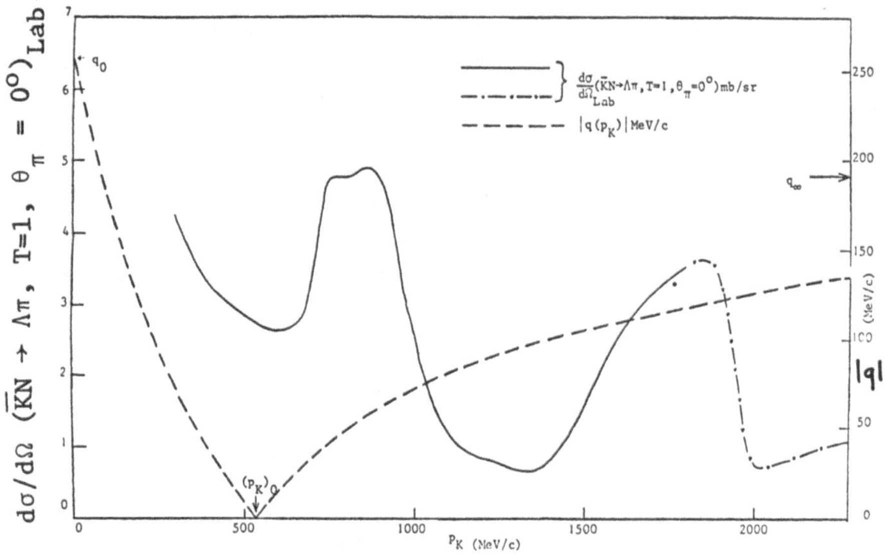

Fig. 4. The data available on $d\sigma/d\Omega$ ($\overline{K}N \rightarrow \Lambda\pi$, T=1, $\theta_\pi = 0^\circ)_{Lab}$ (in mb) is displayed as function of the incident $\overline{K}$ laboratory momentum p_K. The momentum transfer q given to the final Λ particle in this reaction, when it is projected backward relative to the $\overline{K}$ meson in the c.m. frame, is also plotted vs. p_K, showing the special momenta $(p_K)_0$, q_0 and q_∞ referred to in the text.

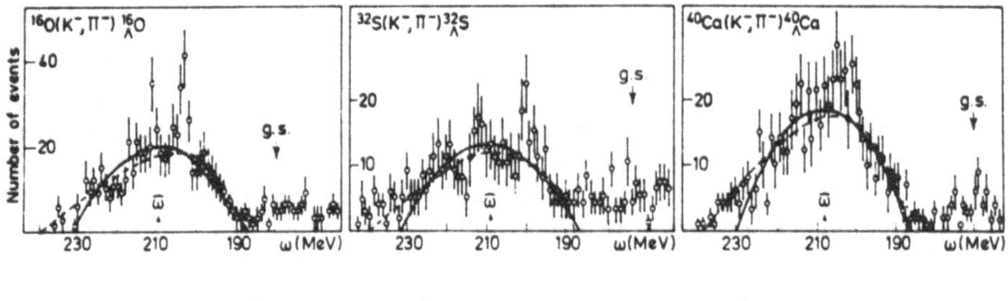

Fig. 5. Hypernuclear excitation spectra for (K^-,π^-) reaction on ^{16}O, ^{32}S and ^{40}Ca. The solid line gives the QF shape, eq. (15), for a constant momentum transfer q and a linear dependence of the energy transfer ω on k_z. The dashed line gives the QF shape with the full k_z dependence for ω and q. The location of (normalized) shapes has been fitted (by eye) to the observed broad hump. $\overline{\omega}$ divides the area of the spectra into two equal parts (from Ref. 17).

Bouyssy[19] and Bouyssy and Hüfner[20] performed a Distorted Wave Impulse Approximation (DWIA) particle-hole calculation for the p_K = 900 MeV/c data[6]. From their fit to the observed strongest coherent excitations, the latter always involving a coherent substitution of a valent neutron by a Λ, the value U_Λ = (27 ± 3) MeV is derived[20], in agreement with the discussion above. On the whole, the shapes of the observed spectra are reproduced[19]. The incoherent excitations which in their microscopic calculation give rise to the observed QF hump correspond to creation of the Λ in unbound single-particle states. The overall normalization in their calculation is still debatable, a point to which we shall come back below.

2.2. Strangeness Analog Resonance

Kerman and Lipkin[21] suggested that since the Λ single-particle wave-functions may closely resemble the corresponding nucleon single particle wavefunctions, this resemblance being roughly confirmed within the framework of a Hartree-Fock calculation[22], an approximate SU(3) symmetry is expected to generalize the isospin SU(2) symmetry into hypernuclei. If a^+_α, b^+_α and c^+_α denote creation operators for proton, neutron and lambda in a quantum state α, then the SU(3) generators are defined by

$$T_+ = \sum_\alpha a^+_\alpha b_\alpha, \quad T_- = \sum_\alpha b^+_\alpha a_\alpha, \quad T_0 = 1/2 \sum_\alpha (a^+_\alpha a_\alpha - b^+_\alpha b_\alpha) \tag{17a}$$

$$U_+ = \sum_\alpha b^+_\alpha c_\alpha, \quad U_- = \sum_\alpha c^+_\alpha b_\alpha, \quad U_0 = 1/2 \sum_\alpha (b^+_\alpha b_\alpha - c^+_\alpha c_\alpha) \tag{17b}$$

$$V_+ = \sum_\alpha c^+_\alpha a_\alpha, \quad V_- = \sum_\alpha a^+_\alpha c_\alpha, \quad V_0 = 1/2 \sum_\alpha (c^+_\alpha c_\alpha - a^+_\alpha a_\alpha) \tag{17c}$$

(Note that $T_0 + U_0 + V_0 = 0$.) In the limit of this symmetry the above generators commute with the Hamiltonian; nuclear and hypernuclear states of the same space-spin structure are then classified into uniquely prescribed SU(3) multiplets, which are constructed in the Sakata triplet version. Since for $q \to 0$, and within the Plane Wave Impulse Approximation (PWIA), the transition operator for (K^-,π^-) is given by the generator U_-, the strength of this strangeness exchange reaction is expected to concentrate on the Strangeness Analog Resonance (SAR) defined by

$$|S_n\rangle = \frac{1}{\sqrt{N}} U_- |\pi\rangle \quad , \tag{18}$$

where $|\pi\rangle$ denotes the parent state and N is the total neutron number in this state. It is evident that this analog state is obtained from the parent by substituting coherently each of the neutrons by Λ in the same space-spin quantum state. For $N > Z$ nuclei the U-spin analog state is not an eigenstate of isospin, and two partial-analogs may be projected out of it, with $T_< = T-1/2$ and $T_> = T+1/2$. The situation is somewhat simpler for the V-spin analog state defined by

$$|S_p> = \frac{1}{\sqrt{Z}} V_+|\pi> \quad , \tag{19}$$

which has, for $N \geq Z$, a unique isospin value of $T_>$. This is the state on which the strength of the (K^-,π^0) reaction is expected to concentrate for $q \to 0$ and within the PWIA. The excitation energy (in the sense of minus the appropriate Q-value) of the V-spin analog is estimated by:

$$\Delta E_V = <S_p|H|S_p> - <\pi|H|\pi>$$

$$= \frac{1}{Z} <\pi|[V_-[H,V_+]]|\pi> \tag{20}$$

which, within the single particle framework, is evaluated to be given by the difference in the Hartree potentials for Λ and proton. This is the same as found above ($\sim$ 30 MeV) for the shift of the quasi-free peak for $q \to 0$. Indeed, in this symmetry limit and for $q \to 0$ the whole quasi-free spectrum shrinks to just one state, the analog state.

It is now simple matter to estimate the location of the two isospin components of the U-spin analog state. The one with larger value of isospin, $T_> = T + 1/2$ is the isospin analog of $|S_p>$:

$$|S_> > = \frac{1}{\sqrt{2T+1}} T_+|S_p> \tag{21}$$

and it is expected to lie higher than $|S_p>$ by the usual Coulomb energy. $|S_< >$ is obtained by orthogonalization and is expected to lie below $|S_> >$ by about the usual symmetry energy. The situation for heavy nuclear targets, where the nuclear symmetry energy and the Coulomb energy approximately cancel each other, is depicted in Fig. 6. The appropriate transition amplitudes f are easily evaluated to be

$$f_p = <S_p|V_+|\pi> = \sqrt{Z} \tag{22a}$$

$$f_{>} = \langle S_{>}|U_{-}|\pi\rangle = \sqrt{\frac{Z}{2T+1}} \tag{22b}$$

$$f_{<} = \langle S_{<}|U_{-}|\pi\rangle = \sqrt{\frac{2T}{2T+1}\,(N+1)} \quad , \tag{22c}$$

so that only the lower U-spin analog is expected to be strongly excited for (N-Z) >> 1.

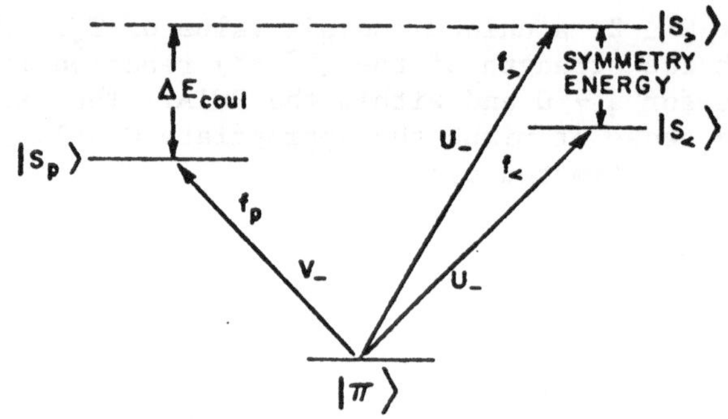

Fig. 6. Schematic representation of the strangeness analog states and the transitions from the parent.

The SU(3) symmetry is expected to be badly broken, at least for light nuclei. The simplest example[23] for this breaking is provided by ${}^{12}_{\Lambda}C$, for which the single-particle estimate for the energies of the coherent $(p^{-1}_{3/2})_n(p_{3/2})_\Lambda$ and $(s^{-1}_{1/2})_n(s_{1/2})_\Lambda$ 0^+ states gives a difference of about 10 MeV for these states. The residual ΛN interaction is incapable of appreciably mixing the two 0^+ states, so it appears very likely that the strong ${}^{12}_{\Lambda}C^*$ excitation, with $B_\Lambda \sim 0$, corresponds basically to the $(p^{-1}_{3/2})_n(p_{3/2})_\Lambda$ 0^+ state.

The situation for heavy nuclei is not yet experimentally resolved. It has been suggested[24] that the close level spacings increase the degeneracy of particle-hole excitations for large values of A and enhances the formation of at least one coherent excitation based on deeply lying neutron hole states. However, the width of these hole states may be prohibitively large and be reflected in the physical spreading widths of these theoretical constructs.

2.3 Is the DWIA Applicable to (K^-,π^-)?

This question arises in the context of the summed hypernuclear formation rate. As discussed in the literature[17], the summed rate at 0° is evaluated within the DWIA to be given by:

$$d\sigma(0^o)/d\Omega_L = \sum_f |\int \chi^{(-)*}_{p_{\pi^-}}(\tilde{r})\langle f|\sum_j U_-(j)\delta(\tilde{r}-\tilde{r}_j)|i\rangle \chi^{(+)}_{p_{K^-}}(\tilde{r})d\tilde{r}|^2\delta(\omega-E_f+E_i)$$

$$\times\ d\sigma(0^o)/d\Omega_L(N\!\rightarrow\!\Lambda) \quad , \qquad (23)$$

where $\chi^{(\pm)}$ denote the appropriate distorted waves. Integrating over ω and neglecting the weak variation of p_{π^-} along the excitation spectrum (corrections to which variation can readily be estimated to involve several percent for the present case), closure in the summation on final hypernuclear states is encountered. The ultimate result is expressed by

$$d\sigma(0^o)/d\Omega_L = N_{eff}\, d\sigma(0^o)/d\Omega_L(N\!\rightarrow\!\Lambda) \quad , \qquad (24)$$

where the effective neutron number

$$N_{eff} = \int \rho_N(\tilde{r})|\chi^{(-)}_{p_{\pi^-}}(\tilde{r})|^2|\chi^{(+)}_{p_{K^-}}(\tilde{r})|^2 d\tilde{r} \qquad (25)$$

reduces to the actual neutron number N for no distortion.

Expression (25) has been recently evaluated[25] under several alternative assumptions for the distorting potentials at $p_{K^-} = 900$ MeV/c and ^{12}C, ^{40}Ca, with the results 1.8 and 3.5, respectively. This is to be compared to the experimental finding of Bruckner et al.[6] of $N_{eff} \sim 0.3$ (± 30%), uniformly over the range $9 \leq A \leq 40$ considered in their experiment. A discrepancy of about order-of-magnitude between experiment and theory is seen to arise for ^{40}Ca. The origin of this discrepancy is not yet resolved, but on the theoretical side it may be that the one-step DWIA is insufficient for describing all the strangeness exchange transitions included along the excitation spectrum. Conventional two-step processes which involve meson rescattering are estimated, however, to be quite unimportant. It is, of course, necessary to give realistic estimates for the two-step processes involving $\Sigma \rightarrow \Lambda$ conversion

$$K^-N_1 \rightarrow \pi^-\Sigma \quad , \qquad \Sigma N_2 \rightarrow \Lambda N \quad , \qquad (26)$$

but it is difficult to envisage a situation where these processes would play as important role for the summed formation rate as the one-step process, without appreciably modifying the QF shape discussed above.

3. SPECTROSCOPY

In this lecture we shall give brief examples of what type of information on the ΛN interaction parameters may be deduced from (K^-,π^-) experiments at reach, with resolution somewhat greater than 1 MeV. Future experiments, with improved resolution, will enable one of deducing hypernuclear spectroscopic information almost matching that achieved by strangeness-conserving nuclear reactions.

3.1. $(K^-,\pi^-\gamma)$

A systematic discussion of the γ-yield expected for the electromagnetic decay of hypernuclear low-lying particle-stable states in the 1p shell, following their formation with the (K^-,π^-) forward reaction, has been recently given[16]. The case of ^{7}Li is considered fundamental enough to warrant display[26], since $^7_\Lambda$Li is the lightest p-shell hypernucleus from which γ emission will definitely occur[27]. The particle-stable states expected for $^7_\Lambda$Li are shown in Fig. 7, together with their dominant nuclear parents, for one of the B_Λ-fits[28] performed in the 1p shell. The numbers in the square brackets give the (DWIA) calculated formation rates for the reaction $K^-\ {}^7\text{Li} \to \pi^-\ {}^7_\Lambda\text{Li}^*$ at 0°, relative to the total $1p_N \to 1s_\Lambda$ transition rate. The strongest γ-ray expected would be the E2 $5/2^+ \to 1/2^+$ (g.s.), since the $5/2^+$ state is the one favourably formed. Determination of this γ-ray is supposed to indicate the splitting of the $(5/2^+, 7/2^+)$ hypernuclear doublet built by attaching the 1s Λ to the first excited 3^+ ^{6}Li core state. This splitting is predicted to mostly arise from a term $\ell_N \cdot s_\Lambda$ induced by the ΛN two-body spin-orbit interaction. This is in contrast with the measured[29] doublet splitting for $^4_\Lambda$H which is dominated by the spin-spin term $s_N \cdot s_\Lambda$.

A case of special interest among the p-shell hypernuclei is $^{12}_\Lambda$C for which the observed ($E^* \sim 11$ MeV) coherent excitation, expected to be 0^+ dominantly of the structure $(1p_{3/2}^{-1})_n (1p_{3/2})_\Lambda$, is barely bound. The other J^π states belonging to the latter configuration are shown on Fig. 8. The formation of the 1^+ and 3^+ states is generally forbidden in the forward direction due to angular-momentum and parity conservation, and their excitation at $0^o < \theta \lesssim 30^o$, where present experiments are capable of pion detection, is probably very weak[13]. On the other hand, the 2^+ is expected to be excited as strongly as the 0^+ state away from

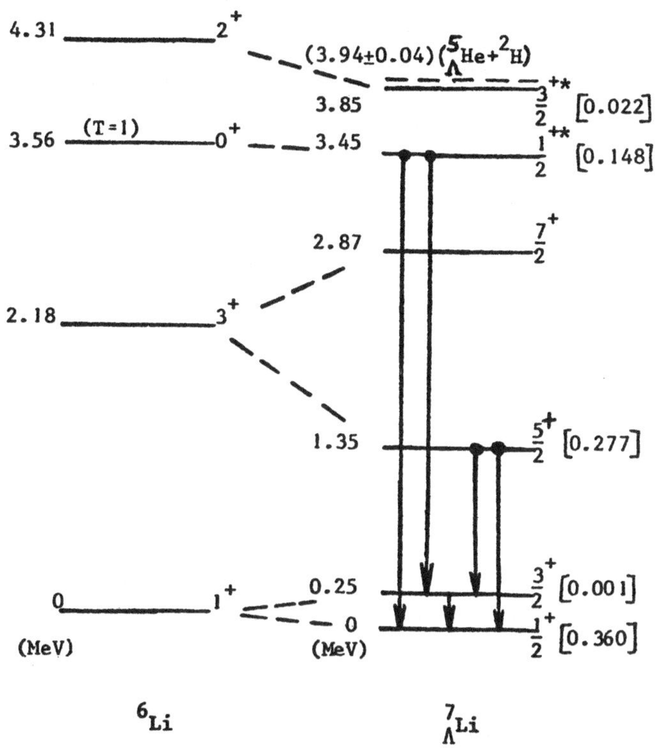

Fig. 7. The energy levels known for ^{6}Li and calculated for $^7_\Lambda$Li using the ΛN parameters of the fit (Δ^+S^+Q;79). The numbers in the square brackets give the formation rates for the reactions $K^-+{}^7Li\to\pi^-+{}^7_\Lambda Li^*$ at 0°, relative to the total $1p_N\to 1s_\Lambda$ transition rate.

the forward direction, although its formation at 0° is probably on the level of 1% of the 0^+ formation[13]. This 2^+ state is generally expected to lie below the 0^+ state of the same configuration, because its structure involves more spatial symmetry than that of the latter, and hence be particle-stable. Starting, for first orientation, from the above pure configuration and representing the ΛN interaction in terms of a central interaction with exchange mixture parameter ε', $V_{\Lambda N} = W_{\Lambda N}\,[(1-\varepsilon') + \varepsilon' P_r]$, the calculational approach adopted in Ref. 30 yields for the $0^+ - 2^+$ energy difference the following values

ε':	0	0.25	0.5	
ΔE(MeV) :	1.9	3.0	4.1	(27)

This estimate will, in real life, be modified by at least two

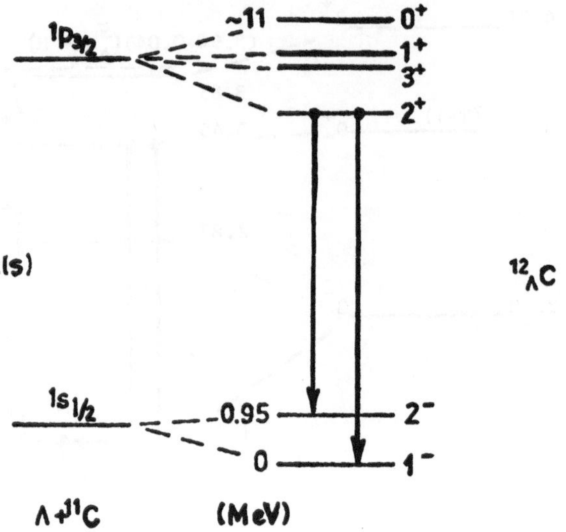

Fig. 8. The lowest energy levels expected for each of the configurations $(1p^{-1})_n(1s)_\Lambda$ and $(1p^{-1})_n(1p)_\Lambda$ in ${}^{12}_{\Lambda}C$. The 0^+ state lies above the particle-stability threshold, which is at 9.8 MeV (for proton emission).

factors: (i) the nuclear core state is not purely $p^{-1}_{3/2}$, so the use of intermediate coupling scheme for the parent nucleus is required, and (ii) if the Λ spin-orbit potential term is quite small, then the mixing of the above 2^+ state with the $[(1p^{-1}_{3/2})_n(1p_{1/2})_\Lambda]_{2^+}$ state should be taken into consideration; this can only push further down (relative to (27)) the lowest 2^+ state of the $(1p^{-1})_n(1p)_\Lambda$ configuration.

If this lowest 2^+ state in ${}^{12}_{\Lambda}C^*$ is indeed particle-stable, as we argued above, it will fastly decay by E1 γ-emission dominantly to the ground-state doublet, as shown in Fig. 8. Observation of any such transition, or of the secondary $2^- \rightarrow 1^-$ M1 spin-flip transition, would be very valuable for putting constraints on the various (not too many!) components of the ΛN interaction.

3.2. ${}^{16}_{\Lambda}O^*$ and the Λ Spin-Orbit

In the ${}^{16}_{\Lambda}O^*$ spectrum shown in Fig. 5, only one coherent excitation clearly sticks out of the QF background. Bouyssy and Hüfner[20], however, argued that a second, although considerably weaker, coherent excitation lies about 5 to 6 MeV below the dominant one and both of these 0^+ states belong to the $(1p^{-1})_n(1p)_\Lambda$

configuration. Since the latter spacing is essentially reproduced by the nuclear $p_{3/2}$-$p_{1/2}$ spin-orbit splitting, there would be not much more room left for an additional Λ spin-orbit effect. Their more involved diagonalization led them then to the estimate of (0 ± 0.3) for the ratio of Λ to N spin-orbit splitting in the ^{16}O region. In their calculation, no exchange mixture for the ΛN interaction was introduced; rather, the less important spin-spin component (which à-priori is not more significant than the other spin-dependent ΛN interaction terms) was included. Here, I shall follow the calculational approach of Ref. 30 for ΛN central interaction of exchange mixture ε', with one-body spin-orbit parameters ζ_N and ζ_Λ for the functional form $\zeta \underset{\sim}{s} \cdot \underset{\sim}{\ell}$. It is convenient, though not necessary, to start from a restricted SU(3) symmetry limit which degenerates only the 1p nucleon and lambda orbitals (with $\zeta_\Lambda = \zeta_N$) and ignores all other orbitals, since the latter do not affect $(1p^{-1})_n(1p)_\Lambda$ excitations. The situation is depicted in Fig. 9. The two 0^+ degenerate basis states $(1p^{-1}_{3/2})_n(1p_{3/2})_\Lambda$ and $(1p^{-1}_{1/2})_n(1p_{1/2})_\Lambda$ are split and mixed by a spin-independent residual ΛN interaction with eigenstates ψ_a and ψ_s, as shown in the figure. The upper state ψ_a is precisely the

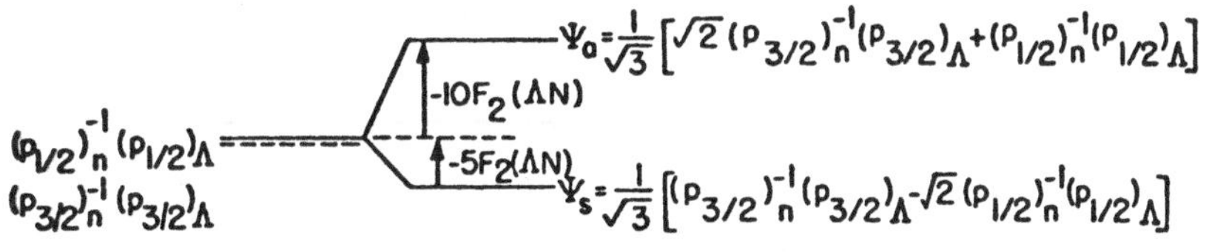

Fig. 9. The $J^\pi = 0^+$ states of $^{16}_{\Lambda}$O which are degenerate in the jj basis for the case $\delta\zeta = 0$, are split into the analog state ψ_a and the supersymmetric state ψ_s by a residual Wigner two body interaction (for a more general interaction, consult Ref. 30). The coincidence of these states with the LS basis states, the analog state being $^1S_0$ and the supersymmetric state being $-^3P_0$, is special for this system and does not hold in general.

strangeness analog state confined to the 1p shell; hence, it is given by the LS assignment $^1S_0$, as for ^{16}O ground state. The lower state ψ_s is then necessarily a $^3P_0$ state with space symmetry higher than that of the analog and provides a simple example of a supersymmetric state[30]. The only spin-dependent ΛN interaction capable of mixing $^3P_0$ and $^1S_0$, in the symmetry limit, is the spin-orbit interaction $\sim(\underset{\sim}{s}_\Lambda-\underset{\sim}{s}_N)\cdot\underset{\sim}{\ell}_{\Lambda N}$. Inasmuch as its effects can be represented by an average Λ spin-orbit potential, we may relate this mixing to $\delta\zeta=\zeta_\Lambda-\zeta_N$. In Table 1 results of diagonalization, for various strengths $\delta\zeta$ and exchange mixture parameters ε' are displayed. ΔE is the spacing (in MeV) between the two levels, the upper one being represented as $\cos\theta|^1S\rangle + \sin\theta|^3P\rangle$ and the lower one by the orthogonal combination; $\tan^2\theta$ then gives the ratio for population

$\delta\zeta$	$\varepsilon' = 0$		$\varepsilon' = 1/4$		$\varepsilon' = 1/2$	
	ΔE	$\tan^2\theta$	ΔE	$\tan^2\theta$	ΔE	$\tan^2\theta$
-2	3.3	0.34	4.3	0.14	5.6	0.07
0	3.6	0	5.3	0	6.9	0
2	5.4	0.08	6.9	0.05	8.3	0.03
4	8.0	0.17	9.2	0.12	10.5	0.09
6	10.8	0.24	11.8	0.18	13.0	0.14

Table 1 - The separation energy ΔE (in MeV) between the upper and the lower 0^+ $^{16}_{\Lambda}O^*$ states belonging to the $(1p^{-1})_n(1p)_\Lambda$ configuration, and the relative intensity $\tan^2\theta$ for (K^-,π^-) direct production at zero momentum transfer of the lower (mostly supersymmetric) state relative to that of the upper (mostly analog) state are given as function of the SU(3) breaking spin-orbit potential $\delta\zeta$ (in MeV) and several values of the Λ exchange-mixture parameter ε'. The Soper (private communication) Slater integrals, reduced by factor 0.6 to go from the NN to the ΛN case, are used.

of the lower state relative to the upper one in the limit $q_\perp\to 0$. In the SU(3) limit for the spin-orbit strength, $\delta\zeta = 0$, only one state is expected to be observed. Note, however, that even for zero strength for the Λ spin-orbit, corresponding in ^{16}O to $\delta\zeta = 4$ MeV, the relative formation rate of the lower state is quite low, ranging between 0.17 and 0.09 as ε' varies from 0 to 0.5. If ΔE is fixed at the "observed" value of $\sim$ 5.5 MeV, a range of $\delta\zeta$ between 2 to -2 MeV is allowed as ε' is varied from 0 to 0.5; however, the relative formation rate for the lower state is smaller than 10%, considerably less than the Heidelberg group[20] would like to extract from the raw data. It seems, therefore, premature to state a value for the Λ spin-orbit coupling.†

3.3. ${}^{9}_{\Lambda}Be^*$ and the ΛN Exchange-Mixture

${}^{9}Be$ is the lightest nucleus exposed to date to (K^-,π^-) and its hypernuclear excitation function at 0°, p_{K^-} = 900 MeV/c, shows two prominent peaks separated by about 11 MeV apart[6]. It is also the only $T_i \neq 0$ nucleus investigated in this experiment, and it is useful to look at the two observed peaks from the point of view of their isospin structure; the predictions of the SAR model, discussed in section 2.2 call for assigning $T_f = T_i - 1/2$ to the lower peak and $T_f = T_i + 1/2$ to the upper one. As we shall see below, the actual picture is somewhat more involved. This is a direct consequence of dealing only with the valent configuration $(1p)^5$ in both ${}^{9}Be$ and ${}^{9}_{\Lambda}Be^*$.

Two coupling schemes appear natural in this context. The most straightforward one is to couple the 1p Λ to the observed core states of ${}^{8}Be$, the latter being taken from a realistic nuclear model calculation[31]. The energy matrices for the ΛN interaction are then set in this basis and diagonalization performed. The other coupling scheme is the symmetry limit, appropriate to the situation $V_{\Lambda N} = V_{NN}$, the latter equality taking place for ΛN and NN states with the same space-spin structure. The SU(2)-isospin multiplets become then embedded in SU(3)-Sakata multiplets. This symmetry limit would be useful even if $V_{\Lambda N}$ differed a lot from V_{NN}, provided that their difference is of relatively long range with small off-diagonal matrix elements. In the particular case of ${}^{8}Be$ and ${}^{9}Be$, the nuclear spin-orbit coupling is known to be weak and the nuclear level schemes are qualitatively interpreted within the SU(4)-Wigner supermultiplet framework, which also provides an approximate basis for the "exact" intermediate-coupling calculation[32]. Considering hypernuclei and nuclei, in this mass region and excitation range, on the same footing means expanding SU(4) to SU(6), the latter group arising from the direct product of SU(2)-spin and SU(3)-Sakata for the discrete degrees of freedom of the five 1p baryons in ${}^{9}Be$ and ${}^{9}_{\Lambda}Be^*$.

We shall outline here the second approach[30], since it is not well known or practised. The lowest states of the $(1p)^5$ configuration are classified according to decreasing degree of space symmetry for the baryonic wave-function. The lowest supermultiplet

---

†See note added at the end of these lectures.

is denoted by (5) and its SU(2) ⊗ SU(3) structure is very simple:

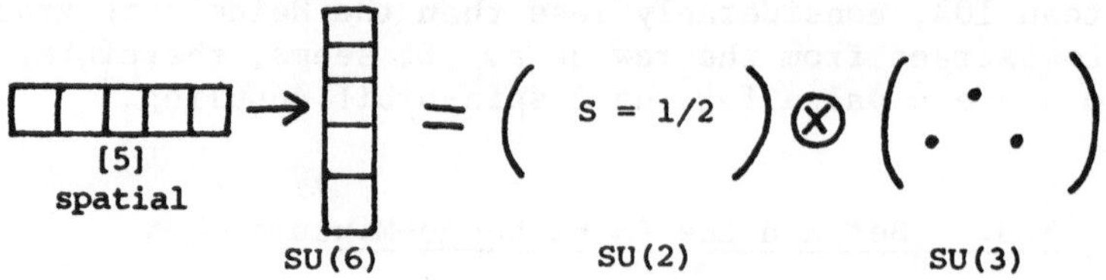
[5]
spatial
S = 1/2
SU(6)
SU(2)
SU(3)

(28)

Another frequent notation for this SU(6) decomposition would be $6 = {}^{2}\overline{3}$, explicitly referring to the dimensionalities involved. The vertical scale for the SU(3) multiplets is given by the appropriate value of strangeness, which in this case is -1 for the upper (${}^{9}_{\Lambda}$Be*, T = 0) member of the SU(3) anti-triplet. This means that the supermultiplet (5) is not realized for ordinary nuclei (zero strangeness), due to the Pauli principle, but is allowed for hypernuclei. ${}^{9}_{\Lambda}$Be* states (generically denoted here by S_o) belonging to the (5)-supersymmetric-representation are not expected to have been seen in (K^-,π^-) low-q experiments, since the dominant components in the ^{9}Be g.s. wave function[32] are of [4,1] symmetry.

Next, in excitation energy, comes the supermultiplet [4,1], with the more complex SU(2) ⊗ SU(3) structure:

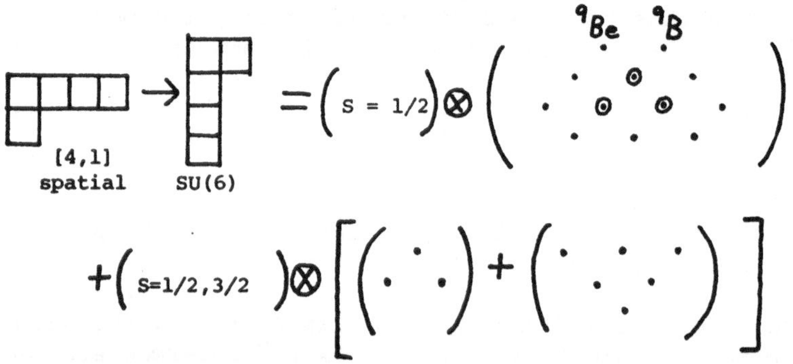
⁹Be ⁹B
[4,1]
spatial
SU(6)
S = 1/2
S=1/2,3/2

(29)

or, with dimensionalities:

$$84 = {}^{2}\overline{15} + {}^{2}\overline{3} + {}^{4}\overline{3} + {}^{2}6 + {}^{4}6 \quad . \tag{30}$$

Of these SU(3) multiplets, only the 15-dimensional representation ([3,2] is one notation, or (1,2) in the more familiar (λ,μ) notation) has states with zero strangeness. Indeed, the upper isodoublet corresponds to ground-state ^{9}Be and ^{9}B, whereas the double-point in the second row corresponds to the T = 0,1 components of the analog ${}^{9}_{\Lambda}$Be* state. The hypernuclear states belonging to the $\overline{3}$ and 6 multiplets are non-analog states. In the SU(6) limit, and for a given value of L, all the states given in Eq. (30) are degenerate with each other. For simplicity, we shall confine attention to $^2P_{3/2}$ states, since this is the dominant component[32] of ^9Be g.s. According to Eq. (30) there are four $^2P[4,1]$ states, two with T = 0 and two with T = 1. These states are partly mixed and split as a result of introducing SU(3) breaking interaction, $\delta V = V_{\Lambda N} - V_{NN}$. We assume this breaking to be spin-independent; it is expected that the unknown spin-dependence of the ΛN interaction, coupled with the known SU(4)-breaking of the NN interaction, will introduce further fine-splittings of the order 1 to 2 MeV. The result of this SU(3) breaking is surprising, at first sight, since the two T = 1 states (of which the $^2\overline{15}$ is denoted below by U_1) do not split at all and remain degenerate with each other, as well as with the upper T = 0 state. The T = 0 states mix and split to give the upper and lower eigenstates:

$$U_0 = -\sqrt{3/8}\ (^2\overline{15}) - \sqrt{5/8}\ (^2\overline{3}) \tag{31a}$$

$$L_0 = \sqrt{5/8}\ (^2\overline{15}) - \sqrt{3/8}\ (^2\overline{3}) \quad , \tag{31b}$$

irrespective of the strength of the SU(3)-breaking central interaction.

The explanation of this degeneracy is quite simple, and may readily be understood in terms of coupling the Λ to ^8Be core states[31]. The [4,1] supermultiplet has parents in ^8Be of the type [4] and [3,1], with (T,S_N) combinations (0,0) for the lower [4] and (0,1) (1,0) (1,1) for the upper [3,1]. Thus, the [4,1] T = 1 $^9_\Lambda Be^*$ states are necessarily obtained by coupling the Λ to the various degenerate (1,S) states of the [3,1] supermultiplet. With spin-independence assumed for the ΛN interaction the two T = 1, S = 1/2 hypernuclear states are labelled by the value of the nuclear spin S_N and, thus, are not mixed, their common shift also equals that for the S = 3/2 hypernuclear state. This same shift also applies to the T = 0,S = 1/2 and S = 3/2 states obtained by coupling the Λ to the (assumed) unique [3,1] core in its S = 1 substate. The other T = 0, S = 1/2 state, that obtained by coupling the Λ to the [4] core, is necessarily decoupled via a central ΛN interaction from the former T = 0, S = 1/2 state since it is based on S_N = 0 nuclear core state.

The situation is summarized in Fig. 10; the upper and lower T = 0 states given by Eq. (31) are precisely those corresponding to coupling the Λ to the [3,1] and [4] ^8Be supermultiplets, respectively. At $q \to 0$, the strangeness exchange strength would be split between these upper and lower states. Applying the calculational procedure of Ref. 30 we find a spacing for these two peaks of 11.2, 9.0 and 6.7 MeV for $\varepsilon' = 0$, 1/4 and 1/2, respectively. This interpretation of the data is novel, in that the upper peak consists of two analog components, at least, with T = 0 and 1, and the lower peak has T = 0 only.

In the above considerations we neglected the mixing induced by a central δV between the [4,1] and [5] $^{9}_{\Lambda}Be^*$ supermultiplets. Since the latter is uniquely based on ^{8}Be [4] states, with $(T,S_N) = (0,0)$, it may be mixed only with the lower T = 0 state L_0. This mixing is relatively unimportant in view of the different space-symmetries involved. Applying again the same calculational procedure as in Ref. 30, the spacing between L_0 and S_0 is found to range from 5.5 to 8.6 MeV for ε' ranging between 0 to 0.5.

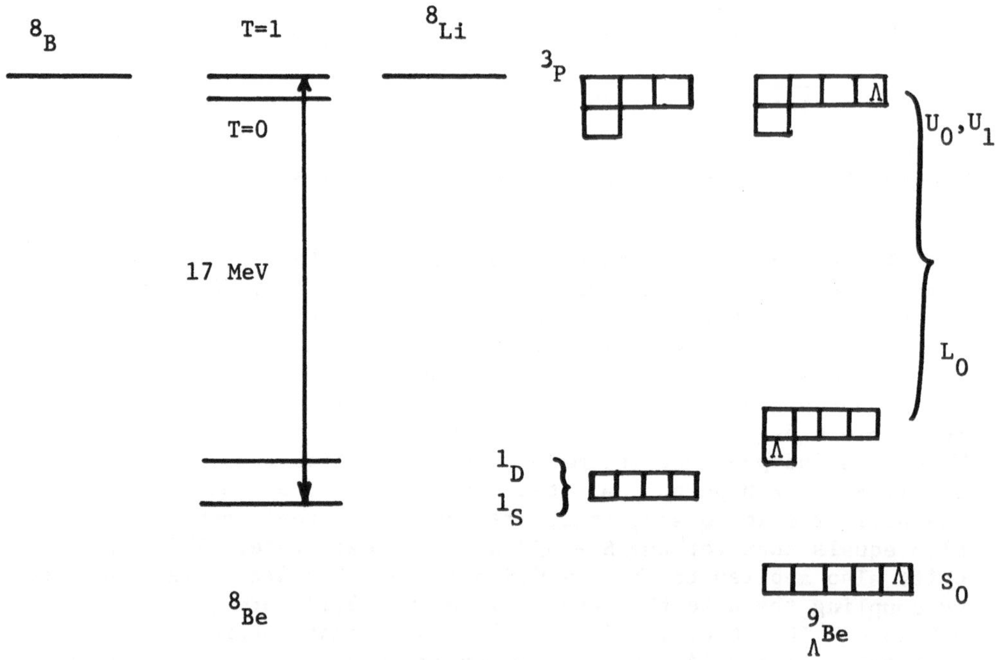

Fig. 10. The lowest levels of ^{8}Be are shown, whose space wavefunctions have permutation symmetry [4] or [3,1]. The $^{9}_{\Lambda}Be$ states (with spin-parity $3/2^-$) made from them by the addition of a 1p Λ particle are then U_0 and U_1, with symmetry [4,1], L_0 with symmetry [4,1] and S_0, the supersymmetric state, with symmetry [5]. The states L_0, U_0 and U_1 will appear as strangeness analogue states in the 0^0 $K^-\to\pi^-$ reaction on target ^{9}Be.

Note added (October, 1978): The discussion of $^{16}_{\Lambda}O^*$ 0^+ states in Section 3.2 must be updated in the light of new CERN experimental evidence, at p_{K^-} = 715 MeV/c and $\theta_{\pi^-} = 0^o$ (W. Brückner et al. "Spin-Orbit Interaction of Lambda Particles in Nuclei", submitted to Phys. Lett. (September, 1978)). Two candidates for 0^+ states are observed with $\Delta E \sim 6$ MeV and ratio of about 1:3 for the formation rate $\tan^2\theta$ of the lower relative to the higher of these states. This suggests extending Table 1 in the direction of negative values of $\delta\zeta$. Thus, for $\delta\zeta$ = -4 MeV, indicating Λ spin-orbit coupling twice as strong as nucleon spin-orbit coupling, the following values are calculated for ΔE and $\tan^2\theta$:

ε':	0	0.25	0.5
ΔE(MeV):	5.7	5.8	6.3
$\tan^2\theta$:	0.87	0.65	0.38

While, for this value of $\delta\zeta$, the values calculated for ΔE are all near the observed value of 6 MeV, the variation in the calculated $\tan^2\theta$ is wide enough to accommodate the measured ratio of about 0.3 to 0.4.

These theoretical considerations are preliminary at best. They should in due course be confronted with the emerging BNL experimental evidence on ^{12}C, at p_{K^-} = 800 MeV/c and $\theta_{\pi^-} = 0^o, 5^o, 10^o$ and 19° (R. Cester, M. May and H. Palevsky, private communication), according to which the two 2^+ states mentioned in Section 3.1 are indirectly seen. As clear from the appropriate discussion, the spacing between these states, as well as their relationship to the dominant 0^+ in $^{12}_{\Lambda}C^*$, are expected to yield useful information on the ΛN parameters ε' and $\delta\zeta$.

I would like to acknowledge stimulating and useful discussions, on topics included in these lectures, with R.H. Dalitz and C.B. Dover. Special thanks are due to R. Cester-Regge, M. May and H. Palevsky for communicating and discussing the preliminary BNL data.

REFERENCES

1. M. Juric et al., Nucl. Phys. B52 (1973), 1.

2. M.I. Podgoretsky, JETP 44 (1963), 695; H. Feshbach and A. Kerman, in "Preludes in Theoretical Physics", edited by A. de Shalit, H. Feshbach and L. Van Hove, North-Holland (Amsterdam 1966), p. 260.

3. M.M. Nagels, T.A. Rijken and J.J. de Swart, Ann. Phys. 79 (1973), 338.

4. F. Eisele et al., Nucl. Phys. B37 (1971), 204.

5. G. Alexander et al., Phys. Rev. 173 (1968), 1452; B. Sechi-Zorn et al., Phys. Rev. 175 (1968), 1735.

6. W. Bruckner et al., Phys. Lett. 62B (1976), 481.

7. D.J. Prowse, Phys. Rev. Lett. 17 (1966), 782.

8. M. Danysz et al., Nucl. Phys. 49 (1963), 121.

9. R.C. Herndon and Y.C. Tang, Phys. Rev. Lett. 14 (1965), 991.

10. H. Feshbach and A. Kerman, ref. 2.

11. H. Feshbach, in "Proceedings of the Summer Study Meeting on Nuclear and Hypernuclear Physics with Kaon Beams", edited by H. Palevsky, BNL 18335 (1973), p. 185; C.B. Dover, in "Proceedings of the Summer Meeting on Kaon Physics and Facilities", edited by H. Palevsky, BNL 50579 (1976), p. 9.

12. C.B. Dover and A. Gal, in preparation (1978).

13. C.B. Dover, private communication.

14. M.M. Nagels, T.A. Rijken and J.J. de Swart, Phys. Rev. D 15, (1977), 2547.

15. H.J. Lipkin, Phys. Rev. Lett. 14 (1965), 18.

16. R.H. Dalitz and A. Gal, Ann. Phys. (1979), in press.

17. R.H. Dalitz and A. Gal, Phys. Lett. 64B (1976), 154.

18. J. Lagnaux et al., Nucl. Phys. 60 (1964), 97; J. Lemonne et al., Phys. Lett. 18 (1965), 354.

19. A. Bouyssy, Nucl. Phys. A290 (1977), 324.

20. A. Bouyssy and J. Hufner, Phys. Lett. 64B (1976), 276.

21. A.K. Kerman and H.J. Lipkin, Ann. Phys. 66, (1971), 738.

22. M. Rayet, Ann. Phys. 102 (1976), 226.

23. J. Hufner, S.Y. Lee and H.A. Weidenmuller, Phys. Lett. 49B (1974), 409.

24. L.S. Kisslinger and N. Van Giai, Phys. Lett. 72B (1977), 19.

25. G.N. Epstein et al., Phys. Rev. C 17 (1978), 1501.

26. R.H. Dalitz and A. Gal, J. Phys. G4 (1978), 889.

27. A candidate for such emission was reported by J.C. Herrera et al., Phys. Rev. Lett. 40 (1978), 158.

28. A. Gal, J.M. Soper and R.H. Dalitz, Ann. Phys. 113 (1978), 79.

29. M. Bedjidian et al., Phys. Lett. 62B (1976), 467.

30. R.H. Dalitz and A. Gal, Phys. Rev. Lett. 36 (1976), 362; erratum, ibid 628.

31. R.H. Dalitz and A. Gal, in preparation (1978).

32. F.C. Barker, Nucl. Phys. 83 (1966), 418.

CURRENT TOPICS IN QUASI-ELASTIC SCATTERING

C.A. Miller

TRIUMF and University of Alberta

TRIUMF, University of British Columbia, Vancouver, B.C.

INTRODUCTION

The study of quasi-elastic scattering of protons or electrons on nucleons bound in nuclei is a relatively mature and well-developed field of nuclear physics. Our purpose herein is to review in a necessarily brief and selective manner some of the recent activities with emphasis on those at TRIUMF. In order to provide a suitable context for the discussion of the implications of recent data, there is first offered a summary of the motivations for studying these reactions and the major assumptions involved in most theoretical analysis. This is followed by a section dealing with a recent innovation in proton quasi-free scattering; namely the use of polarized beam which can enrich the information from this type of reaction as much as it has for others. The comparison with theoretical calculations of such data taken at TRIUMF appears to raise some interesting questions concerning cross sections in spite of considerable qualitative agreement regarding asymmetries. Next is a brief treatment of recent attempts to learn about high momentum components in nuclear wave functions by means of the (p,2p) reaction on a light nucleus. The concluding topic concerns some recent suggestions for future experiments whose specific purpose would be the investigation of off-shell effects in nucleon-nucleon scattering.

GENERAL BACKGROUND

Conceptually, quasi-free scattering is one of the simplest of reactions. The zero'th order picture of the process has an incident particle (proton or electron) suffering a violent interaction with only one nucleon in the target nucleus and knocking it out. The residual or recoil nucleus is considered a spectator whose final state momentum is therefore the same as in the initial state where it is minus the fermi momentum of the struck nucleon. Most experiments involve a measurement of the energies and angles of the two particles in the final state so that it is kinematically completely determined. We therefore know both the separation energy of the knocked out nucleon and the recoil momentum which, in our zero'th order picture, is directly related to the internal momenta of nuclear protons. The discovery that these measured separation energies formed a spectrum of discrete peaks constituted the first direct evidence of the shell structure of the nucleus. Although the recoil momentum distribution is related to the nuclear wave function of the struck nucleon, the relationship is of course complicated by the fact that the presence of the recoil nucleus affects both the incident and final state particles. In fact, at the energies typical of these experiments, the mean free path of nucleons in nuclear matter is of the order of the nuclear radius so it was surprising when early experiments indicated that this very strong absorption and refraction did not qualitatively change the result from what one would expect from the above naive picture. Of course, quantitative interpretation of the data requires the use of some distorted wave theory.

As previously mentioned, the two typical projectiles are protons and electrons. Each has its merits. Electrons are much less strongly absorbed and the electron-proton interaction is much better understood than the nucleon-nucleon interaction, especially since the interaction is in this case off-shell. On the other hand, the (e,e'p) cross section is considerably smaller than the (p,2p) cross section, proton projectiles can knock out neutrons as well as protons from nuclei so that their single particle properties can be compared, and the off-shell nature of the nucleon-nucleon interaction is itself of interest. Both types of experiments continue to be of interest.

For some time now, the theoretical approach usually taken in analyzing quasi-elastic scattering data has undergone only refinements of detail rather than fundamental change. For (e,e'p) and for (p,2p) experiments above about 150 MeV it is believed to be appropriate to use a factorized distorted wave theory called the distorted wave impulse approximation (DWIA). In this theory the cross section can be written in the familiar form[1]

$$\frac{d^3\sigma}{d\Omega_1 d\Omega_2\, d(T_1-T_2)} = K \left.\frac{d\sigma}{d\Omega}\right)_{2\ \text{body}} \sum_m c_m |g_\ell^m|^2$$

where K is a kinematic and phase space factor and $\left.\frac{d\sigma}{d\Omega}\right)_{2\ \text{body}}$ is the cross section for e-p scattering in the case of (e,e'p) or p-p scattering in the case of (p,2p). This cross section is off-shell for more than one reason as detailed in a later section. g_ℓ^m is called the distorted momentum distribution and is usually calculated as

$$g_\ell^m = \int \chi_1^*(\vec{r})\ \chi_2^*(\vec{r})\ \psi_\ell^m(\vec{r})\ \chi_0(\vec{r})d^3\vec{r}$$

where the χ's are optical model scattering wave functions distorted by the presence of the recoil nucleus. Their calculation requires the existence of elastic scattering and total reaction cross section measurements over the energy range of both initial and final state particles. The shortage of such information is still a severe problem for much of the (p,2p) data. For higher energy (p,2p) analysis, the KMT approach[2] has been used to generate the optical potentials directly from the nucleon-nucleon amplitudes. ψ_ℓ^m is a single particle wave function which is supposed to represent the initial bound state of the struck nucleon with orbital angular momentum ℓ.

Some important assumptions implied by the above expressions are as follows:

1. The optical model can represent all important multiple scattering effects of the recoil nucleus. (When the recoil nucleus is left in an excited state, we use the same potential as for the ground state).

2. Factorization is valid. This depends on two things:
   a. The spin-orbit part of the optical model potential can be neglected. Although its effect has been estimated to be small,[3] at least two collaborations are working toward its inclusion in calculations.[4,5]
   b. The ratio of the distorted wave functions to plane waves $D(\vec{r}) = \chi(\vec{r})/e^{\ i\vec{k}\cdot\vec{r}}$ varies sufficiently slowly over the range of the nucleon-nucleon interaction that it may be considered constant (peaking approximation). Its validity has been investigated by Lim & McCarthy[6] and Koshel.[7]

3. The nuclear overlap integral[8]

$$\langle\psi_{A-1}|\psi_A\rangle = \int \psi_{A-1}(\xi)\psi_A^*(r,\xi)d\xi$$

can be approximated by the single particle wave function mentioned above. This can be expected to be reliable only when the final nuclear states have a large fraction of their expected shell model strength.

4. In the case of (p,2p) we can neglect the fact that distortions cause the proton-proton interaction to be fully off-shell. i.e., we can use an (arbitrary) on-shell extrapolation or a half-off-shell cross section. (Redish[9] has suggested a first order correction for this effect of distortion).

5. Pauli effects on the scattering wave functions can be neglected.

6. The iso-spin flip part of the optical potential can be neglected.

7. The recoil nucleus is heavy enough that the static approximation may be used to separate the three body Schroedinger equation for the final state.

POLARIZED (P,2P) EXPERIMENTS

For some time it had been known (but possibly not widely appreciated) that the distortion of the scattering wave functions gives rise in general to an effective initial polarization of the struck nucleon. In 1973 Jacob *et al*[10,11] pointed out that it should be possible to make use of this fact to obtain more information from the reaction. There is a very simple classical picture which illustrates the reason for the polarization of the struck particle. One requires an asymmetrical kinematic situation to see a substantial polarization. For example, assume we have a 200 MeV proton incident, and in the final state, a 140 MeV proton at a small angle (30°) and a 50 MeV proton at a larger angle (∿ 50°) chosen to result in zero recoil momentum. Now as we vary this larger angle in

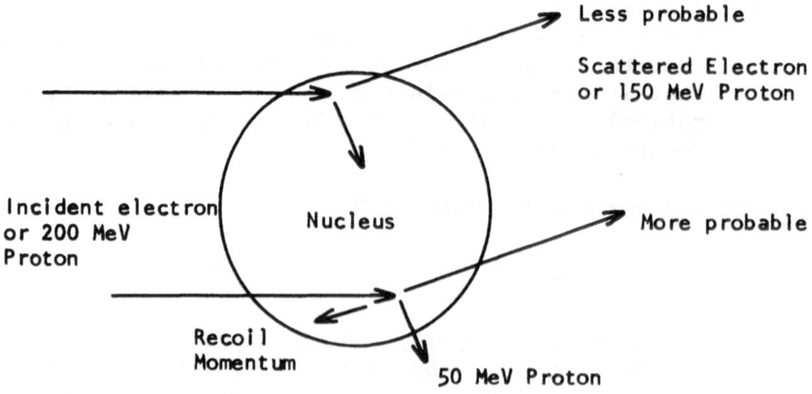

Fig.1. Semi-classical picture of the reason for the polarization of the struck particle.

either direction, the recoil momentum increases in a direction perpendicular to the momentum of the 50 MeV proton. Recall that the initial momentum of the struck particle is equal and opposite to the recoil momentum. Since the 50 MeV proton is more strongly absorbed than the other, one observes a preponderance of events from the side from which it exits the nucleus. This results in a net orbital angular momentum, the direction of which depends on the direction we moved the low energy proton angle. The struck particle's orbital angular momentum is coupled to its spin by the nuclear spin-orbit coupling potential and the sign of this coupling depends on whether $J = \ell+S$ or $J = \ell-S$. The net result is an initial polarization perpendicular to the plane of the figure which depends on both the kinematic situation and the J-value of the shell from which the proton was ejected.

One can measure this polarization of the "target" proton by using a polarized beam and exploiting the large difference in p-p cross sections for spins parallel and anti-parallel. This amounts to saying that the correlation parameter Ayy has substantial values at intermediate energies. In the expression for the (p,2p) cross section, the two body cross section now becomes

$$\left.\frac{d\sigma}{d\Omega}\right)_{2\text{ body}} = \left.\frac{d\sigma}{d\Omega}\right)_{\text{unpol.}} (1 + (\underline{P}_0+\underline{P}_3)\cdot\underline{P} + \underline{P}_0\cdot\underline{P}_3\, A_{yy})$$

where $\underline{P}_0$ is the beam polarization, $\underline{P}_3$ is the struck particle polarization, $\underline{P}$ is a vector proportional to the analyzing power Ay in p-p scattering and is perpendicular to the scattering plane. $\underline{P}_3$ is calculated from the g_ℓ^m's. It is only the third term which is useful in determining $\underline{P}_3$ experimentally.

In the calculations described below, $\left.\frac{d\sigma}{d\Omega}\right)_{\text{unpol}}$, Ay and Ayy are derived from the latest nucleon-nucleon phase shifts.[12]

Herscovitz *et al.* have more recently suggested[13] that it may also be possible to exploit the same phenomenon in a polarized (e,e'p) measurement. In this case, one varies the final state proton direction out of the plane so that its initial polarization is in the plane of the electron trajectories. The beam polarization is longitudinal. The author is unaware of any such experiments planned or in progress.

The first application of this effect to come to mind is the identification of the j-values of unknown hole states in nuclei in a high resolution experiment. However, it was clear that before this was attempted, one should seize the opportunity to test the DWIA theory more comprehensively than previously possible by making a measurement on a good shell model nucleus with well known hole states and comparing the measured asymmetries with the substantial values predicted by the DWIA. The reason why this could be expected to be a good test is that the asymmetry is entirely due to distortion effects. Also, parameter studies[11] showed that these asymmetries should be less sensitive to the choice of bound state and optical model parameters than are cross sections.

Kitching *et al.*, have done such a measurement at TRIUMF on ^{16}O at 200 MeV. Only a small portion of the data have been published at this time.[14] The energy resolution of the experiment was adequate to separate protons ejected from $P_{3/2}$ and $P_{1/2}$ states. Recent (e,e'p) measurements on ^{16}O done at Saclay[15] with higher resolution have shown that the $P_{1/2}$ strength is fragmented not at all and the $P_{3/2}$ only slightly in a way that does not affect the results. With the high duty cycle and relatively intense polarized beam of TRIUMF, it was possible to take data in a very wide range of kinematic conditions. An energy-sharing spectrum was taken at each of 24 angle combinations. This has turned out to be an imporimportant asset in comparing the data with the theory since it is possible to have a reasonable fit as a function of proton energy at one angle pair and a poor fit at another.

The most striking thing about the data, some of which are shown in figure 2, is that the asymmetries are large and, in some, kinematic regions, agree with the theoretical predictions. These calculations were done using the method of partial wave expansion of the optical model wave functions.[16] The solid curves in the figure correspond to a "standard" calculation which will henceforth be called fit A. The two-body observables were calculated using a half-off-shell prescription which is described in some detail in a later section. A factorized t-matrix is used so that experimental phase shifts may be used for on-shell information while the Mongan potential model[30] is used for the off-shell extension. As discussed later, there is reason to believe that, in this situation, there is very little sensitivity to the choice of potential model.

The overall agreement of fit A with the data as seen in Figure 2 is only fair and there appears to be a problem with cross section magnitudes. The theoretical cross sections for fit A have been uniformly normalized downward by a factor of two. There appears to be little doubt concerning the absolute magnitude of the data because at several angle pairs, it was possible to simultaneously record p-p scattering events from hydrogen in the water target. In view of this problem with the cross sections, it is obviously important to reconsider all the information input to the calculation.

The reason such a discrepancy was not reported earlier in reference 14 which showed good fits to both asymmetries and cross sections at 30° - 30° was that those calculations were done using

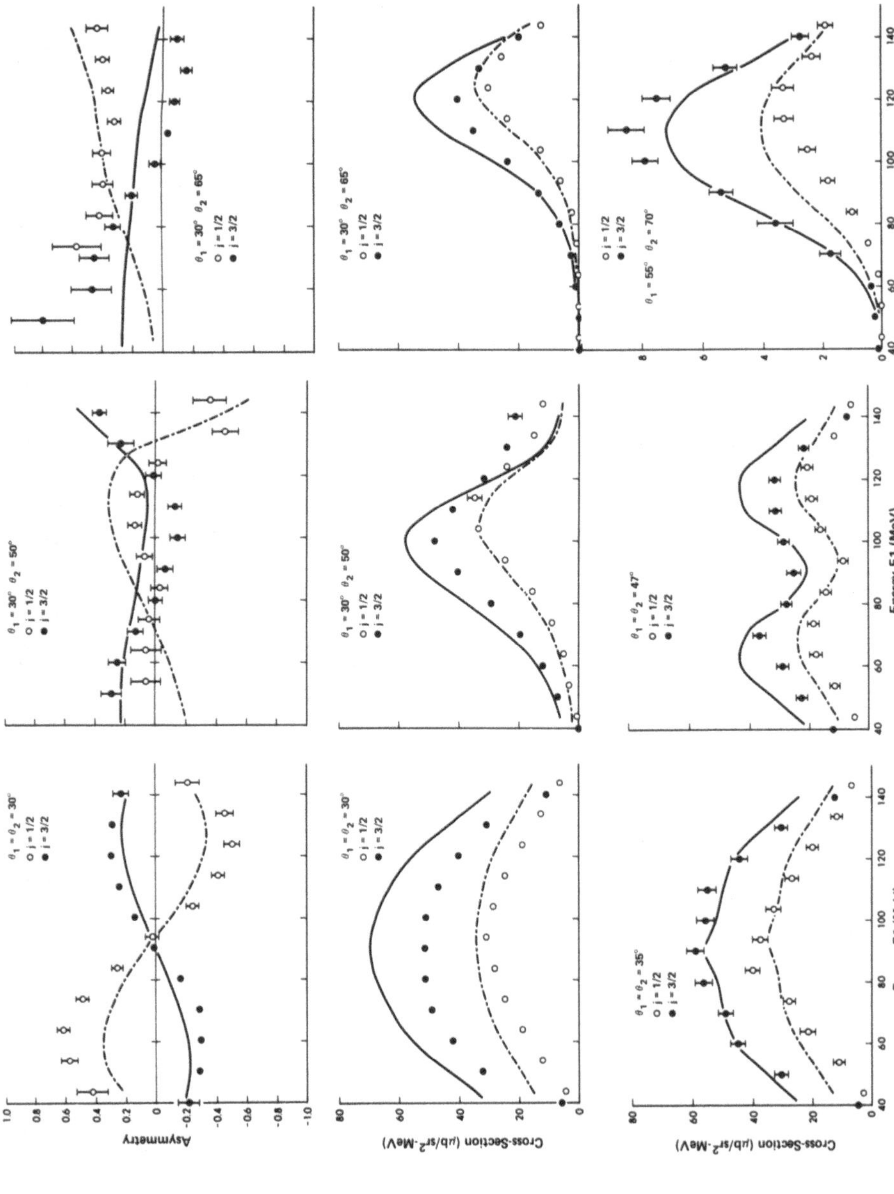

Fig.2. ^{16}O $(\vec{p},2p)$ data at 200 MeV compared with DWIA calculations described as "Fit A" in the text. Theoretical cross sections are normalized downwards by a factor of 2.

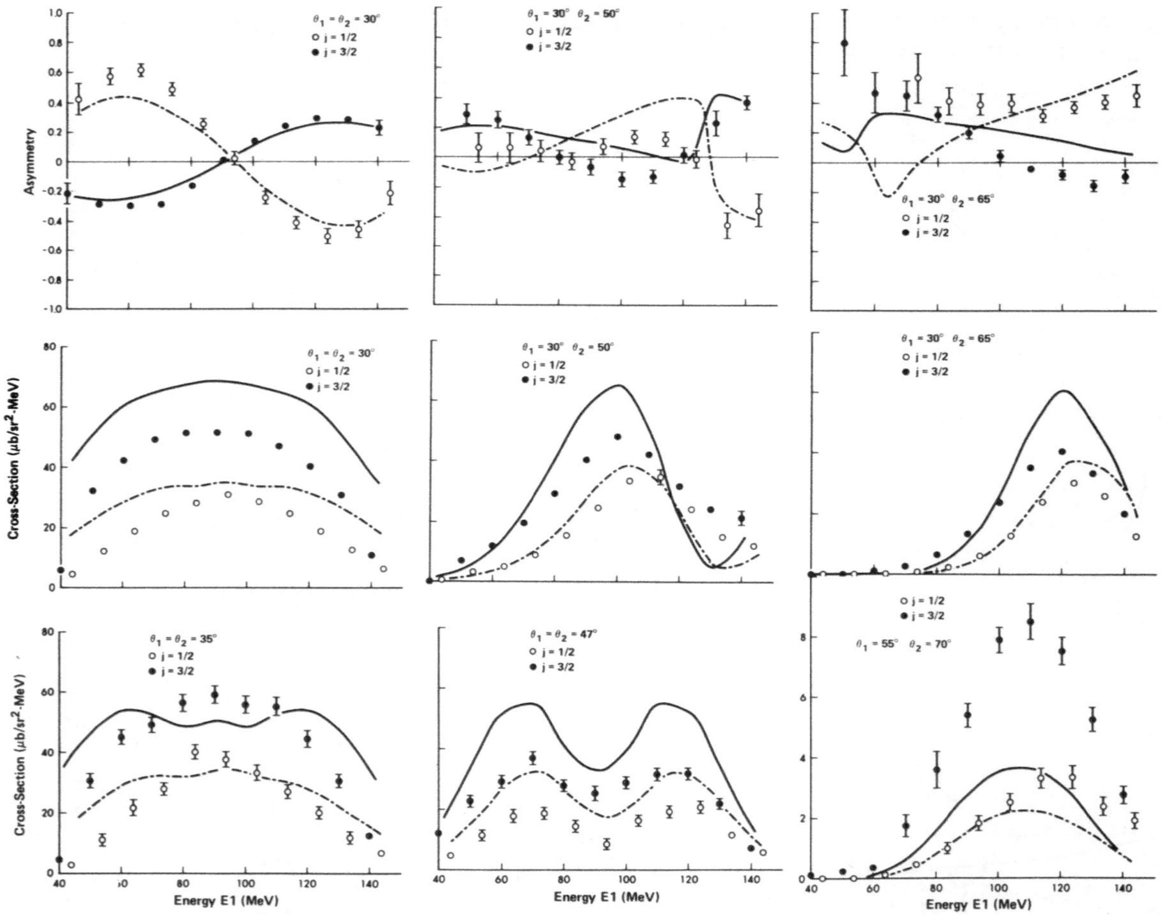

Fig.3. ^{16}O $(\vec{p},2p)$ data at 200 MeV compared with DWIA calculations described as "Fit B" in the text. Theoretical cross sections are normalized downwards by a factor of 1.33.

an optical model parameter set derived for heavier nuclei that was later found to yield proton-nucleus total reaction cross sections (σ_R) for ^{16}O too large by a factor of 2 at 100 MeV. It appears therefore that this good fit was fortuitous. There is, unfortunately, little reliable information about optical model parameters at intermediate energy, especially for light nuclei. The parameter set used for fit A arose from a rough interpolation of previous independent fits for light nuclei including minor adjustments necessary to obtain reasonable values for σ_R over the energy range of importance (50 to 200 MeV). Since these parameters produce less absorbtion, they result in smaller asymmetries than the previous incorrect set but the fit to the experimental asymmetries at forward angles was partially restored by improving the method of calculating the two-body observables. (The previous calculations had employed an on-shell extrapolation using the relative two body energy in the final state). A few typical values of the optical model parameters are given in Table 1.

Table 1. Optical Model Parameters for Fit A (Fit B)

Energy (MeV)	50	100	200
ro (fm.)	1.14 (1.45)	1.29 (1.45)	1.45 (1.45)
V (MeV)	54 (21.7)	22 (16.2)	4.6 (8.3)
ri (fm)	1.3 (.932)	1.3 (.932)	1.3 (.932)
W_v+W_d	5.4 (27)	9.3 (27)	10.0 (27)

It is very probable that this parameter set (A) can be improved upon. Equally important as more experimental elastic scattering data is the information available from nuclear matter calculations.[17] It is even possible that the predicted complex shape acquired by the real potential near the energy at which it passes through zero has a significant influence on (p,2p) calculations. For the present, however, the only further information that can be offered here is the result of trying a rather different parameter set derived by Abdul-Jabil and Jackson from a systematic search using $P + C^{12}$ data at intermediate energies.[18] Their potentials appear to fit some ^{16}O elastic data as well and yield reasonable values for σ_R. The imaginary radius they obtain is much smaller (.93 fm) than that used for fit A and hence the depth is correspondingly greater (27 MeV at all energies). Fit B (figure 3) uses their volume absorption parameter set. Asymmetries are slightly larger but cross section shapes fit the data much less well than does Fit A, although the overall magnitude normalization factor is reduced to approximately 1.33. In short, it seems unlikely that inadequacies in the optical potentials used can entirely account for the problems with cross sections.

The other important input to the calculation is the single particle wave function for the initial bound state. This can affect the shape and especially the magnitude of the cross sections. In contrast, the asymmetries are very insensitive. The wave function used here is calculated in a Wood-Saxon well including the spin-orbit potential. The well parameters are as given by the fits to electron-scattering data by Elton and Swift.[19] Their wave functions also seem to fit the recent Saclay (e,e'p) data for ^{12}C and other nuclei quite well.[20] Therefore. there seems to be little justification for attempting to adjust these parameters. In any case, such adjustment appears to be ineffectual in improving the fit to the cross sections. If the p-state RMS radius is increased from the Elton and Swift value of 2.86 fm to 3.2 fm in order to fit the energy-sharing shapes, then the normalization discrepancy increases to a factor of three.

There is one relatively simple phenomenon which may be relevant to this situation. Nuclear matter calculations indicate that the intrinsic non-locality of the optical potential is even larger than that required to account for the energy dependence of the asymptotically equivalent local potential. (The intrinsic energy dependence of the non-local potential has the opposite sign and has a cancelling effect). The asymptotically equivalent local potential which we are presumably using is not equivalent with regard to the wave function inside the nucleus. It has been noticed by Perry and Buck[21] that wave functions calculated from a non-local potential have amplitudes reduced inside the potential by about 15% at 50 MeV compared with those calculated from the equivalent local potential. The DWIA calculation involves a product of three wave functions, which is then squared to yield the transition amplitude so that such effects are amplified by the sixth power. A crude attempt to determine the importance of this effect was made by multiplying the scattering wave function amplitudes by the nuclear density dependent factor $\left\{\frac{\tilde{m}}{m}\right\}^{\frac{1}{2}}$, values for which were taken from nuclear matter calculations.[22] $\tilde{m}$ is an effective mass which characterizes the non-locality of the potential. The radial dependence of the nuclear density was taken to be that suggested by Negele, also given in reference [22]. Including this effect did not drastically change the overall magnitude of the cross sections except for a reduction at relatively backward angles corresponding to large values of recoil momenta. Although it seems to be worthwhile to take account of the non-locality of the optical model potential (preferably in a more self-consistent manner than that used here), there remain discrepancies with the data that will require further investigation.

INVESTIGATION OF HIGH MOMENTUM COMPONENTS

There have been some (p,2p) measurements specifically directed towards obtaining information about high momentum components of nuclear wave functions. There are uncertainties overshadowing this effort because it is in these kinematic regions that one expects multi-step processes to become important and off-shell effects to become large. However, a recent paper by Zabolitsky and Ey[23] has provided additional motivation. Although they scrupulously avoid comment on how one should measure the momentum distribution of nuclear protons, they point out that such measurements are sensitive to nuclear correlations while the form factor derived from electron scattering is not. Their calculations of the momentum distributions of protons in ^{4}He and ^{16}O employing the generalized Brueckner-Hartree-Fock approximation with various phenomonological nucleon-nucleon potentials show that nuclear correlations can have very dramatic effects on the momentum distribution above approximately 2 fm $^{-1}$. As they suggest, even a rather crude measurement would provide useful information.

One of the motivations for a ^{4}He(p,2p) experiment done at TRIUMF[24] was to make measurements at large recoil momenta. Figure 4 shows some of their data taken at 500 MeV where they varied the recoil momentum up to a maximum of 2.4 fm^{-1} by varying the symmetric angles. Also included is some 600 MeV SREL data which extends out to 1.5 fm^{-1}. There appears to be some evidence of the change in slope at $\sim$ 2 fm^{-1} predicted by Zabolitsky and Ey. It should be born in mind that distortion smears out the momentum at which one is sampling the wave function so that, when the wave function is falling rapidly, one is effectively sampling it at a smaller momentum than the recoil momentum. This is reflected by the fact that in these large angle regions, the DWIA cross section becomes larger than the plane wave prediction. It would be desirable to somehow combine Zabolitsky and Ey's prediction with a distorted wave calculation.

OFF-SHELL EFFECTS

The quest for a clear manifestation of off-shell effects in nucleon-nucleon scattering remains largely unrequited. In principle, p-p bremsstrahlung provides the cleanest situation for this type of investigation. However, (p,pγ) experiments, after the expenditure of great effort, often yield results that differ little from the predictions of the soft photon approximation using on-shell information only.

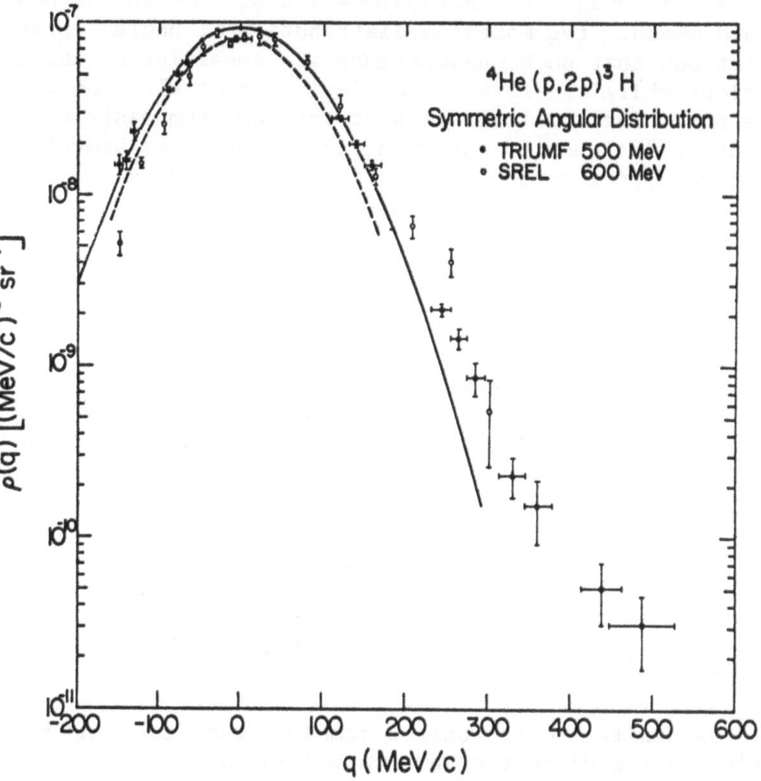

Fig.4. Distorted momentum distributions from the ^{4}He(p,2p) measurements of reference 24 and reference 31. The curves are DWIA calculations of Roos[32].

It is well known [25] that, in some kinematic regions, nucleon knockout cross sections can show sensitivity to off-shell aspects of the nucleon-nucleon interaction. This has not sparked a great deal of specifically related experimental activity up to now presumably because of possible doubts about the DWIA theory and the lack of reliable optical model potentials. If we are to learn anything about the nucleon-nucleon interaction, it is clearly necessary to find ways of minimizing the sensitivity of the theory to such uncertainties while maximizing the differences between predictions arising from various models for the interaction.

Two possible stratagems for achieving this have recently emerged. Jackson and Ionnides[26] have suggested that by a suitable choice of experimental geometry and kinematics, the least known aspects of the reaction may be kept roughly constant while the "off-shellness" as measured by the ratio of the on-shell to the off-shell momenta is varied. They argue that the effects of distortions are sensitive mainly to the energies of the final state particles and not their angles. Consequently they propose keeping these energies constant. Also, in order to reduce the dependence on the shape of the initial bound state wave function, the recoil momentum is also kept constant. Within these constraints, it is still possible to widely vary the parameters of the nucleon-nucleon interaction by simultaneously varying angles of the detected final state nucleons so that the direction θ_3 of the recoil momentum rotates through 180°. They find that, at $\theta_3 = 180^\circ$, there are significant differences among p-p cross sections calculated from the Reid soft core potential using two extreme on-shell extrapolation procedures and the half off shell prescription. One on-shell extrapolation is the so-called "initial state" prescription in which the cross section is evaluated at a laboratory energy corresponding to the energy of the two protons in their centre of mass in the initial state while the other is the similarly-defined "final state" prescription. That these results differ significantly is a necessary but not sufficient condition to be able to learn about off-shell amplitudes.

The other possibility for minimizing the effects of uncertainties in the parameters used in the DWIA is to measure asymmetries with a polarized beam. This has the added possibility of emphasizing certain specific scattering amplitudes. Parameter studies with DWIA calculations show that in sharp contrast with cross sections, the predicted asymmetries are relatively insensitive to both the bound state wave function and to the optical model parameters provided that the latter remain consistent with total reaction cross sections for protons on the target nucleus.

The determination of the on-shell scattering amplitudes has occupied an army of physicists for decades. Of course, one cannot hope to make comparable studies of the off-shell amplitudes. The most that can be attempted is to determine which of the existing models for the nucleon-nucleon interaction best describes the results of a measurement in a situation where the predictions of these models disagree. However, the situation is even further complicated by the fact that these models may differ in their predictions of on-shell amplitudes. The intention is not to determine whether one potential model is "better" than another in an overall sense; this is probably more easily done by elastic scattering experiments. Rather it is to determine which, if any, potential best models the off-shell behaviour of the interaction. We must have some way of disentangling these effects. One method is to compare the prediction of phase-shift-equivalent transformations of a single potential model. This has been done for the Reid soft core potential by Stephenson *et al* .[25] It is also desirable to compare potential models which may not be phase-equivalent. Mongan has suggested that this might be done using the Noyes-Kowalski factorization[27] of the half-off-shell t-matrix into a real off-shell extension function times the on-shell t-matrix:

$$t(q,q_E,E) = f(q,q_E)\ \ t(q_E,q_E,E)$$

The on-shell t-matrix $t(q_E,q_E,E)$ (in the form of partial wave amplitudes) can be calculated from the experimentally determined phase shifts while the off-shell extension function $f(q,q_E)$ may easily be calculated from potential models.[28] In the case of coupled waves, all these quantities take the form of two by two matrices.

A.W. Thomas and the author have investigated the usefullness of this approach for comparing the off-shell properties of two quite different potential models as perceived in hypothetical (p,2p) or (p,pn) experiments. We have attempted to compare a static local potential[29] with a separable non-local one.[30] However, we have encountered practical difficulties at energies where a phase shift passes through zero. In the case of coupled waves, the corresponding condition is that the on-shell two by two t-matrix is singular (has zero discriminant). At such energies, the off-shell function f has a pole. Again in the case of coupled waves, all of the elements of the matrix f (q,qE) are large in the broad vicinity of this energy so that the above matrix multiplication by the on-shell t-matrix involves subtraction of large but similar numbers.

If $f(q,q_E)$ has this "pole" and $t(q_E,q_E,E)$ is not derived from the same potential, the delicate cancellations will not work. This problem exists for the Reid potential in the J=2 coupled wave at energies appropriate for such investigations using the (p,2p) reaction The only way we have been able to achieve a valid comparison is to calculate the on-shell t-matrix from the Reid soft-core potential in all cases and compare the results of calculating $f(q,q_E)$ from either the Reid potential or the Mongan separable non-local potential. The difficulties do not appear in this case because the Mongan potential yields well-behaved off-shell extension functions. Such a comparison is shown in figure 5. Cross sections and asymmetries for the $^{16}O(\vec{p},2p)$ reaction calculated via two on-shell extrapolation prescriptions are compared with half off-shell quantities using the above two potentials. The Mongan " CaseII" fit was used. Even though the on-shell prescriptions differ drastically from each other, the half-shell results show little descrimination between these potential models. This is not encouraging from the point of view of the original goal.

There is another difficulty which must not be overlooked. Within the context of the half-shell prescription, the only quasi-elastic kinematic region where the two-body t-matrix moves significantly off-shell is in the extreme forward direction where the energy in the two-body centre of mass becomes relatively small. This is precisely the region where the factorization approximation becomes suspect. A crude measure of the validity of the "peaking approximation" mentioned earlier as assumption 2 b can be derived from the first order approximations suggested by Redish[9] for the effect of the distorting nuclear potential on the momenta and propogator energy at which the (fully off-shell) two-body t-matrix should be evaluated. His simplest corrections amount to taking the momenta corresponding to the kinetic energies of the protons inside the optical potential well. We have studied the implications of this simple method in the kinematic situation corresponding to figure 5. We used the optical potentials of fit A in table 1. In order to account for the fact that the bound state wave function of the struck particle is only partially within the optical well, we have arbitrarily reduced by 50% the well depths used to calculate the corrected momenta. Even then, the corrections are large enough to reduce the propogator energy from approximately 50 MeV to near zero. This would seem to invalidate the first-order expansions underlying Reddish's approach. We interpret this as an indication that even this modified factorization approximation is inappropriate in this kinematic situation, although at angles such as those of figure 2, there is no such indication.

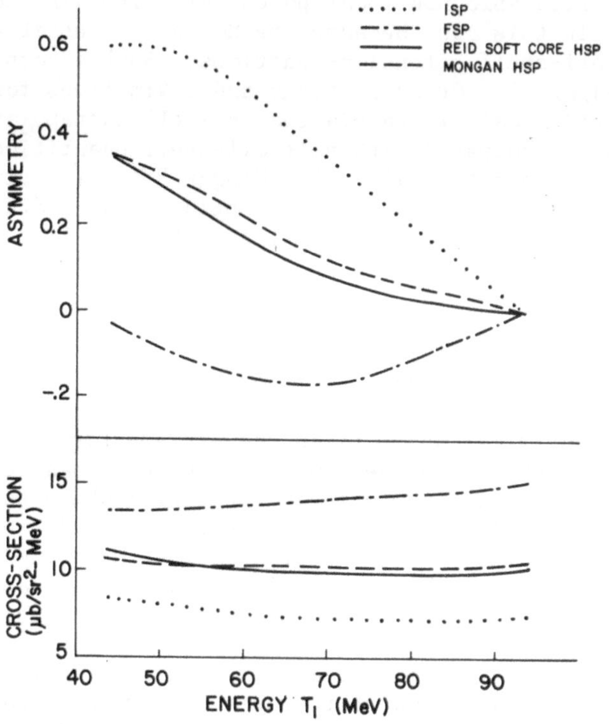

Fig.5. $^{16}O(\vec{p},2p)$ calculations at 200 MeV and symmetric proton angles of 20° plotted as a function of the energy of one proton. The four curves correspond to the initial state prescription (ISP) and final state prescription (FSP) for extrapolating the p-p t-matrix on-shell and half-off-shell prescription (HSP) using two different p-p potentials.

As mentioned previously, others have studied the validity of factorization in more general terms. However, the above simple argument should be sufficient to raise doubts in the present context. This, combined with the apparent insensitivity of the calculations to the choice of nucleon-nucleon potential model, leads us to conclude that hopes of learning about the off-shell nucleon-nucleon interaction using factorized half-shell calculations are naive.

REFERENCES

1. G. Jacob and Th. A.J.Maris, Rev.Mod.Phys. 45(1973)6 and Rev.Mod.Phys. 38(1966)121
2. A.K. Kerman, H. McManus and R.M. Thaler, Ann.of Phys. 8(1959)551
3. C. Schneider, Nucl.Phys. A300(1978)313
4. D.F. Jackson, private communication
5. N.S. Chant and P. Kitching, private communication
6. K.L. Lim and I.E. McCarthy, Nucl.Phys. 88(1966)433
7. R.D. Koshel, private communication
8. W.T. Pinkston and G.R. Satchler, Nucl.Phys. 72(1965)641 T. Berggren, Nucl.Phys. 72(1965)337
9. E.F. Reddish, Phys.Rev.Lett. 31(1973)617
10. G. Jacob, Th. A.J. Maris, C. Schneider and M.R. Teodoro, Phys.Lett. B45(1973)181
11. G. Jacob, Th.A.J. Maris, C. Schneider and M.R. Teodoro, Nucl.Phys. A257(1976)517
12. R.A. Arndt, R.H. Hackman and L.D. Roper, Phys.Rev. C15(1977)1002
13. V.E. Herscovitz, Th.A.J. Maris and M.R. Teodoro, Phys.Lett. 69B (1977)33
14. P. Kitching, C.A. Miller, D.A. Hutcheon, A.N. James, W.J. McDonald, J.M. Cameron, W.C. Olsen and G. Roy, Phys.Rev. Lett. 37(1976)1600
15. J. Mougey, Bull.Am.Phys.Soc. 23(1978)535
16. D.F. Jackson and T.Berggren, Nucl.Phys. 62(1965)353
17. C. Mahaux, ibid
18. I. Abdul-Jalil and D.F. Jackson, preprint (1978)
19. L.R.B. Elton and A. Swift, Nucl.Phys.A94(1967)52

20. J. Mougey, M. Bernheim, A. Bussière, A. Gillebert, Phan. Xuan Hô, M. Priou, D. Royer, I. Sick and G.J. Wagner, Nucl.Phys. A262 (1976)461.

21. F.G. Perey, Phys.Rev.131 (1963)745, F.G. Perey and B. Buck, Nucl.Phys. 32 (1962)353.

22. J.-P. Jeukenne, A. Lejeune and C. Mahaux, Phys.Rev. C16 (1977)80.

23. J.G. Zabolitzky and W. Ey, Phys.Lett. 76B (1978)527.

24. B.K.S. Koene, B.T. Murdoch, C.A. Smith, W.T.H. van Oers, M.B. Epstein, D.J. Margaziotis, J.M. Cameron, L.G. Greeniaus, G.A. Moss, J.G. Rogers and A.W. Stetz, Proc. 3rd Int.Conf. on Clustering Aspects of Nuclear Structure and Nuclear Reactions, Winnipeg, Manitoba (1978), AIP Conf.Proc.No.47,636.

25. G.J. Stephenson Jr., E.F. Reddish, G.M. Lerner and M.I. Haftel, Phys.Rev. C6 (1972)1559.

26. A.A. Ionnides and D.F. Jackson, Nucl.Phys.A308 (1978)305,317.

27. K.L. Kowalski, Phys.Rev.Lett. 15 (1965) 798. H.P. Noyes, Phys.Rev.Lett. 15 (1965)538.

28. T.R. Mongan, Phys.Rev. 184 (1969)1888.

29. R.V. Reid, Jr., Ann.Phys. (N.Y.) 50 (1968)411.

30. T.R. Mongan, Phys.Rev. 178 (1969)1597.

31. C.F. Perdrisat, L.W.Swenson, P.C. Gugelot, E.T. Boschitz, W.K. Roberts, J.S.Vincent and J.R. Priest, Phys.Rev. 187 (1969)1201.

32. P.G. Roos, Phys.Rev. C9 (1974)2437.

Part IV

Pion Physics

PION ABSORPTION BY NUCLEI

Daniel S. Koltun

Department of Physics and Astronomy

University of Rochester, Rochester, New York, U.S.A.

I. General Survey
II. Theoretical Problems

INTRODUCTION

The subject of these lectures has a history of experiments and theory of about thirty years. It is not the present purpose to review the subject in detail. There are by now a number of review articles on the subject[1-3], and regular reports at two year intervals, at the International Conferences on High Energy Physics and Nuclear Structure (of which the most recent is by H.K. Walter[4]), which have up-to-date information (see also Ref. 1a). The purpose here is to give an overall impression of what the subject is about, and how it stands.

I. GENERAL SURVEY

1. What Happens in Pion Absorption?

Commonly nuclear absorption of pions is studied for π kinetic energies (lab) in the range $0 < T_\pi < 300$ MeV. This is, of course, the same range as for π scattering experiments, and is determined by the same two (related) facts: (1) The πN system has a broad resonant state (the $\Delta(1232)$) in the middle of that region, $T_\pi \sim 190$ MeV; and (2) current meson machines (LAMPF, SIN, TRIUMF) have beams in this energy range. Experiments are designed to absorb the beam

*Work supported in part by the U.S. Department of Energy.

directly (capture in flight), or slow the mesons (negative only) in a large target until they are captured first in atomic orbits (mesic atoms), from which they may be absorbed by the nucleus (stopped capture or absorption).

We designate the absorption reaction by

$$\pi + A \rightarrow A^* \rightarrow X \tag{1}$$

where A is the nuclear target, and A* is an excited state of the nucleus, having absorbed the meson. This state is always unstable, and breaks up by emission of one or many particles (X). We shall not specifically talk about radiative absorption, where some of the emission is electromagnetic (photons), or absorption through weak interactions: both are possible, but much less probable than the strong interaction branches which shall interest us. In these last cases, X consists entirely of nuclear particles.

The general facts are:

Many particles are emitted, on the average. A few of these take away most of the available energy—which is given by $\omega_\pi = T_\pi + m_\pi$, where $m_\pi = 140$ MeV (c^{-2}). The kinds of particles seen commonly are: p, n, d, t, ^{3}He, α. The charges of the emitted particles tend to follow that of the π, i.e., more p's, ^{3}He, etc. for π^+; n's, t's for π^-.

This kind of information comes from (often older) photographic studies: emulsions and bubble chambers.

It is useful to rewrite (1) in the form:

$$\pi + A \rightarrow x + B_x^* \tag{2}$$

where x are the emitted particles, and B_x^* a daughter nucleus in an excited state. If we insist that B_x^* be particle stable, e.g., by looking for some particular <u>nuclear</u> γ or β transition (a commonly used experimental technique) then x includes all particles emitted, fast or slow. For stopped π^- absorption, the number of particles increases with target mass number A; the number also increases, as might be expected, with T_π. However, one might also consider x to be a specific set of fast particles (for reasons to be discussed), in which case B_x^* could still emit particles, e.g., thermodynamically (evaporation).

To keep in mind the transfer of quantum numbers: the pion bring in, besides T_π of a few hundred MeV, a linear momentum q of a few hundred MeV/c (200 MeV/c = 1 fm^{-1}), which means an angular momentum

$$\ell \lesssim qR < 10\ (\hbar) \qquad (3)$$

and isospin T = 1, parity = $(-1)^{\ell+1}$ from the odd intrinsic parity of the π.

For contrast, a nucleon of T ∿ 500 MeV, which can also excite a nuclear target by some 150-200 MeV, brings in linear momentum q ∿ 1 GeV/c, and therefore $\ell \lesssim 50\ (\hbar)$ units. A heavy ion of T ∿ 500 MeV has q ∿ $\sqrt{A}$ GeV/c, and can bring in proportionately more angular momentum. A pion, which is much lighter than a nucleon or ion, has little linear or angular momentum, compared to the energy it can deposit--aside from the rest energy, which is also absorbed in the π reaction.

2. What is Measured, and What Can Be Measured, in Absorption Reactions?

First of all, everything can be measured as a function of target size (A) and of meson energy (T_π). Aside from the nuclear structure aspects of looking at different targets, some effects of the variation with A and T_π reflect what is going on with the pion before it is absorbed, i.e., scattering reactions, to which we shall return later. For 150 < T_π < 250 MeV, we expect significant effects of the $\Delta(1232)$ resonance to show up.

Now, what to measure? Photographic experiments (emulsions, bubble-chambers) are, in a sense, "complete": all products are seen (actually only charged particles) and the momenta (including direction) of emitted particles can, in principle, be measured, too, although the accuracy here may be very low both because of statistics and limitations of size of tracks. (Spin polarizations are not detected, however.) If this method were more accurate, and applicable to all targets, energies, etc., the available information would be vast. To use it, we would have to slice it up in all sorts of ways, if only to be able to plot data on graphs! But indeed, it would be useful raw information to have to begin with. Now in fact, we do not generally have such complete experiments, and we rely on slices through the possible data, depending on what we choose to, or are able to measure.

We may display the kind of experiments available on a scale which runs from inclusive to exclusive. These terms, now in fairly general use in particle and nuclear physics, focus on the degree of detail required in a measurement, as follows. The most inclusive data is the total rate or cross section for the reaction--in this case, π absorption--to happen. The most exclusive would be a particular branch of some "complete" measurement, e.g., in the bubble chamber, of

$$\pi^+ + {}^4\text{He} \to p + p + d \tag{4}$$

with all momenta (and spin projections measured). Most experiments lie between, and include some classes of things while excluding others. Let us consider the following partial list:

Inclusive

$\sigma_{abs}(T_\pi, A)$ - total absorption cross section
$\Gamma_{n\ell}$ - absorption width of π-atomic level

The second is a measurement of the absorption rate for a bound ($T_\pi \sim$ zero) π^-, which is absorbed from an orbit with definite angular momentum. (One needs some atomic theory to remove the electromagnetic effects involved here.)

Slightly Exclusive

particle-types emitted: $\langle N_p \rangle$, $\langle N_n \rangle$, $\langle N_d \rangle$, etc. in more detail, energy spectra (singles): $N(E_p)$ ... $j\pi + A \to p +$ anything

final nuclides reached (B_x^*) - e.g. by measurement of nuclear γ or β-decay: $\pi + A \to B_x^* +$ anything

These two examples give rather different information: the first indicates something about how the absorbed energy was distributed among emitted particles, the second tells what final daughter nuclear states are reached, but not how. As an example of the former, we show (Fig. 1) a plot of $d\sigma/d\Omega dE$ vs. E for protons emitted at 45° from absorption of 220 MeV π^+ on various targets (Jackson et al[5]). Here one sees an interesting feature, namely the structure in the distribution at $E_p \sim 200$ MeV, which may be interpreted as a sign of a direct absorption process, since the proton comes out as if the pion were absorbed by a deuteron (or two-nucleon cluster in the target.

One evident difficulty with these inclusive measurements is that they often are unable to distinguish absorption reactions from other inelastic reactions with similar products, e.g. (π,πp) in the case of Jackson et al. One must use other experimental information or some theoretical argument (e.g. on the improbability of knockout of 200 MeV nucleons from He) to make the distinction. (This is usually no problem with stopped π^- absorption.)

More Exclusive

Correlated emissions--like the famous ratio of emitted nn/np pairs--a number quoted in the range 3-10 for light nuclei.

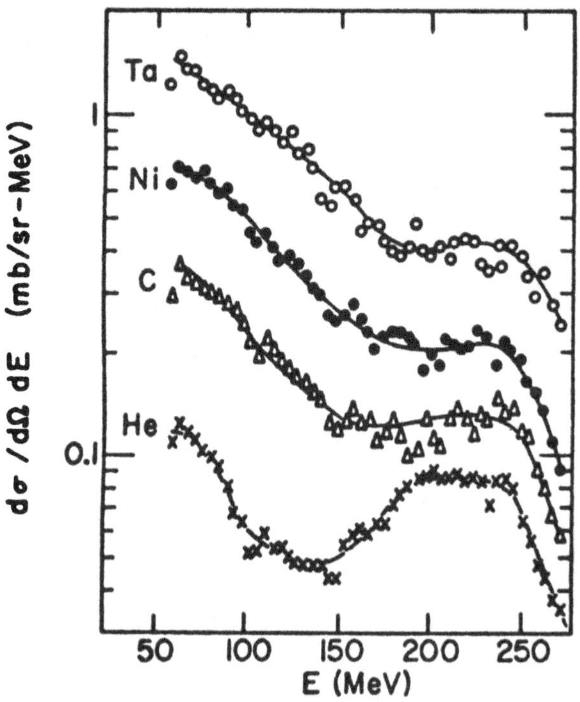

Fig. 1. (from Jackson et al., Ref. 5).

These may have more information: i.e.,

- opening angle (see Fig. 2), perhaps also
- energy of one or both legs: e.g. (p,n),(d,n),(t,n) measurement of Lee at Virginia, E_x with n, x = p,d,t.

The fact that there are correlations between emissions is widely taken as evidence that there is a strongly direct component to the reaction. This makes it interesting from the point of view of nuclear structure studies. For example, there has been considerable interest in the "quasi deuteron" mode of absorption:

$$\pi^{\pm} + A \rightarrow \{{pp \atop nn}\} + B^* \tag{5}$$

which is supposed to proceed like

$$\pi^{\pm} + "d" \rightarrow \{{p+p \atop n+n} \tag{6}$$

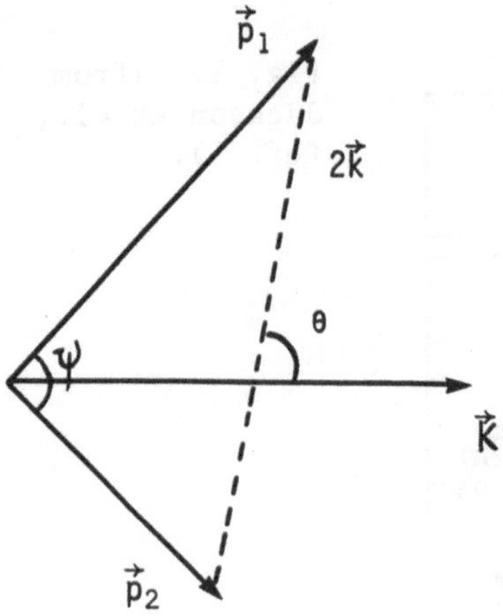

Fig. 2a. Correlated pair emission $A(\pi,\vec{p}_1,\vec{p}_2)B$. Capture at rest - lab frame
ψ - opening angle in lab
θ - angle

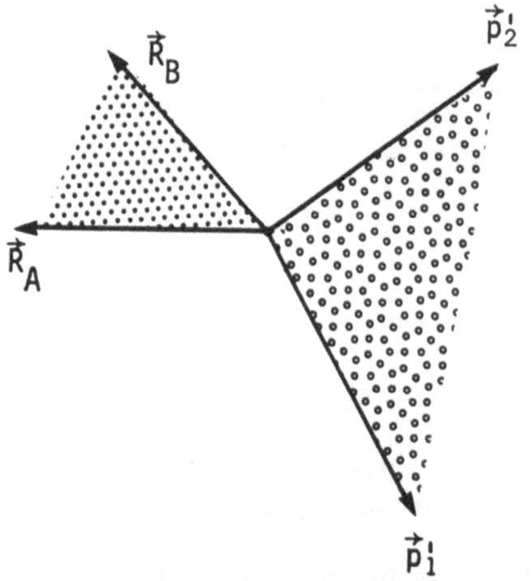

Fig. 2b. Capture in flight
-π rest frame ($\vec{q}_\pi' = 0$)
$\vec{R}_A,\vec{R}_B$ nuclear recoil
T-Y angle - between ($\vec{R}_A \times \vec{R}_B$) and ($\vec{p}_1' \times \vec{p}_2'$) planes

with a spectator residual nucleus (B*) providing the initial binding for the initial n-p pair, which may or may not be like a free deuteron, but which might be expected to have a similar internal structure, dominated by $^3S_1$ and $^3D_1$ relative motion. The evidence from the study of (π^+,pp) or (π^-,nn) on light nuclear targets, partially supports this picture: e.g., for the extremely likely case of ^6Li:

$$\pi^{\pm} + {}^6\text{Li} \rightarrow \{{}^{pp}_{nn}\} + {}^4\text{He} \quad (7)$$

particularly for ground state ^{4}He. But here the nucleus helps us considerably. The case is not so clear for heavier targets, that (6) is the dominant mechanism for reactions like (5), or more generally for the total absorption cross-sections or rates. We shall return to this again. Reaction (7) might fit at the "exclusive" end of our reaction scale, since one can determine the states of all the products in this case. We list a few more:

Exclusive

$A(\pi^+,p)B^*$	to well defined final states of (A-1)
$A(\pi,2N)B^*$	to well defined final states of (A-2)
$A(\pi^+,2pd)B^*$	to well defined final states of (A-4)
(as in (4) above).	
$^{14}N(\pi^+,2p)3\alpha$	(emulsion expt. - Afa-65, see Ref. 1)

These may be more or less complete, depending on whether all or only some of the kinematic variables are measured.

Now the problem is to sort out what to make of all these possibilities: how to interpret the reaction modes, how to apply the interpretation—when applicable—to nuclear structure questions.

3. Description of Absorption Reactions

First we must recognize that there may be several different possible descriptions of what goes on in the dynamics of π-absorption—particularly if we are considering the entire reaction, from start to finish, with all its possible branches. For example, consider the "quasi-deuteron" description of the ^{6}Li reaction (7): as we mentioned, there is evidence that the reaction is direct, and that (6) is the dominant dynamical aspect of the reaction. Here we have the favourable nuclear structure: that the overlap of ^{6}Li with ^{4}He × d is fairly large: i.e., the spectroscopic factor

$$S = |\langle {}^4\text{He} \times d | {}^6\text{Li} \rangle|^2 \sim \frac{1}{2} \ (?) \quad . \quad (8)$$

In addition, the outgoing nucleons may scatter from the residual ^{4}He without major disruption of the direct reaction kinematics.

But what about absorption by a larger target, say ^{40}Ca?

Let us suppose that (6) is the important "elementary" reaction: what would we expect to see? (For recent data, see Ref. 6, and Fig. 3.) On sample DWIA arguments, the strength S for (6) would now be spread over a fairly broad range of excitation of B*, say ^{38}K, for ^{40}Ca(π^-,2n)^{38}K*, at least over $\sim$ 12 MeV available for 2 valence holes (sd)$^{-2}$. The ground state transition might be relatively weak, and not particularly instructive. Integrating after the excitation region given by the (sd)$^{-2}$ strength distribution would presumably show the direct aspect of the reaction more clearly (assuming it were there!). However, here the residual target is larger than for ^{6}Li(π,2N/^{4}He), which means first: that there can be more scattering of the outgoing neutrons than for He--changing the kinematics by diffraction, and second: there can be inelastic scattering of these particles, leading to emission of more than 2 neutrons.

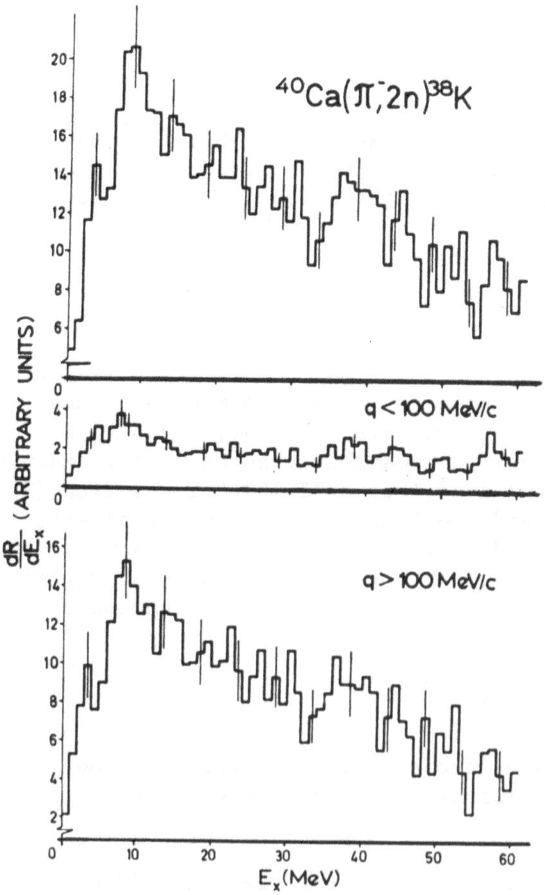

Fig. 3. Excitation energy spectra for the residual ^{38}K nucleus. Upper curve: all data: middle curve: events with q < 100 MeV/c; lower curve: events with q < 100 MeV/c. R is absorption rate. Statistical errors are indicated. (From Ref. 6.)

Now the inclusive measurement $^{40}Ca(\pi^-,2n)^{38}K^*$ might look like quasi-deuteron absorption, but heavily distorted by the interactions in the final state. A series of further measurements of other emissions, e.g.

$$^{40}Ca(\pi^-;2n+xn+yp)Z \tag{9}$$

might be helpful in settling how these final interactions take place. They could be quasi-free knockout collisions, or evaporation from some "equilibrium" state of $^{38}K^*$, or anywhere in between: "pre-equilibrium". If one could make some reasonable description of these later stages, then one might still be able to use the inclusive $(\pi^-,2n)$ measurements to study the two-hole structure in ^{40}Ca. The problems are similar to those of the analysis of (p,2p). We have assumed for the sake of argument, that there is a quasi-deuteron absorption mechanism. There are some theoretical arguments for this, for stopped π^-, based on specific models. Without going into detail, the arguments are that a free pion cannot easily give up both its momentum and energy to a bound, but uncorrelated single nucleon, but it can do so more easily to an interacting pair of nucleons. The relative NN states with ^3S or ^3D have the strongest interactions—hard core and tensor—thus, the "quasi-deuteron". This is oversimplified—and does not strictly eliminate other low-ℓ states: ^1S_o, ^3P. It has not been shown that higher-order correlations, e.g., "quasi-α"—of (2n+2p)—do not also contribute to the possibility of absorption, and such α-modes have also been of considerable interest. The theoretical problem is complicated, because one must separate the question of "quadrupole-correlations" from pair correlations in the same (α) cluster. In fact, the pair-correlation model gives a reasonable (although perhaps big by 2x) prediction of the rate of absorption (Γ_{1s}) for stopped π^- on ^4He[7-9].

What makes the argument for pair-absorption at least possible for stopped π^-, becomes much more problematic for energetic $\pi^{\pm}$, particularly in the energy domain $150 < T_\pi < 250$ of the $\Delta(1232)$. Now the interactions of the incoming π should become more important. This is not just a "distorted wave" problem, since the incoming π can scatter inelastically, exchanging both energy and momentum with various nucleons, thus increasing its "chances" for absorption. For particular exclusive reactions with strong spectroscopic favouring, like the ground state reaction (7), these many-body features could be suppressed, but the need not be in, e.g., $^{40}Ca(\pi,2N)^{38}B$.

This brings up the point that we know very little about the energy-dependence of π-absorption reactions. The total absorption cross-sections are not well known--what is known is that they are large and weakly energy dependent in the Δ region. The reactive cross sections here are almost geometrical (πR^2) here, and the absorption seems to be a good fraction of that. Some recently quoted numbers[4]:

$$
\begin{array}{llll}
^{12}C, & T_\pi = 130\ \text{MeV}, & \sigma_{abs} = 190\ \text{mb} \sim .6\ \pi R^2 \\
^{62}Ni, & T_\pi = 70\ \text{MeV}, & \sigma \qquad 500\ \text{mb} \\
 & T_\pi = 220\ \text{MeV}, & \sigma \qquad 700\ \text{mb} \sim .7\text{-}.8\ \pi R^2
\end{array}
$$

Now there is not at present a good theory for the σ_{abs}. It seems clear that this requires some understanding of the coupling of the scattering processes with the absorption channels, which we shall return to in the second lecture. It also seems clear that the nuclear structure aspects of absorption may be hard to disentangle in the resonance region.

4. Nuclear Structure Aspects

Clearly to use π absorption as a tool for conventional nuclear structure study requires us to be able to identify and isolate direct reaction modes. What can we study? If fast particles are emitted, and if we can make reasonable estimates of the distortions from interactions in the final states (as well as of the incoming pion), then we are looking at hole structure in the target. If we think of final targets in low energy final states (B*), then (π,N), (π,2N), etc. are similar to one, two, etc. nucleon removal experiments done by particle transfer: (p,d), (p,t), etc. (Good energy resolution is then required, e.g., 20 - 100 keV.) But as mentioned before, unless the spectroscopic strengths are very large for the particular initial and final states ($A \rightarrow B^*_x + x$), the assumption of a simple direct mode may fail. Therefore, a better use of these transfer reactions might be to look at deep hole states, with not-so-good energy resolution, as in (p,2p) or (e,e'p) reactions at several hundred MeV. Then one may look at distributions of two-hole, three-hole, etc., states, deep in the target nucleus, which are otherwise inaccessible. The interactions in final states may well be more tractable for the highly excited states reached, than for valence-shell reactions.

Study of the angular distributions of emitted particles yields some information about angular momentum transfers involved, under the assumptions of direct reactions. Two particularly interesting angles are shown in Fig. 2; the θ-angle for stopped π absorption, and the Trieman-Young angle for capture in flight--both of which involve correlations of two emitted particles, e.g., in (π^-,nn) or (π^+,pp).

What if direct modes do not dominate absorption: e.g., for $T_\pi \sim 200$ MeV on a large target? Certainly statistical reaction methods may then be of some assistance in describing the reaction, but not very much nuclear information can emerge. What should be of considerably more interest would be details of the approach to equilibrium: i.e., how the first few particles emitted get out, with what energy, what types, and so on. The theoretical methods here would have to be a compromise between direct and statistical reaction models—perhaps starting with the exciton models used for other pre-equilibrium studies, as for α or heavy-ion induced reactions. A good theory for how an energetic π transfers its energy and momentum in sequence, before it is absorbed by a nuclear target, would be good nuclear physics.

II. THEORETICAL PROBLEMS

1. General Dynamical Formulation

The problem of absorption of mesons is interesting from a theoretical point of view because it forces us to stretch our usual notions of nuclear structure and nuclear reactions to accommodate interactions with meson fields. The point is that we have a well known collection of theoretical tools for dealing with nuclear interactions in terms of the Schroedinger equation, in which the nucleons provide the dynamical degrees of freedom. We know that there are other degrees of freedom in a nucleus: mesons, whose virtual exchange provide the nuclear interactions, and virtual photons, which produce electromagnetic interactions. In normal, low energy nuclear physics, we suppress explicit reference to these virtual excitations of meson or photon fields, and replace their effects by potential interactions—and we are back at the Schroedinger level.

Now when we deal with real mesons, which can be detected at large distance from the nucleus, and which can scatter, be absorbed or emitted by that nucleus, we encounter new problems, since the meson degrees of freedom must now be included in some form, including the possibilities of changing the number of mesons (by absorption or emission). We have a choice of how to deal with this: we can try to reformulate our ideas of nuclear physics in terms of the quantum theory of fields, so that techniques common in that discipline become available to us. Sometimes this is particularly attractive, as when weak or electromagnetic interactions are involved, and we may profitably treat nuclei in particular states as "elementary particles" with quantum numbers and extended structure (i.e., form factors). However it is not clear that this will work for strong interactions with mesons. A more general field theoretic approach to nucleons and mesons is a many, many body problem, and a relativistic one at that, and has not been much advanced. It seems simpler, if less elegant, to try to reduce the field problems to the more familiar terms (to us nuclear physicists) of Schroedinger language.

The simplest form of pion coupling to nuclei is through the Yukawa interaction, which couples $N \leftrightarrow N + \pi$ (see Fig. 4a), and is given by

$$H_I(x) = j\ \phi\ (x) \tag{10}$$

with $\phi(x)$ the pion field and of a nucleon current operator, e.g.

$$j = i\ g\ \sum_i \gamma_5(i)\tau(i) \tag{11a}$$

for pseudoscalar coupling, where the sum is over nucleons. An equivalent nonrelativistic (static nucleon) operator is

$$j = \sqrt{4\pi}\ \frac{if}{m_\pi} \sum_i \tau(i)\ \vec{\sigma}(i)\cdot\vec{\nabla} \quad , \tag{11b}$$

where the coupling constants are related by

$$f^2 = \frac{g^2}{4\pi}\left(\frac{m}{2M}\right)^2 \simeq 0.08 \quad . \tag{11c}$$

Now this interaction (10) is supposed to be responsible for absorption or emission of single pions, much as the electromagnetic interaction $\vec{j}\cdot\vec{A}(x)$ is responsible for emission and absorption of single photons. However, the Yukawa interaction is also a contributor to the nucleon-nucleon interaction, e.g., through one-pion-exchange (OPE) (Fig. 4b). It also is responsible for πN scattering, through

Fig. 4a

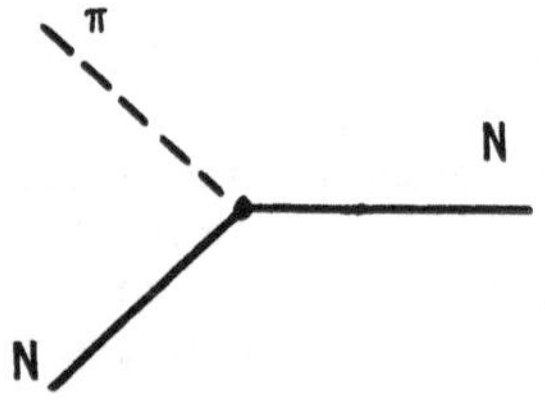

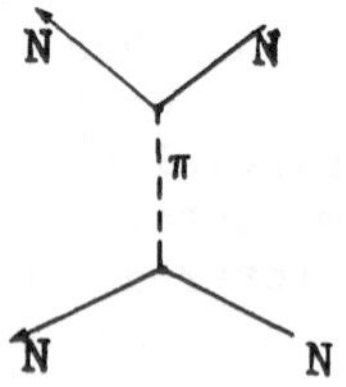

Fig. 4b

processes like those shown in Fig. 4c, d, the direct and crossed Born terms, which are the processes which are supposed to generate the low energy πN p-wave scattering, in the theory of Chew and Low. The behaviour of the low energy πN p-waves, particularly the $T = J = \frac{3}{2}, \frac{3}{2}$ wave (whose resonance is the $\Delta(1232)$ mentioned earlier), and the long-range OPE behaviour of the NN interaction, are the best evidence we have for the Yukawa interaction[10,11].

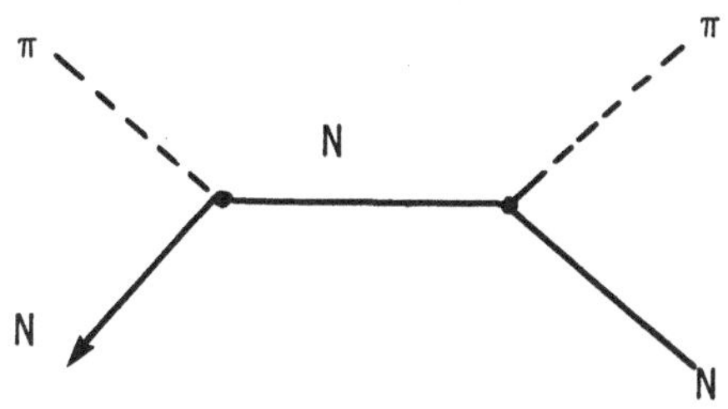

Fig. 4c

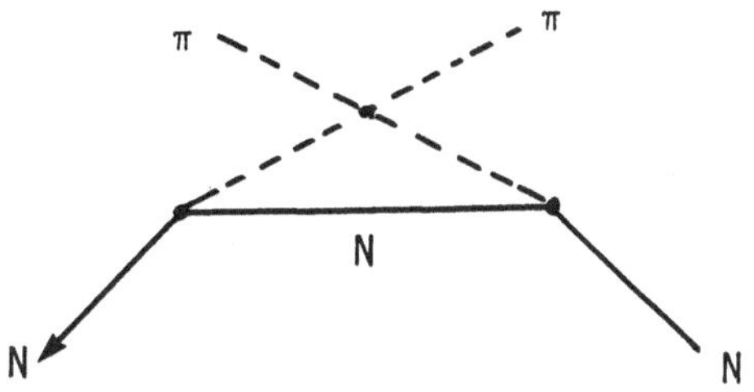

Fig. 4d

Now the nuclear interaction can be treated in terms of potentials, if we wish, but there are problems in trying to treat the πN interaction similarly. It is easy to see that the problem begins right with the Born term (Fig. 4c). Consider the diagram of Fig. 4e, which shows a fourth-order (in 10) process involving two nucleons. This particular diagram can be looked at two different ways: either as a double scattering through the Born term (c), once on each nucleon, or alternatively, as an absorption of the pion by one nucleon, and re-emission by the other nucleon, with a nucleon-nucleon interaction (OPE) intervening. The problem is that with interaction (10), this diagram only appears once, while if we try to treat both (b) and (c) as potential interactions in a Schroedinger equation the diagram will be generated twice: we will not have counted correctly. This illustrates the difficulty of dealing with real pions (i.e., which can scatter, be absorbed and emitted in real reactions) and nucleons in a Schroedinger theory. Any method based on Schroedinger theory will also fail similarly, e.g., the Watson

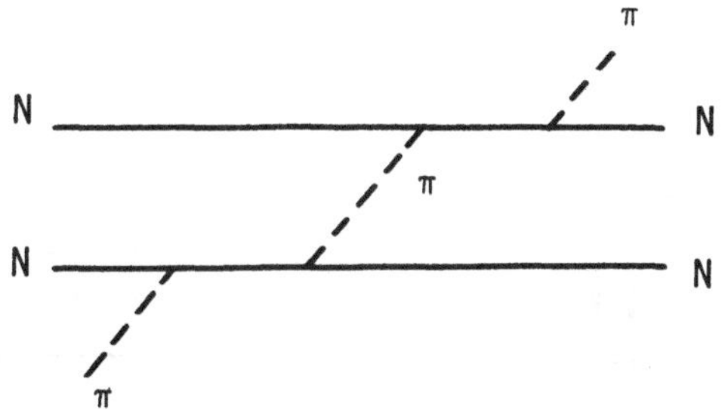

Fig. 4e

multiple scattering expansion, and the conventional optical potential methods derived from it.

There are a number of possible ways out. Some, as we have mentioned, involve non-Schroedinger methods throughout, for example using the Low equation for scattering, following in analogy to the derivation of the Chew-Low theory for πN scattering. Although this can be done for π-nucleus scattering, the non-linear equations of the theory are quite unattractive, and not particularly well suited to calculating pion absorption (which, in fact, must be known as an input in this approach).

A method which Mizutani[10] and I have developed (see also Thomas[11]) sticks much closer to conventional Schroedinger methods as follows. We classify contributions to π-nucleus scattering as in Fig. 5a, b: where we show some 8th order terms. (The "caps" stand for the nuclear ground state.) The separation is according to

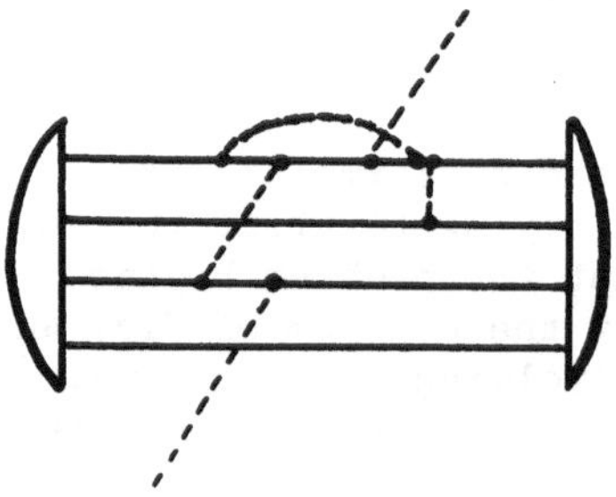

Fig. 5a

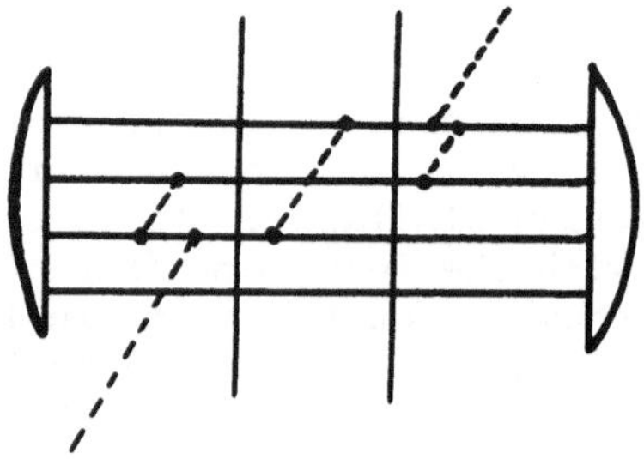

Fig. 5b

whether one can or cannot find a time (vertical lines in (b)) at which no pions are present. These terms contain all the contributions of the Born term (Fig. 4c), and are non-Schroedinger. Terms like (a) have no counting problems: they can in fact be regrouped into an equivalent Schroedinger scattering calculation. The (b) terms may be called absorption-re-emission; since they, in fact, contain intermediate states with no pions, they also contain the transition amplitudes for true absorption reactions, as illustrated in Fig. 5c, d.

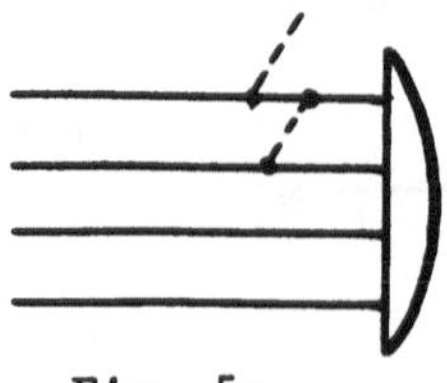

Fig. 5c

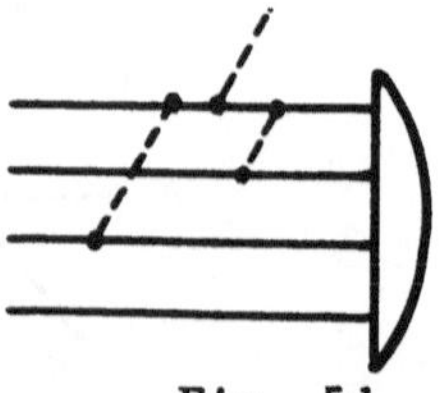

Fig. 5d

Formally, this approach may be summarized as follows. There is an effective Schroedinger equation

$$(E - H_o - v)\psi_1^{\pm(+)} = 0 \tag{12}$$

which is equivalent to class (a) of diagrams: the interaction v includes all nuclear interactions, and all πN interactions <u>except</u> those generated by the direct Born term (Fig. 5c). The t-matrix for, e.g., elastic scattering of a pion by the target takes the form

$$T = T_1 + \langle\psi^{(-)}|a^+ G_N a|\psi^{(+)}\rangle \tag{13}$$

where T_1 and $\psi^{(\pm)}$ refer to solutions of Eq. (12), with no absorption possible, while the second term of (13) gives the contribution of Fig. 5b, in terms of an effective absorption operator a (and a^+), and a propagator G_N for the nuclear system, e.g., between the two vertical lines in (b). The absorption amplitudes takes the form

$$T_{abs} = \langle\phi^{(-)}|a|\psi^{(+)}\rangle \tag{14}$$

where $\phi^{(-)}$ is a <u>nuclear</u> wave function for the final state.

All this is quite formal; it does not tell us what will be important in the reaction. But it does tell us how to use our conventional Schroedinger tools: nuclear wave functions, multiple scattering theory, etc. in the case of coupling to real pions. Secondly, this discussion underlines another important facet of pion dynamics, namely that the scattering and absorption of pions are strongly linked to each other. This was already the case in the Chew-Low πN theory. In fact our formulation follows the structure of the coupled channels theory of nuclear reactions.

2. Detailed Reaction Theory

Now we return to the problem of calculating real absorption processes, as discussed in the first lecture. With the exception of calculations of some special exclusive reactions: e.g., A(π,2N)B* for stopped π^- or at low energy, or for A(π^+,p)B, all for light nuclear targets, this is an unsolved problem. For example, there is no theory of total cross sections for absorption, to apply, for example for $100 < T_\pi < 300$ MeV, and for heavy nuclear targets. Cascade calculations have been applied here, and they may give us a rough idea of what is going on, but we have no idea how accurate they are. Although a many step process may eventually reach the "classical" behaviour of the cascade process, it would be nice to know what happens in the early steps of the process. One might expect, for a π at resonant energies ($T_\pi \sim 200$), where the Δ dominates πN scattering, that several strong collisions, perhaps quite off-shell, precede absorption, and that fast particles are produced in this non-statistical part of the reaction. How many particles? How many will get out? Here the problem is to put together the formalism we have for multiple scattering, which has phase or off-shell information, along with some statistical methods which make it possible to calculate average, rather than detailed data, as applicable to the more inclusive reactions.

The doorway state approach of Kisslinger[12], Lenz[13] and Moniz[14] may give a model which can be developed in this direction. In these theories, a pion must make a Δ-hole state in the nuclear target (doorway state) before anything else can happen, including absorption. Absorption takes place by the Δ-hole state interacting with another nucleon, forming a 2p-2h state. The two particles may then be emitted, giving $A(\pi,2N)B^*$. If more modes of excitation were coupled in, e.g., Δp-2h, Δ2p-3h, etc., we could allow for emission of 2, 3, 4, etc. nucleons at various stages of the reaction.

3. Optical Potential

An interesting question one encounters is where to put absorption into the optical potential for elastic scattering. It is tempting to try to write

$$\upsilon_{opt} = \upsilon_{scatt} + \upsilon_{abs} \tag{15}$$

where the two terms refer to scattering processes and true absorption, respectively, but it may not be easy to specify either potential without some knowledge of both processes, except in a very weak coupling limit.

In the formulation of scattering indicated by Eq. (13), it was natural to separate the scattering amplitudes, not the potentials. Clearly the scattering affects the absorption term, but not vice-versa. An optical potential can be found for Eq. (13), but it is extremely cumbersome. If we assign υ_{scatt} of (15) to the elastic part of $\psi_1^{(+)}$, the scattering solution of the effective Schroedinger equation (12), we may be able to approximate, using the usual (e.g., Watson) low density arguments, by a local potential

$$\upsilon_{scatt} \sim t_o \rho(r) \tag{16}$$

where $\rho(r)$ is the nucleon distribution in the target. But the appropriate υ_{abs} would be highly non-local, partly because of propagation G_N in (13), and partly because of interference of the scattering with the absorption.

On these grounds, I would question the "conventional wisdom" of using a $\rho^2(r)$ form for absorption in an optical potential. This local form was introduced by the Ericsons[15] for zero-energy scattering, where weak coupling and the long wavelength of the pion may justify the use of $\rho^2(r)$, since non-locality would then be irrelevant.

A striking example is given by the doorway model mentioned above. In this theory the optical potential is non-local, with or without absorption, because of Δ-propagation effects in the target. But the nonlocality is <u>changed</u> by the absorption. Eq. (15) does not seem to be a particularly useful representation in this case.

4. How is a Stopped π^- Absorbed?

The last problem we shall mention is somewhat technical, and involves the form of the absorption operator (10) in the limit of very slow pions, e.g., for s-wave orbits of π-atoms. The problem arises because the relativistic forms, e.g., γ_s in (11a), behave "peculiarly" for almost static systems. The static term (11b) is only part of the low energy limit of (11a) given is nonrelativistic (i.e., Pauli spinor) form, so that we can use it with ordinary nonrelativistic nuclear physics. Now (11b) is proportional to the meson momentum (or velocity), because of the gradient which operates on the field $\phi(x)$. As the velocity of the π vanishes, so does the matrix element (11b). Presumably there are terms left out of the limit (11b) which depend, e.g., on the nucleon velocities. These can indeed be generated, by some well-known techniques (see, e.g., Friar[15], but unfortunately the forms depend on a number of things we do not know. One is whether the basic relativistic coupling is γ_s as in (11a), or pseudovector: $\gamma_s\gamma_\mu\partial/\partial x_\mu$. In the static limit, both relativistic forms give the same current (11b), but the correction terms differ. Worse yet, the nonrelativistic forms depend on the nuclear interactions among the absorbing nucleons, and on the behaviour of these potentials relativistically, as well. In other words, in order to do this calculation right, with nonrelativistic nuclear physics, we have to know from what relativistic theory we have taken our limit. It is the pseudoscalar property of the pion which leads to this "unfortunate" result.

REFERENCES

1. D.S. Koltun, Advanced in Nuclear Physics (ed. Baranger and Vogt), 3 (1969), 71; Proc. 4th Int. Conf. on High Energy Physics and Nuclear Structure, Dubna 1971, p. 201.

1a. J.M. Eisenberg and D.S. Koltun, "Theory of Meson Interactions with Nuclei", (Wiley-Interscience, forthcoming).

2. T.I. Kopaleishvili, Particles and Nuclei 2 (1973), 87.

3. J. Hufner, Physics Reports (Phys. Letters C) 21C (1975), 1.

4. H.K. Walter, Proc. 7th Int. Conf. on High Energy Physics and Nuclear Structure, Zurich, 1977, p. 225.

5. H.E. Jackson et al., Phys. Rev. Letters 39 (1977), 1628.

6. B. Bassalleck et al., Z. Physik A 286 (1978), 401.

7. G. Backenstoss et al., Nucl. Phys. A232 (1974), 519.

8. D.S. Koltun and A. Reitan, Nucl. Phys. B4 (1968), 629.

9. G.F. Bertsch and D.O. Riska, Phys. Rev. C18 (1978), 317.

10. T. Mizutani, Ph.D. Dissertation, University of Rochester (1975), Technical Report, UR--53 (1975); T. Mizutani and D.S. Koltun, Annals of Physics 109 (1977), 1; D.S. Koltun and T. Mizutani, to be published.

11. A.W. Thomas, Ph.D. Dissertation, Flinders University (1973).

12. L. Kisslinger and W. Wang, Annals of Phys. 99 (1976), 374.

13. F. Lenz, Annals of Phys. 95 (1976), 348.

14. E. Moniz, Proc. Int. Conf. on Meson Nuclear Physics (Carnegie-Mellon, 1976), p. 105.

15. J.L. Friar, Phys. Rev. C10 (1974), 955.

BEYOND LOWEST-ORDER RESULTS IN PION-NUCLEUS REACTIONS*

J.M. Eisenberg

*Dept. of Physics and Astronomy, Tel-Aviv University, Tel-Aviv, Israel** and Dept. of Physics, University of Virginia, Charlottesville, Virginia 22901, U.S.A.*

In recent years, in large measure due to the advent of the meson factories, there has developed a great interest in using pions for elastic and inelastic scattering on nuclei, in parallel to similar reactions for nucleons. The general objective of such hadronic scattering is to determine nuclear static and dynamic distributions of hadronic matter: proton and--more especially--neutron densities, two-nucleon correlations (with and without spin and isospin vector forms), α-clustering, and so on. Theoretical analysis is generally directed at joining together reliable descriptions of the reaction mechanism and of the pertinent nuclear structure, with sufficient control on each so that their separate influences may be examined. The hadronic processes are very intricate, and the existence of a variety of probes of this nature is to be welcomed as offering greater likelihood for unraveling them. Towards this end, the pion offers several advantages and several disadvantages, as we shall see, at least implicitly, in the following. It also is quite distinct from the nucleon in that it can undergo true absorption in the nucleus, a feature upon which we shall not dwell at much length here where our primary interest will be to examine general features of multiple-scattering theory.

I have been asked to develop in these lectures the basic formalism which is used to discuss elastic and inelastic

---

*Work supported in part by the U.S.-Israel Binational Science Foundation and by the Center for Advanced Studies of the University of Virginia.

**Permanent Address

pion-nucleus scattering reactions at medium energies. I shall try to do this beginning more or less from first principles (the Lippmann-Schwinger equation for scattering on a many-body system) and developing a complete, and, in principle, exact formalism[1]. The main emphasis will be on methods for going beyond lowest-order results in the basic projectile-nucleon free amplitude $t_{\pi N}$ or in the target system density $\rho(\underline{r})$ for calculations of elastic scattering based, say, on an optical potential approach or for inelastic processes evaluated using distorted-wave impulse approximation (DWIA). Such higher-order effects seem to be required by the theory itself for internal consistency and are apparently increasingly demanded by the vastly improved data of the post-meson-factory era. Moreover they incorporate interesting physical effects about which we would like to be better informed. Actual evaluation of these effects is still in very preliminary stages which we shall try to outline briefly in hopes of indicating where future developments are to be expected. Many aspects of the theory, both formal and practical, are treated in greater detail in ref. 2, where there is also extensive discussion of pion absorption reactions, both for their own sake and for their implications in a complete and consistent many-body scattering formalism, and a much more complete list of references.

1. OPTICAL POTENTIAL FORMALISM FOR ELASTIC SCATTERING

We shall start by deriving a general formalism for the projectile-nucleus scattering amplitude in a Watson multiple-scattering series and for the optical potential in a similar series. Elaboration of the analogy between this formalism and physical optics has been given elsewhere[1-3] and will only be touched upon here. We shall instead focus on a reasonably rapid and telegraphic formal development, with stress on the nature of the higher-order features. Towards this end there are two very standard manipulations which we must master. The first of these resums a many-potential Lippmann-Schwinger equation to produce a series involving projectile-nucleon amplitudes closely related to the free amplitudes, while the second provides a formal way to compare two amplitudes which are "closely related". Subsequently, we shall deal with a third standard feature: the expansion in successive degrees of diagonality in matrix elements in the target (nuclear) space. Many of our subsequent developments will depend on these manipulations. Lastly, we note that our initial formalism in the critical stages is quite general and the use of the words "pion" and "nucleus" (π and A in shorthand) is made interchangeably with "projectile" and "target".

1.1. Multiple-Scattering Series for the πA Amplitude

The basic problem which we address is defined in terms of a

hamiltonian

$$H = H_o(\underset{\sim}{r}_1,\underset{\sim}{r}_2, \ldots , \underset{\sim}{r}_A;\underset{\sim}{r}) + V(\underset{\sim}{r}_1,\underset{\sim}{r}_2, \ldots , \underset{\sim}{r}_A;\underset{\sim}{r})$$

$$= H_N(\underset{\sim}{r}_1,\underset{\sim}{r}_2, \ldots , \underset{\sim}{r}_A) + K(\underset{\sim}{r}) + \sum_{i=1}^{A} v_i(\underset{\sim}{r}_i,\underset{\sim}{r}) \tag{1.1}$$

where we shall constantly be distinguishing between the nucleon variables $\underset{\sim}{r}_1,\underset{\sim}{r}_2, \ldots , \underset{\sim}{r}_A$, appearing, for example, in the nuclear hamiltonian H_N with A nucleons, and the projectile space with coordinate $\underset{\sim}{r}$ and free projectile hamiltonian K. We have everywhere suppressed spin and isospin coordinates at this stage since the variables only appear in order to define the relevant spaces. The interaction between the projectile and nucleus is taken as the sum over A pion-nucleon potentials $v_i(\underset{\sim}{r}_i,\underset{\sim}{r})$, which already implies a severe approximation in that we do not allow for true absorption in the case of a boson projectile.

We write the Lippmann-Schwinger equation[1] intended to solve the scattering problem for the hamiltonian of eq. (1.1) as

$$T = V + VG_oT= \sum_{i=1}^{A} v_i + \sum_{i=1}^{A} v_iG_oT, \tag{1.2}$$

where the propagator is

$$G_o \equiv \frac{1}{E - H_o + i\varepsilon} = \frac{1}{E - H_N - K + i\varepsilon} , \qquad \varepsilon \to 0^+ \tag{1.3}$$

and has buried in it all of the purely nuclear dynamics in that it tacitly supposes that a solution is known for the nuclear hamiltonian H_N appearing in the denominator. We shall now recast eq. (1.2) into a form such that the summation of all rescatterings of the projectile on a given, single target nucleon is executed before the full many-body scattering problem is addressed. This we do because the strong interactions involved in our problem make it advantageous to develop a reliable description of the projectile-nucleon two-particle subsystem before going on to the full nuclear problem. In particular, in analogy with the introduction of the G-matrix in the nuclear matter or nuclear effective interaction problem, this avoids difficulties which will otherwise destroy perturbative approaches if, for instance, there is a hard-core in the projectile-nucleon interaction. Moreover, the net result of this line of attack is to replace the potential v_i by the corresponding πN amplitude t_i, which may be better known or more easily parametrized from experiment.

Let us first study the resummation in a diagrammatic language indicated in fig. 1. With time progressing from right to left, we order the interactions according to which took place last (leftmost). By summing over all terms in which the pion interacts only with the first nucleon, we obtain the πN amplitude t_1, but with a propagator for which the other nucleons are always present (G_0). A similar result pertains for each nucleon, yielding $t_1+t_2+...+t_A$. Then we proceed to sum all interactions which end between the pion and nucleon 1, but just before that have an interaction with 2 or 3 or 4.... Eventually we obtain terms for arbitrary numbers of t's in arbitrary order, with propagators G_0 between them, but never with two adjacent t_j's referring to the same nucleon (since that was already summed to produce the pertinent amplitude t_j in the first place).

To obtain this result formally from eq. (1.2), we write [MANIPULATION #1]

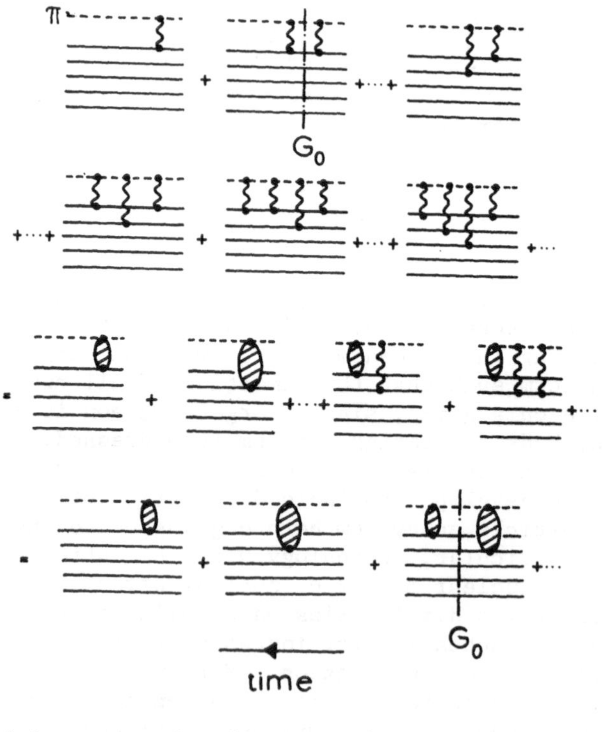

Fig. 1. Multiple-scattering diagrams whose resummation yields the Watson series.

$$T = \sum_{i=1}^{A} T_i \tag{1.4}$$

with

$$T_i = v_i + v_i G_o T \quad , \tag{1.5}$$

which is the sum of all diagrams in which the last interaction is on nucleon i. From eqs. (1.4) and (1.5),

$$T_i = v_i + v_i G_o \sum_{j=1}^{A} T_j \quad , \tag{1.6}$$

or

$$(1 - v_i G_o) T_i = v_i + v_i G_o \sum_{j \neq i} T_j \quad . \tag{1.7}$$

Then defining

$$t_i = (1 - v_i G_o)^{-1} v_i \tag{1.8}$$

we have

$$T_i = t_i + t_i G_o \sum_{j \neq i} T_j \quad , \tag{1.9}$$

which is a set of coupled integral equations for the T_i. The amplitude t_i in eq. (1.8) clearly satisfies

$$t_i = v_i + v_i G_o t_i \quad , \tag{1.10}$$

and thus is the amplitude for pion scattering on the ith nucleon in the presence of the other nucleons which enter through the appearance of H_N in G_o. Lastly, we can use eq. (1.9) in (1.4) to get

$$T = \sum_{i=1} t_i + \sum_{i=1} t_i G_o \sum_{j \neq i} T_j \quad , \tag{1.11}$$

and iterating eq. (1.9) we obtain the Watson multiple-scattering series for the scattering amplitude

$$T = \sum_{i=1}^{A} t_i + \sum_{i=1}^{A} t_i G_o \sum_{j \neq i} t_j + \sum_{i=1}^{A} t_i G_o \sum_{j \neq i} t_j G_o \sum_{k \neq j} t_k + \ldots \tag{1.12}$$

This is an infinite scattering series involving single, double, triple, ...scattering in which no two successive scatterings are on the same nucleon. We note that this result, Manipulation #1, applies for any amplitude containing a sum of interactions and that the same propagator must appear in eq. (1.10) for t_i as in eqs. (1.2) or (1.12) for T.

1.2. Comparing Two Amplitudes with the Same Interaction and Similar Propagators

Eq. (1.10) for t_i involves a propagator which is slightly different from that for the free πN system,

$$t_i^{free} = v_i + v_i g_o t_i^{free} \tag{1.13}$$

with

$$g_o = \frac{1}{E_\pi - K + i\varepsilon} \quad , \tag{1.14}$$

in which the nuclear hamiltonian and energy do not appear. (Previously, we could think of the system energy as involving the nuclear plus pion energies, $E = E_\pi + E_N$. Note also that the single-nucleon recoil kinetic energy should be included in K.) We wish to relate t_i^{free} to t_i, which will, incidentally, provide us with our second general formal trick, MANIPULATION #2. To do this we first note that we can invert eq. (1.13) to read

$$v_i = t_i^{free} (1 + g_o t_i^{free})^{-1} \quad , \tag{1.15}$$

which we substitute into eq. (1.10) to obtain

$$\begin{aligned} t_i &= v_i(1 + G_o t_i) = t_i^{free} (1 + g_o t_i^{free})^{-1} (1 + G_o t_i) \\ &= t_i^{free} (1 + g_o t_i^{free}) (1 + g_o t_i^{free})^{-1} (1 + G_o t_i) \\ &\quad - t_i^{free} g_o t_i^{free} (1 + g_o t_i^{free})^{-1} (1 + G_o t_i) \\ &= t_i^{free} + t_i^{free} G_o t_i - t_i^{free} g_o v_i (1 + G_o t_i) \\ &= t_i^{free} + t_i^{free} (G_o - g_o) t_i \quad , \end{aligned} \tag{1.16}$$

which is again a very general result for two amplitudes with the same driving term in their respective Lippmann-Schwinger equations, but two different propagators. Of course, our hope here is that, under some conditions (high-energy projectiles, for example), G_o and g_o will be very similar, so that t_i may be obtained from t_i^{free} through eq. (1.16) without undue exertion.

1.3. The Optical Potential

As part of our justification for working with an expansion of T in terms of the t_i's rather than the v_i's, we noted that the former might lead to much better convergence properties than the latter, as, for example, in the case of v_i's containing a hard-core part. Our series for T given in eq. (1.12), however, may also not converge rapidly or at all. Its convergence is governed by something like the parameter, $f_{\pi N}/R_{NN}$, i.e. the ratio of the average basic scattering amplitude to the average internucleon distance. For pion-nucleus scattering near the 3,3 region, both of these quantities are near 1 fm and rapid convergence cannot be expected. For this reason we shall try now to develop a scheme for summing over infinite subsets of the infinite series for T. This will yield an expansion for the optical potential, whose convergence is governed by a parameter of the order of $f_{\pi N}\, R_{corr}\rho/k$; where R_{corr} is a typical nuclear correlation length, ρ is the target density and k is the incident pion momentum. This parameter will be relatively small in our applications, and will clearly be reduced for high energies or if only short correlation lengths are involved. The optical potential can therefore be expanded in its series more reliably in what will amount to a cluster expansion, the result then being used in a Schroedinger or Lippmann-Schwinger equation for the amplitude.

The basic idea of the optical potential series rests on the notion of establishing a hierarchy of diagonality in matrix elements of the Σt_i in the nuclear space. We will suppose that the diagonal nuclear matrix elements are generally much larger than the nondiagonal ones and systematically expand in the number of nondiagonal terms appearing, which will lead to a cluster expansion. The nuclear ground state--on the assumption that it is scattering on it which interests us--is then singled out for special consideration at the start, though, more generally, any nuclear state which is reached will also be expected to involve large diagonal matrix elements, that is, to give a preference for the system to return to that state. The hierarchical treatment of the nuclear matrix elements is illustrated in great detail in ref. 2, which affords some insight into the motivations for what otherwise remain abstract manipulations; here in the interests of economy we shall start with a particular formal development of the optical potential and then hint at other possible avenues for generating such potentials.

Our formal tool will require the introduction of a projection operator P_0, for the (preferred) nuclear ground state $|0\rangle$, having properties

$$P_0|0\rangle = |0\rangle, \quad (1-P_0)|0\rangle = 0 \quad , \quad P_0^2 = P_0 \; , \; P_0^+ = P_0 \; , \; P_0 = |0\rangle\langle 0| \tag{1.17}$$

We then define an optical operator $\mathcal{U}$ such that it satisfies a Lippmann-Schwinger equation very similar to that of eq. (1.2) for T,

$$\mathcal{U} = V + V\, G_0\, (1 - P_0)\mathcal{U} \quad , \quad V = \sum_{i=1}^{A} v_i \quad . \tag{1.18}$$

We use Manipulation #2 to compare eqs. (1.2) and (1.18), identifying $\mathcal{U}$ with t_i^{free} and T with t_i in eq. (1.16). Then

$$T = \mathcal{U} + \mathcal{U}\,[G_0 - G_0(1 - P_0)]T = \mathcal{U} + \mathcal{U}\, G_0 P_0 T \quad . \tag{1.19}$$

For elastic scattering, we need the nuclear ground state expectation value of this, namely

$$\begin{aligned} T_c &\equiv \langle 0|T|0\rangle = \langle 0|\mathcal{U}|0\rangle + \langle 0|\mathcal{U} G_0 P_0 T|0\rangle \\ &= \langle 0|\mathcal{U}|0\rangle + \langle 0|\mathcal{U}|0\rangle G_0 \, \langle 0|T|0\rangle \; , \end{aligned} \tag{1.20}$$

but this last is nothing other than the Lippmann-Schwinger equation in the projectile space for the exact πA scattering. Thus the ground state expectation value

$$\mathcal{U}_c \equiv \langle 0|\mathcal{U}|0\rangle \tag{1.21}$$

of the optical operator of eq. (1.18) determines an optical potential such that the wave function which solves the corresponding Schroedinger equation is an exact solution to our problem.

We can now exploit eq. (1.18) and Manipulation #1 leading from eq. (1.2) to eq. (1.12) since $\mathcal{U}$ is also given in terms of a form similar to (1.2). Then we have immediately

$$\mathcal{U} = \sum_{i=1}^{A} t_i' + \sum_{i=1}^{A} \sum_{j\neq i} t_i' G_0 (1 - P_0) t_j' + \ldots \quad , \tag{1.22}$$

where now t_i' solves

$$t_i' = v_i + v_i G_o (1 - P_o) t_i' \qquad (1.23)$$

or

$$t_i' = t_i - t_i G_o P_o t_i' \qquad (1.24)$$

using Manipulation #2 on eqs. (1.10) and (1.23). Thus our pertinent t_i' for the optical potential expansion has moved a bit further away from the original t_i, which in turn was not quite the free πN amplitude t_i^{free}. The crucial new feature in the expansion for $\mathcal{U}$ of eq. (1.22) is the appearance of the projection operator $(1 - P_o)$ which forbids intermediate return to the nuclear ground state, and guarantees that in the optical potential $<0|\mathcal{U}|0>$ there will be at least two nondiagonal nuclear matrix element in terms beyond the lowest order $\Sigma t_i'$ and these nondiagonal factors will reduce the higher-order contributions. As we shall see, this presciption in fact leads to a cluster expansion in two-particle, three-particle, ... correlations.

1.4. The Exploitation of Antisymmetrization: KMT

Before proceeding to a further examination of the properties of the optical potential, it is convenient to exploit a feature of our problem which is accidental from the point of view of multiple scattering, but which leads to a considerable simplification. This is the antisymmetry of the nuclear wave function which describes our target. (The antisymmetrization between a nucleon projectile and the nucleons in the target is ignored here.) This was first exploited systematically for this problem by Kerman, McManus and Thaler[4] whence the rubric KMT for the approach[5]. We start by defining a new πN scattering amplitude through

$$\tau_i = v_i + v_i \frac{1}{E - H_o + i\varepsilon} A\tau_i = v_i + v_i G_o A\tau_i \quad , \qquad (1.25)$$

where the operator A projects on nuclear states which are fully anti-symmetric. (Note that in the analogous eqs. (1.10) and (1.23) the summation on intermediate nuclear eigenstates of H_N is meant to include unphysical states which lack proper Pauli antisymmetry.) Now our original Lippmann-Schwinger equation for the full πA amplitude, eq. (1.2), could just as well have been written

$$T = V + VG_o AT = \sum_{i=1}^{A} v_i + \sum_{i=1}^{A} v_i G_o AT \quad , \qquad (1.26)$$

since it will ultimately be taken between fully antisymmetric nuclear states and the operator $V = \Sigma v_i$ is of necessity fully symmetric in the nucleon coordinates. Thus only fully antisymmetric intermediate states would have survived any way in the spectrum of G_o.

Comparing eqs. (1.25) and (1.26) we see that we have now brought about a situation in which the conditions of Manipulation #1 apply, i.e. identical propagators and a driving term for T which is a sum of those for the t_i. Thus we can immediately write, as in eq. (1.12), that

$$T = \sum_{i=1}^{A} \tau_i + \sum_{i=1}^{A} \tau_i G_o \mathcal{A} \sum_{j\neq i} \tau_j + \sum_{i=1}^{A} \tau_i G_o \mathcal{A} \sum_{j\neq i} \tau_j G_o \mathcal{A} \sum_{k\neq j} \tau_k + \ldots \tag{1.27}$$

Furthermore, since <u>all</u> nuclear matrix elements will now involve fully antisymmetric nuclear states, both initially and finally, the matrix elements of each τ_i, i = 1, 2, ... , A will be identical, so that the summation prescription in this equation can be dealt with by a counting factor, yielding

$$T = \sum_{i=1}^{A} \tau_i + \left(\frac{A-1}{A}\right) \sum_{i=1}^{A} \tau_i G_o \mathcal{A} \sum_{j=1}^{A} \tau_j + \left(\frac{A-1}{A}\right)^2 \sum_{i=1}^{A} \tau_i G_o \mathcal{A} \sum_{j=1}^{A} \tau_j G_o \mathcal{A} \sum_{k=1}^{A} \tau_k + \ldots \tag{1.28}$$

This can be rewritten as

$$T = \sum_{i=1}^{A} \tau_i + \frac{A-1}{A} \sum_{i=1}^{A} \tau_i G_o \mathcal{A} T \quad , \tag{1.29}$$

or, defining

$$T' \equiv \frac{A-1}{A} T, \qquad \mathcal{T}' \equiv \frac{A-1}{A} \sum_{i=1}^{A} \tau_i \quad , \tag{1.30}$$

as

$$T' = \mathcal{T}' + \mathcal{T}' G_o \mathcal{A} T' \quad . \tag{1.31}$$

(Incidentally, we note that at this stage the antisymmetrization instruction $\mathcal{A}$ could be dropped in eqs. (1.29) or (1.31) - though it is to be retained in (1.25) - since T or T' will be taken between fully antisymmetric states and $\tau \equiv \sum_i \tau_i$ or τ' are completely symmetric.) Obviously, T' and T differ by a (trivial) correction of order A^{-1}.

Now we shall explore the construction of an optical potential for T' of eq. (1.31). This time we shall develop the formalism in a more explicit manner*, partly for the added insight and partly to allow for further generalizations of the result of eq. (1.22). We start by recalling that we ultimately require

$$T_c = \langle 0|T|0\rangle = \frac{A}{A-1} T'_c = \frac{A}{A-1} \langle 0|T'|0\rangle \quad , \qquad (1.32)$$

and that the expectation values here refer to the nuclear space only, so that for the ground state

$$|0\rangle = |\Phi_0 (\underset{\sim}{r}_1, \underset{\sim}{r}_2, \ldots , \underset{\sim}{r}_A)\rangle \quad , \qquad (1.33a)$$

with Φ_0 fully antisymmetric, while for excited states, which we shall also take as everywhere antisymmetric,

$$|\alpha\rangle = |\Phi_\alpha (\underset{\sim}{r}_1, \underset{\sim}{r}_2, \ldots , \underset{\sim}{r}_A)\rangle \quad . \qquad (1.33b)$$

Then from eq. (1.31) we can write the exact result

$$\langle 0|T'|0\rangle = \langle 0|\mathcal{T}'|0\rangle + \sum_\nu \langle 0|\mathcal{T}'|\nu\rangle\, G_0(\nu)\, \langle \nu|T'|0\rangle \quad , \qquad (1.34)$$

where the propagator possesses the property

$$\langle \nu'|G_0|\nu\rangle = \langle \nu'|\frac{1}{E-H_N-K+i\varepsilon}|\nu\rangle = \langle \nu|\frac{1}{E-E_N^\nu-K+i\varepsilon}|\nu\rangle \delta_{\nu',\nu} \equiv G_0(\nu)\delta_{\nu',\nu} \quad . \qquad (1.35)$$

Separating out the diagonal nuclear matrix elements, which we expect to be larger than the nondiagonal ones, we have ($E_N^0 = 0$)

$$\langle 0|T'|0\rangle = \langle 0|\mathcal{T}'|0\rangle + \langle 0|\mathcal{T}'|0\rangle \frac{1}{E-K+i\varepsilon} \langle 0|T'|0\rangle$$

$$+ \sum_{\nu\neq 0} \langle 0|\mathcal{T}'|\nu\rangle\, G_0(\nu)\, \langle \nu|T'|0\rangle \quad , \qquad (1.36)$$

where the last term is supposed to be small since it is "doubly nondiagonal". If we drop it and compare with the Lippmann-Schwinger equation for $\langle 0|T'|0\rangle$, from which the optical potential

*Alternatively, we could have proceeded as in eqs. (1.18) to (1.22), leading to yet another single-nucleon amplitude $\tau_i' = v_i + v_i G_0\, A(1-P_0)\tau_i'$.

can be read off [cp. eq. (1.20)], we have immediately the lowest-order result

$$\mathcal{V}_c(1)' = \langle 0|\mathcal{T}'|0\rangle = \frac{A-1}{A} \langle 0|\sum_{i=1}^{A} \tau_i|0\rangle \quad , \tag{1.37}$$

which, to order A^{-1} and ignoring the difference between τ_i and t_i, is identical to the lowest order in eq. (1.22).

Higher orders are now easily generated. We use eq. (1.31) to produce nondiagonal elements of T',

$$\langle\nu|T'|\mu\rangle = \langle\nu|\mathcal{T}'|\mu\rangle + \sum_{\lambda} \langle\nu|\mathcal{T}'|\lambda\rangle\, G_o(\lambda)\, \langle\lambda|T'|\mu\rangle \quad , \tag{1.38}$$

which, in eq. (1.36), leads to

$$\begin{aligned}\langle 0|T'|0\rangle = {} & \langle 0|\mathcal{T}'|0\rangle + \langle 0|\mathcal{T}'|0\rangle\, G_o(0)\, \langle 0|T'|0\rangle \\ & + \sum_{\nu\neq 0} \langle 0|\mathcal{T}'|\nu\rangle\, G_o(\nu)\, \langle\nu|\mathcal{T}'|0\rangle \\ & + \sum_{\nu\neq 0} \langle 0|\mathcal{T}'|\nu\rangle\, G_o(\nu) \sum_{\lambda} \langle\nu|\mathcal{T}'|\lambda\rangle\, G_o(\lambda) \\ & \times \langle\lambda|T'|0\rangle \quad . \end{aligned} \tag{1.39}$$

From this we again separate out the $\lambda=0$ term and read off the next-order optical potential

$$\mathcal{V}_c(2,0)' = \sum_{\nu\neq 0} \langle 0|\mathcal{T}'|\nu\rangle \frac{1}{E-E_N^{\nu}-K+i\varepsilon} \langle\nu|\mathcal{T}'|0\rangle \ , \tag{1.40}$$

again to be compared with eq. (1.22). The correction to this now involves at least three matrix elements which do not contain the nuclear ground state on one side or the other. But, they may be only "doubly nondiagonal" due to terms in eq. (1.39) with $\lambda=\nu$. Such terms will modify eq. (1.40) so that

$$\begin{aligned}\mathcal{V}_c(2,1)' = {} & \sum_{\nu\neq 0} \langle 0|\mathcal{T}'|\nu\rangle\, G_o(\nu) \times [1 + \langle\nu|\mathcal{T}'|\nu\rangle\, G_o(\nu) \\ & + \langle\nu|\mathcal{T}'|\nu\rangle\, G_o(\nu)\, \langle\nu|\mathcal{T}'|\nu\rangle\, G_o(\nu) + \ldots] \times \langle\nu|\mathcal{T}'|0\rangle \\ = {} & \langle 0|\mathcal{T}'|0\rangle + \sum_{\nu\neq 0} \langle 0|\mathcal{T}'|\nu\rangle \frac{1}{E-E_N^{\nu}-K-\langle\nu|\mathcal{T}'|\nu\rangle} \langle\nu|\mathcal{T}'|0\rangle \quad . \end{aligned} \tag{1.41}$$

This version of the second-order result would appear to lead to a more rapidly convergent result than $\mathcal{U}_c^{(2,0)\dagger}$, since, already at this level, it sums over many diagonal nuclear matrix elements which otherwise must enter in higher orders. Of course, this is at the price of a rather more complicated propagator in eq. (1.41). (Note that we have dropped the "+iε" there since $\langle\nu|\mathcal{T}'|\nu\rangle$ is complex--even with negative imaginary part as required.) We refer to the two approaches of eqs. (1.40) and (1.41) as Formalisms 0 and 1, whence the additional superscripts there, and note that these are only the simplest results in a family of formalisms in which the propagator may be corrected by successive orders of the optical potential (not for the ground state, however) appearing in the denominator, with corresponding omission of return to intermediate states in higher order. Each of these formalisms is, in principle, complete and consistent, the differences between them arising presumably in rate of convergence and certainly in complexity of usage since the distortion of intermediate pions by means of the optical potential is a non-negligible complication. Because the second-order results of eqs. (1.40) and (1.41) are, in practice[6], about as far as one can go in a practical calculation, we shall not develop these formalisms further; this is done, for example, in ref. 2, where the implications for computational work are also discussed.

1.5. Some Explicit Forms for the Optical Potential

1.5.1. The Lowest-Order Potential. We shall now translate our expressions for the optical potential into more practical forms, though still not specializing in detail to features peculiar to pions (which we shall do in Chapter 4). The lowest-order result of eq. (1.22), for example, is (we drop the prime on t_i for notational convenience since, at the moment, the distinctions between t_i, τ_i and τ_i' are not of concern to us)

$$\mathcal{U}_c^{(1)} = \langle 0|\sum_{i=1}^{A} t_i|0\rangle \quad . \tag{1.42}$$

Taking the matrix element in momentum space for the projectile this is

$$\langle \underset{\sim}{k}', \underset{\sim}{P}'|\mathcal{U}_c^{(1)}|\underset{\sim}{k},\underset{\sim}{P}\rangle = \sum_{i=1} \int \frac{d\underset{\sim}{p}_1}{(2\pi)^3}\frac{d\underset{\sim}{p}_2}{(2\pi)^3}\cdots\frac{d\underset{\sim}{p}_i}{(2\pi)^3}\frac{d\underset{\sim}{p}_i'}{(2\pi)^3}\frac{d\underset{\sim}{p}_A}{(2\pi)^3}$$

$$\times\ \Phi_0^{+}(\underset{\sim}{p}_1,\underset{\sim}{p}_2,\ \cdots\ ,\underset{\sim}{p}_i',\ \cdots\underset{\sim}{p}_A)\ \langle\underset{\sim}{k}',\underset{\sim}{p}_i'|t_i(E)|\underset{\sim}{k},\underset{\sim}{p}_i\rangle$$

$$\times\ \Phi_0(\underset{\sim}{p}_1,\underset{\sim}{p}_2,\ \cdots\ ,\underset{\sim}{p}_i,\ \cdots\ ,\underset{\sim}{p}_A), \tag{1.43}$$

where $\underset{\sim}{k}$, $\underset{\sim}{P}$, $\underset{\sim}{k}'$ and $\underset{\sim}{P}'$ refer to initial and final pion and nucleus momenta. Momentum conservation for the πN amplitude requires

$$\langle \underset{\sim}{k}',\underset{\sim}{p}_i'|t_i(E)|\underset{\sim}{k},\underset{\sim}{p}_i\rangle = (2\pi)^3\,\delta(\underset{\sim}{k}'+\underset{\sim}{p}_i'-\underset{\sim}{k}-\underset{\sim}{p}_i)$$

$$\times\ t_i(\underset{\sim}{k}',\underset{\sim}{k};\underset{\sim}{p}_i;E), \qquad (1.44)$$

whence

$$\langle \underset{\sim}{k}',\underset{\sim}{P}'|\mathcal{U}_c^{(1)}(E)|\ \underset{\sim}{k},\underset{\sim}{P}\rangle = A\int \rho(\underset{\sim}{p}',\underset{\sim}{p})\ \langle \underset{\sim}{k}',\underset{\sim}{p}'|t(E)|\underset{\sim}{k},\underset{\sim}{p}\rangle\ \frac{d\underset{\sim}{p}'d\underset{\sim}{p}}{(2\pi)^6}\ , \qquad (1.45)$$

where, if we ignore spin and isospin effects,

$$\rho(\underset{\sim}{p}',\underset{\sim}{p}) \equiv \frac{1}{A}\sum_{i=1}^{A}\int \Phi_o^+(\underset{\sim}{p}_1,\ \ldots\ ,\underset{\sim}{p}_{i-1},\underset{\sim}{p}',\underset{\sim}{p}_{i+1},\ \ldots\ ,\underset{\sim}{p}_A)$$

$$\times\ \Phi_o(\underset{\sim}{p}_1,\ \ldots\ ,\underset{\sim}{p}_{i-1},\underset{\sim}{p},\underset{\sim}{p}_{i+1},\ \ldots\ ,\underset{\sim}{p}_A)\ \frac{d\underset{\sim}{p}_1}{(2\pi)^3}\ \cdots\ \frac{d\underset{\sim}{p}_{i-1}}{(2\pi)^3}\,\frac{d\underset{\sim}{p}_{i+1}}{(2\pi)^3}\ \cdots$$

$$\times\ \frac{d\underset{\sim}{p}_A}{(2\pi)^3}\ . \qquad (1.46)$$

Here the interpretation in terms of averaging over the nucleon momenta in the target should be clear. When nucleon spin and isospin features are included, eqs. (1.45) and (1.46) are easily generalized to incorporate spin and isospin densities and amplitudes[2].

If we ignore the dependence of the amplitude in eq. (1.45) on the nucleon momenta, on the grounds that the operative quantity is the Galilei invariant $k - \frac{\omega}{M}p_i\ v_\pi - v_N$ and $v_N/v_\pi \lesssim 1/3$ near the 3,3 region, then

$$\langle \underset{\sim}{k}',\underset{\sim}{P}'|\mathcal{U}_c^{(1)}|\underset{\sim}{k},\underset{\sim}{P}\rangle = A\ \langle \underset{\sim}{k}'|t|\underset{\sim}{k}\rangle\ \rho(\underset{\sim}{k}-\underset{\sim}{k}')\quad , \qquad (1.47)$$

with

$$\rho(\underset{\sim}{k}-\underset{\sim}{k}') \equiv \int e^{i(\underset{\sim}{k}'-\underset{\sim}{k})\cdot\underset{\sim}{r}}\ \rho(\underset{\sim}{r})d\underset{\sim}{r} \qquad (1.48)$$

and

$$\rho(\underset{\sim}{r}) \equiv \frac{1}{A} \sum_i \int d\underset{\sim}{r}_1 \ldots d\underset{\sim}{r}_A \; \Phi_o^+(\underset{\sim}{r}_1, \ldots, \underset{\sim}{r}_A)\; \delta(\underset{\sim}{r}-\underset{\sim}{r}_i)\; \Phi_o(\underset{\sim}{r}, \ldots \underset{\sim}{r}_A) \tag{1.49}$$

(plus spin and isospin generalizations). Since the nuclear size is much greater than the nucleon size, $\rho(\underset{\sim}{k}-\underset{\sim}{k}')$ will tend to be rather forward peaked relative to the πN amplitude and $<\underset{\sim}{k}'|t|\underset{\sim}{k}>$ can be approximated by its forward value t_o, whence

$$\mathcal{V}_c^{(1)}(\underset{\sim}{r}) = At_o\rho(\underset{\sim}{r}) = -\frac{4\pi}{2m} f_o A\rho(\underset{\sim}{r}),\; \int \rho(\underset{\sim}{r})d\underset{\sim}{r} = 1 \; . \tag{1.50}$$

The lowest-order optical potential is then, in this approximation, local, though we shall see that almost any improvement on this simplistic result leads <u>inter alia</u> to nonlocality. Using the optical theorem (and approximating t_o by the free πN amplitude, as is reasonable, say, for high energies), we see that

$$\text{Im}\,\mathcal{V}_c^{(1)} = -\frac{1}{2} v\rho\, \sigma_{\pi N}^{\text{total}} = -\frac{1}{2}\frac{v}{\ell},\; \ell \equiv \frac{1}{\rho\sigma_{\pi N}^{\text{total}}}, \tag{1.51}$$

where v is the pion velocity, $\sigma_{\pi N}^{\text{total}}$ is the total πN cross section and ℓ is the pion mean free path in the nucleus. This leads to the well-known interpretation of the imaginary, non-hermitian part of the optical potential as embodying optical absorption due to scatterings of the pion on nucleons (in which the latter are here thought to be removed from the nucleons in our high-energy approximation, thus taking the nucleus out of the elastic channel). A graph of the mean free path $\ell = (\rho\sigma_{\pi N}^{\text{total}})^{-1}$ is shown in fig. 2, and indicates that, near the 3,3 region, the pion wave is quenched to 1/e of its free-space value within about 0.5 fm, so that the pion sees the surface region almost exclusively--about which more later.

<u>1.5.2. The Second-Order Potential</u>. We shall estimate the next order for the optical potential using the easier Formalism 0 result of eq. (1.22) or (1.40). Referring to the nuclear ground state as $|0>$ and suppressing the nuclear total momenta $\underset{\sim}{P}$ and $\underset{\sim}{P}'$, we have from eq. (1.22),

$$\langle \underset{\sim}{k}'0|\mathcal{V}_c^{(2,0)}|\underset{\sim}{k}\ 0\rangle = \sum_{i=1}^{A} \sum_{j\neq i} \sum_{\alpha\neq 0} \int \langle \underset{\sim}{k}'0|t_i|\underset{\sim}{k}''\ \alpha\rangle$$

$$\times \frac{1}{E-E_\pi(\underset{\sim}{k}'')-E_N^\alpha+i\varepsilon} \langle \underset{\sim}{k}''\ \alpha|t_j|\underset{\sim}{k}0\rangle \frac{dk''}{(2\pi)^3}$$

$$\cong \int \Phi_o^+(r_1, \ldots ,r_A) \sum_{i=1} \sum_{j\neq i} \langle \underset{\sim}{k}'|t_i|\underset{\sim}{k}''\rangle$$

$$\times \frac{e^{i(\underset{\sim}{k}''-\underset{\sim}{k}')\cdot \underset{\sim}{r}_i}\, e^{i(\underset{\sim}{k}-\underset{\sim}{k}'')\cdot \underset{\sim}{r}_j}}{(E-\overline{E}_N) - E_\pi(\underset{\sim}{k}'') + i\varepsilon} (1 - P_o) \langle \underset{\sim}{k}''|t_j|\underset{\sim}{k}\rangle$$

$$\times \Phi_o(\underset{\sim}{r}_1, \ldots ,\underset{\sim}{r}_A)d\underset{\sim}{r}_1 \ldots d\underset{\sim}{r}_A \frac{d\underset{\sim}{k}''}{(2\pi)^3} , \qquad (1.52)$$

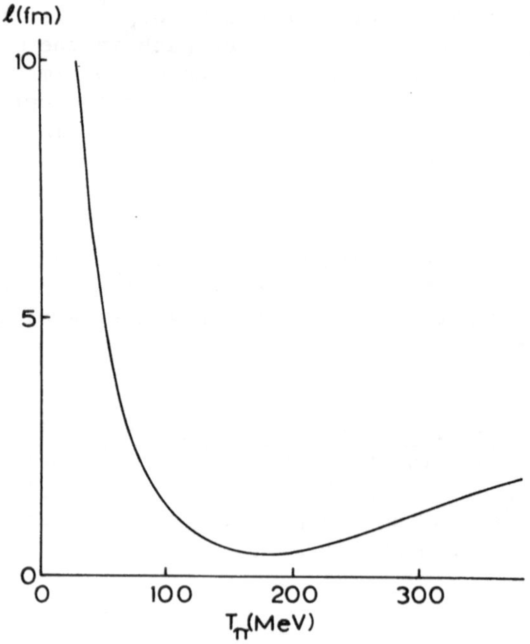

Fig. 2. The pion-nucleus mean free path $\ell=[\rho\cdot\frac{1}{2}(\sigma_{\pi p}+\sigma_{\pi^+ n})]^{-1}$ near the 3,3 resonance for constant nuclear density (N=Z) $\rho=0.17\ fm^{-3}$.

where the second equation follows as in (1.45) - (1.49), and with the use of a closure argument which led to the appearance of an average nuclear excitation energy $\bar{E}_N$, which is presumably much smaller than the high scattering energy E. We shall now ignore spin and isospin parts of the amplitude to rewrite this as

$$\langle \underset{\sim}{k}'|\mathcal{U}_c^{(2,0)}|\underset{\sim}{k}\rangle = A(A-1)\int \langle \underset{\sim}{k}'|t|\underset{\sim}{k}''\rangle \frac{e^{i(\underset{\sim}{k}''-\underset{\sim}{k}')\cdot\underset{\sim}{r}'}e^{i(\underset{\sim}{k}-\underset{\sim}{k}'')\cdot\underset{\sim}{r}}}{(E-\bar{E}_N)-E_\pi(\underset{\sim}{k}'')+i\varepsilon}$$

$$\times \frac{d\underset{\sim}{k}''}{(2\pi)^3}\langle \underset{\sim}{k}''|t|\underset{\sim}{k}\rangle\,[P^{(2)}(\underset{\sim}{r}',\underset{\sim}{r}) - \rho(\underset{\sim}{r}')\rho(\underset{\sim}{r})]\,d\underset{\sim}{r}d\underset{\sim}{r}' \quad , \tag{1.53}$$

where we have introduced the two-nucleon correlation function

$$P^{(2)}(\underset{\sim}{r}',\underset{\sim}{r}) \equiv \frac{1}{A(A-1)}\int \Phi_o^+(\underset{\sim}{r}_1,\ldots,\underset{\sim}{r}_A)\sum_{i=1}^{A}\sum_{j\neq i}\delta(\underset{\sim}{r}'-\underset{\sim}{r}_i)\delta(\underset{\sim}{r}-\underset{\sim}{r}_j)$$

$$\times\, \Phi_o(\underset{\sim}{r}_1,\ldots,\underset{\sim}{r}_A)\,d\underset{\sim}{r}_1,\ldots,d\underset{\sim}{r}_A \quad , \tag{1.54}$$

normalized to

$$\int d\underset{\sim}{r}' \int P^{(2)}(\underset{\sim}{r}',\underset{\sim}{r})d\underset{\sim}{r} = \int \rho(\underset{\sim}{r}')d\underset{\sim}{r}' = 1 \quad . \tag{1.55}$$

The combination in square brackets is referred to as the true correlation function since trivial correlations due to localization in the same volume have been subtracted. This subtraction arises from the projection operator P_o in the optical potential, and would not appear, of course, for the amplitude T_c, which is one way to see why the expansion for $\mathcal{U}_c$ may be more rapidly convergent than that for T_c. It is useful to define another form of correlation function

$$\rho(\underset{\sim}{r}')\rho(\underset{\sim}{r})\,C(\underset{\sim}{r}',\underset{\sim}{r}) = P^{(2)}(\underset{\sim}{r}',\underset{\sim}{r}) - \rho(\underset{\sim}{r}')\rho(\underset{\sim}{r}) \quad ; \tag{1.56}$$

since $P^{(2)}(\underset{\sim}{r},\underset{\sim}{r}) = 0$ for strong, repulsive correlations which prohibit two nucleons at the same point, we have $C(\underset{\sim}{r},\underset{\sim}{r}) = -1$. Then

$$\langle \underset{\sim}{k}'|\mathcal{U}_c^{(2,0)}|\underset{\sim}{k}\rangle \cong A(A-1)t_o^2\int e^{-i\underset{\sim}{k}'\cdot\underset{\sim}{r}'}\Big[\int \frac{e^{i\underset{\sim}{k}''\cdot(\underset{\sim}{r}'-\underset{\sim}{r})}}{\frac{p^2}{2m}-\frac{k''^2}{2m}+i\varepsilon}\frac{d\underset{\sim}{k}''}{(2\pi)^3}\Big]$$

$$\times\, e^{i\underset{\sim}{k}\cdot\underset{\sim}{r}}\rho(\underset{\sim}{r}')\rho(\underset{\sim}{r})C(\underset{\sim}{r}',\underset{\sim}{r})d\underset{\sim}{r}'d\underset{\sim}{r} \quad , \tag{1.57}$$

where we have made a forward approximation for t, assumed nonrelativistic kinematics for the pion, and introduced $p^2/2m = E-\bar{E}_N$.

Thus we see that $\mathcal{U}_c^{(2,0)}$ involves two-particle correlations, and we can anticipate that $\mathcal{U}_c^{(3,0)} \ldots \mathcal{U}_c^{(n,0)}$ will deal with 3-, ... n-particle clusters, so that if multi-particle correlations are small the optical potential series will essentially be truncated at that stage. Higher formalisms have the same property, though the Green functions are more complicated than the square-bracket expression of eq. (1.57) since they involve distorted pion propagation; the pertinent correlation functions beyond second order also assume somewhat different forms.

1.6. Estimates of Higher-Order Features

The multiple-scattering formalism which we have developed thus far is a complete, exact theory within the framework of potential scattering. Its lowest-order result is the simple statement that one should take an optical potential given by $\mathcal{U}_c^{(1)}(\underset{\sim}{r})=t_{\pi N}\rho(\underset{\sim}{r})$, where $t_{\pi N}$ is the free πN amplitude and $\rho(\underset{\sim}{r})$ is the system density, and calculate from it the full πA amplitude. Indeed, most calculations done in the area have been based on just such an approach and have examined the consequences of various parametrizations for $t_{\pi N}$ and for $\rho(\underset{\sim}{r})$, with spin and isospin effects above about 100 MeV and in the forward cone. With more precise data becoming available and a more complete understanding of the lowest-order result--and its limitations--it is becoming increasingly urgent to examine the higher-order effects: correlation contributions to $\mathcal{U}_c$, nuclear corrections to the free amplitude $t_{\pi N}$, and the general treatment of the nuclear dynamics in the multiple-scattering problem. In this subsection we shall try to make some crude estimates for these effects, to which we shall return in Chapter 4.

1.6.1. The Second-Order Optical Potential $\mathcal{U}_c^{(2,0)}$. We start with an estimate of the two-nucleon correlation effect, based on eq. (1.57). Although it is perfectly straightforward to evaluate the propagator appearing in square brackets there to obtain

$$G(\underset{\sim}{r}',\underset{\sim}{r}) = -\frac{2m}{4\pi}\,\frac{e^{ip|\underset{\sim}{r}'-\underset{\sim}{r}|}}{|\underset{\sim}{r}'-\underset{\sim}{r}|}\;, \quad p = \sqrt{2m(E-\bar{E}_N)}\;, \tag{1.58}$$

we shall prefer to estimate $\mathcal{U}_c^{(2,0)}$ at high energies and thus to approximate the Green function in a forward-scattering, eikonal form. We therefore take

$$G(\underset{\sim}{r}',\underset{\sim}{r}) \equiv \int \frac{e^{i\underset{\sim}{k}''\cdot(\underset{\sim}{r}'-\underset{\sim}{r})}}{\frac{p^2}{2m}-\frac{k''^2}{2m}+i^3}\frac{d\underset{\sim}{k}''}{(2\pi)^3} = \int \frac{e^{i\underset{\sim}{k}''\cdot(\underset{\sim}{r}'-\underset{\sim}{r})}}{\frac{1}{2m}(\underset{\sim}{p}+\underset{\sim}{k}'')\cdot(\underset{\sim}{p}-\underset{\sim}{k}'')+i\varepsilon}\frac{d\underset{\sim}{k}''}{(2\pi)^3}$$

$$\cong \int \frac{e^{i\underset{\sim}{k}''\cdot(\underset{\sim}{r}'-\underset{\sim}{r})}}{\underset{\sim}{v}\cdot(\underset{\sim}{p}-\underset{\sim}{k}'')+i\varepsilon}\frac{d\underset{\sim}{k}''}{(2\pi)^3} \qquad (1.59)$$

where in the last step we have approximated $\underset{\sim}{k}'' \approx p = mv$ with the direction of $\underset{\sim}{p}$ or $\underset{\sim}{v}$ given roughly by $^1/_2$ $(\underset{\sim}{k}+\underset{\sim}{k}')$, taken also as the z-axis. Then

$$G(\underset{\sim}{r}',\underset{\sim}{r}) \cong \frac{1}{v}\int \frac{e^{i\underset{\sim}{k}''_{\perp}\cdot(\underset{\sim}{b}'-\underset{\sim}{b})}}{(p-k''_z)+i\varepsilon}\, e^{ik''_z(z'-z)}\,\frac{d^2\underset{\sim}{k}''_{\perp}dk''_z}{(2\pi)^3}$$

$$= -\frac{i}{v}\,\delta^{(2)}(\underset{\sim}{b}'-\underset{\sim}{b})e^{ip(z'-z)}\,\theta(z'-z), \qquad (1.60)$$

where $\underset{\sim}{k}''_{\perp}$, $\underset{\sim}{b}'$ and $\underset{\sim}{b}$ refer to directions perpendicular to z, the two last being impact parameter variables, i.e. $\underset{\sim}{r} = (\underset{\sim}{b},z)$, $r' = (\underset{\sim}{b}',z')$. Eq. (1.60) shows characteristic high-energy, eikonal propagation: the impact parameter variable is preserved $(\underset{\sim}{b}'=\underset{\sim}{b})$, propagation is unidirectional $(z'>z)$, and has a phase $-ie^{ip(z'-z)}$.

We shall now use this result for the propagator at high energies in our estimate for $\mathcal{V}_c^{(2,0)}$, first noting

$$\underset{\sim}{k}\cdot\underset{\sim}{r}-\underset{\sim}{k}'\cdot\underset{\sim}{r}' = \frac{1}{2}(\underset{\sim}{k}+\underset{\sim}{k}')\cdot(\underset{\sim}{r}-\underset{\sim}{r}') + (\underset{\sim}{k}-\underset{\sim}{k}')\cdot\frac{1}{2}(\underset{\sim}{r}+\underset{\sim}{r}')$$

$$= p(z-z')+(\underset{\sim}{k}-\underset{\sim}{k}')\cdot\frac{1}{2}(\underset{\sim}{b}+\underset{\sim}{b}') \quad , \qquad (1.61)$$

having assumed $\overline{E}_N \ll E$ so that $p\approx k=k'$. Then from eq. (1.57) we have

$$\langle \underset{\sim}{k}'|\mathcal{V}_c^{(2,0)}|\underset{\sim}{k}\rangle \approx -\frac{i}{v}A(A-1)t_o^2 \int d^2\underset{\sim}{b} \int_{-\infty}^{\infty} dz\, e^{i(\underset{\sim}{k}-\underset{\sim}{k}')\underset{\sim}{b}}$$

$$\times\, \rho(\underset{\sim}{b},z)\int_z^{\infty} dz'\,\rho(\underset{\sim}{b},z')\,C(\underset{\sim}{b},z';\underset{\sim}{b},z) \quad , \qquad (1.62)$$

and we shall assume the nucleus so large that

$$C(\underset{\sim}{r}',\underset{\sim}{r}) = C(|\underset{\sim}{r}'-\underset{\sim}{r}|) \quad , \tag{1.63}$$

and z' is sufficiently close to z in the correlation function that the density $\rho(\underset{\sim}{b},z') \approx \rho(\underset{\sim}{b},z)$. Then

$$\langle \underset{\sim}{k}' | \mathcal{U}_c^{(2,0)} | \underset{\sim}{k} \rangle = -\frac{i}{v} A(A-1) t_o^2 \int d^2\underset{\sim}{b} \int_{-\infty}^{\infty} dz\, e^{i(\underset{\sim}{k}-\underset{\sim}{k}')\cdot \underset{\sim}{b}}$$

$$\rho^2(\underset{\sim}{b},z)\, R_{corr}, \tag{1.64}$$

where

$$R_{corr} \equiv \int_0^{\infty} C(\rho)d\rho \tag{1.65}$$

is the two-particle correlation length.

From eq. (1.64) we then have

$$\mathcal{U}_c^{(1)}(\underset{\sim}{r}) + \mathcal{U}_c^{(2,0)}(\underset{\sim}{r}) \approx \mathcal{U}_c^{(1)}(\underset{\sim}{r}) \left[1 + \frac{2\pi i}{k} f_o (A-1)\rho(r) R_{corr}\right] \tag{1.66}$$

Note that due to our high-energy simplification $\mathcal{U}_c^{(2,0)}$ is still a local potential.

As an estimate of the fractional second-order effect, we take central nuclear densities and the 3,3 region:

$$\frac{2\pi i}{k} \cdot f_o \cdot (A-1)\rho(\underset{\sim}{r}) \cdot R_{corr} = \frac{2\pi i}{k_{res}} \frac{4i}{3k_{res}} \frac{1}{6fm^3} (-0.4 \text{ fm})$$

$$= 0.23 \quad , \tag{1.67}$$

for $k_{res} \cong 304$ MeV/c and short-range repulsive correlations. This is likely to be an overestimate since at the central densities supposed here the nucleus is highly absorptive ($\ell \sim 0.5$ fm!) so that we primarily see the nuclear surface where $A\rho(r) < 0.17\ fm^{-3}$. In more detailed calculations[6], the second-order potential tends to be negligible for forward scattering and to have its main effect at and just beyond the first minimum. Third-order effects enter at the second minimum and third maximum, and so on.

1.6.2. Corrections to the Free Single-Nucleon Amplitude $t_{\pi N}$. In the previous subsection we examined corrections generated by pion scattering on one nucleon and then another. We now turn to two successive scatterings on the same nucleon with intermediate propagation governed by nuclear dynamics and not just for a free nucleon. Specifically, we shall consider "binding corrections" as arising from H_N in eq. (1.10) for t_i with (1.3) for G_o. For the latter we first note

$$\frac{1}{E-K-H_N+i\varepsilon} = \frac{1}{E_\pi-K+i\varepsilon} + \frac{1}{E_\pi-K+i\varepsilon}\,[H_N-(E-E_\pi)]\,\frac{1}{E-K-H_N+i\varepsilon}\ , \tag{1.68}$$

or

$$G_o = g_o+g_o\,[H_N-(E-E_\pi)]\,G_o\ , \tag{1.69}$$

in terms of the notation of eq. (1.14). Then from (1.16),

$$t_i = t_i^{free} + t_i^{free}\,g_o\,[H_N-(E-E_\pi)]\,G_o t_i = t_i^{free} + \Delta\ . \tag{1.70}$$

To evaluate Δ, we assume mean nuclear excitation energies

$$R \equiv \langle H_N-(E-E_\pi)\rangle \tag{1.71}$$

are small and estimate, very roughly,

$$\Delta \cong R\cdot t_i^{free} g_o^2 t_i^{free} \tag{1.72}$$

or, for $\underset{\sim}{k}' \approx \underset{\sim}{k}$, and p-wave amplitudes,

$$\langle \underset{\sim}{k}'|\Delta|\underset{\sim}{k}\rangle \cong R\int \langle \underset{\sim}{k}'|t_i^{free}|\underset{\sim}{k}''\rangle \frac{d\underset{\sim}{k}''/(2\pi)^3}{[E_\pi-E(k'')+i\varepsilon]^2} \times \langle \underset{\sim}{k}''|t_i^{free}|\,\underset{\sim}{k}\rangle$$

$$\rightarrow \frac{2\pi R}{(2\pi)^3}\int \left[t_i^{free}(E_\pi,E'')\right]^2 \frac{k''^2\left(\frac{dE''}{dk''}\right)^{-1} dE''}{[E_\pi-E''+i\varepsilon]^2}\ . \tag{1.73}$$

Assuming the main contribution to come from the double pole, extending the range of integration and using the residuum theorem gives

$$\frac{\langle\Delta\rangle}{\langle t^{free}(E_\pi)\rangle} \sim \frac{iR}{2\pi t^{free}(E_\pi)}\frac{d}{dE_\pi}\left[\ (t^{free}(E_\pi))^2\, k_\pi^2\, \left(\frac{dE_\pi}{dk_\pi}\right)^{-1}\ \right]$$

$$= \frac{1}{\pi}\, i\, R\, t^{free}\, E_\pi k_\pi \left[\ \frac{1}{2E_\pi}\left(1+\frac{E_\pi^2}{k_\pi^2}\right) + \frac{d}{dE_\pi}\log t^{free}\ \right] \tag{1.74}$$

where we have used relativistic kinematics. In evaluating the mean excitation energy R we shall suppose the nuclear mean potential to be slowly varying and the "constant" part to be immaterial since it mainly serves to change the zero of energy, which we can offset by our choice of effective E_π. Then $R \sim 8$ MeV, the average binding energy, and taking the 3,3 form

$$t^{free}(E_\pi) \sim \frac{1}{E_{res}-E_\pi-\frac{i\Gamma}{2}} , \; E_{res} \sim 335 \text{ MeV}, \; k_{res} \sim 304 \text{ MeV}, \tag{1.75}$$

we get

$$\left|\frac{<\Delta>}{<t^{free}>}\right|_{res} \sim 2|k_{res}f_{res}|\cdot R\cdot\left|\frac{1}{E_\pi}+\frac{2i}{\Gamma}\right| \quad , \tag{1.76}$$

where E_π^{-1} represents a typical projectile time scale and $2\Gamma^{-1}$ the resonance decay time. These times are compared to R^{-1}, the bound state cycle time, the collision being "impulsive" if they are much shorter than R^{-1}. In practice, with $f_{res} \sim \frac{4i}{3k_{res}}$,

$$\left|\frac{<\Delta>}{<t^{free}>}\right| \sim 0.3 \quad , \tag{1.77}$$

or a fairly appreciable correction from the binding effect in this case.

1.6.3. The Fixed-Scatterer Approximation. Already in evaluating $\mathcal{U}_c^{(2,0)}$ we used a closure approach to eliminate details of nuclear structure in favor of a correlation function. In general, the multiple-scattering formalism has most often been applied by taking the nucleons as fixed; calculating the multiple scattering amongst them, and then using the nuclear wave function to average over nucleon positions. To what degree is this justified? We shall examine this by rederiving our previous results incorporating the severe fixed-scatterer assumption essentially from the start[7].

We write the Lippmann-Schwinger equation for the πA wave function in configuration space,

$$\psi^{(+)}(\underset{\sim}{r}_1, \ldots, \underset{\sim}{r}_A;\underset{\sim}{r}) = \Phi_o(\underset{\sim}{r}_1, \ldots, \underset{\sim}{r}_A)\phi_{\underset{\sim}{k}}(\underset{\sim}{r})$$
$$+ \frac{1}{E-H_N(\underset{\sim}{r}_1, \ldots, \underset{\sim}{r}_A)-K(\underset{\sim}{r})+i\varepsilon}\sum_{i=1}^{A} v_i(\underset{\sim}{r},\underset{\sim}{r}_i)$$
$$\times\, \psi^{(+)}(\underset{\sim}{r}_1, \ldots, \underset{\sim}{r}_A;\underset{\sim}{r}) \quad , \tag{1.78}$$

where Φ_o refers to the nuclear ground state, $\phi_k(\underset{\sim}{r})$ to a pion plane wave and the v_i's are taken as local. More explicitly this is

$$\psi^{(+)}(\underset{\sim}{r}_1, \ldots, \underset{\sim}{r}_A; \underset{\sim}{r}) = \Phi_o(\underset{\sim}{r}_1, \ldots, \underset{\sim}{r}_A)\phi_k(\underset{\sim}{r}) - \frac{2m}{4\pi}\sum_\alpha \Phi_\alpha(\underset{\sim}{r}_1, \ldots, \underset{\sim}{r}_A) \times \int d\underset{\sim}{r}_1' \ldots d\underset{\sim}{r}_A'\, d\underset{\sim}{r}' \frac{e^{iK_\alpha|\underset{\sim}{r}-\underset{\sim}{r}'|}}{|\underset{\sim}{r}-\underset{\sim}{r}'|} \times \Phi_\alpha^+(\underset{\sim}{r}_1', \ldots, \underset{\sim}{r}_A') \sum_{i=1}^{A} v_i(\underset{\sim}{r}', \underset{\sim}{r}_i')\psi^{(+)}(\underset{\sim}{r}_1', \ldots, \underset{\sim}{r}_A'; \underset{\sim}{r}'), \tag{1.79}$$

where

$$K_\alpha^2 = k^2 + 2m\,(E_N^o - E_N^\alpha) \tag{1.80}$$

We now assume high-energy projectiles,

$$k^2 \gg 2m(E_N^o - E_N^\alpha) \quad , \tag{1.81}$$

replace K_α by $\overline{K} \simeq k$ and use closure

$$\sum_\alpha \Phi_\alpha(\underset{\sim}{r}_1, \ldots, \underset{\sim}{r}_A)\Phi_\alpha^+(\underset{\sim}{r}_1', \ldots, \underset{\sim}{r}_A') = \delta(\underset{\sim}{r}_1 - \underset{\sim}{r}_1') \ldots \delta(\underset{\sim}{r}_A - \underset{\sim}{r}_A') \tag{1.82}$$

whence

$$\psi^{(+)}(\underset{\sim}{r}_1, \ldots, \underset{\sim}{r}_A; \underset{\sim}{r}) = \Phi_o(\underset{\sim}{r}_1, \ldots, \underset{\sim}{r}_A)\,\phi_{\underset{\sim}{k}}(\underset{\sim}{r}) - \frac{2m}{4\pi}\int d\underset{\sim}{r}' \frac{e^{i\overline{K}|\underset{\sim}{r}'-\underset{\sim}{r}|}}{|\underset{\sim}{r}'-\underset{\sim}{r}|}\sum_{i=1}^{A} v_i(\underset{\sim}{r}', \underset{\sim}{r}_i) \times \psi^{(+)}(\underset{\sim}{r}_1, \ldots, \underset{\sim}{r}_A; \underset{\sim}{r}') \quad , \tag{1.83}$$

which is easily solved by a separable form

$$\psi^{(+)}(\underset{\sim}{r}_1, \ldots, \underset{\sim}{r}_A; \underset{\sim}{r}) = \Phi_o(\underset{\sim}{r}_1, \ldots, \underset{\sim}{r}_A)\,\psi_{\underset{\sim}{k}}^{(+)}(\underset{\sim}{r}) \quad . \tag{1.84}$$

The projectile function satisfies

$$\psi_{\underset{\sim}{k}}^{(+)}(\underset{\sim}{r}) = \phi_{\underset{\sim}{k}}(\underset{\sim}{r}) - \frac{2m}{4\pi} \int d\underset{\sim}{r}' \frac{e^{i\bar{K}|\underset{\sim}{r}'-\underset{\sim}{r}|}}{|\underset{\sim}{r}'-\underset{\sim}{r}|} \sum_{i=1}^{A} v_i(\underset{\sim}{r}',\underset{\sim}{r}_i) \times \psi_{\underset{\sim}{k}}^{(+)}(\underset{\sim}{r}') \quad , \tag{1.85}$$

from which we read off the amplitude

$$<\underset{\sim}{k}'|f|\underset{\sim}{k}> = -\frac{2m}{4\pi} \int d\underset{\sim}{r}_1 \ldots d\underset{\sim}{r}_A \, \Phi_o^+(\underset{\sim}{r}_1, \ldots, \underset{\sim}{r}_A) \times \int d\underset{\sim}{r} \, \phi_{\underset{\sim}{k}'}^+(\underset{\sim}{r}) \sum_{i=1}^{A} v_i(\underset{\sim}{r},\underset{\sim}{r}_i)\psi_{\underset{\sim}{k}}^{(+)}(\underset{\sim}{r}) \times \Phi_o(\underset{\sim}{r}_1, \ldots, \underset{\sim}{r}_A) \tag{1.86}$$

We have therefore succeeded here in separating the nuclear and projectile parts of the problem by holding the nucleons fixed, i.e. by exploiting closure, which will be valid provided

$$\frac{(\underset{\sim}{k}-\underset{\sim}{k}')^2}{2M} << \frac{k^2}{2m} \, , \quad \text{or} \quad \sin\frac{\theta}{2} << \frac{1}{2}\sqrt{\frac{M}{m}} \quad , \tag{1.87}$$

that is, provided no individual nucleon recoil is comparable to the pion energy. This clearly restricts us to forward angles for the fixed-scatterer approximation to be valid. Otherwise we must expect to need the more complete formalism of the "fully dynamic" multiple-scattering series (cluster expansion) for the optical potential.

2. DWIA FORMALISMS FOR INELASTIC SCATTERING

We shall now generalize the procedures of the previous chapter to the case of inelastic scattering. We will make a (somewhat artificial) distinction between two aspects of inelastic scattering: the "soft" distortion of the wave for incident and exiting pion, and the "hard" encounter in which the relevant nuclear transition occurs. Our formalism will be quite general, however, and at the end of this chapter we shall examine corrections to the lowest-order DWIA (distorted wave impulse approximation), in which we shall relax the sharpness of this distinction.

2.1. General Formulation of the DWIA

Our point of departure will be eq. (1.31) which is valid here since we shall again have in mind the use of fully antisymmetrized nuclear states (though now, of course, two different states appear initially and finally). Since the inelastic process may preferentially promote or modify a particular nucleon (e.g. the "outermost" one), it is especially important to treat the formalism consistently, i.e. at least ab initio to use τ_i of eq. (1.25) with the A-prescription. Our equation for the full πA amplitude is now

$$T' = \mathcal{T}' + \mathcal{T}'G_o AT' = \mathcal{T}' + \mathcal{T}'G_o T' \quad , \tag{2.1}$$

where in the last step we exploit the antisymmetry of the final state and the symmetry of $\mathcal{T}'$ of eq. (1.30).

Now for the inelastic process we examine the successive orders of nondiagonality in the nuclear space:

$$\begin{aligned} \langle\alpha|T'|0\rangle &= \langle\alpha|\mathcal{T}'|0\rangle + \langle\alpha|\mathcal{T}'|0\rangle G_o(0)\langle 0|T'|0\rangle \\ &+ \langle\alpha|\mathcal{T}'|\alpha\rangle G_o(\alpha)\langle\alpha|T'|0\rangle + \sum_{\nu\neq 0,\alpha} \langle\alpha|\mathcal{T}'|\nu\rangle G_o(\nu)\langle\nu|T'|0\rangle \ , \end{aligned} \tag{2.2}$$

where at the start we see a significant distinction relative to the elastic case, namely we must single out for special attention both the ground state $|0\rangle$ and the excited state $|\alpha\rangle$. "Diagonality" must be considered with respect to both of them. Moreover, whereas in the elastic case the next order in nondiagonality always involved two nondiagonal elements $\langle 0|\ldots|\nu\rangle\langle\nu|\ldots|0\rangle$, $\nu\neq 0$, here the correction is only one such factor higher, since the lowest-order effect already has $\langle\alpha|\ldots|0\rangle$ and corrections carry $\langle\alpha|\ldots|\nu\rangle\langle\nu|\ldots|0\rangle$, $\nu\neq 0,\alpha$.

To lowest order, we ignore the last term in eq. (2.2) and write the results as

$$\begin{aligned} [1-\langle\alpha|\mathcal{T}'|\alpha\rangle G_o(\alpha)]\langle\alpha|T'|0\rangle &\cong \langle\alpha|\mathcal{T}'|0\rangle \\ &\times [1+G_o(0)\langle 0|T'|0\rangle] \end{aligned} \tag{2.3}$$

or

$$\begin{aligned} \langle\alpha|T'|0\rangle \cong [1-\langle\alpha|\mathcal{T}'|\alpha\rangle G_o(\alpha)]^{-1}\langle\alpha|\mathcal{T}'|0\rangle \\ \times [1+G_o(0)\langle 0|T'|0\rangle] \quad . \end{aligned} \tag{2.4}$$

The rightmost factor here represents distortion of the initial pion wave in the full optical potential, since if it acts on the undistorted pion wave $\phi_{\tilde{k}}$ we have

$$\phi_{\tilde{k}} + G_o(0)\langle 0|T'|0\rangle\phi_{\tilde{k}} \equiv \psi_{\tilde{k}} \quad , \tag{2.5}$$

while from eq. (1.20) - modified for the (A-1)/A factor -

$$\langle 0|T'|0\rangle = \mathcal{V}_c{}' + \mathcal{V}_c{}'\, G_o(0)\langle 0|T'|0\rangle \quad , \tag{2.6}$$

from which

$$\psi_{\tilde{k}} = \phi_{\tilde{k}} + G_o(0)\,\mathcal{V}_c{}'\,\psi_{\tilde{k}} \quad . \tag{2.7}$$

The central factor $\langle\alpha|\mathcal{T}'|0\rangle$ in eq. (2.4) is, of course, the "hard" collision in which the nucleus is kicked into the state α. Then the left-most factor distorts the final pion wave, in this order only with the lowest-order optical potential. This may be seen from the definition

$$\langle\phi_{\tilde{k}'}|[1-\langle\alpha|\mathcal{T}'|\alpha\rangle G_o(\alpha)]^{-1} \equiv \langle\chi_{\tilde{k}'}^{(-)}| \quad , \tag{2.8}$$

where the minus superscript refers to a collapsing spherical wave for the final state, and

$$\langle\phi_{\tilde{k}'}| = \langle\chi_{\tilde{k}'}^{(-)}| - \langle\chi_{\tilde{k}'}^{(-)}|\langle\alpha|\mathcal{T}'|\alpha\rangle G_o(\alpha) \quad ,$$

or

$$\langle\chi_{\tilde{k}'}^{(-)}| = \langle\phi_{\tilde{k}'}| + \langle\chi_{\tilde{k}'}^{(-)}|\langle\alpha|\mathcal{T}'|\alpha\rangle G_o(\alpha) \quad , \tag{2.9}$$

as claimed ($\mathcal{V}_c{}'^{(1)} = \langle\alpha|\mathcal{T}'|\alpha\rangle$). Note that this implies that the ket $|\chi_{\tilde{k}'}^{(-)}\rangle$ is distorted by use of the conjugate of the (lowest-order) optical potential. (Furthermore, since the optical potential is complex and energy-dependent--and appears in different orders to right and left--one must not expect that $|\chi_{\tilde{k}'}(+)\rangle$ and $|\Psi_{\tilde{k}}^{(+)}\rangle$ are orthogonal.) Thus finally we have the usual lowest-order DWIA result

$$\langle\phi_{\tilde{k}'};\alpha|T'|\phi_{\tilde{k}};0\rangle = \langle\chi_{\tilde{k}'}^{(-)};\alpha|\mathcal{T}'|\psi_{\tilde{k}}^{(+)};0\rangle \quad , \tag{2.10}$$

The corrections to this are easily exhibited by retaining the last term in eq. (2.2), iterating for $\langle\nu|T'|0\rangle$ as given by that equation itself, and dropping the last, triply nondiagonal term. This means inserting eq. (2.4), with $\alpha\to\nu$, into eq. (2.2), whence

$$\langle\alpha|T'|0\rangle \cong [1-\langle\alpha|\mathcal{T}'|\alpha\rangle G_o(\alpha)]^{-1}$$

$$\times \{\langle\alpha|\mathcal{T}'|0\rangle + \sum_{\nu\neq 0,\alpha} \langle\alpha|\mathcal{T}'|\nu\rangle \frac{1}{E-E_N^\nu-K-\langle\nu|\mathcal{T}'|\nu\rangle}$$

$$\times \langle\nu|\mathcal{T}'|0\rangle\} [1+G_o(0)\langle 0|T'|0\rangle] \quad , \qquad (2.11)$$

where the final distortion factor is taken as if the same for the state ν and the state α. The second term in curly brackets represents the correction to the lowest-order DWIA in terms of two-step (or correlational) processes to reach the final state α from the initial state $|0\rangle$. Finally, by introducing t_i in terms of τ_i (Manipulation #2) we can write our second-order DWIA result as[2,6]

$$\langle\phi_{\underset{\sim}{k}'};\alpha|T|\phi_{\underset{\sim}{k}};0\rangle \cong \langle\chi_{\underset{\sim}{k}}^{(-)};\alpha|\Big(\sum_{i=1}^{A} t_i + \sum_{i=1}^{A}\sum_{j\neq i} t_i G_o(1-P_o-P_\alpha)t_j\Big)$$

$$\times |\psi_{\underset{\sim}{k}}^{(+)};0\rangle \quad , \qquad (2.12)$$

where the antisymmetrization prescription within the τ_i's has provided us with the missing pieces. Once again we see that we can generate consistent versions of the formalism (completely) with or (completely) without explicit antisymmetrization instructions.

2.2. Reactive Content of the Optical Potential

We shall now note briefly an approach[8] to the multiple-scattering problem which focuses on unitarity, and, in the process, elucidates the role of inelastic reactions in the unitarity balance, finally leading to an alternate view of the DWIA[9]. Our point of departure is eqs. (1.18) and (1.19) for the optical operator $\mathcal{U}$ and the elastic amplitude T:

$$T = \mathcal{U}+\mathcal{U}G_oP_oT \quad , \quad \mathcal{U} = V+VG_oQ_o\mathcal{U} \quad , \quad Q_o \equiv 1 - P_o \qquad (2.13)$$

We shall use these in conjunction with a general theorem[8] pertinent to unitarity which states that, for operators which satisfy

$$A(E) = B(E)+B(E)C(E)A(E) \quad , \qquad (2.14)$$

the difference

$$\Delta A(E) \equiv A(E+i\varepsilon)^{+} - A(E+i\varepsilon) \tag{2.15}$$

is given by

$$\Delta A = A^{+}\Delta CA+(A^{+}C^{+}+1)\Delta B(1+CA) \quad , \tag{2.16}$$

and is proved easily by direct substitution. Applied to T of eq. (2.13), this yields[9]

$$\Delta T = T^{+}\Delta(P_oG_o)T+(T^{+}(P_oG_o)^{+}+1)\Delta\mathcal{V}(1+(P_oG_o)T) \tag{2.17}$$

and

$$\Delta\mathcal{V} = \mathcal{V}^{+}\Delta(Q_oG_o)\mathcal{V} \quad . \tag{2.18}$$

Now at high energies it will be legitimate[8] to take the limit $\varepsilon\to0^+$ within the surrounding operators here, which makes it useful to define the elastic and inelastic phase space factors

$$\Lambda(E) \equiv \text{disc } (P_oG_o) \quad , \tag{2.19a}$$

$$M(E) \equiv \text{disc } (Q_oG_o) \quad , \tag{2.19b}$$

with

$$2\pi i \quad \text{disc } A(E) = \lim_{\varepsilon\to0^+} \; [A(E+i\varepsilon)^{+}-A(E+i\varepsilon)] \quad . \tag{2.20}$$

Then

$$\text{disc } T = T^{+}\Lambda T + (T^{+}(P_oG_o)^{+}+1)\mathcal{V}^{+}M\mathcal{V}(1+(P_oG_o)T) \tag{2.21}$$

expresses unitarity for T as a generalized optical theorem, where the two terms on the right are the elastic and inelastic contributions, respectively. The inelastic amplitude to a particular channel α can be read off from eq. (2.21) directly as

$$T_\alpha = P_\alpha\mathcal{V}(1+(P_oG_o)T)P_o \quad , \tag{2.22}$$

where $P_\alpha = |\alpha\rangle\langle\alpha|$ is the projection operator for the nuclear state α. From eq. (1.22), or the treatment following eq. (1.31), we know that to order A^{-1} the optical operator satisfies

$$\mathcal{U} \cong \sum_{i=1}^{A} t_i' + \sum_{i=1}^{A} t_i'(Q_oG_o)\mathcal{U} = \sum_{i=1}^{A} t_i' + \mathcal{U}(Q_oG_o)\sum_{i=1}^{A} t_i' \ , \tag{2.23}$$

where, as in eq. (1.23)

$$t_i' = v_i + v_i Q_o G_o t_i' \tag{2.24}$$

Thus

$$\text{disc } T = T^+\Lambda T + (T^+(P_oG_o)^+ + 1)\ (\mathcal{U}^+(Q_oG_o)^+ + 1)$$
$$\times \sum_i t_i'^+ M \sum_j t_j'(1+(Q_oG_o)\mathcal{U})\ (1+(P_oG_o)T) \quad . \tag{2.25}$$

If we wish to single out inelastic contributions of a particular nuclear excited state α, we must take into account, for lowest order, the recurring diagonal elements with respect to it, or, from eqs. (2.22) and (2.23),

$$T_\alpha \cong P_\alpha(1+\mathcal{U}(P_\alpha G_o))\ [P_\alpha \sum_{i=1}^{A} t_i' P_o]\ (1+(P_oG_o)T)P_o \quad . \tag{2.26}$$

This is equivalent to our earlier, lowest-order DWIA result, and shows how the latter serves to produce a multiple-scattering theory which embodies unitarity consistently. (Note that the lowest-order $\mathcal{U}_c$ will not satisfy eq. (2.18) and hence will not participate in this consistency[8].)

2.3. Second-Order Corrections to the DWIA for $A(\pi^+,\pi^0)A'$ Reactions

At least one instance, namely pion-nucleus single charge exchange scattering leading to excitation of the isobaric analog of the original ground state, the second-order corrections to the DWIA would appear to be large and may provide a partial resolution to one of the outstanding difficulties in our understanding of pion scattering reactions. The difficulty is perhaps most clear in the case of the reaction $^{13}C(\pi^+,\pi^0)^{13}N_{i.a.r.}$. There the angle-integrated cross section has been measured as a function of energy through the 3,3 resonance region[10,11]. Theoretical expectations based on the ideas of the present and the previous chapter would suggest that, as the 3,3 region is traversed, the increased πN cross section would cause more and more optical absorption, with a subsequent decrease in the (π^+,π^0) channel (and presumably an increase in pion-induced nucleon knock-out). Instead, experiment finds a rather flat cross section, with slight peaking near the resonance, of about 1 mb--some five times larger than even the optimistic theoretical predictions[11].

For a semi-quantitative view of this situation we shall start with the lowest-order DWIA in a rough--but utilitarian and instructive--eikonal approximation. For eq. (2.12) to lowest order, we have

$$<\pi^o;a|T|\pi^+;0> \cong \int \chi_{\underset{\sim}{k}'}^{(-)*}(\underset{\sim}{r})<\alpha| \sum_{i=1}^{A} (t_s+t_v\vec{I}\cdot\vec{\tau}) \times \delta(\underset{\sim}{r}-\underset{\sim}{r}_i)|0>\chi_{\underset{\sim}{k}}^{(+)}(\underset{\sim}{r})d\underset{\sim}{r} \quad , \tag{2.27}$$

where we have taken both the initial and final pion wave functions, $\chi_{\underset{\sim}{k}}^{(+)}$ and $\chi_{\underset{\sim}{k}'}^{(-)}$, to lowest order, t_s and t_v refer to isoscalar and isovector parts of the πN amplitude, with $\vec{t}$ and $\vec{I}$ the nucleon and pion isospin operators, and a is the normalized nucleon isobaric analog state,

$$|a> = \frac{1}{\sqrt{N-Z}} T_+|0> = \frac{1}{\sqrt{N-Z}} \cdot \frac{1}{2} \sum_{j=1}^{A} (\tau_1+i\tau_2)_j|0> \quad . \tag{2.28}$$

Our isospin conventions are such that

$$\frac{1}{2}(\tau_1+i\tau_2)|n> = |p> \ , \quad \frac{1}{2}(\tau_1+i\tau_2)|p> = 0 \quad , \tag{2.29a}$$

$$\tau_3|p> = |p> \quad , \quad \tau_3|n> = -|n> \quad , \tag{2.29b}$$

for neutron and proton states, $|n>$ and $|p>$. Then, since

$$<\pi^o|\vec{I}\cdot\vec{\tau}|\pi^+> = \sqrt{2} \cdot \frac{1}{2}(\tau_1+i\tau_2) \tag{2.30}$$

we have

$$<\pi^o;a|T|\pi^+;0> = \sqrt{2}\, t_v \int \chi_{\underset{\sim}{k}'}^{(-)*}(\underset{\sim}{r}) \frac{1}{\sqrt{N-Z}} (N\rho_n(\underset{\sim}{r})-Z\rho_p(\underset{\sim}{r})) \times \chi_{\underset{\sim}{k}}^{(+)}(\underset{\sim}{r})d\underset{\sim}{r} \quad , \tag{2.31}$$

where $\int \rho_{n,p}(\underset{\sim}{r})d\underset{\sim}{r} = 1$.

The eikonal approximation gives for the wave function

$$\chi_{\underset{\sim}{k}}^{(\pm)}(\underset{\sim}{r}) \cong e^{i\underset{\sim}{k}\cdot\underset{\sim}{r}}\, e^{-\frac{i}{v}\int_{\pm\infty}^{z} \mathcal{V}_c^{(\pm)}(b,\xi)d\xi} \tag{2.32}$$

integration being along a straight-line, classical path, and the pertinent optical potential is taken for the initial and final states,

$$\mathcal{V}_c^{(\pm)} = -\frac{4\pi}{2m} A\rho(\underset{\sim}{r})\, \bar{f}_o\binom{\pi^+}{\pi^o} \quad , \tag{2.33}$$

where the density $\rho(\underset{\sim}{r}) = \frac{1}{A}(N\rho_n(\underset{\sim}{r})+Z\rho_p(\underset{\sim}{r}))$, $\int \rho d\underset{\sim}{r} = 1$, and the forward amplitude and f_o are averaged over neutron and proton numbers,

$$\bar{f}_o = \frac{ik}{4\pi}\frac{1}{A}\{N\sigma_{\pi n}(1-i\alpha_{\pi n}) + Z\sigma_{\pi p}(1-i\alpha_{\pi p})\} \quad . \tag{2.34}$$

Here $\sigma_{\pi N}$ and $\alpha_{\pi N}$ (N = n,p) are total πN cross sections and ratios of real to imaginary part for the forward amplitude.

Gathering these results together gives for the charge exchange amplitude

$$\langle\pi^o;a|T|\pi^+;0 \cong \sqrt{2}\, t_v(0) \int e^{i(\underset{\sim}{k}-\underset{\sim}{k}')\cdot\underset{\sim}{r}} \frac{1}{\sqrt{N-Z}}(N\rho_n(\underset{\sim}{r})-Z\rho_p(\underset{\sim}{r}))$$

$$\times\, e^{-\frac{1}{2}\bar{\sigma}(1-i\bar{\alpha})A\int_{-\infty}^{\infty}\rho(\underset{\sim}{b},\xi)d\xi}\, d\underset{\sim}{r} \quad , \tag{2.35}$$

where $\bar{\sigma} \sim \frac{2}{3}\sigma(\pi^+p\to p\pi^+)$ and $\bar{\alpha}$ have been averaged over N,Z,π^+,π^o, and we have approximated a forward, rectilinear trajectory. The angle-integrated cross section at high energies is easily evaluated as $[\underset{\sim}{\Delta} \equiv (\underset{\sim}{k}-\underset{\sim}{k}')_\perp]$

$$\sigma = (\frac{\omega}{2\pi})^2 \int |T|^2 d\Omega = (\frac{\omega}{2\pi k})^2 \int_{\Delta\leq 2k} |T|^2 d^2\underset{\sim}{\Delta}$$

$$\cong (\frac{\omega}{2\pi})^2\, |\sqrt{2}\, t_v(0)|^2 \frac{1}{k^2(N-Z)} \int d^2\underset{\sim}{\Delta} \int d^2\underset{\sim}{b}\, dz\, e^{i\underset{\sim}{\Delta}\cdot\underset{\sim}{b}}$$

$$\times\, (N\rho_n(\underset{\sim}{r})-Z\rho_p(\underset{\sim}{r}))e^{-\frac{1}{2}\bar{\sigma}(1-i\bar{\alpha})A\int_{-\infty}^{\infty}\rho(b,\xi)d\xi}$$

$$\times \int d^2b'\, dz'\, e^{-i\underset{\sim}{\Delta}\cdot\underset{\sim}{b}'} (N\rho_n(\underset{\sim}{r}')-Z\rho_p(\underset{\sim}{r}'))\, e^{-\frac{1}{2}\bar{\sigma}(1+i\alpha)A\int_{-\infty}^{\infty}\rho(b',\xi)d\xi'}$$

$$\cong \sigma_0(\pi^+n\to p\pi^0)\, (\frac{2\pi}{k})^2 \int d^2\underset{\sim}{b}\, e^{-\bar{\sigma}\, A \int \rho(b,\xi)d\xi}$$

$$\times \frac{1}{N-Z} \left|\int_{-\infty}^{\infty} dz(N\rho_n(\underset{\sim}{b},z)-Z\rho_p(\underset{\sim}{b},z))\right|^2 \quad . \tag{2.36}$$

This result shows very clearly the exponential absorptive effect near resonance in the lowest-order DWIA, an effect which, of course, wins out over the linear dependence on the fundamental charge exchange cross section. Charge exchange to the analog is especially sensitive to absorptive effects since there is no transfer of angular momentum and therefore no suppression of the nuclear interior form factors of r^L.

The second, correction term in eq. (2.12), involves processes in which charge is exchanged in a two-step mechanism such that the nucleus is excited during the exchange. The evaluation of this term can be cast in the language of correlation functions, as in sections 1.5 and 1.6. Due to the charge exchange feature, these will then be isovector correlation functions. Detailed evaluation[11,12] gives increases of as much as 70% over the lowest-order result for isovector correlations arising from the Pauli principle. This is not enough to produce agreement with experiment, but it is plenty to raise worries concerning the convergence of the DWIA. Such worries may be most acute for charge exchange to the analog, where absorption is especially crucial and where the good overlap between initial and final nuclear states may enhance the second-order terms. Even there one has some reason to believe that third-and higher-order corrections are not too important. This is because estimates[13] based on Glauber's multiple-scattering theory (to which we turn in a moment) arrive at similar results to those of second-order DWIA. Glauber's approach has the virtue of summing the multiple-scattering series to all orders, and thus automatically includes the higher-order terms which we seek. On the other hand, it is intended to be applied in the high-energy limit, with many partial waves, so that its application for pion scattering in the 3,3 region is suspect, and we are inclined, therefore, to take the Glauber result for charge exchange as indicative rather than definitive.

3. GLAUBER THEORY

The enormous advantage of being able to sum the full scattering series at high energies makes Glauber theory[14] a very valuable tool for our present purposes, though, as we have already cautioned, one must be suspicious of it at energies pertinent to the 3,3 region, where the πN cross section is certainly not diffractive. As we shall see, Glauber theory accomplishes this feat of summation by cancelling an infinite number of terms of the Watson series for T--leaving only a finite number of terms behind--against all of the off-shell propagation in Watson theory, so that there is yet another significant virtue in the Glauber approach, which will not require difficultly acquired off-shell πN input.

3.1. The Heuristic Approach

We start with a direct, heuristic "derivation" of the Glauber result, based on the usual partial wave expansion for the amplitude (taken as for spinless particles, a limitation easily overcome),

$$f(\theta) = \frac{1}{2ik} \sum_{\ell=0}^{\infty} (2\ell+1)\,(e^{2i\delta_\ell}-1)\,P_\ell(\cos\theta) \quad , \tag{3.1}$$

where $\delta_\ell(E)$ is the phase shift in the partial wave ℓ and $P_\ell(\cos\theta)$ is the Legendre polynomial. At high energies we suppose the scattering to be primarily diffractive, $kR >> 1$, where R is the dimension of any system--nucleus or nucleon--involved. Then many partial waves enter and we replace the discrete variable ℓ with a continuous impact parameter variable b,

$$b = \frac{\ell}{k} \, , \quad db = \frac{1}{k}\,\Delta\ell = \frac{1}{k} \quad , \tag{3.2}$$

and

$$2\delta_\ell(E) \rightarrow \chi(b) \tag{3.3}$$

(note that the energy dependence must conform with this assignment).

The small angles involved in scattering at high energies allow us to approximate

$$P_\ell(\cos\theta) = J_0((2\ell+1)\sin\frac{\theta}{2}) + \text{Order}\,(\sin^2\frac{\theta}{2}) \quad , \tag{3.4}$$

so that

$$f(\theta) = -ik \int_0^\infty J_0(2kb\sin\frac{\theta}{2})\,(e^{i\chi(b)}-1)\,b\,db$$

$$= -ik \int_0^\infty J_o(\Delta b)\ (e^{i\chi(b)} - 1) b\, db \quad , \tag{3.5}$$

with $\Delta = 2k \sin\frac{\theta}{2}$. Eq. (3.5) is the usual diffractive scattering result.

Let us suppose that the scatterer is made up of A constituents, located at $\underset{\sim}{r}_1, \ldots ; \underset{\sim}{r}_A = \underset{\sim}{b}_1, z_1; \ldots ; \underset{\sim}{b}_A, z_A$ and each fulfilling the diffractivity condition. Then Glauber makes a further assumption of additivity of phase parameters,

$$\chi(\underset{\sim}{b}) = \sum_{j=1}^{A} \chi_j(\underset{\sim}{b} - \underset{\sim}{b}_j) \quad . \tag{3.6}$$

Using this in eq. (3.5) and defining the profile functions

$$\Gamma_j(\underset{\sim}{b}) = e^{i\chi_j(\underset{\sim}{b})} - 1 \quad , \tag{3.7}$$

we get

$$F(\theta) = -ik \int_0^\infty J_o(\ b)\ [e^{i\sum_{j=1}^{A} \chi_j(\underset{\sim}{b} - \underset{\sim}{b}_j)} - 1] b\, db$$

$$= -ik \int_0^\infty J_o(\Delta b)\ [\prod_{j=1}^{A} [\Gamma_j(\underset{\sim}{b} - \underset{\sim}{b}_j) + 1] - 1] b\, db$$

$$= -ik \int_0^\infty J_o(\Delta b)\ [\sum_{j=1}^{A} \Gamma_j(\underset{\sim}{b} - \underset{\sim}{b}_j) + \sum_{j=1}^{A} \sum_{k<j} \Gamma_j(\underset{\sim}{b} - \underset{\sim}{b}_j)\Gamma_k(\underset{\sim}{b} - \underset{\sim}{b}_k)$$

$$+ \ldots + \sum_{j=1}^{A} \sum_{k<j} \ldots \sum_{q<p} \Gamma_j(\underset{\sim}{b} - \underset{\sim}{b}_j)\ \Gamma_k(\underset{\sim}{b} - \underset{\sim}{b}_k) \ldots \Gamma_q(\underset{\sim}{b} - \underset{\sim}{b}_q)]$$

$$\times b\, db \quad . \tag{3.8}$$

The fundamental profile functions are identified from the πN amplitudes through

$$\Gamma_j(\underset{\sim}{b}) = \frac{i}{2\pi k} \int e^{-i\underset{\sim}{\delta}\cdot\underset{\sim}{b}}\ f_j(\underset{\sim}{\delta}) d^2\underset{\sim}{\delta} \quad , \tag{3.9}$$

where $f(\delta)$ is the on-shell πN amplitude, so that eq. (3.8) is indeed a finite series involving only on-shell information. Note that we have not had to assume potential scattering in arriving at this result.

3.2. Relation to the Watson Series

The Glauber result can also be obtained directly from the Lippmann-Schwinger equation if we are prepared to assume that the basic scattering arises from a local potential. Then we have

$$t = V + V G_o t \quad , \quad \langle \underset{\sim}{r}' | V | \underset{\sim}{r} \rangle = V(\underset{\sim}{r}) \delta(\underset{\sim}{r}' - \underset{\sim}{r}) \quad , \tag{3.10}$$

and from eq. (1.60) we can immediately write the propagator appropriate for use at high energies, namely

$$\langle \underset{\sim}{r}' | G_o | \underset{\sim}{r} \rangle \underset{=}{\sim} -\frac{i}{v} \delta^{(2)}(\underset{\sim}{b}' - \underset{\sim}{b}) \, e^{ip(z'-z)} \, \theta(z'-z) \quad . \tag{3.11}$$

The Lippmann-Schwinger equation is then easily seen to be solved by

$$\langle \underset{\sim}{r}' | t | \underset{\sim}{r} \rangle = \delta^{(2)}(\underset{\sim}{b}' - \underset{\sim}{b}) e^{ip(z'-z)} \frac{d}{dz'} \theta(z'-z) e^{-\frac{i}{v} \int_z^{z'} V(\underset{\sim}{b},\xi) d\xi} V(\underset{\sim}{b},z)$$

$$= \delta^{(2)}(\underset{\sim}{b}' - \underset{\sim}{b}) e^{ip(z'-z)} \Big[\delta(z'-z) - \frac{i}{v} \theta(z'-z) V(\underset{\sim}{b},z') e^{-\frac{i}{v} \int_z^{z'} V(\underset{\sim}{b},\xi) d\xi} \Big] \times V(\underset{\sim}{b},z) \quad , \tag{3.12}$$

The on-shell $[k_z = k'_z = p]$ amplitude is then

$$_{on}\langle \underset{\sim}{k}' | t | \underset{\sim}{k} \rangle_{on} = \int d^2\underset{\sim}{b} \, e^{i(\underset{\sim}{k} - \underset{\sim}{k}')_{\perp} \cdot \underset{\sim}{b}} \int_{-\infty}^{\infty} dz \, e^{-\frac{i}{v} \int_z^{\infty} V(\underset{\sim}{b},\xi) d\xi} V(\underset{\sim}{b},z)$$

$$= iv \int d^2\underset{\sim}{b} \, e^{i\underset{\sim}{\Delta} \cdot \underset{\sim}{b}} \Big[e^{-\frac{i}{v} \int_{-\infty}^{-\infty} V(\underset{\sim}{b},\xi) d\xi} - 1 \Big] \quad , \tag{3.13}$$

to be compared with eq. (3.5) of the heuristic approach, with the identification of the phase parameter

$$\chi(\underset{\sim}{b}) = -\frac{1}{v} \int_{-\infty}^{\infty} V(\underset{\sim}{b},\xi) d\xi \quad . \tag{3.14}$$

For the local potential case, the additivity of phases is immediately guaranteed by the additivity of potentials, and we obtain for the πA amplitude

$$
{}_{on}\langle \underset{\sim}{k}'|T|\underset{\sim}{k}\rangle_{on} = iv\int d^2b\, e^{i\underset{\sim}{\Delta}\cdot\underset{\sim}{b}}\left[e^{-\frac{i}{v}\int_{-\infty}^{\infty}\sum_{j=1}^{A}V_j(\underset{\sim}{b}-\underset{\sim}{b}_j,\xi-z_j)d\xi}-1\right]
$$

$$
= iv\int d^2\underset{\sim}{b}\, e^{i\underset{\sim}{\Delta}\cdot\underset{\sim}{b}}\Big[\sum_{j=1}^{A}\Big[e^{-\frac{i}{v}\int_{-\infty}^{\infty}V_j d\xi}-1\Big]
$$

$$
+ \ldots + \sum_{j=1}^{A}\sum_{k<j}\ldots\sum_{q<p}\Big[e^{-\frac{i}{v}\int_{-\infty}^{\infty}V_j d\xi}-1\Big]
$$

$$
\Big[e^{-\frac{i}{v}\int_{-\infty}^{\infty}V_k d\xi}-1\Big]\times\Big[e^{-\frac{i}{v}\int_{-\infty}^{\infty}V_q d\xi}-1\Big]\Big]\ , \qquad (3.15)
$$

as in the heuristic case. One can explicitly exhibit[15] the cancellations in the Watson series which produce this finite, on-shell series. The Glauber series has an obvious interpretation in terms of single-, double-, A-scatterings, but now with no return to previously struck nucleons.

4. APPLICATIONS, REFINEMENTS AND NONLOCALITIES

We shall now turn to a very brief survey of applications of the general multiple-scattering formalism to pion-nucleus scattering from zero energy up through the 3,3 resonance region. The consequences of higher-order effects there are only beginning to be studied and we shall look at these as yet tentative results with a view towards guessing what lies ahead. A common aspect of the corrections from the higher-order features is that they lead to generalizations of our previous simple forms for the optical potential [for examples eqs. (1.50) and (1.66)] which are nonlocal, sometimes in a more or less trivial way, sometimes more profoundly. Indeed, it is only after the most extreme simplifications that a local optical potential is encountered.

4.1. The Kisslinger Potential

One of the first features which must be incorporated into the πA optical potential for use at kinetic energies between 0 and 300 MeV, or so, is the dominance of p-wave scattering due to the 3,3 p-wave resonance. The πN amplitude can there be parametrized as

$$\langle \underset{\sim}{k}'|t(E_\pi)|\underset{\sim}{k}\rangle = b(E_\pi)+c(E_\pi)\underset{\sim}{k}'\cdot\underset{\sim}{k}+id(E_\pi)\underset{\sim}{\sigma}\cdot\underset{\sim}{k}'\times\underset{\sim}{k} \quad , \tag{4.1}$$

where $\underset{\sim}{k}$ and $\underset{\sim}{k}'$ are the initial and final pion momenta in the c.m. system for πN, $\underset{\sim}{\sigma}$ is the nucleon Pauli spin matrix, and b,c,d are energy-dependent coefficients related to s- and p-wave πN phase shifts. The s-wave part, $b(E_\pi)$, has no resonant behaviour, while $c(E_\pi)$, $d(E_\pi)$ do. Assuming our nuclear target to be approximately spin saturated, we may ignore $d(E_\pi)$.

In the forward direction, eq. (4.1) then yields for the optical potential [see eq. (1.50)]

$$\mathcal{U}_c(\underset{\sim}{r}) \cong -\frac{4\pi}{2m} f_o A\rho(\underset{\sim}{r}) = \left[b(E_\pi)+k^2c(E_\pi)\right]A\rho(\underset{\sim}{r}) \tag{4.2}$$

with c resonating at the 3,3 peak. (There, the nucleus will appear very black and details of nuclear structure will be obscured, the surface region, at best, being visible. This may militate for work above or below the resonance as being more revealing.) To take the p-wave nature of eq. (4.1) more fully into account, we retain the general form of t in evaluating the optical potential as

$$\begin{aligned}\langle \underset{\sim}{k}'|\mathcal{U}_c|\underset{\sim}{k}\rangle &= A\, b(E_\pi) \int e^{-i\underset{\sim}{k}'\cdot\underset{\sim}{r}}\, \rho(\underset{\sim}{r})\, e^{i\underset{\sim}{k}\cdot\underset{\sim}{r}}\, d\underset{\sim}{r} \\ &+ A\, c(E_\pi) \int (\underset{\sim}{\nabla} e^{-i\underset{\sim}{k}'\cdot\underset{\sim}{r}}) \cdot \rho(\underset{\sim}{r})(\underset{\sim}{\nabla} e^{i\underset{\sim}{k}\cdot\underset{\sim}{r}})\, d\underset{\sim}{r} \\ &= A\, b(E_\pi) \int e^{-i\underset{\sim}{k}'\cdot\underset{\sim}{r}}\, \rho(\underset{\sim}{r}) e^{i\underset{\sim}{k}\cdot\underset{\sim}{r}}\, d\underset{\sim}{r} \\ &- A\, c(E_\pi) \int e^{-i\underset{\sim}{k}'\cdot\underset{\sim}{r}}\, [\underset{\sim}{\nabla}\cdot\rho(\underset{\sim}{r})\underset{\sim}{\nabla}]\, e^{i\underset{\sim}{k}\cdot\underset{\sim}{r}}\, d\underset{\sim}{r} \quad ,\end{aligned} \tag{4.3}$$

whence the Kisslinger potential[16]

$$\mathcal{U}_c(\underset{\sim}{r}) = A\, b(E_\pi)\rho(\underset{\sim}{r}) - A\, c(E_\pi)\underset{\sim}{\nabla}\cdot\rho(\underset{\sim}{r})\underset{\sim}{\nabla} \quad , \tag{4.4}$$

where the p-wave piece exhibits a "minimal" nolocality. (Eq. (4.4) may be transformed into an "apparently local" form by the substitution $\psi(\underset{\sim}{r}) = \phi(\underset{\sim}{r})/\sqrt{1+2mAc(E_\pi)\rho(\underset{\sim}{r})}$, and the reader is cautioned then to recall that only ψ--not ϕ--has meaning as a wave function.)

Had we assigned the off-shell behaviour in eq. (4.1) differently, we could have written

$$\langle \underset{\sim}{k}'|t(E_\pi)|\underset{\sim}{k}\rangle = [b(E_\pi)+k^2c(E_\pi)] - \frac{1}{2}c(E_\pi)q^2 \quad , \qquad (4.5)$$

where $\underset{\sim}{q} = \underset{\sim}{k}-\underset{\sim}{k}'$ is the momentum transfer and we now take k^2 in the first term at its on-shell value. Since we now have a function of $\underset{\sim}{k}-\underset{\sim}{k}'$, this leads to a local potential

$$\mathcal{U}_c^{loc}(\underset{\sim}{r}) = A\,[b(E_\pi)+k^2c(E_\pi)]\,\rho(\underset{\sim}{r}) + \frac{1}{2}A\,c(E_\pi)(\nabla^2\rho(\underset{\sim}{r})) \qquad (4.6)$$

If d-waves enter, we could imagine extending this argument to include $f(q^2) = f_o+f_1q^2+f_2q^4+ \ldots$ leading to $\nabla^4\rho(r)$, and so forth. The results of eqs. (4.4) and (4.6) represent two extremes in handing the off-shell aspects of the p-wave feature of N scattering near the 3,3 resonance; a more reasonable approach may well be to take the form of eq. (4.1) with phenomenological cutoff factors in the off-shell behaviour of $\underset{\sim}{k}$ and $\underset{\sim}{k}'$, which will lead to the necessity of handling the optical potential in momentum space as thoroughly nonlocal with consequent numerical complications.

Other consequences of the p-wave resonance (inter alia in inducing nonlocal features) are discussed in ref. 2. We here raise--and at that very briefly--only one other such feature, which is the nature of the transformation which takes us from the πN c.m. system to the appropriate frame for πA scattering. This is most easily discussed if we use the manifestly (nonrelativistic) covariant form for the amplitude (in a self-evident notation)

$$\langle \underset{\sim}{k}'_\pi,\underset{\sim}{p}'_N|t(E_\pi)|\underset{\sim}{k}_\pi,\underset{\sim}{p}_N\rangle = b(E_\pi)+c(E_\pi)$$

$$\left[\frac{M}{M+m}\underset{\sim}{k}'_\pi - \frac{m}{M+m}\underset{\sim}{p}'_N\right]\cdot\left[\frac{M}{M+m}\underset{\sim}{k}_\pi - \frac{m}{M+m}\underset{\sim}{p}_N\right] \quad , \qquad (4.7)$$

the covariance being manifest since the quantities in square brackets involve $\underset{\sim}{v}_\pi-\underset{\sim}{v}_N \xrightarrow[\text{transformation}]{\text{galilean}} (\underset{\sim}{v}_\pi-\underset{\sim}{V})-(\underset{\sim}{v}_N-\underset{\sim}{V}) = \underset{\sim}{v}_\pi-\underset{\sim}{v}_N$.

In the c.m. frame, $\underset{\sim}{k}_\pi = \underset{\sim}{k}, \underset{\sim}{p}_N = -\underset{\sim}{k}$ and this reduces to eq. (4.1), while in the lab frame, appropriate to πA scattering, this is

$$\langle \underset{\sim}{k}'_\pi,\underset{\sim}{p}'_N = \underset{\sim}{k}_\pi-\underset{\sim}{k}'_\pi|t(E_\pi)|\underset{\sim}{k}_\pi,\underset{\sim}{p}_N = 0\rangle = \left[b(E_\pi)-c(E_\pi)\frac{mM}{(M+m)^2}k_\pi^2\right]$$

$$+\ c(E_\pi)\frac{M}{M+m}\underset{\sim}{k}'_\pi\cdot\underset{\sim}{k}_\pi \quad , \qquad (4.8)$$

involving changes of order $\frac{m}{M} \sim \frac{1}{6.7}$; relativistically these will be $\frac{\omega}{M} \sim \frac{1}{3}$. Eq. (4.8) then represents a presumably improved point of departure for calculation based on Kisslinger-like potentials. Such calculations have had reasonable success from 0 to 300 MeV in representing--at least phenomenologically--the A scattering data. Especially in the range below 100 MeV it would appear to be important to carry out more careful averaging over the nuclear distribution of nucleon momenta in using eq. (4.7). The inclusion of the second-order optical potential would also seem to be of some moment[17], especially near the second maximum in the angular distribution.

Thus far, only a very few calculations of this sort have been reported. Second-order effects may also arise from the consideration of binding corrections, i.e. the binding effects arising from the appearance of the nuclear hamiltonian in the multiple-scattering propagators. Both of these effects* also produce nonlocalities in $\mathcal{U}_c$. For our present purposes we shall sketch very briefly some specific features arising at zero energy (pionic atoms) or in the 3,3 region due to the p-wave resonance or Δ isobar.

4.2. Pionic Atoms and the Lorentz-Lorenz Effect

The formalism we have developed may be applied at zero pion kinetic energy to generate optical potentials for pionic atoms, which have been extensively reviewed[18] in recent years. In that context, there occurs a rather dramatic higher-order nonlinear effect--though in fact, leading still to a local optical potential --analogous to the Lorentz-Lorenz (LL) effect of electromagnetic theory. This was first pointed out by N. Kroll and studied extensively by the Ericsons[19]. Since it has been reviewed and discussed extensively, we here note only some recent developments[20-25], the recent papers on which can be consulted for details and extensive references to the extensive literature.

A derivation of the Lorentz-Lorenz optical potential,

$$\mathcal{U}_c(\underset{\sim}{r}) = \frac{1}{2m} \underset{\sim}{\nabla} \cdot \frac{4\pi c A\rho(\underset{\sim}{r})}{1+\frac{4\pi}{3} cA\rho(\underset{\sim}{r})} \underset{\sim}{\nabla} \tag{4.9}$$

in the formalism of chapter 1 has been given[20], which also lead

*See also ref. 32 for a comprehensive study of πA scattering incorporating these.

to a realization of the crucial role of the range of the πN interaction in possibly vitiating the effect. Subsequently, the possibility that ρ-exchange might also contribute to the LL phenomenon was raised[21] and the role of Pauli correlations more fully studied[22]. The nature of quadrupole terms in the LL effect was examined and various formal questions, such as its origin in higher formalisms, were explored[23]. Lastly, a geometrical manifestation of the LL effect has been noted[24], and the validity of the effect, or its analog, has been considered[25] up through 300 MeV. Because of the problems of πN range effects, and the like, the theoretical status of the LL effect remains uncertain, while clear experimental indicators are obscured by uncertainty in the various parameters required to fix the pionic-atom optical potential. It therefore seems fair to say that, at this time; the Lorentz-Lorenz effect should not be singled out for preferred treatment, but rather should be considered on the same footing as other higher-order effects in the optical potential.

4.3. The Delta-Hole Doorway State Approach

We close this very brief exposition of modern efforts to study higher-order effects with a theoretical viewpoint which tries to incorporate some of the dynamic features particular to the Δ isobar which dominates πN effects in our kinematic region. This is the Δ-hole doorway state model[26-28], which attempts to consider Δ dynamics in the πA scattering process.

The free πN amplitude in the 3,3 region, with spin saturation, is taken as

$$\langle \underset{\sim}{k}'_\pi \underset{\sim}{p}'_N | t(E_\pi) | \underset{\sim}{k}_\pi, \underset{\sim}{p}_N \rangle \propto (2\pi)^3 \, \delta(\underset{\sim}{k}'_\pi + \underset{\sim}{p}'_N - \underset{\sim}{k}_\pi - \underset{\sim}{p}_N)$$

$$\times \, t_o \, \frac{\Gamma(E_\pi)}{2} \, \frac{1}{E_\pi - E_{res} + \frac{1}{2} i\Gamma(E_\pi)} \, , \tag{4.10}$$

where we have made explicit the conservation of momentum in the fundamental process, and the normalization factor is fixed by the optical theorem,

$$t_o = \frac{|\underset{\sim}{v}_\pi - \underset{\sim}{v}_N|}{2\chi^2} \, \sigma_{total} \Big|_{E_\pi = E_{res}} \, , \tag{4.11}$$

with $\underset{\sim}{\chi} = (M\underset{\sim}{k}_\pi - m\underset{\sim}{p}_N)/(M+m)$. In eq. (4.10), we fix the pion energy at

$$E_\pi = E - H_N(A-1) - \frac{(\underset{\sim}{k}_\pi + \underset{\sim}{p}_N)^2}{2M^*} \quad , \quad M^* = M+m, \tag{4.12}$$

where the need to average over nuclear dynamics is clear. Here $H_N(A-1)$ refers to the energy of passive, spectator nucleons, the active participant being that with momentum $\underset{\sim}{p}_N$, expected to form a Δ isobar with the incident pion.

Using a shell model description, and denoting the single-particle states by ϕ_N, the removal energy for the pertinent hole state is

$$\langle\phi_n^{-1}|H_N(A-1)|\phi_n^{-1}\rangle \equiv -\varepsilon_n \quad , \tag{4.13}$$

and the first-order optical potential becomes

$$\langle\underset{\sim}{k}_\pi'|\mathcal{V}_c^{(1)}(E)|\underset{\sim}{k}_\pi\rangle \propto \frac{1}{2}\, t_0 \Gamma \sum_{n=1}^{A} \int \frac{d\underset{\sim}{p}_N}{(2\pi)^3}$$

$$\times\ [\phi_n^*(\underset{\sim}{p}_N') \frac{1}{E+\varepsilon_n - E_{res} - \frac{(\underset{\sim}{k}_\pi+\underset{\sim}{p}_N)^2}{2M^*} + \frac{i\Gamma}{2}} \phi_n(\underset{\sim}{p}_N)] \ , \tag{4.14}$$

where $\underset{\sim}{p}_N' = \underset{\sim}{p}_N + \underset{\sim}{k}_\pi - \underset{\sim}{k}'$ and we will take the Δ width as constant, Γ = 110 MeV. The resonating πN subsystem, whose recoil energy appears in the resonance denominator, has an important effect as we see from integrating in eq. (4.14) to obtain

$$\langle\underset{\sim}{r}'|\mathcal{V}_c^{(1)}|\underset{\sim}{r}\rangle = -\frac{2M^*}{4\pi}\left(\frac{1}{2}\,\bar{\chi}^2 t_0 \Gamma\right) \sum_{n=1}^{A} \phi_n^*(\underset{\sim}{r}') \frac{e^{iK_n|\underset{\sim}{r}'-\underset{\sim}{r}|}}{|\underset{\sim}{r}'-\underset{\sim}{r}|} \phi_n(\underset{\sim}{r}) \ , \tag{4.15}$$

with

$$K_n^2 = 2M^*(E+\varepsilon_n - E_{res} + \frac{i\Gamma}{2}) \quad . \tag{4.16}$$

In eq. (4.15) the nonlocality of the optical potential due to Δ recoil is self-evident. For a very short-lived Δ, the propagation again becomes local,

$$\frac{1}{4\pi i} M^*\Gamma \frac{e^{iK_n|\underset{\sim}{r}'-\underset{\sim}{r}|}}{|\underset{\sim}{r}'-\underset{\sim}{r}|} \xrightarrow[\Gamma\to\infty]{} \delta(\underset{\sim}{r}'-\underset{\sim}{r}) \quad , \tag{4.17}$$

and

$$\langle \underset{\sim}{r}' | \mathcal{U}_c^{(1)} | \underset{\sim}{r} \rangle \underset{\Gamma\to\infty}{\longrightarrow} -i\bar{\chi}^2 t_o \sum_{n=1}^{A} |\phi_n(\underset{\sim}{r})|^2 = -\frac{1}{2}\sigma_{\text{total}} A\rho(r) \quad , \quad (4.18)$$

as in our most primitive result. The nonlocality spreads over distances

$$\ell_\Delta \sim \frac{1}{\sqrt{M^*\Gamma}} \sim 0.6 \quad \text{fm} \quad , \quad (4.19)$$

and arises from the energy-dependence of the transition amplitude and the nonvanishing lifetime of the Δ. To the degree to which details of the interaction of the Δ with the residual nucleus of (A-1) nucleons can be incorporated, much more refined features of the dynamics, as they are manifested in the multiple-scattering process, can be studied (see refs. 27-31).

There remains the question as to whether the refined features of the Δ-hole approach can be observed through elastic scattering in the 3,3 region where absorption and geometry seem to carry the day. It may prove necessary to seek nuclear structure details in pion scattering below the 3,3 resonance, or to take much more seriously than hitherto the large-angle elastic scattering or inelastic pion reactions. In all these approaches, the multiple-scattering language seems to provide a well-based, systematic point of departure from which to develop studies of important effects beyond the lowest order $\mathcal{U}_c = t^{free} A\rho(r)$.

It is a pleasure to acknowledge many valuable exchanges and discussions over some years with Messrs. G.E. Brown, T.E.O. Ericson, H. Feshbach, A. Gal, J. Hüfner, A.K. Kerman, D.S. Koltun, E.J. Moniz, J.V. Noble and H.J. Weber which have helped to shape some of the viewpoints presented here.

REFERENCES

1. See, for example, M.L. Goldberger and K.M. Watson, Collision Theory (Wiley, New York, 1964) or A.L. Fetter and K.M. Watson, in Advances in Theoretical Physics, vol. 1, K.A. Brueckner, ed. (Academic, New York, 1965), p. 115.

2. J. M. Eisenberg and D. S. Koltun, The Theory of Meson Interactions with Nuclei (Wiley-Interscience, New York, 1980).

3. J.M. Eisenberg, in Lectures from the LAMPF Summer School on the Theory of Pion-Nucleus Scattering, W.R. Gibbs and

B.F. Gibson, V, eds., LA-5443-C (National Technical Information Service, Washington, 1973).

4. A.K. Kerman, H. McManus and R.M. Thaler, Ann. Phys. (New York) 8 (1959) 551.

5. See also N. Austern, F. Tabakin and M. Silver, Am. J. Phys. 45 (1977) 361.

6. See, for example, H. Feshbach, A. Gal and J. Hüfner, Ann. Phys. (New York) 66 (1971), 20, and subsequent work of H. Feshbach and collaborators, including E. Lambert and H. Feshbach, Ann. Phys. (New York) 76 (1973) 80 and J.J. Ullo and H. Feshbach, Ann. Phys. (New York) 82 (1974) 156, and references cited therein.

7. L.L. Foldy and J.D. Walecka, Ann. Phys. (New York) 54 (1969) 447.

8. P.C. Tandy, E.F. Redish and D. Bollé, Phys. Rev. Letters 35 (1975) 921; Phys. Rev. C16 (1977) 1924. For a specifically pionic application, see A.W. Thomas and R.H. Landau, Phys. Lett. 77B (1978) 155.

9. J.M. Eisenberg, Phys. Rev. C19 (1979) 559.

10. The pion charge exchange process has recently been thoroughly reviewed, especially in its experimental aspects, by J. Alster and J. Warszawski, Phys. Repts., to be published. See also ref. 11 for a general assessment of theoretical aspects of the problem with reference to the earlier literature.

11. J. Warszawski, A. Gal and J.M. Eisenberg, Nucl. Phys. A294 (1978) 321.

12. J.M. Eisenberg and A. Gal, Phys. Lett. 53B (1975) 390; A. Gal and J.M. Eisenberg, Phys. Rev. C14 (1976) 1273.

13. E. Oset, Phys. Lett. 65B (1976) 46.

14. R.J. Glauber, in Lectures in Theoretical Physics, vol. 1, W.E. Brittin *et al.*, eds. (Interscience, New York, 1959), p. 315.

15. D.R. Harrington, Phys. Rev. 184 (1969) 1745. J.M. Eisenberg, Ann. Phys. (New York) 71 (1972) 542.

16. L.S. Kisslinger, Phys. Rev. 98 (1955) 761.

17. T.-S.H. Lee and S. Chakravarti, Phys. Rev. C16 (1977) 273.

18. G. Backenstoss, Ann. Rev. Nucl. Sci. 20 (1970), 467. L. Tauscher, in High-Energy Physics and Nuclear Structure, Santa Fe and Los Alamos, 1975, D.E. Nagle et al., eds. (A.I.P. New York, 1975), p. 541; J.Hüfner, Phys. Repts. 21C (1975), 1; L. Tauscher, in Exotic Atoms, Int. School of Physics of Exotic Atoms, Erice, 1977, G. Fiorentini and G. Torelli, eds. (Frascati, 1977), p. 145.

19. M. Ericson and T.E.O. Ericson, Ann. Phys. (New York) 36 (1966) 323.

20. J.M. Eisenberg, J. Hüfner and E.J. Moniz, Phys. Lett. 47B (1973) 381.

21. G. Baym and G.E. Brown, Nucl. Phys. A247 (1975), 395 and erratum A262 (1976) 539.

22. J. Delorme and M. Ericson, Phys. Lett. 60B (1976), 451; J. Delorme, M. Ericson, A. Figureau and C. Thevenet, Ann. Phys. (New York), 102 (1976) 273.

23. J. Warszawski, J.M. Eisenberg and A. Gal, Nucl. Phys. A312 (1978) 253.

24. W.R. Gibbs, B.F. Gibson and G.J. Stephenson, Jr., Phys. Rev. Lett. 39 (1977) 1316.

25. H. Garcilazo, Nucl. Phys. A302 (1978) 493.

26. F. Lenz, Ann. Phys. (New York) 95 (1975) 348.

27. M. Hirata, F. Lenz and K. Yazaki, Ann. Phys. (New York) 108 (1977) 116.

28. M. Hirata, J.H. Koch, F. Lenz and E.J. Moniz, Phys. Lett. 70B (1977) 281.

29. E.J. Moniz, in Meson-Nuclear Physics - 1976, P.D. Barnes et al., eds. (A.I.P., New York, 1976) p. 105. See also ref. 33.

30. M. Dillig, R. Händel and M.G. Huber, to be published.

31. K. Klingenbeck, M. Dillig, H.M. Hofmann and M.G. Huber, to be published.

32. J.P. Maillet, J.P. Dedonder and C. Schmit, to be published.

33. E.J. Moniz, lectures at the NATO Advanced Study Institute on Theoretical Methods in Intermediate Energy and Heavy-Ion Physics, Madison, Wisconsin, June, 1978 (M.I.T. preprint).

THE PION-NUCLEUS OPTICAL POTENTIAL[+]

W. R. Gibbs[++]

Los Alamos Scientific Laboratory, T-5

Los Alamos, New Mexico, 87545, U.S.A.

ABSTRACT

A review of the pionic optical model is given. Some recent calculations and fits to data are discussed briefly.

I. MOTIVATION

It may seem strange to start out by discussing the reasons for wanting a pion-nucleus optical model since we have had a nucleon-nucleus optical model for many years and we know its usefulness. However the pionic case is sufficiently different to consider on its own. Following is a series of considerations that have been addressed in developing the pionic optical model.

i) Testing our Understanding of Multiple Scattering Theory

There are several reasons for thinking we might be able to develop such an optical potential from first principles more easily than the nucleon-nucleus optical potential.

One significant reason is the pion's small mass. Even at moderate energies it moves much faster than the nucleons in the

---

[+]Work supported by the U.S. Department of Energy

[++]New work reported on in this talk was done in collaboration with B.F. Gibson and G.J. Stephenson, Jr.

nucleus. Even with these greater velocities its wave number is such that only $\ell = 0$ and $\ell = 1$ are necessary in the basic interaction. Thus we have simple amplitudes and fixed scatterers at the same time: a situation which is not possible with nucleons because of the larger mass.

Another simplification is the lack of spin of the pion. This again reduces the degree of complexity of the basic interaction.

Pions are available in both positive and negative states for scattering (neutral pions may be used in the final state). This allows the known Coulomb potential to be used to test the short range (internal to the nucleus) properties of the potential, or in other words--test the off-shell properties of the scattering.

The probe in this case is not identical with the scatterers so the effects of antisymmetrization are smaller.

The on-shell strength of the interaction can be varied. Because of the existence of the very sharp T=3/2, J=3/2 resonance

at $T_\pi \sim 180$ MeV a large change in interaction strength can be achived with a moderate change in energy and momentum. Since some effects drastically change their importance with interaction strength this can be a valuable tool in checking parts of the theory.

ii) Learn About the Off-Shell Pion-Nucleon Interaction

In order to learn about the pion-nucleon off-shell t-matrix there must be (at least) a third strongly interacting body present. In the deuteron the multiple scattering effects are not very important so that π-d scattering is not very sensitive to off-shell effects. Scattering on a nucleus is entirely different. Here multiple scattering is very important. The situation here is similar to calculating the binding energy in nuclear matter to test nucleon-nucleon potentials.

Knowledge of the off-shell t-matrix allows comparison with more fundamental theories of the pion-nucleon interaction.

iii) Examine Relative Neutron-Proton Radii in Nuclei[1]

Since positive pions see mostly protons and negative pions see mostly neutrons (π^+p and π^-n are pure T=3/2 systems and hence are coupled to the 3-3 resonance most strongly), one should observe a difference in the diffraction minima positions corresponding to neutron and proton radii.

There is also a difference in the apparent radii due to Coulomb effects alone. Since the negative pion is bent into the nucleus and the positive pion is bent away, the π^- has a larger impact parameter for a grazing collision than a π^+. This is illustrated schematically in Fig. 1.

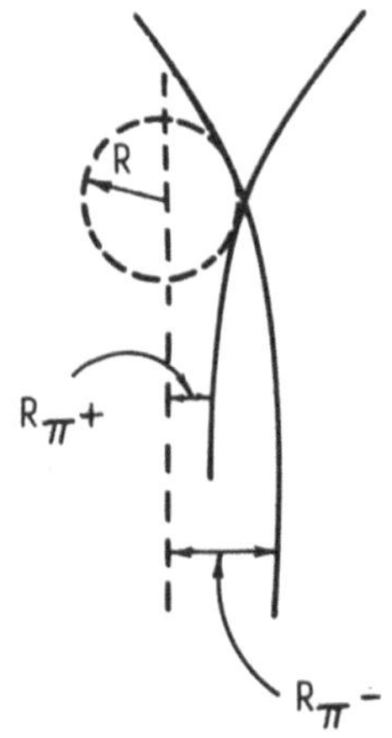

Fig. 1. Schematic representation of how the Coulomb interaction can give an apparent $\pi^+ - \pi^-$ radius difference.

If one assumes that this Coulomb effective radius difference is the same for ^{40}Ca and ^{48}Ca one may hope that the correction may be made using the data. Such an experiment was performed by Egger et al[1] at S.I.N. and a representation of their data is shown in Fig. 2.

The effects just discussed are apparent in the data. Using a "black disk" model they find "ΔR" = 0.21 for ^{40}Ca and "ΔR" = 0.51 for ^{48}Ca. Assuming that the neutron and proton radii are equal in ^{40}Ca we may infer that 0.51 - .21 = .30 is the neutron-proton radius in ^{48}Ca. This is very nice--one has determined the radius difference rather directly from the data. However there remains one serious problem, i.e. what radius difference has been measured? Since the experiment was done near the resonance the absorption is very strong. It would seem that the nucleus is effectively black well beyond the half density point. Present estimates are that the ΔR measured is at $\rho \sim 0.2\ \rho_0$. This is illustrated in Figure 3.

I have shown the situation in which the half-density radii are equal but, because of a difference in diffuseness, a substantial

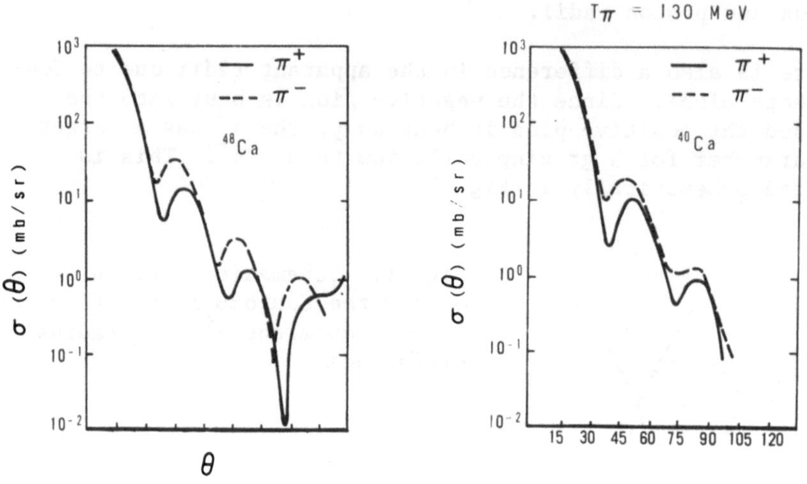

Fig. 2. Representation of the data of ref. 1 showing how the neutron-proton radius difference in ^{48}Ca can be measured using ^{40}Ca to remove the Coulomb effects.

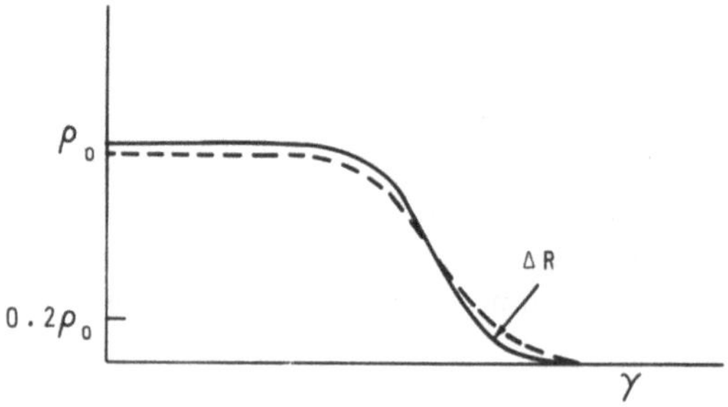

Fig. 3. Relative proton (solid line) and neutron (dotted line) densities.

ΔR is measured at 0.2 ρ_0. Clearly the degree of blackness is crucial. An optical model is an excellent method for determining this property. If one does these experiments at lower energies different considerations will enter. I will discuss this in detail at the end of the lecture.

iv) Provide Wave Functions for DWIA Calculations of Reactions

This is the classical use in nuclear physics so not much needs to be said. However I might remark that there are a number of exotic reactions for which these wave functions are essential, e.g., double charge exchange, single nucleon absorption and two nucleon absorption.

v) Other Specialized Uses

There seem to be a number of reasons to want an optical model that were not apparent in the beginning. No doubt there will continue to be new applications arriving on the scene as time passes. I list three uses below.

a) Analyse Angular Distributions to Obtain Total Cross Sections and Re f(0). For low energies and/or heavy nuclei the problem of untangling the forward Coulomb-nuclear interference to obtain total cross sections from transmission experiments is severe. In doing optical model fits to data we have noticed that the forward amplitude of the fit is relatively insensitive to the ambiguities in the parameters used to obtain the fit. This method may supply an alternate procedure for obtaining these important quantities in these difficult cases.

b) Obtain Amplitudes to Check Isospin Predictions. If one assumes that iso-spin is a good quantum number in the nuclear and pion system then certain amplitudes can be related. For example, if A^+ is the amplitude for π^+ elastic scattering on ^{13}C, and A^- is the amplitude for π^- scattering on ^{13}C, the analogue charge exchange cross section is given by

$$\sigma_{(\theta)}^{CEX} = \frac{1}{2} \left| A^+(\theta) - A^-(\theta) \right|^2 .$$

Since we seem to be having considerable difficulty understanding single (and double) charge exchange, it might be very useful to check iso-spin on this reaction. A violation might be present if the π^0 did not behave as it "should" in a hadronic environment e.g., if it decayed before leaving the nucleus.

To carry out this program one must have the amplitudes. These may reasonably be obtained by fitting an optical model to the data.

c) Investigate the Possibility of Pion-nucleus Bound States. Ericson and Myhrer[2] have pointed out that the pion-nucleus potentials are very close to supporting a bound state. If one could find a particularly favourable case these bound states might exist. To get a reliable estimate of these possibilities one needs a realistic optical potential. I will give some results on this point later.

II. DEVELOPMENT OF THE PION-NUCLEUS OPTICAL MODEL

I would like to discuss some of the important features of the pionic optical model in order to lay a foundation for a particular formulation to be given. I will do this in a more-or-less chronological order. Not all effects will be considered--probably not all important effects will be considered.

i) Simple (Nuclear Matter) Optical Model

One often argues (from nuclear matter considerations) that the expression for the first order optical potential can be simplified:

$$V(\bar{k},\bar{k}') = f(\bar{k},\bar{k}')\rho(\bar{k}-\bar{k}') \approx f(\bar{k},\bar{k})\rho(\bar{k}-\bar{k}') \quad .$$

This is reasonable for the nucleon-nucleus optical potential but does not work in the case of pions because the pion-nucleon amplitude $f(\bar{k},\bar{k}')$ is a strong function of its variables (and, in fact, is somewhat backward peaked below the 3-3 resonance).

ii) Kisslinger Optical Model

For this reason Kisslinger[3] included the p-wave part explicitly. This gives a term of the type $\bar{k}\cdot\bar{k}'$ in momentum space which becomes a non-local operator

$$V\psi \sim \nabla\cdot(\rho\nabla\psi)$$

in coordinate space. Note that this potential is non-local only in the derivative sense. We will consider more general non-local forms later.

Auerbach, Fleming and Sternheim[4] showed that the comparison with data strongly suggested that this form was needed.

iii) Lorentz-Lorenz Effect

It was pointed out by the Ericsons[5] that the p-wave nature of the π-nucleon interaction causes a modification of the π-nucleus optical potential similar to the Lorentz-Lorenz effect in electromagnetism. This idea was applied to pionic atoms by Krell and Ericson[6]. This phenomenon is being discussed by both Ericson and Eisenberg at this meeting so I will not discuss the derivation.

It leads to a "saturation" of the potential since it is most important at high densities. Thus the effect is to replace the density with an effective density

$$\rho \to \frac{\rho}{1+b\rho} \quad , \quad b = \frac{4\pi}{3}(A-1)a$$

where "a" is the average pion-nucleon scattering volume.

iv) Finite Range Pion-Nucleon Interaction

We may note that the Kisslinger form of the t-matrix,

$$b_o + b_1 \bar{k}\cdot\bar{k}' \quad ,$$

corresponds to a zero range separable potential in coordinate space. Since we believe that both the pion and the nucleon are composite particles, and hence have finite size, the zero range approximation may be rather poor. We may introduce this size by using a t-matrix corresponding to a separable potential of finite range, i.e.

$$b_o v_o(k) v_o(k') + b_1 v_1(k) v_1(k') \bar{k}\cdot\bar{k}' \quad .$$

Using a Kisslinger form for the pion-nucleus optical potential Jones and Eisenberg[7] found that the magnitude of the cross section for pion production was greatly increased over the plane wave result.

We[8] noticed that the cross section for $\pi^+ + d \to p + p$ was calculated to be about a factor of 5 too large with the Kisslinger form. The introduction of a finite range brought the total cross section into agreement with data.

Miller and Phatak[9] showed that a similar replacement in the optical potential corrects the magnitude for the p,π reaction on nuclei.

One method for determining the functions v_o and v_1 is to construct a model for pion-nucleon scattering using them, and to use the energy dependence of the π-N phase shifts to get the v's as a function of momentum. This has been done by several groups[10] with similar results.

This finite range has a profound effect on the Lorentz-Lorenz effect. Since the latter is equivalent to short range correlations[11,12] a very large range of the pion-nucleon interaction would destroy the effect. This is generally expressed quantitatively by replacing the quantity "b" with

$$b \to \xi\hat{b}$$

where $\xi = 1$ corresponds to the full L-L effect as proposed by the Ericsons. The expression for ξ is[11,12]

$$\xi = \frac{1}{2\pi^2} \int_0^\alpha k^4 dk \frac{G(k)v^2(k)}{\omega_0^2-\omega^2(k)+i\varepsilon}$$

where $v(k)$ is the finite range function and $G(k)$ is the short range correlation function between the nucleons.

Clearly ξ depends on the range of $v(k)$. For ranges of the order given by Refs. 10 $\xi \sim 0.2$[11]. However ξ can increase slightly with energy[12].

Pauli correlations can lead to an increase of ξ as well.

Baym and Brown showed[13] that ρ exchange could lead to a similar form and could once again increase ξ considerably.

Thus we see that the value of ξ is not well determined theoretically at the present time. We shall see what can be learned from the elastic scattering data shortly.

v) The "Angle Transform"

Since the nucleons in the nucleus are moving, the scattering angle (and relative momentum and energy) are not the same as in a free scattering. These effects were included[14] in the nucleon-nucleus optical model but were not very important because the form factor forces the nucleon-nucleon amplitude on-shell and to 0°.

For the π-nucleus system the effect is very large and clearly seen, especially for π-^{4}He scattering[15].

One can see the important features of this correction in a simple non-relativistic picture which, in fact, is adequate for low pion energies.

If we write the off-shell t-matrix in Galilean invariant form,

$$b_0 \mp b_1 (\bar{k} - \frac{\mu}{m}\bar{p}_i)\cdot(\bar{k}' - \frac{\mu}{m}\bar{p}_f)$$

where $\bar{p}_i$ and $\bar{p}_f$ are the initial and final nucleon momenta, and use the conservation of momentum

$$\bar{k} + \bar{p}_i = \bar{k}' + \bar{p}_f$$

we have $b_o + b_1(k - \frac{\mu}{m} p_i)\cdot[k' - \frac{\mu}{m}(k + p_i - k')]$.

To get a first approximation drop the terms of order $(\frac{\mu}{m})^2$ and those linear in nucleon momenta (they tend to average to zero). Thus we have

$$b_o - b_1 \frac{\mu}{m} k^2 + b_1(1 + \frac{\mu}{m})\, \bar{k} \cdot \bar{k}' \quad .$$

In a single scattering approximation the momenta would be on shell. For ^{4}He the single scattering approximation is not toally wrong so we might get away with using this for the on-shell t-matrix. We notice that for $k = \mu$, since $\mu^2 b_1 \sim -7\, b_o$, the magnitude of the s-wave part of the t-matrix is about doubled. Notice that in the single scattering approximation

$$F(\theta) = [b_o - \frac{\mu}{m} b_1 k^2 + b_1(1 + \frac{\mu}{m})k^2\cos\theta]S(\theta) \quad .$$

Thus there is a zero in the cross section whose position is given by

$$\cos\theta = \frac{-b_o + b_1 k^2}{k^2 b_1 (1+\frac{\mu}{m})} ,$$

Even after multiple scattering this zero remains as a minimum (for light nuclei). It is referred to as the "s-p interference minimum". Note that it has a weak energy dependence, unlike the diffraction minima, and that, as energy increases, it actually moves out toward 90°, just opposite to the behaviour of diffraction minima.

In the resonance region this correction must be done relativistically, a non-trivial problem. There is a unique solution[16] in a relativistic potential model. This solution is apparently being adopted in recent calculations, e.g. Ref. 17.

vi) "True" Pion Absorption

I will not say much about this topic since it is being covered by both Rinat and Koltun.

It has generally been assumed that the two nucleon absorption dominates the true absorption. This is not necessarily the case.

For example after a number of scatterings within the nucleus, the pion may find itself quite far off-shell and with the proper momentum to absorb on a nucleon. This type of absorption is linear in the density (the higher orders in scattering and density being provided by the solution of the Schrodinger equation--equivalent to a Born expansion) unlike the terms usually inserted in the optical potential which are quadratic in the density. The size of this effect can be calculated and is not terribly small[18,19]. Figure 4 shows the effect on the elastic angular distribution as calculated in Ref. 19. Note that it may be misleading to think of this as strictly a one nucleon mechanism. It is really more like a (A-1)-nucleon-one-nucleon mechanism.

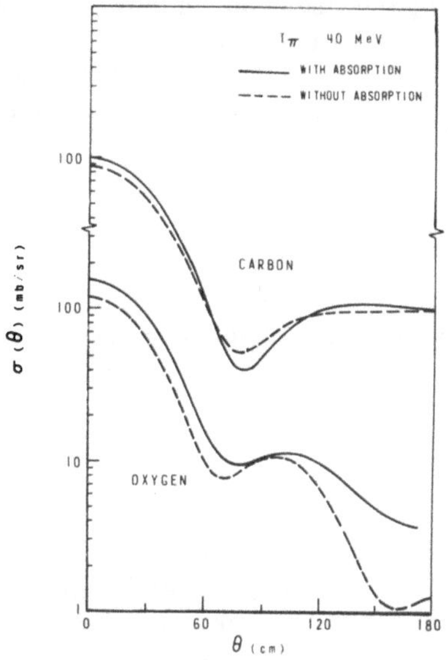

Fig. 4. The effect of "one nucleon absorption" on elastic scattering on carbon and oxygen at 40 MeV.

Other multi-nucleon absorption mechanisms are possible but have been considered very little.

vii) Energy Subtraction

Since the imaginary part of the simple first order optical potential comes from the imaginary part of the π-nucleon t-matrix, this absorptive term is due (primarily) to quasi-elastic knock-out of nucleons. Since this cannot occur for $T_\pi < E_B$ a strong modification of the t-matrix is to be expected at low energies. This usually takes the form of an energy subtraction, i.e., one uses the free t-matrix at a lower energy or an average over lower energies. The residual nucleus (after knock-out) must have some kinetic energy and this further reduces the effective energy. Since the initial nucleon is moving, a certain amount of energy goes into the total centre of mass energy of the π-nucleon system, reducing that available in the centre of mass. All of these effects lower the effective energy of the collision. This tends to make the nucleus more transparent below the resonance. These effects have been treated most carefully in Refs. 17 and 20.

viii) Δ-hole States and the Lifetime of the Delta

One may (perhaps must?) treat the delta as a different particle so that when the π-nucleon system forms a delta, a hole is created in the nucleon shell. Such pictures were introduced by Kisslinger and Wang[21] and may be treated in an optical model framework.

ix) Soft Pion Approach

The use of current algebra techniques allows one to calculate the off-shell t-matrix for the π-nucleon system. Cammarata and Banerjee[22] have used this method and obtain a somewhat different form than that usually assumed. It is not yet clear if these differences are observable in practical cases.

x) Three Body Effects

I have already mentioned this in the discussion of the Lorentz-Lorenz effect but it is separate. The pion can be absorbed on a nucleon to form a delta. This delta can emit an ρ--this ρ can be absorbed by another nucleon to become a second delta and this delta can de-excite by emitting a pion. This provides an explicit three body force and hence is outside the usual optical model and must be added by hand. Besides Ref. 13 this has also been considered by Refs. 23 and 24.

As I said at the beginning I have doubtless left out several effects even, perhaps, important ones.

III. GENERAL FEATURES OF PION-NUCLEUS ELASTIC SCATTERING

I would like to briefly review the general features of pion-nucleus scattering so that you may recognize some of the features directly. The energy regimes are naturally separated into a strongly absorbing and a weakly absorbing one.

i) Strongly Absorbing $-T_\pi \sim 160$ MeV

Here the angular distribution can be characterized as black disk scattering. Of course the nucleus has a diffuse surface so if it is truly black it is also infinite in extent. What one has found[25,26,27] is that with "free" optical parameters the effective radius is too large. This may be because the optical parameters are too strongly absorbing. We may ask how much too strong would they have to be to give this effect. Using a simple volume absorption and assuming there is some value of the imaginary potential for which the nucleus is effectively "black", W_c. Then using the amount the radius must be decreased to give a fit as given by Ref. 25 in the eq.

$$\frac{W}{1 + e^{\frac{r-c}{a}}} = W_c$$

We may estimate the factor by which the nucleus is too strongly absorbing. We obtain

^{9}Be	1.57
^{28}Si	1.41
^{58}Ni	1.32
^{208}Pb	1.20

These are not unexpected as order of magnitude estimates and hence this may be the explanation of the apparent small radius.

ii) $T_\pi \sim 30$-80 MeV

The most prominent feature in this energy region (at least for light nuclei) is the s-p interference minimum. This dip occurs in around 60-70° and is easily identifiable in nuclei as heavy as Calcium.

In the forward direction the interference with the Coulomb amplitude is clearly visible. The signs are such that this interference is destructive for π^+ and constructive for π^-.

As the atomic number is increased diffraction minima appear and for Pb they dominate the angular distribution. It is useful to remember that as the incident pion energy is increased diffraction minima move to smaller angle while the s-p minimum is essentially fixed.

IV. THE LOS ALAMOS OPTICAL MODEL

I would like now to describe an optical model code which we (Jerry Stephenson, Ben Gibson and myself) have developed and used at Los Alamos. We solve the equation

$$(T+V)\psi = E\psi$$

with relativistic kinematics for the pion. This is equivalent to the usual truncated Klein-Gordon equation.

However here V is to be interpreted as a non-local operator

$$V\psi \to \int d\bar{r}'\; V(\bar{r},\bar{r}')\psi(\bar{r}').$$

In momentum space the first order optical potential is given by:

$$V_E(\bar{q},\bar{q}') = t_E(\bar{q},\bar{q}')S(\bar{q}-\bar{q}')$$

which we take as (using a separable form for the pion-nucleon t-matrix)

$$[\lambda_0(E)\,\frac{(\alpha_0^2+K^2)^2}{(\alpha_0^2+q^2)(\alpha_0+q'^2)} + \lambda_1(E)\,\frac{(\alpha_1^2+K^2)^2}{(\alpha_1^2+q^2)(\alpha_1^2+q'^2)}\,\frac{\bar{q}\cdot\bar{q}'}{K^2}\,]\;S(\bar{q}-\bar{q}')$$

$$\equiv V_{E,0}(\bar{q},\bar{q}') + V_{E,1}(\bar{q},\bar{q}')$$

where K is the on-shell momentum, i.e.

$$E + \mu = \sqrt{\mu^2+K^2} = \omega \quad .$$

Since we wish to work in coordinate space it is necessary to make a transformation. This requires a little bit of algebra which I won't go through. If we define

$$f_\alpha(\bar{r}-\bar{s}) = \frac{\alpha^2+K^2}{4\pi}\,\frac{e^{-\alpha|\bar{r}-\bar{s}|}}{|\bar{r}-\bar{s}|}$$

then by using

$$\nabla^2 f_\alpha(\bar{r}-\bar{s}) = \alpha^2 f_\alpha(\bar{r}-\bar{s}) - (\alpha^2+K^2)\delta(\bar{r}-\bar{s})$$

and a number of integrations by parts we find for the p-wave piece

$$V_{E,1}\psi = -\frac{\lambda_1(E)(4\pi)^2}{K^2} \int d\bar{r}'\ \psi(\bar{r}')$$

$$\{\alpha^2 \int d\bar{s} f_\alpha(\bar{r}-\bar{s})\rho(\bar{s})f_\alpha(\bar{s}-\bar{r}') - \frac{1}{2}(\alpha^2+K^2)[\rho(\bar{r}')+\rho(\bar{r})]f_\alpha(\bar{r}-\bar{r}')$$

$$-\frac{1}{2}\int d\bar{s} f(\bar{r}-\bar{s})[\nabla^2\rho(\bar{s})]f_\alpha(\bar{s}-\bar{r}')\} \quad .$$

Note that there are no gradients acting on the wave function. However as $\alpha \to \infty$ this becomes the Kisslinger form as can be explicitly demonstrated. This expression can be easily broken into partial waves and we can solve the equation for each partial wave by a general technique for integro-differential equations to be discussed now.

There is one technical point. Since we wish to be able to calculate for moderately large values of α in order to compare with Kisslinger codes and the integrals must be able to represent derivatives in this case, it was necessary to develop special quadrature weighting schemes adjusted to a particular form of the nuclear density and to the energy of the pion.

For a given pion-nucleus partial wave we may write a discrete form of our equation as a matrix equation

$$\begin{pmatrix} H_{11}-E & H_{12} & H_{13} & \cdots & H_{1H} \\ H_{21} & H_{22}-E & H_{23} & \cdots & H_{2H} \\ \vdots & \vdots & & & \\ H_{N1} & & & & H_{NN}-E \end{pmatrix} \begin{pmatrix} \phi_1 \\ \phi_2 \\ \vdots \\ \phi_N \end{pmatrix} = 0$$

We generally use N = 80 and the derivatives are represented by a 20 point formula. For boundary conditions we have $\phi_1=\phi(0)=0$ and $\phi_N=1$, the second is an arbitrary normalization to be corrected when the s-matrix element is obtained.

The equations are:

$$(H_{11}-E)\phi_1 + H_{12}\phi_2 + \ldots H_{1N}\phi_N = 0$$

$$H_{21}\phi_1 + (H_{22}-E)\phi_2 + \ldots H_{2N}\phi_N = 0$$

$$\vdots$$

$$H_{N1}\phi_1 + H_{N2}\phi_2 + \ldots (H_{NN}-E)\phi_N = 0$$

The boundary conditions make the first term in each equation zero and the last term a known quantity. Thus we have N equations in N-2 unknowns. Dropping the first and last equations the matrix system for the new equations is

$$\begin{pmatrix} H_{22}-E & H_{21} & \cdots & H_{2N-1} \\ H_{23} & H_{33}-E & \cdots & H_{3N-1} \\ & & & \vdots \\ H_{2N-1} & & & H_{N-1N-1}-E \end{pmatrix} \begin{pmatrix} \phi_2 \\ \phi_3 \\ \vdots \\ \phi_{N-1} \end{pmatrix} = \begin{pmatrix} H_{2N} \\ H_{3N} \\ \vdots \\ H_{N-1,N} \end{pmatrix}$$

The matching conditions to be applied at large r are

$$c\phi = \frac{1}{2} [h^-(kr) + Sh^+(Kr)] \quad .$$

These give the S-matrix elements and hence the cross section.

We have checked this code against Abacus for a Kisslinger solution and a momentum space code.

Note that the bound state problem has the boundary conditions $\phi(0)=\phi_1=0=\phi(\infty) \approx \phi_N$ so that:

$$\begin{pmatrix} H_{22}-E & H_{23} & \cdots & H_{1N-1} \\ H_{23} & & & \vdots \\ \vdots & & & \\ H_{2N-1} & \cdots & & H_{N-1,N-1}-E \end{pmatrix} \begin{pmatrix} \phi_2 \\ \phi_3 \\ \vdots \\ \phi_{N-1} \end{pmatrix} = 0$$

Thus the eigenvalues of the same matrix used to solve the

scattering problem give the energies of the bound states and the eigenvectors give the wave functions.

While one could calculate the cross section with various assumptions relating the effective parameters to the pion-nucleon free t-matrix we have taken the point of view that we fit the data to determine the best effective parameters. The idea is to see if geometric corrections (Lorentz-Lorenz, ρ^2 corrections from true absorption, etc.) can be separated from strength modifications.

We might try to classify generally our qualitative understanding of the strengths. The "b" parameters are in the notation of ref. 4.

$\mathrm{Re}\,b_1$ - This is changed very little from the free values by various corrections. It is the best determined.

$\mathrm{Re}\,b_o$ - Can be altered greatly by the "angle transformation".

$\mathrm{Im}\,b_1$ - Since we have no explicit ρ^2 term to represent true absorption that effect must be mocked up by this parameter.

$\mathrm{Im}\,b_o$ - Affected by both the angle transform and true absorption.

We shall let the code search for the best value of these four parameters while we "control" the geometry by hand. We start from the body density given by electron scattering with the finite size of the proton removed. The data is from the work of the Los Alamos - South Carolina - Oak Ridge - V.P.I. Collaboration at Los Alamos. The ^{16}O data at 50 MeV is published[28] and the rest is in press.

Even though we have only been discussing a first order optical potential we can include the Lorentz-Lorenz effect for a Woods-Saxon potential by a trick. Recall that the replacement is

$$b_1\rho \rightarrow \frac{b_1\rho}{1+b\rho} \qquad \text{with} \qquad b = \xi\hat{b} \quad .$$

Thus we have for a Woods-Saxon shape

$$\frac{b_1\rho}{1+b\rho} = \left(\frac{b_1 N(c,a)}{1+e^{(r-c)/a}}\right)\left(1+\frac{bN(c,a)}{1+e^{(r-c)/a}}\right)^{-1}$$

$$= \frac{b_1 N(c,a)}{1 + bN(c,a) + e^{\frac{r-c}{a}}}$$

$$= \frac{b_1 N(c,a)}{1+bN(c,a)} \frac{1}{1 + e^{\frac{r-c'}{a}}}$$

where

$$c' \equiv c + a\ln\,(1 + bN(c,a))$$

and N(c,a) is the normalization constant required to make

$$4\pi \int_0^\infty r^2 dr\; \rho(r) = 1$$

Thus the effect of this correction is simply a renormalization and a change in the radius leaving the diffuseness fixed. Since we know c and a, if we also know the c' needed to fit the data we may calculate b and hence the effective fraction ξ.

The search procedure will be as follows: For a given α a curve of minima on b_o, b_1 will be traced as a function of c'. These will form a sequence of parabolas for a number of values of α. Such a collection of curves is shown in figure 5, taken from reference 29. We note that it is possible to get an excellent fit for a number of different values of α. However if we take α = 3000 MeV/c (corresponding to a very short range for the pion-nucleon interaction range) we must take $\xi = 0$. For such a value of α we expect $\xi \gtrsim 1$. If we take the fit for α = 300 MeV/c we would be forced to take a value of $\xi \gtrsim 0.6$ while we would expect $\xi \lesssim 0.2$ for this long a range. Thus a solution must lie in the intermediate values.

Figure 6 shows a similar, but less extensive, analysis for ^{40}Ca. Here we see a value of $\xi \sim 0.5$ is indicated, not inconsistent with the first result. Table 1 summarizes the parameters and compares them with "expected" values. The value of $\mathrm{Re}b_1$ tends to favour a value of $\alpha \gtrsim 400$ MeV/c. The other parameters are consistent with our qualitative expectations.

Figure 7 shows a similar result for ^{16}O at 40 MeV where again α = 400 MeV/c corresponds to $\xi \sim 0.5$, and Figure 8 from the

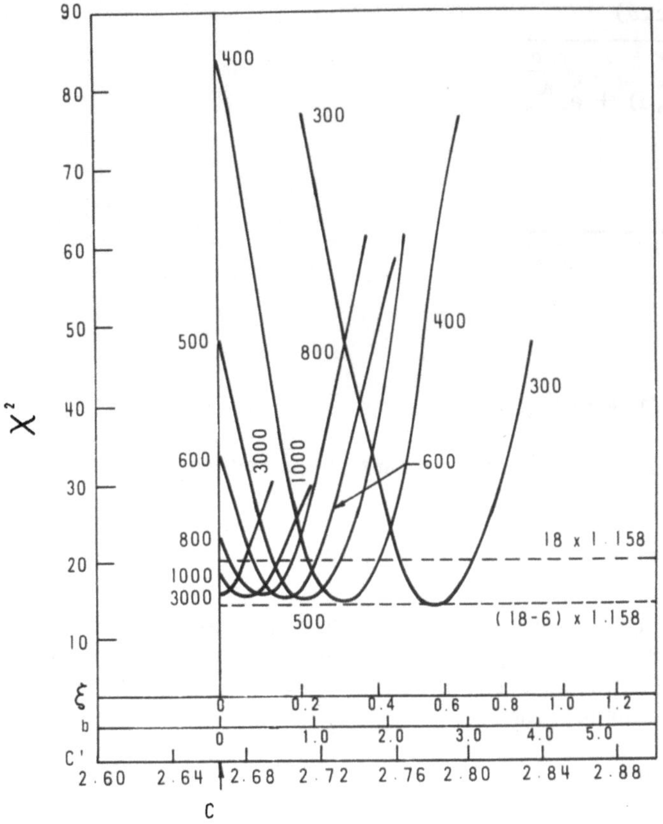

Fig. 5. χ^2 vs the triple parameter [c',b,ξ] for various values of the off shell parameter α. The nucleus is ^{16}O and the pion energy is 50 MeV. The dotted lines are drawn at the number of data points minus 6 (the number of parameters being varied) times the χ^2/η coming from a phase shift analysis. This provides a calibration for experimental errors.

TABLE 1

Parameters corresponding to the minima of parabolas (Fig. 4 & 5)

	α(MeV/c)	ξ	Reb_1	Imb_1	Reb_o	Imb_o
^{16}O	300	0.58	8.71	1.98	-4.06	-0.01
	400	0.32	7.83	1.65	-3.76	-0.01
	500	0.21	7.43	1.48	-3.62	-0.02
	600	0.15	7.19	1.38	-3.53	-0.06
	800	0.10	6.87	1.30	-3.43	-0.12
	1000	0.05	6.61	1.19	-3.34	-0.14
	3000	0.00	5.67	0.96	-3.09	-0.19
	Free		6.85	1.02	-1.01	0.79
	A.T. (q=k)		7.56	0.91	-2.55	0.50
	A.T. (q=500)		7.56	0.91	-6.76	-0.01
^{40}Ca	400	.48	7.96	1.24	-3.99	.02
	800	.56	7.60	1.22	-3.69	- .19

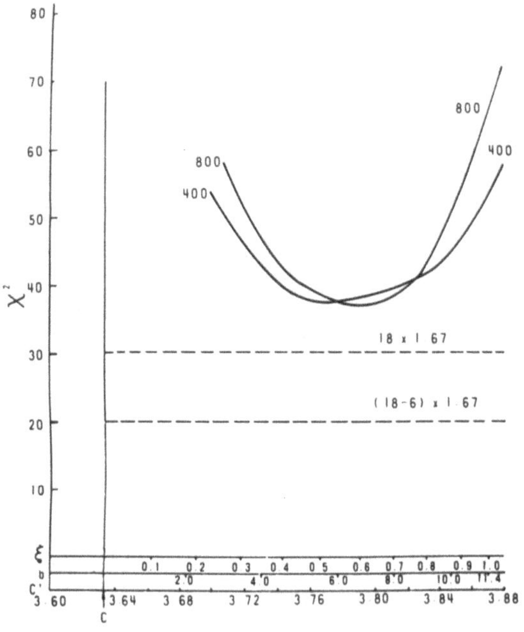

Fig. 6. Same as Figure 5 for ^{40}Ca at 50 MeV.

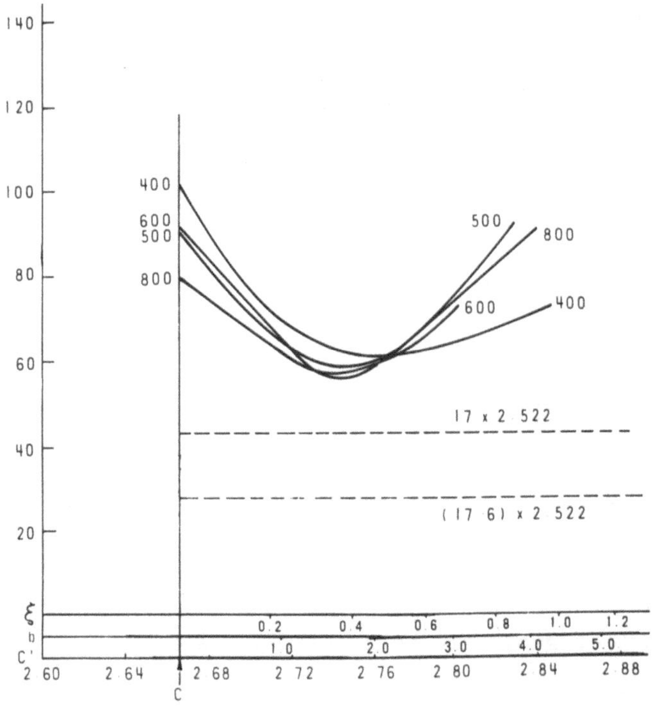

Fig. 7. Same as Figure 5 for ^{16}O at 40 MeV.

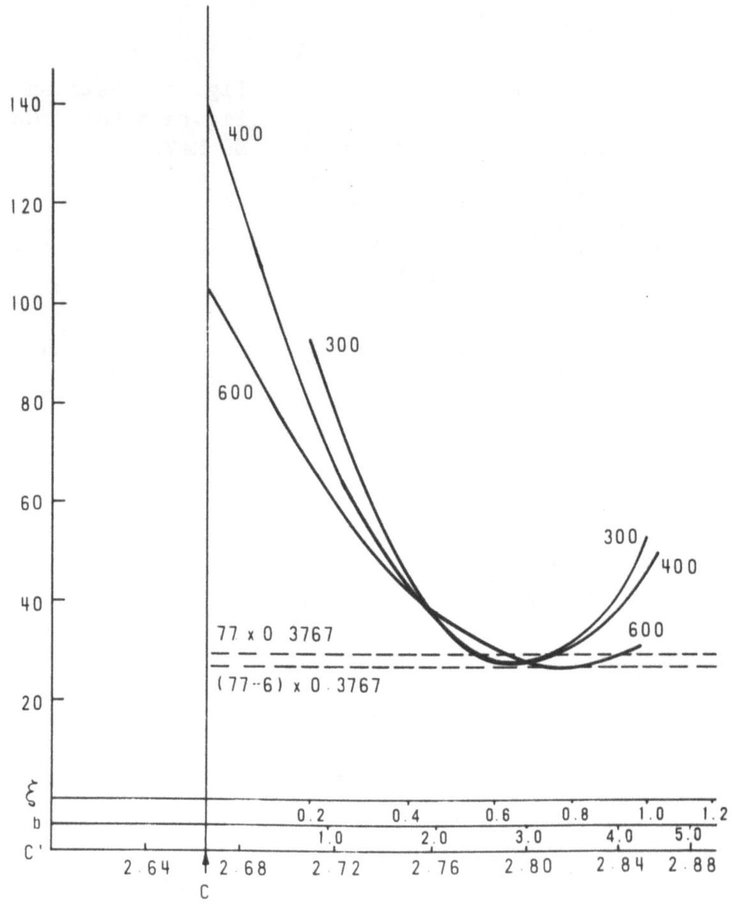

Fig. 8. Same as Figure 5 for ^{16}O at 80 MeV.

data of ref. 27 shows a slightly larger value of $\xi \sim .65 - .8$.

These fits show a consistent picture. We must remember that the picture is one of a larger c'. We have chosen to express this in the language of the Lorentz-Lorenz effect. Mixed in with this are other ρ^2 effects as well. Thus we have shown that while an interpretation in terms of the Lorentz-Lorenz effect alone is consistent with the data it is certainly not required by the data--although some sort of geometric correction is required.

We have made a small study of the bound states discussed by Dr. Ericson. The new element that we are including is the finite range of the pion-nucleon interaction in the off-shell t-matrix. This has the effect of always giving a finite number of bound states. It also requires a large value of the strength before binding can take place, reducing considerably the chances of these states existing. We also find that the bound states with the most structure seem to be least bound so that the order is reversed from the zero range case. How one case limits to the other is not well understood at the present.

I will now present the results of a short study to determine how the α-ξ ambiguity will affect the attempt to determine the neutron-proton radius difference by $\pi^{\pm}$ scattering at low energies. That is to say, if we are not able to tell with certainty the right value of α will there be an appreciable error introduced in the radius difference?

We start by finding equivalent fits to real data with α = 800 MeV/c and α = 400 MeV/c to get the corresponding c'.

We find for these values of c':

		^{16}O	^{40}Ca
π^+	800	2.687	3.82
	400	2.735	3.84

We now generate π^- "data" by assuming that 800 MeV/c is the true value and that the neutron radius is the same as the proton radius. We then fit this "data" insisting that α = 400 MeV/c. We find

		^{16}O	^{40}Ca
π^-	400	2.653	3.78

Thus the error in c' due to using the wrong value of α is

For ^{16}O 2.735 - 2.653 = 0.082 fm

For ^{40}Ca 3.84 - 3.78 = 0.06 fm ,

This would be a substantial error indeed. If this comparison is indeed so sensitive to α, might we use a case in which $r_n = r_p$ to

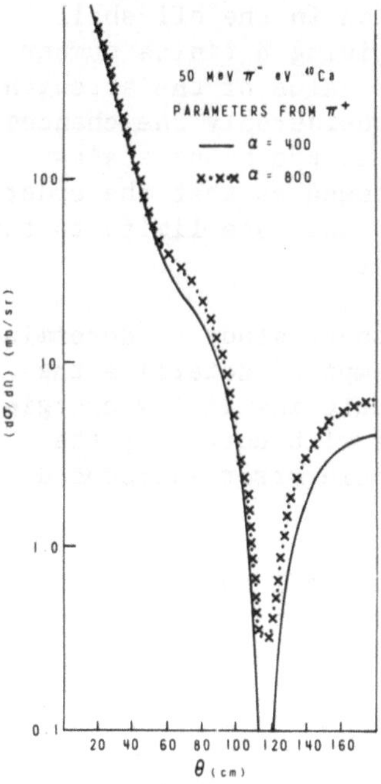

Fig. 9. Two predictions of π^- scattering on ^{40}Ca at 50 MeV. The two curves correspond to strengths which give identical fits to the π^+ data.

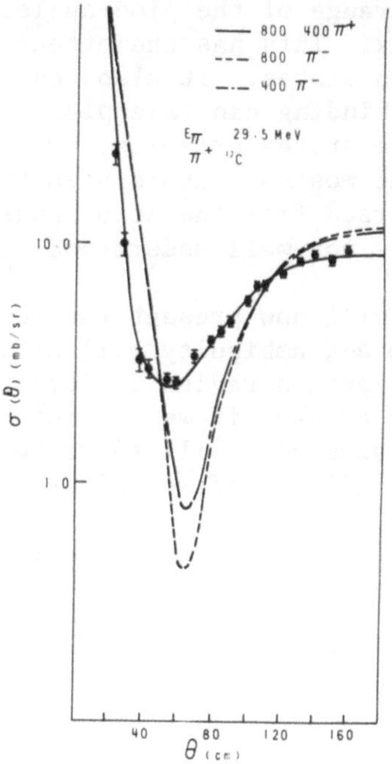

Fig. 10. Elastic scattering fit for π^+ on ^{12}C at 30 MeV and the two π^- predictions corresponding to two different values of the range parameter α. The points are preliminary data from the Los Alamos - South Carolina - Oak Ridge - V.P.I. collaboration.

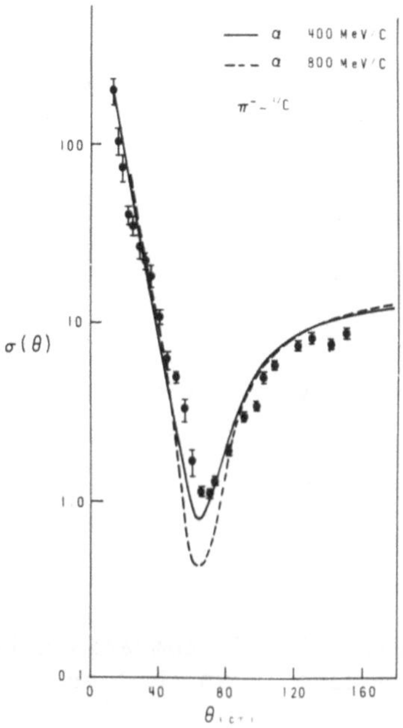

Fig. 11. Comparison of the two predictions in Figure 10 with preliminary data from TRIUMF of π^- elastic scattering on ^{12}C at 30 MeV (T. Masterson, private communication).

measure α?

If we determine the strong parameters from a fit to π^+ data with two different values of α and use those parameters to predict π^- scattering by simply changing the sign of the charge we get two different predictions. Figure 9 shows the result of this process for 50 MeV scattering on ^{40}Ca. As may be seen there is considerable sensitivity. To actually determine α by this procedure is a little more difficult since we have assumed that the strong parameters do not change while the effective energy of the pion in the nucleus will change with coulomb effects and this will, in general, change the strong parameters. However this may not be such a difficult complication since we are near the scattering length-scattering volume limit so that the strengths are slowly changing and their variation can be established empirically.

Figure 10 shows the fits to the π^+ ^{12}C data as well as the two predictions for the π^- scattering. While the differences between the two curves are not as great this case has the advantage that there exists π^- scattering data.

Figure 11 shows the comparison. While the value α = 400 MeV/c clearly shows better agreement in the minimum region neither curve is a good enough representation of the data to instill confidence that we have at last determined the elusive range α. Note that we are attempting to find the short range part of the pion-nucleus (and pion-nucleon) interaction by comparing it with the known Coulomb potential.

We are continuing studies of the type discussed here and we look forward to the continued arrival of new data to test various aspects of the pion-nucleus optical model.

REFERENCES

1. J.-P. Egger et al., Phys. Rev. Lett. 39 (1977), 1608.

2. T.E.O. Ericson and F. Myhrer, Phys. Lett. 74B (1978), 163.

3. L.S. Kisslinger, Phys. Rev. 98 (1955), 761.

4. Auerbach, Fleming and Sternheim, Phys. Rev. 162 (1967), 1683.

5. M. Ericson and T.E.O. Ericson, Ann. Phys. (N.Y.) 36 (1966), 323.

6. M. Krell and T.E.O. Ericson, Nucl. Phys. B11 (1969), 521.

7. W.R. Jones and J.M. Eisenberg, Nucl. Phys. A154 (1970), 49.

8. B. Goplen, W.R. Gibbs and E.L. Lomon, Phys. Rev. Lett. 32 (1974), 1012.

9. G.A. Miller and S.C. Phatak, Phys. Lett. B51 (1974), 129.

10. R. Landau and F. Tabakin, Phys. Rev. D5 (1972), 2746; J.T. Londergan, K.W. McVoy and E.J. Moniz, Ann. Phys. 86 (1974), 147; M. Reiner, Phys. Rev. Lett. 38 (1977), 1467.

11. J.M. Eisenberg, J. Hufner and E.J. Monitz, Phys. Lett. 47B (1973), 381.

12. H. Garcilazo, Nucl. Phys. A302 (1978), 493.

13. G. Baym and G.E. Brown, Nucl. Phys. A247 (1975), 395.

14. M.L. Adelberg and A.M. Saperstein, Phys. Rev. C5 (1972), 1180.

15. R. Landau, Phys. Lett. 57B (1975), 13; W.R. Gibbs, B.F. Gibson, A.T. Hess and G.J. Stephenson, Jr., Phys. Rev. C13 (1976), 2433.

16. L. Heller, G.E. Bohannon and F. Tabakin, Phys. Rev. C13 (1976), 742.

17. R.H. Landau and A.W. Thomas, Nucl. Phys. A302 (1978), 461.

18. G.A. Miller, Phys. Rev. C14 (1976), 361.

19. H. Garcilazo and W.R. Gibbs, Los Alamos preprint LA-UR-78-897.

20. L.C. Liu and C.M. Shakin, Phys. Rev. C16 (1977), 333.

21. L.S. Kisslinger and W.L. Wang, Ann. Phys. (N.Y.) 99 (1976), 374.

22. J.B. Cammarata and M.K. Banerjee, Phys. Rev. C13 (1976), 299; Phys. Rev. D16 (1977), 1334.

23. M. Theis, Phys. Lett. 63B (1976), 43.

24. Digiacomo, Rosenthal, Rost and Sparrow, Phys. Lett. 66B (1977), 421.

25. Zeidman et al., Phys. Rev. Lett. 40 (1978), 1316.

26. Ingram et al., Phys. Lett. 76B (1978), 173.

27. Albanese et al., Phys. Lett. 73B (1978), 119.

28. Malbrough et al., Phys. Rev. C17 (1978), 1395.

29. W.R. Gibbs, B.F. Gibson and G.J. Stephenson, Jr., Phys. Rev. Lett. 39 (1977), 1316.

RECENT DEVELOPMENTS IN THE DESCRIPTION OF THE NNπ SYSTEM

A.S. Rinat

Department of Physics, Weizmann Institute of Physics

Rehovot, Israel

ABSTRACT

A survey is given of recent developments in the description of the NNπ system. Starting from a non-relativistic potential model for πd-induced reactions we discuss covariant potential models, field theories where pion-absorption channels are included and finally the role of extraneous degrees of freedom, such as vector mesons.

1. Preamble

In spite of the fact that the first treatments of the NNπ system are well over 20 years old, there is a constant stream of publications on the topic. At least two major incentives seem to explain this interest.

1) Experimental facts on the NNπ system, or more precisely, on πd-induced reactions leading to πd, NN, NNπ - final states, are best known in the region $T_\pi \lesssim 300$ MeV[1,2]. In that region pions are relativistic but nuclear energies are still comfortably below the nuclear mass. Compared with a description of the Nd system at comparable energies one will have to cope with some complications due to relativity, not met till now.

Again different from the 3N system, one deals with pions which can be absorbed and emitted even at zero energy and a description of that feature will require a framework beyond a simple potential model which suffices for the 3-nucleon system. Finally, while the latter can be well described by phenomenological pair interactions without a microscopic description of its origin, this is not the case for the NNπ system where the pion is not only projectile, but is also in part responsible for the NN force. It is then not a priori clear whether the π and N explicit degrees of freedom will suffice. All these aspects make the NNπ system in itself an intriguing, theoretical laboratory worth being explored in its own right.

2) In a description of nucleon-induced reactions, one usually tries to express amplitudes for (p,p), (p,p'), (p,2p) etc. processes in the simplest effective interactions, viz. the NN (off-shell) scattering matrix.

The very aspects mentioned under 1) and in particular the fact that pions can be scattered as well as absorbed, renders the πN scattering matrix insufficient as an effective interaction. The absorption process is more likely to occur on two nucleons and πNN scattering matrices play the role of minimal effective interactions.

Since π-nucleus scattering will be discussed in other lectures, only the NNπ system itself will be treated. Regarding its scope we shall stay far from an exhaustive and complete review, and instead present ideas and developments with minimal technical complication. The few cited references should suffice as a guide to the literature.

2. Non-relativistic three-body potential model

Consider πd-induced reactions with pion energies low enough to neglect relativistic effects. For the moment we also disregard the possibility of π absorption (π+d→N+N). It is then attractive to figure NNπ as a non-relativistic (NR) three-body system interacting through pair forces. Complete descriptions have been given for these systems which we shall not reproduce in full here. Below we only give an outline in order to introduce a widely used nomenclature: for details we refer to an extensive literature (see, for instance, ref 3).

Consider the Hamiltonian of a non-relativistic three-body system $H = H_o + \Sigma_\alpha v_\alpha$. The index α refers to a pair, the particle not present in the pair or a three-body channel formed by a pair α and the remaining spectator particle.

We start with the two-body system. For any given interaction v, the Lippmann-Schwinger equation ($G_o(z) = (z-H_o)^{-1}$)

$$\langle\vec{p}|t(E)|\vec{p}'\rangle = \langle\vec{p}|v|\vec{p}'\rangle + \int\frac{d\vec{p}''}{(2\pi)^3} \langle\vec{p}|v|\vec{p}''\rangle G_o(E,p'')\langle\vec{p}''|t(E)|\vec{p}'\rangle \quad (1)$$

provides the corresponding scattering matrix t. The latter is in principle off-shell, i.e. $E \neq p^2/2\mu, p'^2/2\mu$, with μ, the reduced mass of the pair.

The scattering, say 3+(12) → 1+(23) and in general $\alpha \rightarrow \beta$, is then computed in a standard fashion as[3]

$$<\vec{q}_\alpha|T_{\alpha\beta}|\vec{q}\beta> = <\vec{q}_\alpha\phi_\alpha|U_{\alpha\beta}|\vec{q}_\beta\phi_\beta> \tag{2a}$$

$$U_{\alpha\beta}(z) = (1-\delta_{\alpha\beta})(z-H_o-v_\alpha)+\sum_{\gamma\neq\alpha} v_\lambda[1+G_0(z)\sum_{\delta\neq\beta} v_\delta] \tag{2b}$$

$U_{\alpha\beta}$ in (2a) is a transition operator, of which matrix elements have to be taken by product eigenstates of $\vec{q}_\alpha$, the relative momentum of pair and spectator and of ϕ_α, the wave function of the pair. The latter is in momentum space as follows related to the so-called form factor g and binding energy $-|\varepsilon_\alpha|$

$$\phi_\alpha(p_\alpha) = -(p_\alpha{}^2/2\mu + \varepsilon_\alpha)\ g_\alpha(p_\alpha) \tag{3}$$

Next one eliminates by means of (1) the potentials in [2b] and this results in coupled integral equations for $U_{\alpha\beta}$ in two vector variables $\vec{p}_\gamma\vec{q}_\gamma$,

$$U_{\alpha\beta}(z) = (1-\delta_{\alpha\beta})G_o^{-1}(z) + \sum_{\gamma\neq\alpha} t_\gamma(z)U_{\gamma\beta}(z) \tag{4}$$

Eq. (4) features the t matrix (1) as effective interaction replacing v itself. The enormous difficulties encountered in a numerical solution of (4) are considerably relieved if an assumption is made on v. If these interactions are separable, i.e. if

$$v(p,p') = \lambda g(p)\ g(p') \tag{5}$$

so is the solution t of eq. (1)

$$<\vec{p}|t(z)|\vec{p}'> = g(p)\ G(z)\ g(p'), \tag{6}$$

where G is the propagator of the interacting pair

$$G^{-1}(z) = \lambda^{-1} - \int \frac{d\vec{p}}{(2\pi)^3}\, g^2(p)\ G_0(z,p) \tag{7a}$$

$$= (z+|\varepsilon|) \int \frac{d\vec{p}}{(2\pi)^3}\, g^2(p)\ G_0(z,p)\ G_0(-|\varepsilon|,p) \tag{7b}$$

The second form holds if the potential v supports a bound state of energy $-|\varepsilon|$. In that case one can show that the potential component g(p), eq.(5) is just the form factor defined in (3). It should further be noted that form factors (but not wave functions) remain defined for resonances.

The representation (6), exact for a potential of the form (5), is approximately valid for any v provided the energy z is in the neighbourhood of a bound state or resonance. This fact sometimes justifies the use of (6). Also, the phase shift of a separable potential resembles the one for short-range forces. Substituting (6) into (4) one easily succeeds to turn (4) into coupled integral equations directly for amplitudes, viz

$$\langle\vec{q}_\alpha|T_{\alpha\beta}(E)|\vec{q}_\beta\rangle = \langle\vec{q}_\alpha|B_{\alpha\beta}(E)|\vec{q}_\beta\rangle + \sum_{\gamma\neq\alpha}\int\frac{d\vec{q}_\gamma}{(2\pi)^3}\langle\vec{q}_\alpha|B_{\alpha\gamma}(E)|\vec{q}_\gamma\rangle$$

$$\times G_\gamma(E,q_\gamma)\langle\vec{q}_\gamma|T_{\gamma\beta}(E)|\vec{q}_\beta\rangle \qquad (8)$$

$G_\gamma(E,q_\gamma)$ is the interacting two-body propagator (7) in the presence of a non-interacting spectator, i.e. with $E\rightarrow E - q_\gamma^2/2\mu_\gamma$, with μ_γ the reduced mass of the pair γ and the spectator. $B_{\alpha\beta}$ appears as a driving or Born term and can be shown to be

$$B_{\alpha\beta}(z) = (1-\delta_{\alpha\beta})\langle g_\alpha|G_o(z)|g_\beta\rangle \qquad (9)$$

and is thus made up of form factors in the two channels it connects. Figure 1 shows graphically its interpretation as a single-particle exchange graph: pair (12) dissociates and (23) receives the exchanged particle 2 after propagation.

The so-called Mitra-Amado-Lovelace eq. (8) are coupled channels equations of a generalized Lippmann-Schwinger type in one vector variable[4]. Its iterative solution is readily seen (Fig. 2) to amount to the summation of any number of repeated, single particle exchanges B (Fig 1) between pairs. In between exchanges $B_{\alpha\gamma}, B_{\gamma\beta}$, the formed pair and non-interacting spectator γ propagate as discussed after eqs. (8). A partial wave analysis is possible for any conceivable spin configuration[5], but amplitudes are generally diagonal only in J^π,I, i.e. in total angular momentum, parity and isospin. When not conserved, eqs (8) become after partial wave analysis coupled integral equations in $|\vec{q}_\gamma|$, channel angular momentum and spin (LS) and whatever quantum numbers left in 'γ' characterizing the 3-body γ channel.

At this point we make an interlude. It is well-known that the

Lippmann-Schwinger eq. (1) is the most general equation satisfied by a non-relativistic two-body amplitude which is generated by a real potential v. One can also show that NR elastic and rearrangement amplitudes $T_{\alpha\beta}$ satisfying three-body unitarity must be coupled as in eq (8). provided one deals again with real, and now separable pair forces. In different words, the requirement

$$T_{\alpha\beta} - T^{+}_{\alpha\beta} = i\sum_{\gamma}\int T_{\alpha\gamma} \text{ (phase space)} T^{+}_{\gamma\beta} \qquad (10)$$

(where γ contains also open break-up channels) essentially implies eq. (8), provided the same arbitrary formfactors which enter the Born terms $B_{\alpha\beta}$, eq. (9) build the propagators G_{γ} as prescribed in eq (7). We shall return to the embodiment of unitarity requirements. For the time being we apply the theory just outlined to the NR NNπ system disregarding for a moment the observed NN channel. A solution of (8) is then possible once the input two-body channels have been selected.

Since the deuteron is the only stable target possible one needs at least the $^3S_1-^3D_1$ NN channel ('d'). Other NN partial waves do have appreciable phases in the interesting energy region, but appear to be of secondary importance and will be disregarded.

By far the most important πN wave is of course P_{33} ('Δ') with, in particular S_{31} and S_{11} as next important perturbations. Again for simplicity we shall only retain the Δ although its dominance occurs in the region $250 \geq T_{\pi}(\text{MeV}) \gtrsim 70$ where the π actually is relativistic. The set $(\underset{\sim}{8})$ then becomes ($T_{d\Delta} = T_{\pi d,N\Delta}$ etc.).

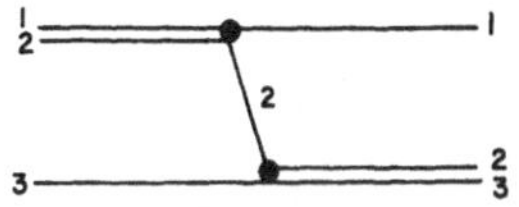

Fig. 1. Graph representing exchange of particle 2 in a Born or driving term B[3+(12) → 1+(23)].

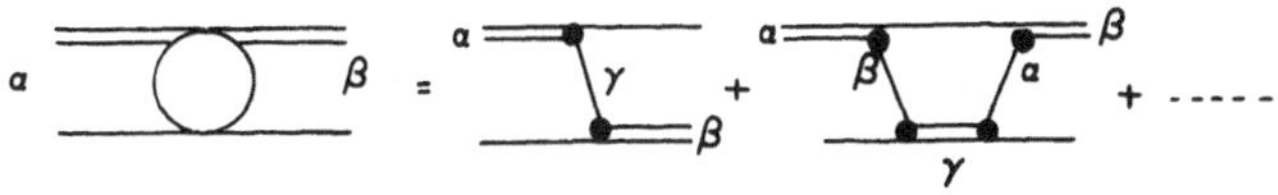

Fig. 2. Iteration of Mitra-Amado-Lovelace coupled equation for amplitude $T_{\alpha\beta}$.

$$T_{dd} = 2B_{d\Delta}\, G_\Delta\, T_{\Delta d}$$

$$T_{\Delta d} = B_{\Delta d} + B_{\Delta d}\, G_d\, T_{dd} + B_{\Delta\Delta}\, G_\Delta\, T_{\Delta d} \tag{11}$$

with the factor 2 accounting for the equivalence of the two nucleons. Notice further the absence of $B_{dd} = B_{\pi d,\pi d}$ since no π or N can be exchanged between the two πd configurations. Eq. (11), when necessary, generalized to encompass other relevant πN waves, should indeed give a reasonable description of elastic πd scattering.

Were it not for the fact that one deals with an exact solution for composite target scattering where the major two-body channel is governed by an outspoken resonance, πd scattering at this stage would hardly be of greater interest than low-energy Nd scattering at comparable energies. However:

i) Around the interesting Δ resonance energy region, kinetic energies are of order m_π. Is it possible to formulate a proper relativistic framework for πd scattering?
ii) The pion can be absorbed and emitted on a single nucleon and the NN system will thus occur as a possible intermediate or final state. Can one incorporate these states in a three-body system?
iii) Are the pionic and nucleonic degrees of freedom sufficient in order to describe the πd reactions?

We shall now enter these questions and outline what insight the answers have been provided in the past few years.

3. A relativistic potential model for the NNπ system

The only framework permits a proper relativistic treatment of interacting particles is field theory or utilization of demonstrable analytic properties of amplitudes in dispersion relations. In field theories particles may be emitted and absorbed and it often makes no sense to consider a fixed number of particles.

In our case it is of course the number of pions which is not conserved. It is, however, meaningful to ask whether a relativistic theory can be formulated, if again the NN channel resulting from the possibility of π absorption or emission, is disregarded.

There are, of course, other aspects of relativity, namely the creation of $N\bar{N}$ pairs and the like. At this point one should emphasize that in the energy region on which we focus no real pair creation is possible. Since thresholds are relatively remote it seems reasonable to neglect also virtual pairs. This definitely is some violation of relativistic aspects, but under the conditions of interest, probably a minor one.

Neglect of $N\bar{N}$ pair formation and, for the moment disregard for changes in the number of pions brings us back to a theory where particles colliding will also emerge, i.e. characteristic features of a potential theory. We now recall the observation on the sufficiency of unitarity, which provides a frame for a non-relativistic scattering theory. Is the requirement of two and three particle unitarity (10) sufficiently restrictive to lead to a unique covariant formulation?

Consideration of a two-body channel alone shows that the answer is negative, because contrary to a non-relativistic theory, relativistic particles may propagate with positive as well as negative energies: unitarity can only restrict positive energy amplitudes. Yet it has been shown that classes of two-body equations exist which are both covariant and satisfy two-body elastic unitarity, and thus meet minimal requirements for energies below production thresholds. The one used most frequently is the Blankenbecker-Sugar[6] equation, which in the 2-body CM system reads ($\varepsilon_{1,2} = (\vec{k}''^2 + m_{1,2}^2)^{\frac{1}{2}}$, s = total energy squared)

$$\langle\vec{k}|T(s)|\vec{k}'\rangle = \langle\vec{k}|V|\vec{k}'\rangle + \int \frac{d^3\vec{k}''}{8\pi^3} \frac{\varepsilon_1+\varepsilon_2}{2\varepsilon_1\varepsilon_2} \frac{\langle\vec{k}|V|\vec{k}''\rangle\langle\vec{k}|T(s)|\vec{k}'\rangle}{s-(\varepsilon_1+\varepsilon_2)^2} \qquad (12)$$

The above is obviously a Lippmann-Schwinger type of equation, to which it of course reduces in the non-relativistic limit. Likewise it has been shown [7-9] that if v in (11) is separable there are classes of three-body equations like (8) from which three-body unitarity can be derived. Thus, in spite of indeterminancies and ambiguities,minimal requirements can be satisfied. The effect of these ambiguities can of course not be assessed in a general fashion. In the few cases studied, they seem to be minimal[9,10].

We shall not enter here a detailed discussion of the three-body Blankenbecker-Sugar equations, which are indeed of the form (8). We only wish to warn the uninitiated that there is more to it than the replacement of $d\vec{k}$ by its covariant analogue $d\vec{k}/2E$ as in the two-body case (cf. eq (11)). The latter equation holds in a two-body CM frame, and the, generally off-shell, solutions have to be embedded in a three-body CM system[7,8]. A galilean transformation suffices in the NR case, but Lorentz boosts in a relativistic theory are well-defined only for on-mass shell particles. The crux of a covariant three-body theory is thus a proper definition of relative momenta, which for separable forces enter in (8) as arguments of form factors g and as a consequence driving terms $B_{\alpha\beta}$. Again, ambiguous prescriptions can be given all leading to equations of the type (8), but we forego a discussion of the accompanying technical difficulties[9]. In conclusion, one can point at possible covariant descriptions of the NNπ system restricted as explained above. This was to be expected since the spelled-out simplifications actually bring us back to a strict three-body potential model, for which a, now, covariant description is demanded.

4. The NN absorption channel and the breakdown of potential theories.

Genuine field theoretical models.

Since the NN channel is already open at the πd threshold we are first interested in a NR NNπ theory, which now also incorporates NN states.

Afnan and Thomas[11] argued that one can still use a potential model, provided one includes the $P_{11}\pi N$ channel, which carries the N as a direct-channel pole. When interpreted as a πN bond state, NN, together with NΔ states will emerge.

Let us disregard for the moment the questionable point whether the dressing of the nucleon can be mocked-up by πN scattering data as is required by the model. The latter, however attractive, suffers then from at least one serious drawback. In intermediate NN states one N is elementary while one is dressed. The resulting violation of the Pauli principle is illustrated in figs. 3. The graph shown in fig. 3a represents second order πd scattering through a NN like it would for NΔ. Figure 3b is the part obtained from 3a by interchange of two equivalent N but it is absent in the Afnan-Thomas model. It also clearly depicts the absorption on one and reemission by a different nucleon: no potential model can generate such a mechanism, which belongs to the realm of a genuine field theory.

Since we cannot hope to construct exact solutions starting from elementary fields, one is led to short-cuts. Can one, for instance, just add the NNπ vertex to the potentials already used? The answer is negative as inspection of fig. 4 shows. The graph can be read as i) a pion rescattered in a P_{11} interaction, ii) a NN interaction through exchange after initial π absorption and final emission. Application of both mechanisms leads to overcounting of contributions.

By far the most thorough treatment of a theory free of overcounting is found in a Ph.D. thesis by Mizutani[12] and in a subsequent paper[13]. Consider a Hamiltonian

$$H = \bar{H}_o + U_{NN\pi} + V'_{NN} + W'_{\pi N} \tag{13}$$

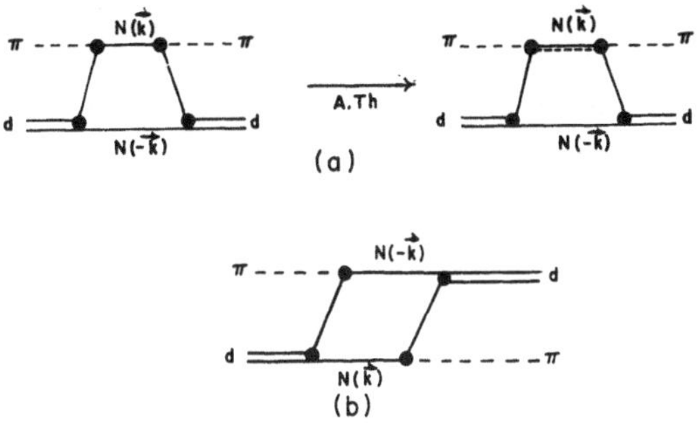

Fig. 3. a) Graph showing a second order πd elastic amplitude proceeding through a NN intermediate state. The upper N is in the model of Afnan and Thomas[11] a πN bound state. (b) The counterpart of the graph in fig. 3a, required by the Pauli principle, but which is missing in the Afnan-Thomas model.

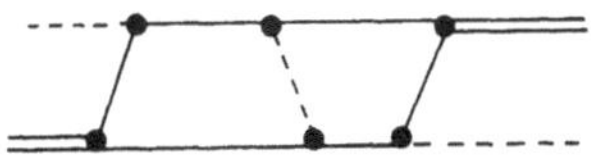

Fig. 4. A third order πd elastic scattering graph which can be interpreted as pion rescattering through the P_{11} Nπ states, or alternatively as a NN interaction through a one-pion exchange potential preceded by pion absorption and followed by pion emission.

$U_{NN\pi}$ describes π absorption (emission), thus in terms of nuclear and pion creation and destruction operators

$$U_{NN\pi} = \int [g(k,k'q)a^{+}_{k}a_{k'}b_{q} + h.c.]\, d\vec{k}\, d\vec{h}'d\vec{q} \tag{14}$$

V'_{NN} is the NN interaction with parts generated by U removed. The latter are for instant the one-pion-exchange, crossed two-pion-exchange contributions, etc.

Likewise is $W'_{\pi N}$, the πN interaction similarly truncated, and one has, for instance, to remove the direct N pole in the P_{11} πN partial wave (Fig.5a). The vertex part $U_{NN\pi}$ of course also generates a corssed N pole which contributes to all direct-channel partial waves. From Chew-Low theory we know for instance that the Δ in the P_{33} channel heavily draws on that crossed pole (Fig.5b) and its projection onto the P_{33} channel is thus absent in $W'_{\pi N}$.

Mizutani and Koltun then show how in principle one can calculate all desired amplitudes in perturbation theory. A second order elastic πd amplitude will now contain both contributions shown in figs. 3a,b. Quite generally one recovers in spite of a non-appealing initial division of interactions as in (13) all potential model results,* now properly corrected for genuine absorption and emission options.

Fig. 5. a) The direct pole in the P_{11} πN partial wave which is removed from $W_{\pi N}$. (b) The crossed nucleon pole, which contributes to all partial waves and for instance builds the Δ in the P_{33} partial wave.

A disadvantage of the Mizutani approach is that, contrary to the 3-body Afnan-Thomas model, one obtains series expansion and not closed, coupled channel equations for the relevant amplitudes.

That result does emerge if one assumes the dominance of one-pion intermediate states. In that case one can solve the field theoretical model defined by (13) and two sets of coupled integral equations have been derived[19]

$$2^{-\frac{1}{2}} T_{dN} = B_{dN}(1+G_o t_{NN}) + B_{d\Delta} G_\Delta T_{\Delta N}$$

$$T_{\Delta N} = B_{\Delta N}(1+G_o t_{NN}) + B_{\Delta\Delta} G_\Delta T_{\Delta N} + 2^{\frac{1}{2}} B_{\Delta d} G_d T_{dN} \tag{15}$$

and

$$2^{-\frac{1}{2}} T_{dd} = T_{dN} G_o B_{Nd} + T_{d\Delta} G_\Delta B_{\Delta d}$$

$$T_{d\Delta} = 2^{\frac{1}{2}} B_{d\Delta} + T_{dN} G_o B_{N\Delta} + T_{d\Delta} G_\Delta B_{\Delta\Delta} + 2^{\frac{1}{2}} T_{dd} G_d B_{d\Delta} \tag{16}$$

In the first, one solves for the genuine absorption amplitude $T_{dN} = T_{\pi d,NN}$ and it, together with 'normal' Born terms drives the second set (16). Notice further the presence of G_o, the propagator of two free nucleons (cf. Fig.3a).

*With two equivalent elementary nucleons!

Alternatively, one may derive explicit expressions for T_{dN}, T_{dd} in terms of solutions $T^{o}_{\alpha\beta}$ of the potential model discussed lacking the NNπ vertex and thus NN states. The resulting equations read

$$T_{dN} = \{\sqrt{2}(1+T^{o}_{dd}G_{d})B_{dN} + T^{o}_{d\Delta}G_{\Delta}B_{\Delta N}\}(1+G_{o}t_{NN}) \tag{17}$$

$$T_{dd} = T^{o}_{dd} + T_{dN}G_{o}\ \{B_{N\Delta}G_{\Delta}T^{o}_{\Delta d}+\sqrt{2}B_{Nd}(1+G_{d}T^{o}_{dd})\} \tag{18}$$

and have been used in actual computations to be discussed below. Fig. 6 shows the lowest order πd→NN graphs and fig. 7 does the same for the absorption correction to πd elastic scattering.

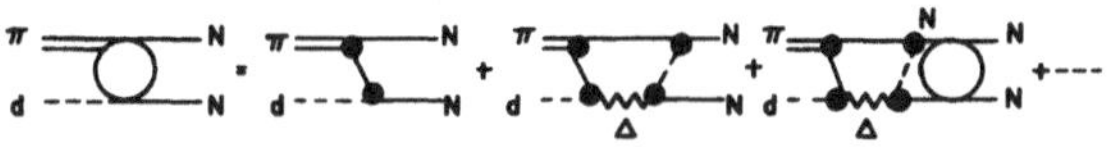

Fig. 6. Lowest πd → NN graphs: Direct absorption, rescattering through Δ(β), same with NN final state interactions, etc.

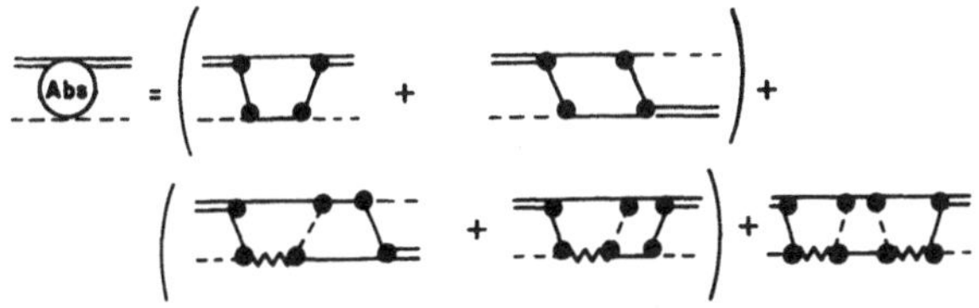

Fig. 7. Graphs showing lowest order absorption corrections to the πd elastic scattering amplitude.

We close this section with two remarks. First we stress that the model used in this section is a NR field theory. The relativistic version can be guessed comparing a NR with a covariant three-body potential theory. The same relation can be used to transform either set (15), (16) or (17), (18) in a covariant manner.

A more delicate point regards the one-pion approximation and the appearance in (15)-(18) of amplitudes, like $T_{d\Delta}$. We recall that the P_{33} πN channel, truncated as discussed does not lead to a Δ and that for instance a crossed two-pion graph as in Fig 5b is needed for its description. When using 'real' Δ's we obviously went beyond the one-pion approximation. Also when eventually correcting for crossing one should be aware that some terms have already been included (see ref. 14a for details).

5. The role of vector meson exchange

In this section we wish to elaborate on some selected multipion contributions, notably those embodied in correlated exchanges, like ρ exchange. From the onset one ought to emphasize that from this point on results are no more rigorous and we can only use plausibility arguments.

In particular the Stony Brook School vigorously advocated inclusion of ρ meson effects, since vector mesons, although much heavier than pions, couple much more strongly to N and Δ than do π. One has for instance $\frac{f^2_{NN\pi}}{4\pi} = 0.081$, $\frac{f^2_{\Delta N\pi}}{4\pi} = 0.34$ compared with $12.5 \gtrsim \frac{f^2_{\Delta N\rho}}{4\pi} \sim 2.9\ \frac{f^2_{NN\rho}}{4\pi} \gtrsim 10$ (ref. 15).

A basic theory of ρ-exchange effects would presumably have to start with a modification of the Mizutani hamiltonian (13), where ρNN vertices are added to π vertices and contributions to V'_{NN} and $W'_{N\pi}$ are subsequently substracted. Such a procedure is much more hazardous than in the Mizutani model, because the ρ is part of the 2π continuum.

Rather than looking for a selfconsistent theory, the advocates of a prominent ρ role of course point at its proven role in building the medium range NN interaction[16]. Further evidence comes from calculation of the total cross-section for $\pi d \to NN$ which has been computed up to second order (see fig. 6). Using realistic πNN vertex functions agreement with experimental values of $\sigma^{total}_{\pi d,NN}$ is only obtained if in addition to $\pi N \underset{\Delta}{\to} \pi N$ also $\pi N \underset{\Delta}{\to} \rho N$, i.e. $\pi\rho$ conversion is included in the second rescattering graph of fig. 6[17,18]. If indeed by inclusion of ρ no overcounting occurs, one can in principle extend also our formulation and add ρ to π exchange wherever possible, i.e. $B^{\pi}_{\Delta N} = B^{\pi}_{\Delta N,NN} \to B^{\pi+\rho}_{\Delta N,NN}$ and $B^{\pi}_{\Delta\Delta} = B^{\pi}_{\Delta N,\Delta N} \to B^{\pi+\rho}_{\Delta N,\Delta N}$.

Use of eqs. (15), (16) then guaranties inclusion of ρ exchange to any (and not only to second) order.

6. Numerical results

A. Input. The input of our theory consists in essence of form factors. Some of these like for instance the $\Delta N\pi$ ($\beta N\Delta$) or dNN vertices may be related to (near) elastic πN and NN scattering. One parametrizes the formfactor as appearing in (5) and solves the Blankenbecker-Sugar eq. (12). On-shell information, that is, the observed phases then provide best-fit values for the parameters in g(q).

In contradiction to the examples above, the NNπ, NNρ and $\Delta N\rho$ matrices cannot be related to elastic scattering data. Information on coupling constants comes from dispersion relations or from relations based on strong interaction symmetries. Regarding cut-off momenta one can only rely on theoretical models[15,17]. The three vertices above were all chosen to be of the form

$$g(q) = Nq/(q^2 + \beta^2) \tag{19}$$

with N fitted to the coupling constants $\frac{f^2}{4\pi} = 0.081$, 3.45 and 10, respectively. The cut-off momenta used are 1.0, 2.0 and 2.0 GeV/c[17].

It is characteristic of the discussed field theoretical models that $T_{\pi d,NN}$ and $T_{\pi d,\pi d}$ are not directly coupled as they would be in a potential model[11] (see however ref. 14a). One rather has to first solve for T_{dN}, which amplitude subsequently serves as input for T_{dd} (see for instance eqs. (17) and (18)). It thus seems natural to first discuss T_{dN}. Agreement with π absorption data on d, will give faith in the results for absorption corrections computed with (half off-shell) absorption amplitudes.

B. $\pi d \to NN$. The kind of data which exist throughout the resonance region are total cross-sections for numerous energies, about 10 angular distributions and a limited number of polarization measurements[2,9,20,21]. In by far the largest number of calculations, $T_{\pi d,NN}$ has been approximated by the first two terms in the expansion of (17), i.e. by the direct capture and the rescattering terms, with or without NN final state interaction[17,22,23]. Not all of these computations are comparable, because even within the so-called rescattering approximation different input can and has been used. A detailed comparison[24] lies outside the scope of these lectures and we therefore only list findings of several groups.

a) The major part of the integrated $\pi d \to NN$ cross-section is given by direct capture, and lowest order π rescattering including

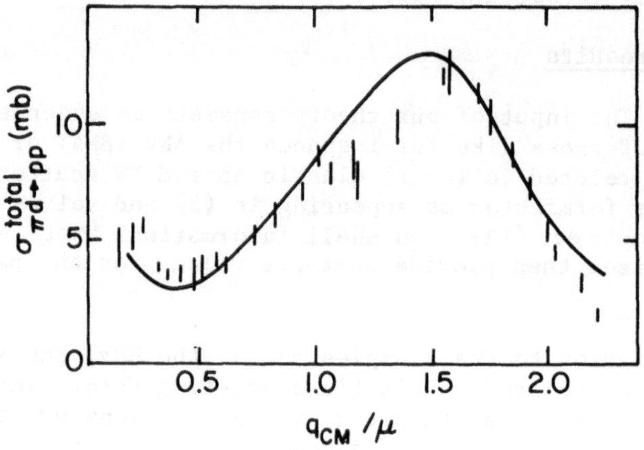

Fig. 8. Integrated cross-section for $\pi d \to NN$ as function of q_{CM}/m_π.

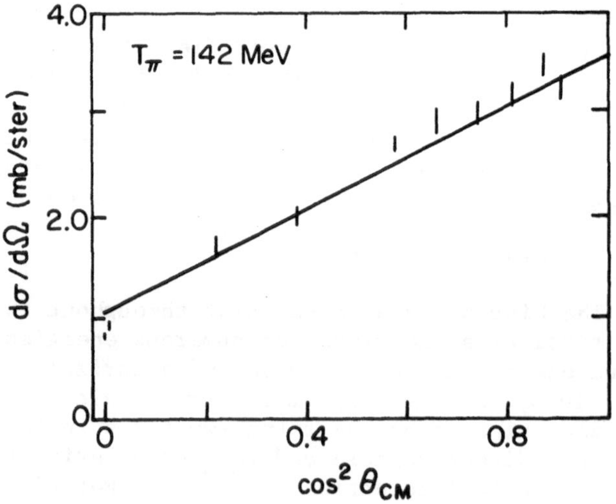

Fig. 9. Differential $\pi d \to NN$ cross-section for T_π = 142 MeV and prediction based on Eq. 18.

π-ρ conversion (Fig. 8). The influence of changes in parameters has extensively been studied in ref. 17.

b) Rescattering effects of higher than second order may amount to 20-30% changes in partial wave amplitudes[24].

c) NN final state interactions about influence as the on-shell phases suggest. Their effect can cause appreciable changes in both directions depending on the partial wave under study.

d) Whereas it is fairly easy to get agreement for total cross-sections differential cross-sections (essentially the slope of $d\sigma/d\Omega$ as function of $\cos^2\theta$) depends sensitively on the smaller S πN waves and even on the chosen parametrizations for those.

e) Some hitherto neglected high partial waves (NN F waves) contribute significantly to cross-sections.

The paramount feature of $\sigma^{total}_{\pi d,NN}$ is of course the peak which reflects the off-shell resonant P_{33} amplitude in the scattering terms. The downward shift of the peak and its precise form (including the upward trend towards $T_\pi \to 0$) are then caused by the embedding of the πN amplitude in the rescattering terms and by interference with the relatively small direct capture term B_{dN}.

In Fig. 9 we give the πd→NN differential cross-section for T_π = 142 MeV; similar results have been obtained for other energies[24]. The measure of agreement should be judged with caution, because several parameters are only approximately known. In particular $\beta_{NN\rho}$ ($\sim\beta_{\Delta N\rho}$) is thought to be uncertain by maybe 20% and the same is true for the parametrization of the off-shell S_{11}, S_{31} πN partial waves.

One should look forward to polarization data (mostly for $\vec{p}+\vec{p}\to\pi^+ +d$) which, when sufficiently accurate, may correlate uncertainties in the input. However, it seems that the original hope of indirectly extracting, what was once thought to be the major uncertainty namely the D-state probability of the d groundstate wave function[23] appears now to be somewhat naive.

C. πd elastic scattering

Elastic πd observables are from (18) seen to consist of two parts. Those which can be computed from a potential model ("$T^{(0)}$") and further absorption corrections $\propto T_{dN}$. Calculations of the former type with an ever more extended input regarding πN (NN) partial waves and relativistic effects have been performed in the past few years[9,25], and reasonable agreement with experiment has been obtained regarding differential and total cross sections. The outstanding exception is the minimum in the differential cross section for T_π = 256MeV where predictions overshoot the data by a factor ∿3-5*

*No such extreme discrepancy exists for the neighbouring energy T_π = 230MeV[26]!

Addition of absorption corrections, Eq (18), is off-hand thought to be of minor importance, since, say integrated $\pi d \to NN$ cross sections, are only a fraction of πd total cross-sections[2]. Yet the following argument invites for caution: The deuteron is a loosely bound system and for not too strong elementary forces multiple scattering is not very probable. This means that by and large the πd elastic amplitude is given by the πN amplitude, dominated by the Δ. The presence of the nucleons in a bound d causes smearing of the πN P_{33} amplitude over several πd partial waves and in general one will find $|\mathrm{Re}\ f^o_{\pi d}| \lesssim |\mathrm{Im}\ f^o_{\pi d}|$ as is the case for $f_{\pi N}$. Re $f^o_{\pi d}$ actually changes sign around $T_\pi \sim 160$MeV and is thus sensitive to any change of input. Indeed adding to the dominant P_{33} πN wave less important πN S and P waves sometimes causes relatively large changes in πd partial wave amplitudes. The same phenomenon is observed for absorption corrections. Their effects on elastic cross section is relatively minor and as a rule away from the agreement obtained without the corrections (Fig 10). As is the case for $\pi d \to NN$, integrated elastic and total cross-sections are in reasonable agreements with experiment.

The relatively violent changes in Re $f_{\pi d}$ will affect in particular polarizations as can be seen from Figs 11, where we display

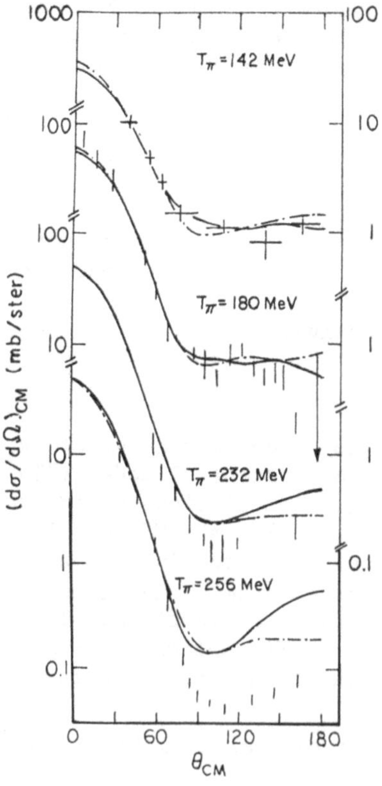

Fig. 10. Elastic πd cross-section for T_π = 142, 180, 232 and 256 MeV compared with predictions (P_D = 6.7%). $T^{(o)}$ (diag + non-diag --abs corrections included.)

tensor and vector polarizations computed for T_π = 142MeV (P_D=4, 6.7% respectively). One should look forward to planned experiments[27] for additional tests on the theory. One may expect that uncertainties in $T_{\pi d,NN}$ and πd polarizations or analyzing powers will become even more restrained.

7. Conclusion

There can be no doubt that great progress has recently been made in the understanding of πd induced reactions. We mentioned a number of problems which emerged and have been tackled. Yet it cannot be denied that the refinement of the description, although of course unavoidable, has not only been beneficial, because accurate assessment of their numerical influences is limited by insufficient knowledge of the input. For the time being it may well be that partial ignorance of the particular input parameters more than technical difficulties threaten to stall further progress. Yet we emphasize that even at this stage the ideas developed above already have found applications to a sounder description of absorption corrections to the pion-nucleus potential[14,28,29,30,31]. There is little doubt that for that purpose alone, the efforts spent on understanding the πNN system were both useful and necessary.

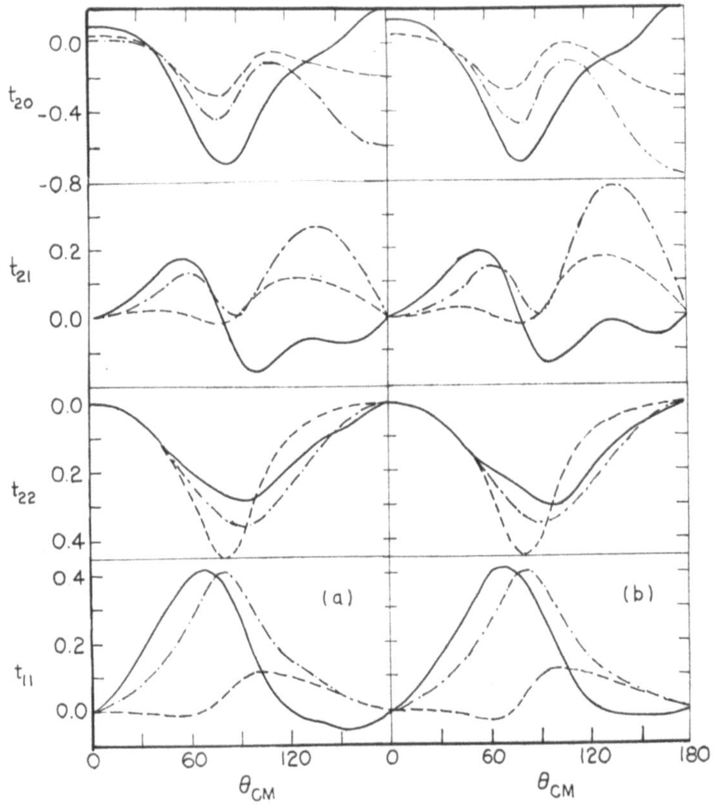

Fig. 11. Predicted tensor and vector polarizations (T_π = 142 MeV) in their dependence on P_D (4 and 6.7%). Legend as for Fig. 10 — T^o, only diagonal.

REFERENCES

1. E.G. Pewitt et al., Phys. Rev. 131 (1963), 1826; J.H. Norem, Nucl. Phys. B33 (1971), 512; R.H. Cole et al., Phys. Rev. C17 (1978), 681; K. Gabathuler et al., Nucl. Phys. B55 (1973), 397.

2. C. Richard-Serre et al., Nucl. Phys. B20 (1970), 413.

3. Modern Three-Hadron Physics, Editor A.W. Thomas, Springer-Verlag, 1977.

4. E.G.C. Lovelace, Phys. Rev. 135B (1964), 1225.

5. M. Stingl and A.S. Rinat, Nucl. Phys. A154 (1970), 613.

6. R. Blankenbecler and R. Sugar, Phys. Rev. 142 (1966), 1051.

7. D. Freedman, C. Lovelace and J.M. Namystowski, Nuovo Cim. 43A (1966), 258.

8. R. Aaron, R.D. Amado and J.E. Young, Phys. Rev. 174 (1968), 2022.

9. A.S. Rinat and A.W. Thomas, Nucl. Phys. A282 (1977), 365.

10. T.I. Kopaleishvili, A.I. Machavariani and G.A. Emelyanenko, Nucl. Phys. A302 (1978), 423.

11. I.R. Afnan and A.W. Thomas, Phys. Rev. C10 (1974), 109.

12. T. Mizutani, Ph.D. thesis, Univ. of Rochester 1975; Univ. of Rochester technical report UR-553, COO 2171-55.

13. T. Mizutani and D. Koltun, Ann. Phys. 109 (1977), 1.

14. A.S. Rinat, Nucl. Phys. A287 (1977), 399.

14a. A.W. Thomas and A.S. Rinat, Phys. Rev., to be published.

15. E.G. Brown and W. Weise, Physics Reports C22 (1975), 279.

16. J.W. Durso, M. Saarela, G.E. Brown and A.D. Jackson, Nucl. Phys. A278 (1977), 445.

17. M. Brack, D.O. Riska and W.W. Weise, Nucl. Phys. A287 (1977), 425.

18. J.A. Niskavnen, Nucl. Phys. A290 (1978), 417.

19. B.M. Freedom et al., Phys. Rev. C17 (1978), 1042.

20. D. Axen et al., Nucl. Phys. A256 (1976), 387.

21. G. Jones, 2nd Int. Conf. on the NN Interactions, Vancouver, BC, 1977.

22. C. Lazard, J.L. Ballot and F. Becker, Nuovo Cim. 65 (1970), 117.

23. B. Goplen, W.R. Gibbs and E.L. Lomon, Phys. Rev. Lett. 32 (1974), 1012; R. Gibbs, B.F. Gibson and G.J. Stephenson, Meson-Nuclear Phys. ed., P.D. Barnes et al., AIP Conf. Proc., Vol. 33 (1976), 464.

24. A.S. Rinat, E. Hammel and Y. Starkand, in preparation.

25. V.B. Mandelzweig, H. Garcilazo and J.M. Eisenberg, Nucl. Phys. A256 (1976), 461 (Err. ibid. A285 (1977), 505); N. Giraud, Y. Avishai, C. Fa yard and G.H. Lamot, preprint LYCEN/7833.

26. A.S. Rinat, E. Hammel, Y. Starkand and A.W. Thomas, in preparation.

27. W. Gruebler, private communication.

28. F. Lenz, Zuoz Spring School Lectures 1976.

29. L.C. Liu and C.M. Shakin, Phys. Rev. C16 (1977), 333, and several preprints.

30. R. Landau and A.W. Thomas, Nucl. Phys. A302 (1978).

31. H. Hofmann, Erlangen preprint; E. Oset and W. Weise, Regensburg preprints; G.E. Brown, B.K. Jennings and V. Rostokin, Stony Brook preprint; A.S. Rinat, 7th Intern. Conf. on High-Energy Physics and Nuclear Structure, Zurich 1977 and mss. in preparation.

THE $\lambda D \rightarrow NN\pi$ REACTION AND THE THREE-BODY KINEMATICS GAME

G. Tamas

DPh-N, CEN Saclay, BP 2

91190 Gif-sur-Yvette, France

Is it possible to use the electromagnetic probe to check nuclear processes involving pions, exchange current in the Δ resonance region? This question was somehow discussed during this meeting. I would like to illustrate it, and concentrate myself on the most simple case namely the two nucleons-one pion system, experimentally studied at the Saclay 600 MeV electron linac.

Let us consider the reaction:

$$\gamma + D \rightarrow p + p + \pi^- .$$

Experimentally it is a very favourable case: the three final state particles can be easily detected and their momentum measured, if their energy allows them to escape the target. It is very likely that at least two of them can be detected, and so even if a bremsstrahlung photon beam with a continuous spectrum is at your disposal, you will be able to determine the energy of the interacting photon.[1] If now you have a monochromatic photon beam, by detecting one of the particles you can measure directly the missing mass spectrum of the two undetected particles. What can be said on this reaction? First the most likely to occur is the absorption of the photon by the neutron and emission of a π nucleon pair, which are very often coupled to a Δ resonance, if the γ energy is close to 300 MeV, the proton remaining spectator (Fig. 1). This quasi free photo-production is characterized by the invariant mass Q of the produced π^-p pair, the angles ω between the π and the photon in the

---

[1]The only restriction is to set the end-point energy of the bremsstrahlung spectrum in order to avoid an other pion emission.

rest frame of the πp system (Fig. 2) and the momentum of the spectator nucleus.

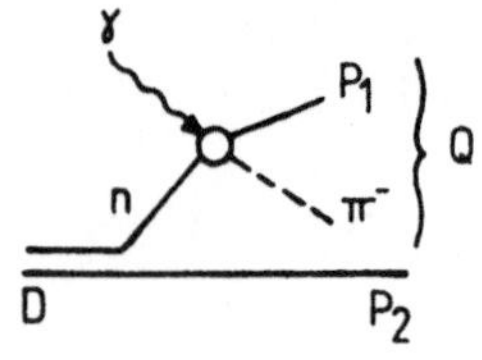

Fig. 1

In this quasi free process, if $\vec{p}_0$ is the momentum of the neutron before the reaction we have obviously

$$\vec{p}_2 = -\vec{p}_0.$$

We can describe this reaction by the impulse approximation and the differential cross-section is then proportional to the elementary cross-section multiplied by the momentum distribution $\rho_D(p_0)$ in the deuterium

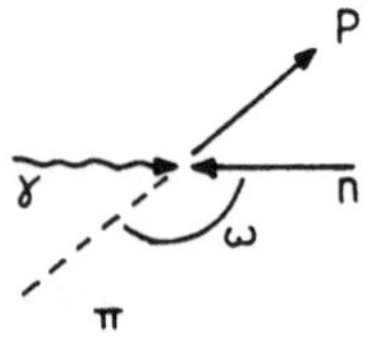

Fig. 2

$$d\sigma \propto \frac{d\sigma_n}{dn}(Q,\omega)\ \rho_D(|\vec{p}_2|).$$

How is it possible to have a signature of this process? If one keeps all the parameters, except one, constant, the variation of the cross-section as a function of this last quantity will give the answer. So, in a first experiment, we set $\vec{p}_0$, ω constant and we measured the cross-section as a function of Q. In a second one, Q, ω and $|\vec{p}_2|$ were constant and only θ_2, the angle between the photon and the second proton was varying. The quasi-free model predicts of course that the cross-section is invariant in respect of θ_2.

As we are working on relatively low values of $|p_2|$, where the deuteron wave function is precisely known, we detected the proton p_1 and the π^-. We have also to take in account diagrams in which p_1 is the spectator nucleon. But these diagrams give negligible contributions because:

i) $|p_1|$ is large, so $\rho_D(p_1)$ is small;
ii) the kinematics is chosen in such a way that the proton p_2 is emitted on the same side as the pion, and their invariant mass small and not far from threshold. Another consequence of this kinematics is to minimize the Pauli principle effects as the two protons direction is very large.

For $|p_2|$ = 50 MeV/c, we found a very good agreement with the quasi-free model qualitatively and quantitatively[1] for the Q variation (better than 5%) and for the θ_2 angular distribution (within 2% for $(d\sigma_{mes} - d\sigma_{QF})/d\sigma_{QF})$). But if $|p_2|$ increases, the picture changes. This was of course expected because then $\rho(|p_2|)$ decreases strongly and more complicated processes may appear. For instance (Fig. 3), at $|p_2|$ = 150 MeV/c, the θ_2 angular distribution of $(d\sigma_{mes} - d\sigma_{QF})/d\sigma_{QF}$ is far of being isotropic. The uncertainties on $d\sigma_{QF}$, mainly due to the deuteron wave function, are less than 10%, and affect only the position of the abscissa axis, not the shape.

In order to interpret these data, we included some second order processes. The pion rescattering is responsible of the backward bump, and the pp rescattering term enhances the cross-section because their relative kinetic energy decreases. This gives a qualitative agreement with the data but before trying more complicated processes, we wanted to check the π rescattering contribution. Considering this process, it is dominated by two diagrams shown in Fig. 4, because the πN scattering amplitude is dominated by the Δ, and is a way of measuring the exchange part of the Δ-N interaction.

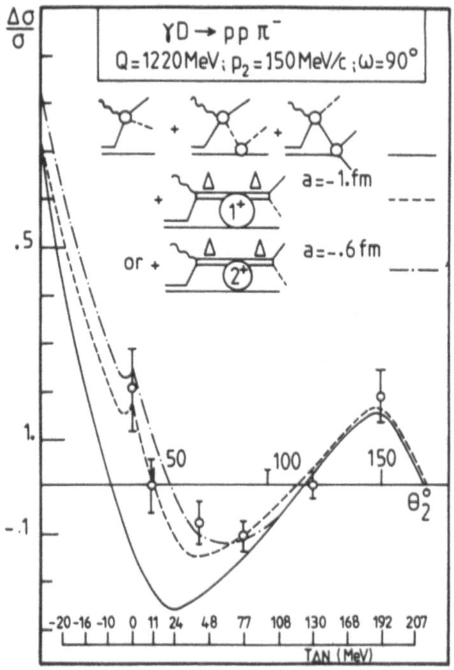

Fig. 3

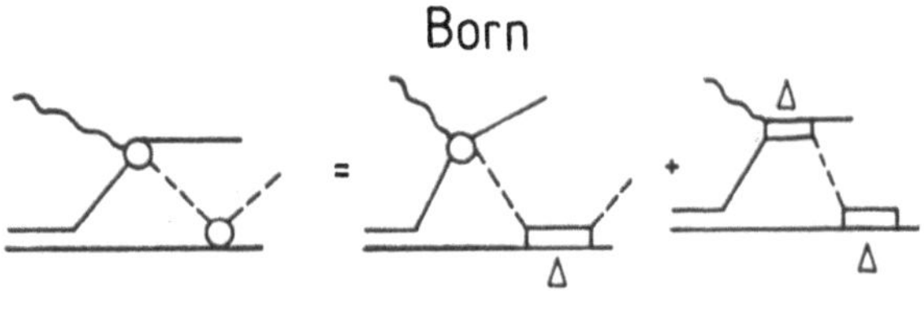

Fig. 4

The corresponding amplitude must be large when the intermediate pion is on mass shell. As we have to perform an integration over momentum in the triangular part of the diagram, the larger the kinematical region allowing an intermediate on-mass shell pion is, the maximum the effect will be. So we have chosen, in a third experiment, to measure the θ_2 angular distribution in order:

i) to minimize the quasi-free process, which means a large value of $|p_2|$;
ii) by a large relative energy of these two particles to suppress the proton final state interaction;
iii) to favourize the Δ production.

As p_2 became large, it was possible then to detect the two protons. The result[2], presented on Fig. 5, shows the ratio of the measured yield for Q = 1200 MeV, $\omega = 90^{\circ}$ and $|p_2|$ = 400 MeV/c to the

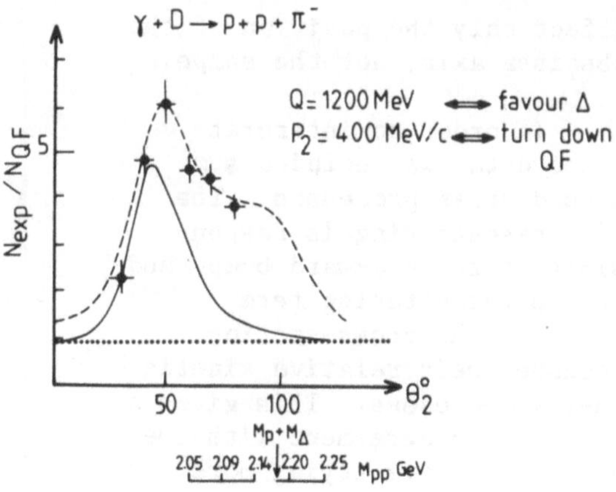

Fig. 5

quasi-free prediction (using the Reid soft core wave function). The experimental points present very large deviation with this model, represented by the dotted line. The solid line, strongly peaked around $\theta_2 = 45^o$, is the result of the calculation including now the π rescattering term. It gives qualitatively the description but there is still something missing specially for large angles. I give on Fig. 5, the mass scale of the two protons system and the discrepancy between the experience and the solid line increases with this mass. This was a guide to consider the contribution of a diagram (Fig. 6) similar to the double pion photoproduction on a nucleon followed by the reabsorption of one of the pions (like in the deuterium photodisintegration). Adding this new process leads to the dashed curve, in good agreement with the data.

Fig. 6

Then we tried to have a direct proof of the validity of such a diagram. We have changed the kinematical conditions:

i) increasing again $|p_2|$ to minimize again more strongly the quasi-free contribution $|p_2| = 500$ MeV/c;
ii) going far from the Δ resonance region to minimize also the π-rescattering process Q = 1100 MeV. This also vanishes the region for on-shell rescattering. The corresponding angular distribution is again referred to the quasi-free photoproduction. The solid line represents the quasi-free plus rescattering contribution, and the dashed one the result obtained by adding the double pion production term.

The agreement with the experiment is excellent (Fig. 7) (note that it was necessary to introduce the exchange of the ρ meson and a form factor at each pion-baryon vertex to take in account the data).

Returning now to the 150 MeV/c angular distribution, assuming that the photon is producing a Δ, we can notice that the relative kinetic energy of the Δ and the other nucleon becomes small around $\theta_2 = 40^o$ where the discrepancy with the previous calculation is

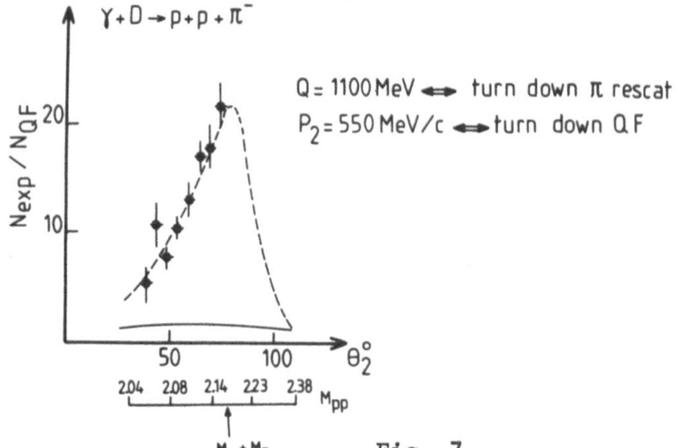

Fig. 7

large. So it is tempting to attribute this fact to a Δ-N scattering length. The curves, dashed and mixed lines, on Fig. 3 are fitted on the data for the two possible angular momenta of the ΔN system in a S-wave $J = 1^+$ and $J = 2^+$. This calculation done by J. M. Laget, as most of those used in this work[3], leads to those values of the scattering length: $a = -1$ fm for $J = 1^+$ and $a = -.6$ fm for $J = 2^+$.

As we can dispose of a monochromatic photon beam using the position annihilation in flight technique, we measured very recently the missing-mass spectrum of the two undetected protons.

The Fig. 8 shows what one can expect looking at the pion spectrum. The dotted curve is the quasi-free contribution and the solid one represents what happens if one turns on the pp final state interaction. The effect of the ΔN interaction is presented by the dashed ($J = 1^+$) and mixed curves ($J = 2^+$).

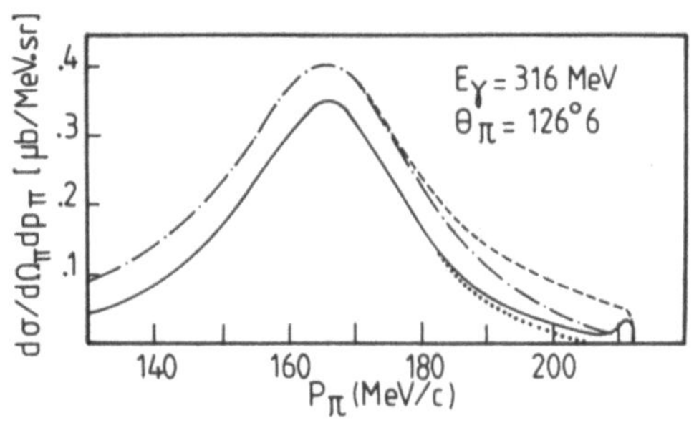

Fig. 8

On Fig. 9, I represent very preliminary and rough data taken in July to indicate how the measurement is going on: the quasi-free photoproduction is nicely shown and after deconvolution of the γ spectrum ($\Delta E_\gamma/E_\gamma \sim 1\%$) the (pp) interaction will very likely be clearly visible.

Returning to the (Δ-N) interaction, the best place to check it is to study the variation of the cross-section around $T_{\Delta N} = 0$ as a function of energy. The predictions and the preliminary results are shown on Fig. 10, with the same caption as the previous one. This

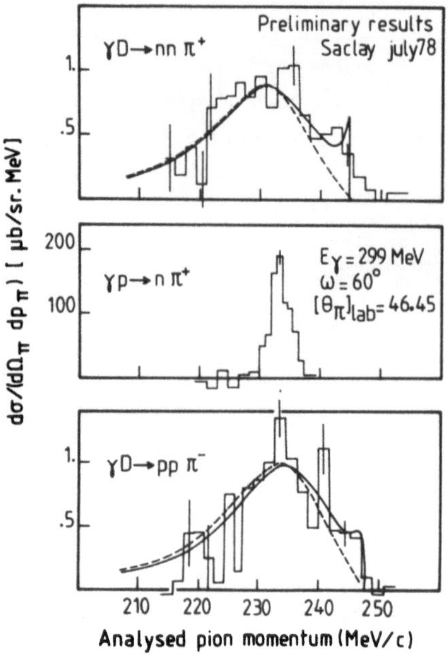

Fig. 9

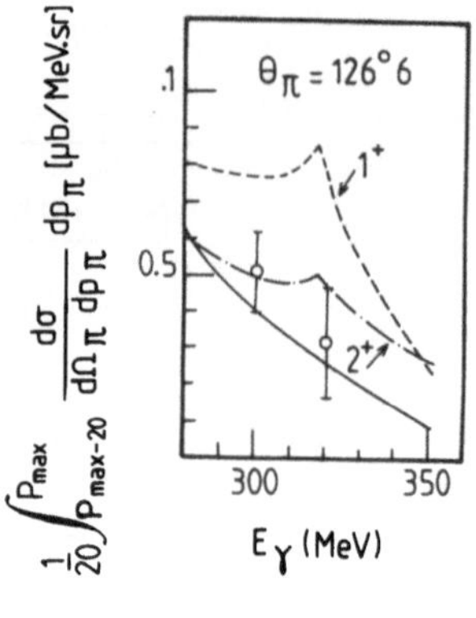

Fig. 10

excludes the J = 1^+ interaction and is not precise enough to distinguish between J = 2^+ and no Δ-N interaction.

This is of course not the end of the story, but I think it was worth showing that it is possible to understand the pion photoproduction on deuterium pretty well now, and at least to use this reaction as a laboratory for learning something about the Δ-N interaction.

REFERENCES

1. P.E. Argan et al., Nucl. Phys. A296 (1978), 373.
2. P.E. Argon et al., Phys. Rev. Lett. 41 (1978), 86.
3. See for instance: J.M. Laget, Phys. Rev. Lett. 41 (1978), 89; J.M. Laget, Nucl. Phys. A296 (1978), 413.

THRESHOLD PION PHOTOPRODUCTION AND THE NUCLEAR STRUCTURE

C. Tzara

Departement de Physique Nucléaire, CEN Saclay

91190 Gif-sur-Yvette, France

Instead of a standard review of the electro and photoproduction of pion, I have decided to present a selection of data and to discuss some of the problems encountered at this occasion, and which I feel of particular interest. It is thus a rather personal view of the field that will emerge, and many aspects will be missing. For instance, I deliberately ignore the "elementary particle" approach to these processes, and adhere completely to the microscopic analysis, the reason being that I want to know how the nucleus is made of "elementary particles". Let me recall at this point the conceptual framework in which the following discussion takes place.

The basic theory of strong interaction is not yet known. This forces us to proceed by successive approximations. The first step is to assume that the building block of a nucleus is the nucleon. The nucleons are unaffected by the vicinity of others nucleons, they merely interact through a 2N potential. Consider all sorts of probes whose interactions with a free nucleon are well parametrized. Their interactions with a nucleus are the resulting interaction with the constituent nucleons. By increasing the data gathered with a variety of probes, an increasingly precise picture of the nucleus is obtained.

Now, in practice, difficulties arise when the elementary interactions at work are strong enough to necessitate the account of complicated multiple reaction expansion. Hence the prejudice in favour of mild interactions as a faithful tool for exploring the nucleus.

The finest results on nuclei are certainly obtained with

photons and electrons. Indeed, it is the electromagnetic interaction which has provided the first sign of a breakdown of the simple image of a nucleus, outlined above. I refer to the thermal neutron radiative capture on the proton: $np \to d\gamma$ and the backward electrodisintigration of the deuteron near threshold: $e + d \to e' + n + p$ and equally to the magnetic moment of light nuclei.

These data cannot be explained in the first order nuclear model, using "purely nucleonic" wave function or one-nucleon currents. Hence the necessity, revealed in these particular situations, to enlarge the description of the nucleus.

One knows that in the true nuclear state, to the nucleon state and mesons and pairs vacuum is superposed an unending series of states with mesons and pairs. By a convenient procedure one can construct, in principle, a purely nucleonic state (the equivalent of the simple first order picture). But in that process, the N-N interaction generates, in principle, many body forces, one body currents become many-body currents... Confronted with this complicated situation, we must select properly the target nuclei and the probes, so as to minimize the calculational difficulties which would obscure the main object of the investigation.

As for the target nuclei, our first choice is obviously the 2N and 3N nuclei, their states being accurately calculable with a variety of realistic NN potentials. The 4N system should soon join them.

Among the probes, the electro-magnetic one provides the ideal tool. Whereas the electric interaction is not, or less[1], sensitive to the non-nucleonic current, the magnetic part is sensitive to others currents than the orbital, as noticed above.

It happens that the low-energy pion photoproduction amplitude, because of the parity of the produced (absorbed) pion, flips the nucleon spin and bears resemblance with the spin magnetic interaction. Furthermore, the π-N scattering lengths are small compared to the distance between nucleons in a nucleus, the masses ratio m_π/M_N are small, enabling simplification in the treatment of the multiple scattering, and their interaction is isoscalar. Thus, the pion radiative capture and the pion photoproduction at threshold are described in first order by an amplitude proportional to the axial form factor:

$$\langle\phi|\sum_N E_{0+}(i)\vec{\sigma}_i\,\vec{\varepsilon}\,e^{i\vec{k}\vec{r}_i}|\phi\rangle$$

Let us summarize the present knowledge on the γ-Nπ and π-N interaction at low energy.

Table 1

Channel	Decomposition in isospin space	measured E_{0+}^{th} in 10^{-16} cm
$\gamma p \to n\,\pi^+$	$\sqrt{2}\,S + V_3 + \sqrt{2}\,V_1$	40.0 ± 0.7
$\gamma n \to p\,\pi^-$	$-\sqrt{2}\,S + V_3 + \sqrt{2}\,V_1$	44.9 ± 2.8 45.6 ± 0.7*
$\gamma p \to p\pi^o$	$-\,S + \sqrt{2}\,V_3 - V_1$	3.2 ± 0.3
$\gamma n \to n\pi^o$	$S + \sqrt{2}\,V_3 - V_1$	not measured -0.8 ± 0.6*

*These values are obtained from the measured channels ($n\pi^o$) and ($p\pi^o$) by using only the following general laws:

- Gauge invariance of the ε.M. interaction
- PCAC hypothesis
- the conventional isotensorial rank of the e-m and strong interactions

More definite amplitudes can be obtained from the theory, but at the expense of model dependence[2,3]. At any rate, these approaches assume always the mentioned symmetry in isospin space. This assumption can be questioned in the vicinity of threshold. This is one reason for trying to fill the gaps in Table 2. We shall return to this point later.

As for the π-N scattering lengths, the more recent set is as follows[4] in Table 2.

These quantities are not sufficient to work out a complete microscopic calculation. Because of the momentum acquired by a bound nucleon, it is necessary to consider the momentum dependent interaction operators. For instance $|M_{1+}| \simeq 0.08\ q$ in 10^{-16} cm, q in MeV/c, irrespectively of the pion charge. For charged pion interaction, it contributes as a small correction. For neutral pion, it is crucial even at a few MeV above threshold.

The p-wave π-N scattering amplitude has less influence in the

Table 2

Channel	Scattering length in 10^{-13} cm
$\pi^+ p \to \pi^+ p$ $\pi^- n \to \pi^- n$	-0.130 ± 0.003
$\pi^- p \to \pi^- p$ $\pi^+ n \to \pi^+ n$	0.117 ± 0.003
$\pi^- p \to \pi^o n$ $\pi^+ n \to \pi^o p$	-0.175 ± 0.003
$\pi^o p \to \pi^o n$ $\pi^o n \to \pi^o n$	-0.006 ± 0.002

multiple scattering expansion.

Now, once one considers the nucleon momenta, one begins to accumulate difficulties related to the fact that the participating particles are off-mass-shell. Sophisticated treatments have been proposed[5]; even those neglect the quantum nature of the system in the intermediate states. Some people think that it is sufficient, and perhaps even safer, to treat correctly the impulse approximation step and to assume frozen nucleons in the next orders of multiple scattering. At any rate, these ambiguities are not dangerous for the charged pions production or absorption, where the rescattering correction is small.

I. CHARGED PION PHOTOPRODUCTION

At this point, the prospect is, for the interpretation of the data, rather encouraging.

We shall now describe the status of the experiments, which indeed need a good standard of precision, say a few percent.

I insist on the necessity to reach this level, otherwise, at the 10% - 20% level, one cannot learn more than the overall agreement between the values of similar form factors*.

The experimental approaches are:

$$\left.\begin{array}{l}\pi^- A \to \gamma B^- \\ \pi^- A \to \pi^0 B^-\end{array}\right\} \quad \text{from electromagnetically bound pions}$$

$$\left.\begin{array}{l}\gamma A \to \pi^{\pm} B^{\mp} \\ \gamma A \to \pi^0 A \\ \varepsilon A \to e' \pi^{\pm} B^{\mp}\end{array}\right\} \quad \text{near threshold}$$

The radiative capture experiments, otherwise a powerful tool, are difficult to interpret because one needs to measure the width of the 1S level and the population of that state when the radiative capture takes place.

The Panofsky ratio $\pi^-\pi^0/\pi^-\gamma^0$ bypasses the difficulty but mixes two types of operators, an axial one and a scalar one. As for the cancelling of the pion distortion effect, often invoked as an advantage, we shall see, in the case of an ^{3}He target, that it does not hold automatically.

The π^- photoproduction was investigated by activation techniques, but proved much too imprecise , and applicable in few cases only.

The π^+ photoproduction was shown to be precise at the level of a few percent[8]. One of the advantages is that the detecting apparatus and the beam parameters can be calibrated by the reaction:

$$\gamma\ p \to n\ \pi^+ \tag{1}$$

which is the underlying elementary process for the reaction:

$$\gamma\ A \to \pi^+\ B \tag{2}$$

Thus the comparison of the yields of these two reactions enable to obtain with a negligible uncertainty the total cross section of (2) relatively to that of (1), which is just the quantity needed for our purpose.

(The (γ,π^0) reactions posing quite different problems, their study will be postponed at the end of the talk.)

---

* This is true in the perspective described here, and not in cases where nuclear structure problems can be solved with less precise measurement. An example is the reaction $\gamma\text{-}N^{14} \to \pi^+ C^{14}$

The 2N System

The cross section for the reaction:

$$\gamma\, d \rightarrow nn\pi^{+} \tag{3}$$

has been measured at the Bates Linac[9] and at Saclay[10]. The data of the latter laboratory, more precise, agree well with those of MIT, but they were analyzed with more scrutiny. The resulting cross section is obtained, relatively to the reference reaction (1), with an accuracy of ± 2.5%.

The best published calculation of the cross section[11] uses the following ingredient:

- Elementary production amplitudes including momentum dependent terms.
- Hulthen bound state wave function.
- n-n scattering wave functions obtained from a separable potential.
  The phase shifts of the higher partial waves are neglected.

The resulting cross section is in agreement with the experimental one (Fig. 1).

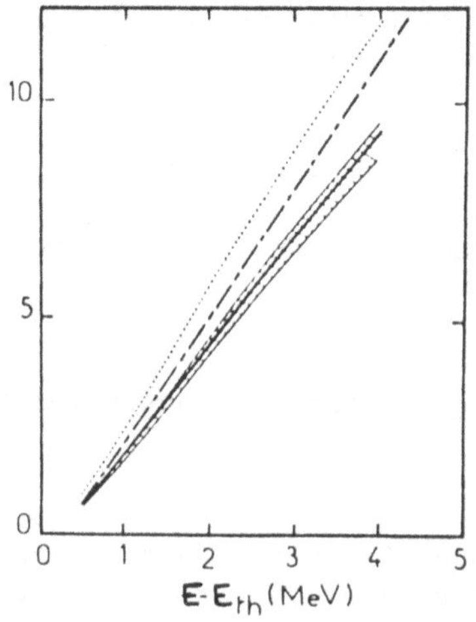

Fig. 1. Cross section of the (γ,π^{+}) reaction on deuterium deduced from this experiment. $E-E_{th}$ is the excess energy above threshold in the laboratory system. The shaded area corresponds to the experimental determination: the uncertainty on $a_p = 4\pi\, E_{0'}{}^{2}$ is included in addition to the uncertainty on a_d. Dotted (Ref. 4), dash-dotted (Ref. 7), and solid (Ref. 11) lines are the theoretical predictions.

We have verified that the use of more realistic wave functions, such as those generated by the Reid-Soft-Core potential, does not change appreciably the matrix element of $E_0 + \vec{\sigma}\cdot\vec{\varepsilon}$. As this is the dominant amplitude, it is sound to assume that changing to more realistic wave functions would not modify appreciably the cross section computed by Noble.

As for the effect of the π^+ rescattering, it has a negligible effect ($\sim$ 1%) only if the central densities of the wave functions are depressed by a soft core potential. This correction increases the theoretical cross section, because the $\pi^+ n \rightarrow \pi^+$ scattering length is positive.

Finally as the d-state does not contribute significantly to the cross section, one should decrease the Noble cross section by the d-state percentage $\sim$ 5%.

In conclusion, the theoretical cross section agrees with the experimental one at the level of better than 5%.

The electro-disintegration of the deuteron near threshold:

$$e\ d \rightarrow e'\ n\ p \qquad (4)$$

is a similar process, driven by a magnetic operator, essentially a spin-flip one, and leading to the same isospin triplet, spin singlet final state as for the pion photoproduction (3).

The measured cross section[12] is substantially higher than the one computed by using the impulse approximation and realistic wave functions[13,14]. At $q_\mu^2 = 0$, the process is reduced to the inverse of the celebrated thermal neutron-proton radiative capture, and the defect of the I.A. calculation is 10%. At $q_\mu^2 \sim m_\mu^2$, the defect reaches more than 20% and increases when the momentum transfer increases.

Thus the magnetic amplitude is sensitive to the non-nucleonic degrees of freedom. In contrast, it seems that the threshold pion production operator is almost insensitive to these mesonic effects.

Is this still true in the case of the 3N system?

The 3N System

^{3}He and ^{3}H are more tightly bound than the deuteron. Moreover, 3N forces could be present. These are characteristics of a genuine nucleus.

- The magnetic form factors of these two nuclei have been

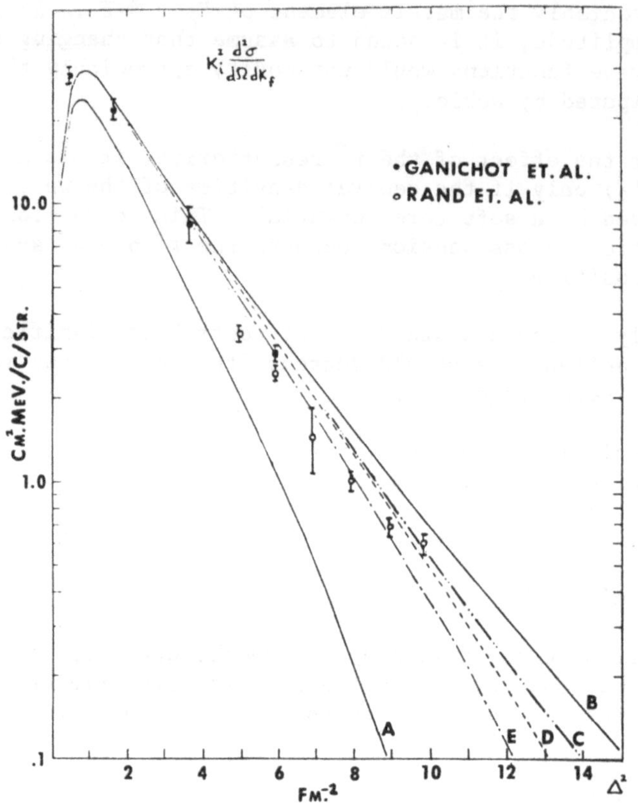

Fig. 2. Deuteron electrodisintegration cross section near threshold from back-scattered electrons whose four-momentum transfer is Δ with the Reid hard core parametrization of the n-p potential. Final n-p relative energy is E_{cm} = 3.0 MeV. Curve A is the impulse cross section to the final $^1S_0$ state. Curves B, C, D and E are the impulse plus meson interaction cross section. In curve B, the catastrophic process is proportional to $E_1^V(\Delta^2)$ and the pion exchange process is proportional to $E_\pi(\Delta^2)$. In curve C, the catastrophic process is proportional to $F_1^V(\Delta^2)$ and the pion exchange process is proportional to $F_\pi(\Delta^2)F_{\pi\pi}(\kappa_1^2\kappa_2^2)$. In curve D, the catastrophic process is proportional to $F_A(\Delta^2)$ with m_A = 0.81 GeV and the pion exchange process is proportional to $F_\pi(\Delta^2)$. In curve E, the catastrophic process is proportional to $F_A(\Delta^2)$ with m_A = 0.81 GeV and the pion exchange process is proportional to $F_\pi(\Delta^2)F_{\pi\pi}(\kappa_1^2,\kappa_2^2)$.

measured accurately[15].

- The Panofsky ratio has recently been measured with a 3% precision[16].

- The radiative capture partial rate and the width of the 1S state in $\{\pi^- \, ^3He\}$ atom have been measured[17,18].

Let us express these data in terms of an axial form factor. Applying the impulse approximation to the charge exchange and the radiative capture, one obtains

- from the Panofsky ratios for ^3He and ^1H:

$$|M|^2 \left(\frac{\pi-\pi^o}{\pi-\gamma} \right) = (0.59 \pm 0.02) \frac{C_{\pi-\pi^o}}{C_{\pi-\gamma}} \tag{5}$$

- from the radiative partial rate and the average 1S state width (55 eV) and assuming that the capture takes place only from s states, a lower limit:

$$|M|^2 \, (\pi-\gamma) = (0.97 \pm 0.20) \frac{1}{C_{\pi-\gamma}} \tag{6}$$

The coefficients $C_{\pi-\pi^o}$ and $C_{\pi-\gamma}$ account for the strong interaction experienced by the pion. We have remarked above that the equality $C_{\pi-\pi^o} = C_{\pi-\gamma}$ is generally taken for granted. This impression comes from the use of the concept of a potential interaction between the pion and the nucleus. In the present case, where we deal with an absorption of the incoming pion, and describe it by a phenomenological elementary amplitude, this concept is not applicable, and one must appeal to a multiple scattering expansion. For a light nucleus like ^3He, the two approaches yield a completely different result.

As remarked above, the lightness and the small momenta of the participating mesons enable us to use the fixed scatterer approximation and to neglect the p wave scattering.

At first order in the scattering expansion, the π^- before being radiatively absorbed by one proton is scattered by an (n-p) pair, but only very weakly because $a(\pi^- n) + a(\pi^- p) \simeq 0$. On the other hand, the $\pi^- \, ^3He$ potential is in first order generated by the scattering of the π^- on the whole ^3He nucleus, leaving thus practically only the $\pi^- p$ scattering which is attractive and enhances the central density of the π^- wave by $\simeq$ 8%.

Consequently, we have estimated $C_{\pi^- \pi^o}$, $C_{\pi^- \gamma}$ and $C_{\gamma\pi^+}$ up to the second order scattering.

We computed

$$<\frac{1}{r}>_k = [\int|\phi|^2 \, e^{ikr_i}/r_{ij}]\cdot[\int|\phi|^2 \, e^{ikr_i}]^{-1}$$

and

$$<\frac{1}{r^2}>_k = [\int|\phi|^2 \, e^{ikr_i}/|r_{ij}|^2]\cdot[\int|\phi|^2 \, e^{ikr_i}]^{-1}$$

assuming that

$$<\frac{1}{r_{ij}r_{jk}}> = <\frac{1}{r_{ij}r_{ij}}>$$

at k=0 (for the $\pi^-\pi^\circ$ process) and $k = m_\pi$ (for the radiative processes), and using directly the measured ^{3}He form factor as suggested by Fabre de la Ripelle[19].

For the radiative processes, we find that not only the first scattering but also the second one are subjected to a large cancellation, leading to:

$$C_{\pi^-} \sim 0.98$$

$$C_{\gamma\pi^+} \sim 0.99$$

For the charge exchange process, the second order scattering is not cancelled out, and yields a fairly large correction

$$C_{\pi^-\pi^\circ} \sim 0.88$$

contrary to the common belief.

The π^- ^{3}He scattering length has been recently calculated with the same method and agrees reasonably well with the measured energy shift of the 1S level in the π^- ^{3}He atom. Thomas gives argument for the neglect of the π-N p wave scattering.

Thus values given above for the C's are certainly of the right order of magnitude. Inserted in (4) and (5) they give

$$|M|^2 \left(\frac{\pi^-\pi^\circ}{\pi-\gamma}\right) = 0.53 \pm 0.02 \qquad (7)$$

and

$$|M|^2\ (\pi-\gamma) = 0.99 \pm 0.20 \tag{8}$$

This latter value of the axial form factor squared is difficult to understand. The value of $|M_a|^2$ computed with realistic wave functions[21] does not exceed 0.55, in agreement with the experimental value (7). Either the experimental widths are badly in error, or there is a physical effect modifying equally the charge exchange and the radiative capture of a π^-.

In this confusing situation, it was necessary to measure the cross section for the reaction

$$\gamma\,^3He \rightarrow \pi + {}^3H \tag{9}$$

with the precise method devised at Saclay. Great care was given to

$\gamma d \rightarrow d\pi^0$

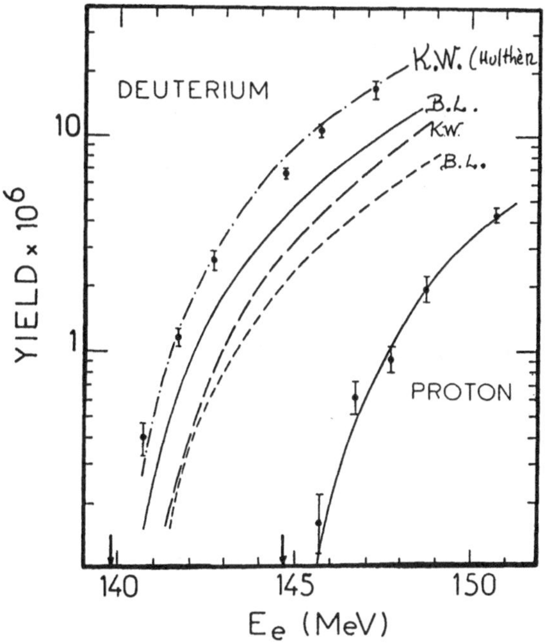

Fig. 3

determination of the densities of the hydrogen and helium liquefied targets. Their vapor pressure was recorded and transmission measurements performed with a 0.511 MeV γ-ray beam. They yielded densities in agreement at the level of 2%. The resulting cross section relative to the reference cross section (1) yields:

$$\frac{a_{^3He}}{a_p} = \frac{\frac{k}{9}\sigma_{^3He}S^{-1}}{\frac{k}{9}\sigma_p} = 0.62 \pm 0.02$$

from which we obtain:

$$|M|^2_{(\gamma M^+)} = (0.52 \pm 0.02)\,\frac{1}{C_{\gamma\pi^+}} = 0.53 \pm 0.02 \qquad (10)$$

in striking agreement with the value (5) extracted from the Panofsky ratio and comparable with an impulse approximation prediction[21].

Under the reservation that the neglected nucleon Fermi momentum has indeed a negligible influence on the process, our main and important conclusion is that the π^+ photoproduction at threshold is insensitive to extra-nucleonic degrees of freedom in ^3He. In other words, this process is a faithful tool for the exploration of the body axial form factor, whereas the magnetic spin dependent form factor reveals the presence of exchange currents.

Further, considering that the radiative capture rate is the mirror process of the π^+ photoproduction, it is likely that the large experimental value (8) of $|M|^2_{r^-\gamma}$ is explained by an overestimation of the 1S level width. It is interesting to note that by using our estimate (9) of $|M|^2$, are obtained instead of Γ_{1S} = 55 eV the value Γ_{1S} = 30 eV, in agreement with Phillips calculation and invalidating the conclusions drawn from the large experimental width[18].

Other nuclei has been studied with the same probes: ^6Li [8], ^{12}C [23]. Only is the ^6Li photopion cross section precise enough for one purpose. As we lack fundamental wave functions for ^6Li and ^6He, we can only compare the photopion cross section to the magnetic electron form factor. Unfortunately its spin part cannot be extracted without ambiguities from the data[24]. Furthermore, no direct information is available on the radial difference between the analogue $(0^+,1)$ states in ^6Li and ^6He. Thus no firm confirmation can be drawn from the ^6Li of the effect found in the case of the 2N and 3N nuclei, namely the insensitivity of the π^+ threshold photoproduction to the mesonic degrees of freedom.

Let us summarize our conclusions.

- The π^+ photoproduction at threshold on very light nuclei seems to be correctly described by an impulse approximation amplitude, computed with accurate conventional wave functions generated by an N-N potential. It appears thus as a faithful tool for investigating the conventional nuclear structure.

- Some improvements must be made on the side of the theory, namely i) in the case of the 2N system, better wave functions, ii) in the case of the 3N system, inclusion of pion and nucleon momentum dependent terms.

- The experimental precision attained seems difficult to improve, and is sufficient for the present purpose.

- For heavier nuclei, the pion photoproduction decreases rapidly, due to the Coulomb repulsion and the smallness of the form factor, yielding poor precision. The comparable process, magnetic electron scattering, in plagued with ambiguity in the respective contributions of the spin and orbital moment.

In view of the importance of the issue, the following paths should be explored.

i) Threshold π^+ photoproduction measurements on heavier nuclei than ^6Li with the technique initiated at Saclay. This is possible with a very high duty cycle accelerator, providing an intense photon beam without the inconvenience of blinking the detectors with too intense pulses.

Correspondingly, improvement should be made in magnetic electron scattering with the aim to reduce the ambiguity in the extraction of the spin part from the magnetic form factor.

ii) The investigation of the axial form factor at higher momentum transfer through the reaction:

$$eA \rightarrow e' \; B \; \pi^{\pm} \qquad (11)$$

following the same methods as have been used for the proton[25]. The obvious candidates are A = ^3He and ^3H. With the present moderate duty cycle accelerators, it seems a difficult experiment. Higher duty cycles would lift the main experimental obstacle.

Extension of these techniques to heavier nuclei is dubious:

i) either one performs a coincidence experiment between the scattered electron and the recoiling residual nucleus; the latter require an energy which, with increasing mass M_B, is unable to win the absorption in the target.

ii) either one searches on the radiative tail of the scattered electron the signature of the process. For π^+ emission, the Coulomb repulsion depresses the cross section to such an extent that it will not be visible. π^- emission could be seen thanks to the step shape of the threshold cross section or virtual π^- atomic S states as bumps superimposed on the tail[26,27]. The main interest of this last process would be to extend the spectroscopy of the s levels to heavy nuclei where the competing capture process forbid the π^- to reach low <u>n</u> levels. In such a hypothetical state, the Coulomb wave function is inside the nuclear volume, and the strong interaction is no more a perturbation. See in connection with that process, Ref. 28.

II. THE π° PHOTO-PRODUCTION

As mentioned above, the π° photoproduction at threshold has been barely explored experimentally for two principal reasons:

1) The smallness of the cross section

$$\frac{d\sigma}{d\Omega}(p\pi^\circ) \simeq z/k\ 10^{-31}\ \mathrm{cm}^2$$

2) The low π° detection efficiency.

However, the neutral pion channel are of great interest especially in the thredhold region. The π° photoproduction amplitude is much more sensitive to the details of theoretical models than the charged pion photoproduction amplitude. For instance, the pseudo-scalar π-N coupling and the pseudo-vector coupling give completely different Born-amplitudes for the neutral pion channel, whereas the charged pion channels are almost insensitive to that choice. The application of the P.C.A.C. hypothesis to this process, completed by the Gauge invariance of the electromagnetic interaction and the accepted isotensorial rank of the strong and electromagnetic interaction, allows a prediction of $E_0 + (p\pi^\circ)$ and $E_0 + (n\pi^\circ)$ without any model dependence, but with a limited precision. In the case of purely electromagnetic interaction the corresponding low energy theorems are exact at zero energy of the photon. In meson-nucleon physics the low energy theorems amount to determine only the first terms in a power expansion of $\mu = m_\pi/M_N \sim 0.15$[29]. Model dependent predictions are available.

All these theoretical investigations assume the isoscalar character of the strong interaction. However this symmetry is not exact, as shown by the $\pi^\circ - \pi^\pm$ mass difference. Could the threshold values of E_{0+} be affected by the breaking of this symmetry?

These considerations reinforce the interest of direct measurements of $E_{0+}(N\pi^\circ)$ as near the threshold as possible. Indeed, the measurements on the proton are made much too far from the threshold (at energies exceeding 13 MeV above the threshold), and the uncertainty quoted[30] is considered as underestimated.

We have embarked on an experimental program which should provide some information in this field.

It consists in measuring the π° photoproduction near threshold on the proton, the deuteron, the Helium-3 and the Helium-4. In impulse approximation and right at threshold, the amplitudes would be proportional to:

$E_{0+}(p\pi^\circ)$	on the proton target.
$E_{0+}(p\pi^\circ) + E_{0+}(n\pi^\circ)$	on the deuteron target.
$E_{0+}(n\pi^\circ)$	on the Helium-3 target (neglecting the mixed symmetry state).
0	on the Helium-4 target.

After we started the preliminary measurements, complications due to the pion rescattering were pointed out[31]. In addition to the various problems related to the reaction mechanism, they obscure the simple original picture, and it is not clear at the moment if we shall be able to extract any reliable value for $E_{0+}(N\pi^\circ)$.

The experimental situation is the following. The measurements have been performed on the four targets mentioned above from 1 to 10 MeV above threshold. What is measured is essentially the total cross section folded with the Bremsstrahlung spectrum and the π° detection efficiency, as a function of the end point energy of the photon spectrum. The absolute efficiency is known with poor accuracy, so that the measurement on the proton target is used to calibrate the apparatus. The total cross section on the proton is:

$$\sigma = \frac{k}{q} \left|E_0 + (p\pi^\circ)\right|^2 + 2\, M_1(p\pi^\circ)\,|^2$$

$M_{1+}(p\pi^\circ)$ is reasonably known, and the ratio $|E_{0+}(p\pi^\circ)|/|M_{1+}(p\pi^\circ)$ is adapted to the shape of the measured yield. One gets thus together the efficiency and an experimental value of $E_{0+}(p\pi^\circ)$ which, admittedly, is not very precise, but confirm the value extracted from higher energies[30] and from the theories most in favor nowadays.

This is used in turn to interpret the data obtained on the other targets.

Up to now, the deuteron case has been analyzed and yields a cross section more than two times larger than the one obtained in impulse approximation. It is likely that the mechanism proposed in[31] is at work. It is simply the effect of the photoproduction of a charged pion which is scattered with charge exchange, yielding an outgoing $\pi°$. The direct amplitude is proportional to

$$A_d = \{E_{0+}(n\pi°) + E_{0+}(p\pi°)\}\ F(m_\pi^2)$$

and the rescattering one to:

$$A_r = \alpha_{exch}\ F^-(m_\pi^2)\ \{E_{0+}(p\pi^-) + E_{0+}(n\pi^+)\}$$

where

$$F(m_\pi{}^2) = \int |\phi_d|^2\ e^{ik-q\cdot\frac{r}{2}}\ d^3r$$

and

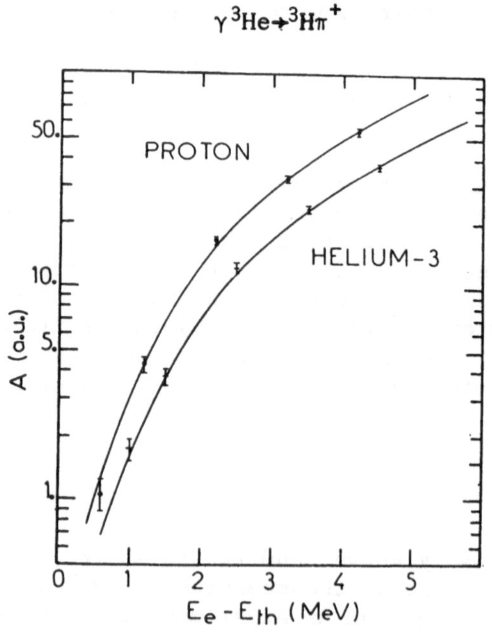

Fig. 4

$$F^{-}(m_\pi^2) = \int |\phi_d|^2 \frac{1}{r} e^{ik-q\frac{v}{2}} d^3r = \frac{F(m_\pi^2)}{\langle r \rangle} \sim 0.5\ F(m_\pi^2)\ (\text{in fm}^{-1})$$

The ratio of these two amplitudes is thus:

$$\frac{A_r}{A_d} = \frac{a}{\langle r \rangle} \frac{E_{0+}(p\pi^-) + E_{0+}(n\pi^+)}{E_{0+}(n\pi^\circ) + E_{0+}(p\pi^\circ)} = 0.18 \times 0.5 \frac{85}{2.4} \sim 3.$$

This particular mechanism produces thus an enormous enhancement over the impulse approximation prediction. The next orders in the multiple scattering expansion do not present the same character, and are of no more importance than in the charged pion photoproduction.

Detailed calculations with a realistic deuteron wave function, various degrees of sophistication in the treatment of the elementary amplitudes and of the pion propagator have been published[32], but the agreement is not satisfactory. It must be recalled that in these theories, the photoproduction amplitudes are directly computed with an effective Lagrangian and are not free. Thus, a different type of approach, where the elementary amplitudes enter as free parameters, must be devised. This will be done in the final stage of our work, in order to extract all the possible content of the four measured cross sections.

The future of such investigations can be foreseen easily.

1. It is the next urging task to measure the absolute cross section of $\gamma p \to p\pi^\circ$ by simply calibrating the π° detection system. This additional effort is considered at the A.L.S.

2. An order of magnitude should be gained in the precision of the experimental yields. This is impossible with the Bates or the A.L.S. linac and require a high duty cycle beam.

3. Investigation of heavier nuclei are performed at Bates. The process is then dominated by the coherent π° production, which goes as $A^2\ F^2(Q^2)\,|M_{1+}|^2$. Matter form factors could be obtained, as shown in an old work[33].

In conclusion, I would like to insist on the interest to improve the calculations in the frame of the impulse approximation and the estimates of the multiple scattering series.

Furthermore, predictions have been published on the large influence of the exchange currents on the threshold photoproduction. These claims should be reexamined in the light of the experimental evidence presented above.

REFERENCES

1. H. Jyuga, H. Ohtsubo, Nucl. Phys. A294 (1978), 348.

2. R.D. Peccei, Phys. Rev. 181 (1969), 1902.

3. I. Blomquist, J.M. Laget, Nucl. Phys. A280 (1977), 405.

4. M.M. Nagels et al., Nucl. Phys. B109 (1976), 1.

5. J.M. Laget, Nucl. Phys. A296 (1978), 388.

6. A. Figureau, N.C. Mukhopadyay in "Meson nuclear physics", editors P.D. Banees et al. (1976), 616.

7. A.M. Bernstein et al., 6th International Conf. on high energy physics and nuclear structure, Santa Fe, New Mexico, June 1975.

8. G. Audit et al., Phys. Rev. C15 (1977), 1415.

9. E.C. Booth et al., Phys. Rev. C16 (1977), 236.

10. G. Audit et al., Phys. Rev. C16 (1977), 1517.

11. J.V. Noble, Phys. Lett. 67B (1977), 39.

12. D. Ganichot, B. Grossetête, D.B. Isabelle, Nucl. Phys. A178 (1972), 545.

13. J. Hockert et al., Nucl. Phys. A217 (1973), 14.

14. J.A. Lock, L.L. Foldy, Phys. Lett. 51B (1974), 212.

15. J. McCarthy et al., Phys. Rev. Lett. 25 (1970), 884.

16. M.D. Hasinoff et al., 7th International conference on high energy physics and nuclear structure, Zurich (1977).

17. R. Abela et al., Phys. Lett. 68B (1977), 429.

18. G.R. Mason et al., Phys. Lett. 68B (1977), 429.

19. M. Fabre de la Ripelle, Phys. Lett. 8 (1964), 340 and private communication.

20. A.W. Thomas, preprint TRI PP 782, Triumf, U.B.C. Canada.

21. B. Goulard, A. Laverne, J.D. Vergados to be published in Phys. Rev.

22. A.C. Phillips, Report on Progress in Physics 40 (1977), 905.

23. A. Bernstein et al., BAPS 23 (1978), 4, 611.

24. J.C. Bergstrom, I.P. Auer, R.S. Hicks, Nucl. Phys. A251 (1975), 401.

25. E.D. Blom et al., Phys. Rev. Lett. 30 (1973), 1186; P. Brauel et al., Phys. Lett. 45B (1973), 389; D.R. Botterill et al., Phys. Lett. 45B (1973), 405.

26. C. Tzara, Nucl. Phys. B18 (1970), 246.

27. G.T. Emery, Phys. Lett. 60B (1976), 351.

28. Private Communication.

29. De Baenst, Nucl. Phys.

30. B.B. Govorkov and E.V. Minarik, Sov. J. Nucl. Phys. 14 (1972), 245.

31. J.H. Koch and R.M. Woloskyn, Phys. Lett. 60B (1976), 221.

32. P. Bosted and J.M. Laget, Nucl. Phys. A296 (1978), 413.

33. R.A. Schrack, Phys. Rev. 140 (1965), 897.

PION DOUBLE CHARGE EXCHANGE

Martin D. Cooper

University of California, Los Alamos Scientific Laboratory, Los Alamos, N.M. 87515, U.S.A.

ABSTRACT

The pion double charge exchange data on the oxygen isotopes is reviewed and new data on ^{9}Be, ^{12}C, ^{24}Mg, ^{26}Mg and ^{28}Si are presented. Where theoretical calculations exist, they are compared to the data.

Pion double charge exchange to definite nuclear states is a new field of experimental investigation. At the time of this lecture, there are only three groups which have observed this reaction. The measurements include 0° cross sections on a variety of nuclei[1], 18° measurements[2] on ^{18}O, and two angles[3] on ^{9}Be. A collation of all of the data on this reaction is given in Table I, and the scarcity of data makes most of the conclusions that one can draw somewhat qualitative. Nevertheless, some interesting trends do seem to be appearing.

The motivations for doing pion double charge exchange have been recognised[4] for some time. Under the assumption that the reactions mechanism is dominated by the process involving two sequential single charge exchanges, the reaction is expected to be more sensitive to two nucleon effects than reactions which do not necessarily involve two nucleons in the first order. The simplicity of the projectile and final particles raises the hope of separating reaction mechanism issues from those of nuclear structure.

TABLE I

A COLLATION OF ALL DOUBLE CHARGE EXCHANGE RESULTS*

		0°	13°	18°
Target	Energy (MeV)	cross sections (μb/sr)		
^{9}Be	145	0.08 ± 0.05	0.20 ± 0.05	0.14 ± 0.04
^{12}C	145	0.65 ± 0.2		
^{16}O	145	0.87 ± 0.21		
^{18}O	95	1.67 ± 0.38		
	126	2.19 ± 0.44		
	139	2.00 ± 0.34		
	148			0.30 ± 0.10
	187			0.21 ± 0.08
^{24}Mg	145	0.67 ± 0.2		
^{26}Mg	145	0.18 ± 0.1		
^{28}Si	145	0.35 ± 0.1		

*Results on all nuclei except the oxygen isotopes should be regarded as preliminary.

The leading candidate for a simple reaction was thought to be the double analog transition. Such reactions were thought to take the initial state into its mirror state by sequential actions of the isospin raising or lowering operator and without any substantial change to the radial part of the wave functions; an example of a double analog transition would carry ^{18}O into ^{18}Ne. The simplicity of the reaction mechanism and the similarity of the initial and final states suggests that a simple understanding of them might evolve. As we shall see, a richer spectrum of theoretical ideas appears to be necessary to explain the experimental results.

The matrix element for this two-step mechanism may be written as

$$T_{FI} = \langle \Phi_F^A(Z+2)\phi^{(-)} | \sum_{\lambda} \sum_{i \neq j} t_i(\vec{\tau}\cdot\vec{I}) \frac{|\Phi_\lambda^A(Z+1)\rangle\langle\Phi_\lambda^A(Z+1)|}{E-E_\lambda-k_\pi-U_\lambda}$$

$$t_j(\vec{\tau}\cdot\vec{I})|\Phi_I^A(Z)\phi^{(+)}\rangle \quad . \qquad (1)$$

It turns out that the places for theoretical interpretation of Eq. (1) are deciding what the structure of the nuclear wave function Φ_I and Φ_F is which contributes to the reaction and determining what set of intermediate states Φ_λ are important. To date, the data appear to address these questions separately. Excitation functions and angular distributions bear on reaction mechanism issues, while the A dependence of the cross section deals with the structure of the initial and final state.

It should be noted that the experiments are difficult and the field is young. Hence, the experimental results are largely preliminary, and some of the conclusions one now draws may evolve as the experiments become more refined.

The first observation of pion double charge exchange to discrete states[5] was made to the double analog state of ^{18}O. The spectrum obtained is shown in Fig. 1, and a clear peak is observed at the energy of the ground state of ^{18}Ne. The peak has a substantial background whose shape is not perfectly known. The systematic error in estimating this background dominates the uncertainty in the cross section measurement; systematic errors dominate all the cross sections measured to date.

An excitation function[6] of the 0° cross sections is shown in Fig. 2. The predictions for several theoretical approaches are also given (for the details, see Ref. 6). None give the proper shape except the fixed scatterer, which has had its normalization somewhat arbitrarily adjusted. The optical model calculations which contain short range correlations fit the magnitude of the cross section in the region of the 3-3 resonance much better than those without the correlations. This makes one hopeful that the reaction will eventually shed detailed information on the nature of short range correlations in nuclei.

The two data points near 140 MeV make a preliminary angular distribution. These points are plotted in Fig. 3 along with the theory of Sparrow and Rosenthal[7]. Their theory includes only a few intermediary states in Eq. (1), but appears to give some agreement with the data. The dashed curve shows the theory divided by two. It appears that in order to get the rate of angular

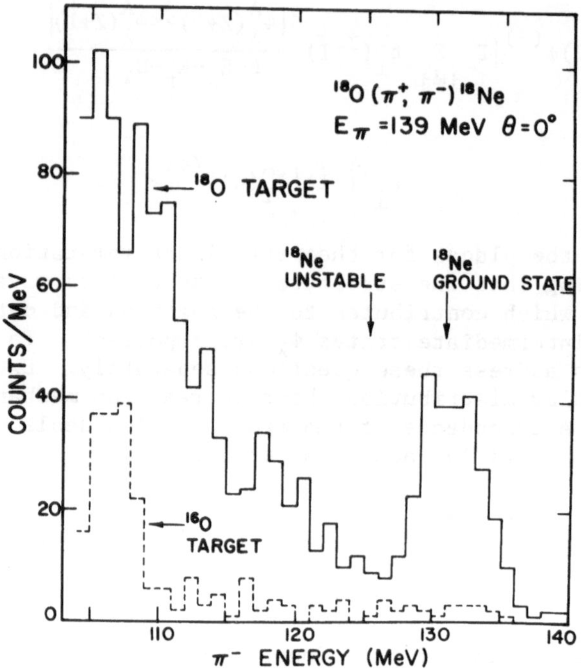

Fig. 1. $^{18}O(\pi^+,\pi^-)^{18}Ne$ spectrum at T_π = 139 MeV.

fall-off required by the data points, higher values of the intermediate angular momenta will have to be important.

In addition to observations of double charge exchange on ^{18}O at 0° with a 2 μb cross section, it has been seen with a 0.9 μb cross section on ^{16}O. The large value for a non-analog transition suggests that both the details of the nuclear wave functions and cross shell charge exchange are important in determining the size of the cross section. This suggestion was made plausible by Lee, Kurath and Zeidman[8]. They consider both the particle-hole structure of the nuclei, the analog transitions (type I) and the non-analog transitions (type II) as exemplified in Fig. 4. Their calculations are done in a $1d_{5/2}$, $2S_{1/2}$, $1p_{1/2}$ space and use the plane wave impulse approximation (they only address the question of the ratio of the cross sections). They are able to get the correct ratio for ^{18}O to ^{16}O by having cross shell strengths which are about 25% of the analog transitions and particle-hole admixtures of about 25%.

Recently, measurements have been made on a number of other nuclei. The new 0° data is shown in Figs. 5 and 6 for T_π = 145 MeV. These nuclei do not show as clean peaks as the oxygen

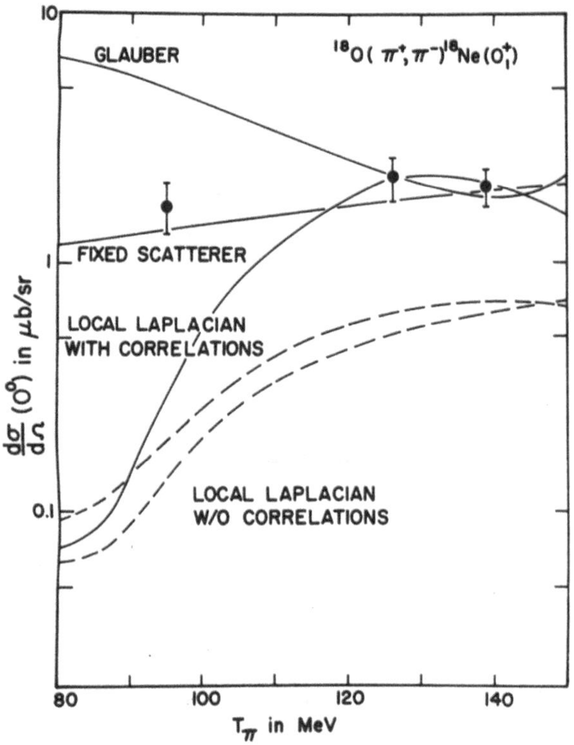

Fig. 2. The double charge exchange excitation function on ^{18}O. The theoretical curves are discussed in Ref. 6.

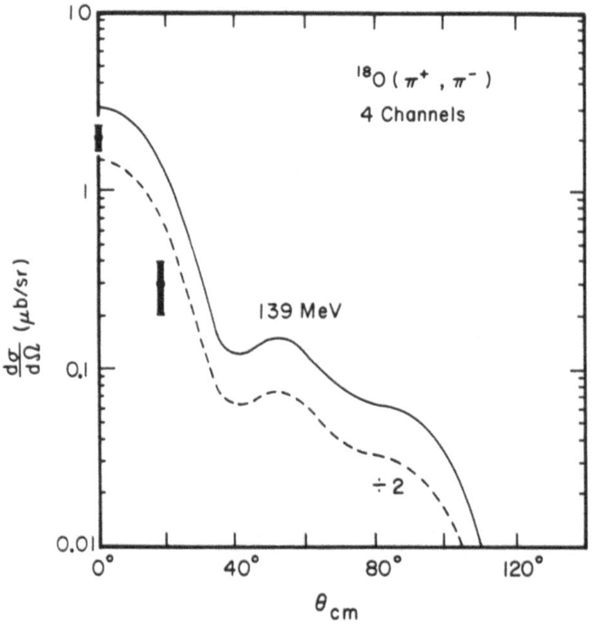

Fig. 3. A two point angular distribution for ^{18}O double charge exchange.

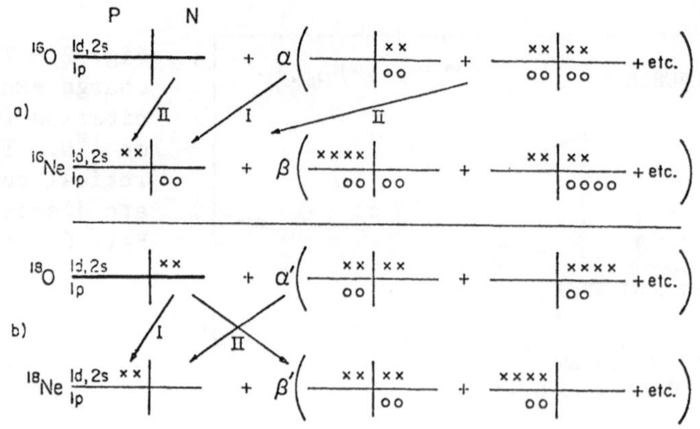

Fig. 4. Symbolic representation of the particle-hole structure of the oxygen and neon nuclei. Also shown are analog (I) and non-analog (II) transitions.

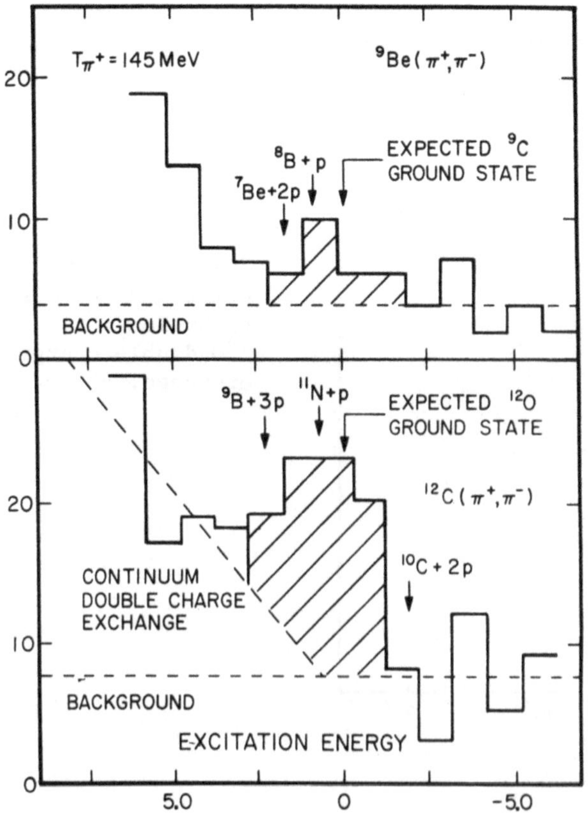

Fig. 5. (π^+,π^-) spectra at T_π = 145 MeV on ^{9}Be and ^{12}C.

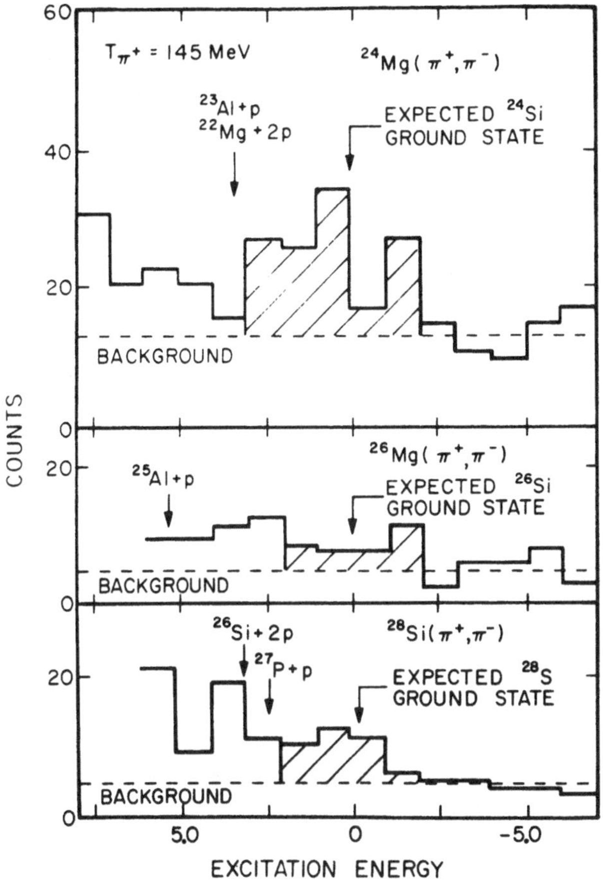

Fig. 6. (π^+,π^-) spectra at T_π = 145 MeV on ^{24}Mg, ^{26}Mg and ^{28}Si.

isotopes because the breakup channels lie lower in excitation energy and the cross sections are smaller. Therefore, the measurements necessarily have large errors. The three-point angular distribution[9] on ^{9}Be is shown in Fig. 7. These points are consistent with a flat angular distribution, and it is a mystery to explain this shape theoretically.

Another mystery is the ratio of the 0° cross sections for the s-d shell nuclei. ^{26}Mg to ^{26}Si should be a double analog transition, but it has the smallest cross section. In this light, it will require an unusual conspiracy of reaction mechanism and nuclear structure to explain the large cross section on ^{24}Mg. No one has calculated the double charge exchange t-matrix for wave functions

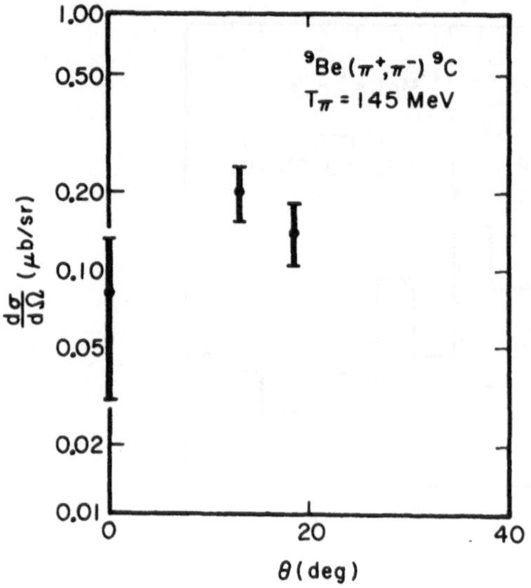

Fig. 7. The T_π = 145 MeV angular distribution on ^{9}Be.

appropriate to highly rotational nuclei.

In the near future, we can expect more complete angular distributions, cross sections at lower energies, and data from the (π^-,π^+) reaction. In addition, extensive data on single charge exchange is soon to be forthcoming, and that data should shed more light on the pion charge exchange process. Unraveling the mysteries of double charge exchange is certain to be exciting.

REFERENCES

1. The experiment group includes: M.P. Baker, R.L. Burman, M.D. Cooper, R.H. Heffner, R.J. Holt, D.M. Lee, D.J. Malbrough, T. Marks, B.M. Preedom, R.P. Redwine, J.E. Spencer, M. Yates-Williams, B. Zeidman.

2. C. Perrin, J.P. Albanese, R. Corfu, J.P. Egger, P. Gretillat, C. Lunke, J. Piffaretti, E. Schwarz, J. Jansen, B.M. Preedom, Phys. Lett. 69B, (1977), 301.

3. The experimental group includes: S. Iverson, H. Nann, A. Obst, K.K. Seth and H.A. Thiessen.

4. S.D. Drell, H.J. Lipkin and A. de Shalit, as cited by R.G. Parsons, J.S. Trefil and S.D. Drell, Phys. Rev. 138 (1965), B847.

5. T. Marks, M.P. Baker, R.L. Burman, M.D. Cooper, R.H. Heffner, R.J. Holt, D.M. Lee, D.J. Malbrough, B.M. Preedom, R.P. Redwine, J.E. Spencer and B. Zeidman, Phys. Rev. Lett. 38 (1977), 149.

6. R.L. Burman, M.P. Baker, M.D. Cooper, R.H. Heffner, D.M. Lee, R.P. Redwine, J.E. Spencer, T. Marks, D.J. Malbrough, B.M. Preedom, R.J. Holt and B. Zeidman, Phys. Rev. C17 (1978), 1774.

7. D.A. Sparrow and A.S. Rosenthal, Univ. of Maryland Technical Report 78-067, Preprint 78-155, College Park, MD (1978).

8. T.S. Lee, D. Kurath and B. Zeidman, Phys. Rev. Lett. 39 (1977), 1307.

9. The 13° and 18° points come from K.K. Seth, private communication (1978).

4. S.J. Greene, W.J. Braithwaite and A. de Sueljée, as cited by A.B.
    Laragne, J.B. Frankle and J.N. Ginocchio, Phys. Rev. C 18 (1965),
    984.

5. E. Marks, M.P. Baker, R.L. Burman, M.D. Cooper, R.H. Heffner,
    R.J. Holt, W.R. Lee, O.L. Halbresen, R.M. Kieson, M.P. Rede-
    mann, J.B. Spencer and B. Zeidmann, Phys. Lett. A (1971),
    185.

6. M.L. Barlett, R.P. Baker, M.D. Cooper, L.V. Cluet, R.H. Hef-
    fner, R.V. Poelman, J.R. Spencer, N. Marks, D., Malbrough, S.M.
    Nieedna, R.D. Holt and B. Zeidmann, Phys. Rev. C (1979), 1774.

7. D.A. Sparrow and A.S. Rosenthal, Univ. of Maryland Technical
    Report DOE/ER, Preprint 79-125, College Park, MD (1979).

INDEX